AVIAN VIRUSES

FUNCTION AND CONTROL

BRANSON W. RITCHIE, DVM, PhD
Associate Professor of Avian and Zoologic Medicine
The University of Georgia
College of Veterinary Medicine
Department of Small Animal Medicine
Athens, Georgia

KIP CARTER, CMI
Medical Illustrator
The University of Georgia
College of Veterinary Medicine
Department of Educational Resources
Athens, Georgia

© 1995 Wingers Publishing, Inc., Lake Worth, Florida

WINGERS PUBLISHING, INC.

Post Office Box 6863, Lake Worth, Florida 33466-6863

Library of Congress Catalog Card Number: 95-060930
ISBN: 0-9636996-3-6

Editorial Director: Linda R. Harrison
Art Director: Kari W. McCormick
Electronic Production: Roy Faircloth
Medical Editors: Janice Matthews, K. Storm Hudelson, Paul Hudelson

AVIAN VIRUSES: FUNCTION AND CONTROL   ISBN: 0-9636996-3-6

10 9 8 7 6 5 4 3 2 1

Avian Viruses: Function and Control
and the original research findings it contains are dedicated to
Phil Dean Lukert, DVM, PhD:
a wise, respected and cherished mentor,
colleague, advisor and friend.

Contents

Acknowledgements

Compiling *Avian Viruses: Function and Control* would not have been possible without the contributions of numerous individuals, the greatest of which have been from my colleagues who performed the original research summarized in this book. The research findings on PBFD virus, avian polyomavirus, EEE virus and proventricular dilatation disease conducted by the Psittacine Disease Research Group at the University of Georgia, College of Veterinary Medicine were made possible by the support of hundreds of private aviculturists, avicultural organizations, bird clubs, bird product companies and veterinarians. To all those who have given: thank you for helping us help birds. I would particularly like to thank Irving Cowan, Terry Clyne, Joe and Sue Still, Richard and Luanne Porter, Allen Berk, John Vaughn, the International Avian Research Foundation, Knick Enterprises, Midwest Avian Research Exposition, the International Aviculturists Society, Avian Research Associates, Lafeber Co., and Zeigler Brothers, Inc. for their past and continued support. Problems faced by the avicultural community regarding the health of companion birds can only be resolved by the community itself.

I am most appreciative of the support provided to our research group and our clinical program by the administration of the University of Georgia. Particularly, Drs. Joe Key, Janice Kimple and John Ingel have provided us an unencumbered environment in which to study viral diseases of companion birds; without their interest in these projects, much of the work chronicled in this book would not exist.

I would also like to thank the members of the Psittacine Disease Research Group: Frank Niagro, Kenneth Latimer, Cheryl Greenacre, Chris Gregory, W.L. Steffens, Denise Pesti, Ray Campagnoli and Phil Lukert for their long-standing commitment to improving the health of companion birds and their dedication to problem solving and research excellence. Without the perseverance and dedication of these professionals, none of the new findings on PBFD virus, avian polyomavirus, EEE virus or proventricular dilatation disease would have been possible.

Early insights into the tone and direction for this book were provided by Terry Clyne, Luanne and Richard Porter, Steve and Judy Hartman, Linda Pietroski, Greg Harrison and Heather Wilson. I appreciate the contribution of each of these individuals to this book. The exceptional illustrations that bring this book to life were produced by Kip Carter, Hursh Jain and Thel Melton. I think the reader will especially appreciate the artistic genius of Kip Carter, who can visually simplify even the most complex concepts.

As was the case for *Avian Medicine: Principles and Application*, Linda Harrison and the staff of Wingers Publishing, Inc. provided me with an opportunity to write an easy-to-read, visually stimulating book, unhindered by restrictive budgets and unyielding guidelines. I am appreciative of their commitment to excellence in teaching through innovative publishing. Particularly, I would like to thank Kari McCormick and Roy Faircloth for a design that masterfully integrates the text, illustrations and photographs into an easy-to-follow work.

The excellent photographs used in this book were provided by Avian Research Associates, Biomune, Gilles Berneir, Bob Dahlhausen, Richard Davis, Josh Dein, Cs.N. Dren, Mark Goodwin, R.E. Gough, Cheryl Greenacre, Chris Gregory, S. Robert Hilfer, Brett Hopkins, Horizon Micro-Environments, S.W. Jack, E.F. Kaleta, Kenneth Latimer, Doug Mader, Scott McDonald, M.S.

McNulty, Barbara Oglesbee, David Pass, B. Perelman, Sarah Plant, Richard Porter, Stuart Porter, Amy Raines, Antonio Ramis, Joey Rodgers, Jean Sander, Deanna Shafar, W.L. Steffens. S.S. Tsai, Tom Tully and D.B. Warheit. A special thanks to W.L. Steffens and Mary Ard for the excellent electron micrographs used in the book.

Finally, I would like to acknowledge Drs. Cheryl Greenacre, Don Harris, Sam Vaughn, Jamie Lindstrom, Barbara Oglesbee, Greg Harrison, Rita McManamon, James Harris, Greg Rich, Bob Dahlhausen, Louise Bauck, Tom Tully, Brian Speer, Rich Nye, Scott McDonald, Mike Taylor and Susan Brown for their years of unwavering support as well as their positive impact on me and our profession.

Branson W. Ritchie, DVM, PhD

Foreword

Phil D. Lukert, DVM, PhD

Professor of Medical Microbiology

The University of Georgia

College of Veterinary Medicine

Athens, Georgia

vian Viruses: Function and Control provides a basis for understanding the nature of viruses and how they are capable of causing disease, as well as giving the general methods used to diagnose these infections. The clinical features, epizootiology, pathology, diagnosis and control of each specific viral disease is dealt with in great depth. I would speculate that at this time *Avian Viruses: Function and Control* is the most complete treatise on viruses of companion, aviary and free-ranging birds that has been undertaken. I am confident that Dr. Ritchie's efforts will be appreciated by veterinarians, aviculturists and wildlife biologists.

The concern for disease is of utmost importance to aviculturists, and their survival as successful breeders will be determined by their ability to meet the challenge of disease management and control. Professional aviculture is a relatively new industry and, as such, our knowledge of the diseases of birds is still in its infancy. Disease problems will become amplified in the breeding aviaries, and the need for disease prevention through biosecurity, quarantine and immunization is of paramount importance to the aviculturist.

The future will most surely bring about the recognition of more and more diseases of birds. I have often made the analogy of disease as a whole being similar to a giant onion. As we peel away each layer of disease there is another layer of disease lying underneath. We also must realize that new diseases arise from time to time, which indicates that evolution is still going on.

I would predict that this book will be current only for a relatively short time and Dr. Ritchie will, hopefully, follow this edition with periodic new editions that will represent the rapid accumulation of knowledge about viral diseases of companion birds. If we as researchers are doing our jobs, that will be the case.

Foreword

Terry Clyne

Apalachee River Aviary

Farmington, Georgia

Not so long ago, and not too far away, a pair of parrots bred in captivity for the first time and aviculture was born. Medicine men were drawn into these events, and avian medicine began a sort of creaking and groaning existence, like some rusty squeeze-box, waiting for the right pairs of hands to make music from it. Time passed, and more parrots were raised. The medicine men pressed the psittacine into the mold of their knowledge of other animals, and it was, at best, a difficult fit.

Aviculture, too, underwent incredible change. No longer was the fortuitous coupling of a single pair of birds an event extraordinaire. Many aviculturists sought the help of the medicine men, in order to temper their own intuitions and feelings with the knowledge of healing. Some of the medicine men shared freely what they knew, and progress was made.

Constantly lurking in the near distance were the unknowns — viruses, bacteria and other disease-causing agents — ready to wreak uncontrollable havoc upon finding the least chink in the wall. Where were these chinks destined to appear? How wide would they be? How long before they could be plugged? The lives of many psittacines hung in the balance of these and other questions.

Then, a miracle happened. A group of scientists at the University of Georgia began to examine these chinks and ways to repair or prevent them; it was brutally scientific stuff, yet stuff with a heart. The heart was that of each bird the scientists saw suffer and die. The science itself was impeccable. Nevertheless, it was neither esoteric nor arcane; it was goal-oriented and meaningful — not a single motion that was not very carefully thought out, not a moment to spare, not another life to lose. Abandoning traditional academic arrogance, which behaves as if science should exist for its own sake, these researchers labored to construct useful tools — tools that would help us prevent the loss of life in the avian flocks over which we have been granted control.

Understanding that science in a vacuum benefits no creature, Dr. Ritchie began work on this book. Making the results of scientific progress understandable to the masses is no small task; yet throughout the years, I have often seen Bran Ritchie struggle to do just that. He has written innumerable articles and delivered countless talks in an effort to show aviculturists that chinks in the wall do exist, and to illustrate to us, and to his fellow avian veterinarians, what bricks and mortar can be applied to fill those crevasses — or better still: how these holes can be prevented from cracking in the first place.

This book is about the wall itself, and the chinks in it, as well as about the bricks and mortar available to seal those holes. It is about what will happen if we ignore the holes, as well as the search for new building materials and new tools. It is a remarkable book: user-friendly and full of information. It is a treatise about the impact of viruses on other life forms and how those of us responsible for other beings might prevent or lessen that impact. The book, like all tools, is only as good as our willingness to use it. For those who are willing to read and learn, the information is here. Hopefully, that information will change — become outdated, obsolete — by the time you reach the end of this book. I know that the book's author would especially wish it so.

So, as the caterpillar in Alice Through the Looking Glass said: "...begin at the beginning, go on until you reach the end, and then stop." Let us all hope that the end of this book is just the beginning. There is still a very long way to go — the speed and direction of which will depend on continued communcation between aviculturists and avian veterinarians.

Over the last century, viral infections have been associated with overt signs of disease in most species of living organisms, including some of the most severe and devastating of all diseases in humans and other animals. However, the study of viruses is a relatively new science. In the 1950s, scientists first developed techniques to successfully grow some viruses in the laboratory in plates of living cells. The introduction of these cell culture techniques dramatically changed a virologist's ability to study many viruses, and understand how they replicate and cause disease. Much progress has been made since.

Understanding in general what viruses are and how they function will help individuals develop a perspective of how viruses can affect birds. It will also help them reduce the impact of viral-induced disease. This book was conceived and written for veterinarians, aviculturists, companion bird enthusiasts, ornithologists, zoological personnel and wildlife management officials. It was my intent to provide these readers with an easy-to-understand, easy-to-access overview of the viruses that infect companion, aviary and free-ranging birds.

The book begins with a simplified explanation of how viruses work: what they are, where they come from, where they hide and how they attack their victim/host. The initial chapters were designed to provide the veterinary student or aviculturist with a general understanding of how viruses function and how we can stop them from adversely affecting individuals as well as populations of birds. Given the basic unity of all life forms, this treatise has implications far beyond aviculture and avian medicine.

Information also is provided on specific viruses affecting avian species: how these viruses function and why certain species of psittacine birds, in particular, are susceptible to these pathogens. *Avian Viruses* explores what is available to prevent the spread of viral disease within a flock, as well as what is in the works for the future data.

For viral diseases that have been documented in poultry but not in other avian species, an attempt has been made to provide some perspective as to how a disease process might occur if a related virus was found to infect companion, aviary or free-ranging birds. For in-depth information on viruses of poultry, the reader is referred to *Virus Infections of Birds*, McFerran JB, McNulty MS, Elsevier Science Publishers, 1993.

A warning for the future. This book is a compilation of information generated by disease investigators from every continent. Understanding what are today considered "foreign pathogens" is important. Given continued smuggling, indiscriminate movement of wildlife from continent to continent and renewed discussions of expanding exportation and importation of companion birds, it is likely that diseases that are now restricted to a single continent will not be so isolated tomorrow.

Branson W. Ritchie, DVM, PhD

An Overview of Viruses

A CLINICAL CHALLENGE

In evaluating and treating diseases of suspected viral etiology, the avian clinician works under many handicaps, including: 1) a lack of information on the complex interaction between the avian host, a suspected viral agent, secondary invaders and the resulting disease process; 2) the relative absence of effective anti-viral drugs; 3) the difficulty in diagnosing many viral diseases in living birds; 4) the fact that the duration and routes of viral shedding for many common avian viruses are not fully understood; 5) the inability to detect and remove subclinical virus carriers; and 6) the scarcity of appropriate vaccines.

Viruses continue to cause high levels of morbidity and mortality in companion and aviary birds. In addition to the readily detectable changes that have been associated with viral infections, improved techniques for detecting viral proteins or viral genetic material (DNA or RNA) are implicating subclinical viral infections in debilitating diseases, neoplasia, permanently damaged immune systems and as a predisposing factor in recurring bacterial, fungal and parasitic infections. In many of these cases, the damage to the infected host caused by concurrent infections with bacterial, chlamydial, parasitic and fungal agents may disguise any initial pathology induced by a primary viral pathogen. The degree and type of clinical disease that develops in an infected host is governed by the host's age, condition, nutritional status and immunologic competency, as well as by the virulence of the virus and route of viral exposure. These variables further complicate a virologist's ability to determine what role a virus is playing in a disease process. The risk of an epornitic is greatest for those viruses that are stable outside of the host and can be transported on fomites.

Integral to the increasing occurrence of viral diseases is the mixing of avian species from numerous sources, the shift in aviculture to high-density confined breeding operations, and the frequent relocation of young birds with immature immune systems. The global movement of numerous avian species to meet the demands of the various bird markets results in the exposure of immunologically naive animals to widely varying endogenous microflora that may be endemic to geographically isolated avian populations. The ease of intercontinental mass transportation through airline travel will continue to favor the dissemination of highly pathogenic viruses into naive populations of all types of animals, including humans and birds. Mixing birds from various aviaries can create a similar infectious disease nightmare.

The actual incidence of viral infections in quarantine stations probably is much higher than reported.[4] The diagnostic

Documented virus activity by family of birds

	1	2	3	4	5	6	7	8	9
Adenoviridae	X	-	X	X	X	X	X	X	-
Arenaviridae	?	-	-	-	-	-	-	-	-
Astroviridae	-	-	X	-	-	-	X	X	-
Birnaviridae	-	X	X	-	-	X	X	X	-
Bunyaviridae	-	-	-	-	-	-	-	-	-
Circoviridae	X	-	-	X	?	-	X	-	-
Coronaviridae	X	-	-	X	-	X	X	X	-
Caliciviridae	-	-	-	-	-	-	-	-	-
Filoviridae	-	-	-	-	-	-	-	-	-
Flaviviridae	-	X	-	-	-	-	X	-	-
Herpesviridae	X	X	X	X	X	X	X	X	X
Hepadnaviridae	-	-	X	-	-	-	-	-	-
Iridoviridae	-	-	-	-	-	-	-	-	-
Orthomyxoviridae	X	X	X	X	-	X	X	X	X
Papovaviridae									
Papillomavirus	X	X	X						cranes,herons
Polyomavirus	X	X	-	-	?	-	X	?	-
Paramyxoviridae	X	X	X	X	X	X	X	X	X
Pneumovirus							X	X	
Parvoviridae	?	-	X	-	-	-	?	-	-
Picornaviridae	X	-	X	-	-	-	X	X	-
Poxviridae	X	X	X	X	X	X	X	X	X
Reoviridae	X	X	X	X	X	X	X	X	X
Rotavirus	-	-	-	X	-	-	-	-	-
Retroviridae	?	?	X	X	?	X	X	X	quail
Rhabdoviridae	-	X	-	-	X	-	X	-	-
Togaviridae	X	X	X	X	X	X	X	X	X
Toroviridae	-	-	-	-	-	-	-	-	-

procedures that are used in these facilities are designed to detect only those birds excreting one of the few viruses (including influenza A virus and paramyxoviruses) with hemagglutinating activity.

The prevalence of subclinical virus carriers was dramatically demonstrated by one study that found 17 of 500 (3.4%) clinically normal companion birds to be shedding a virus that could be recovered in cell culture. It was not determined how many of these birds were shedding viruses that could not be recovered in cell culture.[4]

Twenty-four families of animal viruses have currently been described. Fifteen of these families have been associated with some level of disease in domestic fowl; 11 families have been associated with diseases in psittacine birds; 10 families have been associated with diseases in pigeons and doves; and only 11 families have been demonstrated in ratites *Table 1.1.* Either commonly maintained species of companion and aviary birds are tremendously resistant to viral infections, or the veterinary community has an alarmingly limited understanding of the association between viruses and disease in these avian species. The latter is certainly more likely. As investigations into avian diseases of suspected viral etiology increase, many families of animal viruses, which have not previously been associated with companion and aviary birds, will undoubtedly emerge.

AVIAN VIROLOGY - PAST, PRESENT AND FUTURE

The improved health and breeding performance of companion animals, livestock and poultry during the past century has been made possible by the veterinary community's ability to prevent

TABLE 1.2

Information available concerning diseases of dogs and cats in comparison to those that affect companion and aviary birds

Dogs and Cats Disease	Etiology	Management Technique
Canine distemper	Paramyxovirus	Vaccination
Canine parvovirus	Parvovirus	Vaccination
Feline cardiomyopathy	Nutritional	Special diets
Heartworms	Parasite	Monthly preventive
Feline panleukopenia	Parvovirus	Vaccination
Avian Disease	Etiology	Management Technique
Proventricular dilatation disease (Psittaciformes, geese)	Unknown	Unknown, terminal
Avian circovirus (PBFD) (Psittaciformes, Columbiformes)	Circovirus	DNA probe testing
Papillomatosis	Unknown	Unknown
Iron storage hepatopathy	Unknown	Unknown, terminal
Avian polyomavirus	Polyomavirus	Vaccination
Incubation failures (Psittaciformes, ratites)	Unknown	Unknown
Hypocalcemia syndrome (African Grey Parrots)	Unknown	Unknown

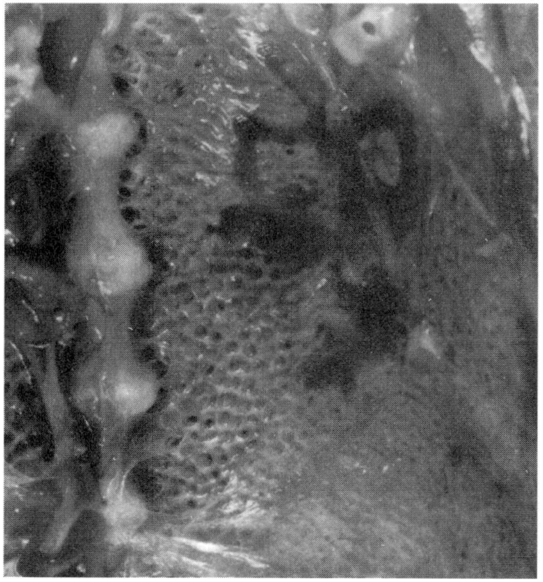

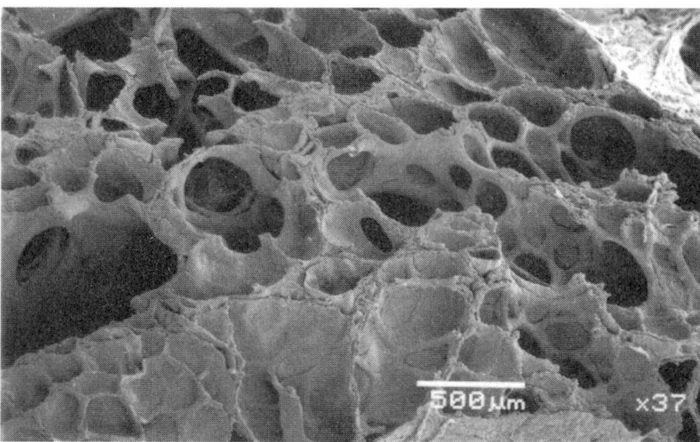

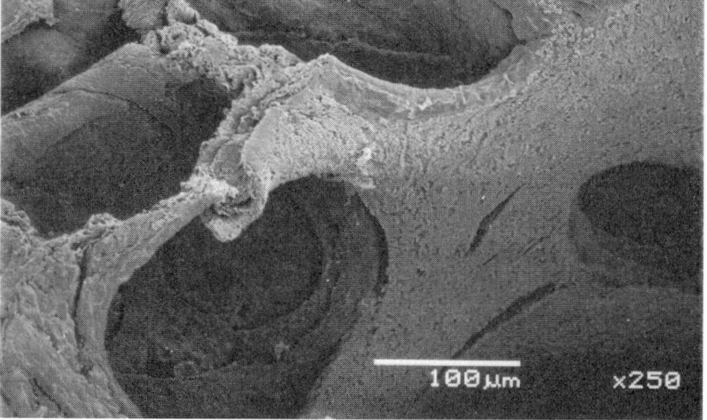

FIG 1.1

Each organ within the body is composed of thousands of individual cells that function in concert to support life. The complexity of the interaction of these cells is clear when viewing increasingly magnified photographs of the lung.
Top left *The gross appearance of the lung.*

Top right *A view of the lung magnified 37 times showing its multiple divisions and small holes where oxygen and carbon dioxide exchange occurs between the inspired air and blood.*

Same view magnified 250 times.

Same view magnified 650 times. The irregularity of the surface becomes visible.

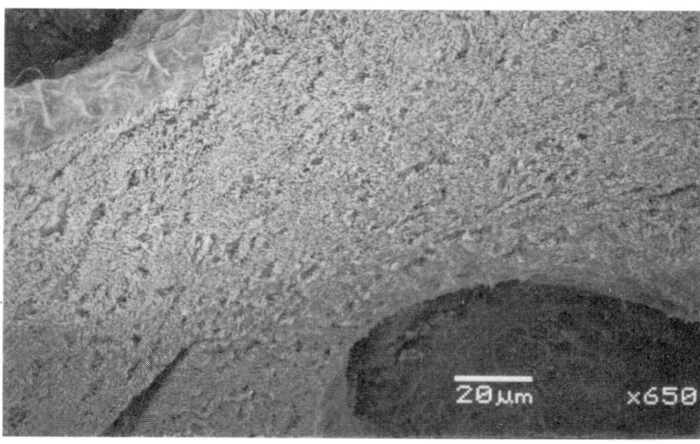

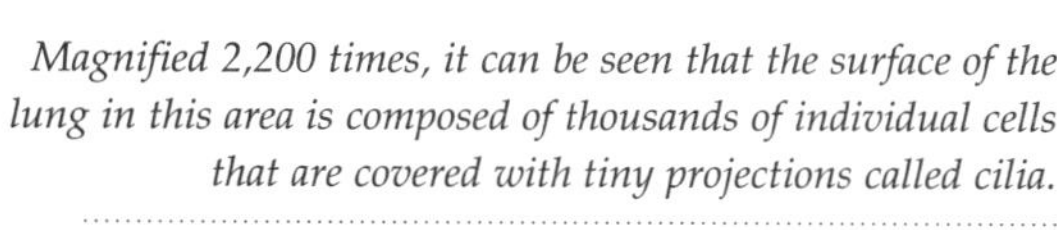

Magnified 2,200 times, it can be seen that the surface of the lung in this area is composed of thousands of individual cells that are covered with tiny projections called cilia.

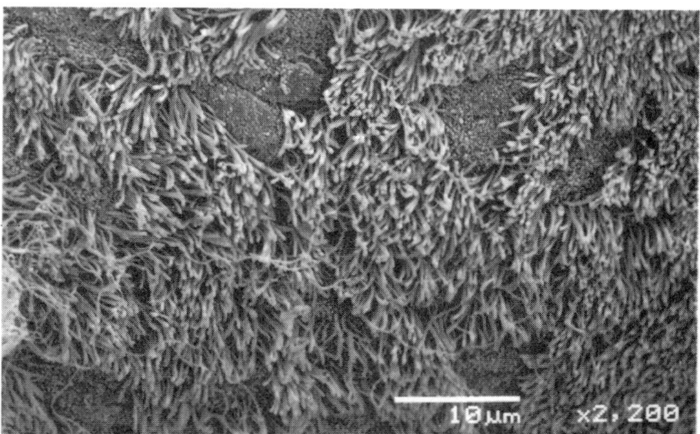

photographs courtesy of W.L. Steffens

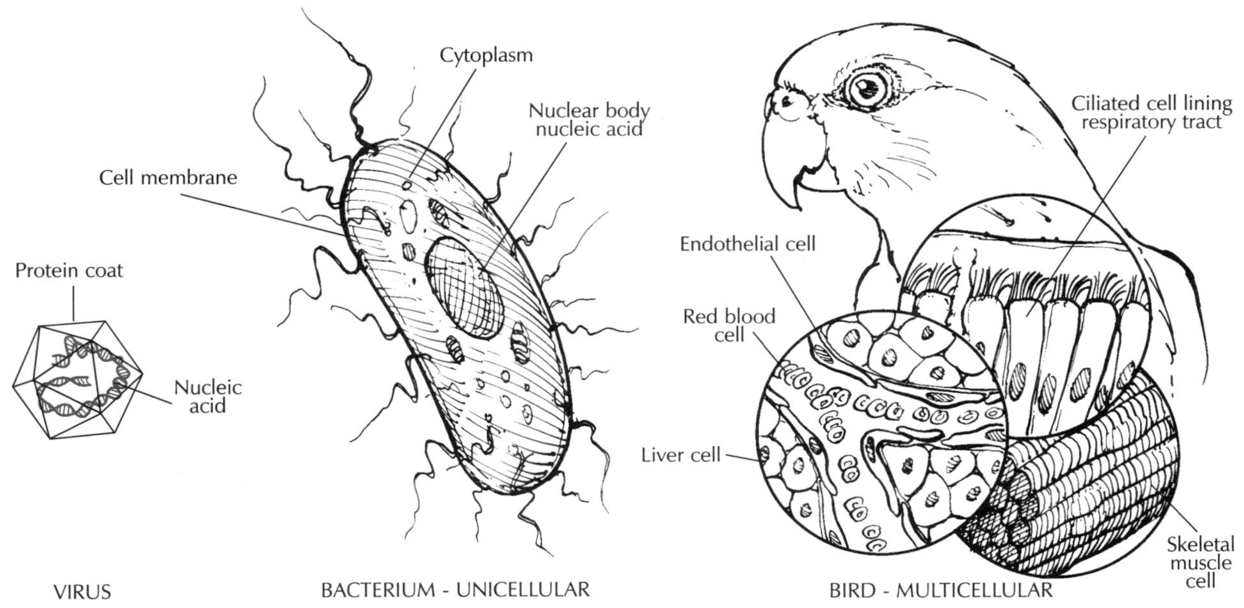

or control infectious diseases, particularly those caused by viruses. Historically, veterinary care for each of these animal groups advanced through a roughly similar set of "achievement zones." The "achievement zone" which most closely mirrors the current state of avian medicine occurred in the middle part of this century with dogs and cats. During that time, dogs and cats became increasingly popular as human companions, and the veterinary community was challenged to evaluate canine and feline problems and diseases. However, veterinary practitioners primarily had been trained to manage diseases of large farm animals. Many of the clinical syndromes that were recognized quickly in dogs and cats had not been studied and were of unknown etiology.

Companion bird medicine is currently experiencing a similar period. Numerous disease syndromes are being recognized. Unfortunately, the etiologies for many of these syndromes have not yet been determined. For many of those disease problems for which a causative agent is known, preventive techniques have not yet been developed. If one compares several commonly recognized diseases of both dogs and cats with some of the more prevalent diseases of companion and aviary birds, the difference in quality of preventive medicine that is available for avian patients is dramatically illustrated *Table 1.2*.

Avian health care is poised to enter a new "achievement zone." The application of rapidly expanding information on the molecular biology of viruses to the investigation of avian viral diseases will undoubtedly improve the veterinary community's ability to confirm what role a virus plays in a disease process, and to diagnose, prevent or possibly treat viral infections in both captive and free-ranging birds. The development and widespread availability of improved diagnostic techniques and prophylactic measures will

FIG 1.2

Viruses are the most rudimentary of all single-celled organisms. In their simplest form, they contain only nucleic acid surrounded by a protein coat.

Bacteria, fungi, mycoplasma, rickettsiae and chlamydia are also single-celled organisms, but unlike viruses have a nuclear body (made of nucleic acid) and a cytoplasm that contains components that convert nutrients into energy to drive functions of the cell.

Each individual organ within a bird is constructed of its own type of cells. There are complex differences among the cells that form skeletal muscle, line the respiratory tract or collectively are found in the liver. Endothelial cells are those that line the blood vessels.

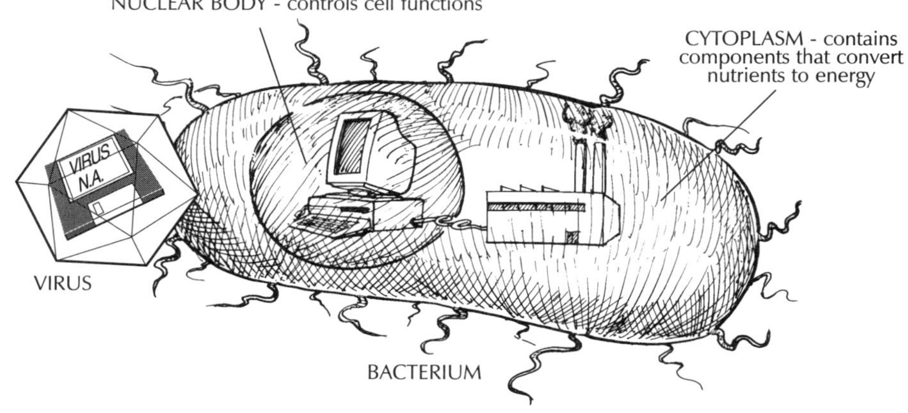

Because bacteria contain their own machinery to convert nutrients to energy, they are capable of reproducing when placed in any environment that contains their required nutrients. Thus, many bacteria can grow in unpurified, nutrient-rich water. The nucleus of a cell can be viewed as the control center; the cytoplasm as the factory. Viruses are completely dependent on the cell they infect for all the machinery they need to produce energy and replicate themselves. In this manner, a virus can be viewed as having a computer program (nucleic acid, N.A.) without a computer or factory.

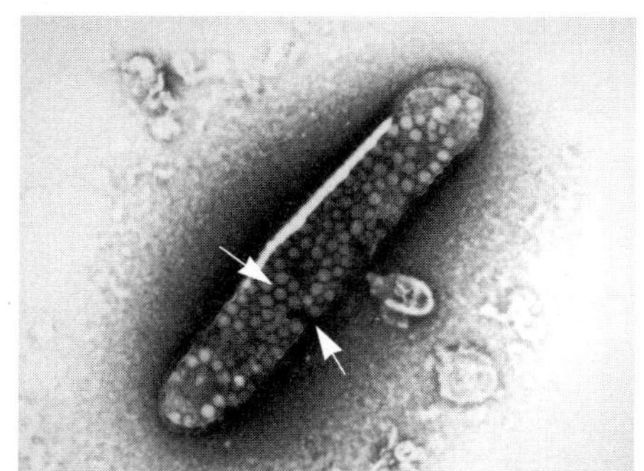

FIG 1.4

Viruses are so small that there are actually viruses that infect and destroy bacteria. In this photograph, a bacterium is packed full of individual virus particles (arrows). When the bacterial cell ruptures, the virus particles will be released and can then infect new bacteria. Magnification 26,000 times. photograph courtesy of W.L. Steffens

provide the avian clinician with methods to decrease morbidity and mortality, extend longevity, improve reproductive success and protect endemic bird species from foreign viral pathogens.

The avian community can reach the same heightened capacity to care for companion and aviary birds that is enjoyed by individuals concerned with dogs, cats, cattle, poultry and horses. Aviculture will reach its full potential when the principles of disease prevention, through sound hygiene and maintenance of closed breeding populations, are combined with an increased knowledge of the physiologic systems, diseases and management techniques necessary to maintain birds in a state of health.

CHARACTERISTICS OF VIRUSES

Birds are composed of millions of individual cells, each with a specific function within a bird's body. Each individual organ within a bird (lung, spleen, brain, blood, intestinal tract) is constructed of its own type of cells *Figure 1.1*. For example, the cells that form the liver have a specific form and function that are related to metabolism and to detoxifying the body. In contrast, the cells of the brain have a specific form and function that are designed to send and receive electrical and chemical signals that control the functions and movements of the body. The body of a bird is, then, a complex mixture of different cells that, when healthy, functions in concert to support life.

By comparison, microorganisms such as protozoa, fungi, bacteria, mycoplasmas and rickettsiae are all single-celled organisms *Figure 1.2*. These organisms can survive without the assistance of other cells because each individual organism contains all the nucleic acid and metabolic pathways it needs to survive. Nucleic acid is the term that is collectively used to describe all of a cell's DNA and RNA — the "computer software" that controls how a cell functions and replicates. Metabolic pathways can be viewed as the "machinery" that an organism uses to take up necessary nutrients and produce the macromolecules necessary to sustain life. The ability of unicellular organisms to produce their own macromolecules is important because antimicrobial agents can be artificially designed to attack the metabolic pathways of the invading organism while causing minimal damage to the bird they infect *Table 1.3.*

Viruses also are single-celled. Smaller than other unicellular microorganisms, viruses are so simple in structure that they must enter a living organism (host) and use the metabolic machinery of an infected cell in order to replicate *Figure 1.3.* Unlike bacteria and fungi that may reproduce outside an organism, viruses are completely dependent on the cell they infect for the machinery they need to produce the components of new virus particles. In other words, viruses are not metabolically active ("alive") until they infect (parasitize) a living cell.

The nucleic acid of viruses is either DNA or RNA, but not both. The RNA-containing viruses are currently the only known organisms that exclusively use RNA to transfer genetic material to their progeny. All other known organisms use DNA as their genetic material.

In some ways, chlamydia can be viewed as a biologic intermediate between viruses and bacteria. Like viruses, chlamydia cannot replicate

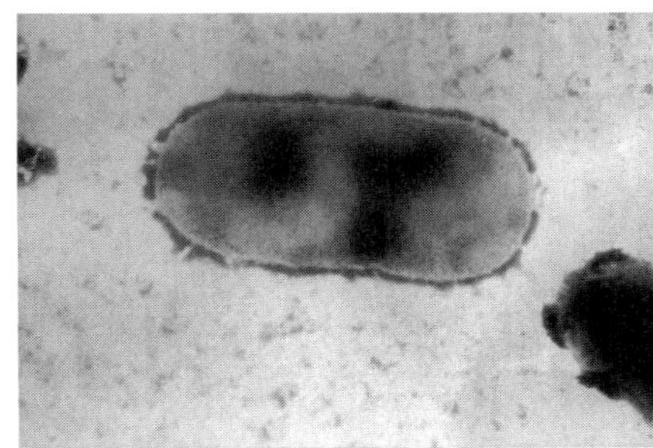

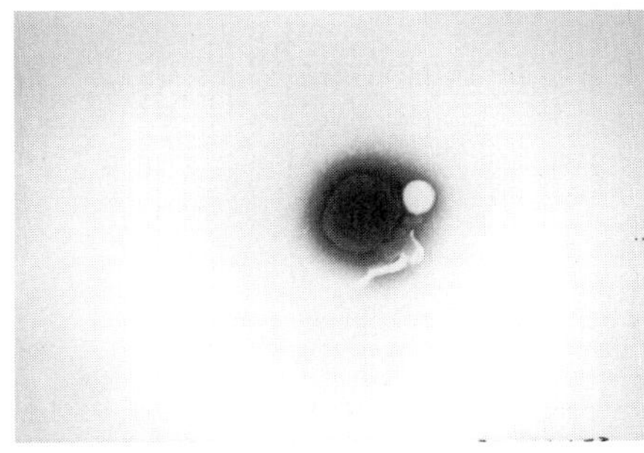

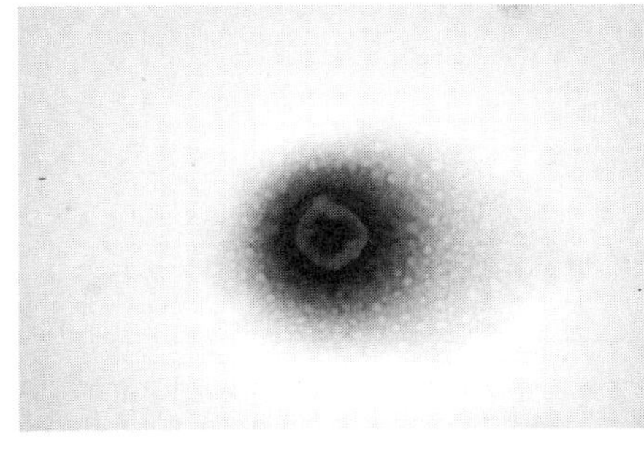

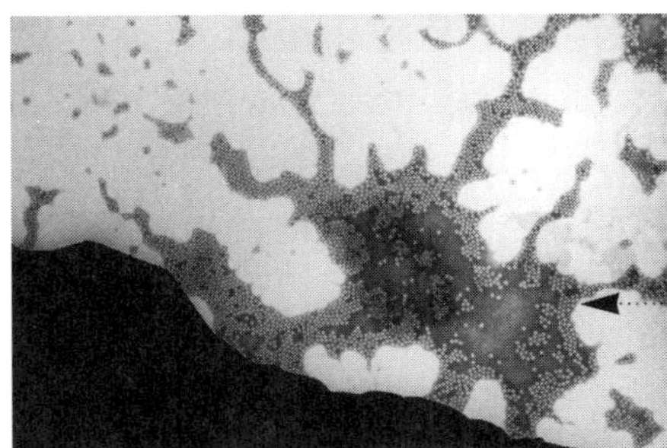

FIG 1.5

To demonstrate their relative sizes, some common microbial organisms were magnified 55,000 times: (Note that all of these organisms are composed of a single cell.)

Pseudomonas, a gram-negative bacteria

Poxvirus, the largest virus that infects birds

Chlamydia psittaci

PBFD virus, the smallest pathogenic animal virus. Each tiny, white dot is a virus particle.

photographs courtesy of W.L. Steffens

TABLE 1.3

Comparative characteristics of bacteria, chlamydia, rickettsia, mycoplasma and viruses

	Viruses	Chlamydia	Rickettsia	Mycoplasma	Bacteria
Grow outside host	no	no	no	no	yes
Contain both DNA and RNA	no	yes	yes	yes	yes
Sensitive to interferon	yes	yes	no	no	no
Sensitive to antibiotics	no	yes	yes	yes	yes
Replicate by cell division	no	yes	yes	yes	yes

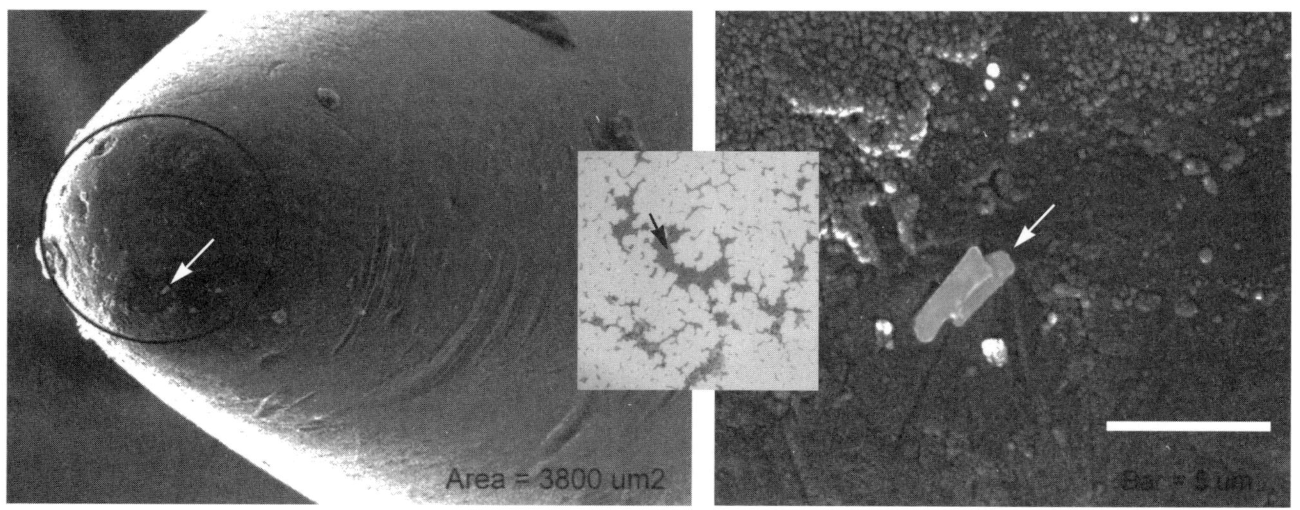

except in living cells; in having both DNA and RNA, being susceptible to antibiotics and multiplying by binary fission, they are like bacteria *Table 1.3*.

SIZE

Viruses are among the smallest of all organisms *Figure 1.4*. The companion animal viruses that have been identified vary in diameter from 14 nanometers (PBFD virus) up to 450 nanometers (poxvirus). Other viruses undoubtedly exist but have yet to be identified or characterized. The poxvirus is comparable in size to some of the smallest bacteria *Figure 1.5*, but it would take 25 PBFD virus particles to equal the size of the same bacteria *Figure 1.6*. With such a small size, viruses can easily be concealed or transported by dust particles *Figure 1.7*.

Filters designed to remove aerosolized virus particles must be capable of trapping viruses that range in diameter from 14 to 450 nanometers, but filters with larger pore sizes are frequently effective in stopping virus particles that are simply floating in the air, because these viruses are usually attached to larger debris and dust. In contrast, virus particles that are in a liquid solution can be washed free of the dust particles to which they may be attached and can then easily pass through filters with larger pores.

STRUCTURE OF THE VIRAL CAPSID AND ENVELOPE

Viruses are all similar in that they are structurally composed of genetic material (DNA or RNA) that is surrounded by a protein coat (capsid), but each family of viruses has a different and characteristic gross appearance *Figure 1.8*.

The capsid can be viewed as a protective, protein shell. The simplest viruses have only this single protein coat surrounding the nucleic acid. Some more complex viruses have an additional lipoprotein covering called an envelope *Figure 1.9*. A complete infectious virus particle that contains nucleic acid, a protein coat and an envelope (if present in a particularly family of viruses) is referred to as a virion.

The protein coat that surrounds the nucleic acid of the virus is produced by an infected cell in response to the command of the infecting virus. The protein coat can be viewed as being viral in origin, because its production was controlled by the virus. In contrast, the envelope that surrounds some viruses can be acquired as the virus is passing through an infected cell's membrane in a process called budding. Thus, the envelope can be derived from the infected cell of a bird *Figure 1.10*.

A virus particle is a naturally designed package for carrying nucleic acid (the DNA or RNA that controls replication) from one cell to another, or from one bird to another. The capsid that surrounds the nucleic acid is composed of many different chains of folded proteins. Each individual protein chain is called a subunit. Collectively these subunit proteins form a protective covering for the nucleic acid contained within *see Figure 1.9*. The contact and interaction among all the subunit proteins determines the structure of a virus.

The protein coat and the lipoprotein envelope, if present, control the way that a virus survives and functions in and out of a host. These structures connect the virus particle to a cell, control the bird's defensive response to the virus and determine how susceptible the virus is to inactivation when outside of the bird's body. In general, viruses without an envelope are more durable than enveloped viruses when outside of an infected bird and are more difficult to inactivate with disinfectants. A list of the enveloped and nonenveloped viruses is provided in *Table 1.4*.

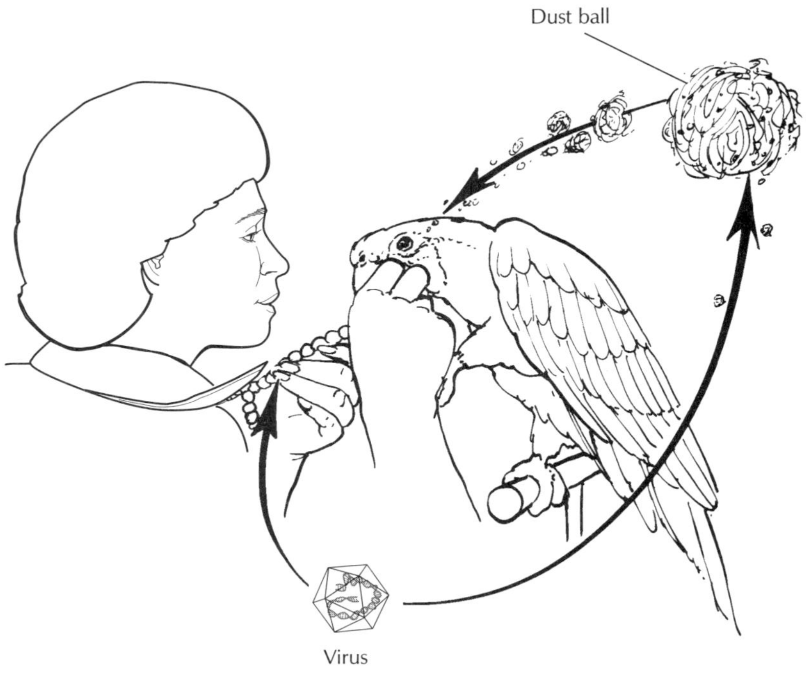

FIG 1.7
Virus particles originating from an infected bird can contaminate fingernails, skin, hair, jewelry or clothes and be transported to other birds in a home or aviary.

TABLE 1.4

By family, the relative resistance of enveloped and nonenveloped viruses to inactivation

Enveloped Viruses

Arenaviridae	unstable
Bunyaviridae	unstable
Coronaviridae	unstable
Filoviridae	unstable
Flaviviridae	unstable
Herpesviridae	varies with strain, most unstable
Hepadnaviridae	varies with strain, most unstable
Iridoviridae	unstable
Orthomyxoviridae	unstable
Poxviridae	very stable
Paramyxoviridae	varies with strain, most unstable
Rhabdoviridae	unstable
Togaviridae	unstable
Toroviridae	—

Nonenveloped Viruses

Adenoviridae	stable
Astroviridae	stable
Birnaviridae	stable
Caliciviridae	stable
Circoviridae	extremely stable
Papovaviridae	very stable
Parvoviridae	extremely stable
Picornaviridae	very stable
Reoviridae	stable

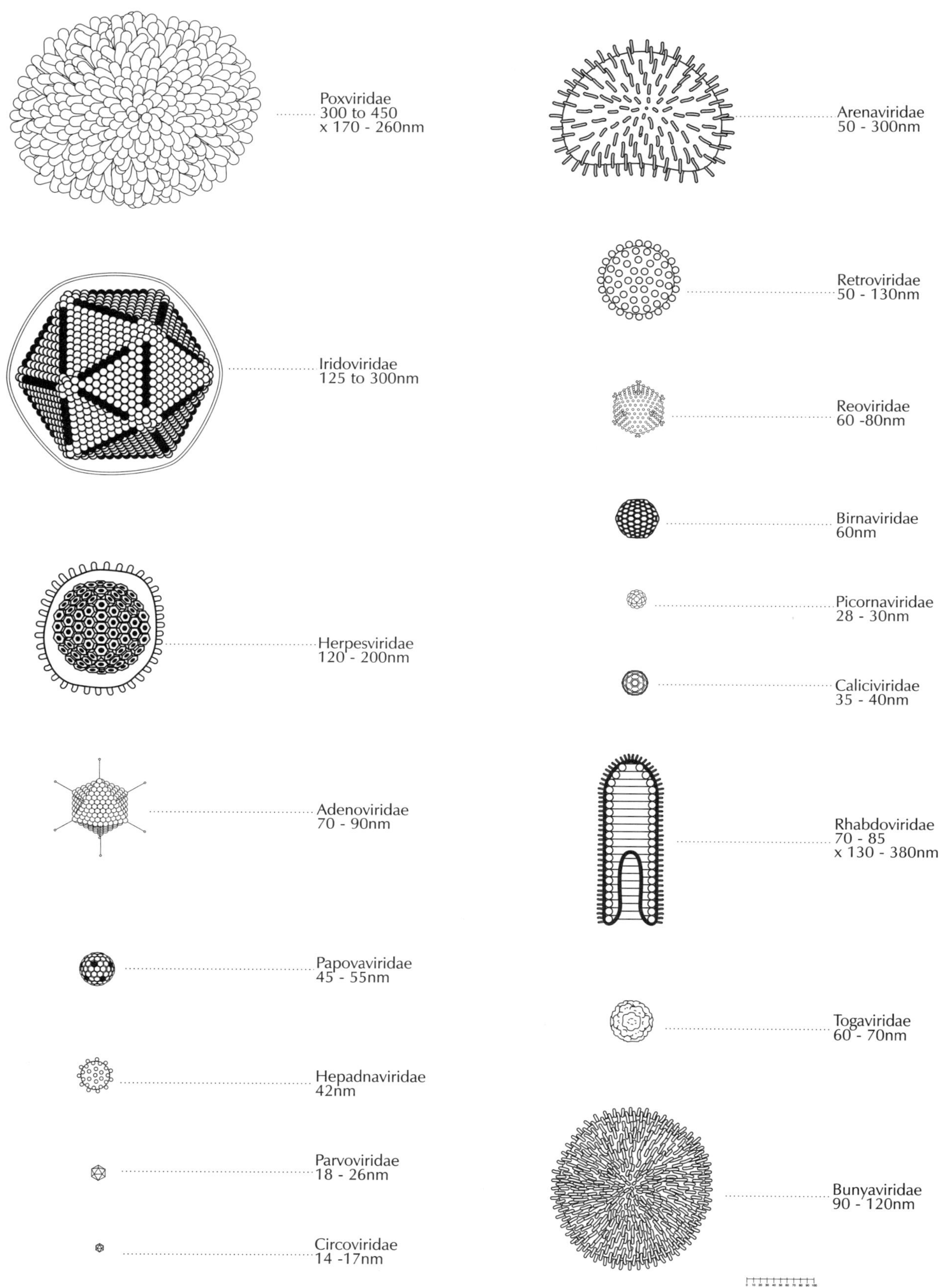

Poxviridae
300 to 450
x 170 - 260nm
Arenaviridae
50 - 300nm
Iridoviridae
125 to 300nm
Retroviridae
50 - 130nm
Reoviridae
60 -80nm
Birnaviridae
60nm
Herpesviridae
120 - 200nm
Picornaviridae
28 - 30nm
Caliciviridae
35 - 40nm
Adenoviridae
70 - 90nm
Rhabdoviridae
70 - 85
x 130 - 380nm
Papovaviridae
45 - 55nm
Togaviridae
60 - 70nm
Hepadnaviridae
42nm
Parvoviridae
18 - 26nm
Bunyaviridae
90 - 120nm
Circoviridae
14 -17nm

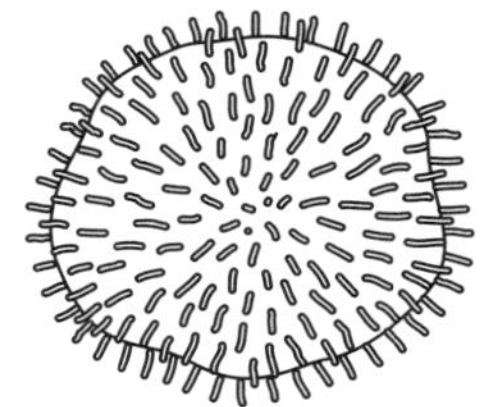

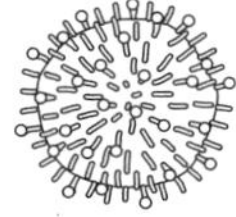

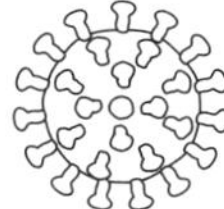

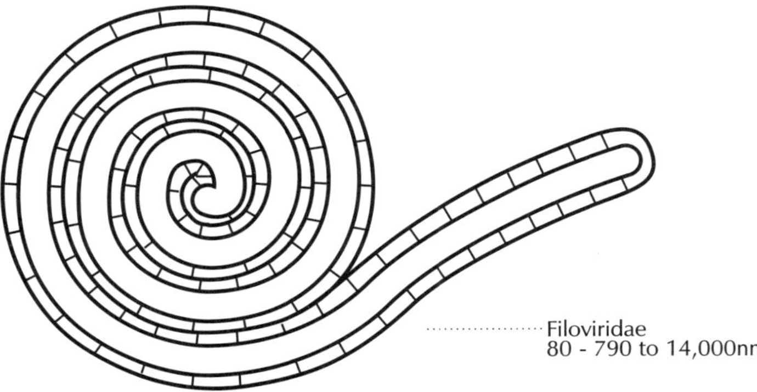

FIG 1.8

Examples of mammalian diseases caused by some virus families

POXVIRIDAE

Smallpox

HERPESVIRIDAE

Herpes simplex virus
Epstein-Barr virus

ADENOVIRIDAE

Infectious canine hepatitis virus

PAPOVAVIRIDAE

Human papillomavirus

HEPADNAVIRIDAE

Human hepatitis B virus

PARVOVIRIDAE

Feline panleukopenia virus

RETROVIRIDAE

Human immunodeficiency virus (HIV)

PICORNAVIRIDAE

Polio virus

CALICIVIRIDAE

San Miguel sea lion virus
Feline calicivirus

RHABDOVIRIDAE

Rabies virus

TOGAVIRIDAE

Hog cholera virus

BUNYAVIRIDAE

California encephalitis virus
Hantavirus

PARAMYXOVIRIDAE

Mumps virus
Measles virus

ORTHOMYXOVIRIDAE

Influenza A virus

CORONAVIRIDAE

Common cold virus

FILOVIRIDAE

Ebola virus

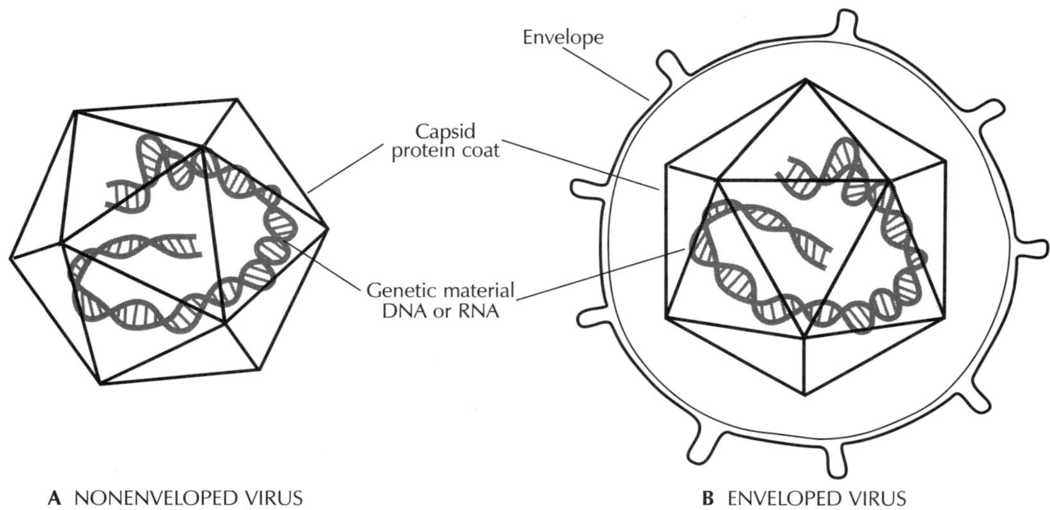

FIG 1.9
A The simplest viruses contain nucleic acid surrounded by a capsid.
B More complex viruses also contain an envelope.

The capsid or envelope, depending on the type of virus, serves as a binding site for attachment of a virus to an appropriate cell. This attachment is the initial step in the infection process. Thus, the capsid or envelope is responsible for determining which type of birds are susceptible to infection *Figure 1.11*.

To resist infection, the bird's defense systems must recognize the protein coat and/or lipoprotein envelope (if present) of the virus as a foreign invader and target it for destruction. In this manner, the protein coat or envelope serves to initiate the immune response. The body's reaction to the protein coat or lipoprotein envelope of the virus particle results in antibody production and stimulation of special defensive cells designed to destroy the virus and protect the bird from future infections by the same virus.

VIRAL TAXONOMY

It is easiest to develop an understanding of the relationship of different viruses to one another — and consequently, their classification by scientists — by using the animal kingdom as an analogy. Ostriches, Amazon parrots and swans belong to the same class, Aves, to reflect their basic similarities, but to different families, Struthionidae, Psittacidae and Anatidae, respectively, to reflect their unique differences *Figure 1.12*.

Viruses are placed in various taxa, or classification groupings, on the basis of a number of specific characteristics including the type of organism they infect, the type of nucleic acid they contain, the presence or absence of a lipoprotein envelope, the diameter of the virus particle, the symmetry of the capsid and the specific type of antibodies they induce.

Viruses can be initially divided into large groups based on whether they infect bacteria, plants, invertebrates or vertebrate animals. Within these large groups, viruses are divided into specific families based on common shared characteristics, including the size and shape of the virus particle, whether it contains DNA or RNA and whether this nucleic acid material is single-stranded or double-stranded, circular or linear. A viral family is indicated by the suffix viridae. For example, the avian polyomaviruses have been placed in the family Papovaviridae; its members are 45 nm, nonenveloped virus particles containing a double-stranded circular DNA molecule. New-castle disease virus, on the other hand, has been placed in the family Paramyx-oviridae, which contains 150 to 300 nm, enveloped virus particles with a sin-gle-stranded linear RNA molecule. It is theorized that the members of a viral family share an evolutionary history. Thus, all of the herpesviruses are thought to have a common ancestor that coevolved with the specific host they were infecting.

Virus families may be divided further into subfamilies based on refined dif-ferences in the structure of their protein coats and based on the host they infect. Subfamilies are indicated by the suffix virinae.

Subfamilies of viruses are divided into genera and species. Each genus is given the suffix virus. A genus is a population of viruses within a family that shows certain characteristics distinct from oth-er genera in the same family. Species are generally determined on the basis of antigenic differences (the antibody response that the virus induces in an infected animal).

Species of viruses can be further divid-ed into subspecies, strains and variants. As an example, consider herpesviruses

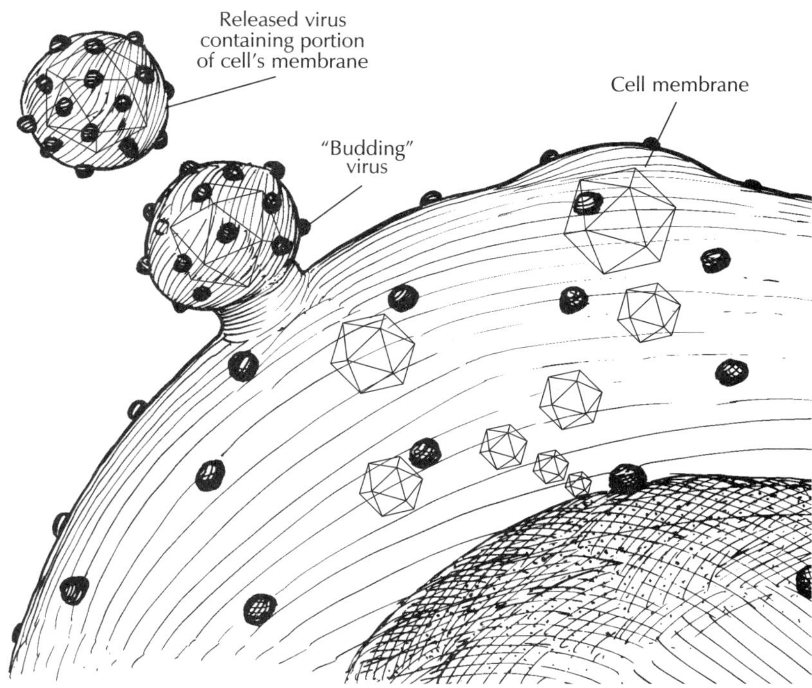

FIG 1.10

The envelope that surrounds some more complex viruses can be obtained as the virus is passing through an infected cell's membrane. When this occurs, the envelope of the virus contains portions of the infected bird's cell membrane.

that infect birds and mammals *Figure 1.13*. These herpesviruses share some basic characteristics, and thus all are placed in the same family, Herpesviri-dae. However, the herpesviruses that infect birds are not known to infect mammals, and the herpesviruses that infect mammals are not known to infect birds. Among the avian Herpesviridae, some infect falcons while others infect owls, cranes, chickens, turkeys or par-rots; thus they are divided into a num-ber of genera. Within the genera that infect parrots, more subdivisions are possible. For example, at least three antigenically distinct herpesviruses cause lesions that are described fre-quently as Pacheco's disease. These are classified as different species. When the actual Latin or specific names of a genus or species are used, they are underlined or italicized.

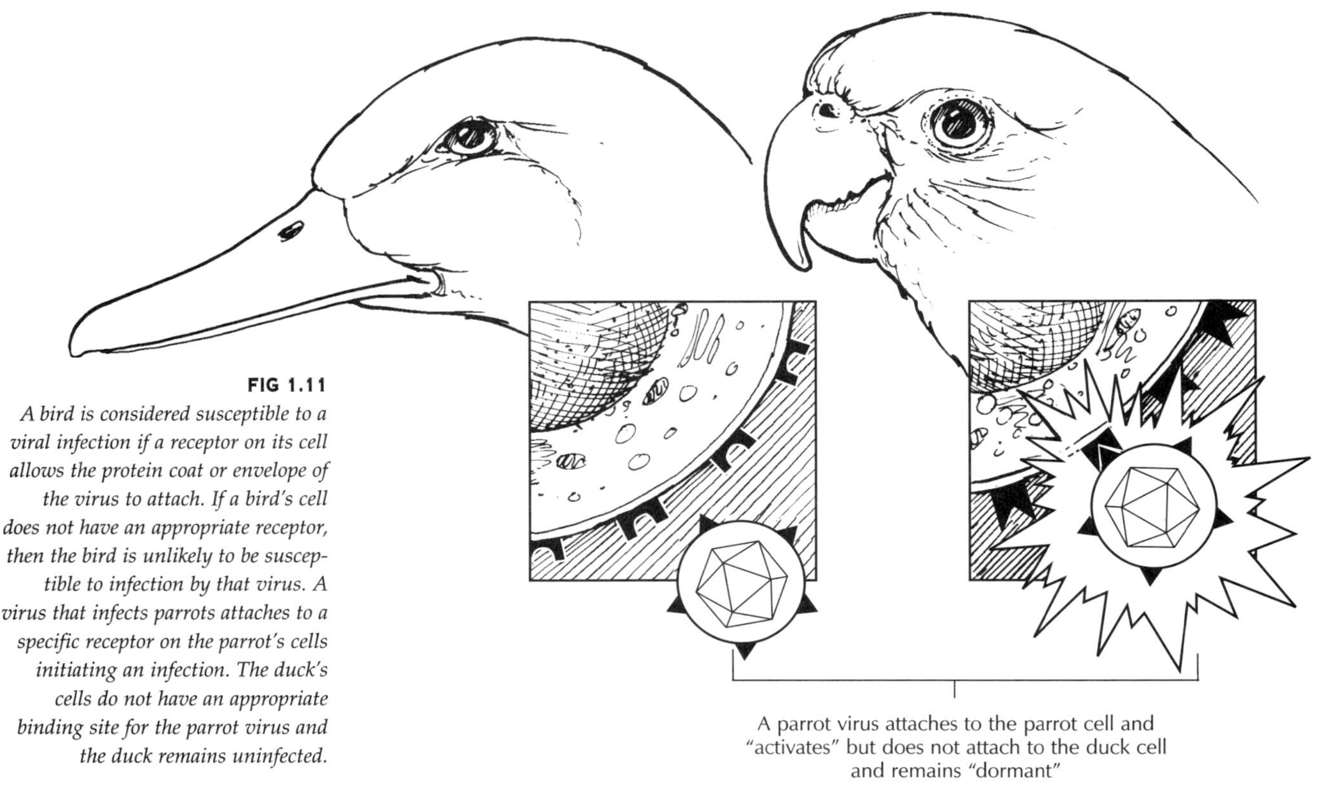

FIG 1.11

A bird is considered susceptible to a viral infection if a receptor on its cell allows the protein coat or envelope of the virus to attach. If a bird's cell does not have an appropriate receptor, then the bird is unlikely to be susceptible to infection by that virus. A virus that infects parrots attaches to a specific receptor on the parrot's cells initiating an infection. The duck's cells do not have an appropriate binding site for the parrot virus and the duck remains uninfected.

A parrot virus attaches to the parrot cell and "activates" but does not attach to the duck cell and remains "dormant"

VIRUSES IN THE ENVIRONMENT

Viruses are generally classified as being active or inactive, rather than alive or dead. A virion would be capable of entering a bird's cell and causing an infection, whereas an inactive virus particle would not be capable of infecting a bird's cell.

Some types of viruses can survive for prolonged periods in such a state of suspended animation outside the bird's body, while other types of viruses are inactivated quickly once they are released from an infected cell *Figure 1.14*. Many viruses with an envelope are quickly and easily destroyed by disinfectants, ultraviolet light and drying. Many nonenveloped viruses are resistant to harsh environmental conditions and many disinfectants; in addition, they are generally more stable than enveloped viruses and can remain infectious for longer periods when outside of a bird. For example, Pacheco's disease virus has a lipoprotein envelope, is unstable in the environment and is easy to destroy with common disinfectants; the avian polyomavirus does not have an envelope, is stable in the environment and is resistant to some commonly used disinfectants.

How do viruses compare with other microorganisms for stability? In general, bacterial and fungal spores are the most resistant to inactivation, followed in decreasing order of resistance by fungi, bacteria, nonenveloped viruses and enveloped viruses. While some types of nonenveloped viruses are considered relatively durable when outside the host, most viruses are less stable than are bacteria and other microor-

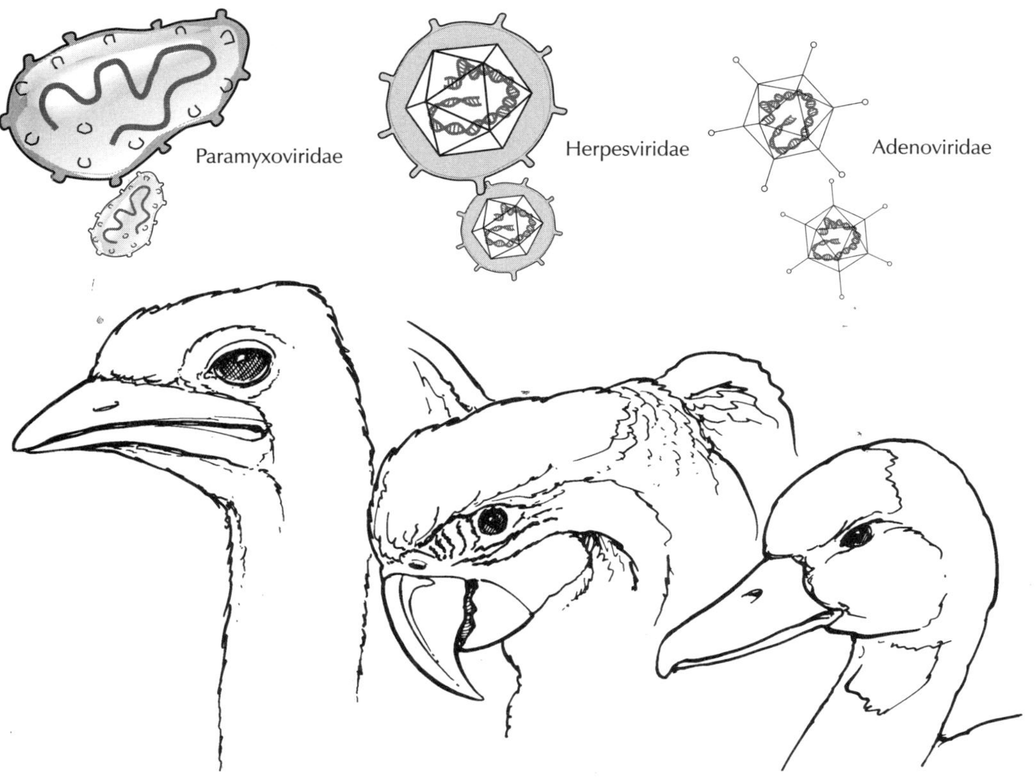

FIG 1.12

Birds that have shared characteristics are placed in the same family. Viruses are classified into families using a similar scheme. For comparison purposes, one could say that there is as little similarity among an ostrich, a macaw and a duck as there is among a paramyxovirus, a herpesvirus and an adenovirus.

ganisms. Any organic matter (soil, food, feces, blood, bedding material, mucus) that is in contact with these organisms can potentially serve as a protective matrix that increases their survival time outside the bird.

Most viruses are stable when frozen. Only a few very sensitive viruses are destroyed in the freezing process. Many viruses are destroyed when exposed to a low pH (acidic environment). However, some viruses that infect the cells that line the digestive tract (for example, picornaviruses and parvoviruses) must pass through the acidic stomach (pH 2 to 3) before they can infect a cell. Not surprisingly, these

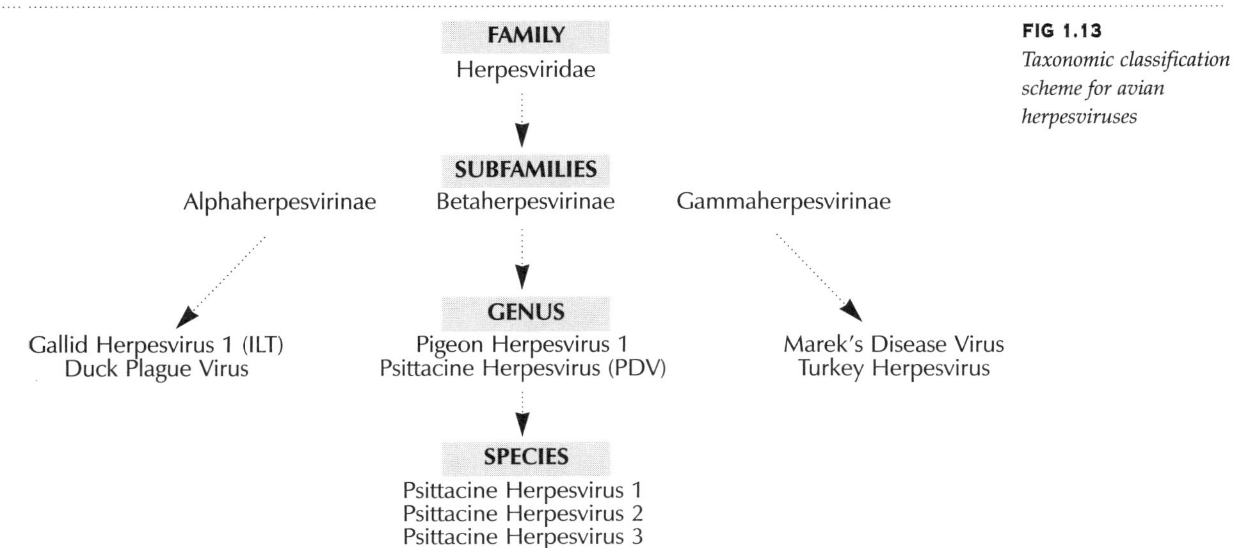

FIG 1.13

Taxonomic classification scheme for avian herpesviruses

FIG 1.14

The length of time a virus can remain infectious outside of a bird generally is determined by whether or not it has a lipoprotein envelope. Nonenveloped viruses are generally more stable, and can remain infectious for longer periods when outside a bird. Many enveloped viruses survive only days to weeks outside of a host, but some nonenveloped viruses can remain viable for over a year. Over time, even the most durable virus particle will eventually be inactivated.

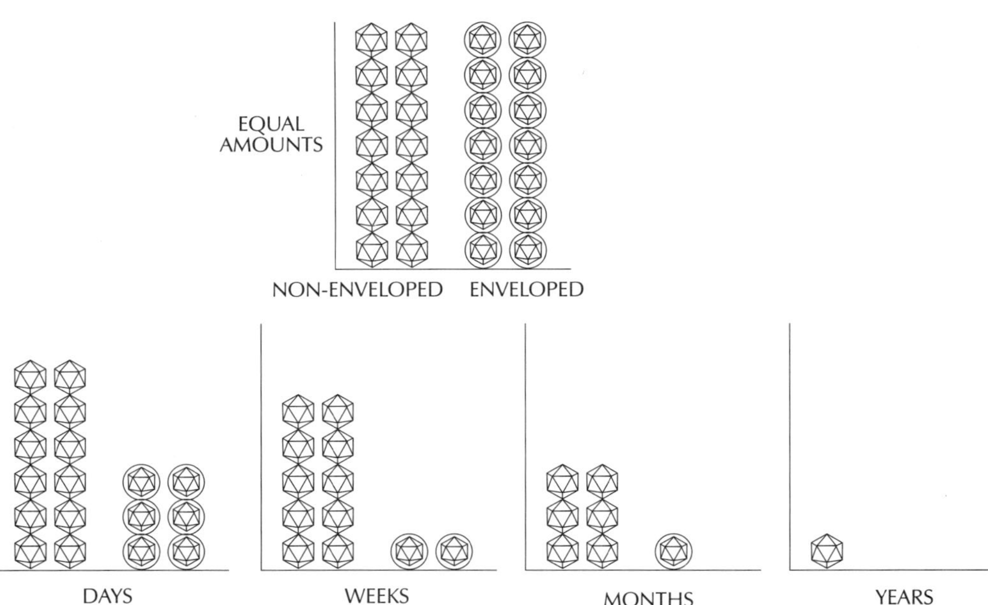

viruses are resistant to the destructive effects of stomach acids.

VIRAL INFECTION AND REPLICATION

Replication is the viral equivalent of reproduction. To survive, a virus must come in contact with a susceptible host, enter a cell and replicate to produce progeny that are released to infect new hosts. A virus' scheme for replicating may govern the way an infected bird responds to the invading virus *Figure 1.15*. Some types of viruses can complete the entire replication cycle in as little as 6 hours; others may take days to weeks. Depending on the type of virus, an infected cell may produce only a few copies of the virus, or can produce over 100,000 copies of the infecting virus.

ATTACHMENT

Attachment of a virus to a cell, the initial step in the infection process, is typically a highly specific reaction. A particular type of virus has specialized binding sites that must come in contact with appropriate receptors on a bird's cell *see Figure 1.11*. If the binding site of the virus and the receptor on the bird's cell do not meet and match, many viruses cannot enter the cell and no infection will occur.

Some types of viruses have multiple binding sites, allowing them to attach to more than one type of cell through different receptors *Figure 1.16*. These viruses frequently infect multiple tissues (such as blood cells, brain, liver, kidney and spleen) and many different species of animals. Other types of viruses have very specific binding sites that attach only to certain types of cells. These viruses usually cause infections in specific tissues (only in the liver or only in the brain, for example) and may infect only a few species of animals. Effective cleaning and disinfecting procedures can disrupt this attachment step by removing the virus from a bird's environment and damaging the

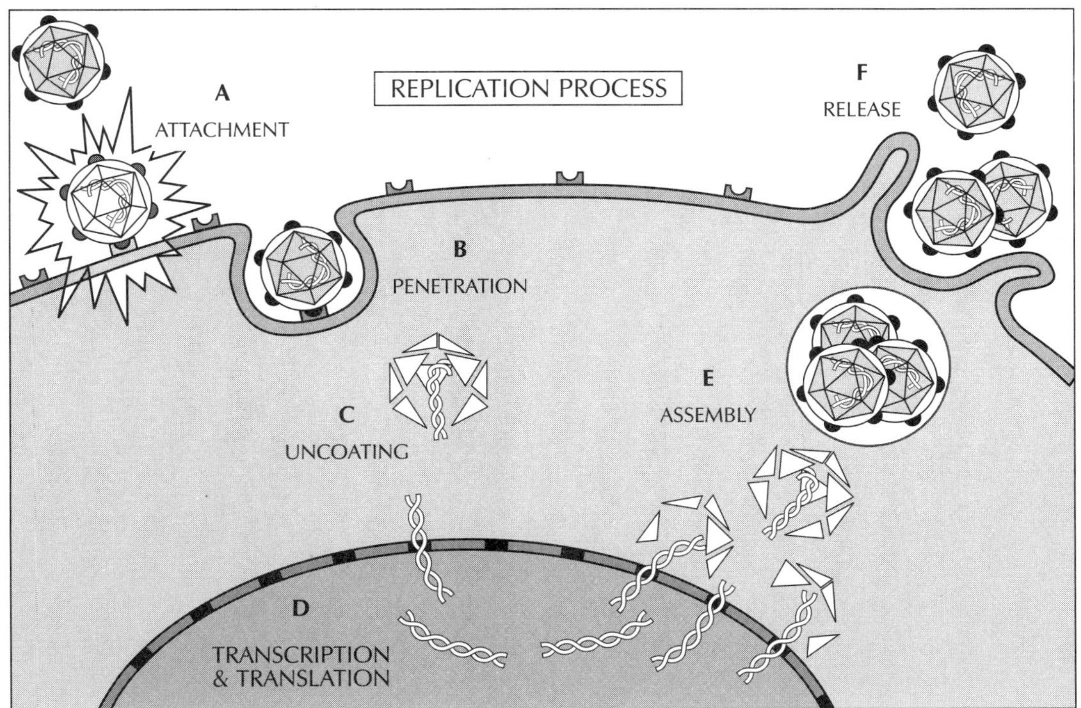

protein coat of the virus, thus damaging the binding site *Figure 1.17*.

PENETRATION

A virus attached to the appropriate receptor on the surface of an avian cell can be taken inside the cell by several methods. The most common method is called endocytosis. The process of endocytosis can be viewed as a cell "swallowing" the material that is attached to its surface.

UNCOATING, TRANSCRIPTION AND TRANSLATION, AND ASSEMBLY

Once inside the cell, the virus takes over the cell's machinery, induces the cell to produce energy for the virus and uses the nucleic acid found inside the host's cell to replicate and produce more virus particles *Figure 1.18*. The nucleic acid of the virus achieves this by altering the function of the metabolic machinery of the cell sufficiently to cause the cell to cease its normal function of supporting the life of the cell, and switch its function to producing new virus particles. Under the direction of the viral DNA or RNA, the infected bird's cell makes more copies of the virus' nucleic acid and protein coat. The newly synthesized nucleic acid molecules are then packaged into the newly created protein coats *see Figure 1.15*.

RELEASE

Depending on the family of virus, the new progeny contained within the infected cell can be released in several ways. Highly pathogenic viruses typically release their progeny by causing the infected cell to die and rupture. Other viruses that infect rapidly dividing cells that are replaced frequently, such as the cells lining the intestinal tract, may accumulate in the cell to be released when the cell undergoes a natural cyclic death. Still other viruses, particularly

FIG 1.15
Each family of viruses has a unique scheme for replicating; however, the general steps in the replication process are similar irrespective of the type of virus. These include:
A attachment of a virus to a receptor on a bird's cell;
B penetration into the interior of a bird's cell;
C uncoating or removal of the protein coat releasing the nucleic acid into the infected cell;
D transcription and translation of the virus' nucleic acid, resulting in the infected cell's producing new copies of the protein coat and nucleic acid;
E assembly of the newly formed nucleic acid and protein coat into a virus particle; and
F release of the newly formed virus particles from the infected cell.

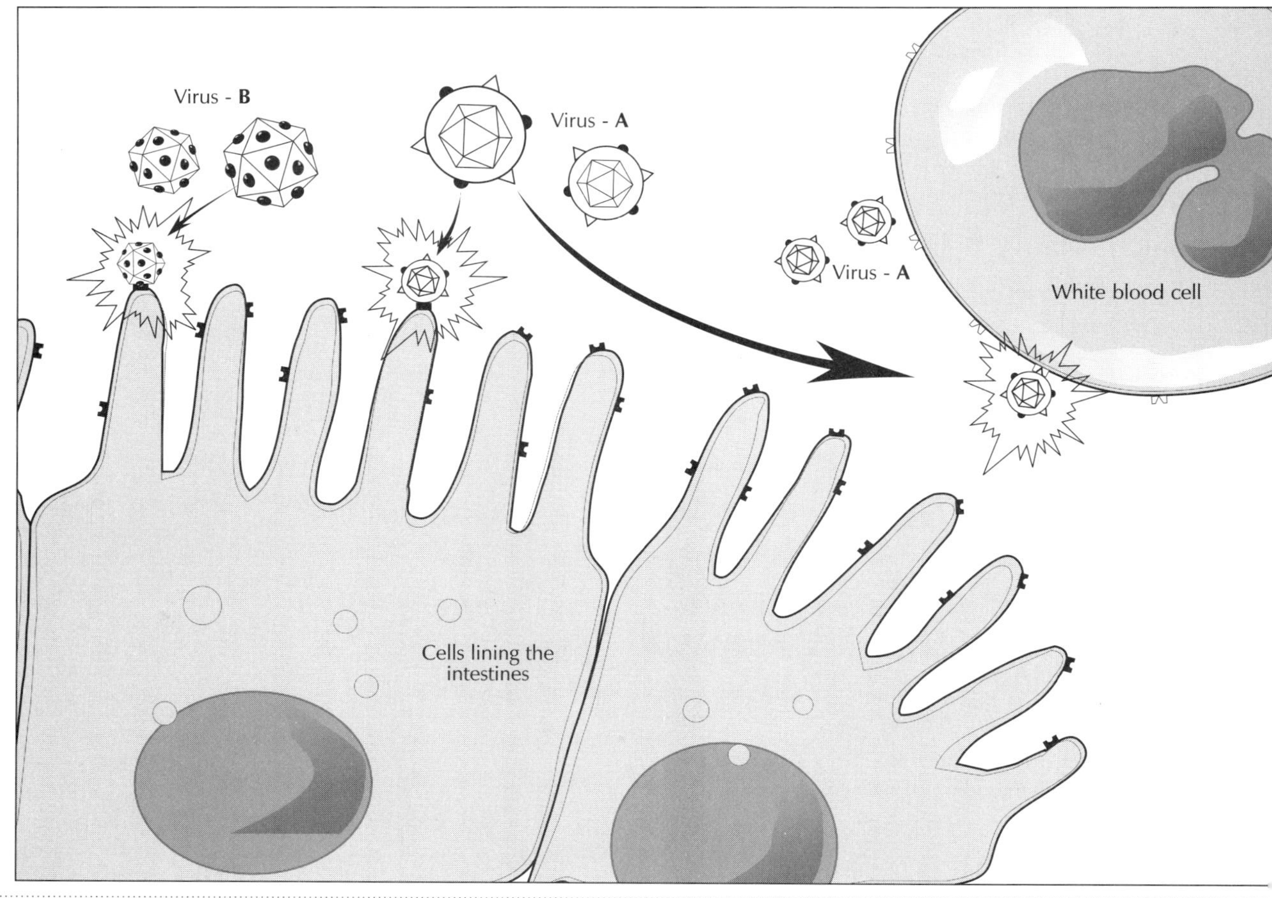

those that cause persistent infections, will be released by budding. Budding can be viewed as a cell's "regurgitating" a virus. The new progeny are released from an infected cell without fatally damaging the cell *Figure 1.19*.

Enveloped viruses obtain their envelope as they pass through the membranes of an infected cell. Some enveloped viruses, like togaviruses, can cause infected cells to lyse. Others, like retroviruses, are less damaging to the infected cell and are released without lysis. Many nonenveloped viruses are highly lytic and are released when the infected cells rupture.

Recently released virus particles either infect new cells at the site of the initial infection or are transported in the blood stream to other organs. To continue the cycle of infectivity, progeny virus produced by the infected cells must then be released to the environment so that other susceptible birds or animals will be exposed to the virus.

DIFFERENCES IN REPLICATION OF RNA AND DNA VIRUSES

The replication of RNA viruses is much simpler than the replication of DNA viruses; thus, RNA viruses frequently infect a wide variety of different hosts whereas DNA viruses are frequently host-specific. Because many RNA viruses can easily infect varying hosts, they typically cause the most severe pandemics, or outbreaks of disease that may affect a great many species of animals at the same time.

GROWING VIRUSES IN THE LABORATORY

Viruses are comparatively difficult to work with in a laboratory setting. Bacteria grow in a test tube on specially prepared medias that are easy to make, but viruses will grow only in living cells. Some viruses will grow in many types of cells; others are very selective. For example, a virus may replicate in a kidney cell but not in a spleen cell. This means that in the laboratory, and for research purposes, some viruses must be grown in specific types of living cells. Tissues derived from embryonated hens' eggs are most frequently used to establish cell cultures for the growth of viruses found in birds.

Even if a virus will grow in cell culture, the type of bird from which the virus was recovered must be used for any studies designed to determine how the virus induces disease, what lesions or clinical signs the virus causes, how the immune system responds to the virus, what diagnostic methods can be used to determine whether a bird is infected by a virus, and for the development and testing of vaccines to prevent infections.

WHERE DID VIRUSES COME FROM?

Three theories are currently put forth to explain the origin of viruses: the regressive theory,[3] the discard theory and the coevolution theory.

The regressive theory of virus origin suggests that viruses started as self-

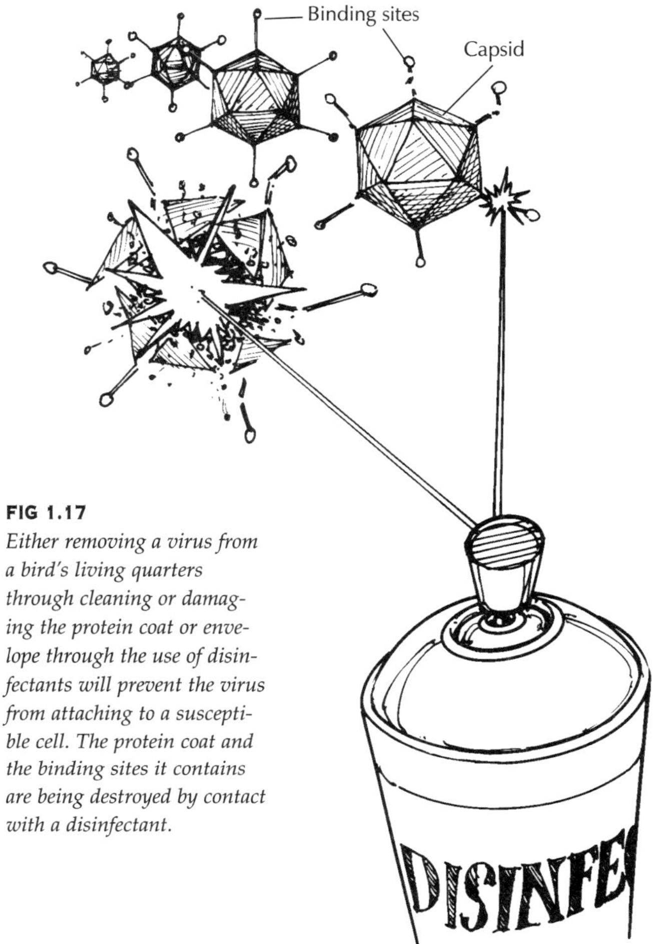

FIG 1.17
Either removing a virus from a bird's living quarters through cleaning or damaging the protein coat or envelope through the use of disinfectants will prevent the virus from attaching to a susceptible cell. The protein coat and the binding sites it contains are being destroyed by contact with a disinfectant.

contained microorganisms that evolved to become intracellular parasites. As they adapted to being intracellular parasites, viruses lost more and more of the metabolic machinery that was necessary for them to survive independently. It is difficult to explain the origin of all types of viruses using this theoretical model, and most virologists believe that it is more likely that viruses evolved by one of the other schemes.

The discard theory of virus origin is that viruses evolved from portions of host cell nucleic acid that became separated from the cell but retained an ability to replicate.

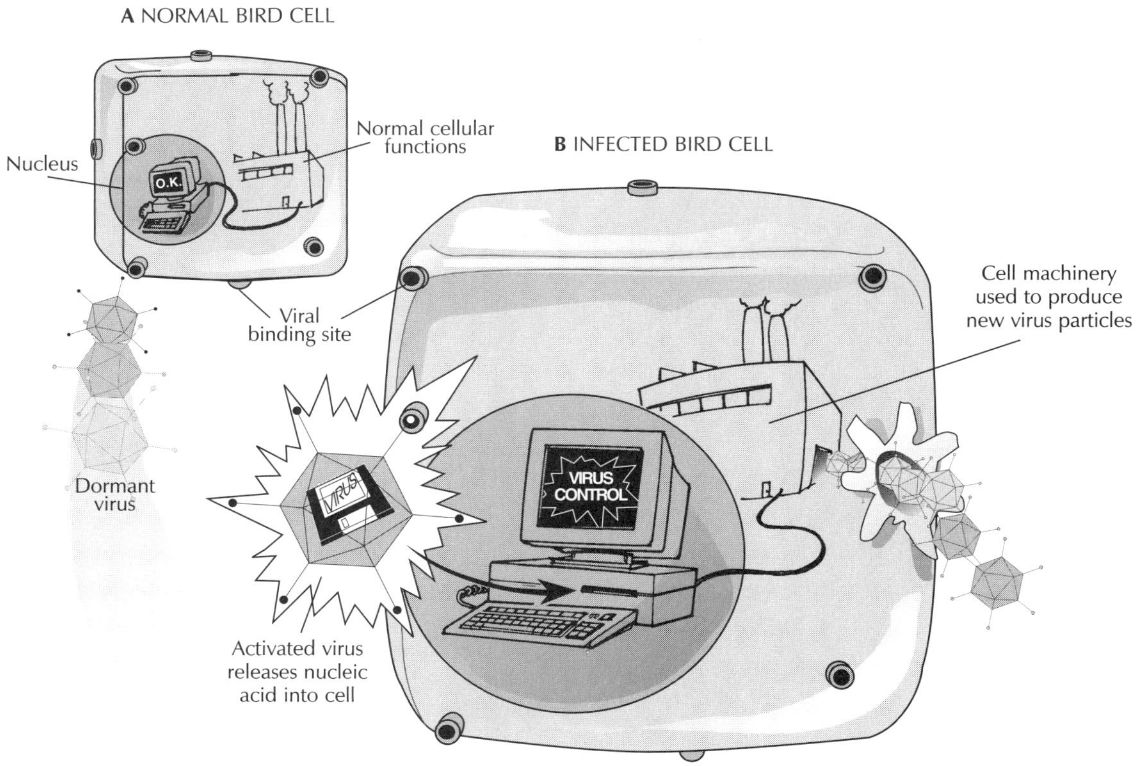

FIG 1.18

A When functioning normally, a bird's cell performs functions as determined by its nucleic acid. B When a virus enters a cell, the nucleic acid from the virus takes over the cell and instructs the cell's machinery to produce the components of new virus particles.

The coevolution theory is that both the virus and the host which it infects started an evolutionary process together as simple, self-replicating organisms. These simple forms of life then evolved in different directions, with some becoming more complex (bacteria, amebae, plants, animals) and others becoming less complex (chlamydia, rickettsiae and viruses). The viruses then evolved with the host they infected to become increasingly dependent intracellular parasites, which eventually became completely dependent on the cell of a host for their metabolic machinery.

EVOLUTION AND VIRAL ADAPTATION

It is clear that like the living organisms they infect, viruses are in a constant state of evolutionary change. Like all evolutionary changes, the adaptations that occur to viruses are driven by selection pressures that affect the survival of the virus. Because viruses are intracellular parasites, the selection pressures that affect the survivability of a population of susceptible cells or a susceptible host also affect the survival of the virus.

Virologists speculate that natural selection favors evolutionary changes that increase the transmissibility of a virus, improve survival of an infected cell, increase susceptibility of multiple types of cells, increase susceptibility of new hosts, increase survivability of a host, increase duration of virus shedding or increase virus durability outside of a host. Some viruses have evolved specific ways to avoid the host's defense system or to be unaffected by the antibodies or other viral-specific defenses produced by the host.

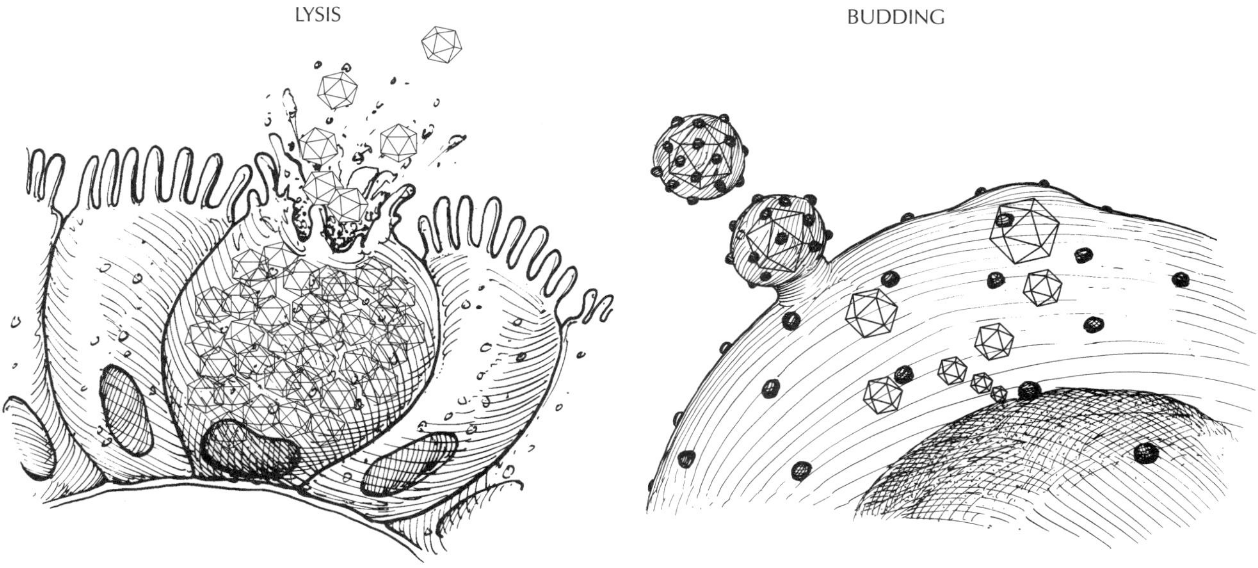

In the evolution of a virus, several structural alterations favoring survival would be expected to occur frequently. These include:

- changes in the capsid or envelope that allow a virus to infect a new species of bird.[7]

- changes in the capsid or envelope that allow a virus to reinfect a bird that has developed immunity against the original unevolved virus.[7]

- changes in the structure of a virus that allow it to more easily infect a cell or to infect a new type of cell.[6]

- changes that increase the number of infectious particles produced by an infected cell.

- changes that increase the number of progeny released from an infected bird.

- changes that improve the resistance of a virus to inactivation in the environment.

Changes in the virus that allow it to infect a new host would most drastically change the evolutionary direction of that virus. These changes could also alter the evolutionary future of the host, which becomes suddenly susceptible to a virus to which it was previously resistant. The most devastating viral disease outbreaks generally occur when a virus enters a large population of highly susceptible individuals. A virus that moves into a new host generally encounters a highly susceptible population of immunologically naive individuals, in which transmission from host-to-host is rapid and occurs unimpeded by specific defense mechanisms.

Because of their rapid replication cycle, some viruses can produce as many as 17 generations (250,000 copies) of DNA in a single infected cell during one cycle of replication.[2] The speed and efficiency of this replication cycle becomes clear when one considers that it would take at least 250 years to pro-

FIG 1.19
Some viruses are released when an infected cell lyses or undergoes a natural death. For example, the cells that line the intestinal tract are being constantly replaced. Other viruses are released from an infected cell through a process called budding.

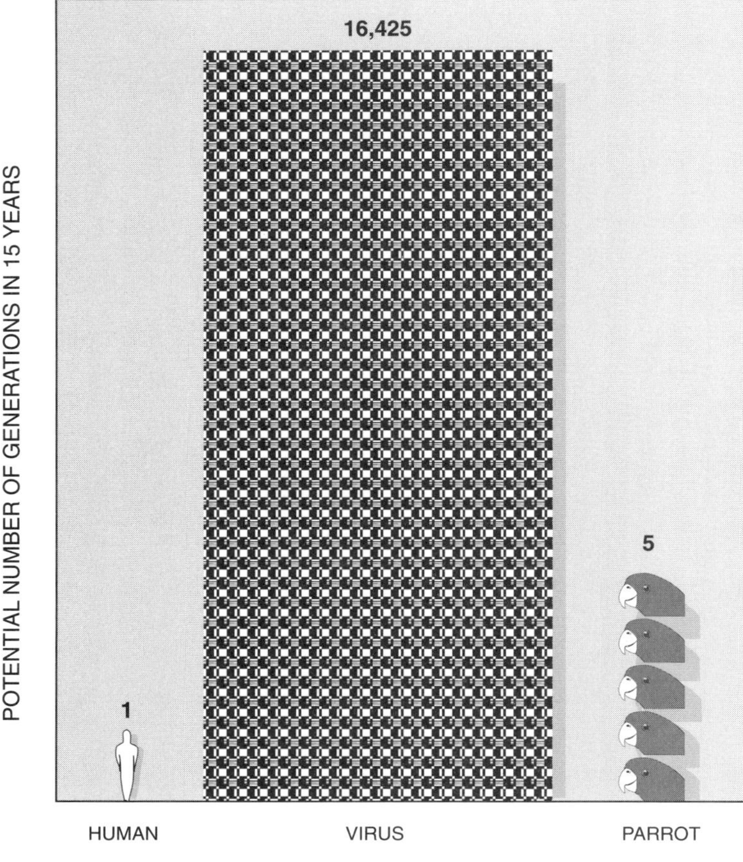

FIG 1.20

The capacity of a virus to quickly produce numerous generations of progeny allows them to evolve much more rapidly than the animals they infect. Assume that some viruses can produce 3 generations per day; some parrots, a new generation every 3 years; and humans, a new generation every 15 years. Over a 15-year period, one generation of humans, 5 generations of parrots and 16,425 generations of a virus could be produced.

duce 17 generations of humans or 68 years to produce 17 generations of African Grey Parrots. The ability of viruses to quickly produce numerous generations of progeny allows them to evolve much more rapidly than the animals they infect *Figure 1.20*.

RNA viruses have been found to be capable of changing one million times faster than higher forms of life.[1] Because of the rate of mutations that occur in RNA viruses, the progeny produced from a single infected cell can vary dramatically in their genome sequences when compared to DNA viruses *Figure 1.21*.

RECOMBINATION

Alterations that favor the survival of some viruses can occur through a process known as recombination. Recombination occurs when a cell is infected with two different types of the same virus and the progeny that the infected cell produces have portions of each of the original viruses *Figure 1.22*. This combining of the genome from two different but related viruses results in the production of a new virus that an animal's immune system has not previously encountered.

Recombination is one of the events that leads to the changes that frequently occur, for example, in the influenza A virus, which allows it to infect repeatedly a population of animals on an annual basis. Influenza viruses change frequently when the genome from one influenza virus is shared with that of another to form a completely new virus.

TREATING VIRAL INFECTIONS

Ideally, the effective antimicrobial agent damages metabolic machinery of the unicellular organism without damaging the cell membrane or metabolic machinery of the host *Figure 1.23*. Once damaged, the microbial organisms are unable to replicate and consequently die. The defense systems of the healthy bird must then remove the dead or dying invaders from the body.

Few drugs are available to destroy viruses once they have entered a bird. Because viruses depend on the metabolic pathways of the infected cell to replicate *see Figure 1.18*, any therapeutic

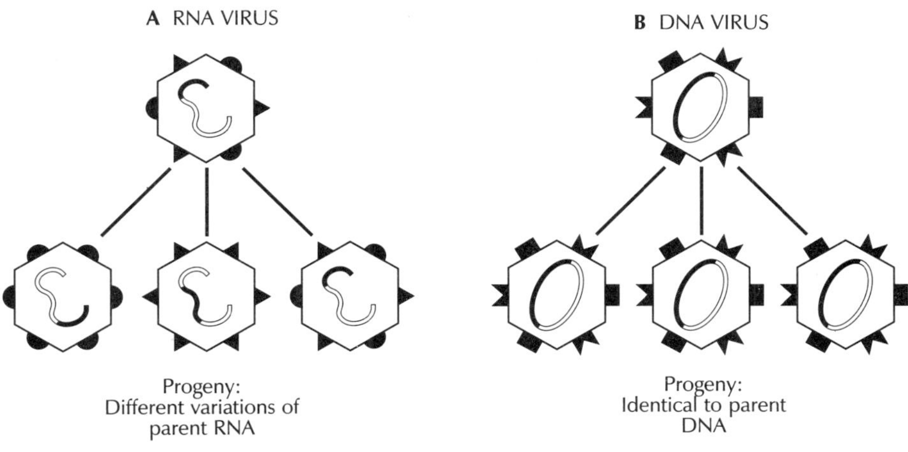

FIG 1.21

Viruses that contain RNA can mutate more quickly than viruses that contain DNA. Because of these differences in mutation rates: A the new virus particles produced from a cell infected by an RNA virus can have a different nucleic acid sequence or surface receptors from the virus particle that initiated the infection; while B the new virus particles produced from a cell infected by a DNA virus are usually the same as the virus particle that initiated the infection.

agent that damages or destroys the virus is also likely to damage or destroy the infected cell. In general, viral infections are controlled by preventing a virus from entering the body, by preventing it from attaching to a cell or through vaccination. Development of vaccines to prevent viral infections will probably continue to be the best technique for combatting this group of infectious agents.

Supportive treatment may help some birds recover from specific types of viral infections. Nonspecific therapy may include the use of immunostimulants, nutritional supplementation (including the addition of vitamin C), and, when necessary, antimicrobial agents to prevent secondary bacterial agents from taking advantage of an immunosuppressed bird.

The few antiviral agents that are being used (acyclovir for some but not all herpesviruses and AZT for some retroviruses) suppress viral replication but do not eliminate the virus from the infected host. These drugs only slow the virus' progression in an infected host.

Ongoing advances in the application of molecular biology undoubtedly will improve the clinician's ability to treat viral infections. In theory, once the specific receptor on the cell surface to which a virus binds is determined, then a competitive binder could be produced to block the receptor site and prevent a virus from attaching to a cell. This blocking action would prevent the initial attachment of the virus to the cell — the initial critical step in the infection process.

INTERFERONS

Interferons are a group of molecules produced by some virus-infected cells that function as chemical signals to notify adjacent cells to produce antiviral compounds. Because of this characteristic, they are frequently discussed as a therapeutic agent for viral infections. Interferons are being actively studied and may prove to be of value in helping cells naturally resolve some viral infections. The interferons produced by animals appear to be largely

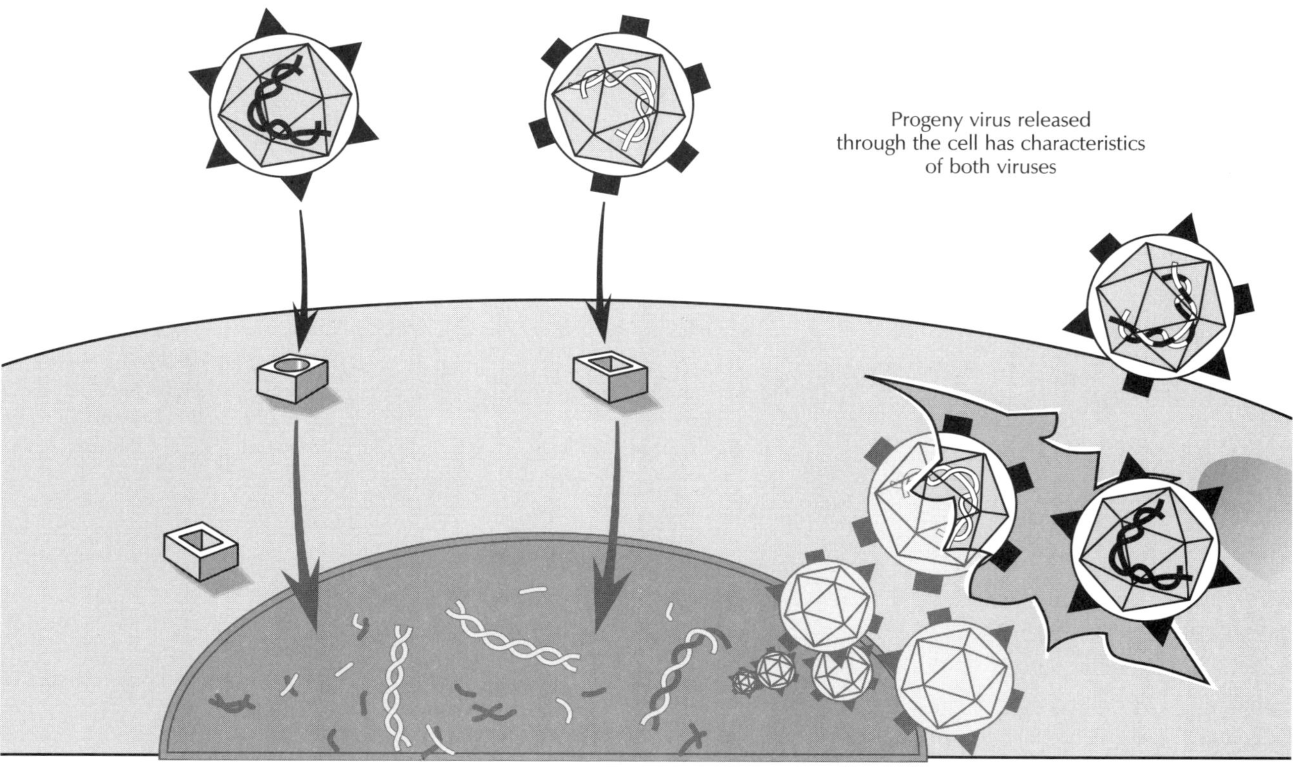

FIG 1.22

Some types of viruses can be changed through a process called recombination. This process can be viewed as a type of "viral breeding." If two similar viruses that are slightly different infect a cell at the same time, the progeny virus produced by the infected cell may have characteristics of both viruses that entered the cell.

host-specific. For example, interferons produced by humans might not be functional in other species.

Normal, nonviral-infected cells do not produce many interferons, while virus-infected cells may produce large quantities. The compounds produced in response to interferon stimulation have been shown to inhibit further replication of some viruses, stop the production of additional viral progeny and improve the function of the immune system. Combined, these effects slow the progression of virus transfer from cell to cell within the infected host. Interferons may prevent some viruses from penetrating the cell, thus stopping an infection before it has a chance to start. Interferons have also been shown

to reduce the neoplastic effects of some cancer-producing viruses.

Because interferons are produced a few hours following a viral infection, it is believed that they may initiate a primary defense response by providing the body with an opportunity to combat a viral infection before the immune system becomes fully functional, which may require several days.

Viruses vary in their capacity to stimulate the production of interferons. While RNA viruses are good inducers of interferons, DNA viruses (except for poxviruses) are poor inducers of them. Additionally, the sensitivity of viruses to interferons varies widely from one individual family of viruses to another. For example, influenza A virus is very sensitive to interferon-stimulated defense

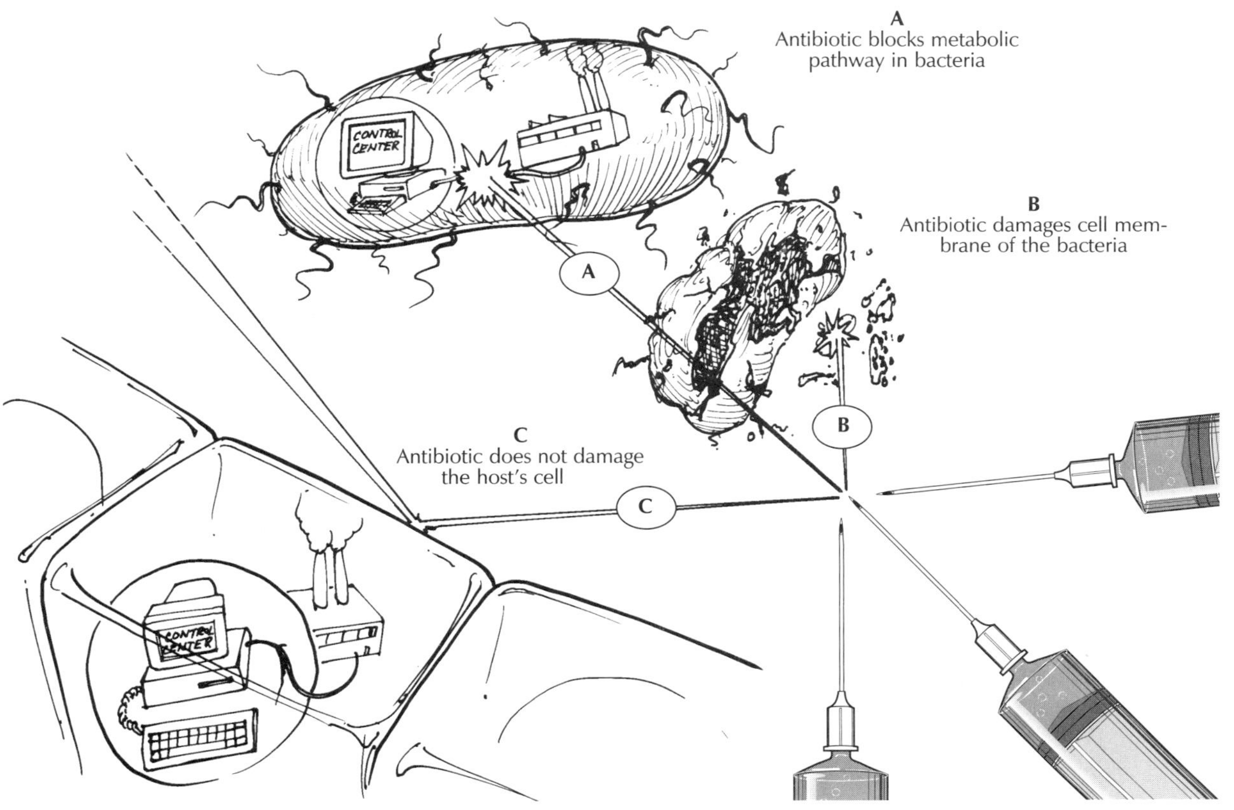

FIG 1.23

Most antimicrobials function by:
A inhibiting a specific metabolic pathway that is unique to the organism they are designed to affect; or
B damaging components of the cell wall that are unique to a microbial organism.
C In either case, the end result is that the target microorganism is damaged while minimal changes occur to the cells within the host. Because viruses depend on the cells they infect for the machinery necessary to replicate, agents that damage the metabolic pathways of the virus can also damage the metabolic pathways of normal host cells.

mechanisms, whereas poxviruses are relatively resistant to interferon-stimulated defenses.

Interferons may prove to be most helpful in preventing or eliminating persistent viral infections in some individuals, particularly those persistent viral infections associated with tumor formation. Unfortunately, interferons can also cause numerous side effects associated with organ toxicity.[5]

REFERENCES

1. Britten RJ: Rates of DNA sequence evolution differ between taxonomic groups. Science 231:1393-1398, 1986.

2. Doerfler W: The molecular biology of adenoviruses, 30 years of research 1953-1983. Curr Top Microbiol Immunol 3:109-111, 1983.

3. Lwoff A: Factors influencing the evolution of viral diseases of the cellular level and in the organism. Bacteriol Rev 23:109-24, 1959.

4. Pieper K, Kaleta EF: Virus isolations from psittacine birds. Proc Europ Chap Assoc Avian Vet, 1991 pp 199-201.

5. Quesada JR, Talpez M, Rios A, et al: Clinical toxicity of interferons in cancer patients: A review. J Clin Oncol 4:234-243, 1986.

6. Scheid A, Choppin PW: Protease activation mutants of Sendai virus: Activation of biological properties by specific proteases. Virology 69:265-277, 1976.

7. Weiner HL, Powers ML, Fields BN: Absolute linkage of virulence and central nervous system tropism of reoviruses to viral hemagglutinin. J Infect Dis 141:609-614, 1980.

Infection
and
Disease

Infection can be defined as the process of a virus entering a bird's cell and replicating. Disease includes the abnormal changes that occur in a bird's tissues as a result of a virus infecting a cell. The type of disease that occurs depends upon a bird's age, nutritional level, environmental conditions and level of stress in conjunction with the types and quantity of infectious agents to which a bird is exposed. Because a disease process is rarely caused by a single factor, it is prudent to evaluate the general health, nutritional status and environmental conditions of any diseased bird to identify factors that may be contributing to any noted abnormalities. Some of the ways a virus can cause disease within a bird are illustrated *Figure 2.1*.

DISEASE IN BIRD POPULATIONS

Epizootiology is the study of the occurrence of a disease in a population of animals. This science attempts to define how an infectious agent like a virus interacts with its host and external environmental factors to cause a disease process. An epizootic occurs when there is a recognized increase in the frequency or number of animals with a specific disease. An epornitic is an increased occurrence of disease in a single population of birds — whether it be in a pet shop, an aviary or a free-ranging flock.

Some types of viruses establish themselves in a population of birds and cause few disease-related problems but persist over long periods of time. These viruses are said to be enzootic within the affected population. A panzootic occurs if multiple populations of animals from different regions of the world experience an increased occurrence of a disease at the same time. For example, proventricular dilatation disease can be said to be panzootic because this disease is occurring with increased frequency in bird populations in North America, South America and Europe.

OUTCOME OF VIRUS EXPOSURE

Just because a bird is exposed to a virus does not mean that it will either be infected or develop disease. In fact, given the low frequency with which viruses are implicated in diseases of companion and aviary birds, it is of interest to note that the majority of avian viral infections are subclinical. Subclinical infections occur when an infectious agent enters a bird, but the infected bird develops no observable signs of having been infected. If the infected bird develops clinical signs, then it is said to be diseased *see Types of Infections and Figure 2.6.*

The fact that most viral infections are subclinical can be demonstrated by comparing the large number of birds

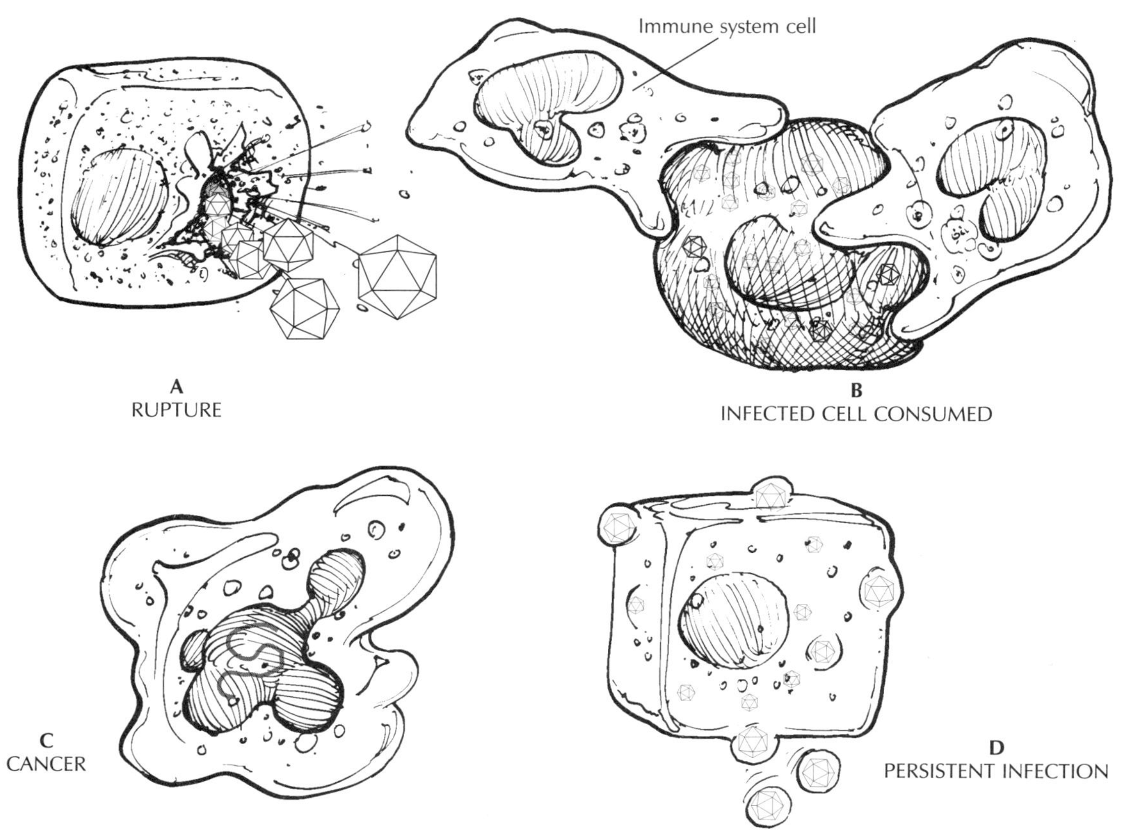

FIG 2.1

A viral infection can cause disease within a bird in several ways: A The virus can directly induce lesions in an organ by causing rupture of infected cells; B The virus may stimulate the bird's immune response, which then destroys the virus-infected cells. (If overstimulated, the immune response can severely damage the tissues of the bird leading to autoimmune disease.); C The virus may damage the infected cells, causing them to become neoplastic; D The virus may establish itself in the bird such that the infected bird becomes a persistently infected carrier of the virus.

that have antibodies to a particular virus to the small number of birds that actually develop disease. Detecting antibodies in a bird indicates that at some time in its life it has been exposed to a specific infectious agent. For example, testing of psittacine birds indicates that many individuals have antibody titers against avian polyomavirus, yet the vast majority of birds with antibodies to this virus do not show clinical signs of disease; that is, their infection remains subclinical. In comparison, young psittacine birds often do show signs of an avian polyomavirus infection and frequently die.

Several factors affiliated with a virus, its host and the environment determine whether a bird exposed to a virus will remain asymptomatic or will develop

signs of disease. The outcome of an infection depends upon a number of factors, including the virulence of the infecting virus and the health of the bird's defense system. If a given group of birds is naturally exposed to a particular virus, varying outcomes could be expected: some birds will not become infected, some birds will develop asymptomatic infections, some birds will develop clinical signs of disease followed by recovery and some birds will develop clinical signs of disease and will not recover.

Depending on the condition of the bird and the characteristics of a particular virus, either the virus will kill the bird; the bird will mount an effective immunologic response to destroy the virus and provide protection from

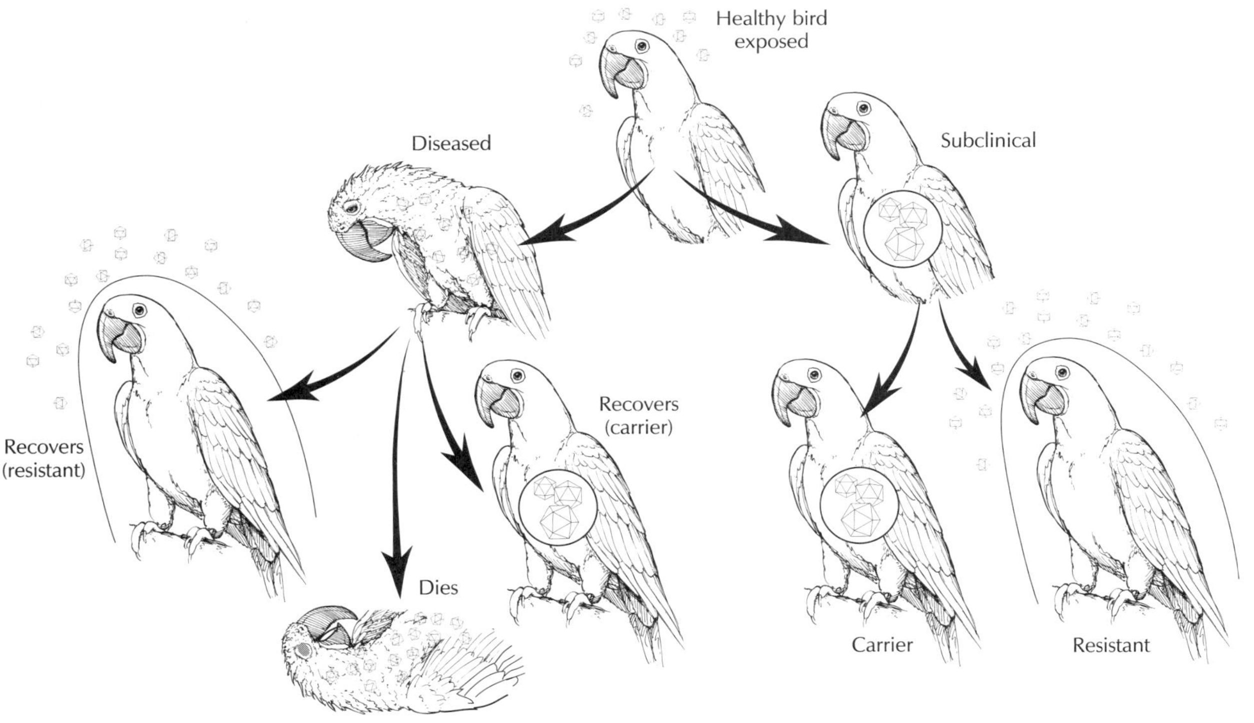

future infections; the bird will mount an immune response that results in recovery from the infection but does not protect it from future infections; or the virus may persist in the bird's body and produce a clinical disease years later. Some viruses may not be eliminated from the body following an infection, even though the bird recovers from any induced disease. These birds are considered to be persistently infected, and may experience periodic episodes of disease resurgence and virus shedding *Figure 2.2*.

Infections by various types of viruses can have different outcomes:

1) A complete productive infection, in which the virus replicates in an infected cell and causes the destruction of the infected cell when progeny virus are released.

2) A complete productive infection in which the virus replicates and progeny are released, but the infected cell is not damaged by the infection. In some cases, a persistent infection may develop.

3) A nonproductive infection in which the virus infects the cell but is unable to replicate. In this case, the viral genome that has entered the cell may be destroyed or the genome can persist in the cell. This type of infection can cause transformation in the infected cell or can cause a latent infection.

Virus-related factors that govern the changes that might occur when a virus has entered a bird include: the number of virions to which the bird is exposed, the virulence of the invading virus and the route of viral exposure *Figure 2.3*. Additionally, the exposure route must be appropriate; a virus that typically

FIG 2.2
The result of exposure to a virus depends on the age, species and condition of the bird and the characteristics of the particular virus. A healthy bird that is exposed to a virus to which it is susceptible can become infected. This infected bird can remain clinically normal or it can become obviously diseased. If the bird develops a subclinical infection, it may destroy the virus and be protected from future infections, or it can remain persistently infected (a carrier state). Likewise, a clinically infected bird may mount an effective immunologic response to destroy the virus and develop protection from future infections, or it may recover from the disease but remain persistently infected, or it may die.

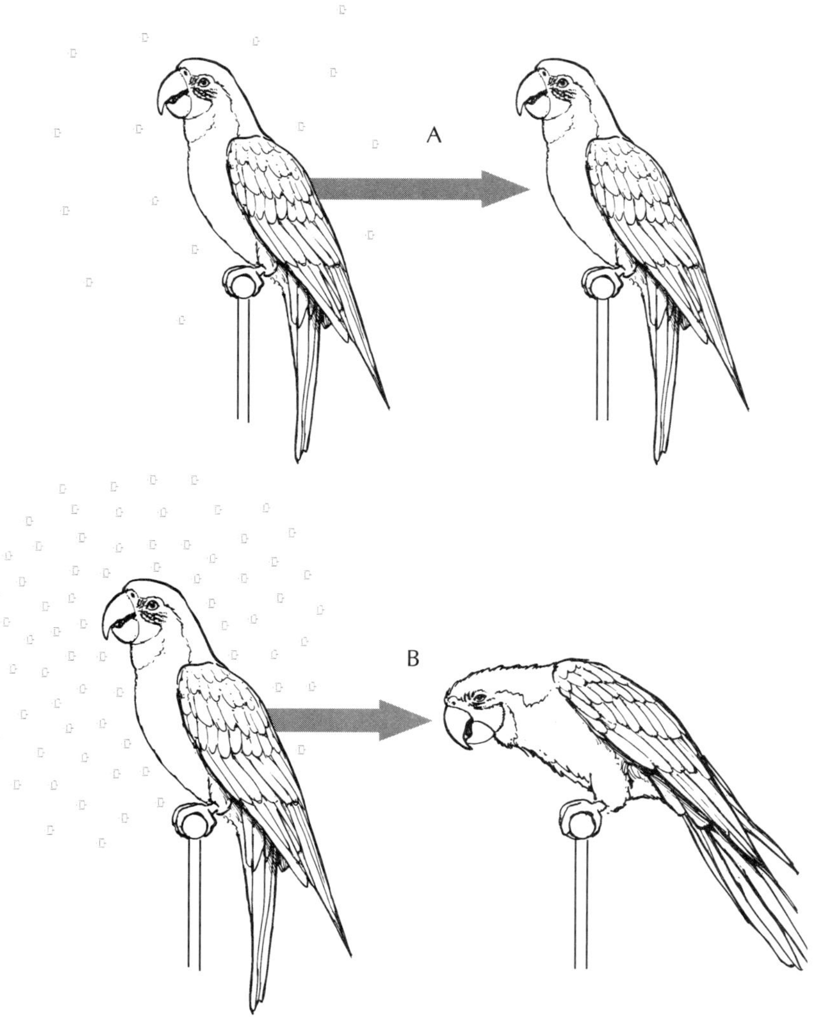

FIG 2.3

*There are numerous factors that govern what changes, if any, will occur when a bird is exposed to a virus. The most important factors are the condition of the bird, the amount of virus to which the bird is exposed and the virulence of the virus. With all other factors being equal: **A** a healthy bird that is exposed to only a few viral particles is less likely to develop disease than **B** a healthy bird exposed to a large concentration of a virus.*

enters the body through the mouth is not likely to cause an infection in a bird that inhales the virus into the lungs.

Many families of viruses are primary pathogens, capable of passing a bird's defense systems and causing disease in an otherwise healthy bird. If these primary pathogens damage the immune system, they can persist unimpeded and induce further damage within the bird's body. Other viruses do not normally cause disease in a healthy bird, but can cause abnormal changes in an immunosuppressed bird; these are called secondary pathogens.

Virulence is a term used to describe the relative ability of a virus to cause disease in an infected bird. A particular virus that is capable of producing disease is said to be virulent. A member of the same group of viruses that does not produce disease would be classified as avirulent. A small quantity of an extremely virulent virus could infect and produce disease in the healthiest of birds. By comparison, a virus that is of low virulence might not cause disease in a healthy bird, even if the bird were exposed to large concentrations of that virus.

Many viruses are selective in the type of birds they infect. A virus that infects only a particular genus or family of birds is said to be host-specific. For example, in a mixed group of birds including parrots, pheasants and finches, the Pacheco's disease virus (a parrot herpesvirus) would be expected to cause disease only in the parrots. In the same group of birds, Newcastle disease virus (an avian paramyxovirus) might infect all the birds. The host specificity of a virus is generally determined by possession of a receptor which allows it to attach to and subsequently infect a particular cell. A bird that has the receptor to a virus on its cell may be susceptible, whereas a bird that does not have the receptor on its cell could be resistant *see Figure 1.12.*

Host factors — a bird's age, species, nutritional condition, level of stress and whether the bird has been previously exposed to the same virus — play an important role in the outcome of a viral exposure. Very young birds and

FIG 2.4
A bird's age may determine whether an invading virus causes an infection or disease.
A Young birds are particularly susceptible to infections because their defense systems are incompletely developed.
B Birds are the most resistant to infections from the time they reach sexual maturity through their reproductively active years.
C As birds reach an older age (7 to 10 years for a Budgerigar or over 50 years for a macaw), they again become more susceptible to infectious agents because of a reduction in the functional capacity of the immune system.

very old birds are more likely to be infected by viruses and other microorganisms than are birds in middle age. These periods of increased susceptibility to disease are a result of the immaturity of the immune system in young birds and a poorly functioning immune system in older birds *Figure 2.4*.

In any given bird, the better the plane of nutrition (proper and sufficient diet, no deficiencies or excesses), the better the hygiene and the less stress that the bird is experiencing, the less likely it is to develop disease following viral exposure. For example, a two-year-old bird that is fed a high quality formulated diet, has a clean, dust-free environment and is secure in a household is less likely to develop disease following exposure to a virus than is a two-year-old bird that is fed an all-seed diet, frequently exposed to cigarette smoke and constantly harassed by young children *Figure 2.5*.

A virus that would normally induce a subclinical infection can cause a more severe, clinically recognizable disease in a bird that is stressed, malnourished or in suboptimum environmental conditions. It should be stressed that malnutrition can occur not only when a bird is fed a diet that is deficient in nutrients, but also if the diet has an excess of nutrients or has an improper ratio of nutrients. Malnutrition may also occur if a bird does not properly absorb ingested nutrients, if the absorbed nutrients are not properly used by the body, or if a food stuff or water supply contains detrimental contaminants such as fungal toxins, bacterial toxins, pesticides or fertilizer residues.

TYPES OF INFECTIONS

To complete a cycle of infection, a virus must be capable of surviving a sufficient amount of time outside the initially infected bird to come in contact with a new susceptible bird. The virus parti-

cle must then enter the susceptible bird, attach to the correct type of cell, take over the infected cell and produce progeny virus that are released into the environment to infect new susceptible birds.

Once a virus particle has infected a cell, it may cause a subclinical, peracute, acute or persistent infection. Persistent infections are generally divided into three groups: chronic infections, latent infections and slow infections *Figure 2.6*. It is likely that persistent viral infections will be shown to be involved in many of the diseases that affect the immune, nervous or endocrine systems, are currently of unknown etiology or appear to develop over a long period of time.

Most viral infections in birds are subclinical. The infected birds appear to be

healthy and the complete viral cycle, from the initial entrance of the virus into the bird to the shedding of new virus particles from the bird, remains unrecognized.

Peracute and acute viral infections, on the other hand, are relatively easy to characterize: they are frequently associated with a rapid, recognizable onset of abnormal changes. Peracute infections are characterized by an extremely rapid onset of disease that in many cases is fatal, as in the case of a psittacine bird infected with a virulent strain of the Pacheco's disease virus. Acute infections also are characterized by a fast onset of disease, such as occurs when poxvirus causes an acute infection in a psittacine bird. In many cases, the birds survive and the acute viral infection may be documented by demonstrating an increasing level of viral-specific antibodies in a bird with characteristic clinical signs of a particular disease.

With most viruses, the end of an infection is signaled by the removal of the virus from the body, a situation which can be detected by a cessation of clinical signs or a rise in viral-specific antibodies. However, with some viruses the end of the initial cycle of infection does not necessarily correlate with a clearance of the inciting virus from the body. These viruses remain in the body for prolonged periods of time and cause ongoing chronic infections or recurring latent infections.

Birds with chronic infections may or may not have clinical signs of disease, but virus can be consistently demonstrated in the bird or recovered from

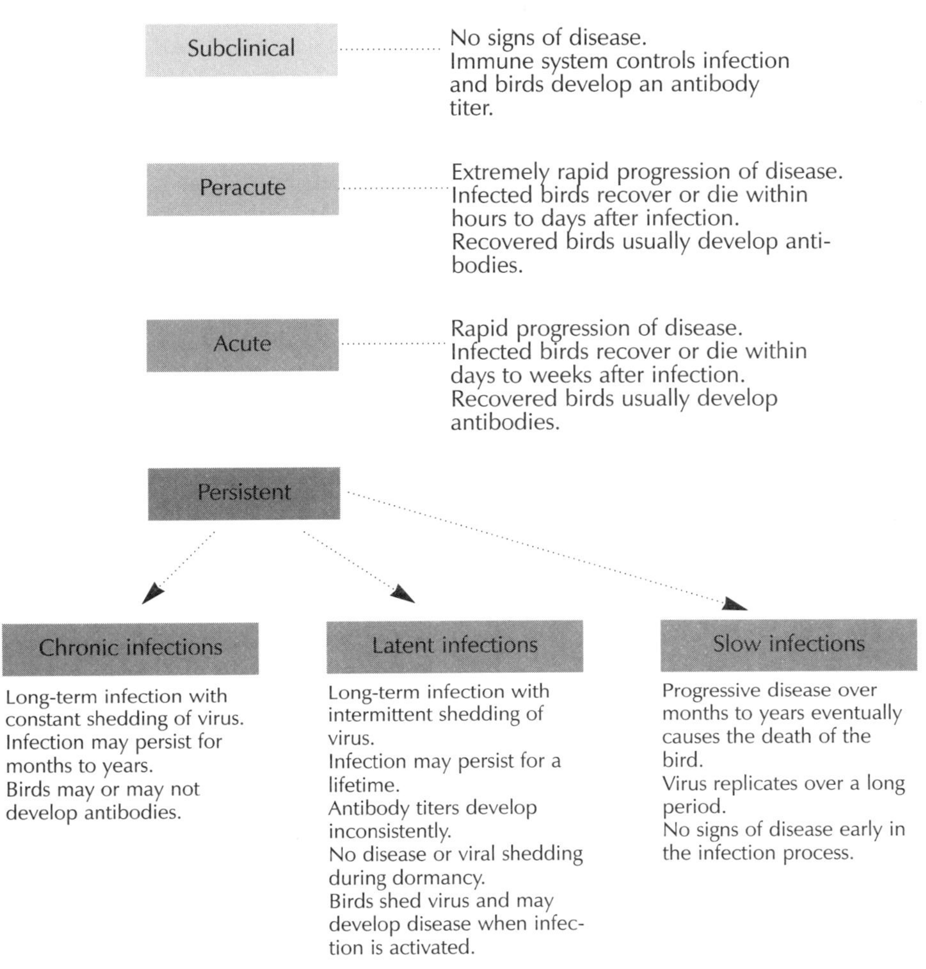

body excretions or secretions for a finite period of time. Chronically infected birds usually shed virus for periods of months to years but eventually may stop shedding virus.

With latent infections, a bird is infected but does not show any signs of disease, nor can virus be recovered from the bird during periods of virus dormancy. Virus dormancy can be viewed as a period when a virus is "sleeping" within an infected bird, and new viral particles are not being produced. At varying intervals when the viral infection becomes active ("awakens"), the bird may develop lesions or may shed virus. Some latent viral infections may cause severe disease or death in birds whose immune systems are suppressed as a result of concomitant infections, malnutrition, administration of drugs, breeding activity or environmental stresses.

With a chronic infection, virus can be consistently recovered from the infected bird, whereas with a latent infection, virus is intermittently recovered from the infected bird. For example, some birds that recover from herpesvirus infections develop lifelong latent infections. These latently infected birds can intermittently shed virus and may be responsible for infecting susceptible birds with which they come in contact while shedding virus.

Some viruses can cause both a chronic and a latent infection. This situation occurs when an infected bird consistently sheds virus for a period of time (frequently months) and the virus then enters a period of dormancy in which virus shedding stops, only to resume at some future time.

Slow infections are characterized by a long period (months to years) in which a virus is replicating but does not cause a detectable disease. As the infection progresses, the bird slowly develops clinical changes which frequently are fatal. The apparent slow progression of proventricular dilatation disease in some birds suggests that it might be caused by a slow viral infection.

Because of its capacity to induce widely differing clinical changes among various infected birds, avian polyomavirus is an interesting model for discussing the types of infections that can be caused by a virus. In some young birds, polyomavirus causes a peracute type of infection characterized by death a few hours after a bird develops clinical signs. In other birds, avian polyomavirus infections are acute, and birds die several days after developing clinical signs or slowly recover over several weeks. In Budgerigars, avian polyomavirus can cause a latent infec-

tion with periodic episodes of virus shedding by clinically normal birds.

EFFECTS OF PERSISTENT INFECTIONS

Birds with persistent viral infections are particularly dangerous, because they can serve as a source of continued virus exposure in a flock. Birds with chronic infections can frequently be detected by using a diagnostic test which demonstrates that the bird is shedding virus (eg, DNA probes, electron microscopy, virus isolation). Birds latently infected with a virus create a diagnostic and management dilemma within a flock. These latently infected birds may develop resistance to the disease induced by a virus and appear clinically normal while shedding infectious virus in their feces, urine, respiratory secretions or exfoliated cells from the feathers or skin. As these carriers shed the virus, susceptible birds in the collection develop disease. When a bird persistently infected with a particular virus is exposed to a group of immunologically naive birds, such as a nursery full of chicks, a severe disease outbreak can occur.

Many viruses that are not stable outside of a bird's body have evolved to cause long-term, latent infections. This is the mechanism of survival for most herpesviruses. The latent infections caused by various herpesviruses can be recognized clinically as an occasional outbreak of Pacheco's disease virus in a flock of captive, psittacine birds or massive die-offs of waterfowl on lakes or ponds contaminated with duck plague virus (a waterfowl herpesvirus).

Viruses that cause persistent, and particularly chronic, infections are favored from a survival standpoint *Table 2.1*. A bird with an acute viral infection may be clinically ill and shed virus for a period of days to weeks depending on the virus. There is, of course, a grave potential for the number of new birds that could be exposed to this bird while it is shedding virus, but when it dies, the number of other individuals that can come in contact with it is dramatically reduced. In contrast, a chronically infected bird will shed virus for prolonged periods of time, in some cases years. During this period of prolonged shedding, the chronically infected bird may expose hundreds of new birds or new geographic areas to infectious virus.

TABLE 2.1

Viruses that have been shown to establish persistent infections in some host

Adenoviruses
Arenaviriuses
Hepadnaviruses
Herpesviruses
Lentiviruses (Retroviridae)
Papillomaviruses
Paramyxoviruses
Parvovirus
Picornaviridae
Polyomaviruses
Psittacine beak and feather disease virus

Viruses that establish chronic or latent infections are generally well adapted, indicating a long history of virus-host interactions. The fact that Pacheco's disease virus causes rapidly fatal infections in some infected birds suggests that this virus is infecting an immunologically naive host. It is possible that various genera of psittacine birds have their own host-adapted herpesvirus,

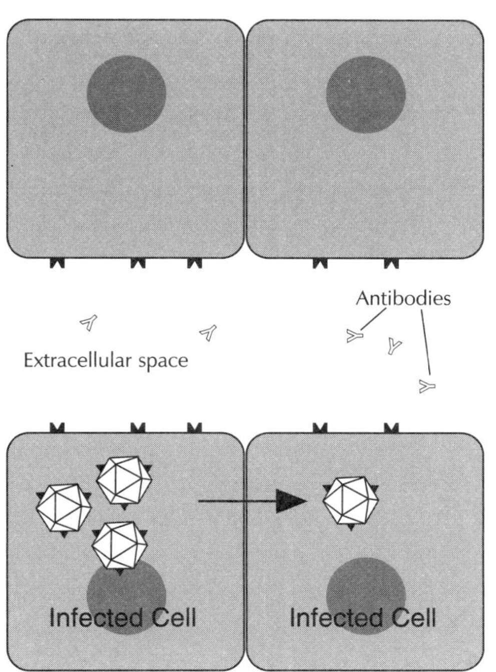

FIG 2.7
Antibodies generated in response to a viral infection help clear the invading organism and protect the body from future infections by the same virus. Some viruses have developed mechanisms that allow them to persist even though the infected bird develops antibodies to the virus. These viruses persist by causing minimal or no damage to infected cells and by spreading directly from cell to cell without being exposed to circulating antibodies found in fluids outside the cell.

and infected birds develop severe disease only when they are exposed to a herpesvirus that normally infects a different genus of birds.

Persistent infections are generally induced by viruses that cause minimal or no damage to infected cells and are spread from cell to cell without being exposed to circulating antibodies *Figure 2.7*. Viruses that cause death of the cells they infect can establish persistent infections by infecting only a small number of cells within an organ, by affecting cells with a high capacity for rejuvenation, such as skin, or by infecting cells in organs that can withstand a substantial amount of damage, such as the liver or digestive tract.

INDUCTION OF PERSISTENT INFECTIONS

Birds may develop persistent viral infections if they produce an incomplete antibody response, or have a weak or partially damaged immune system, if the virus is able to persist inside certain cells of the immune system (macrophages, B-lymphocytes, T-lymphocytes) or if the virus infects cells that are in an area protected from the immune system. The latter is called an immunologically privileged site and includes nervous tissue, epithelial surfaces of the kidney and possibly the developing feather. Some types of viruses are able to persist by inserting their own nucleic acid into the nucleic acid of the bird. This is how some types of viruses are able to remain dormant for years and then suddenly cause disease in a bird that appeared normal.

Viruses that cause persistent infections have two important factors in common. First, they have evolved mechanisms that allow them to avoid the bird's immune system. Second, they cause minimal or no damage to infected cells, so that the infected bird does not die. The latter can occur if a virus adapts to an infected cell in such a way that the cell is not damaged by the virus, or if the cell adapts to the virus in such a way that it is not adversely affected by the virus.

Some viruses able to persist inside the cells of the immune system may severely weaken the bird's defense mechanisms, causing the bird to die from bacterial, fungal or other viral infections that they would otherwise be able to fight off.

In general, viruses cause changes on the surface of the cells they infect, sending a signal to the immune system that the cell is infected and should be destroyed. Some viruses are able to persist by not causing these changes on the surface of the cells they infect. Because there is no change on the cell surface, the bird's immune system does not know that the cell is infected.

Viruses that replicate in or persist within macrophages are able to pass one of the major defense mechanisms of the immune system. Macrophages are the cells responsible for consuming viruses and activating the immune response. Viruses that are known to persist in macrophages include psittacine beak and feather disease virus, some togaviruses, some poxviruses, some lentiviruses, some coronaviruses and reoviruses.

Depending on the type of virus, some birds with latent infections may have a partially damaged immune system that causes them to be constantly ill, or they may develop tumors that originate in the persistently infected cells. Tumor-inducing viruses in birds include retrovirus, herpesvirus, poxvirus and papillomavirus. Depending on the inciting virus, induced tumors may be benign and localized, or they can be invasive and metastatic.

THE BIRD-VIRUS INTERACTION

HOST SUSCEPTIBILITY

The frequency of problems caused by viral infections in companion and aviary birds is due to the mixing of birds from numerous sources, the shift in aviculture to high-density, confined breeding operations, the inability to screen for subclinical carriers of many viruses and the relative lack of vaccines *Figure 2.8*.

The global movement of numerous avian species will continue to result in the mixing of birds carrying different indigenous microbial flora. Mixing birds from different breeders can cause a similar exposure, on a smaller scale, to previously unencountered infectious organisms.

Viruses that are endemic to birds from other continents are particularly dangerous when introduced to a new area because free-ranging species of birds in a related genera may also be susceptible to infection. For example, viruses indigenous to Passeriformes that are normally found in Africa may also infect Passeriformes that are found in North America. If the Passeriformes in North America were immunologically naive, these birds would be particularly susceptible to infections, and high levels of morbidity or mortality might occur when these birds were exposed to the foreign virus. A similar scenario occurred in humans when Europeans first introduced the smallpox virus to the American Indians. The smallpox epidemic that ensued decimated the native human population of North America, while the Europeans who had experienced this virus over time "lived with it."

The introduction of a new virus to a highly susceptible population of birds is frequently characterized by severe disease with rapid spread among the exposed birds. In some cases, the high frequency of disease is also associated with a high level of death. As the pop-

FIG 2.8
Viral infections are common in avicultural settings because birds from varying sources are frequently mixed together and many aviculturists use high-density, confined breeding operations. Maintaining numerous birds in close proximity will increase the likelihood that a virus can be transmitted from bird to bird.

ulation of susceptible birds evolves with the newly introduced virus, the character of an infection may change and the severity of disease may decrease. This can be recognized clinically by a reduction in the number of infected birds that are either sick or die. In any population, neonates and older birds should be considered the most susceptible to infection.

HOST SPECIFICITY

Many viruses are restricted in the species of animals they infect. For example, the herpesviruses that infect mammals are not known to infect birds, and the herpesviruses that infect birds are not known to infect mammals. Newcastle disease virus and influenza A virus are examples of viruses that may be capable of infecting numerous species of birds and mammals, including humans.

Viruses that are capable of being transmitted between animals and humans are called zoonotic viruses. The few viruses that infect both birds and humans are listed in *Table 2.2*. These viral diseases are of particular concern to the aviculturist because of their potential negative impact on public health.

TABLE 2.2

Viruses that have been shown to infect birds and humans

Bunyaviridae - Birds serve as virus reservoir for insect vector that can infect mammals

Flaviviruses - Birds serve as virus reservoir for insect vector that can infect mammals

Influenza A virus - Causes "flu" in humans and birds

Newcastle disease virus - Causes mild flu-like signs in humans

Togaviruses - Birds serve as a virus reservoir, mosquito transmitted, causes encephalomyelitis in humans

Host specificity frequently influences the type of disease that develops when two birds from related but different genera are exposed. For example, Blue-fronted Amazon Parrots exposed to poxvirus often develop severe upper respiratory disease and may die, while other species of *Amazona* exposed to the same virus may develop only mild lesions (papules) involving the skin around the eyes, face and mouth.

HOST RESPONSE TO A VIRUS INFECTION

The ideal host-parasite relationship is one in which the parasite infects the host and causes little or no damage. This allows the parasite to replicate freely in the host, producing an abundance of progeny. If a virus destroys the cell it is infecting and subsequently causes the death of the bird, then the virus loses its life support for replication. If the progeny virions that have already been produced are not able to infect new living cells within a specified period of time, the new virions may become inactivated. Thus, viruses that kill the birds they are infecting are less likely to be spread to a new bird than viruses that cause chronic infections and are shed from the affected bird for long periods of time.

When a virus and a bird meet, a battle ensues between the virus and the bird, with the virus attempting to enter the body and establish and maintain an infection, and the bird's defense system attempting to clear the infection. In most cases, a naive, susceptible bird that is exposed to a virus through the proper route, such as Pacheco's disease virus through the gastrointestinal tract, will become infected. However, depending on the encounter and the type of virus, a bird may be:

- exposed to the virus and remain uninfected; these birds will not develop a detectable antibody response.

- exposed to the virus and become infected, but remain clinically normal. These birds may become persistently infected or develop resistance to future infections by the same virus. These birds usually develop a detectable antibody response.

- exposed to the virus, become infected and develop abnormalities associated with the infection. These birds may recover and be resistant to future infections by the same virus or they may become persistently infected or may die. These birds usually develop a detectable antibody titer if they live, but they may not develop antibody titers if they die rapidly.

If the immune response mounted against a viral infection is slow, the

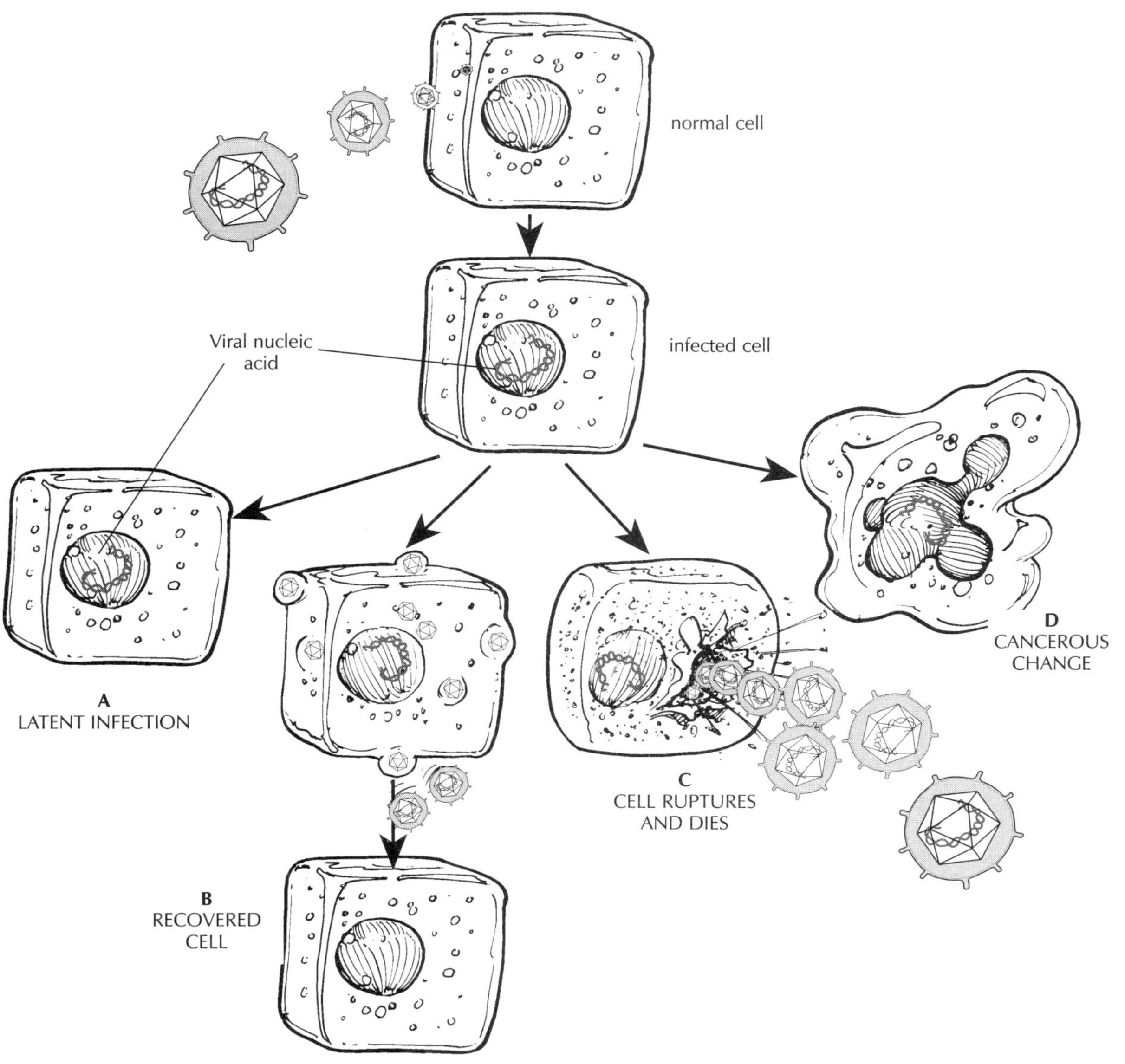

FIG 2.9

A viral infection can have multiple effects on a cell. A virus can:
A cause a latent infection in which the virus is dormant for extended periods of time;
B infect a cell and be released by budding so that the cell remains intact;
C infect a cell and be released by causing lysis of the cell; or
D cause the infected cell to become cancerous.

virus may damage the body so severely that the bird dies. A virus may also infect a bird and cause temporary or permanent damage to the immune system, which prevents the immune system from protecting the bird from normally encountered bacteria, fungi and parasites. These opportunistic or secondary pathogens can take advantage of the compromised immune system and induce disease. A poor immune response may also allow the virus to persist so that the bird becomes a carrier. Viral infections in the respiratory tract, intestinal tract and skin are generally characterized by a rapid disease onset in which the bird either mounts a sufficient, rapid immune response and recovers, or mounts a slow, poor immune response and dies.

The interactions that occur between a virus and a bird are in a constant state of evolution. Viruses that cause the most severe damage to a bird are generally thought to be those that have been recently introduced to a new pop-

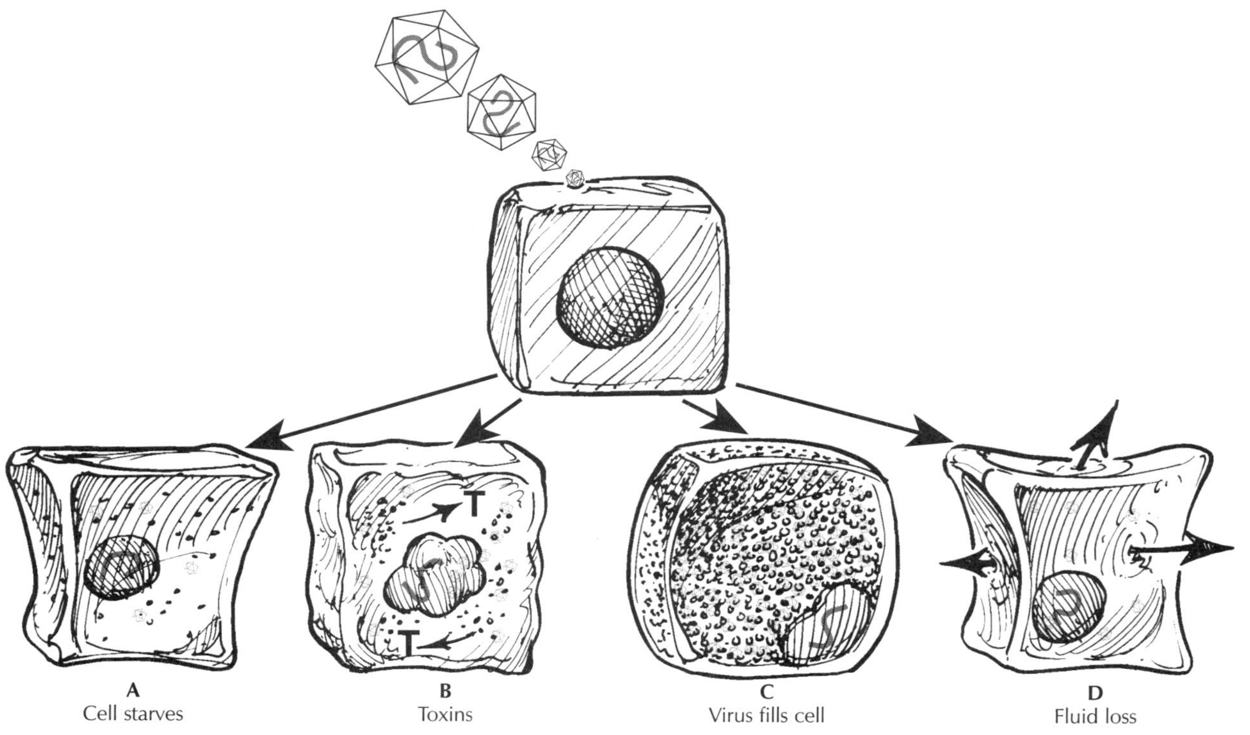

FIG 2.10

Viruses can damage the cells they infect in several ways. These include: A using the cells' metabolic pathways for virus functions (effect: the cell basically starves); B producing virus components which are toxic to the cell (the virus protein coat may be particularly toxic); C producing large numbers of virus particles that occupy too much space in the cell (effect: the cell cannot function or ruptures); and D altering a cell's membrane, which causes it to be more permeable (effect: the cell membrane develops leaks and loses vital fluids).

ulation of highly susceptible birds. Viruses that cause less severe or no damage to a bird are thought to have been infecting a population of susceptible birds for eons. This long-term association between the virus and the bird has resulted in a favorable compromise in which the infected bird lives and the well adapted virus is able to replicate and produce abundant progeny that are released to infect new, susceptible birds.

In mammals and poultry, viruses that establish persistent infections have been found to be responsible for a number of reproductive problems. It is likely that the same situation is occurring in companion and aviary birds, and that reproductive failures may be warning signs of previous or ongoing infectious diseases.

EFFECTS OF VIRAL INFECTIONS ON CELLS

Viral infections can cause a number of changes in an infected cell. These include death by rupture or lysis, transformation, formation of inclusion bodies, hyperplasia or persistence of the virus *Figure 2.9*. These changes may occur alone or in combination. It is the changes that a virus causes in a cell's structure that are searched for when necropsy or biopsy samples of a diseased bird are sent to a laboratory for histologic evaluation. Some, but by no means all, viruses produce inclusion bodies within an infected cell. In general, these inclusion bodies can be viewed as "factories" for the production or storage of large quantities of new virus components or virus particles. Some inclusion bodies, particularly those in the nucleus, represent visual damage to the infected cell. The viruses that cause inclusion bodies are the easiest to detect using a microscopic examination of infected tissues *Table 2.3*.

Viruses can damage the cells they infect in several ways *Figure 2.10*. Bacteria, fungi and protozoa frequently produce toxins that adversely affect the cells and tissues of infected animals. Viruses per se do not produce any specific toxins; however, some of the proteins that are components of the virus can be toxic to infected cells. In general, viruses cause problems in an infected host by adversely changing the function of the metabolic machinery of the infected cell. As a virus takes over the machinery of an infected cell, the changes that occur can cause the cell to stop functioning properly, a situation which leads to death of the cell.

The process of leaving the cell also can be damaging. Viruses that have an envelope typically leave an infected cell through a process called budding that may not destroy the cell. Viruses that do not have an envelope generally leave an infected cell by causing the cell to lyse. Obviously, the viruses that cause an infected cell to lyse would be expected to cause more damage to the bird than viruses that exit the bird's cell by budding.

Viruses may further damage a bird by initiating an autodestruction process that tells the body's defense system to eliminate virus-infected cells. A bird can die as an outcome of its own immune system attempting to eliminate a virus from the body. When a large number of cells in a critical organ (brain, liver, heart or kidney) are determined by the immune system to be infected and are subsequently targeted for destruction by the immune system, the result can be lethal *see Figure 2.13*.

Viruses that destroy the cells they infect generally cause acute diseases with clinical changes associated with the organ system that is affected. For example, the rupture of liver cells caused by Pacheco's disease virus causes clinical changes associated with liver disease. Viruses that cause more chronic infections generally alter the function of cells and create a slowly progressive disease process. The papillomaviruses are an example of a group of viruses that cause slow, progressive changes (warts) in infected hosts.

Viruses that do not destroy the cells they infect can still have a negative effect on the cell. These negative effects, which typically occur over time, include pre-

TABLE 2.3

Virus families that cause inclusion bodies and their cellular location

Virus	Cellular Location of Inclusion Bodies
Adenoviridae	intranuclear inclusions contain virus, intracytoplasmic inclusions do not

Circoviridae
  PBFDV
  CAV
	
	intranuclear and intracytoplasmic
	intranuclear, identified inconsistently

Herpesviridae	intranuclear
Papovaviridae Polyomavirus	 intranuclear
Paramyxoviridae	intranuclear and intracytoplasmic
Parvoviridae	intranuclear
Poxviridae	intracytoplasmic
Reoviridae	intracytoplasmic
Rhabdoviridae	intracytoplasmic

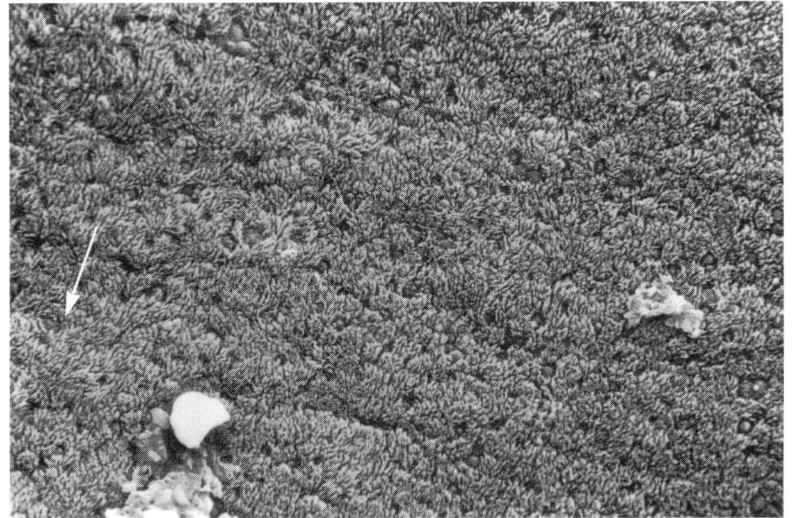

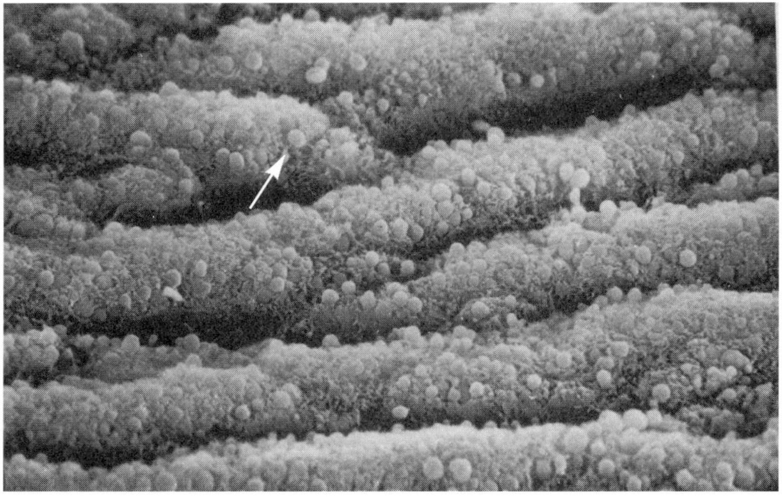

Another example of virus-induced damage occurs with some viruses that infect the cells that line the trachea. This specialized group of cells secretes mucus and contains cilia on the surface *Figure 2.11*. The mucus secreted by these cells is responsible for trapping inhaled foreign material that is then transported up the trachea by the cilia and expelled from the body as a bird coughs. Some viruses that infect the ciliated cells that line the trachea will cause the cilia to stop functioning, decreasing a bird's ability to remove inhaled material, such as bacteria, fungal spores and dust. Until the tracheal lining is able to regenerate, the bird has increased susceptibility to infectious agents. Exposing a bird to aerosolized chemicals, particularly some disinfectant fumes, cigarette smoke or any aerosolized toxin could also damage the ciliated cells that line the trachea and make a bird more susceptible to infectious agents that attack or enter the body through the respiratory system.

CANCER

Cancer can be viewed as a process in which a cell begins to grow progressively and persistently without regulation. Some viruses can cause cells to be transformed (become cancerous); these specific types of viruses are potent inducers of cancer *Table 2.4*. Viruses induce cancer by altering the structure of the nucleic acid within the infected cell. With most cancer-causing viruses, the nucleic acid of the infected cell is damaged when the virus' nucleic acid becomes a part of the nucleic acid of the infected cell *Figure 2.12*.

venting a cell from secreting products that are important to the bird or altering the mechanical function of a cell. Either of these changes can have dramatic effects on the bird, particularly when the damaged cell is part of the immune system. For example, lymphocytes, the blood cells responsible for secreting antibodies and for destroying viruses once they are inside the body, are critical components of the immune response *see Figure 3.18*. A virus that infects lymphocytes may cause these cells to be destroyed or to stop functioning properly. Once these cells are damaged, the immunosuppressed bird is more susceptible to other infectious agents.

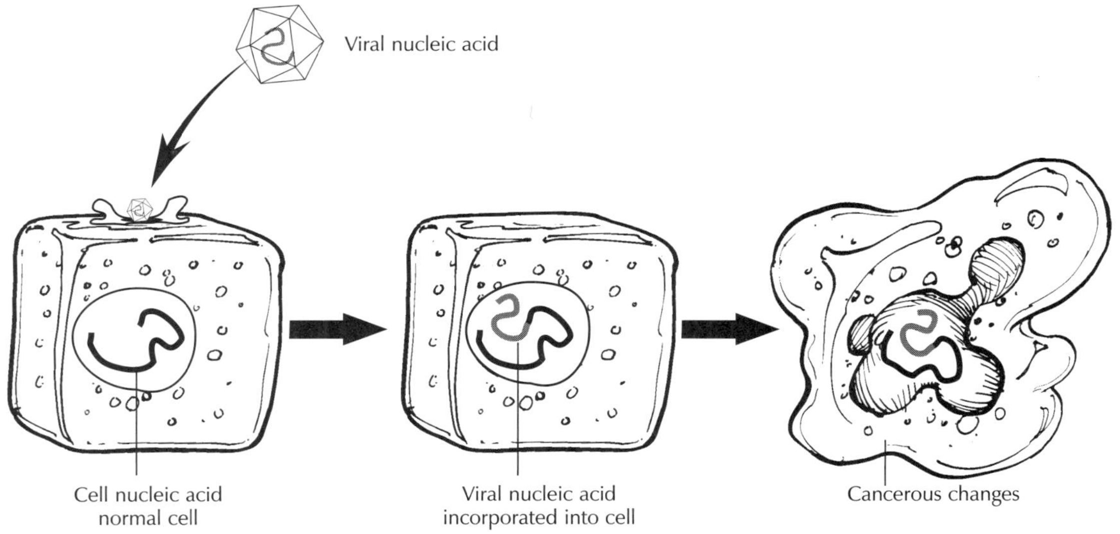

In some cases, viruses may induce benign tumors that become malignant because of the environment in which the tumor cell is growing. Papillomaviruses, for example, induce a benign tumor that may become malignant, particularly when present in the female reproductive tract.

The viral nucleic acid present in the affected cell is usually noninfectious at the time the cell becomes cancerous. Once the virus has caused the genetic change that induces tumor formation, the alteration is irreversible and the cell continues to be neoplastic even though the virus is no longer replicating. Retroviruses are the most intriguing of the cancer-inducing viruses because these viruses can cause cells to become neoplastic, irrespective of tumor growth, and conversely, tumor growth can occur irrespective of viral replication.[2] Most cancerous changes that are caused by exposure to toxins, such as cigarette smoke, occur over a prolonged period of time. In contrast, many viral-induced cancers can occur rapidly, while others are not noted for years following the initial viral infection that caused the cancerous change.

The immune system attempts to destroy viral-altered cells that become cancerous. Specialized T-lymphocytes attempt to identify and destroy cancerous cells. The proper function of these immune system cells may partially explain why the polyomaviruses and adenoviruses that commonly cause cancer in experimental situations have not been associated with tumors in animals with a functional immune system.

VIRUS MOVEMENT THROUGH AN INFECTED BIRD

Viruses that cause systemic infections must spread from their site of initial entrance into the bird to other susceptible cells in the body. Viruses move from one area of a bird's body to another principally in the blood or via the nerve tracts. Birds that have virus in their blood stream are said to be viremic.

FIG 2.12
Most viruses that cause cancer insert a portion of their nucleic acid into the nucleic acid of a host cell. The inserted viral nucleic acid substantially alters the normal function of the host cell, causing cancerous changes.

TABLE 2.4

Viruses that have been shown to cause cancerous changes in some infected animals

Virus	Type of Cancer
Adenoviruses*	solid tumors in various organs
Hepadnaviruses	liver tumors
Herpesviruses	lymphomas and some carcinomas
Papillomaviruses	tumors (warts) on epithelial surfaces
Polyomaviruses*	solid tumors in various organs
Poxvirus	myxomas and fibromas, particularly in pigeons and canaries
Retroviruses	generally blood cell tumors like lymphomas and erythroblastomas

** Polyomaviruses and adenoviruses have been associated only with tumors in experimentally infected animals. The other viruses have been associated with naturally occurring tumors.*

Some viruses are transported in the blood after they have been injected into the body by biting insects. Other types of viruses first replicate at the site of initial infection and the progeny virus enter the blood stream when the initially infected cells release new viral particles. The concentration of virus particles in the blood is particularly high with viruses that are transmitted by insects. Many types of viruses do not enter the blood stream at all. Others are associated with a low titer viremia or cause a viremia that lasts for only a brief period of time (hours to days).

VIRAL-INDUCED CHANGES IN A BIRD

Viruses can induce clinical abnormalities that involve different organ systems and vary from mild changes to death. Some viruses cause localized lesions only at the site of initial infection. Others enter the bird, spread throughout the body and cause changes in numerous tissues. Viruses may also infect cells at the site in which they enter the body and the progeny produced at this entry site are carried to other locations in the body, causing more severe disease. Such is the case in mammals infected with rabies virus. This virus initially multiples in the cells adjacent to a bite wound. Progeny virus are carried to the central nervous system and cause the most serious, usually fatal changes associated with the infection.

Many of the organs that viruses infect must be severely damaged before clinical changes occur that would indicate the presence of a disease. For example, 70% of the liver in most animals must be damaged before clinical signs of liver disease are present. The intestinal tract also must be severely damaged before clinical signs are noted. Some tissues affected by viruses, such as the skin and epithelial linings of the intestinal and respiratory tracts, have a tremendous regenerative capacity, and viral infections of these tissues are frequently characterized by an acute onset of clinical signs followed by a progressive, and often rapid, recovery. Additionally, these tissues can frequently withstand severe damage with few long-term side effects. In comparison, minor damage to the cells that form the heart or brain can be rapidly fatal *Figure 2.13*.

The progression of clinically recognizable signs that occurs following damage to many organ systems is usually similar, irrespective of whether it is caused by a virus, bacteria, parasite, fungus or toxin, or some combination thereof. Thus, specific diagnostic tests are often needed to determine the cause of diseases, particularly those of the gastrointestinal or respiratory systems. For example, a bird may develop diarrhea. Microscopic examination of feces indicates that the bird has a gram-neg-

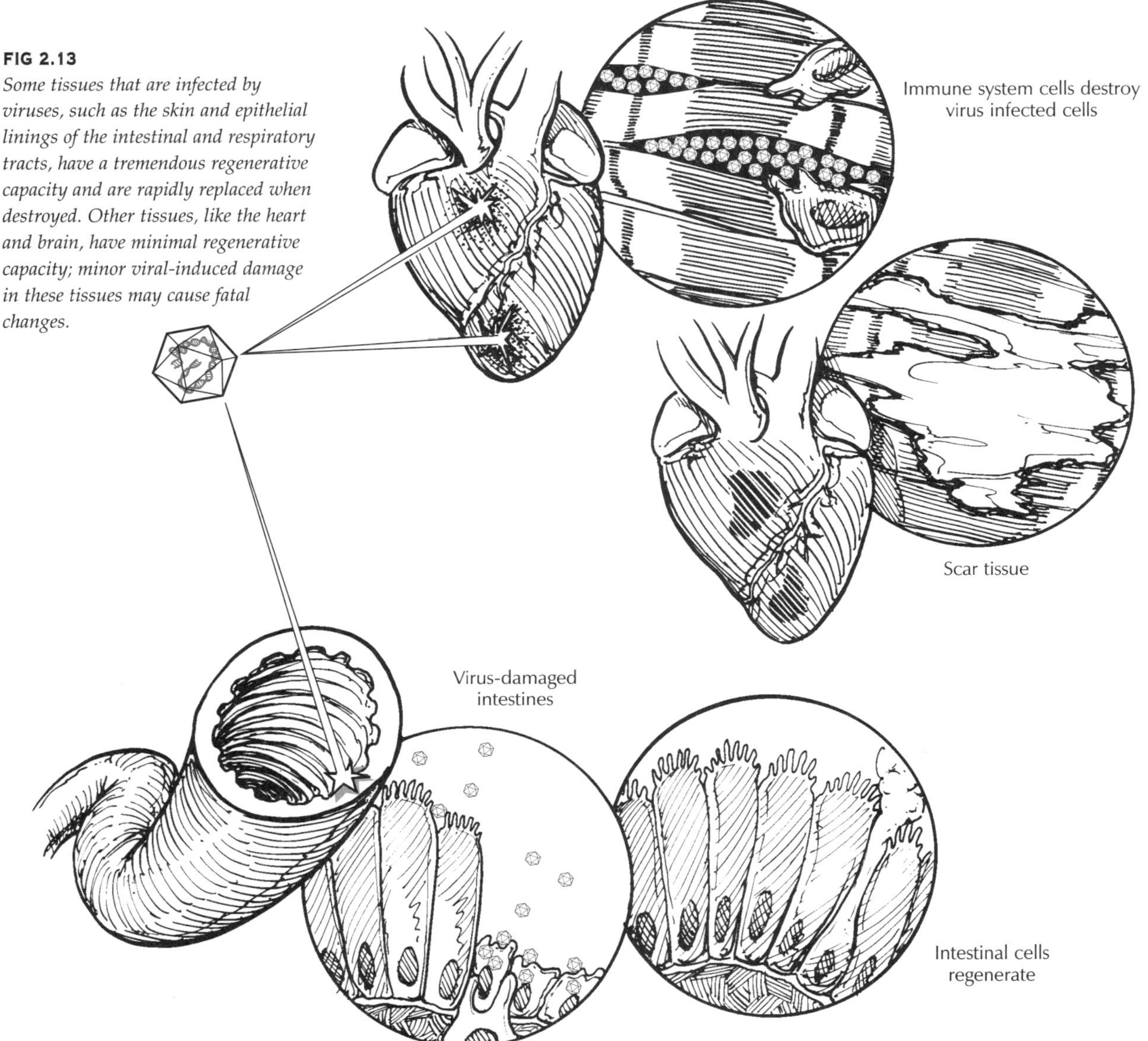

FIG 2.13

Some tissues that are infected by viruses, such as the skin and epithelial linings of the intestinal and respiratory tracts, have a tremendous regenerative capacity and are rapidly replaced when destroyed. Other tissues, like the heart and brain, have minimal regenerative capacity; minor viral-induced damage in these tissues may cause fatal changes.

ative bacteria, an adenovirus and *Giardia*. Adenovirus, gram-negative bacteria and *Giardia* can all be demonstrated in clinically normal birds, but can also be recovered from birds that are ill. It becomes the clinician's responsibility to determine which of these organisms, if any, may be causing the diarrhea and how best to treat the patient.

Reductions in the number of circulating heterophils or lymphocytes can occur with some viral diseases. It should be noted, however, that the type of changes that occur in the white blood cells of an infected bird vary dramatically with the type of viral infection, the stage of viral infection (acute vs recovery) and the age and condition of the host. Like any diagnostic test, a white blood cell count should be used only as a method of collecting data about a bird's overall state of being, and it can rarely be interpreted as an absolute indication of a particular disease process.

Some types of chronic infections may be associated with hyperglobulinemia. Globulins are specialized proteins that are involved in the immune response. Hyperglobulinemia is recognized in a routine blood examination by an increase in the total amount of protein and can be specifically identified using a specialized test called serum electrophoresis.

In mammals, some viral infections have been associated with early death of an embryo or genetic malformations. Specific references to viral-induced changes in the embryos of companion or aviary birds are lacking. However, they would be expected to occur. To monitor effectively the health of a flock,

TABLE 2.5

Viruses that specifically damage components of the immune system

Virus	Damage
IBD Virus	Destroys the cloacal bursa
PBFD Virus	Damages thymus and bursa Virus circulates in white blood cells and persists in macrophages
CAV	Damages bursa and thymus

it is important that complete necropsies, including histopathology, be performed on embryos that die.

IMMUNOSUPPRESSION

Immunosuppression occurs when the defense systems of a bird are damaged, rendering the bird less capable of mounting an immune response against an invading microorganism. Immunosuppression may: 1) cause a bird that would otherwise be resistant to a viral infection to become susceptible; 2) cause a bird to develop severe disease following infection with a normally avirulent strain of virus; or 3) cause an infected bird to become persistently infected when it would normally eliminate the virus.

Some viral infections may be immunosuppressive by causing damage to a bird's defense systems so the bird is more susceptible to other viral infections or to secondary bacterial, fungal or parasitic invaders. A bird will be particularly susceptible to secondary infectious agents if a virus damages the skin or linings of the respiratory, intestinal or genitourinary tracts. Any viral infection that causes disease can be expected to be immunosuppressive to some degree. However, the viral agents that damage the thymus, bursa, bone marrow and spleen should be considered particularly immunosuppressive because of the importance of these organs to the immune system *Table 2.5.*

REFERENCES

1. Ahmed R, Canning WM, Kauffman RS, et al: Role of the host cell in persistent viral infection. Cell 25:325-332, 1981.

2. Temin HM: Separation of morphological conversion and virus production in Rous sarcoma virus infection. Cold Spring Harbor Symp Quant Biol 27:407-414, 1963.

Viral Attack and Avian Reponse

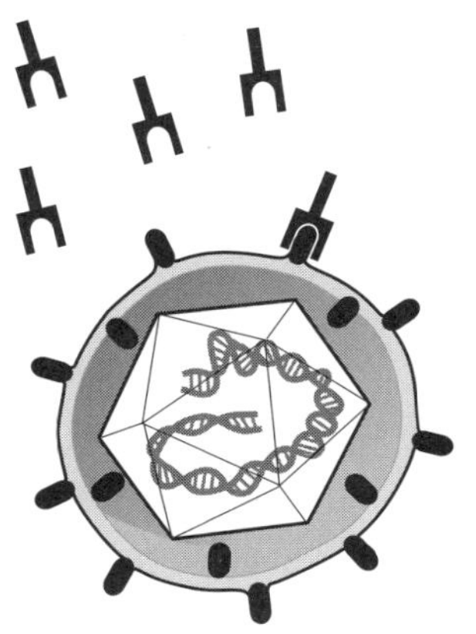

For a virus species to survive, it must replicate in a host and then be transmitted to a new host where it establishes an infection and replicates again to produce progeny. Birds that are clinically or subclinically infected with a virus can release progeny produced in an infected cell into the environment in which the birds live. While the virus does not replicate when outside the bird, it can remain in an infectious state for a defined period of time where it may contact and infect a new susceptible bird. The likelihood that a virus will be transmitted to a new host is directly related to the durability of the progeny virus when outside the originally infected bird.

The initial step of the infection process requires that a virus enter the living cell of a bird. A virus, depending on its type, may gain access to the cells of a bird through one of several routes including inhalation, ingestion, injection or skin-to-skin contact. These methods of transfer are collectively termed horizontal transmission *Table 3.1*.

Some viruses can be transmitted from an infected hen directly into an egg and then into the developing chick. Transmission of a virus from a parent to an egg during fertilization, during development in the oviduct or immediately after laying is called vertical transmission. Once the egg is in the nest, some viruses can pass through the eggshell

TABLE 3.1

Some potential methods of virus transmission in birds

Horizontal
    Preening
    Rubbing
    Inhalation of aerosols
      (usually short distances)
    Coitus
    Insect or animal bites
    Contamination of egg
    Ingestion of contaminated feces
    Contact with contaminated
      fomites
    Ingestion of contaminated food
      or water
    Ingestion by neonates of regur-
      gitated food from parent
    Combat-related injuries
    Insects - mechanical vector
    Insects - biologic vector
Vertical
    Hen transmission to the egg
    Contamination of egg
      immediately after laying

upon contact with contaminated feces, urates or bedding.

Many viruses can enter a bird through more than one route *Table 3.2*. In general, however, viruses that infect the alimentary tract and liver enter a bird through ingestion and are shed to the environment in the feces. Viruses that infect the respiratory tract, sinuses, trachea and lungs usually enter a bird through inhalation of contaminated air; virus is shed back to the environment in respiratory aerosols when the bird exhales, coughs or sneezes. Viruses that affect the genitourinary tract are generally shed in urine or from secretions produced by the male or female reproductive tract.

Typical shedding routes and potential modes of virus transmission

Skin-to-skin Contact
- Herpesvirus
- Papillomavirus
- Poxvirus
- Retroviridae
- Rhabdoviridae (animal bites more common)

Ingestion - Fecal/Oral
- Adenovirus
- Arenovirus
- Birnaviridae
- Calicivirus
- Circoviridae
- Coronavirus
- Enterovirus
- Herpesvirus
- Orthomyxovirus
- Paramyxovirus
- Parvovirus
- Polyomavirus
- Poxvirus
- Reoviridae
- Togavirus

Inhalation - Contaminated Aerosols
- Adenovirus
- Arenovirus
- Calicivirus
- Circoviridae (possible, unproven)
- Coronavirus
- Enterovirus
- Herpesvirus
- Orthomyxoviridae (some strains)
- Paramyxovirus
- Parvovirus
- Picornaviridae (genus *Rhinovirus*)
- Polyomavirus
- Poxvirus

Arthropods - Mechanical or Biologic Vector
- Bunyaviridae
- Circoviridae (mechanical vector possible, unproven)
- Flaviviridae
- Poxvirus
- Reoviridae (genus *Orbivirus*)
- Rhabdoviridae (some strains)
- Togavirus

The most common method of virus transmission is direct contact between an infected bird that is shedding virus and a susceptible bird. This direct transmission of a virus from one bird to another may occur through rubbing, preening, feeding, sexual contact, combat-related injuries or by passing from the hen directly into the developing egg. In some respects, virus-contaminated aerosols that are produced when a bird coughs or sneezes also can be viewed as a type of direct transmission because the aerosols produced rarely travel more than four feet.

Indirect transmission may occur if a susceptible bird is exposed to any object that is contaminated with a virus (fomite). Fomites can be contaminated with virus that is present in aerosolized feather dust, feces, urine, saliva or aerosolized respiratory secretions that are generated when a bird coughs or sneezes *Figure 3.1*. Any object or being that serves to transport a virus from one bird to another can be called a mechanical vector.

To be transmitted on a fomite, a virus must be sufficiently stable outside of a bird's body to remain infectious for the period between shedding from the initial bird and contact with a new susceptible bird. A virus particle that is outside of a host is in a state of suspended animation. Even the hardiest of viruses will eventually become inactivated. Viruses that survive for prolonged periods in the environment are more likely to be spread by indirect contact. Viruses that do not survive for long in the environment are almost always spread through direct contact *see Table 1.4.*

Environmentally stable viruses can be transmitted among birds on contaminated bedding, examination tables, enclosures, clothing, nets, gloves, perches, toys, towels, food and water containers, nest boxes or foods obtained from open containers. A common food source or feeding utensil can spread viruses among neonates in a nursery *Figure 3.2.* This single, major gap in nursery hygiene frequently results in devastating outbreaks of viral, bacterial or fungal diseases among neonates.

Veterinary medical professionals, pet shop personnel and aviary caretakers must use special care to avoid serving as a source of viral exposure. Individuals who have companion birds, or are in contact with companion birds, can transport viruses from one location to another on their clothes, shoes, hair or hands. It is advisable to maintain birds within a secluded home or closed aviary with no new birds or visitors to reduce the chances of a bird being exposed to a virus.

Insects also may be responsible for indirect transmission of some viruses from bird to bird. The term arbovirus collectively describes all of the viruses transmitted in nature by an arthropod vector. These viruses survive in nature by causing cyclic infections in free-ranging mammals and birds. In their natural hosts, many arboviruses cause minimal disease because of evolutionary pressures that drive survival of the host and the virus. In unnatural hosts, many of

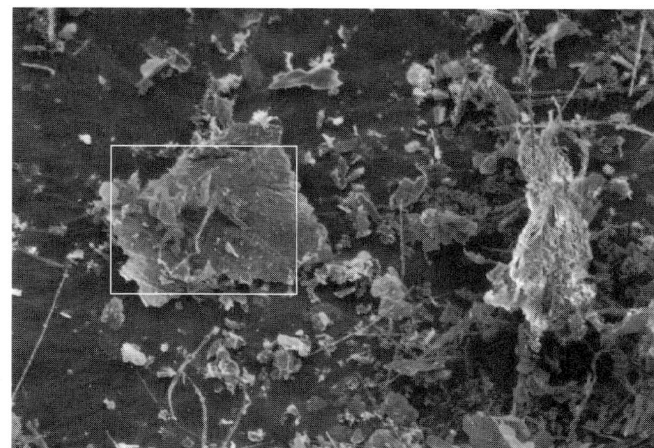

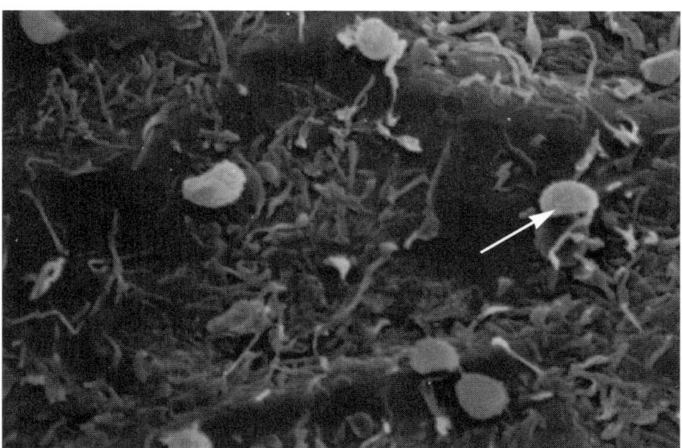

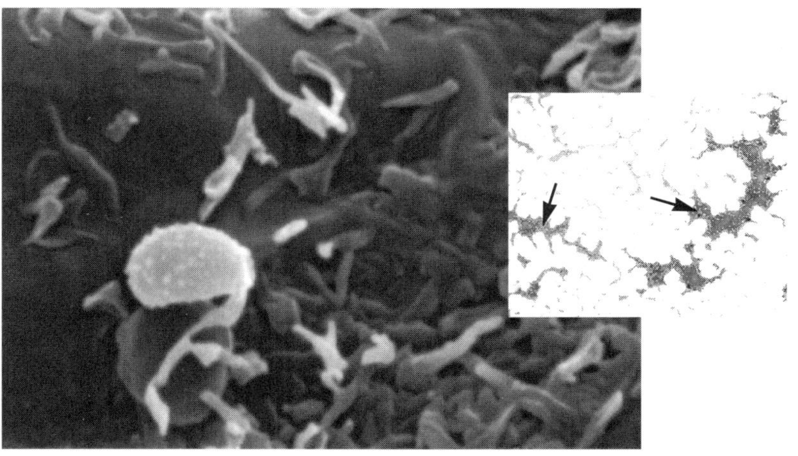

FIG 3.1

A Photomicrograph of dust collected from the wall in a 20' by 20' room where 18 birds were housed. The dust has been magnified 75 times.
B The area marked in A has been magnified 2,000 times. Eight bacteria that were floating on one small area of the dust particle can be clearly visualized (white arrow).
*C A bacterium demonstrated in B has been magnified 7,000 times. PBFD virus magnified 7,000 times **inset** illustrates how many PBFD virus particles could be located on the surface of one particle of contaminated dust. At this magnification, the PBFD virus particles are barely visible as minute white dots in a grey background (arrows).*
photographs courtesy of W.L. Steffens

these viruses can cause severe, life threatening infections. Many of the arboviruses have zoonotic potential.

Most arboviruses are transmitted when an arthropod ingests a blood meal from an infected animal and then moves to an uninfected animal and injects the virus during feeding. Insects may serve as vectors for viruses in two ways: as mechanical vectors that land on contaminated fomites, become contaminated and then carry the virus to another bird; or as biological vectors. In the latter case, the insect ingests the virus and the virus replicates in the insect. The insect then transmits the virus to a susceptible bird during subsequent feedings. Mosquitoes, ticks and other blood-eating insects are the principal biologic vectors of viruses *Table 3.2*.

The insect-transmitted viruses generally cause disease in multiple tissues. A characteristic feature of insect-vectored viruses is the presence of viremia in an infected bird. This high concentration of virus in the blood increases the likelihood that a biting insect will become contaminated or infected by the virus and subsequently transmit it to other susceptible animals. Insect-transmitted viruses are extremely dangerous because many of these viruses can be

transmitted to many species of animals, including birds, mammals and humans.

FACTORS IN VIRAL TRANSMISSION

The species of bird or animal that a virus can infect plays a critical role in controlling how the transmission cycle progresses. Some viruses have a wide host range and can infect a variety of different animals and tissues. Other viruses have a narrow host range and infect only certain animals or types of cells. For example, the paramyxovirus that causes Newcastle disease has a wide host range and can infect most species of birds as well as some mammals. In contrast, the canary poxvirus infects only canaries and other closely related avian species.

Highly transmissible viruses are those that are likely to be spread from a shedding bird and cause an infection in an exposed bird. Viruses with a low transmissibility are not likely to be spread from one bird to another. Transmission of most viruses occurs during an active infection. A bird that has a viral-induced disease is most likely to shed virus in the highest quantities, and thus be the greatest threat to other birds, just before and during the period that they are clinically ill.

The transmission of a virus from bird to bird is most likely to occur in large groups of birds that are frequently in contact with birds from other large groups. The congregation of large numbers of potentially virus-infected birds clearly occurs within current trading procedures used for companion and aviary birds, including holding stations, bird shows, airplanes or import stations. The exposure of birds in these locations to numerous infectious agents is followed by the agents' disbursement to a variety of new locations where they may expose other birds to recently acquired infectious disease-causing organisms. Many of the confined, indoor, high-density breeding facilities used in the captive production of companion and aviary birds create an ideal environment for the transmission of viruses, particularly those that are shed from the respiratory or gastrointestinal tract.

Increasing the number of birds in any area or the movement of birds from location to location favors the amplification and dissemination of a virus to multiple flocks or locations. Thus, outbreaks of infectious diseases would be expected to be most common in quarantine stations, pet shops that obtain birds from numerous sources, nurseries that acquire birds from numerous sources, nurseries that practice poor hygiene, breeding facilities that obtain birds from numerous sources or breeding facilities that frequently allow visitors. Stacking enclosures increases the likelihood that an infectious organism can spread from bird to bird. This arrangement creates a dangerous health hazard because the birds in the lower enclosures can be easily exposed to the contaminated excrement of birds higher in the stack.

Birds that are housed indoors year-round should be considered more susceptible to infectious diseases because of decreased air quality, the accumula-

tion of pathogens in a restricted environment and the lack of exposure to sunlight. These factors function collectively to decrease a bird's natural resistance to disease. The potential for a virus to be transmitted among a group of birds can be reduced by preventing overcrowding, providing maximum separation between birds, providing frequent exposure to sunlight and ensuring a constant supply of fresh uncontaminated air. Sunlight will destroy many viruses that are free in the environment. In birds that are maintained in ideal conditions, the viruses that are vectored by insects become more important than the viruses that are transmitted through direct contact.

Many types of viruses have developed mechanisms for establishing persistent infections that serve to ensure their frequent excretion from an infected bird. The long-term shedding of virus that can occur with persistent infections increases the likelihood that a virus will be able to successfully infect numerous new birds. If a bird is slowly damaged by a virus, higher numbers of progeny will be shed from the bird and released to the environment over a longer period of time. A perfect example of this type of virus transmission occurs with

FIG 3.3
Some birds infected with the PBFD virus can survive for years and shed large concentrations of this stable virus in their feather dust and feces. These two Scarlet Macaws were infected with the PBFD virus when an affected Eclectus Parrot was added to the aviary.
photograph courtesy of Joey Rodgers

psittacine beak and feather disease (PBFD) virus *Figure 3.3.*

Some viral infections occur most commonly at certain times of the year. For example, viruses that are transmitted by mosquitoes, such as eastern equine encephalitis virus or poxvirus, increase in incidence in the late spring, summer and early fall when mosquito populations are highest. Viruses that routinely affect neonates, such as PBFD virus or polyomavirus, are more common during the breeding season when there are more neonates in the nursery. At this time, thorough cleaning is more difficult and the combining of neonates from multiple clutches into one area frequently occurs.

The process of shipping eggs for artificial incubation has been suggested as a method of preventing the introduction of viral infections to immunologically naive avian populations. However, a hen can pass both antibodies and some viruses to her offspring while the egg is developing. These infected neonates can cause severe outbreaks with high mortality in a nursery setting where they expose other neonates with no antibodies to a particular virus.

NOSOCOMIAL INFECTIONS

Nosocomial infections are those that are derived from an animal's visit to a hospital. These infections may occur as a result of a bird's contact with an infected animal or from contact with contaminated equipment or supplies within the hospital. Infectious organisms that enter a bird through the respiratory or gastrointestinal tract would be expected to be the most common causes of nosocomial infections in hospitalized birds.

Viruses that are difficult to inactivate, such as poxvirus, PBFD virus and polyomavirus *see Table 1.4*, are most likely to cause nosocomial infections in birds. Because of the persistence of the PBFD virus, it is recommended that birds with this disease not be hospitalized in the same airspace where other birds will be housed, and particularly not in the same airspace with neonates. Viruses that are easy to inactivate, such as herpesvirus, paramyxovirus and togavirus, are not likely to cause nosocomial infections unless affected and susceptible birds are in direct contact with each other. Poor sanitation practices will increase the likelihood of any virus causing a nosocomial infection.

Given the lack of vaccines for companion and aviary birds, it is best for hospital stays to be of short duration and, where possible, treatment for noninfectious diseases should be administered at home. The use of sound hygiene, a consistent disinfectant program and segregated airflow systems will decrease the risk of nosocomial infections. Contaminated hands are a common source of nosocomial infections. All veterinary personnel should wash their hands after contact with each animal.

Any bird that is hospitalized for illness is in a weakened state and would be expected to be more susceptible to infections. Breeding birds or neonates that have clinical signs of disease (diarrhea, sneezing, coughing, delayed crop emptying or regurgitation) should not be returned to an aviary where they may serve as a reservoir for the spread of an infectious disease-causing organism. To prevent serious viral-induced disease outbreaks, a neonate that is taken from the nursery and has contact with other birds must not be returned to the nursery.

RELEASE OF CAPTIVE-PRODUCED BIRDS

A bird, like any living being, can be viewed as a self-contained environment for a distinct group of bacteria, viruses, fungi and parasites that infect that particular bird. The individual bird is then a member of a population of birds, each of which has its own bacteria, viruses, fungi and parasites. This population of birds lives in association with other animals that are infected also with a distinct group of bacteria, viruses, fungi and parasites. The infectious organisms found in any individual bird or animal interact with each other and with the host they inhabit to prevent or cause disease.

Many articles have been written discussing the release of captive-raised birds produced on a continent to which the birds to be released are not indigenous. This practice could easily result in the introduction of previously not encountered, infectious disease-causing organisms that could have disastrous effects. Programs intended to protect and increase populations of companion birds are best instigated through habitat protection, as well as breeding and release programs that are performed in a bird's natural habitat. Any release program that uses birds that have been exposed to non-indige-

nous species should be discouraged. It should be noted that the human immunodeficiency virus (HIV) was originally isolated to distinct regions of Africa; the transcontinental movement of infected people resulted in the widespread, intercontinental dissemination of this virus.

The introduction of a virus to a new geographic location may occur from the translocation of the virus, a host infected with the virus or a vector that is infected or contaminated with the virus.[5]

Populations of birds that have been isolated to islands with restricted endogenous microbial flora should be considered particularly naive with respect to infectious disease-causing agents. Such is the case with the pink pigeon indigenous to Hawaii. Over time, these pigeons co-evolved with the infectious agents found on the island and developed some resistance to the agents normally found in Hawaii. However, when pigeons were transferred to the mainland United States for captive breeding, with the intention of releasing progeny to Hawaii, the birds died after exposure to a herpesvirus they had not previously encountered.[6]

Polyomavirus antibodies were not detected in any Puerto Rican or Hispanolan Amazon parrots maintained in Puerto Rico for breeding and release purposes. In contrast, the seroprevalence of avian polyomavirus in a group of captive psittacine birds in the United States was 63%. The movement of birds from Puerto Rico (where they had not been exposed to avian polyomavirus) to the United States (where they are likely to be exposed to polyomavirus) and then back to Puerto Rico for release could easily result in the dissemination of this virus into a population of birds where it did not previously exist.

It has been suggested, and there is substantial supporting evidence, that many species of Hawaiian birds were driven to extinction because of the introduction of avian poxvirus and malaria carried to these islands in non-native mosquitoes.[5] It should also be noted that an infectious agent or parasite is not required to cause clinically distinguishable disease in order to adversely affect a free-ranging bird. A viral infection that causes no observable clinical signs of disease can be responsible for the death of a free-ranging bird by causing minor weakness or lethargy that causes the bird to consume inadequate quantities of food or water, to change its social position in a flock or to render it less efficient at avoiding predators.[3]

Quarantine and testing programs have been suggested as methods of preventing the introduction of pathogens to a new geographic area. However, quarantine, monitoring or testing procedures should not be considered adequate for preventing the introduction of disease-causing agents. Specific-pathogen-free (SPF) chickens are free from certain infectious organisms for which they are screened; however, even these birds maintain a population of infectious disease-causing organisms for which testing procedures have not been developed. Flocks of SPF chickens have been selectively bred for decades in an attempt to reduce the background of infectious agents in offspring. The result of these years of expensive, intensive efforts has been a reduction only in the number of infectious organisms that these birds harbor.

The use of vaccines in birds intended for release programs may also adversely affect survival of the species. While these vaccines may protect the recipient from a specific pathogen, they may also alter naturally occurring selection pressures and incorrectly favor the survival of birds that are actually weaker than nonvaccinated birds.

The most prudent way for avian professionals concerned about preservation of avian species to help in global conservation is to work for the protection of habitat and to do as little self-induced destruction as possible. Self-induced destruction can be minimized by reusing resources, recycling, conserving energy, avoiding the use of disposable products (particularly plastics) and avoiding the use of any pesticides and toxic chemicals.

VIRAL ENTRY INTO A HOST

A bird has numerous surfaces that are exposed to the environment through which a particular virus may enter. Generally, a specific family of viruses has its own preferred site for entering a bird *see Table 3.2*. A particular type of virus may enter a bird through the skin, alimentary tract, respiratory tract, conjunctivae, urinary tract or genital tract. Embryos may be infected by the hen when the egg is forming. Some types of viruses may be injected into a bird by

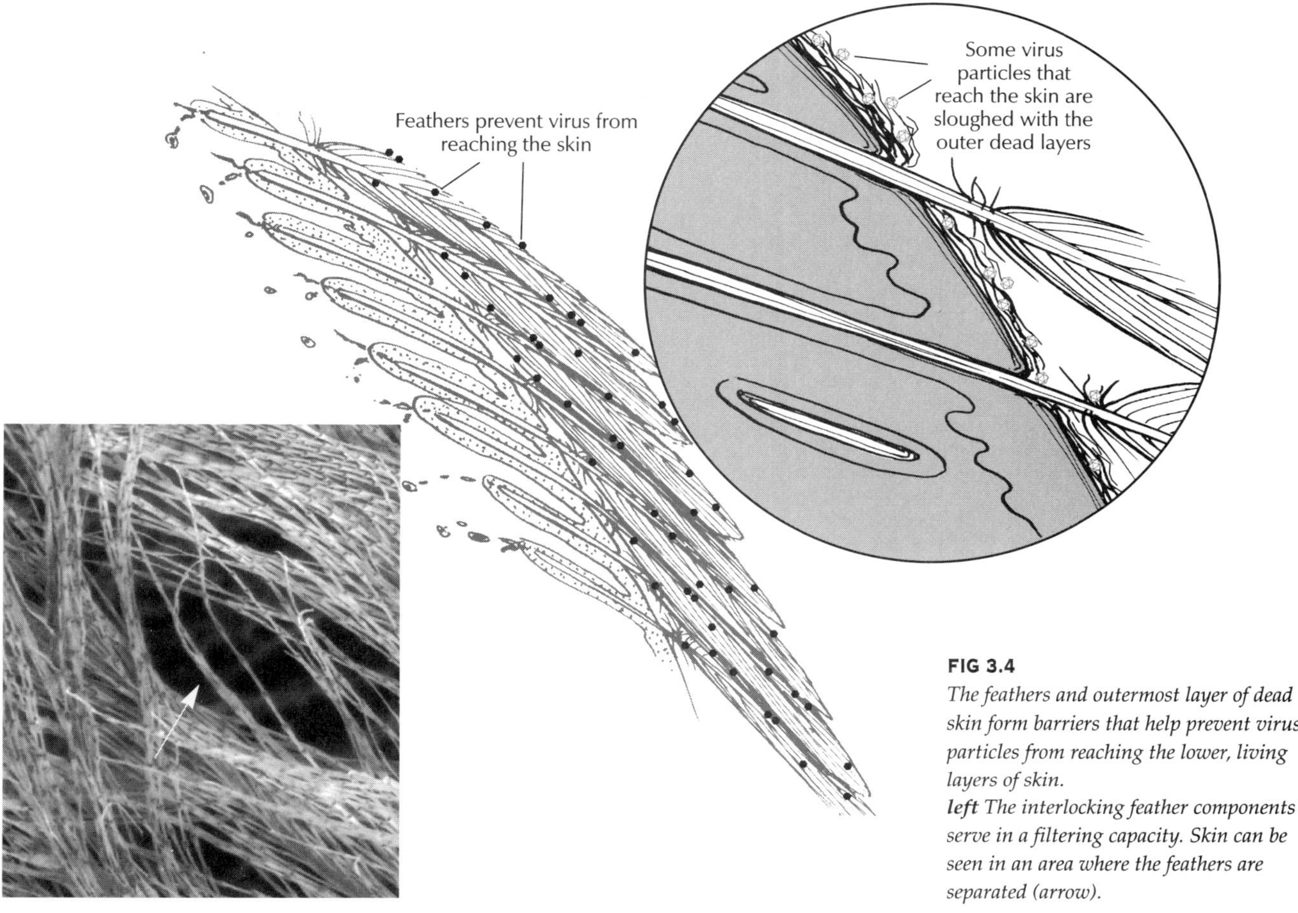

FIG 3.4

The feathers and outermost layer of dead skin form barriers that help prevent virus particles from reaching the lower, living layers of skin.
left The interlocking feather components serve in a filtering capacity. Skin can be seen in an area where the feathers are separated (arrow).

biting insects or can enter through cuts or scratches in the skin. Anything that damages the skin or mucosal lining of the intestinal, respiratory, genital or urinary tract predisposes a bird to infection by microbial pathogens, including viruses.

Once a virus has attached to cells at the primary entry site in the skin, intestinal tract, respiratory tract, eye, genital tract or urinary tract, the virus can remain there and cause localized signs of disease, or it can be spread throughout the body and cause abnormal changes in many organs. For example, papillomaviruses enter the skin of a bird, induce a localized infection in the cells that form the skin and then cause disease (a wart) that is principally limited to the skin. Viruses that target the respiratory and gastrointestinal tracts typically remain localized to the cells lining these areas of the body and do not progress to infect other cells. The mechanisms that restrict the systemic spread of these viruses are not fully understood. They probably involve the absence of viral receptors on other cells as well as host cell defense mechanisms.

By comparison, polyomavirus may enter a bird through numerous routes. Once the virus infects the bird, it can be spread (presumably in the blood) to many tissues, causing disease in numerous organ systems from the liver to the heart and brain.

SKIN

Most viruses that come in contact with the feathers or outermost layer of dead skin become entrapped in the debris and are discarded with the outer layer of dead skin. Viruses that are trapped in the feathers or skin debris can be ingested when a bird preens.

A few types of viruses can enter the lower, viable layers of the skin through abrasions or can be injected past the outer dead layer of skin by biting insects *Figure 3.4*. Virus-contaminated hypodermic needles, endoscopic equipment or surgical instruments also could function as methods for the injection of viruses or other microorganisms past the outer feather and skin barriers and into deeper body tissues.

Even though a few avian viruses are shed from the skin or feathers of some infected birds (poxvirus, papillomavirus, polyomavirus, Marek's disease virus, reticuloendotheliosis virus and PBFD virus), other routes by which virions are released from a bird may be more important sources of viral contamination in the environment. The shedding of PBFD virus in feather dust may be the major route by which this virus is disseminated in the environment.

RESPIRATORY TRACT

Respiratory viruses may enter the body of a bird when it inhales contaminated air or when a viral-contaminated body part (feathers, toes, skin) comes in contact with the nose, mouth or eyes. Viruses typically enter the respiratory tract attached to minuscule drops (smaller than 5 µm) of fluid. Viruses that infect the respiratory tract general-ly remain localized to the cells that line this organ system, and progeny virus are shed to the environment during expiration, sneezing or coughing. Viral families that could be considered to principally involve the respiratory tract include Orthomyxoviridae, Paramyxoviridae, Coronaviridae, Adenoviridae, Picornaviridae (genus *Rhinovirus*) and Caliciviridae.

Many of the viruses that cause diseases of the respiratory tract enter a bird through the conjunctiva. In some cases, these viruses cause changes that remain localized to the eye or upper respiratory tract, while in other cases, the virus spreads to cause systemic changes. The careless veterinarian or aviculturist may be responsible for transmission of viruses through evaluating or touching the eyes of an infected bird and then touching the eyes of a susceptible bird.

DIGESTIVE TRACT

Virus particles originating from several anatomic areas can enter the digestive system. They can be swallowed with contaminated food or water, taken into the oral cavity through preening activities, or passed from adult to offspring or mate during routine feeding activity. Inhaled particles can be coughed up from the respiratory tract and swallowed. Intestinal viruses often infect cells only at the site where they enter the host and do not spread to other tissues.

Enteric viruses usually enter a bird by ingestion; they replicate in the cells that line the gastrointestinal tract and are shed from the gastrointestinal tract in regurgitated material or feces. The passage of a virus from the contaminated

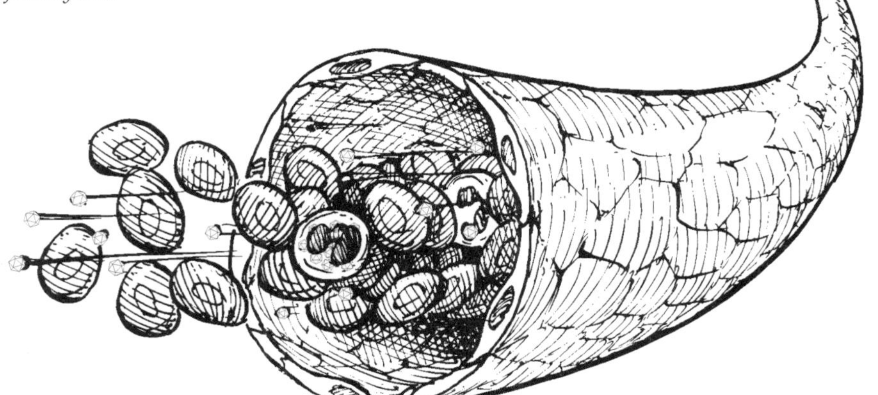

Viruses that circulate in the blood can be transmitted from bird to bird by biting insects, through breeding activity, through preening that involves ingestion of dried blood and through preening-induced damage of blood feathers. Virus that is present in the bloodstream can be contained on or within red blood cells, within the plasma or within the white blood cells that form the body's internal defense system.

feces of one bird to the oral cavity of another bird is commonly referred to us as a fecal-oral transmission cycle. Viral families that could be considered to be primarily enteric pathogens include Reoviridae, Coronaviridae, Picornaviridae (genus *Enterovirus*), Adenoviridae and Caliciviridae.

Many other viruses that enter a bird through the digestive tract will cause a localized infection that remains confined to the cells that line the digestive tract; however, some viruses that enter through the digestive tract cause disease in other organs, particularly the liver.

Genitourinary Tract

Some viruses enter the body through the genitourinary tract during breeding. These viruses may replicate in the cells that line the reproductive tract or kidneys, and can be shed to the environment in reproductive fluids (semen, vaginal secretions) or urine. Some viruses that persist in the cells that line the kidney produce lifelong infections with intermittent shedding. In humans, polyomaviruses have been associated with viruria, indicating that this is a

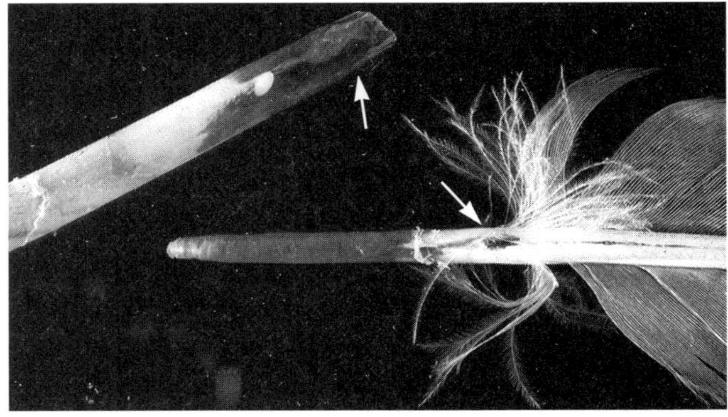

method by which the virus may be shed into the environment. Polyomavirus can be detected in the cells that line the kidney in some infected Budgerigars, suggesting that this virus may be shed in the urine from these birds.

Blood-borne Viruses

If a virus is able to bypass all of a bird's defense mechanisms, it may enter the

FIG 3.6

Dried blood in the shaft of feathers (arrows) can be contaminated with virus particles that were circulating in a bird's blood when the feather was developing. Virus particles that are sufficiently durable outside of a bird (like the PBFD virus, polyomavirus or adenovirus) could be easily transferred from bird to bird within a contaminated feather shaft.

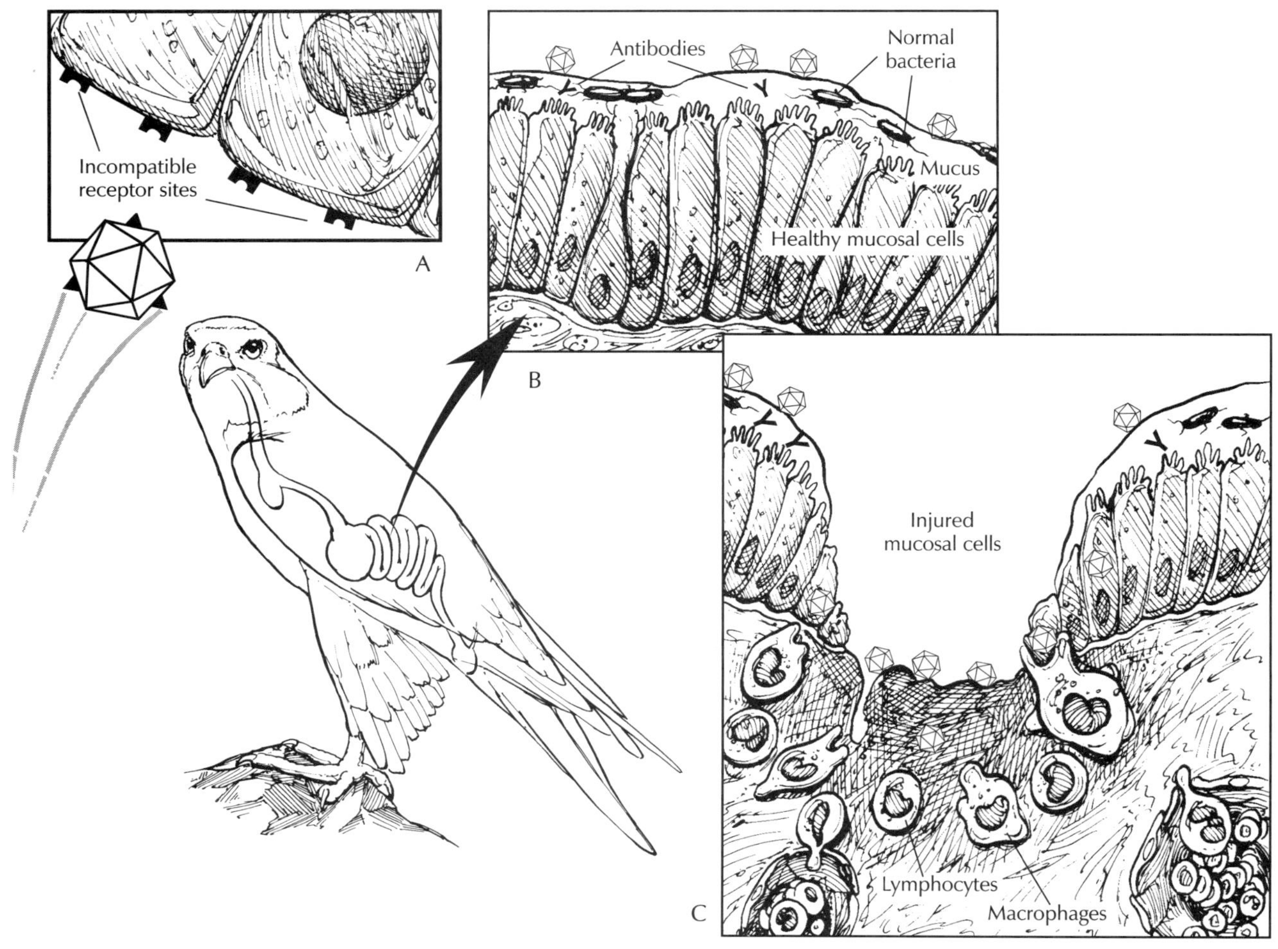

blood stream. A bird that has virus circulating in the blood is said to be
viremic *Figure 3.5*. Viruses that are
spread through contact with blood
(particularly those spread by biting
insects) usually cause a viremia that
persists for an extended period of time.
During this long viremic period, new
virus particles are continuously shed
into the blood stream *Figure 3.6*. Once in
the blood, a virus can be spread to any
tissue and may infect the cells in any
organ that has specific receptors for that
particular virus. While in the blood
stream, a virus is exposed to circulating
macrophages and other specialized
defense cells that attempt to destroy the
virus. Virus circulating in the blood can
enter the egg while it is developing
within the hen.

DEFENSE MECHANISMS

The progression of a viral infection in a
given bird can be viewed as an ongoing
battle between the bird and the invading virus. Every exposure that a bird
has to a particular virus involves a con-
flict between the virus' attempt to
bypass the defense systems and infect
the bird, and the bird's defense systems' attempt to isolate and destroy the
virus. When the bird wins, a virus exposure results in no infection or a subclinical infection. When the virus wins, a
virus exposure results in an infection
that may be associated with disease.

A highly virulent virus that enters a bird with a fully functional defense system is likely to win the battle and cause disease in the bird. A less virulent virus that enters a bird with a fully functional defense system may or may not cause disease in the infected bird. A virulent virus that enters a bird with lowered resistance is likely to cause disease in the infected bird. A less virulent virus that enters a bird with a damaged defense system may or may not cause disease in the infected bird, but in a bird with a fully functioning defense system, it will probably be unable to cause disease. Irrespective of the route by which a virus may enter a bird or the condition of a bird's defense systems, the more virus a bird is exposed to, the more likely the virus is to bypass the bird's defense mechanisms and cause an infection.

A bird has both external and internal defense mechanisms that help protect it from invading organisms *Figure 3.7*. The bird's external defense systems are designed to prevent a virus from entering the bird. If a virus bypasses these defenses and enters the bird, internal defenses are designed to contain the infection by destroying virus-infected cells and released virus particles. In general, the functional capacity of the defense systems is dependent on the nutritional status, genetics, age and environmental conditions of a bird.

In a healthy bird, if a virus bypasses the external defense systems and enters the body, the immune system should respond. The immune response that is activated is designed to identify and

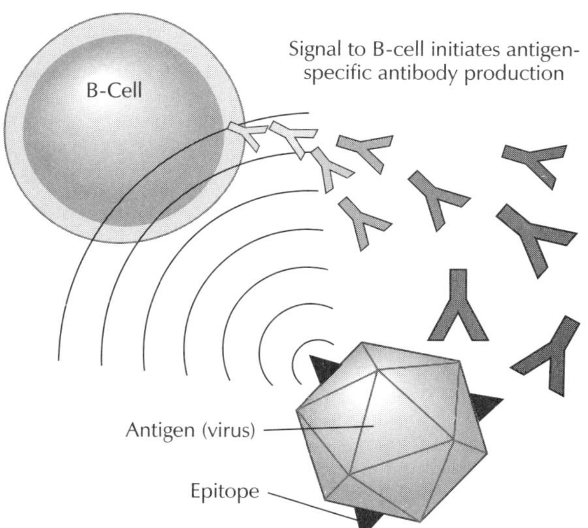

FIG 3.8
A molecule that stimulates the immune system is called an antigen. The specific molecule that the body makes in response to an antigen is called an antibody. Antibodies are generated specifically against a small portion of a virus particle called an epitope.

destroy infectious agents that have bypassed other defense systems *Figure 3.8*.

When an infection is treated with any type of drug, the actual goal is to retard the progress or damage caused by the invading organism for a sufficient peri-

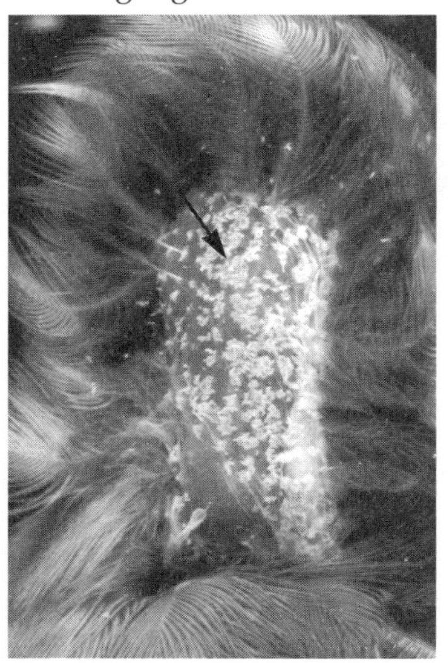
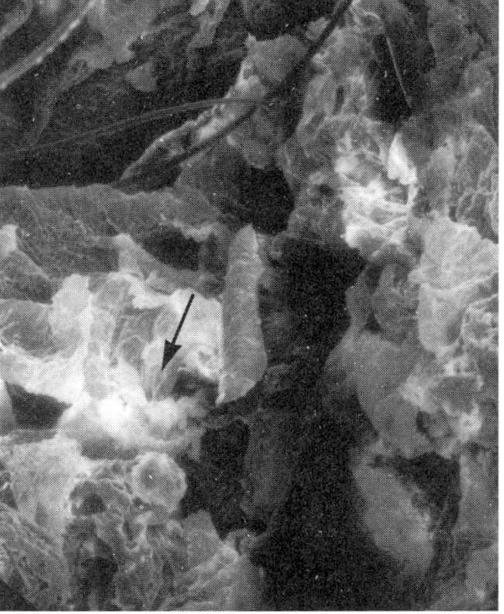

od of time in order to allow the immune system to destroy the intruder. If, for any reason, the immune system is not functioning properly, the patient can relapse when drug therapy is stopped. Any factor that damages the defense systems makes a bird more susceptible to all types of infectious agents. These factors may include poor nutrition,

FIG 3.9
The sloughing, outer layers of dead skin (arrow) can be easily visualized when the feathers are removed.
right *The outer layer of dead skin is magnified 50 times (arrow).*
photographs courtesy of W.L. Steffens

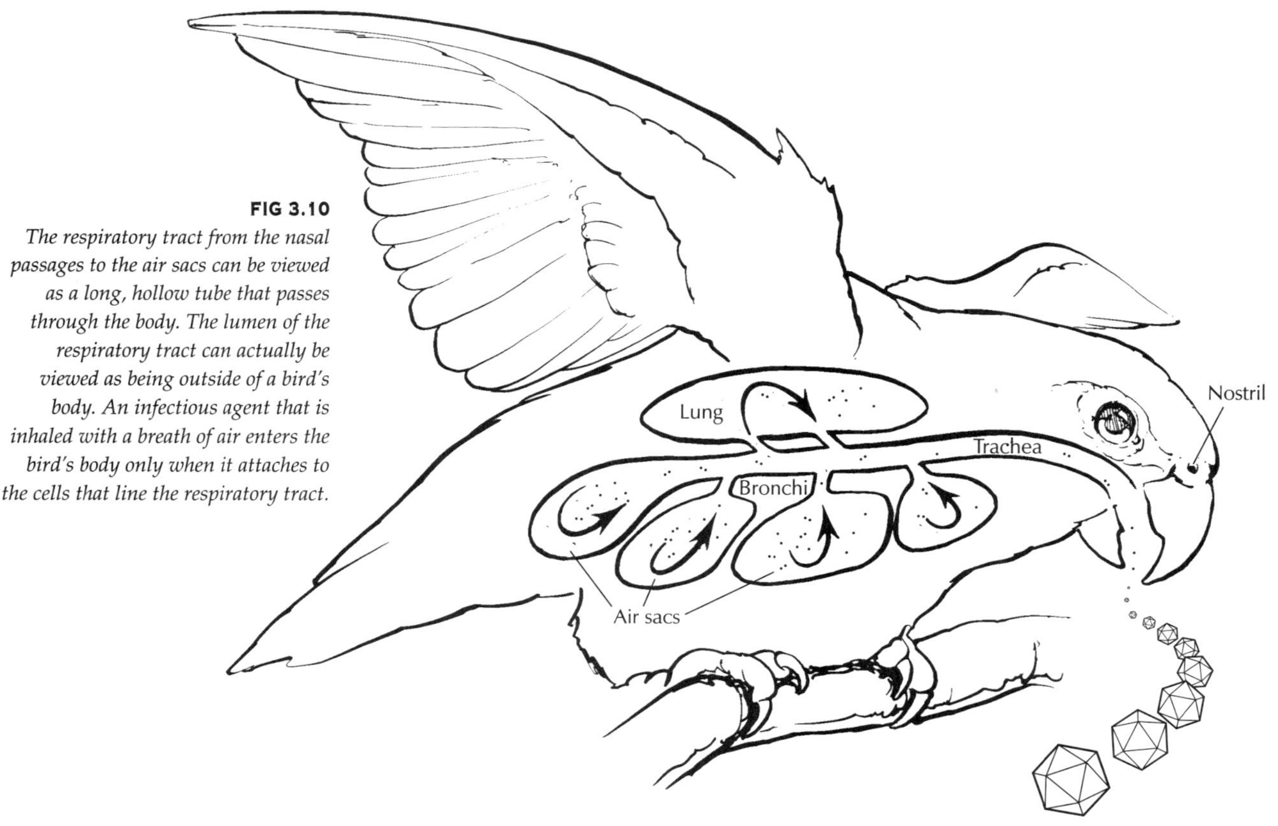

FIG 3.10
The respiratory tract from the nasal passages to the air sacs can be viewed as a long, hollow tube that passes through the body. The lumen of the respiratory tract can actually be viewed as being outside of a bird's body. An infectious agent that is inhaled with a breath of air enters the bird's body only when it attaches to the cells that line the respiratory tract.

poor air quality, exposure to toxins or cigarette smoke, or poor hygiene. The unnecessary administration of antibiotics to birds also can damage the defense systems, making a bird more susceptible to some infectious agents, particularly viruses and fungi.

EXTERNAL DEFENSES: PROTECTIVE BARRIERS

Viruses and other infectious disease-causing organisms are common in the environment. The initial defenses that a bird has for protecting itself against viruses include skin and mucosal barriers that prevent these organisms from entering the body, the interactions of normal microbial flora found on the skin and mucous membranes, and specific cells and their secretions that attack invaders before they can attach to cells.

The cells that line the gastrointestinal, respiratory and urogenital tracts function as physical barriers, preventing many organisms from entering the body. Barriers that prevent organisms from entering the body are generally non-specific and help protect a bird from all types of infectious agents including viruses, parasites, bacteria and fungi. Any factor that damages the integrity of these protective barriers (such as trauma, ingestion or inhalation of toxins, improper use of antibiotics, improper use of disinfectants or poor nutrition) may weaken a bird's defense mechanisms.

As an example of the type of damage that inhaled toxins can cause, it has been demonstrated in humans that the inhalation of cigarette smoke damages the cells that line the trachea, making

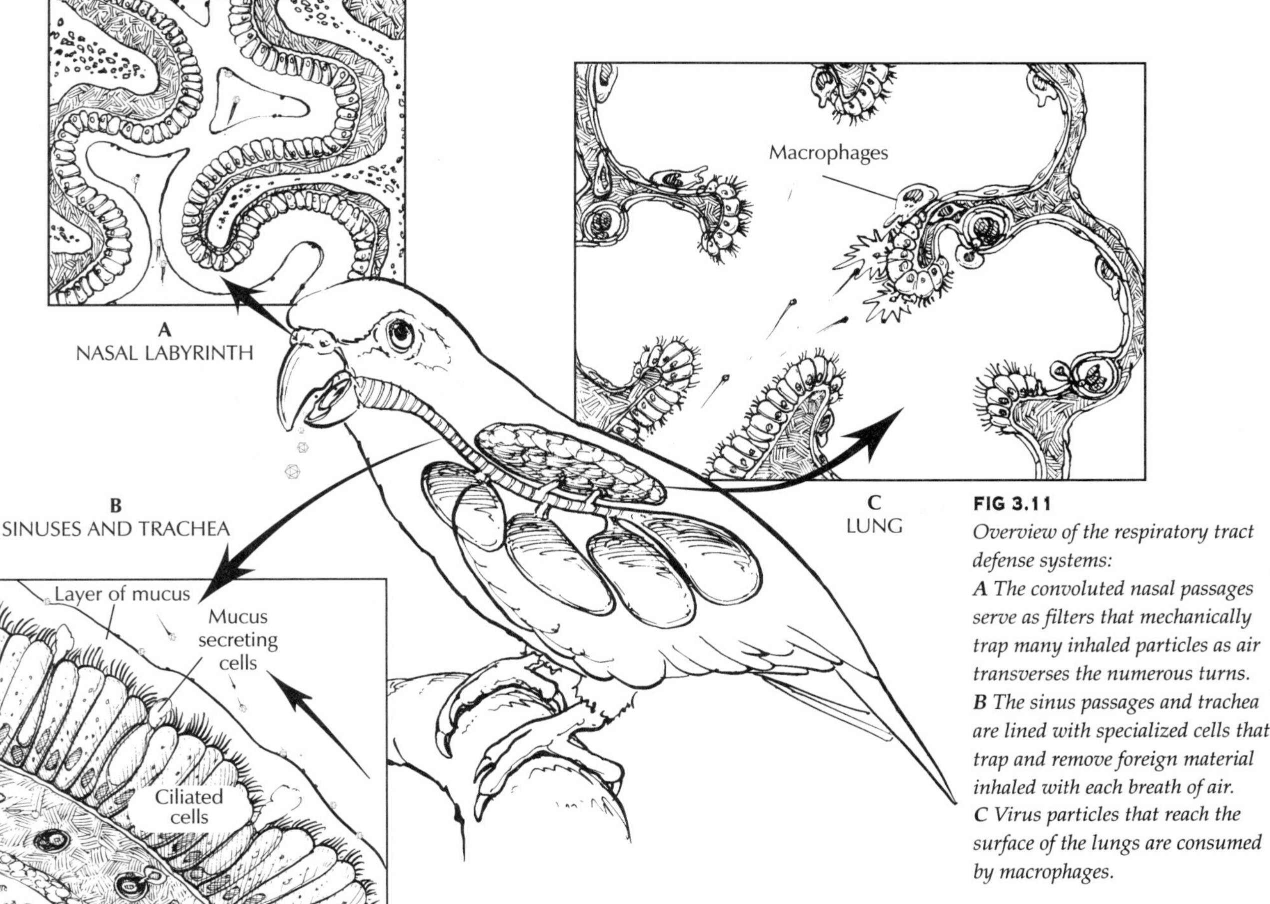

FIG 3.11
Overview of the respiratory tract defense systems:
A The convoluted nasal passages serve as filters that mechanically trap many inhaled particles as air transverses the numerous turns.
B The sinus passages and trachea are lined with specialized cells that trap and remove foreign material inhaled with each breath of air.
C Virus particles that reach the surface of the lungs are consumed by macrophages.

exposed individuals more susceptible to infection by influenza virus. It could be hypothesized that a bird's being exposed to cigarette smoke would result in a greater risk of its being infected by influenza A virus and other respiratory tract pathogens.

The microbial flora normally found on the skin and mucous membranes help prevent pathogens from causing infections by mechanically blocking their access. When these "beneficial" bacteria or fungi are destroyed by careless use of disinfectants or unnecessary antimicrobial agents, the normal flora can be destroyed, making a bird more susceptible to viruses and pathogenic forms of bacteria and fungi.

SKIN DEFENSES

Healthy feathers and skin have several components that help make them effective barriers against many invading microorganisms. The viruses that can enter the body through the skin have developed mechanisms to by-pass the skin's natural defense mechanisms. The developed feathers and outermost surface of the skin are dead and will not support the growth and replication of viruses. Most viruses that come in contact with the feathers or outermost layer of nonviable skin become entrapped in the debris and are discarded with the outer layer of constantly sloughing, dead skin *Figure 3.9.*

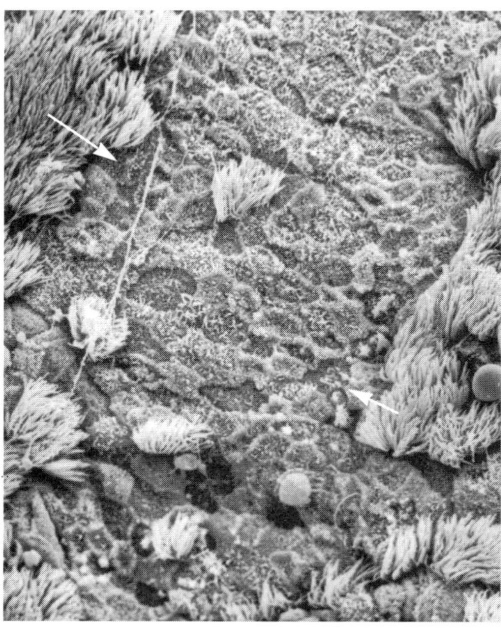

The trachea is lined with two types of specialized cells: goblet cells and another type that is covered with cilia.

Mucus (arrow) secreted by the goblet cells will trap inhaled materials, which are then carried out of the respiratory tract by the cilia (white arrow).

Inhalation of toxic fumes can cause damage to the mucociliary blanket, making a bird more susceptible to infectious agents. A severely damaged area of the trachea is highlighted with white arrows.

photographs courtesy of Jean Sander

The healthier the skin and feathers, the more effective they are as physical barriers. Specialized cells in the skin secret fatty substances that help maintain the health and integrity of the living layers of the skin. These fatty acid substances also have properties that support the growth of "beneficial" bacteria and fungi and may retard the growth of pathogenic bacteria and fungi. Traumatic injuries and insect bites are two impor-

tant ways in which the skin barrier can be breached, allowing some viruses access to a bird.

RESPIRATORY DEFENSES

The respiratory system is constantly exposed to viruses, fungi and bacteria through the lumen of the trachea, bronchi and lungs *Figure 3.10.*

A properly functioning respiratory tract has an arsenal of defenses that prevent viruses inhaled in contaminated air from passing out of the lumen of the respiratory tract and into the body *Figures 3.11, 3.12.* The mucus-entrapped material is expelled from the sinuses and nasal passages when a bird sneezes, or is passed to the back of the throat and swallowed. The mucus-entrapped material is expelled from the primary bronchi and trachea when a bird coughs or is carried by cilia up to the throat and swallowed. A bird also produces viral-specific antibodies, which are secreted into the mucus that lines the trachea. These antibodies will bind to and neutralize the specific viruses to which they have been produced.

Any damage to the delicate mucus-secreting or ciliary-containing cells that line the upper respiratory tract would be expected to make a bird more susceptible to infectious agents that are present in inhaled air. Some types of viruses can infect and damage the cells that secrete mucus and contain cilia. The damage that these viruses cause to the bird's respiratory defense system allows other inhaled virus particles, bacteria and fungi freer access to the deeper tissues in the respiratory tract. Severe damage to the respiratory

AVIAN VIRUSES: FUNCTION AND CONTROL

defense mechanisms can occur from inhalation of stale air, large amounts of dust, or toxic fumes, such as disinfectant fumes, cigarette smoke, aerosols, paint, varnish, or insecticides.

The upper respiratory tract generally contains a number of normal bacteria that help prevent other pathogens from reaching the cells that line the respiratory tract. The trachea, lungs and air sacs of the lower respiratory system should be considered sterile environments.

If inhaled virus particles are not removed by the mucociliary defense mechanism, they can move through the trachea and may be deposited on the surface of the lungs. Virus particles that are deposited on the surface of the lungs are generally consumed by macrophages that line the interior surface of this organ *Figure 3.13*. Macrophages can be viewed as cellular "vacuum cleaners" that engulf foreign debris and carry it out of the body. Many viral particles that pass by the mucosal defenses are consumed and subsequently destroyed by macrophages before the virus has an opportunity to initiate an infection.

GASTROINTESTINAL TRACT DEFENSES

The lumen of the oral cavity, esophagus, proventriculus, ventriculus, intestinal tract and cloaca can be collectively viewed as a long, tubular processing plant for ingested materials *Figure 3.14*. The digestive tract has a number of defenses designed to prevent an infectious agent from attaching to the cells that line the gastrointestinal tract *Figure 3.15*.

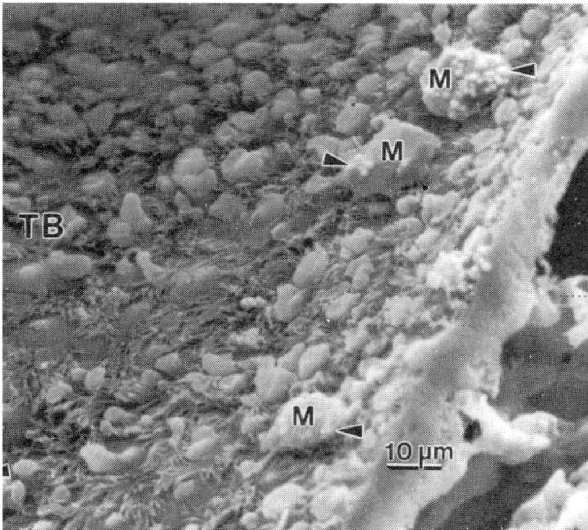

FIG 3.13
Macrophages (M) can be seen on the surface of the passages through the lungs (TB).

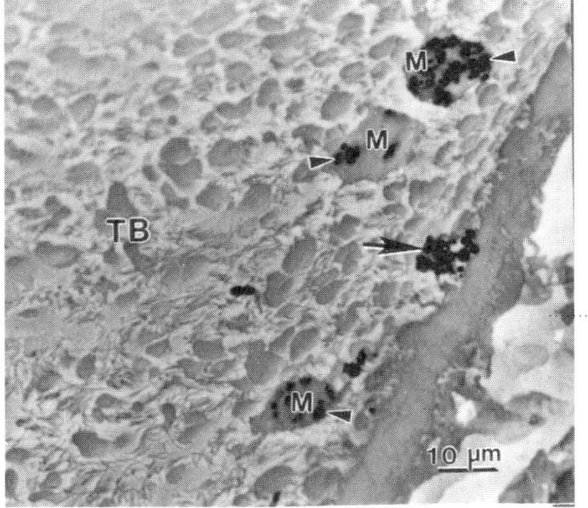

Iron particles (arrows) inhaled by an animal have been consumed by the macrophages and are being carried out of the body by the mucociliary escalator.

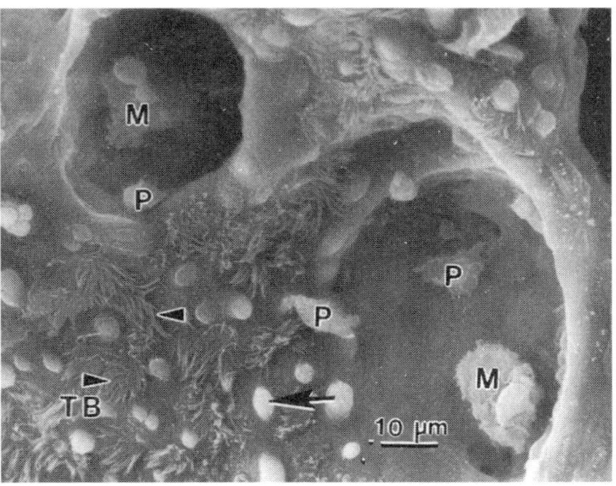

Six hours after an animal inhaled silica both macrophages (M) and other white blood cells (P) are accumulating on the surface of the lungs. Collectively, macrophages and other white blood cells will attempt to remove the inhaled silica and protect the damaged mucosa of the lung from infectious agents. photographs courtesy of D.B. Warheit, reprinted with permission of the Anatomical Record 231:107, 1991.

Virus particles can enter the lumen of the digestive system from many external sources: they can be swallowed with contaminated food or water, picked up on feathers and taken into the oral cavity through preening activities, passed from other birds during

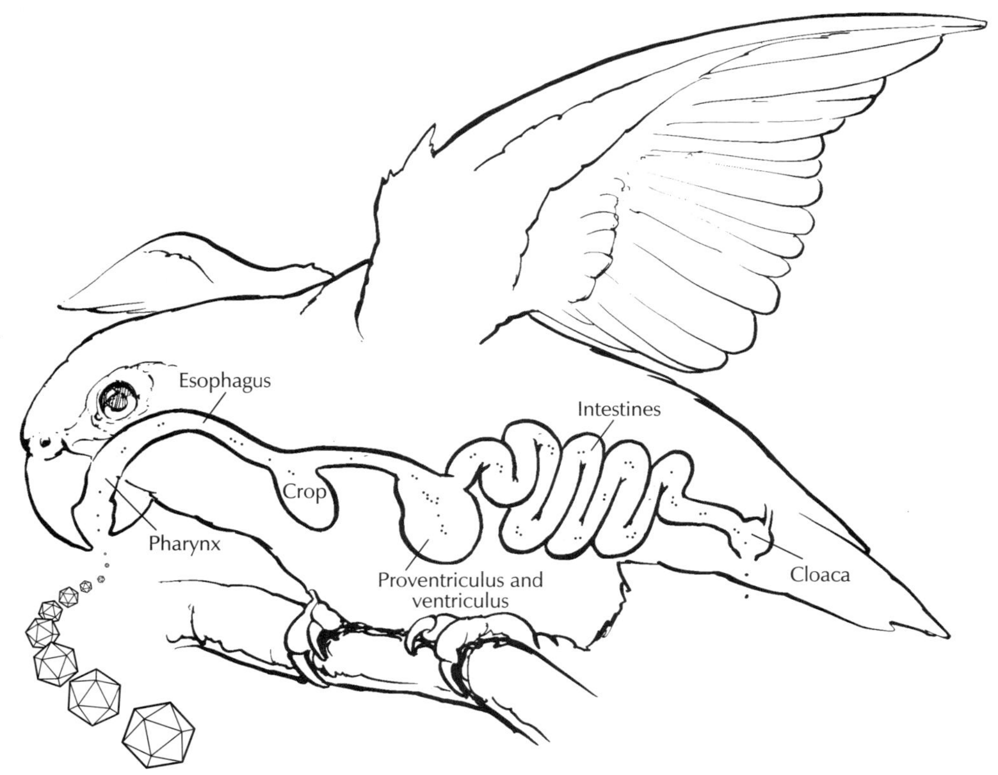

The alimentary tract from the oral cavity to the cloaca can be viewed as a long, hollow tube that passes through the body. The lumen of the alimentary tract can actually be viewed as being outside of a bird's body. An infectious agent that is ingested with a bird's food or water enters the bird's body only when it attaches to the cells that line the gastrointestinal tract.

routine feeding activity, or inhaled and later coughed up and swallowed. Saliva and mucus that are present in the oral cavity and throughout the upper gastrointestinal tract bind to and inactivate some types of ingested viruses. Hydrochloric acid in the proventriculus of a normal bird rapidly destroys most types of ingested viruses. Many other viruses that are ingested are destroyed by the protein- and fat-digesting compounds that are produced by the liver (bile acids) or pancreas (pancreatic enzymes) and secreted into the upper intestinal tract.

The enveloped viruses are particularly susceptible to bile acids, but the protein coats of nonenveloped viruses are relatively resistant to bile acids *see Table 1.4.* Currently, the only enveloped viruses that are known to enter a bird through the digestive tract are members of the Coronaviridae family. The factors that allow this enveloped virus to survive bile acid degradation have not been defined. The infectivity of many enteric viruses is actually enhanced by the changes that the protein-destroying enzymes in the intestinal tract cause to the outer covering of the virus particle. Other intestinal viruses are able to avoid the inactivating effects of these acids and enzymes by blending with food, which then serves as a protective matrix.

The normal bacteria found in the gastrointestinal tract can play an important role in preventing viral infections. These nonpathogenic bacteria compete for nutrients, inhibit organisms from reaching the lining of the alimentary tract and secrete compounds that prevent other bacteria, fungi and some viruses from gaining access to the cells that line the gastrointestinal tract. When the normal flora is destroyed or weakened by the indiscriminate use of antibiotics, by malnutrition or by the ingestion of disinfectant residues, opportunistic organisms that would not normally infect a bird can begin to grow at a rapid rate, resulting in an infection. As an example, it is common for birds that are receiving certain types of tetracyclines to develop secondary yeast infections. This occurs because the tetracyclines destroy the normal bacteria that are preventing the yeast from proliferating in large numbers.

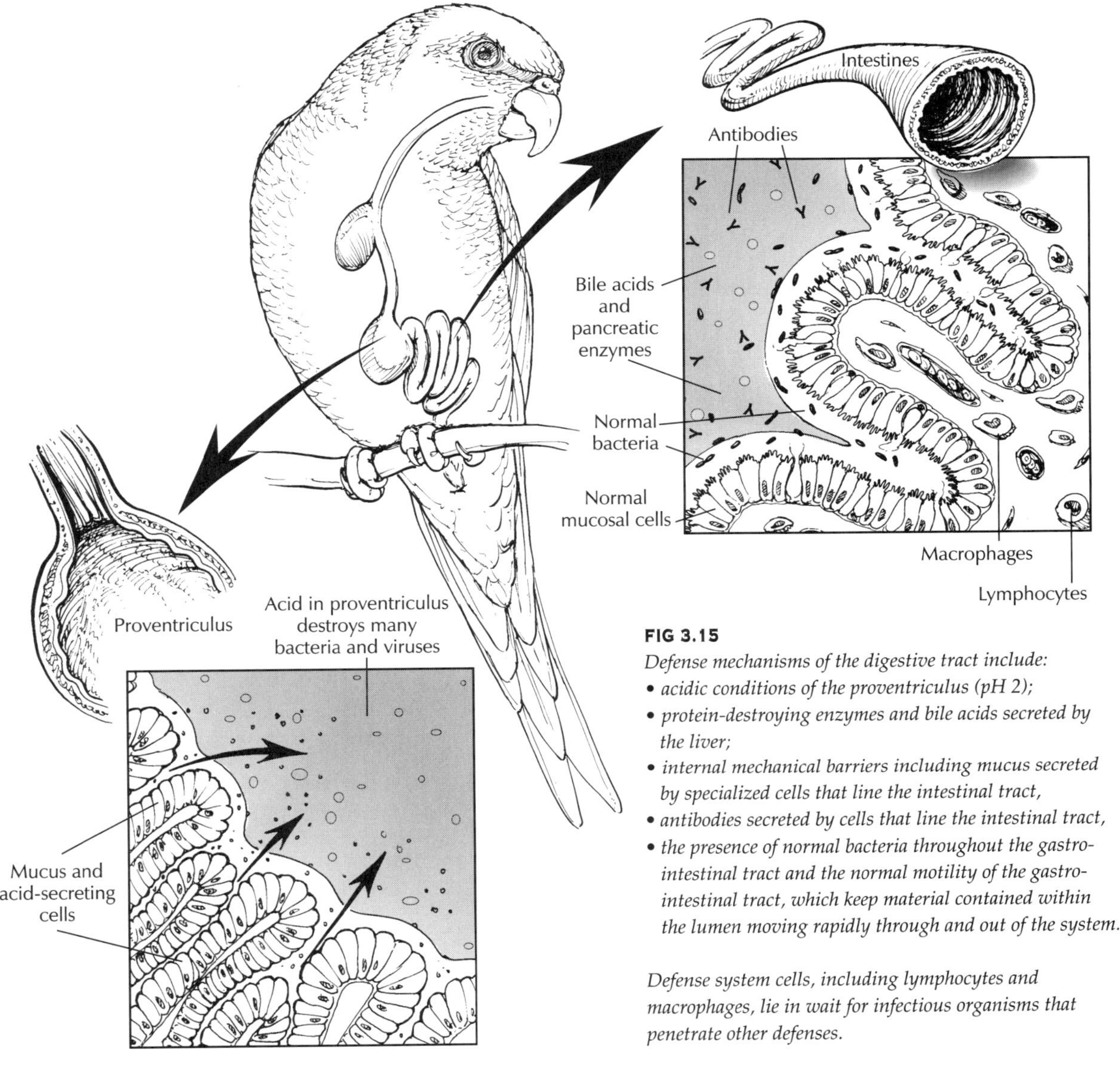

FIG 3.15

Defense mechanisms of the digestive tract include:
- *acidic conditions of the proventriculus (pH 2);*
- *protein-destroying enzymes and bile acids secreted by the liver;*
- *internal mechanical barriers including mucus secreted by specialized cells that line the intestinal tract,*
- *antibodies secreted by cells that line the intestinal tract,*
- *the presence of normal bacteria throughout the gastrointestinal tract and the normal motility of the gastrointestinal tract, which keep material contained within the lumen moving rapidly through and out of the system.*

Defense system cells, including lymphocytes and macrophages, lie in wait for infectious organisms that penetrate other defenses.

When the bacteria are destroyed, the yeast are able to replicate freely, overwhelm the bird's defense mechanisms and induce an infection.

The continuous replacement of the cells that line the gastrointestinal tract and the constant movement of material through the lumen also function to eliminate microbial organisms from the intestinal tract. Any factor that reduces gastrointestinal movement would be expected to make a bird more susceptible to infections.

Enteric viruses replicate in and sometimes destroy the cells that line the intestinal tract. These cells are responsible for protecting the body from infectious organisms found in the digestive tract, for secreting enzymes involved in digestion and for absorbing fluids and digested foodstuffs. When the cells that line the digestive tract are damaged, fluid and material in the lumen are not properly digested or absorbed, resulting in diarrhea. Diarrhea is often considered a problem that should be stopped with medications; however, viewing diar-

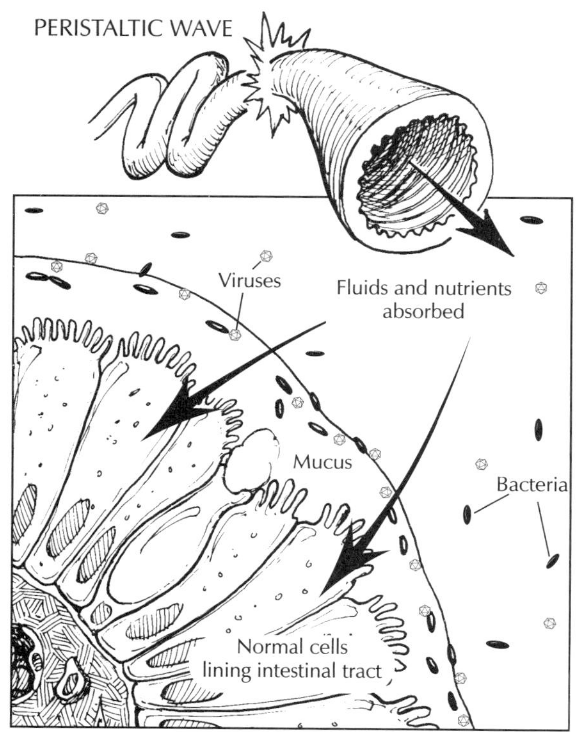

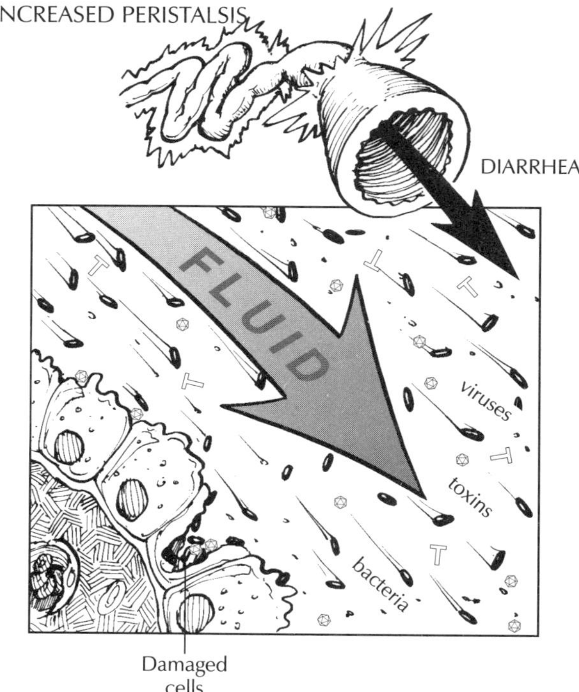

FIG 3.16

A Cells that line the proventriculus, ventriculus and intestines are responsible for; protecting the body from infectious organisms, secreting enzymes involved in digestion and absorbing fluids and digested foodstuffs in the lower digestive tract.

B When the cells that line the digestive tract are damaged, fluid and material in the lumen are not properly digested or absorbed, and the intestinal tract motility increases, resulting in diarrhea.

C Preventing diarrhea causes a stagnation of the material contained within the lumen of the intestinal tract and may actually reduce the body's ability to eliminate undesirable microorganisms.

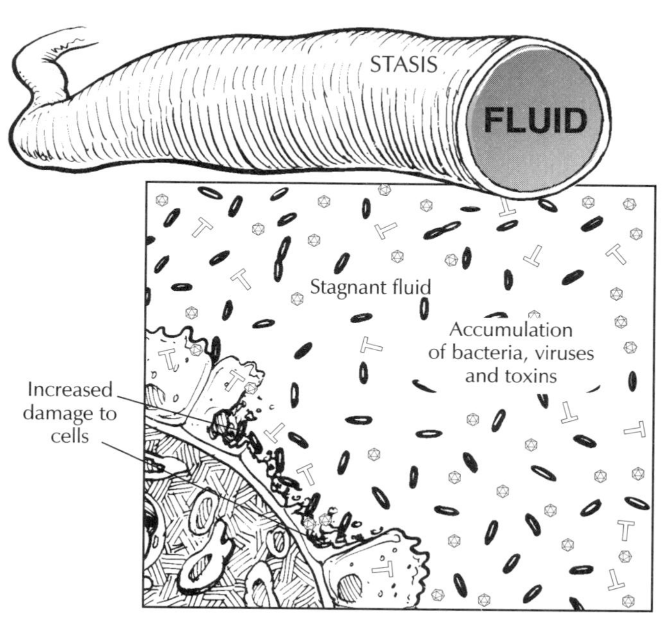

rhea from a different perspective, it could be considered a mechanism that the body uses to remove toxins, irritating substances or microorganisms from the gastrointestinal tract. Diarrhea can be viewed as a natural defense mechanism that the body uses to flush the gastrointestinal tract *Figure 3.16*.

Preventing diarrhea may actually reduce the body's ability to eliminate undesirable microorganisms from the alimentary tract. Furthermore, the reduced motility of the gastrointestinal tract would favor the growth and accumulation of pathogenic organisms and their toxins. In some cases, it may be best for the health of the bird to treat the initiating cause of the diarrhea, ensure that the bird is well hydrated with abundant fluids and allow the diarrhea that is occurring to flush undesirable materials out of the digestive tract.

INTERNAL DEFENSES: THE IMMUNE SYSTEM

The immune system can be viewed as a network of body features, including specific types of cells and their secretions, that collectively form a "search, find and destroy" unit. Functioning in concert, the components of the immune system attempt to identify, destroy and remove any foreign material (including viruses) that bypass other defense mechanisms and enter the body. The ability of a bird to clear a viral infection and develop viral-specific resistance that will protect it from future infections by the same virus requires a functional immune system. The immune system reacts against the viral proteins that compose the capsid (nonenveloped viruses) or envelope (enveloped viruses) *see Figure 1.9*. If the immune response to a virus that enters the body is rapid and sufficient, a viral infection may be controlled before signs of disease are noted; if the immune response is delayed or insufficient, clinical changes that represent damage to the body may occur.

The principal organs that form a bird's immune system are the bone marrow, the cloacal bursa, the thymus and the spleen *Figure 3.17*. The bone marrow is the site where white blood cells (heterophils, macrophages, monocytes, T-lymphocytes and B-lymphocytes) are produced; it can be viewed as a manufacturing plant for critical components of an elaborate defense network. The cloacal bursa and thymus are sites where the lymphocytes produced in the bone marrow undergo maturation and become functional members of the defense system. The cloacal bursa and thymus can be viewed as assembly plants for some of the weapons used by a bird's immune system. Once fully functional, the white blood cells that have been produced in the bone marrow and have matured in the cloacal bursa or thymus begin to circulate through the body where they initiate and control the body's response to invading viruses. The spleen functions as a filter for the circulating blood. Virus particles and other foreign materials that are present in the blood are trapped in the small vessels that course through the spleen.

MACROPHAGES

Macrophages and monocytes are related and have a similar function, which is to identify and consume damaged cells and foreign materials that have entered the body. In this respect, these cells can be viewed as the immune system's radar and initial assault team.

As a radar system, these cells are found either circulating in the blood (in the form of monocytes) or fixed in tissues (macrophages) where viruses may enter or attack a bird. The monocytes that are circulating in the blood stream are attracted to sites of inflammation by special chemicals secreted by damaged cells. Macrophages are found along small blood vessels in the liver, lungs and various body cavities. They occur in the highest concentration in critical

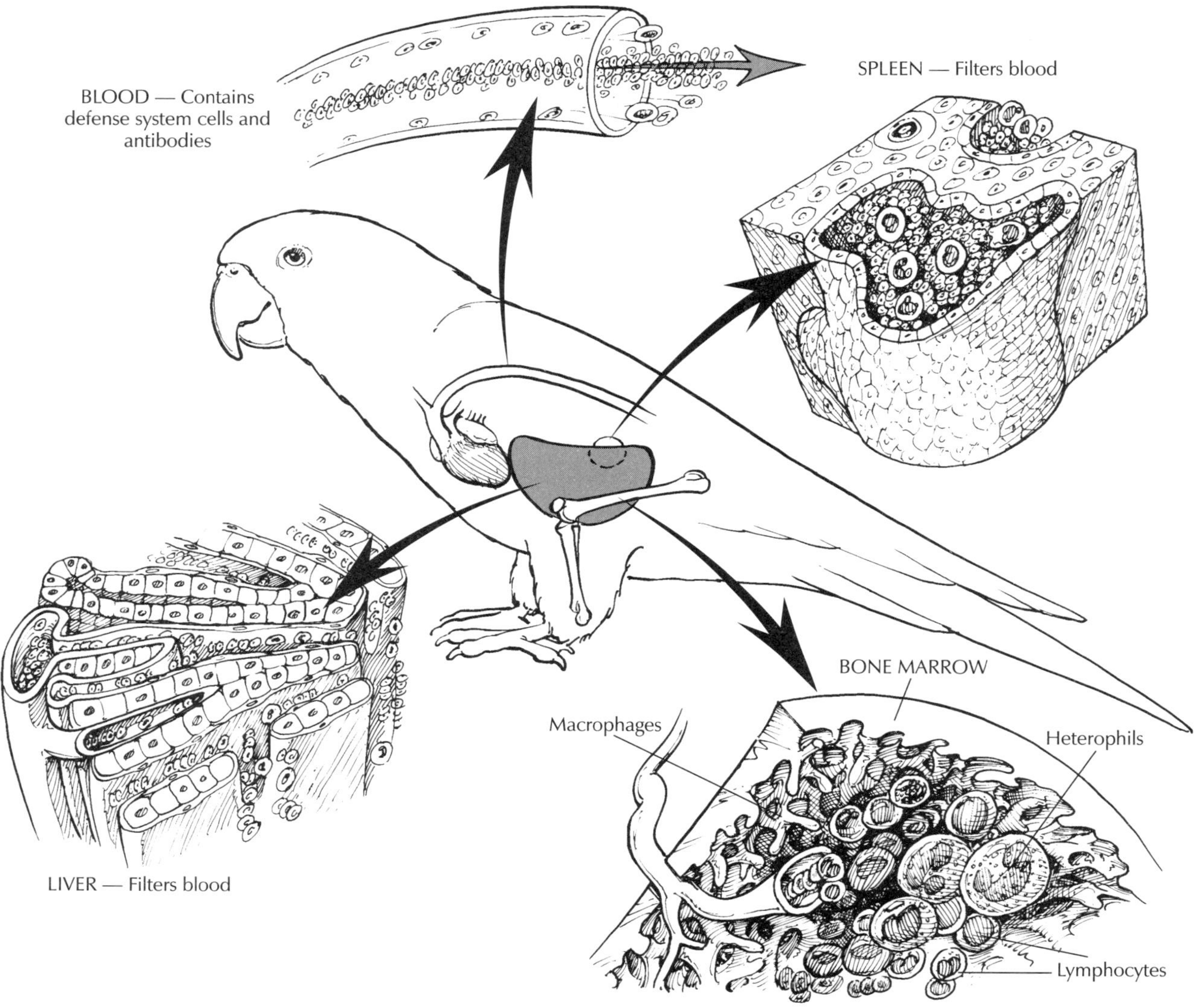

FIG 3.17

The bone marrow, blood, spleen and liver are all involved in defending a bird against infectious agents. White blood cells are produced in the bone marrow and are carried throughout the body in the blood. The liver and spleen function as filters that cleanse the blood. Virus particles and other foreign materials that are present in the blood are trapped in large numbers in the small vessels that course through the spleen.

sites where a virus may enter a bird's body: in the linings of the respiratory, intestinal and urinary tracts *see Figure 3.13*. If undamaged, macrophages can remain functional for several months.

If an invading virus passes the mucosal defense system and enters a bird's body, macrophages will attempt to ingest and then destroy the virus. A macrophage that consumes a virus should become activated, sending chemical signals to alert the body that a virus has entered *Table 3.3*. Thus alerted, the properly functioning immune system will then send other macrophages and immune system cells (B-lymphocytes and T-lym-

TABLE 3.3

Effects on the body of secretions released by macrophages[1]

Increase the body temperature.

Induce sleep to conserve energy and aid the healing process.

Increase the metabolic rate.

Stimulate T-cells, which destroy virus-infected cells and coordinate the immune system.

Stimulate B-cells to produce antibodies.

Stimulate fibroblasts to improve healing of damaged tissues.

phocytes) to try to stop the virus from further invading the bird. This process initiates the production of virus-specific

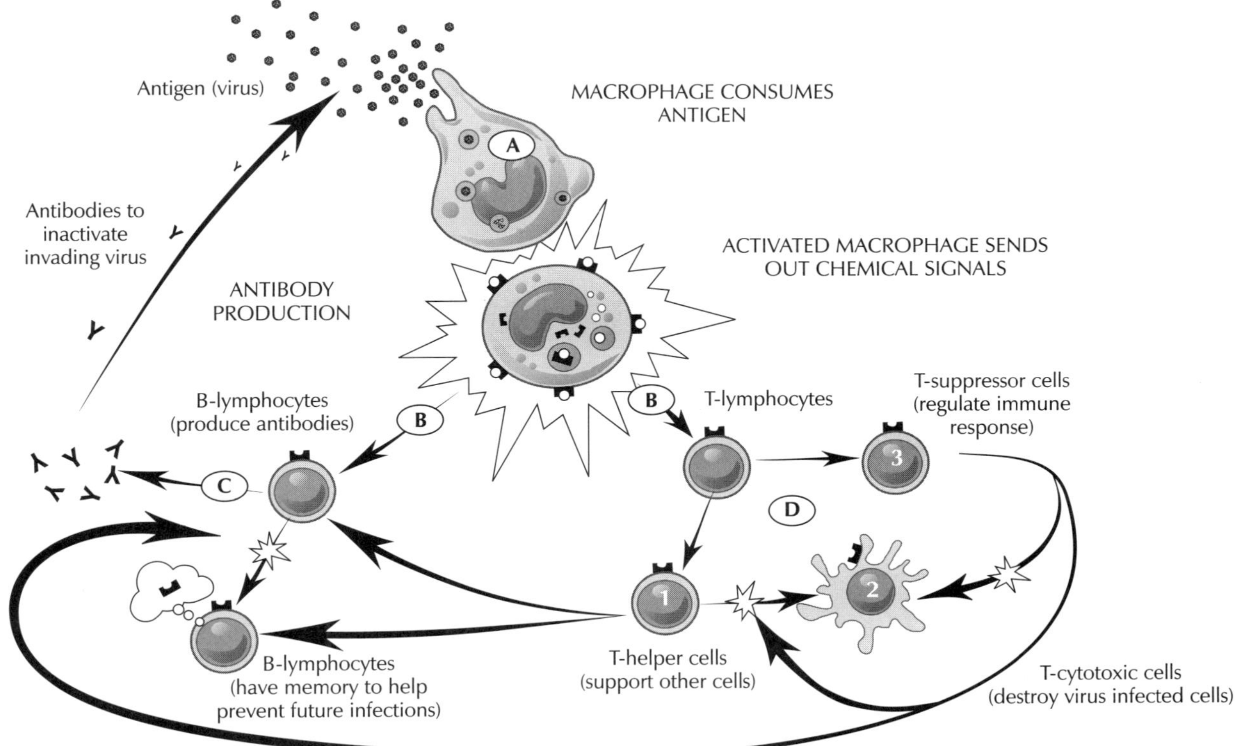

FIG 3.18

A A viral-specific immune response is initiated when a macrophage consumes an invading virus particle. A macrophage that consumes a virus should become activated and send chemical signals to alert the body that a virus has entered. The macrophage in this capacity is responsible for stimulating the immune response.

B Once alerted, the properly functioning immune system will then send other macrophages and immune system cells (B-lymphocytes and T-lymphocytes) to try to stop the virus from further invading the bird.

C This process stimulates B-lymphocytes to produce virus-specific antibodies that are designed to eliminate the invading virus and help protect the bird from future infections by the same virus.

D An activated macrophage also will stimulate:
1) T-helper cells to improve the functional capacity of the immune system;
2) T-cytotoxic cells to seek out and destroy virus infected cells; and
3) T-suppressor cells to govern the overall immune system and keep it from becoming overstimulated.

antibodies which are designed to eliminate the invading virus and help protect the bird from future infections by the same virus *Figure 3.18*.

Just as diarrhea may be viewed as a defense mechanism rather than a problem *Figure 3.16*, fever is not always a condition to be avoided or prevented. The fever that accompanies many viral infections may be one of the body's defenses against a virus. Activated macrophages secrete chemicals that cause an increase in the body temperature. It is theorized that this increase in body temperature has natural antiviral activity.

Experimentally virus-infected mammals given drugs (aspirin) that decrease the fever response typically develop a more prolonged or more severe disease than mammals infected by the same virus that are not given aspirin.[2]

In most cases, a virus particle that comes in contact with a macrophage is engulfed and destroyed. However, some types of viruses have developed methods of protecting themselves from the destructive effects of macrophages, including methods that prevent a virus from being engulfed by macrophages and prevent engulfed viruses from being destroyed. Some viruses, like PBFD virus, adenovirus, some paramyxoviruses and some herpesviruses, may actually persist or replicate inside macrophages or monocytes. These viruses have adapted a method that renders a major portion of a bird's defense system inoperable. Additionally, the viruses that persist in monocytes are particularly troublesome because these infected cells can easily spread the virus throughout a bird's body because these cells circulate in the blood stream *Figure 3.19*.

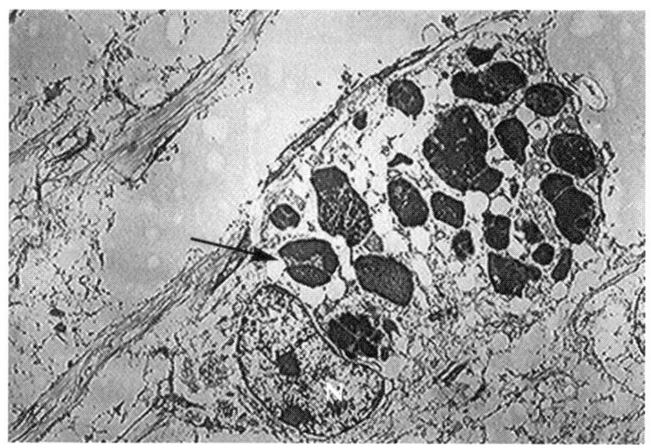

FIG 3.19

The cytoplasm of a macrophage is filled with inclusion bodies containing PBFD virus (arrow). The nucleus (N) of the cell is clearly visible. Because PBFD virus is able to persist in macrophages, the immune system of most affected birds is unable to remove the virus from the body.

B-LYMPHOCYTES

Lymphocytes produced in the bone marrow are carried to either the cloacal bursa or thymus where they undergo maturation to become fully functional components of the immune system. The lymphocytes that mature under the control of the cloacal bursa are called B-lymphocytes. Once mature, these cells produce chemical weapons called antibodies that are specifically designed to bind to and destroy a particular antigen *see Figures 3.8, 3.18.*

Antibodies are highly specific weapons that are designed to destroy the microbial organism that stimulated their production. Like an antitank weapon engineered to specifically destroy tanks and not airplanes, anti-polyomavirus antibodies are designed to destroy polyomavirus and not herpesvirus. The specific antibodies produced by the B-lymphocytes coat and help neutralize virus particles that are outside of cells (see Antibodies). In general, antibodies that are circulating outside of a cell have little effect on viruses that are inside of a cell. To help destroy viruses that are inside of cells, B-lymphocytes also perform surveillance functions by "marking" viral-infected cells to alert other members of the immune system that they should be destroyed *Figure 3.20.*

T-LYMPHOCYTES

Those lymphocytes that mature under control of the thymus are called T-lymphocytes. Based on their function in the immune response, these cells have been divided into three main types: T-helper cells, T-cytotoxic cells and T-suppressor cells. Collectively, these cells function as a "special forces hit team" to seek out and destroy circulating virus particles and virus-infected cells, support other cells involved in the immune response and control the overall response of the immune system to an invading virus.

T-helper lymphocytes release chemical messages that pass instructions to other cells involved in the immune response *see Figure 3.18.* T-cytotoxic lymphocytes seek out and destroy cells that contain virus particles or have been damaged by a virus. The surface of virus-infected cells becomes altered. It is these changes in the surface of an infected cell that signal T-cytotoxic lymphocytes circulating in the blood stream to attack and destroy the cells altered by viruses.

The body's ability to defend itself against many viral infections and certain types of cancer depends on the functional capacity of the T-cytotoxic lymphocytes.[1] Viruses that are released into an extracellular space are primarily controlled by antibodies. Viruses that remain intracellular during their replication cycle are primarily controlled by T-cytotoxic lymphocytes *see Figure 3.22.*

T-suppressor lymphocytes provide a modulating effect on the immune sys-

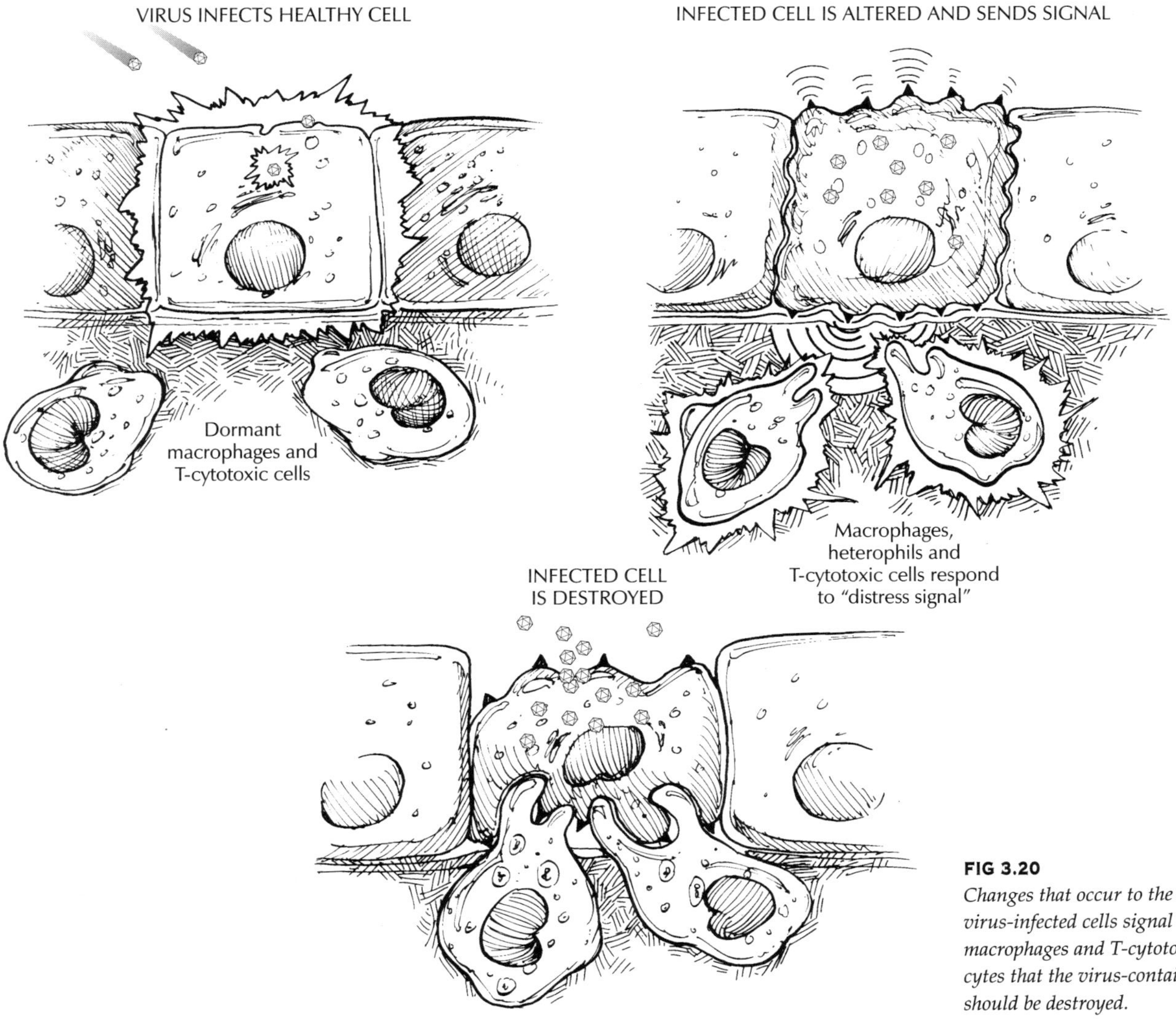

FIG 3.20

Changes that occur to the surface of virus-infected cells signal macrophages and T-cytotoxic lymphocytes that the virus-containing cell should be destroyed.

tem. The secretions from these cells prevent the immune system from becoming overstimulated and causing autoimmune damage to the bird *see Figure 3.18.*

HETEROPHILS

Heterophils, another group of circulating defenders, are primarily responsible for fighting bacterial and fungal invaders and have less of a role in defending against viral infections. However, they are the most numerous phagocytic cells in the body. Heterophils engulf and destroy dust, pollen, bacteria, fungi, protozoa and some viruses and virus-damaged cells.

Attracted to inflammation by special chemical messages that are sent out by damaged cells, heterophils quickly respond. They move out of the circulation, accumulate at the site of inflammation and act like cellular vacuum cleaners to remove injured cells and foreign material *see Figure 3.20.* Pus is an accumulation of degenerated heterophils and the dead cells and debris they have consumed. Heterophils generally have a short life-span, remaining active only a few days.

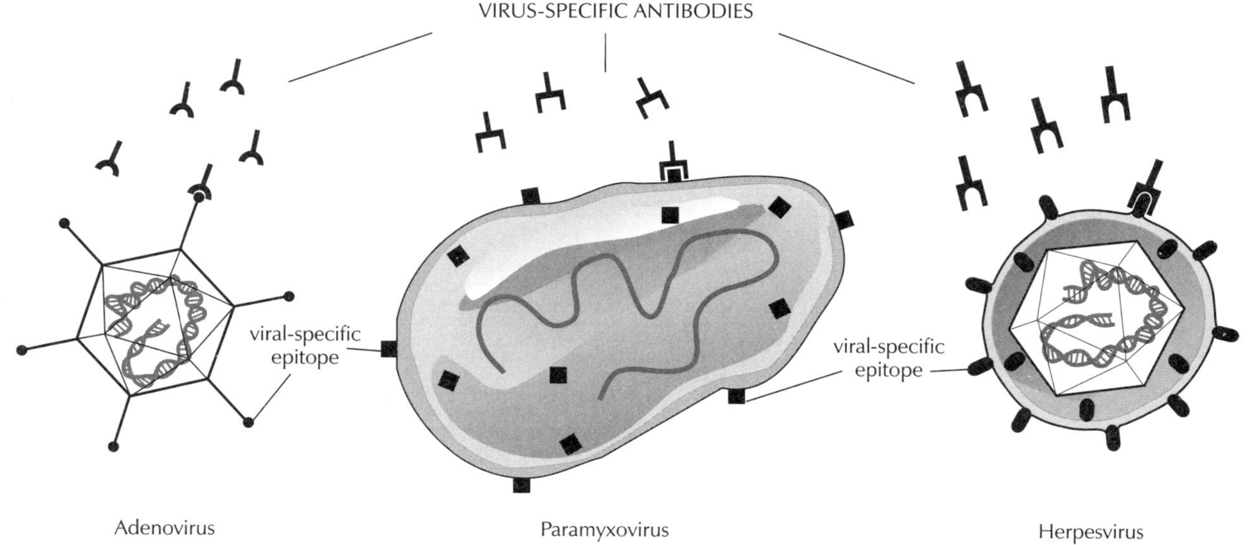

Antibodies produced against adenovirus would not be expected to bind to paramyxovirus or herpesvirus; antibodies produced against paramyxovirus would not be expected to bind to adenovirus or herpesvirus and antibodies produced against herpesvirus would not be expected to bind to adenovirus or paramyxovirus.

ANTIBODIES

Antibodies are generated by the immune system in response to an antigen; are specific for a particular type of virus; and are produced to react with a small portion of a virus particle called an epitope *see Figure 3.8.*

When a virus enters the body, it should be engulfed by a macrophage and then destroyed. During the destruction process, the macrophage sends signals to the immune system that the body is being attacked. These signals from the macrophage should initiate an immune response against the invading virus. Specifically, this process should stimulate B-lymphocytes to produce virus-specific antibodies *see Figure 3.18.* Antibodies are called viral-specific because they generally work against only the virus to which they were produced *Figure 3.21.* Viral-specific antibodies are key components of the immune system. They are involved in destroying and eliminating free virus particles, and helping protect the bird from future infections by the same virus. In some cases they help oth-

er components of the immune system destroy viral-infected cells.

Antibodies circulate in the blood of a bird and, as such, constantly bathe the outside of cells that they contact. Thus, for a virus particle to come in direct contact with antibodies it must be extracellular *Figure 3.22.* When viral-specific antibodies encounter the virus to which they were produced, they bind to the virus. A virus particle that is coated with antibodies is easier for other components of the immune system to prevent from attaching to susceptible cells and destroy. This effectively stops an infection from progressing. Antibodies are relatively ineffective in controlling viral infections in locations where virus particles can be transmitted between susceptible cells without coming in contact with viral-specific antibodies. These locations include the central nervous system and the mucosal surface of the respiratory or gastrointestinal tract.

The antibody response that a bird generates when exposed to a virus can be divided into two parts: a primary

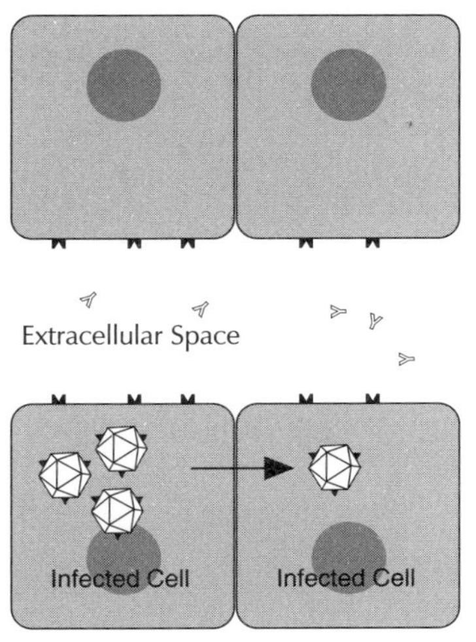

CELL TO CELL TRANSFER
Not affected by antibodies

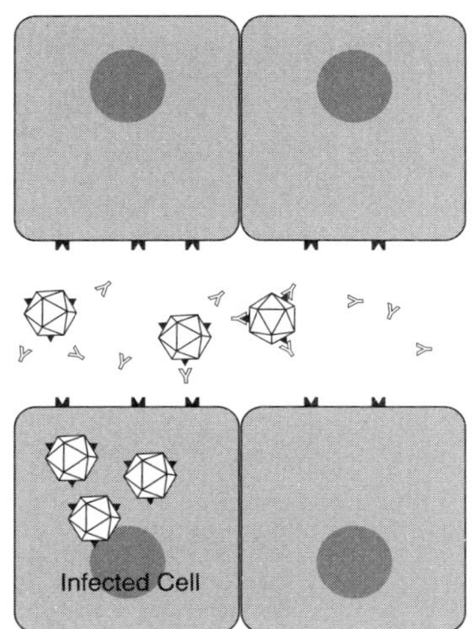

VIRUS RELEASE
Antibodies prevent new cell infection

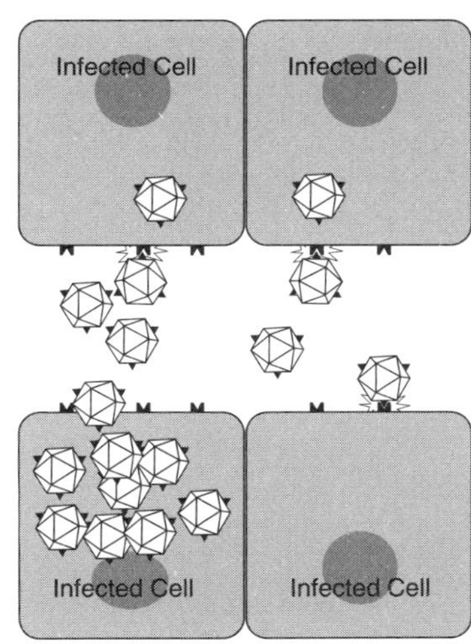

VIRUS RELEASE
No antibodies to prevent new cell infection

response and a secondary response *Figure 3.23*. The primary reponse occurs the first time the immune system is exposed to a pathogen. The secondary response occurs when the immune system is subsequently exposed to the same organism. In some cases, a bird can produce antibodies to a virus so rapidly that the viral infection does not result in disease. For example, this type of fast response occurs when a bird is infected with rabies virus. Instead of developing a severe neurologic disease, such as that which occurs in mammals infected with rabies virus, birds exposed to this virus develop a rapid antibody response. This neutralizes the virus and prevents the infected bird from developing a rabies virus-induced disease.

IMMUNE SYSTEM COMPETENCE

Much of the information on immune system function in companion and aviary birds is based on data derived from studying other avian species such as chickens and ducks. Applying con-cepts derived from these species to companion birds may not be fully accurate, particularly when evaluating neonates. Pheasants, ducks and geese produce precocial chicks that have their eyes open and are fully feathered at the time of hatching. In comparison, young psittacine, passerine, pigeon, ratite and raptor chicks are altricial: they are naked and their eyes are closed at the time of hatching. The inherent variance in growth between these two groups of birds creates numerous differences in the rate of development and functional capacity of the immune system at various ages.

Assays in mammals and chickens allow specific identification of problems in various portions of the immune system. These same tests have not been developed for evaluating the immune system of companion birds, and an understanding of the interaction between the bird's body and infectious organisms is limited. Thus, an avian veterinarian has minimal ability to effectively evaluate the functional capacity of the various

FIG 3.22

Viruses that are released into an extracellular space are primarily controlled by antibodies, while viruses that remain intracellular are primarily controlled by T-cytotoxic lymphocytes. Some viruses can transfer from cell to cell without coming in contact with antibodies contained in the extracellular spaces.

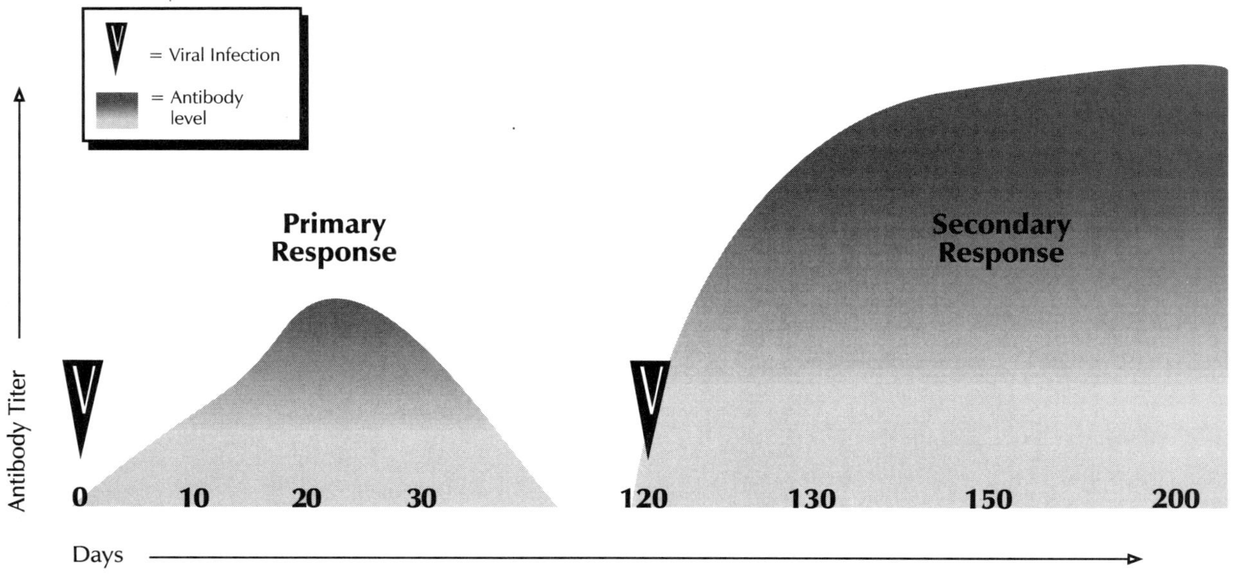

components of the immune system and is primarily limited to determining the type of circulating white blood cells present in a bird's blood. Unfortunately, specific information on the types of viral-induced changes that occur in the white blood cells of companion and aviary birds is scarce.

GENETIC INFLUENCES

The genetics of a bird undoubtedly influence its natural resistance to a particular virus. With many types of viruses, the more ancient the relationship between the bird and the virus, the more likely the virus is to produce a subclinical infection. Conversely, the more recent the relationship between the bird and the virus, the more likely the virus is to cause a severe disease.

Pheasants from North America, where eastern equine encephalitis (EEE) virus is endemic, develop subclinical infections associated with an active, protective immune response following exposure to this virus. In contrast, pheasants from other continents, where EEE virus is not endemic, frequently develop severe neurologic disease and may die when infected by this virus. These differences in the progression of EEE virus infections in these two groups of birds suggest that the pheasants from North America have developed some natural resistance to this disease because of years of coevolution between the virus and the pheasant, whereas the pheasants from other continents have not co-evolved with this virus and have minimal resistance to disease.

The ancestry of a bird may also influence the strength of a bird's immune response. Birds from parental lines that have a strong immune system are more likely to have a greater resistance to disease. Birds from parental lines that have a weak immune system are more likely to have a greater susceptibility to disease. Any breeding of birds that involves the mating of relatives (line-breeding, inbreeding) to alter specific physical characteristics such as size, color or shape would be expected to produce a bird with a weakened immune system and a shortened life span *Figure 3.24.*

Populations of animals can be bred specifically to enhance resistance to certain types of viruses. Genetic selection to increase natural resistance to disease can be practical for animals such as mice, rats and chickens that produce a large number of young quickly. It is of less value in companion birds, which may produce young that are not reproductively active for two to ten years. However, it would be best for aviculturists to breed second and third generation birds from parents that have been shown to produce strong, healthy chicks with personalities that are best suited for life as a companion bird.

AGE EFFECTS

Neonatal birds are particularly susceptible to infectious agents. Unless contaminated by the hen when the egg is developing, a young bird is in essence hatched germ-free. However, immediately upon hatching, a neonate is exposed to an environment that is full of bacteria, fungi and viruses *see Figure 3.1.* As neonates mature, they develop a "normal" microbial flora in their gastrointestinal and upper respiratory tracts that is derived from the organisms they frequently ingest or inhale. It should be noted that the normal microbial flora for a Hyacinth Macaw living in the rain forest may not be the same as the normal flora for a Hyacinth Macaw living in an outdoor flight in the eastern United States nor would it be the same as the flora of a Hyacinth Macaw living totally indoors in the western United States. Additionally, the normal flora of a psittacine bird is different from the normal flora of a pigeon, pheasant, hawk or emu.

A chick's immune system is not fully developed at birth. As the neonate matures, its immune system becomes increasingly functional *Figure 3.25.* When a chick absorbs the yolk sac from the egg, it obtains temporary protection against some of the infectious organisms to which its mother has developed immunity. This yolk contains a "cocktail of antibodies" that will protect the chick during the first weeks of life until its own immune system becomes functional. The antibodies that are transferred to the chick in the yolk have a defined half-life. As the activity of these antibodies decreases, a chick becomes increasingly susceptible to viruses to which it had been initially resistant. At this point, a chick's immune system must begin to function and protect the chick from invaders. Chicks are most susceptible to infections during this transition period when the level of protection derived from the hen begins to decay and the neonate's immune system begins to function. Generally, a neonate cannot be effectively vaccinated against a particular virus until the maternally derived antibodies to that virus decrease to a certain level *Figure 3.26.*

FIG 3.24

*When selecting a companion, breeding or aviary bird, it is advisable to select an individual that most closely resembles a bird seen in nature. The wild-type Budgerigar in **A** would be expected to live a longer life with fewer problems that the lutino Budgerigar in **B**. The lutino Budgerigar has been produced by breeding relatives to each other, a condition that would be expected to weaken the immune system of the progeny.*

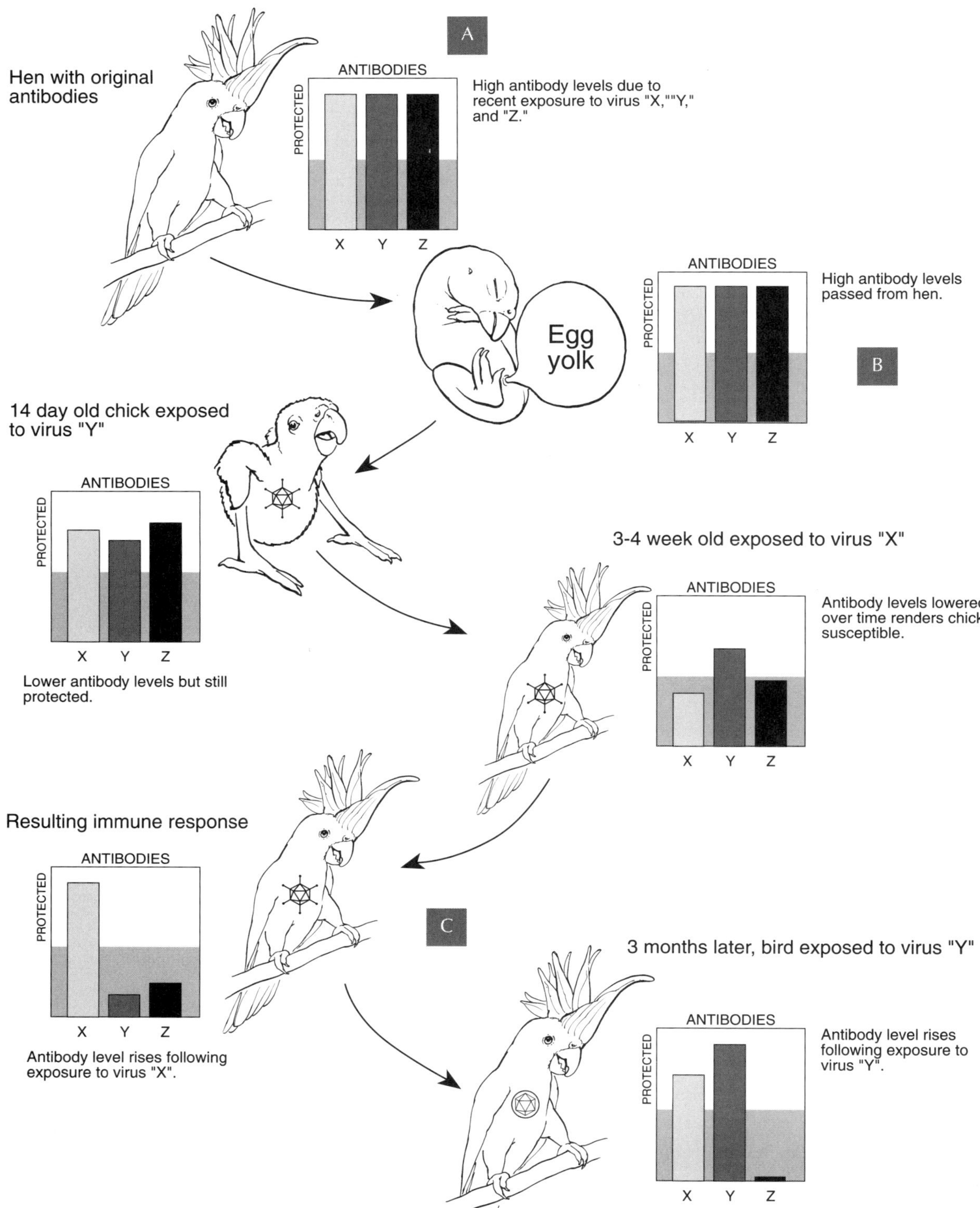

FIG 3.25

A Some antibodies that a hen has developed against infectious agents are transferred to her chicks through the yolk.

B The "cocktail" of viral-specific antibodies the chick obtains from the yolk may protect it from a virus during the first several weeks after hatching when its immune system is not yet fully functional. As the chick matures, the antibodies derived from the yolk begin to decline. The bird becomes susceptible to viruses to which it was previously resistant and the chick's immune system becomes increasingly functional.

C When the maturing chick is infected with a virus, its immune system should produce antibodies to protect it against future infections by the same virus. In this manner, antibody titers to a virus will increase and decrease throughout a bird's life, based on when the bird is infected with a particular virus.

AVIAN VIRUSES: FUNCTION AND CONTROL

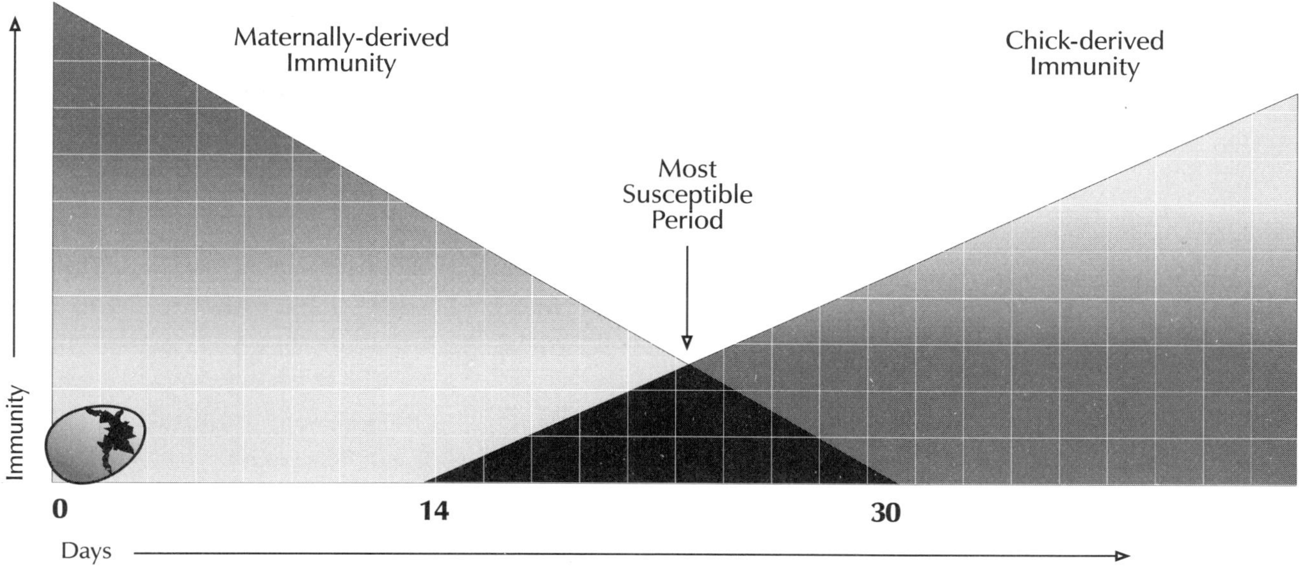

Some types of viruses produce minor disease in neonates and severe disease in adults, while other types of viruses are devastating to neonates and rarely produce disease in adults. The increase in resistance to disease as a bird matures is thought to depend on the presence or absence of maternally derived antibodies and on the progressive development of the bird's immune system. In general, neonates without maternally derived antibodies would be expected to be more susceptible to disease than neonates with maternally derived antibodies. It is for this reason that chicks from a vaccinated hen (hen with antibodies) should be resistant on hatching to the virus against which the hen was vaccinated.

Additionally, a young bird with an immature immune system would be expected to be more susceptible to disease than an adult bird with a fully functional immune system. In mammals, some viruses that infect neonates early in life have been associated with stunting and an increased susceptibility to normally nonpathogenic organisms (yeast, gram-positive bacteria, avirulent viruses). As a bird reaches old age, its immune system begins to fail, and the bird becomes increasingly susceptible to infectious diseases.

It has been demonstrated that the immune system of healthy, young chickens becomes fully functional by six weeks of age *Figure 3.27*. Two studies suggest that at least portions of a psittacine bird's immune system are functional as early as two to four weeks after hatching. In one study to evaluate the response of the immune system in young Blue and Gold Macaws, it was shown that vaccinated hens transferred antibodies to their chicks in the yolk that could still be detected at 14 days of age.[4]

The level of antibodies in the chicks decreased over a 42-day period to a point where they were virtually undetectable.[4] In another study, African Grey Parrot and Umbrella Cockatoo

FIG 3.26

Chicks are most susceptible to infections during the transition period when the level of protection derived from the hen begins to decline and the neonate's immune system becomes increasingly functional. This transition period varies with the type of bird. In general, the chick's immune system will begin to function around 14 days of age and the antibodies to most viruses that are derived from the hen will be gone by 30 days of age.

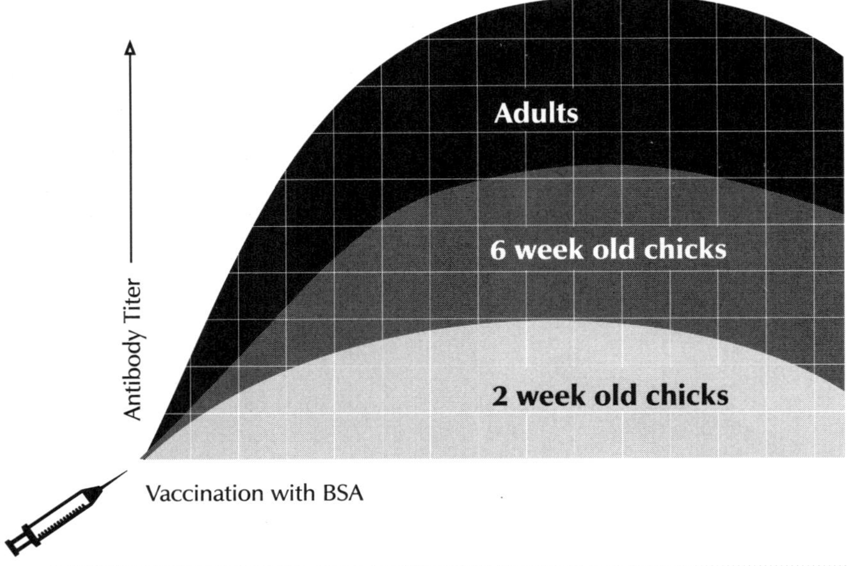

Chicks that were vaccinated with bovine serum albumin (BSA) from two to ten weeks of age produced antibodies to BSA; however, the response in chicks was weaker than the response in adults, and the response in two-week-old chicks was weaker than the response in six-week-old chicks.[4]

chicks from hens vaccinated against PBFD virus had detectable levels of anti-PBFD virus antibodies at 20 days of age, and the antibody levels decreased from 20 to 40 days of age. Adult psittacine birds and chicks as young as 30 days of age have been shown to develop antibodies to the PBFD virus following vaccination. The serum half-life of immunoglobulins in Budgerigars has been described as three to five days.

IMMUNODEFICIENCY

If the defense systems that protect a bird from infectious agents are not functioning properly, the affected bird is defined as immunodeficient. A wide variety of immunodeficiencies have been described in poultry, mammals and humans. It is safe to assume that similar problems occur in companion, aviary and free-ranging birds; however, the capacity to test for problems involving the immune system of these species of birds is minimal.

Immunodeficiencies may occur if any component of the defense system is improperly functioning or damaged. These components include the skin or cells lining the respiratory or gastrointestinal tract, the heterophils and macrophages, or the B-lymphocytes or T-lymphocytes.

Immunodeficiencies can be caused by genetic abnormalities, pathogen-induced damage, malnutrition, exposure to toxins, stress factors or drug administration. Because many immunodeficiency disorders are genetic in origin, breeding of birds with a weakened immune system would be expected to produce young with a weakened immune system. These chicks would be highly susceptible to recurring infectious diseases.

Viruses can be involved with immunodeficiencies in two ways: they can directly damage components of the immune system, or they can more easily infect a bird with a weakened immune system. Any viral infection that causes disease is considered immunosuppressive to some degree. The viruses that replicate in the immune system cells or cause damage to the thymus, bursa, bone marrow and spleen should be considered particularly immunosuppressive because of the importance of these systems to the immune response. Viruses that damage the skin or linings of the respiratory, intestinal or genitourinary tract will make a bird particularly susceptible to secondary bacterial, fungal or parasitic invaders. Immunodeficiencies may cause a bird to develop severe disease following infection with a strain of virus that is normally considered avirulent or can cause a bird to become latently infected by a virus that would be readily destroyed in a healthy bird.

Immunodeficiencies can be temporary or long-term. Viruses that damage the white blood cells circulating in the blood stream frequently cause a temporary immunodeficiency. A bird that has a reduced white blood cell count, particularly lymphopenia, should be considered to be at least temporarily immunodeficient. This deficiency usually resolves when the bird's body responds to the infection by increasing the production of lymphocytes that help the body eliminate the virus. Viruses that damage the bursa, spleen, thymus or bone marrow can cause long-lasting or permanent immunodeficiencies, particularly in young birds. Birds with a weakened immune system commonly develop recurring infections, particularly yeast infections.

Hormones, stress factors, some drugs and the plane of nutrition also influence the immune response. Various reproductive hormones can cause some viral infections to be more severe in pregnant females when compared to non-pregnant females. Other viral infections are more common in males when compared to females. The egg-laying process, which is a particularly stressful event, may weaken a hen's defense systems and make her more susceptible to infectious organisms. Additionally, hens that have latent viral infections may begin to shed virus during the stressful breeding period.

The stresses associated with shipping have been shown to predispose many species of mammals to infectious diseases, particularly those caused by viruses. While the negative effects of shipping on the immune system of companion and aviary birds have not been confirmed, it would be expected that the stresses associated with this process would make a bird more susceptible to infectious agents. Using shipping containers that keep a bird calm, quiet and protected from airborne pathogens may reduce the number of post-shipment infectious diseases *Figure 3.28*. Additionally, shipment of weaned birds with a more fully developed immune system would be considered safer than shipping unweaned birds with an incompletely developed immune system.

The administration of steroids will decrease the response of the immune system and predispose a bird to viral infections. In mammals, aspirin and tetracycline have been shown to be immunosuppressive. Many antibiotics,

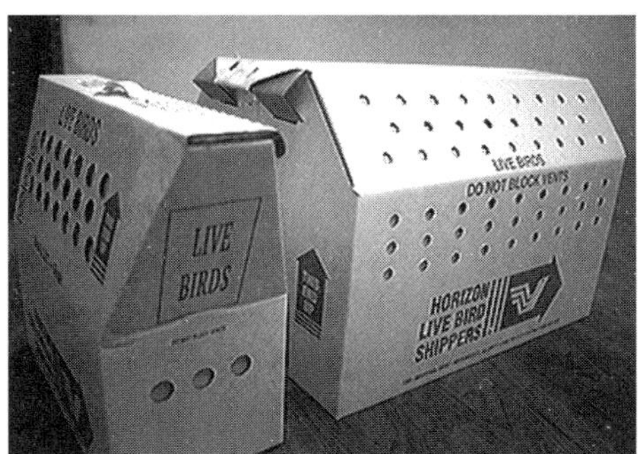

FIG 3.28

Special shipping containers designed for companion birds help keep them calm and protected from airborne pathogens during transit. When sealed, the container provides a barrier. Aerosolized pathogens are unable to pass through the surgical mask-like material which covers the air vents. The interior of the container is lined with a durable material to keep a bird secure. photograph courtesy of Horizon Micro-Environments

including tetracyclines, tylosin and aminoglycosides, are known to be immunosuppressive in chickens. Subjectively, it could be assumed that these agents are also immunosuppressive in other avian species, although their effects remain undetermined.

Malnutrition may alter a bird's defense systems by damaging the linings of the intestinal, respiratory or genitourinary tract or by specifically handicapping the functional capacity of macrophages and lymphocytes. Frequently, birds that are malnourished are also exposed to less-than-ideal husbandry practices, which further predispose them to viral-induced diseases.

VIRUS AVOIDANCE OF THE IMMUNE SYSTEM

The immune system defends a bird from viral infections by producing specialized B-lymphocytes, T-lymphocytes and antibodies that react to the protein coat or envelope proteins *see Figure 3.18*. The reaction of the immune system to these proteins determines if, when and how an infected bird is able to control and resolve a viral infection. Viruses can avoid the immune system by changing themselves so that they appear different to the immune system than they did in the past; by preventing the activation of the immune system; or by damaging the immune system and making it less effective.

When a bird recovers from a viral infection, it should develop viral-specific antibodies that will help protect it from future infections by the same virus (see Antibodies). These viral-specific antibodies are formed to react with specific proteins on the surface of the virus *see Figure 3.21*. The surface proteins of a virus can change over time through a process called antigenic drift *Figure 3.29*. In this manner, a bird that had developed resistance to one strain of virus may be susceptible to a different but related strain of virus that it encounters several months after the initial infection.

An alteration in the proteins on the surface of a virus that cause it to appear different to the immune system is called antigenic variance.

Antigenic variance is an extremely efficient method for a virus to avoid the immune system and occurs commonly with some types of enveloped viruses. Antigenic variation in the envelope proteins of the influenza A virus is responsible for the annual flu epidemics that are common in humans. Each year, the influenza virus that is transmitted through a population can be slightly different than the virus that was being transmitted the previous years.

Viruses can alter the function of the defense system by stopping the manufacture of immune system cells (damaged bone marrow), preventing the proper development of immune system cells (damaged bursa or thymus) or living inside the defense system cells (macrophages and lymphocytes), rendering them inoperable. Viruses that replicate or survive inside macrophages are particularly effective at avoiding the immune system. If the macrophage is viewed as the "radar" for the immune system, then a virus that survives within a macrophage is living inside the system that is responsible for alerting the body that it is being invaded. The ability of PBFD virus to persist inside macrophages may be one mechanism that prevents many diseased birds from clearing the PBFD virus from the body *see Figure 3.19*.

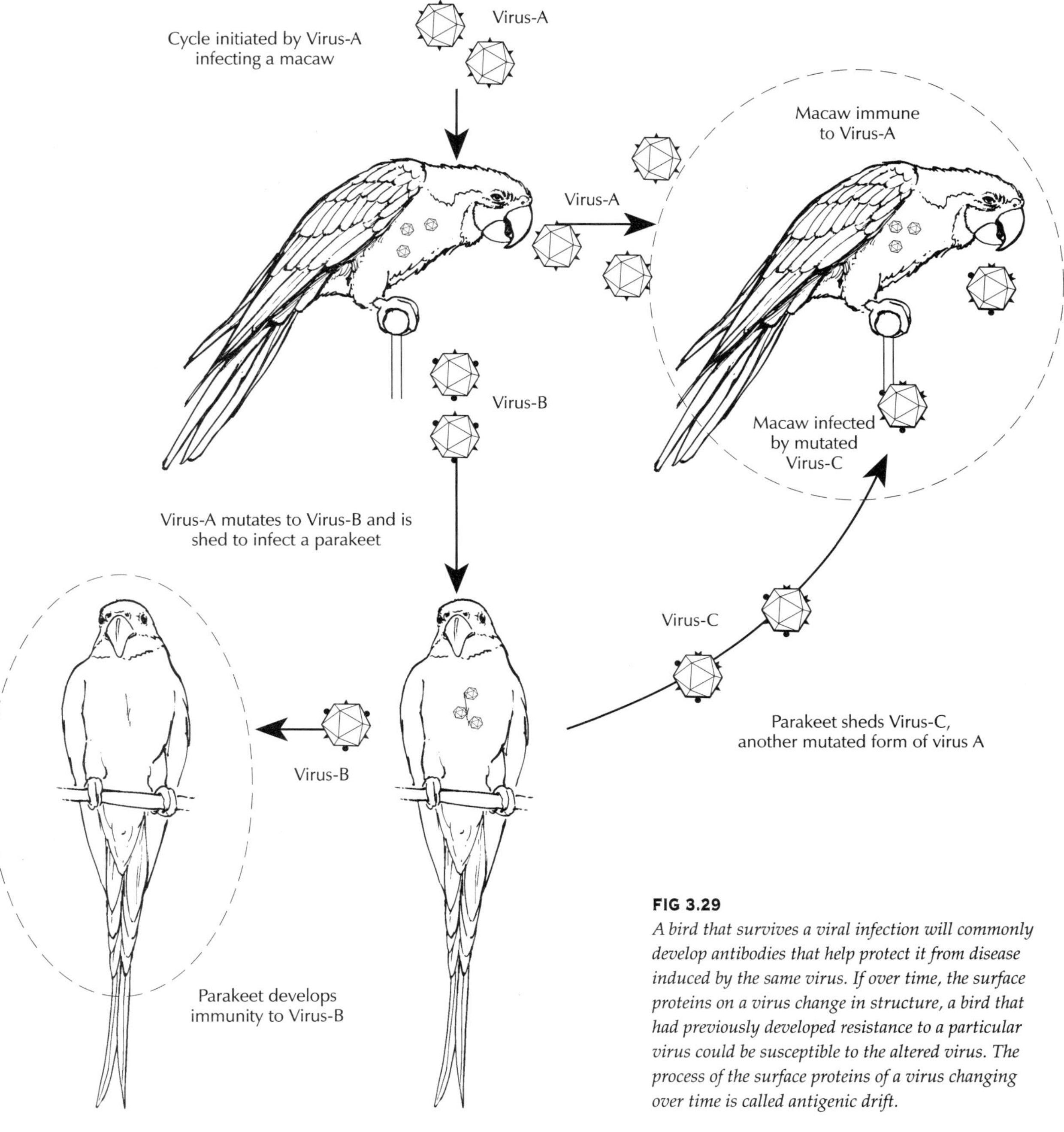

FIG 3.29

A bird that survives a viral infection will commonly develop antibodies that help protect it from disease induced by the same virus. If over time, the surface proteins on a virus change in structure, a bird that had previously developed resistance to a particular virus could be susceptible to the altered virus. The process of the surface proteins of a virus changing over time is called antigenic drift.

As viruses are released from an infected cell, they are susceptible to inactivation by viral-specific antibodies. Some viruses are able to avoid the neutralizing affects of antibodies by remaining inside the protected environment of a cell *see Figure 3.22.* Additionally, the viruses that replicate in the outer cells that line the respiratory, gastrointestinal and genitourinary tracts can cause long-term infections in these cells by moving from cell to cell without coming in contact with extracellular spaces that contain antibodies.

REFERENCES

1. Dinarello CA: Biology of interleukin 1. FASEB J 2:108-115, 1988.

2. Husseini R, Sweet C, Collie M, et al: Elevation of nasal viral levels by suppression of fever in ferrets infected with influenza viruses of differing virulence. J Infect Dis 145:520-524, 1982.

3. Loye JE, Zuk M: Bird-Parasite Interactions. Oxford, England, Oxford Univ Press, 1991.

4. Lung NP, Klein PA, Thompson JP, et al: Passive transfer of immunity and the development of antibody responses in the blue and gold macaw. Proc Assoc Zoo Vet, 1993, pp 25-26.

5. Olsen GH: Introduced avian disease and its effects on the Hawaiian ecosystem. Proc Assoc Avian Vet, 1992, pp 279-289.

6. Snyder B, Thilsted J, Burgess B, et al: Pigeon herpesvirus mortalities in foster reared Mauritus pink pigeons. Proc Am Assoc Zoo Vet, 1985, pp 69-72.)

Diagnosing Viral Infections

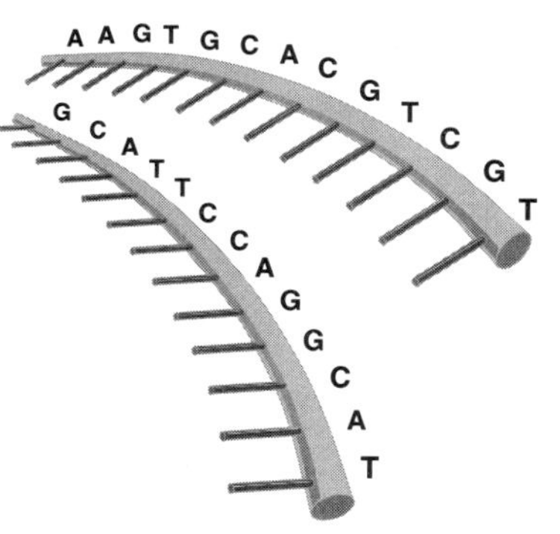

Diseases are frequently caused by the interaction of more than one type of infectious agent, including bacteria, fungi, viruses or parasites. To determine which portion of a disease process, if any, may be caused by each type of infectious agent that can be present in an abnormal bird, extensive investigations may be needed. In many cases, a virus may cause damage to a bird that allows other pathogens (particularly bacteria and fungi) to colonize unhealthy tissues. The bird may clear the viral infection that initiated the disease process, and it is a bacteria or fungus that is recovered from the abnormal tissues of the clinically ill bird. The virus that started the disease process may not be detected, because the immune system removed the virus from the body before diagnostic samples were collected. Thus, many birds that are diagnosed with bacterial and fungal infections may have initially had a viral infection.

The occurrence of any abnormality in a bird should be viewed as a warning to carefully evaluate the entire bird for underlying problems. Detecting an abnormality in a bird that is a member of an aviary should alert the attendant to scrutinize each management technique that may predispose other birds to the same problem. Aggressively pursuing the factors that have contributed to a viral-induced disease when the first affected bird is identified

can help prevent the virus from spreading to multiple birds within the aviary. Waiting to determine why a bird is sick or has died until a second or third bird is also affected gives a virus time to infect multiple birds, making control of a viral disease outbreak more difficult *Figure 4.1.*

DIAGNOSTIC PROCEDURES

Clinical abnormalities, necropsy findings and microscopic abnormalities may suggest that a bird was infected by a virus; however, pathognomonic clinical signs or lesions that absolutely confirm that a particular type of viral infection occurred are rare *Table 4.1.* Therefore, it is imperative for the health of other birds in a household or aviary that a thorough necropsy with collection of necessary diagnostic samples be performed on any bird that dies. Diagnostic samples may include portions of tissues placed in formalin for microscopic evaluation, samples of blood for bacterial or viral isolation, swabs of abnormal areas for bacterial, fungal or viral isolation, and portions of tissues that are frozen for virus isolation or toxicologic studies. Generally, the unequivocal determination that a particular viral infection has occurred can be made only through the use of specific laboratory tests.

As an example, a 70-day-old Moluccan Cockatoo, one of 40 babies in a nursery, dies less than 24 hours after becoming

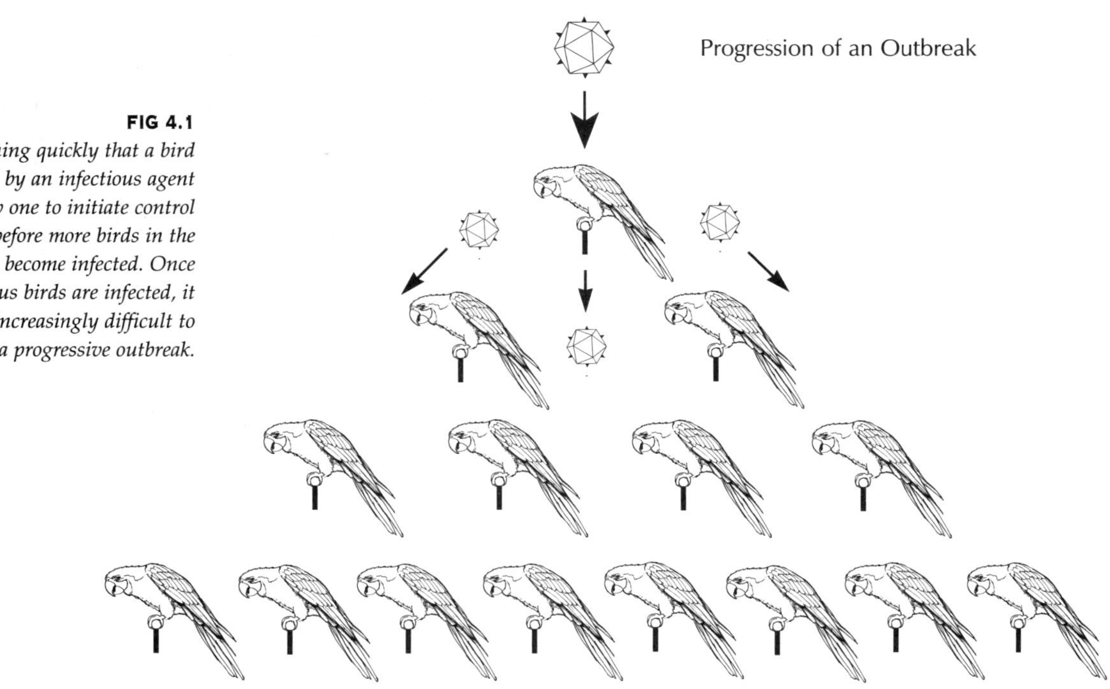

FIG 4.1

Determining quickly that a bird was killed by an infectious agent will allow one to initiate control measures before more birds in the flock become infected. Once numerous birds are infected, it becomes increasingly difficult to prevent a progressive outbreak.

slightly lethargic. The bird is from a well managed aviary that takes all available precautions to prevent infectious disease outbreaks. At necropsy, the bird is in excellent overall condition and has ingesta in the crop, proventriculus and intestinal tract, indicating that the bird died acutely. Hemorrhage is common under the skin and in the liver, heart and abdominal cavity *Figure 4.2*. The acute death in this young bird and the abnormal necropsy findings are highly suggestive of avian polyomavirus infection. However, these same abnormalities can be caused by bacterial septicemia, Pacheco's disease virus and liver disease. In this case, microscopic changes in the liver and polyomavirus-specific DNA probes indicate that the bird died from bacterial septicemia and not an avian polyomavirus infection. If the cause of death in this bird had been subjectively diagnosed as polyomavirus based on the clinical changes and necropsy findings, the wrong management decisions would have been instigated to prevent other problems within the nursery.

SAMPLE SUBMISSION

Some viruses are present in a bird for only a brief period, and they are easiest to detect if diagnostic samples are collected early in the infectious process. If samples are not collected during the correct period, it may not be possible to

TABLE 4.1

Nonspecific clinical signs that may suggest the presence of an infectious disease-causing organism, including a virus, in a bird

Drowsiness, excessive sleeping
Frequent blinking, tucking the head under the wing
Labored breathing
Sitting on the bottom of the enclosure
Coarse, ruffled or moist feathers
Scratching and excessive preening
Lameness, wing droop, standing on one leg
Regurgitation or vomiting
Change in talking or singing
Change in food or water consumption
Change in color, texture, consistency or volume of droppings

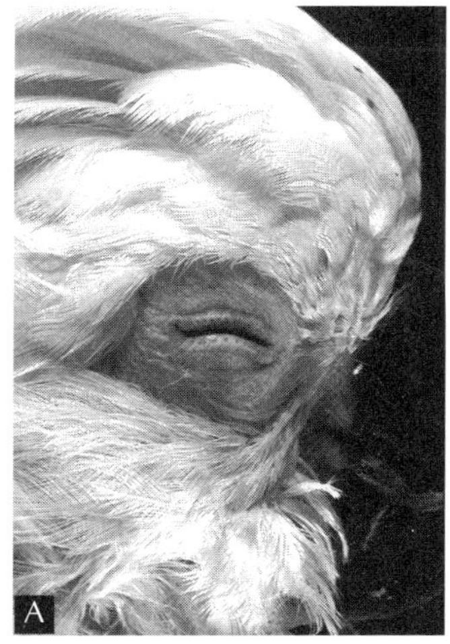

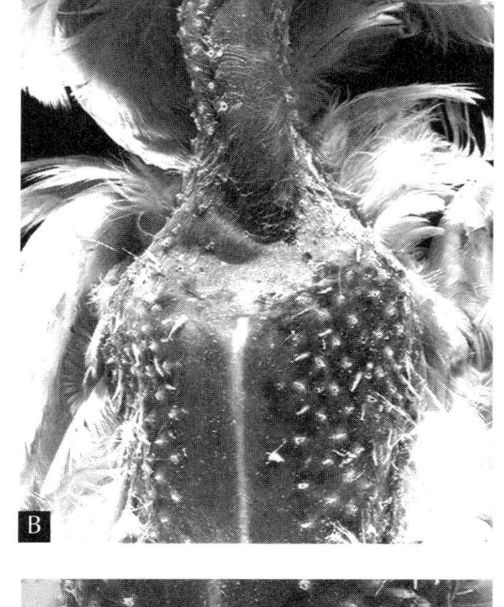

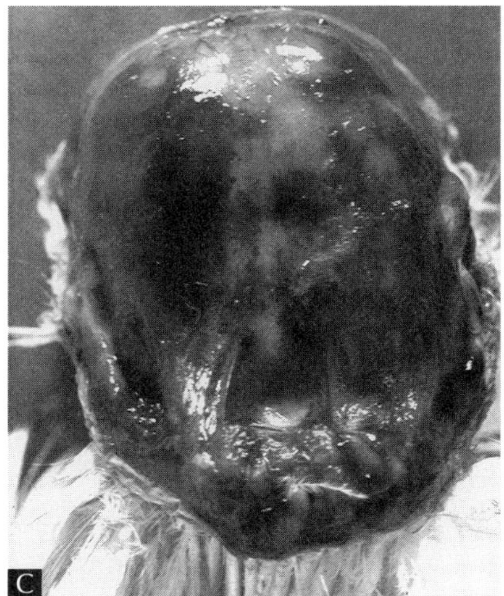

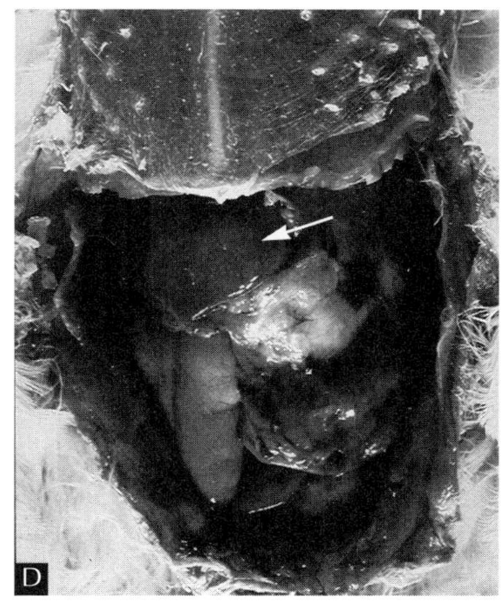

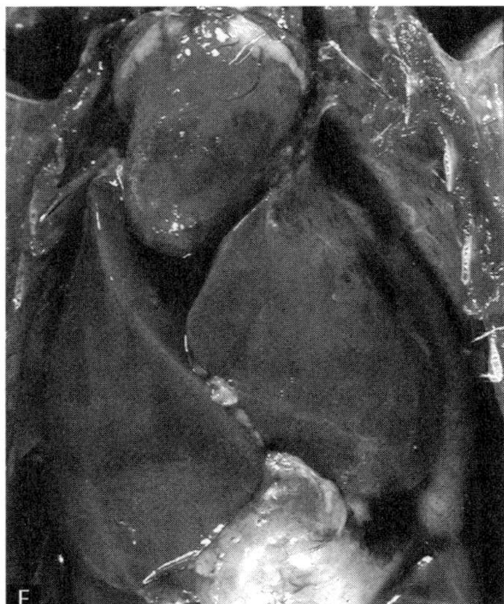

FIG 4.2

A 70-day-old Moluccan cockatoo died less than 24 hours after becoming slightly lethargic. Hemorrhage was evident:

A under the skin of the face;

B pectoral area and neck; and

C skull.

D The liver (arrow) was enlarged and

E both the liver and the heart had areas of hemorrhage. The acute death in this young bird and the necropsy findings were highly suggestive of an avian polyomavirus infection. However, this bird died from a bacterial septicemia that caused changes similar to those that occur with avian polyomavirus. This case emphasizes the importance of histopathology and specialized testing for determining the cause of a bird's death.

demonstrate directly (by cell culture, microscopic evaluation, DNA probes or electron microscopy) that a virus is present. However, in these cases, it still may be possible to document that a viral infection has occurred using indirect techniques (ie, antibody detection methods).

Samples collected from living birds for detection of viruses will vary with the type of virus and the type of test. In most cases, the collection of multiple samples will increase the chances of successfully demonstrating the presence of a virus. Depending on the virus, samples used to document an infection may include feces, tracheal swabs, skin biopsies, feather samples, organ biopsies or blood.

The better the quality of any sample that is sent to a laboratory, the more likely the laboratory is to provide useful information. Incorrectly collected or submitted samples are generally useless. A laboratory that performs a specific viral diagnostic test can advise the submitter on the correct method for collecting and transporting the sample. For accurate results, the submission recommendations of a laboratory should be followed precisely.

Diagnostic tests used to detect viral infections require varying amounts of time. From the time of submission, most tests for diagnosing viral infections will require from seven days to four weeks to complete.

Necropsy

Annual losses in a properly managed group of birds should be minimal. Any bird that dies, whether it is a neonate or an adult, should be immediately moistened with water, refrigerated and necropsied as soon as possible. Birds should not be frozen prior to necropsy. Many of the microscopic changes that will differentiate certain diseases are subtle. The damage that freezing causes to tissues may handicap the ability of a pathologist to make a proper diagnosis.

Complete necropsies are the avian community's most important tool for evaluating problems and developing programs to prevent infectious diseases. Properly performed necropsies serve as a mirror for the general health of a group of birds and allow specific problems that may be occurring to be identified and resolved.

For diseases in which an etiologic agent has not been confirmed, the clinician is encouraged to collect, process and store serum, whole blood, formalin-fixed tissues and fresh-frozen tissues. These archival samples may prove valuable in determining the etiologies of common syndromes as more advanced investigative techniques are applied to resolving viral diseases of birds.

Accuracy of Diagnostic Tests

The value of any diagnostic test is based on its sensitivity and specificity. A virus test with a high sensitivity is able to correctly find small quantities of virus. It is unlikely to provide false negative results, ie, the test is unlikely to indicate that a bird is not infected when it actually is infected. A virus test with a high specificity is able to differentiate

TABLE 4.2

Some direct and indirect techniques for detecting a virus

DIRECT TECHNIQUES
  Isolation of the virus from the test material
  Demonstration of viral particles by electron microscopy
  Demonstration of viral antigen (Ag) using viral-specific antibodies (Ab)
  Demonstration of viral nucleic acid using viral-specific-nucleic acid probes

INDIRECT TECHNIQUES
  Detection of antibodies (previous infection)
  Demonstrating an increase in antibody titer in paired serum samples (an active infection)

between two related viruses. Because of this characteristic, it is unlikely to provide false positive results: the test is unlikely to indicate that a bird is infected when it actually is not infected. The higher the sensitivity and specificity of the test, the more valuable it is as a diagnostic technique.

Each particular type of test has its own levels of sensitivity and specificity. For example, for a virus to be detectable by electron microscopy, there must be at least one million viral particles per milliliter of sample. This test would be considered to have a low sensitivity. However, it is easy with an electron microscope to differentiate between the physical characteristics of a herpesvirus and an adenovirus, which illustrates that electron microscopy has a high specificity *Figure 4.3*.

DNA probe-based tests that are designed to detect the nucleic acid of a virus are the most sensitive and specific of all the virus detection methods. When properly performed, these tests can theoretically detect the nucleic acid from a single viral particle (highly sensitive) and can differentiate between two closely related strains of the same virus (highly specific).

TYPES OF DIAGNOSTIC TESTS

The ideal diagnostic test not only is highly sensitive and specific, but also is quick, inexpensive and easy to perform. There are two general types of tests for detecting viral infections: those that demonstrate the presence of the virus or a portion of the virus, and those that demonstrate that a bird's immune system has responded to a viral infection through the production of antibodies. Techniques that demonstrate the presence of a virus or a component of the virus are called direct techniques. Those that detect the presence of antibodies are called indirect techniques *Table 4.2.*

Depending on the type of virus, direct testing techniques can be used to determine whether tissues or samples of secretions or excretions (blood, saliva, feces, respiratory discharge, semen, urine) contain a particular virus. Tests that directly demonstrate the presence of a virus are best used to document an infection in an individual bird that is experiencing clinical problems. Tests that indirectly demonstrate that an infection has occurred through the detection of viral-specific antibodies in the blood are best for establishing the incidence or prevalence of viral activity

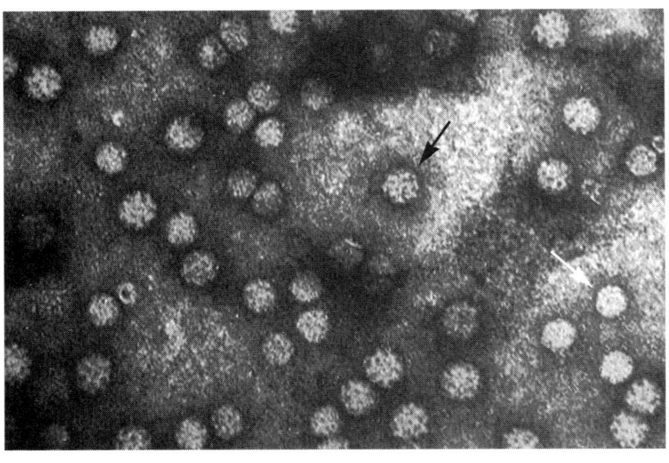

FIG 4.3

When enough virus particles are present in a sample, electron microscopy is a sensitive method to differentiate between viruses. The chicken anemia virus, or CAV, (arrow) is slightly larger that the PBFD virus (white arrow). Additionally, the CAV has a surface structure that makes it look like a small soccer ball, while the PBFD virus has an indistinct surface structure.

photograph courtesy of M.S. McNulty

within a group of birds. For example, if a bird is having clinical signs that are consistent with those caused by Pacheco's disease virus, a test that directly demonstrates the presence of the avian herpesvirus that causes this disease is going to provide the most diagnostic information. On the other hand, if one wishes to determine whether any bird in an aviary has recently been infected with eastern equine encephalitis virus, a test that demonstrates the presence of antibodies to this virus would be best.

The direct and indirect tests that can be used for diagnosing various viral infections are listed in *Table 4.3*. Some of these tests can be used for either application, depending on how they have been engineered. For example, an enzyme-linked immunosorbent assay (ELISA) could be designed so that it detects the presence of avian polyomavirus in the feces of a bird; a different ELISA could be designed to detect the presence of polyomavirus antibodies in the serum of a bird. The ELISA designed to detect antibodies to a polyomavirus cannot be used to detect the virus; the ELISA designed to detect the virus in fecal samples cannot be used to detect antibodies.

Some diagnostic techniques have been developed for certain avian viruses. With others, the avian community does not yet have the techniques needed to effectively determine whether a bird is or has been infected by a particular virus. For example, there is a DNA probe test to determine whether a bird is infected with the PBFD virus. A test to detect antibody levels can be used to determine whether a bird has been previously infected by avian polyomavirus. But currently, there is no way to reliably determine whether a bird has been infected previously with Pacheco's disease virus or to determine whether a bird will develop proventricular dilatation disease.

A particular type of direct or indirect test generally proves to be most valuable for diagnostic work with an individual virus. The type of test that is best suited for detecting a viral infection

TABLE 4.3

Various assays for direct or indirect demonstration of viral infections

DIRECT TESTS - detect presence of virus or portion of virus
        Agar-gel immunodiffusion (AGID)
        Cell culture
        Electron microscopy
        Enzyme-linked immunosorbent assay (ELISA)
        Hemagglutination (HA)
        Indirect fluorescent antibody (IFA)
        Nucleic acid probes (DNA probes)
        Radioimmunoassay (RIA)

INDIRECT TESTS - detect antibodies to a particular virus
        Agar-gel immunodiffusion (AGID)
        Complement fixation (CF)
        Enzyme-linked immunosorbent assay (ELISA)
        Hemagglutination-inhibition (HI)
        Indirect fluorescent antibody (IFA)
        Radioimmunoassay (RIA)
        Virus neutralization (VN)

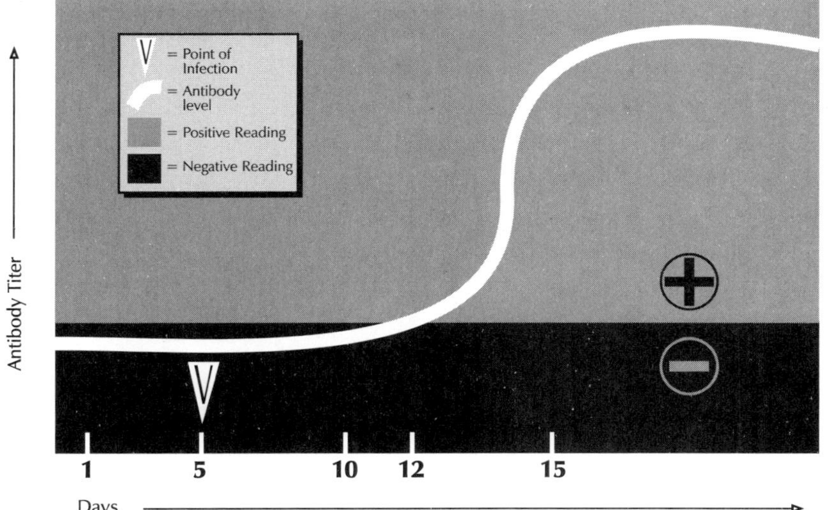

FIG 4.4

An antibody titer may not be detectable for several weeks after a bird is infected. In this case, blood collected from a bird from days 5 through 12 would be interpreted as negative, but blood collected from the bird after day 12 would be considered positive.

depends on how the virus functions in a bird's body, what type of disease the virus may induce and how and when the virus is released from the infected bird. For example, a DNA probe-based test is better than other virus detection techniques for demonstrating PBFD virus in the blood of birds that have feather abnormalities or that are incubating the virus and have not yet developed feather abnormalities. A virus-neutralization (VN) assay has proven most beneficial for detecting antibodies to avian polyomavirus, while a hemagglutination-inhibition (HI) assay is most valuable for detecting antibodies to the PBFD virus.

ANTIBODY TESTS

Antibodies can be viewed as footprints that a virus leaves in the body of some infected birds. Antibody levels may decline over time to a point where they are undetectable. If a bird is again exposed to the same virus, the antibody titer may increase. The virus then leaves a new footprint that can be detected with specific antibody detection assays. The detection of viral-specific antibodies to indicate that an infection has occurred previously requires that the bird survive the infection, that the immune system of the bird produce a detectable quantity of antibodies and that the antibody titer remain high for a sufficient period of time for it to be detected by testing *Figure 4.4.* The types of tests that can be used to demonstrate antibodies to a virus are listed in Table 4.3.

All antibody detection assays share basic characteristics that influence their diagnostic value:

- For an antibody test to provide meaningful data, the test must be run twice at a two- to four-week interval. Demonstrating a four-fold increase in antibody titer indicates an immunologic response is occurring in response to an active infection. The detection of antibodies in a single test indicates only that a bird has been exposed to a virus; this test

FIG 4.5

Blood samples were collected on the same day from a group of birds labeled A through K, and the serum from each bird was tested for antibodies to avian polyomavirus. For this particular assay, an antibody titer of 8 or less is considered negative, while an antibody titer of 16 or greater is considered positive. Birds B, E, F and K have positive titers, while the remainder of the birds have negative titers. This would suggest that avian polyomavirus has been active in the aviary from which the samples were collected; the results provide minimal information on the health status with respect to avian polyomavirus of each individual bird.

generally is of little value in determining if a bird is currently infected with a virus. In some cases, an antibody titer may not be detectable for several weeks after a bird is infected. If a blood sample for a single antibody assay were collected prior to this point, the resulting low titer would suggest that the bird had not been exposed to a particular virus, when in fact it was undergoing an active infection *Figure 4.4*.

- Some antibodies generated against certain types of viruses will cross-react with the antibodies produced against other viruses or microorganisms, causing a false positive result. A false positive result occurs when a test indicates that a bird has antibodies to a virus when it actually does not.

- A bird's immune system must be properly functioning to produce

detectable levels of antibodies against an invading virus.

For diagnostic purposes, tests that demonstrate that a bird has produced an antibody response to a virus are best used in combination with tests that are designed to directly demonstrate the presence of a virus. In general, tests to detect antibodies are best used for screening a population of birds for evidence of previous viral activity, and are of limited value for diagnosing the state of health in an individual bird *Figure 4.5*. Some viral pathogens are present in the bird for a brief period of time. If samples are not collected during this period, it will not be possible to recover a virus. In these cases, demonstrating antibodies to the virus may be the only way to detect that a viral infection has occurred.

A single antibody titer that is sufficiently high to be considered positive indicates that a bird has been previously

AVIAN VIRUSES: FUNCTION AND CONTROL

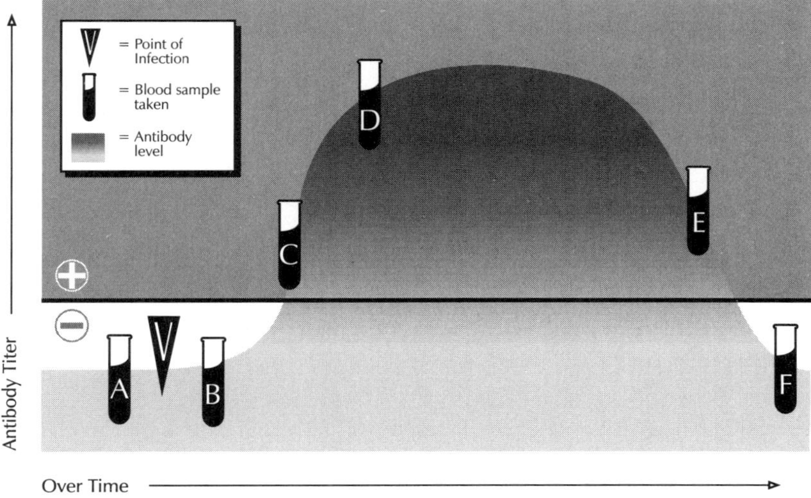

FIG 4.6

A bird that survives a viral infection should produce antibodies to that virus. The highest concentrations of antibodies are present in the blood shortly after the bird recovers from infection. The antibody titer then progressively decreases over the months following the bird's recovery.

A blood sample is drawn from a bird to test for antibodies at points A through F. The point at which the bird is infected with the virus is labeled. Interpretation of the results from each blood sample would be:

A Antibody assay is negative, suggesting that the bird is not or has not been infected.

B Antibody assay is negative; however, the bird is actually infected but the antibody titer has not yet increased to a detectable level.

C Antibody assay is positive, but the titer is low. This indicates that the bird has been previously infected by the virus. It is not possible from this sample to determine if the infection is recent or has occurred in the distant past.

D Antibody assay is positive and the titer is high. This suggests a recent exposure to the virus.

E Antibody assay is positive, but the titer is low. This indicates that the bird has been previously infected by the virus. It is not possible from this sample to determine if the infection is recent or occurred in the distant past.

F Antibody assay is negative, suggesting that the bird is not or has not been infected.

exposed to a particular virus. A single antibody titer that is sufficiently low to be considered negative indicates that a bird has not recently been exposed to a particular virus or that a measurable antibody response has not occurred *Figure 4.6*. Finding that a bird does not have a titer to a particular virus can indicate that the bird has not been infected by that virus or that it was infected so long ago that antibodies can no longer be detected. In general, a single antibody test indicates only if a bird has been previously exposed to a particular virus; these assays should not be expected nor used to provide additional information. For viruses that are common in populations of companion and aviary birds, like polyomavirus and PBFD virus, finding birds with antibodies is of limited diagnostic value. If a group of birds has been vaccinated for a particular virus, antibody tests can be used to screen these birds and determine which ones developed an appropriate antibody response following vaccination.

It may be possible to use paired serum samples to demonstrate how recently a bird has been infected by a particular virus. If an initial antibody titer is low and a second antibody titer that is performed on blood collected several weeks later is at least four times higher than the first sample, it indicates that the bird's immune system is actively involved in fighting a viral infection. If the titer is decreasing, it may indicate that the bird was recently infected and the titers are starting to wane *Figure 4.7*. As an example, if a disease was theorized to be caused by eastern equine encephalitis (EEE) virus, then screening a group of birds and finding EEE virus antibodies would at least indicate that the seropositive birds had been previously infected. Finding that none of the birds had EEE virus antibodies would suggest that the virus had not recently infected any of the birds in the tested group.

Yolk contains antibodies that were present in the hen and that are transferred to the developing egg. Yolk from unfer-

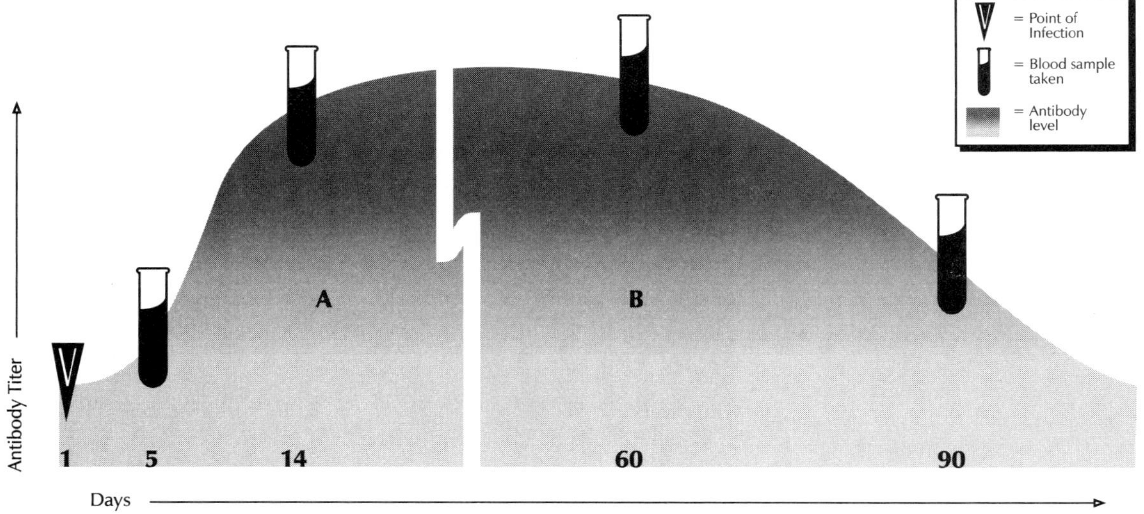

tilized eggs can be used to determine whether a hen transferred antibodies to a particular virus to the egg. It should be noted that the eggs from early in a laying period will generally have a higher level of antibodies than eggs that are laid later in the laying period.

CELL CULTURE

Cell culture is a powerful tool used by virologists for the isolation, identification and propagation of viruses. This laboratory technique involves replicating a virus in plates containing living cells. For practical reasons, cells derived from chicken, duck or goose embryos are most frequently used for isolating viruses that infect birds *Figure 4.8*.

The recovery of a virus in cell culture is a definitive way to demonstrate that a virus is present in the excretions, secretions or tissues of a bird. While virus isolation by cell culture is an important research tool for virologists, it is not routinely performed as a diagnostic technique in companion and aviary birds. When compared to other virus

detection methods, cell culture procedures are inefficient, labor-intensive, expensive and may require weeks to months to complete. The elapsed time between the submission of a sample and the reporting of reliable data may render the findings of little value in managing disease in an individual bird or in a flock. Additionally, many of the viruses that infect companion and aviary birds are difficult to recover in cell culture, and the failure to recover a virus from a bird with a disease suspected to be of viral origin does not mean the bird was not infected. It should also be noted that the isolation of a virus from a bird does not necessarily mean that the recovered virus was the cause of a disease process.

The viruses that grow in cell culture are the easiest to study. However, many viruses that infect companion, aviary and free-ranging birds have very specific requirements for replication and are difficult to grow in the laboratory. For example, the virus that causes papillomas on the skin of birds has not been successfully grown in cell culture, making it much more difficult to study than the Pacheco's disease virus, which is easy to grow in cell culture.

Difficulties in growing viruses in the laboratory occur because many viruses have adapted to replicate only in certain types of cells in a specific species of bird. For example, a virus that infects only psittacine birds may be more difficult to grow in cells derived from chickens than a virus that normally infects chickens. The viruses that infect companion and aviary birds could best be isolated in the laboratory by placing them on cells derived from the species of bird that the virus is likely to infect. However, the cell culture process generally requires the sacrifice of a large number of embryos. It is unlikely that cell cultures derived from companion or aviary birds will be used in North America or Europe where these birds are cherished.

ELECTRON MICROSCOPY

Electron microscopy can be used to demonstrate a virus in excretions or secretions if the virus is present in a sufficiently high concentration (greater than one million viral particles per milliliter of sample). Samples used for electron microscopy must be relatively free of extraneous debris to facilitate the visualization of virus particles. Electron microscopy has been used to demonstrate virus particles in feces, urine, feather pulp, skin scrapings, respiratory secretions, serum, vomitus and organ samples.

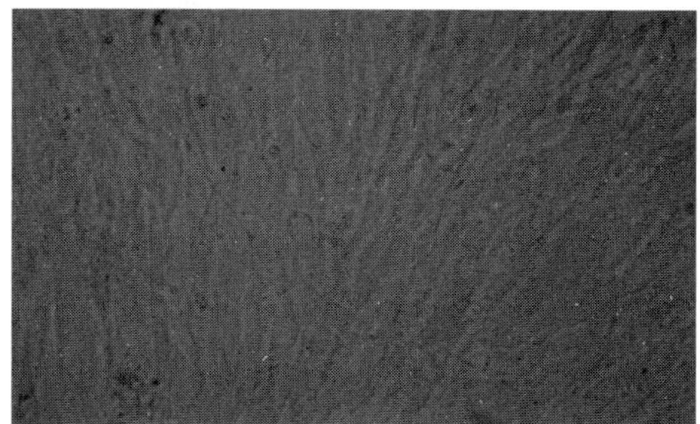

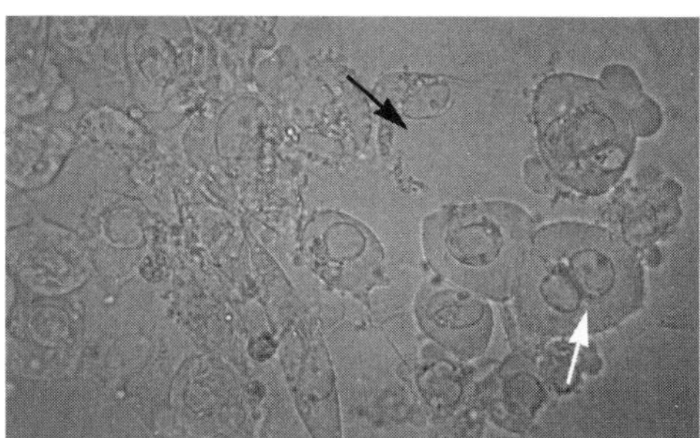

FIG 4.8
A Uninfected cell culture derived from chick fibroblasts. Note the consistent pattern of the cells.

B Cell culture infected with avian polyomavirus. The polyomavirus has killed cells in some areas, causing "holes" in the once consistent layer of cells (arrow). Other infected cells are swollen (white arrow) and are in the early stages of cell death.

photographs courtesy of Scott McDonald

Many viruses that infect the digestive tract cause asymptomatic infections — that is, the virus is infecting cells within the digestive tract but does not cause any signs of disease. Frequently, virus particles are detected by electron microscopy in the feces of normal-appearing birds. Virus particles that are identified in the feces may be the cause of a disease process, may be causing an asymptomatic infection or may merely be passing through the bird and not contributing to a disease process *Figure 4.9*.

Electron microscopy can also be used on organ biopsies or on samples of organs collected at necropsy to demonstrate the presence of virus in infected tissues. Electron microscopy is particularly beneficial for demonstrating the size and structural characteristics of virus particles contained in inclusion bodies *Figure 4.10*.

HISTOPATHOLOGY

Biopsy samples collected from live birds or tissues collected at necropsy

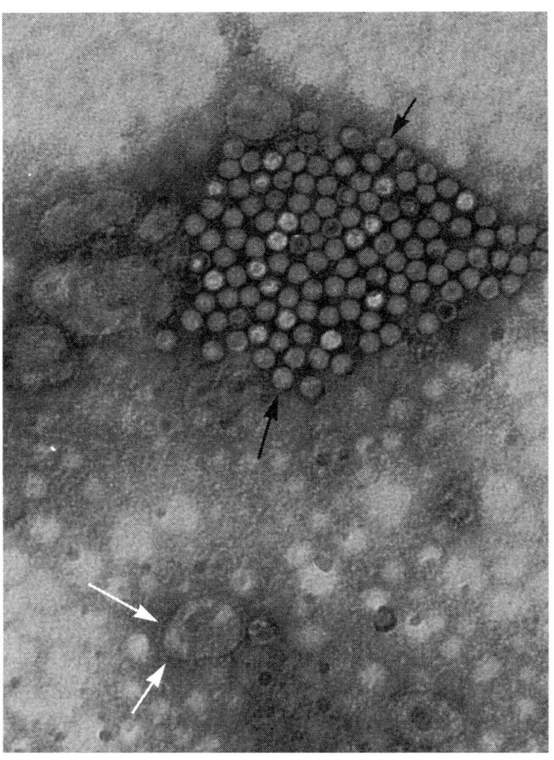

can be examined under a microscope to determine whether inclusion bodies are present. Not all viruses induce inclusion bodies, but those that do are easier to diagnose by microscopic examination than those that do not. If inclusion bodies of a particular type (intranuclear, intracytoplasmic or both) are identified, the pathologist can provide a presumptive diagnosis that a viral infection has occurred. A confirmed diagnosis requires that viral-specific nucleic acid probes, viral-specific antibodies or electron microscopy be used to demonstrate the presence of a virus in the suspect inclusion bodies. Many intranuclear inclusion bodies can appear similar. Specific tests must be used to confirm which particular virus may be causing these inclusion bodies *Figure 4.10*. An unconfirmed diagnosis based on microscopic changes alone can result in initiation of incorrect man-

agement techniques or can result in the unnecessary death of a bird.

DNA PROBES

In the early 1980s, techniques were developed that allowed nucleic acid to be amplified and detected with a high degree of reliability. The DNA or RNA present in each genus or strain of a virus is unique in its composition. For simplicity, DNA can be envisioned as a long chain made of four different types of segments — adenine, guanine, cytosine and thymidine — that repeat themselves in an unique sequence. RNA can be envisioned similarly, but its segments include adenine, guanine, cytosine and uracil *Figure 4.11*.

When placed in the proper conditions, two chains of DNA may bind to each other in a sequence-specific manner so that adenine from one strand binds to thymidine on the other strand, and cytosine on one strand binds to guanine on the other strand *Figure 4.12*. This hybridization process is the basis of using viral-specific DNA probes to detect the presence of a virus' nucleic acid in samples collected from a bird. The hybridization that occurs can be detected by using a number of visual indicator systems based on color changes or radiographic sensitivity. These natural behavioral characteristics of DNA have been used by scientists to develop testing techniques that reliably help to detect the presence of virus nucleic acid in diagnostic samples *Figure 4.13*.

The ideal diagnostic test would provide maximum sensitivity, maximum specificity and rapid results. DNA probe-based tests most completely meet these requirements. They are more sensitive

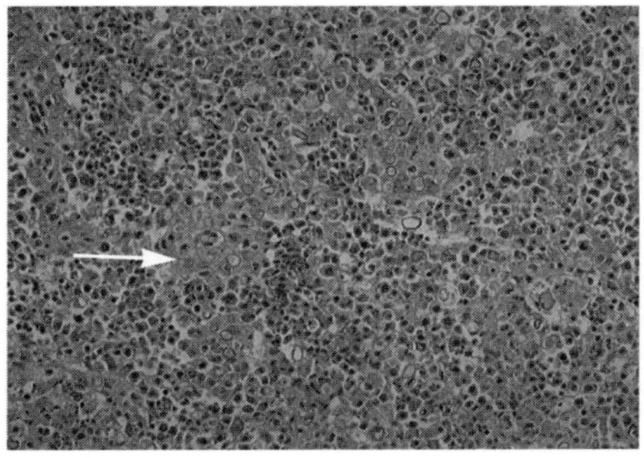

POLYOMAVIRUS

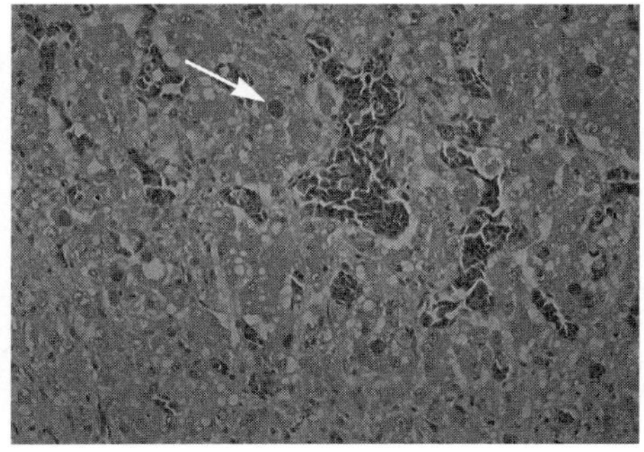

PACHECO'S DISEASE VIRUS

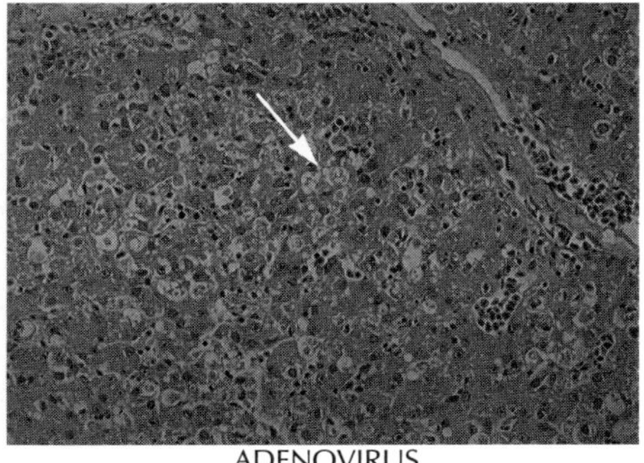

ADENOVIRUS

DNA PROBE STAINING

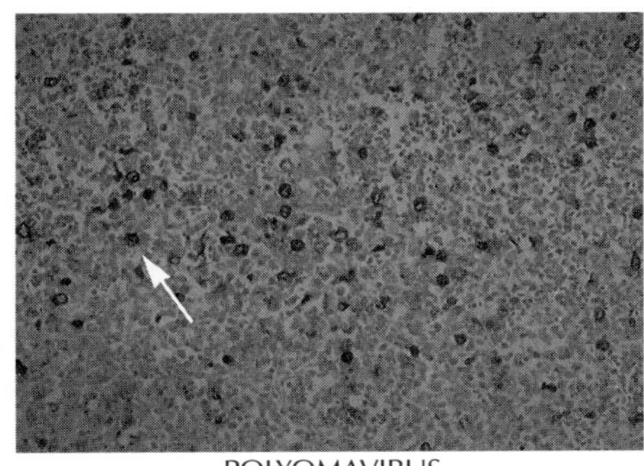

POLYOMAVIRUS

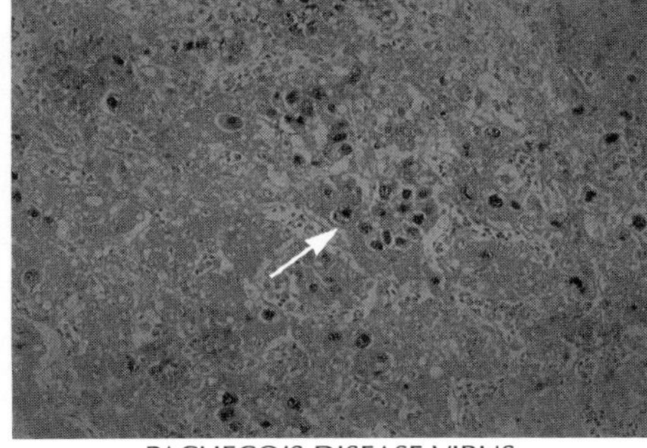

PACHECO'S DISEASE VIRUS

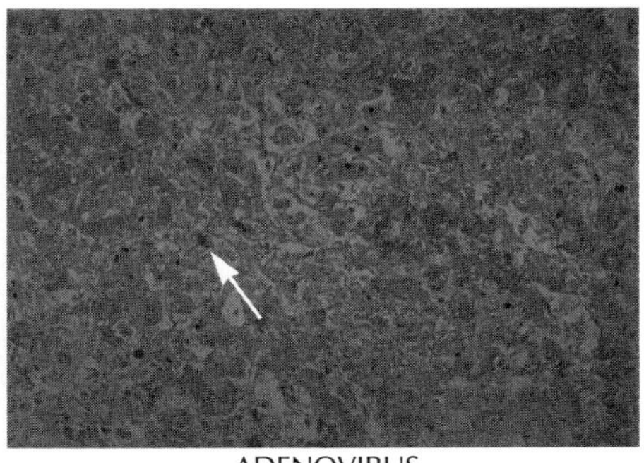

ADENOVIRUS

ELECTRON MICROSCOPY

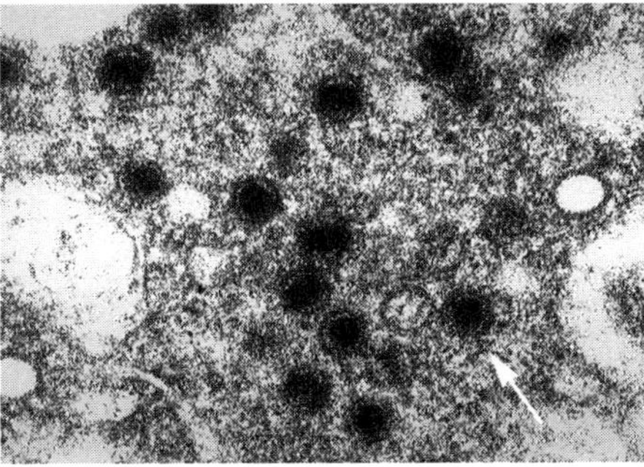

POLYOMAVIRUS

PACHECO'S DISEASE VIRUS

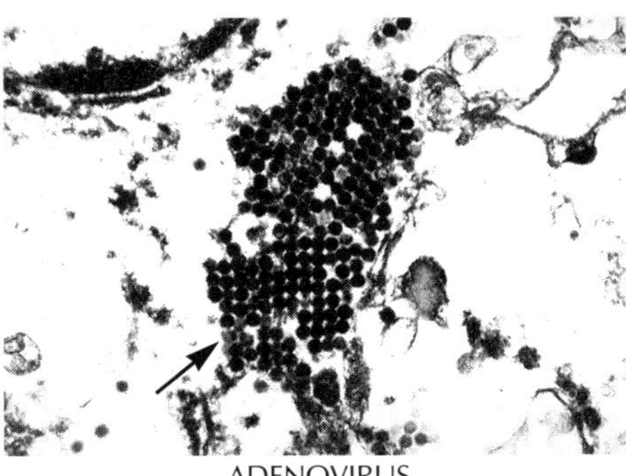

ADENOVIRUS

FIG 4.10

Many of the microscopic changes induced by various types of viruses can appear similar.

Row 1 The intranuclear inclusion bodies (arrows) caused by avian polyomavirus, Pacheco's disease virus and adenovirus may be difficult to differentiate based on routine microscopic examination. In these cases, confirming which virus is present in affected tissues requires the use of viral-specific

nucleic acid probes, viral-specific antibodies or electron microscopic examination.

Row 2 Specialized staining of the tissues from the birds in Row 1 using viral-specific DNA probes indicates by dark staining (arrows) that the viral-specific nucleic acid has been detected in specific cells.

Row 3 Use of an electron microscope to demonstrate the presence of avian polyomavirus, Pacheco's disease virus or adenovirus (arrows) in inclusion bodies in the tissues of infected birds.

photographs courtesy of Mark Goodwin and Kenneth Latimer

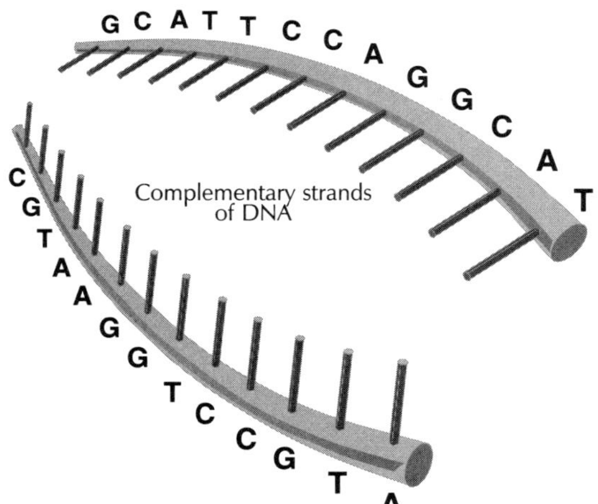

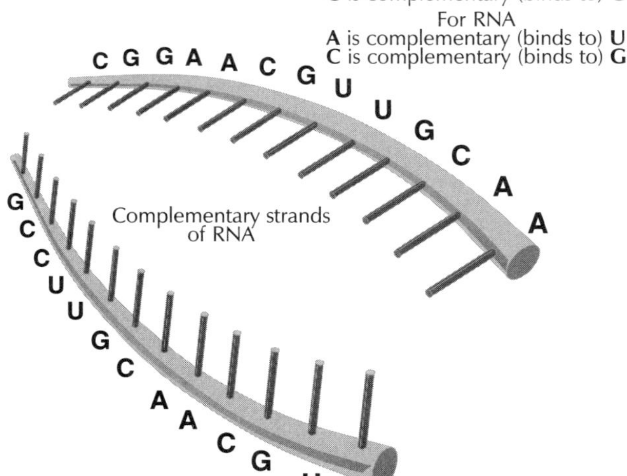

FIG 4.11

DNA can be viewed as a long chain made of four different types of segments that repeat themselves in a unique sequence; RNA can be viewed as a long chain made of four different types of segments that repeat themselves in a unique sequence.

than other virus detection techniques and have the added advantage of detecting small concentrations of virus. In many cases, DNA probes can detect the presence of a virus in samples before recognizable changes have occurred in infected tissues. Their high sensitivity and specificity, the speed with which results are available and the fact that they can be used to detect viruses that will not replicate in cell culture make DNA probes extremely valuable diagnostic techniques.

Viral-specific DNA probes can be developed to detect a small number of virus particles in crop washings, feces, environmental samples, blood, saliva, respiratory secretions, urine or semen. They may also be used to confirm the presence of viral nucleic acid in tissues like liver, spleen, kidney or brain collected from a bird suspected to have a particular viral infection *Figure 4.13*.

With some viruses, DNA probes can be used to detect subclinical, active or persistent infections. The identification of persistently infected birds requires that the site for virus persistence in the body be known, and that samples be collected from the appropriate site at the correct stage of the infection. Depending on the virus and the bird, some common reservoir sites for viral persistence are blood cells, hepatocytes, enterocytes or skin cells. Samples that could be effective in detecting persistently infected birds might then include whole blood, liver biopsies, excrement or skin biopsies, respectively.

For a DNA probe to be specific, it must be developed from the nucleic acid sequence that is unique to the virus in question *Figure 4.14*. The specific nature of the probe prevents cross-reactions with other viruses, imparts specificity, and reduces false-negative results. A probe designed from an unknown sequence can lead to errors in the interpretation of results.

Once a viral-specific probe has been developed, it can be used to detect nucleic acid that is extracted from a sample and attached to a membrane, or it can be used to detect viral-specific

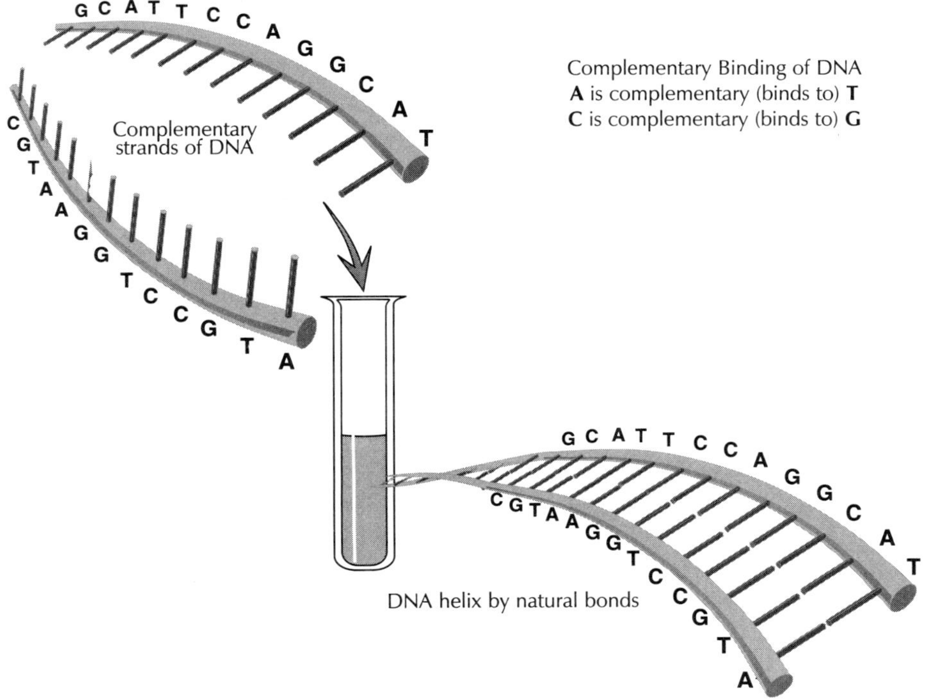

Two individual strands of DNA with complementary sequences (strand 1 = 3′ GCATTCCAGGCAT 5′ and strand 2 = 5′ CGTAAGGTCCGTA 3′).

Mixed together under the correct conditions they will bind together (hybridize) to form a double-stranded molecule.

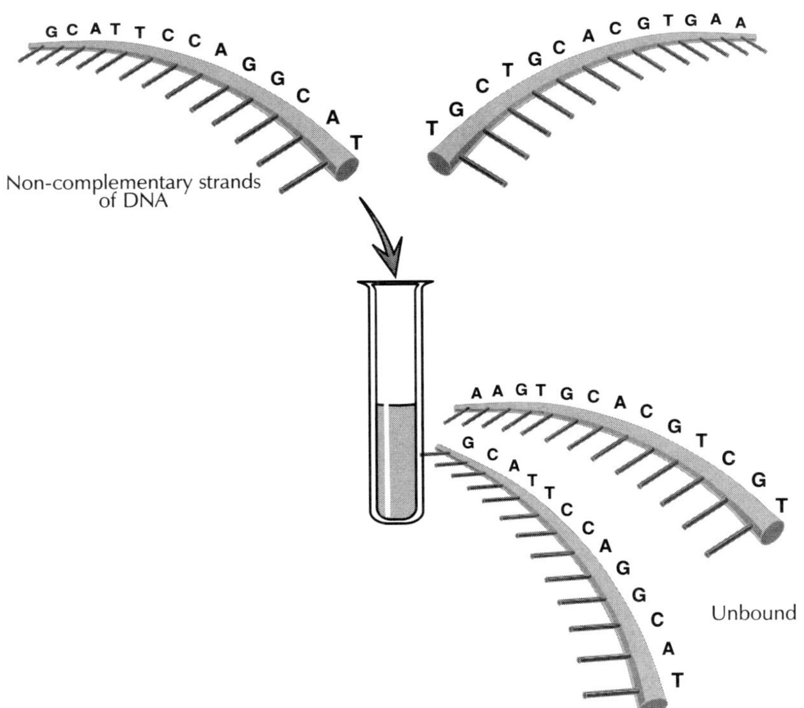

Two individual strands of DNA that do not have complementary sequences (strand 1 = 3′ GCATTCCAGGCAT 5′ and strand 2 = 5′ TGCTGCACGTGAA3′). They will not bind together and will remain as two single-stranded molecules.

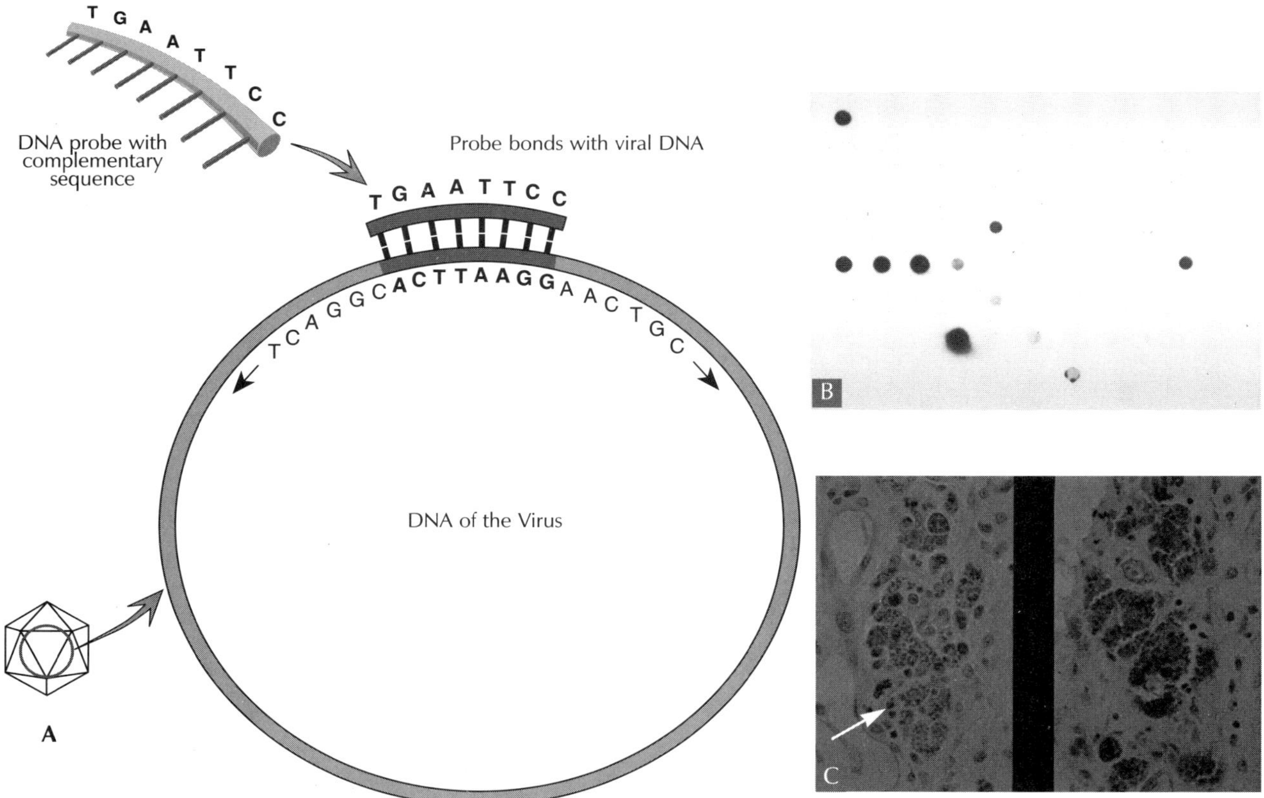

FIG 4.13

A Two single strands of DNA (for example, a DNA probe produced in the laboratory and a target sequence of virus DNA) can bind together in a process called hybridization. In this case, a viral-specific DNA probe with the sequence 3' TGAATTCC 5' will bind specifically to its target 5' ACTTAAGG 3'. B Viral-specific DNA probes can be developed to detect a small number of virus particles in crop washings, feces, environmental samples, blood, saliva, respiratory secretions, urine or semen. In this case, PBFD virus-specific DNA probes have been used to screen a group of blood samples for the presence of PBFD virus nucleic acid. A black spot indicates a positive sample (photograph courtesy of Bob Dahlhausen). C DNA probes may be used also to confirm the presence of viral nucleic acid in tissues (liver, spleen, kidney, brain). In this case, inclusion bodies suspected to contain PBFD virus are shown on the left (arrow). On the right, PBFD virus-specific DNA probes have been used to confirm that these inclusion bodies do contain PBFD virus nucleic acid. photograph courtesy of Kenneth Latimer

nucleic acid in a section of formalin-fixed tissue that has been processed for histopathologic evaluation. The process of using nucleic acid probes to detect viral-specific nucleic acid sequences in tissues is called in situ hybridization.

In situ hybridization using viral-specific nucleic acid probes is particularly valuable in diagnosing a viral infection when a virus is present in small numbers or produces lesions that microscopically resemble those induced by other viruses. For example, the intranuclear inclusion bodies caused by polyomavirus can appear morphologically similar to the intranuclear inclusion bodies caused by PBFD virus, adenovirus or herpesvirus. In situ hybridization using viral-specific DNA probes can determine quickly and correctly which of these viruses induced the detected inclusion bodies. Viral-specific antibodies have also been used to determine which virus is present in an inclusion body; however, when compared to antibody staining techniques for the

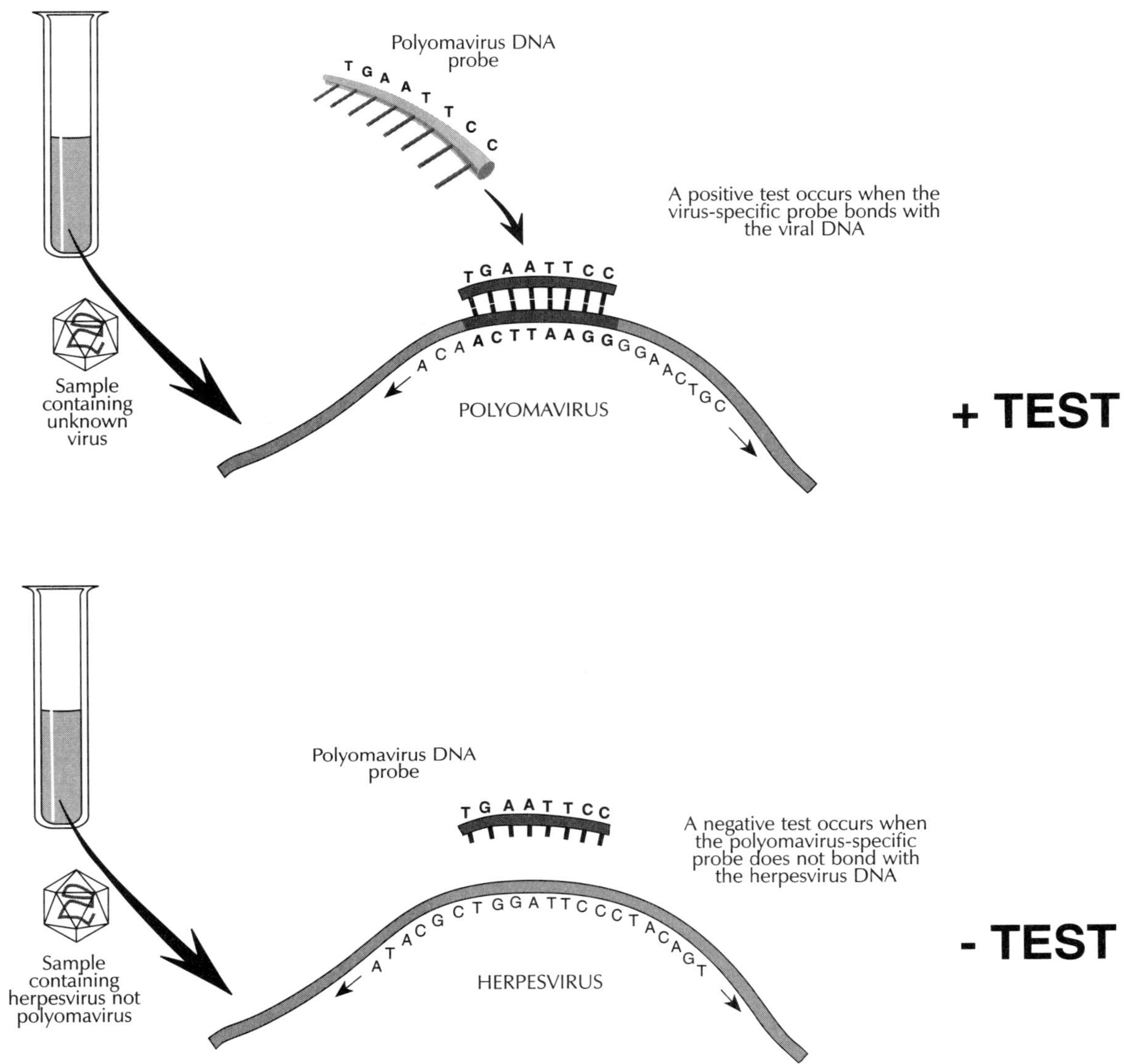

FIG 4.14

*A viral-specific DNA probe can be used to detect the presence of virus nucleic acid in a sample. **A** For example, if a specific portion of the polyomavirus genome had the sequence: 5'...ACAACTTAAGGGGAA...3', a DNA probe with the sequence 3' - TGAATTCC- 5' would be complementary to a portion of this sequence and would bind when the two nucleic acid strands were incubated in the proper conditions. This probe could be used to detect the presence of the polyomavirus nucleic acid in infected liver tissue, saliva, urine or in a contaminated environment as long as the virus was present in the sample. **B** If the DNA probe designed to detect polyomavirus (3' - TGAATTCC- 5') were mixed into a processed sample containing the herpesvirus nucleic acid sequence .... 3' - ATACGCTGGATTCCCTACAGT.... 5 ' then the probe would not bind to the necessary target and would be washed away.*

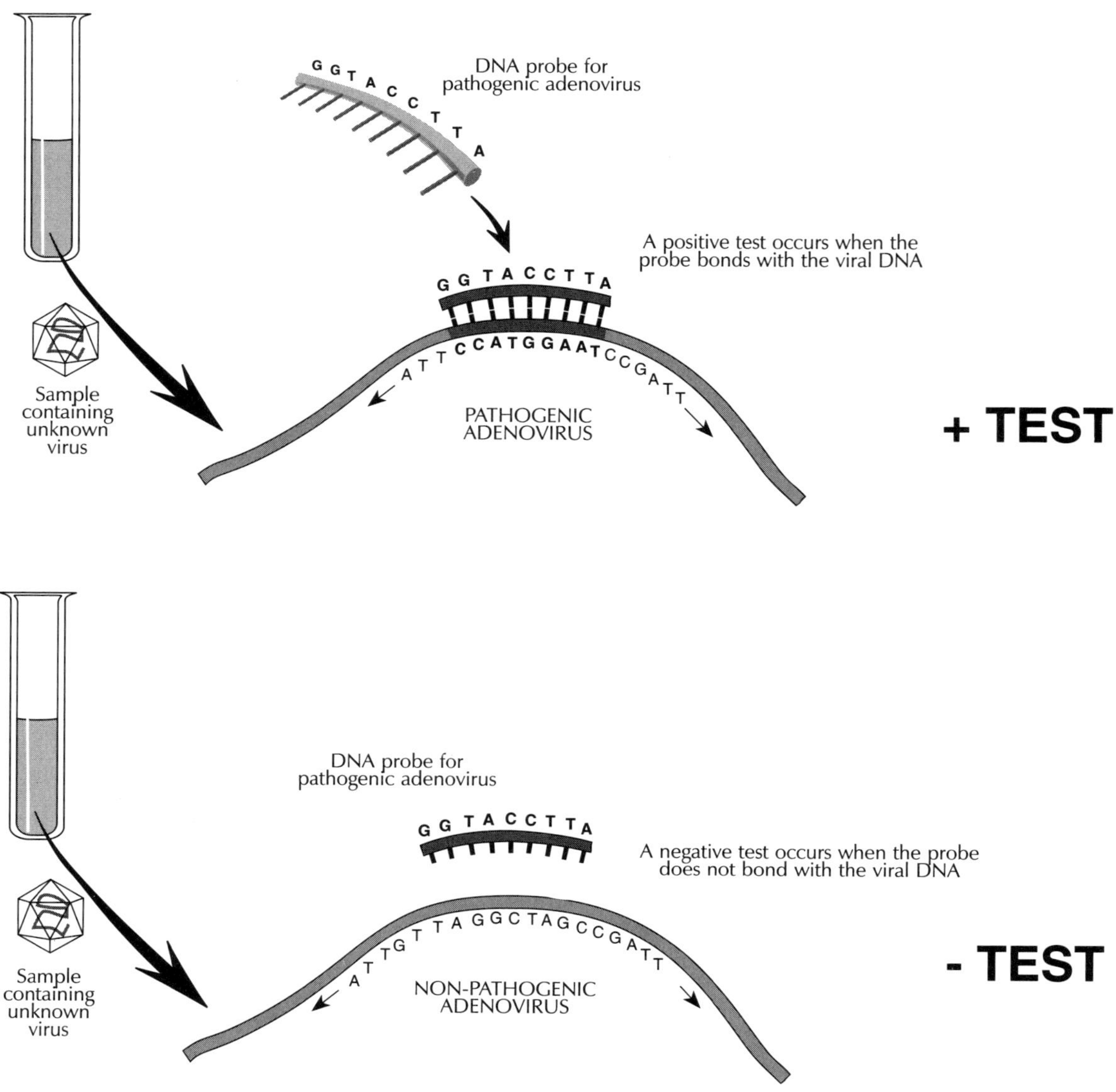

FIG 4.15

DNA probes can be used to differenti-ate between closely related strains of the same virus that have minor differences in their nucleic acid sequence. For example, an adenovirus that causes high mortality might have the nucleic acid sequence 5' ...ATTC-CATGGAATCCGATT... 3', while a strain of adenovirus that causes subclinical infections might have a different nucleic acid sequence 5' ...ATTGTTAGGCTAGCCGATT...3'. A DNA probe with the sequence 3'-GGTACCTTA - 5' could then be used to specifically detect the pathogenic strain of adenovirus.

identification of viruses in tissues, DNA probes are more specific, more sensitive and can detect viruses that may have been altered during processing.

In addition to confirming the presence of a virus in tissue, in situ hybridization can also be used to detect the specific type of cell infected, and whether the nucleic acid of the virus is present in the cytoplasm or nucleus of the cell. This latter finding is of particular impor-tance in understanding the replication

scheme of many viruses, which can be critical to understanding how viral infections can be prevented.

Nucleic acid probes are currently avail-able in only a few diagnostic laborato-ries and have been developed for only a few viral diseases. However, this advanced testing system will likely be the future standard for demonstrating the presence of a virus.

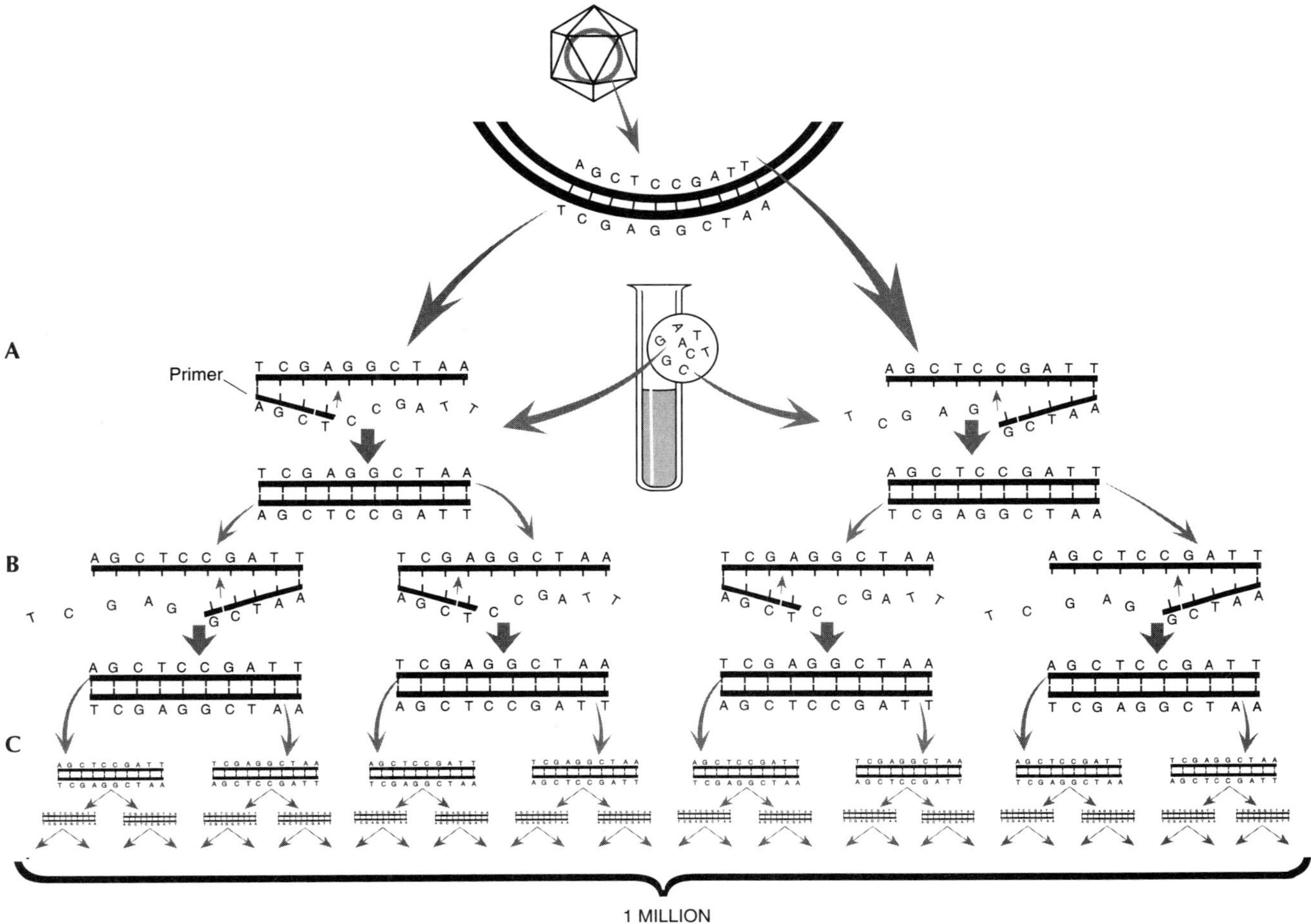

Nucleic acid probes can be designed to be so specific that they can differentiate between two viruses that are antigenically similar but have differences in nucleic acid sequence that alter the pathogenicity of the disease process *Figure 4.15*. As a hypothetical example, two adenoviruses that are antigenically similar could occur in a bird population. Being antigenically similar, these adenoviruses would be difficult to differentiate using an antibody-based diagnostic assay. Clinical evidence might suggest that in some cases, these adenoviruses are highly pathogenic, inducing high levels of mortality, while in other cases, infected birds develop an immune response and remain clinically normal. By determining the DNA sequence of virus recovered from different birds, it might be discovered that the adenovirus that causes high mortality has a different nucleic acid sequence than the virus that induces a subclinical infection.

NUCLEIC ACID AMPLIFICATION

Use of DNA probes to detect the presence of a virus' nucleic acid in infected tissues where high numbers of the virus are present is fairly straight forward. In contrast, detection of nucleic acid in excretions or secretions where numbers of the virus may be small requires addi-

FIG 4.16
*Theoretically, nucleic acid amplification procedures can use one copy of a nucleic acid sequence from a virus to produce one million replicates of the same sequence. **A** This procedure is accomplished by using primers and a mixture containing the building blocks of nucleic acid (A,T,C,G) to construct new strands of desired sequence. **B** The newly constructed sequence (copies of the original) are now processed in the same manner to produce four new strands. **C** These four strands are then processed into eight strands and so on until up to one million strands have been generated from the single initial target sequence.*

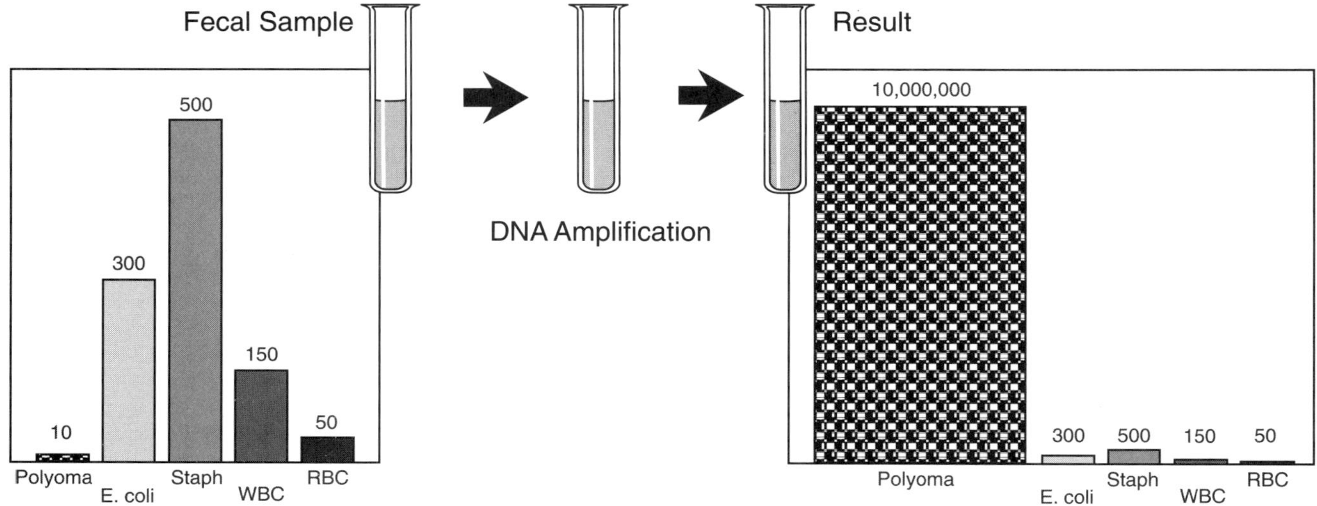

FIG 4.17

DNA amplification procedures are used to preferentially increase the number of target nucleic acid sequences (here the target is polyomavirus), while making the nucleic acid from other sources dilutionally unimportant. As an example, a fecal sample collected for polyomavirus testing might contain 10 polyomavirus particles, 300 E. coli, 500 Staphylococcus sp, 150 host-derived white blood cells (nucleated and containing DNA) and 50 host-derived red blood cells (also nucleated in birds and containing DNA).

It would not be possible to detect only 10 copies of the target polyomavirus DNA. By using primers designed specifically for the polyomavirus nucleic acid, the amount of target sequence (portion of polyomavirus DNA to be detected) can be increased from 10 copies to 10,000,000 copies, while the contaminating DNA from the E. coli, Staphylococcus, white and red blood cells remains the same. The 10,000,000 synthesized copies constitute a quantity that can be easily detected.

tional processing. To increase the likelihood of finding a virus in a diagnostic specimen, a sample to be tested is often subjected to a group of reactions that will amplify the number of DNA molecules in the sample that originated from the virus. Increasing the number of target molecules improves the ability of the probe to detect the presence of the virus.

Theoretically, amplification procedures can use one copy of a nucleic acid sequence from a virus to produce one million replicates of the same sequence. It is obviously easier for a DNA probe to detect one million copies of the target sequence than it is for the DNA probe to detect one copy of the target sequence

Figure 4.16. The most important portion of this process is the development of virus-specific primers. These primers allow the process to preferentially increase the number of virus-specific nucleic acid molecules without increasing the number of all other contaminating nucleic acid molecules that would be present in a sample *Figure 4.17*.

From a simplified perspective, the two most critical components for the amplification and detection of nucleic acid from a virus are the virus-specific amplification procedures used to increase the sensitivity of the test and the virus-specific nucleic acid probes used to ensure the specificity of the test.

If DNA probe testing has a shortcoming, it is the fact that the technology provides such a sensitive test that a small number of virus particles that inadvertently contaminate a sample can result in a positive test in a bird that does not have a viral infection. For example, if a clinician were testing a bird to determine if PBFD virus were present in the blood, and the blood sample was collected from a toenail, a positive result might indicate that PBFD virus was present in the blood, or it could indicate that the bird's toenail was contaminated with PBFD virus that was introduced into the sample tube during the collection process. Washing the bird's nail before collection would not be expected to reduce the potential for contaminating the sample because of the deep crevices in a nail in which PBFD virus (or any virus, bacteria or fungus) can easily hide *Figure 4.18*. A blood sample properly collected into a sterile syringe by venipuncture would be less likely to be contaminated.

If samples are properly collected and submitted to appropriate laboratories, and the results are interpreted by a qualified avian veterinarian, DNA probes provide the best method for detecting the presence of a virus in a selected sample. A knowledgeable clinician can minimize contamination by using proper technique to collect any sample intended for DNA probe testing.

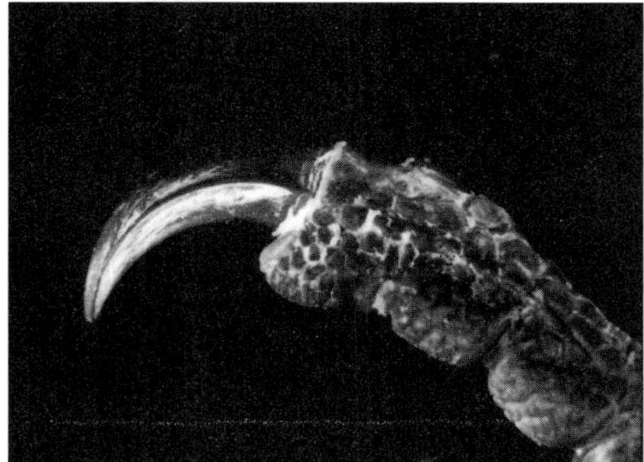

FIG 4.18

Blood samples collected from a toenail can be easily contaminated with bacteria, fungus or virus.

Toenail of an Amazon parrot.

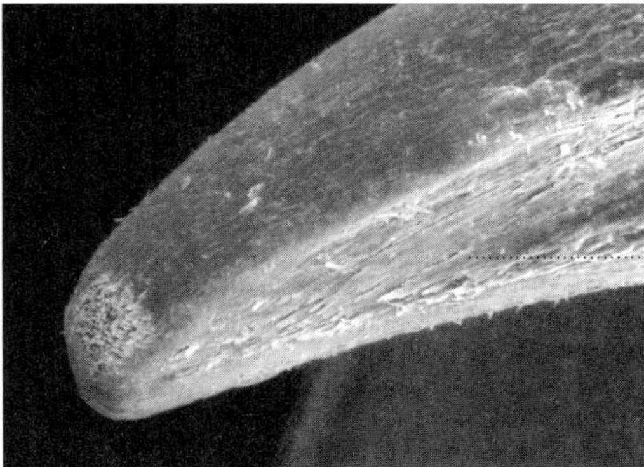

Toenail magnified 40 times; area of magnification.

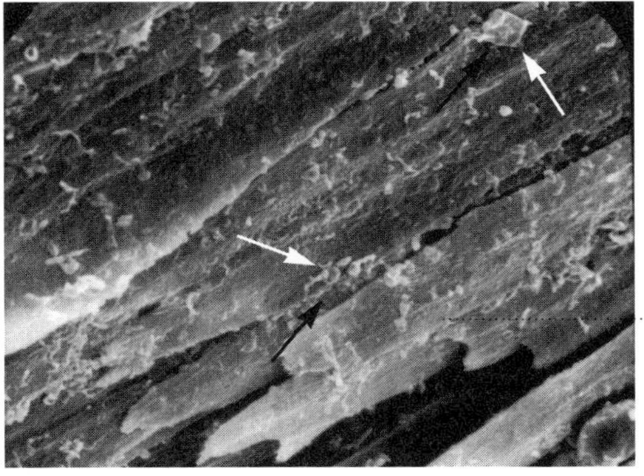

Toenail magnified 1000 times. Note the surface of the nail is covered with fungal hyphae, dirt and debris (arrows). The deep crevices could contain thousands of virus particles that are not visible at this low magnification

photographs courtesy of W.L. Steffens

Preventing Viral Infections

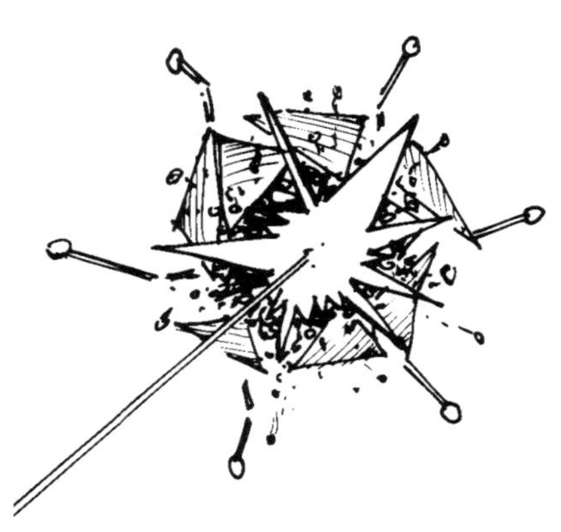

The goal of maintaining any bird in captivity, whether it is a single companion bird or a functional member of a large breeding aviary, is to ensure that the bird is in the best possible condition. Elsewhere in veterinary medicine, vaccination plays a major role in preventing viral infections. Because few effective vaccines exist against the viruses that infect companion and aviary birds, careless exposure of a single bird within a flock to a virus can initiate a cascade of transmission that severely impacts the group. Catastrophic outbreaks with high levels of morbidity and mortality can occur if highly infectious viruses like PBFD virus, avian herpesvirus, paramyxovirus or polyomavirus are introduced into young or highly susceptible birds. This is particularly true in the avian nursery, breeding aviary or pet shop, where numerous immunologically naive birds are congregated into a relatively confined, fixed airspace. Until more effective vaccines are available, birds can be maintained in the best health by applying the principles of quarantine, avoiding direct or indirect contact with birds outside the established flock, providing birds an excellent environment, paying close attention to sanitation methods, maintaining a high plane of nutrition and maintaining birds in a low-stress environment.

Any bird that exhibits signs of illness should be immediately isolated to prevent it from exposing other birds to an infectious disease-causing organism. It should be noted that the common practice of placing a hospital or "sick" room in the same building or air space with a nursery is contrary to sound preventive medical practices. The last place one would choose to place a clinically ill bird is in the same airspace with highly susceptible neonates.

Any bird that dies should be necropsied to determine the cause of death. The type of virus that is causing a problem and its source should be established when the first affected bird is identified. By the time a second bird within a group dies from a viral-induced disease, it may be too late to prevent the widespread transmission of the virus within a susceptible flock.

Birds maintained in less crowded areas with sufficient air circulation, reduced exposure to excrement and frequent exposure to sunlight are less likely to be exposed to, or to succumb to, viral infections than are birds maintained in crowded indoor areas with poor hygiene and poor ventilation. Every procedure that reduces indirect contact between birds will help reduce the possibility of a viral disease outbreak *Table 5.1*. When combined, these procedures reduce a bird's exposure to viruses and help maintain a bird in peak condition to resist infection by many viruses.

A CLEAN ENVIRONMENT

The cleaner a bird's environment, the healthier it is likely to be. Viral infections can move fastest through groups of birds maintained in closed, crowded, unsanitary conditions. Viral infections are less likely to cause substantial disease outbreaks in smaller groups of birds that are maintained in clean, less crowded areas with adequate air circulation.

A key to maintaining the health of a bird is to prevent it from coming in contact with infectious agents that may be present in accumulated excrement or discarded food. For the individual bird, this can be accomplished by using commercial enclosures that have a wire mesh floor that allows food and excrement to pass through. In the breeding aviary, hanging welded-wire enclosures of appropriate size that have both an indoor and outdoor area are ideal. Drop-through enclosures have the added advantage over walk-in type arrangements in reducing the possibility of aviary personnel functioning as fomites in the transmission of infectious agents, as they walk from one enclosure to the next *Figure 5.1*.

CLEAN AIR

The quality of a bird's air could be expected to influence its susceptibility to respiratory infections. Clean, fresh, frequently exchanged air helps to maintain the health of a bird's respiratory system and reduces the accumulation and spread of infectious agents *see Figure 3.1*. It may be prudent for breeders to concentrate on maintaining birds that can adapt to the climatic variations that occur in an aviculturist's area. This allows the breeding birds to have year round access to an indoor/outdoor environment. Electronically filtering air will reduce but rarely eliminate airborne pathogens. Just as people who smoke are more susceptible to infection by influenza virus, a bird exposed to cigarette smoke would be expected to damage its respiratory tract and make it more susceptible to inhaled infectious agents.

AVIARY VISITORS

In established breeding facilities, aviary visitors should be discouraged. Visitors disrupt the daily routine and may indirectly expose birds in the aviary to a virus or other infectious agents. Visitors can carry viruses into a facility on skin, hair or clothing, and should not enter the nursery or touch neonates. An avian veterinarian should not visit different flocks of birds on the same day unless a shower and complete change of clothes are available.

MIXING BIRDS

Mixing birds from different sources (multiple breeding facilities, multiple pet retailers or multiple overseas sources) will increase the risk of a viral-induced disease. The mixing of neonates origi-

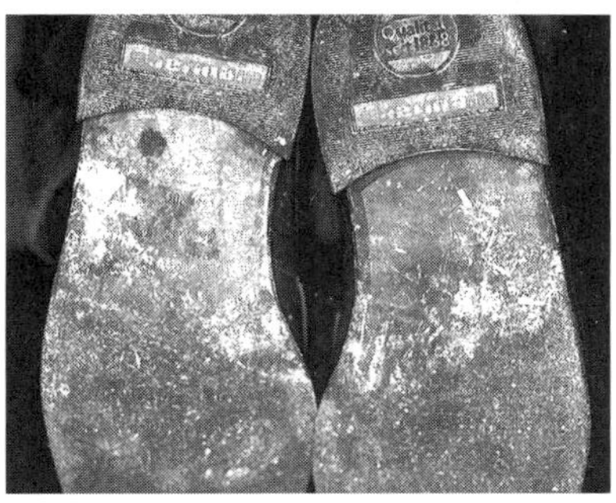

nating from several different aviaries is also flirting with a viral-induced disaster. Any situation where numerous aviculturists bring birds to one location is potentially detrimental to the birds' health. No dog or cat breeder would consider taking their companion into a show ring without the animal being fully vaccinated.

Aviculturists should avoid the addition of imported birds to an established, closed flock. If a bird must be added to an established aviary or companion bird household, it is extremely important to carefully and critically scrutinize the source of that new bird. Care in choosing the source for a new bird is just as important for the individual who has only one or two other birds as it is for the aviculturist with a large facility. It only takes one bird from a "questionable source" that is carrying an infectious disease to cause multiple deaths within a flock.

QUARANTINE

One of the most important techniques for preventing the introduction of an infectious disease to an aviary is to separate new birds from the established group. This period of physical separation is called quarantine.

Many aviculturists want to believe that when they purchase a bird that has been through the USDA quarantine system, they are receiving a bird that is free of disease. The USDA quarantine system is designed only to protect poultry from infectious diseases of economic importance. The system does not insure that a bird is otherwise healthy. It is therefore important for companion and aviary bird enthusiasts to design and implement their own quarantine program for new arrivals. While a quarantine program does not eliminate the possibility of introducing an infectious agent to other birds, it does provide an evaluation period that may prevent some sick birds from being added to a group. A newly obtained bird that was raised in captivity should be quarantined for a minimum of 30 days. All other birds should be quarantined for a minimum of 90 days.

Ideally, birds in quarantine should be maintained in an area that is completely separate from the established group, and quarantined birds should not be cared for by the same individuals that care for the remainder of the group. If a separate facility is not available, quarantined birds should be placed in an area where they are physically separated from the established group with different air, supplies and feeding utensils; the quarantined birds should be fed and cleaned last. Once daily contact has occurred with the quarantined birds, the aviculturist must not approach the area where the established group is housed without taking a shower and changing all garments including shoes. Any bird that has been removed from a group and exposed to any other bird that might be shedding an infectious disease-causing organism should be quarantined before being reintroduced to the group.

New birds should be given thorough physical examinations by a knowledgeable avian veterinarian prior to being placed in quarantine. The examination should include evaluation of the feces for abnormal types or quantities of bacteria, fungi and parasites; a complete blood count (CBC) and selective clinical chemistries; blood testing to determine if a bird is latently infected with PBFD virus; vaccination for avian polyomavirus: a fecal antigen test or culture to detect if a bird is shedding chlamydia; and a serum antibody test to detect chlamydial antibodies.

Aviculturists and conservationists from every geographic area of the world should be concerned about the movement of companion or aviary birds between continents. These birds that can be easily moved to or from remote regions of the earth can be subclinically infected with indigenous microbial flora to which avian species in the importing country may have no immunity. It is not currently possible, nor is it likely to be possible, to ensure that birds being imported from other continents are free of infectious disease-causing organisms.

BOARDING

For a traveling family that has a companion bird, a care provider that comes into a home is preferable to boarding a bird in another location where it may be exposed to infectious disease-causing organisms. Additionally, having a companion bird cared for at home would be expected to be less stressful than having it moved to a new location. Unless a bird must be hospitalized for medical observation or therapy, it is best for birds to stay at home.

The ideal avian hospital or boarding facility would be segmented into small rooms that hold two to three enclosures per room and have separate air flow systems. This arrangement would allow patients that may have infectious diseases to be separated from those that do not have infectious diseases. Staff caring for these birds must not move from hospitalized patients to the areas where "normal" birds are being boarded. Aviaries that are producing neonates should not board birds.

CLEANING AND DISINFECTION

Because of the frequent mixing of birds that are highly susceptible to viral infec-

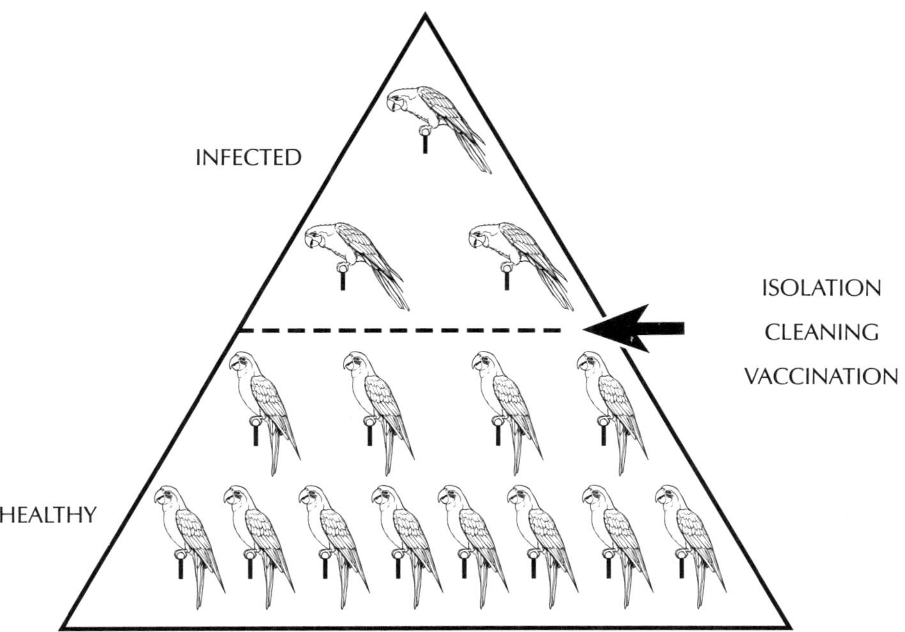

FIG 5.2
Early detection of the presence of a virus in a flock increases the chances that isolation of affected and exposed individuals, cleaning and disinfection procedures, and vaccination (where applicable) can be used to stop further progression of an outbreak.

tions, a lack of vaccines to prevent many viral diseases and the difficulty in detecting subclinical infections, exceptional hygiene and well-designed disinfecting techniques are critical to controlling the spread of viruses. Young, elderly and immunocompromised birds should be considered particularly susceptible to infectious diseases. Proper hygiene is especially crucial to maintain health in these groups of birds.

Understanding the general requirements for a virus to replicate provides insights into how viral infections may be prevented. For a virus to infect a bird, it must enter the bird's body, attach to a specific type of target cell, penetrate the cell and be uncoated to release its nucleic acid into the cell. The virus then takes over the machinery of the cell, reproduces the viral nucleic acid, assembles into a new viral particle and is released from the cell to infect other cells *see Figure 1.15*. To then spread from bird to bird, a virus must be shed into the environment in secre-

tions or excretions and be capable of surviving a sufficient time outside the originally infected bird for direct or indirect contact with a new susceptible bird to occur. If this cycle of transmission is not interrupted, a virus will continue to infect increasing numbers of new, susceptible birds within the exposed population.

If a virus particle does not enter a bird or come in contact with a target cell within the bird, the virus will not be able to replicate and will eventually become inactivated. Viruses are susceptible to inactivation when they are outside of a bird's body. Their destruction or removal from a bird's environment is one of the best methods for breaking a viral transmission cycle *Figure 5.2*. The transmission cycle can be stopped by removing virus particles from the environment (washing), permanently fixing them to a surface (painting), destroying the receptors that allow them to bind to a cell or destroying their nucleic acid (use of disinfectants). (It should be noted

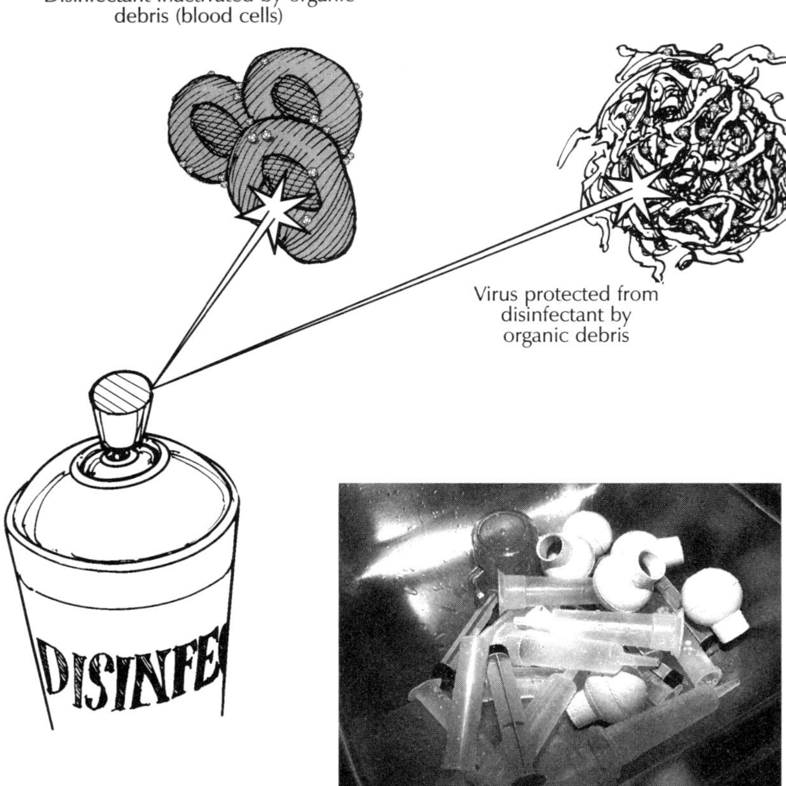

The presence of soil, food, feces, blood, bedding material or mucus can interfere with the disinfection process in two ways: it can directly inactivate the disinfectant, or it can protect infectious agents from coming in contact with the disinfectant.

Feeding utensils used in the nursery must be cleaned, disinfected, rinsed and allowed to dry after each use.

that paints release toxic fumes for extended periods of time. Birds should not be placed in the same airspace with recently painted surfaces for at least one month.) The viruses that can be best controlled by cleaning and disinfecting are those that are transmitted from bird to bird on fomites. Viruses that are transmitted by direct contact are less effectively controlled with cleaning and disinfecting procedures. These viruses will best be controlled through vaccination.

PRINCIPLES OF DISINFECTION

The terminology frequently used when discussing the destruction of microorganisms is summarized in *Table 5.2*. Disinfectants are a valuable part of any disease management program. However, they do not replace the use of sound hygiene in preventing infectious diseases. Numerous disease outbreaks have been linked to veterinarians, pet retailers and aviculturists who have become careless in daily husbandry and have attempted to replace thorough cleaning with the use of disinfectants and drugs to prevent disease problems.

TABLE 5.2

Terminology used to describe control of microorganisms

DISINFECTANT - An agent that will destroy many of the disease-causing microorganisms present on the surface of an inanimate object. These agents may or may not be effective against some viruses, mycobacteria, protozoa or heat-resistant bacterial spores.

STERILANT - An agent that destroys all microbial organisms including heat-resistant bacterial spores. Sterilization can be achieved by boiling, autoclaving or exposure to toxic chemicals. Solutions that contain chlorine or glutaraldehyde are frequently labeled as chemical sterilants.

GERMICIDE - An agent, that when used as directed, will kill a specific group of organisms listed on the label.

SANITIZER - An agent that reduces microbial contamination on the surface of an object to an acceptable level. Sanitizers must not leave a harmful residue.

ANTISEPTIC - An agent that can be used as directed to reduce the microbial population found on skin.

The ideal disinfectant would rapidly inactivate a wide variety of microorganisms, would be safe to use on inanimate objects and would be safe and nontoxic to humans or other animals. Unfortunately, no safe disinfectant is equally effective against all types of infectious organisms. Additionally, few disinfectants work effectively when large quantities of organic debris are present *Figure 5.3*.

Enclosures, bowls, towels, nets, toys, perches, shipping containers, feeding dishes and instrumentation should be thoroughly cleaned and disinfected before use between birds. To maximize the effects of a disinfectant, these objects must be thoroughly cleaned to remove any food, excrement or secretions. The easier an object is to clean, the more likely it is that it can be adequately disinfected. Flaming and steam-cleaning are extremely effective for cleaning the surface of durable objects. Porous objects, such as woods, fibers or ropes, are virtually impossible to clean and disinfect, and require sterilization to destroy contaminating infectious agents. Wooden perches or nest boxes should be replaced if an infectious agent has contaminated the aviary environment *Figure 5.4*. Even surfaces that are considered to be smooth, such as plastics and metals, have irregularities in which virus particles can accumulate *Figure 5.5*. In one study, the surface of a stainless steel table was shown to have 1-micron ridges sufficient to hide individual or clumps of virus particles.[15] Materials that are cleaned should be rinsed before they are placed in disinfectant solution to prevent the cleaning agent from reacting with, and possibly inactivating, the disinfectant.

The capacity of a disinfectant to destroy a virus varies with the type of virus, the type of disinfectant, the concentration of the disinfectant, the quantity of the virus and the material in which the virus is contained. For disinfection to occur, there must be direct contact for the necessary time between an effective disinfectant and the agent to be

FIG 5.4

Extensive, deep crevices and surface irregularities on porous objects provide numerous locations where infectious agents can accumulate and remain protected from the inactivating effects of disinfectants: End of a perch. Arrow marks area of magnification.

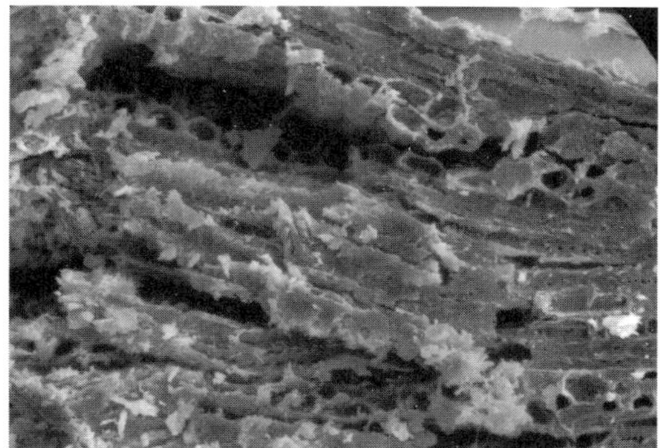

End of a perch showing the extensive, deep cracks and crevices that are visible, even when magnified only 60 times.

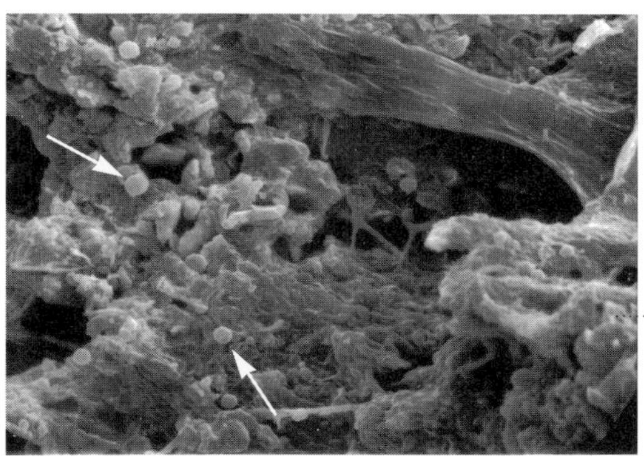

An area of the perch contaminated with feces contains numerous bacteria (arrows). Magnification 1,000x.

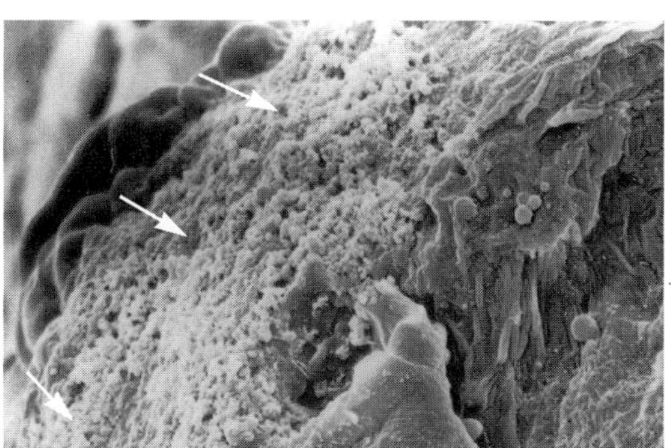

On another area, a bacterial colony has created a "lawn" (arrows) that is covering a larger area of the perch. Magnification 1,500x.

photographs courtesy of W.L. Steffens

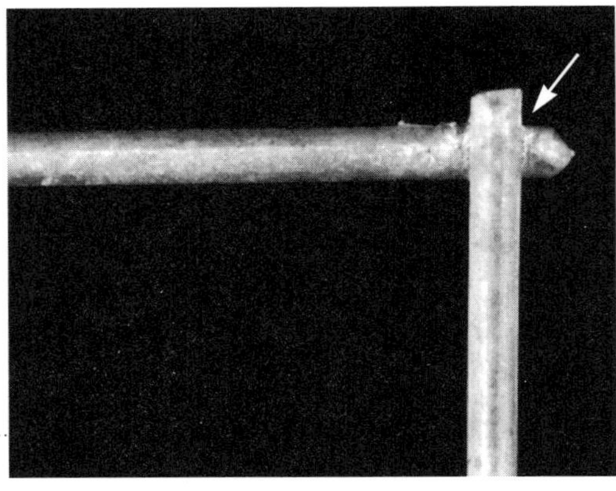

Metal and plastic surfaces that appear relatively smooth actually contain extensive irregularities in which virus particles can accumulate.

End of a galvanized welded-wire enclosure. Arrow marks area for magnification.

End of same enclosure demonstrating irregularities in the surface of the metal that are visible when magnified 48 times.

Irregularities in galvanized welded-wire that are visible when magnified 1,000 times

photographs courtesy of W.L. Steffens

in a disinfectant provides longer contact time, but many disinfectants are corrosive and may damage metals or plastics.

The effects of chemical disinfectants are generally dependent on the pH, temperature and the concentration of the solution.[15] In general, disinfectants can be considered to be more effective at higher temperatures. At one concentration, a chemical may sterilize an object it is in contact with, while at a lower concentration the same chemical may function only as a disinfectant. A safe balance with the use of chemicals designed to destroy microorganisms must be reached in which the target organisms are destroyed without creating a health hazard to the humans, animals or materials that they contact.

If a disinfectant reacts with other substances before it contacts the target agent, it may have insufficient residual "killing capacity" to destroy the infectious agent. The inactivation of a disinfectant is particularly common when materials containing large quantities of organic debris are placed in containers of disinfectants. The presence of these organic materials at best will increase the contact time necessary for a disinfectant to inactivate a microorganism.

Viruses generally have a tendency to stick to other objects such as cellular debris, dust or mucus, or to aggregate together in clumps. Any clumping or accumulation of organic material around a group of virus particles makes them particularly difficult to inactivate *Figure 5.6*.

destroyed.[15] Disinfectants may require minutes to hours of contact time to be effective. However, in many applications, disinfectants may come in contact with an agent for only seconds. For example, wiping an inanimate surface with a disinfectant-laden cloth provides a particularly short time of interaction between the disinfectant and contaminating viral particles. Soaking an object

The susceptibility of viruses to inactivation varies among families, and different virus particles of the same type can exhibit varying susceptibility to inactivation. Most (but not all) viruses that have a lipoprotein envelope are unstable and are inactivated by most disinfectants.[1] Viruses that do not have a lipoprotein envelope generally are resistant to harsh environmental conditions and many disinfectants *see Figure 1.14.* As an example, avian polyomavirus does not have a lipoprotein envelope and is considered relatively stable under adverse conditions. It could be hypothesized that disinfectants that inactivate avian polyomavirus also would be effective against other viruses of importance in birds, like herpesviruses, paramyxoviruses and togaviruses which have an envelope and are less stable than avian polyomavirus.

A virus particle that is exposed to a disinfectant may react in any of several ways. It can remain unaffected or react with the disinfectant causing temporary, reversible damage. The virus particle can be irreversibly changed and thus be inactivated. The virus particle can be damaged, releasing infectious nucleic acid. The virus particle and the nucleic acid can be damaged, rendering the virus inactivated. The nucleic acid can be damaged, with the protein coat remaining intact *Figure 5.7.*[16] Some enveloped viruses may be inactivated by damaging the fatty protein envelope even though the capsid or the nucleic acid it contains, remains undamaged.

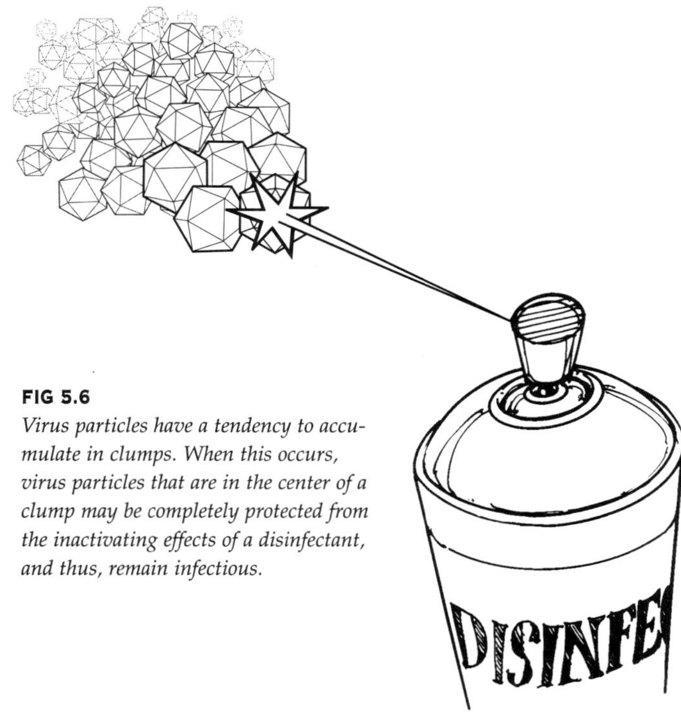

FIG 5.6
Virus particles have a tendency to accumulate in clumps. When this occurs, virus particles that are in the center of a clump may be completely protected from the inactivating effects of a disinfectant, and thus, remain infectious.

Most failures of disinfectants are related to improper use. Repeated exposure of viruses to inadequate disinfection procedures would be expected to favor genetic selection for a population of viruses that are resistant to a particular disinfectant. This phenomenon has been documented for the polioviruses, and is expected to occur with other viruses that are relatively resistant to inactivation.[3] To insure the safe and efficacious use of a disinfectant, the manufacturer's recommendations should always be followed carefully.

Body parts that are contaminated with virus particles are particularly troublesome, because many viruses can remain infectious for prolonged periods and toxic disinfectants cannot be safely used on skin surfaces. Hands contaminated with virus particles may be the most important mechanism for trans-

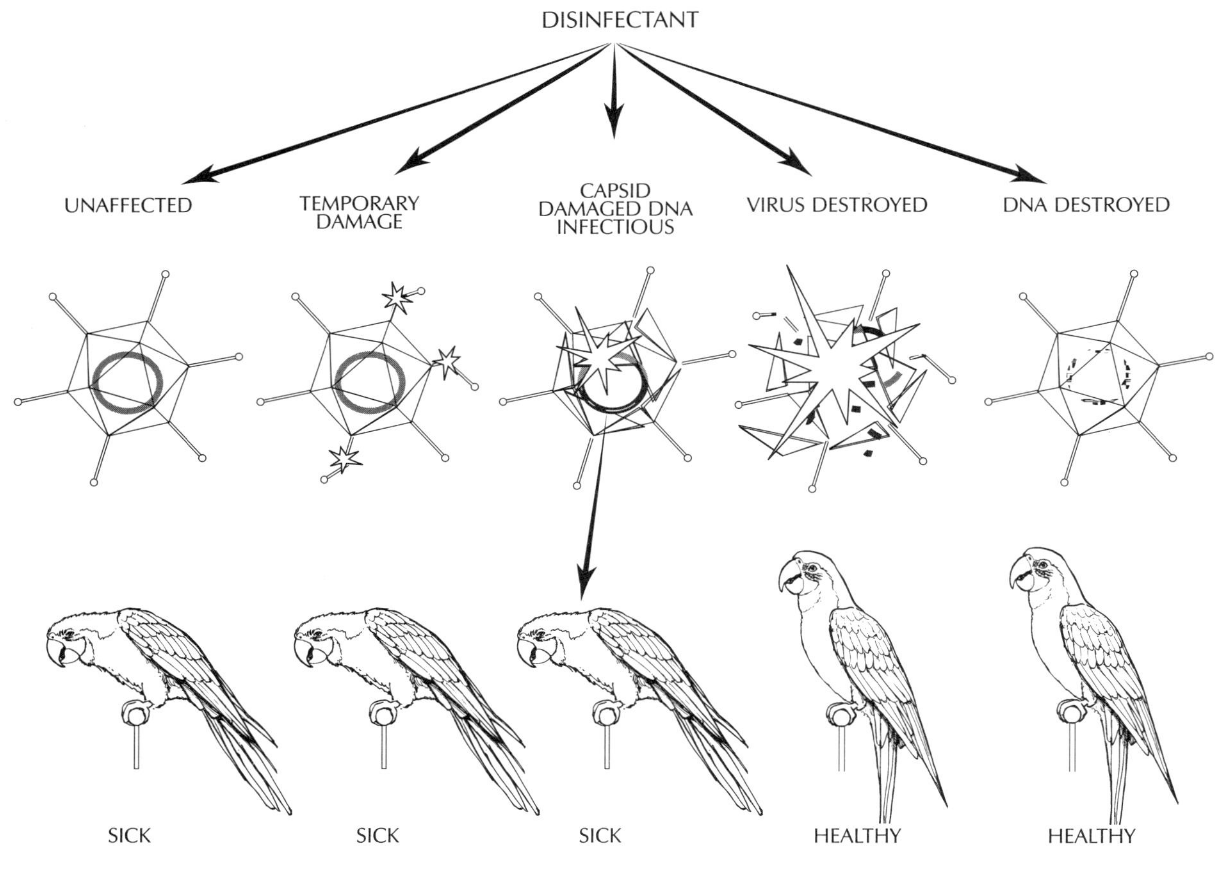

FIG 5.7

The effects that a disinfectant has on a virus will vary with the type of disinfectant, the conditions in which the virus is exposed to the disinfectant and the type of virus.

mission of environmentally stable viruses.[15] Frequently washing one's hands with a soap containing a disinfectant is the best way to reduce the chances of serving as a mechanical vector for virus transmission.

CHARACTERISTICS OF SPECIFIC DISINFECTANTS

A number of disinfectants are widely used in veterinary hospitals, pet retailers and avicultural settings to control pathogens of concern to birds. The active compound and trade name of common disinfectants are listed *Table 5.3*. Sources of the various disinfectants are listed in the Appendix. Individuals who use disinfectants should be aware of the fact that all disinfectants are not effective against all microorganisms.

The use of a disinfectant that is ineffective against a particular pathogen can result in the spread of an infectious disease.

When choosing a disinfectant, the avian professional should consider the ability of a particular product to inactive microorganisms as well as the health risk associated with the use of a particular product. Unfortunately, the health hazards associated with frequent exposure of companion birds (particularly neonates) as well as hospital or aviary personnel, to harsh disinfectants or their fumes are rarely considered when choosing a disinfectant. Any disinfectant that inactivates viruses is toxic, and care should be used to prevent the unnecessary exposure of aviary personnel or birds, particularly neonates, to the residues or fumes of these disinfectants. Perfumes and scents designed to hide the noxious odor of many disinfectants do not decrease the damage that these agents cause if the fumes are inhaled. When any disinfectant solution is used, treated surfaces must be thoroughly rinsed to prevent contact of potentially tissue-toxic solutions with the delicate tissues of birds. Unless suggested by the manufacturer, disinfectants should not be mixed with each other or with other cleaning products.

SOAPS AND DETERGENTS

Soaps and detergents are divided into two groups: anionic soaps (which are negatively charged) and synthetic detergents (which are positively charged). Soaps and detergents reduce the attraction of greases and dirt to an object. In some cases, specific chemical disinfectants are combined with a soap or detergent. These agents are combined to facilitate the cleaning and disinfecting of areas or objects that are contaminated with large quantities of organic debris.

Household detergents are excellent for cleaning bowls, enclosures, perches and nets that may be contaminated with food, excrement or secretions. As is the case with any chemical, surfaces and objects washed with soaps or detergents should be rinsed thoroughly and allowed to dry before being placed in contact with a bird.

CHLORINATED COMPOUNDS

Sodium hypochlorite (bleach) is the most common chlorinated compound used as a disinfectant. It can be purchased as a liquid or as a powder.

TABLE 5.3

The active compound and trade names of some common disinfectants[4]

ACTIVE COMPOUND	PRODUCTS
Chlorinated compounds	Bleach, Clorox, Purex
Chlorine dioxide (stabilized)	Dent-A-Gene
Chlorhexidine gluconates	Hibitane, Hibistat, Nolvasan, Virosan
Glutaraldehydes	Banacide, Cidex, Cybact, MC-25, Sporcide, Sonacide, Sterol, Wavecide
Iodine	Betadyne, Scrubodyne, Povidone, Prepodyne, Virac, Wescodyne
Phenols	Avinol-3, LPH, Lysol, Matar Amerse, One stroke, Environ, O-Syl, Staphhene,
Quaternary ammonium	A-33, Barquat, Cetylcide, Floquat, Hitor, Merquat, Omega, Parvosol, Quintacide, Roccal, Zephiran
Wood tar distillates	Hexol, Pine-Sol

Bleaches are extremely powerful oxidizers that, depending on their concentration, can destroy many if not most microorganisms including bacteria and viruses. Bleach has limited activity against the spores produced by some bacteria and fungi. Most viruses tested have shown some sensitivity to inactivation by chlorine-containing products.[15] However, it should be noted that most virus-inactivation studies with chlorine-containing disinfectants have been performed with the virus suspended in a liquid and in the absence of large quantities of contaminating organic debris. It is well known that organic debris severely reduces the efficacy of oxidizing disinfectants, including chlorine-based products.

Bleach solutions are widely used for disinfecting durable surfaces because they are easy to obtain, inexpensive, and have a wide antimicrobial activity and low residual toxicity. In general, a 1:32 dilution (½ cup of household bleach per gallon of water) is effective in inactivating many infectious agents. Bleach at a ratio of one gallon per 500 gallons of water has been used to cleanse contaminated water pipes.

Bleach solutions are rapidly inactivated by organic debris, extremes in pH, exposure to sunlight (ultraviolet light) or evaporation. They require frequent mixing (every several hours) to maintain an active solution. Unfortunately, bleach solutions and the fumes they produce are toxic to living tissues, including skin, eyes and lungs. Irritation of mucous membranes resulting in watery eyes, nasal discharge and sneezing may be noted in birds exposed to bleach fumes. Bleach is corrosive to metals and produces carcinogenic by-products, including trihalomethanes, chlorophenols or chloramines.[11,12,14] All disinfectants should be used in an area with sufficient ventilation, and this is particularly true for bleach. As is the case with any chemical, surfaces and objects treated with bleach should be rinsed thoroughly and allowed to dry before being placed in contact with a bird.

STABILIZED CHLORINE DIOXIDE

Like bleach, stabilized chlorine dioxide is an extremely powerful oxidizer that can destroy many microorganisms including bacteria, viruses, fungi and protozoa. Some studies suggest that in many applications chlorine dioxide may be a superior disinfectant to sodium hypochlorite.[1] At working dilutions, stabilized chlorine dioxide is considered safe for humans and animals and is used by many municipalities as the principal agent to eliminate potential pathogens from drinking water. In Europe, chlorine dioxide is used to treat drinking water because, unlike chlorine, it does not form carcinogenic trihalomethanes, chlorophenols or chloramines.[11,12] Stabilized chlorine dioxide has been shown to inactive avian polyomavirus despite its relative safety for humans and animals.

Solutions containing stabilized chlorine dioxide and the fumes it produces are toxic to living tissue, including skin, eyes and lungs. Stabilized chlorine dioxide is rapidly inactivated by organic debris and exposure to sunlight. It evaporates quickly and requires frequent mixing to maintain an active solution. All disinfectants should be used in an area with sufficient ventilation. As is the case with any chemical, surfaces and objects treated with stabilized chlorine dioxide should be rinsed thoroughly and allowed to dry before being placed in contact with a bird.

CHLORHEXIDINE GLUCONATES

Chlorhexidine-containing compounds are frequently used as disinfectants for inanimate objects and as antiseptics for cleaning skin and wounds. These disinfectants are relatively nontoxic to skin. Repeated use on wounds is thought to create a layer of residual compound that helps protect wounds from invading microbes. Some chlorhexidine products also contain alcohol. These mixtures have been shown to have superior antimicrobial properties when compared to compounds that contain only chlorhexidine.[15] Chlorhexidine is frequently discussed for use in aviaries and around birds. These disinfectants are relatively nontoxic and noncorrosive, and are considered safe for use on objects that come in direct contact with birds. They have good activity against many bacteria, yeast (particularly *Candida*) and some enveloped viruses.

These disinfectants have limited activity against some bacteria (particularly *Pseudomonas*), spores produced by mycobacteria and nonenveloped viruses. Virosan has been shown to be effective for *Pseudomonas*. In general, chlorhexidines should not be considered reliable viricides.[15] Chlorhexidine is ineffective in the presence of organic debris and has a limited stability. It must be made fresh at least once per day. It has been suggested that hexachlorophene is a potent carcinogen.

GLUTARALDEHYDES

Glutaraldehydes rapidly inactive many microbial agents including most bacteria (including mycobacteria), many viruses and chlamydia. They are effective against many viruses even in the presence of organic debris and are stable as a working solution from two weeks to up to one month. However, glutaraldehydes are infrequently used as general purpose disinfectants because of their widespread side effects in human and animals. These agents may cause irritation to the eyes, respiratory tract and skin (cracking and peeling), particularly following long-term exposure. Some glutaraldehyde-containing products are corrosive to metals; others are not. The product information provided by the manufacturer will list the corrosive properties of a particular glutaraldehyde disinfectant. Aldehydes should not be mixed with ammonia, phenols or oxidizing agents (chlorine or iodine).[15]

IODINES

Iodine-containing solutions are oxidizing agents that are frequently used as antiseptics for cleaning wounds and skin. Most commercially available iodine-containing disinfectants are iodophors (iodines mixed with a detergent) or "tamed" iodines. These agents are less caustic and have an improved antimicrobial activity when compared to the base compound, iodine.

Iodine-containing disinfectants generally produce limited toxic vapors, are available mixed with detergents to both clean and disinfect, and are effective for many bacteria, some viruses and fungi. However, many iodine-containing disinfectants are expensive, require full-strength use, are toxic if ingested, may cause drying and cracking of skin, are not effective against all strains of *Pseudomonas* and are not effective against some viruses.

Iodine-containing disinfectants may be rapidly inactivated by the presence of organic debris. It is important that materials to be disinfected be thoroughly cleaned. Some studies indicate that ionophores are effective in the presence of small quantities of organic debris but are not effective when larger quantities of organic debris are present.[15]

PHENOLS

Sodium orthophenol is the active ingredient in most phenol-containing disinfectants. Phenols can inactivate many bacteria (including *Pseudomonas* and mycobacteria), fungi and some viruses.[15] Phenols vary in activity based on the presence of organic matter and on the temperature, concentration and pH of the disinfectant solution. They are inexpensive and are easy to rinse from inanimate objects so they leave minimal toxic residue. They are toxic to many tissues, irritating to the skin, eyes and respiratory tract, and may be particularly toxic to cats and reptiles.

All disinfectants should be used in an area with sufficient ventilation. As is the case with any chemical, surfaces and objects treated with phenols should be rinsed thoroughly and allowed to dry before being placed in contact with a bird.

QUATERNARY AMMONIUM COMPOUNDS

Quaternary ammonium disinfectants are produced by adding organic compounds to ammonia. Because of their chemical structure, these agents may function as a detergent and help remove organic debris from contaminated objects. Quaternary ammonium disinfectants are inexpensive at working dilution, are relatively safe, and inactivate many types of bacteria, some viruses and chlamydia.

These disinfectants may be inactivated by organic debris and contact with soaps. They are not considered effective for spores, mycobacteria or fungi, and may have reduced efficacy against many nonenveloped viruses and *Pseudomonas*. All disinfectants should be used in an area with sufficient ventilation. Quaternary ammonium compounds are difficult to rinse from some surfaces and frequently leave a slimy residue. Ingestion, and possibly inhalation, of quaternary ammonium compounds can cause respiratory paralysis and death. These agents are not recommended for objects that will be in direct contact with birds, such as anesthesia equipment, food or water bowls, feeding utensils or endoscopic equipment.

WOOD TAR DISTILLATES

Wood tar distillates have a low toxicity but are poor disinfectants. They may be used in some cases to facilitate cleaning of inanimate surfaces prior to disinfecting.

ALCOHOLS

Seventy percent ethyl alcohol (ethanol) inactivates many bacteria and viruses providing that there is sufficient contact time. Many require 20 minutes. Alcohols perform best in the presence of moisture. Parvoviruses and some enteroviruses are resistant to inactivation by ethanol.[15] Alcohols dissolve many plastics, rubbers and glues and must be used cautiously on equipment containing these materials. Alcohol fumes can be irritating to the eyes and mucus membranes.

FORMALIN

Formalin and formaldehyde are extremely dangerous, toxic compounds that should not be used as disinfectants. Formaldehyde will inactivate most viruses, but may require up to one hour of contact time. Formaldehyde can cause dermatitis and is toxic to the eyes and mucous membranes lining the nose, mouth and respiratory tract. This agent is thought to be a powerful carcinogen.[15]

VACCINES

Because drugs to cure viral infections are not available, effective vaccination programs have become indispensable for preventing viral diseases in humans, companion animals and production animals. Vaccines have been responsible for reducing the incidence of (or eradicating in specific populations) some of the most devastating viruses found in humans and animals. While effective vaccines exist for many viral diseases that affect humans, dogs, cats, horses, cattle, pigs and chickens, relatively few vaccines have been developed and tested for use in companion and aviary birds. The development of effective vaccines for preventing viral-induced disease in these avian species will undoubtedly improve the community's ability to prevent viral diseases and should dramatically improve the health and longevity of companion and aviary birds.

HOW VACCINES WORK

When a bird recovers from a viral infection, it does so because the immune system produces specific antibodies and specialized immune system cells that react to the surface proteins that compose a virus' coat or a virus' envelope. In many cases, this immune response will protect a bird from future infections by the same virus. Vaccination procedures are intended to induce a similar immune response.

Depending on the particular virus, a vaccine may initiate an immune response that will prevent a virus from infecting a bird or it may allow a bird to be infected but produce a minor, rather than severe, disease. In either case, a vaccinated bird should develop an active immunologic response following vaccination. In many cases, this immune response can be demonstrated by detecting antibodies in the blood fol-

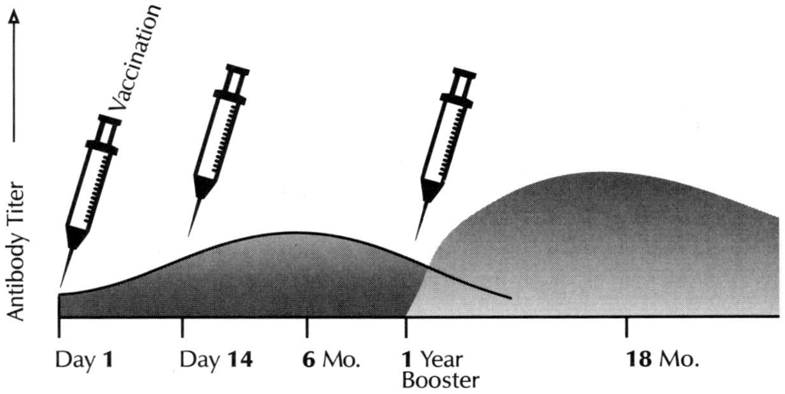

lowing vaccination. Over time, the immunologic response induced by a vaccine will decrease and administration of a booster vaccination will become necessary *Figure 5.8*.

FIG 5.8

Booster vaccines are given on a frequent (usually annual) basis to maintain a strong immunologic response and reduce the chances that a bird will be infected by a virus.

The more birds in a population that are vaccinated, the less likely it is that a virus will find a susceptible host *Figure 5.9*. If very few or none of the birds in a population are susceptible, then a virus shed from an infected bird is unlikely to find a new host to infect and will eventually become inactivated. By using mass vaccination programs, it has been possible to eliminate some viruses from specific populations. For example, smallpox has been eradicated from the

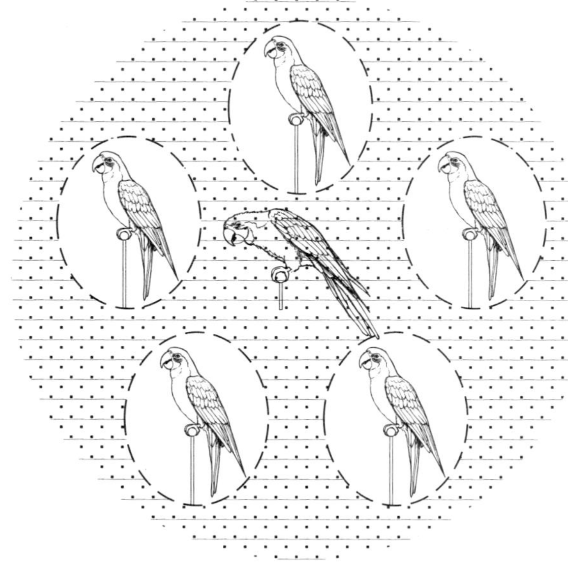

FIG 5.9

With any virus that is spread by bird-to-bird contact, the greater the number of birds in a population that are protected from an infection, the less likely a virus is to find a susceptible host.

human population due to an aggressive worldwide vaccination program.

Because avian polyomavirus is restricted to a limited number of avian species and because a free-ranging host for this virus has not been demonstrated in North America or Europe, it should be possible through vaccination to control polyomavirus infections within specific companion bird populations on these two continents. In contrast, it would be very difficult to eliminate PBFD virus from Australia where free-ranging psittacine birds may be infected and serve as a reservoir for transmission into captive birds.

Vaccination of Hens

Viruses that infect young birds are of particular concern to aviculturists. Virus infections in recently hatched chicks are best prevented by vaccinating hens. The antibodies produced by a vaccinated hen should be passed to the developing chick in the yolk, providing temporary protection to the chick for several weeks after hatching. The process of antibodies being passed from the hen to the chick is called passive immunization. The process of vaccinating the hen, or any other bird, is called active immunization.

Passive immunization provides a chick with temporary immunity against viruses that it may encounter. As the chick matures, the antibodies that the chick received from the vaccinated hen decrease, and the chick becomes susceptible to infection by viruses to which it was previously resistant. At this point, it becomes the responsibility of the chick's maturing immune system to protect it from viral diseases. This transition period from maternally derived immunity to a functional immune system is an interval when young birds should be considered particularly susceptible to infections *see Figure 3.26.*

The speed with which maternally derived protection declines varies widely with the particular type of antibody and the individual bird. In chickens, passively derived maternal antibodies decay very quickly (several weeks), and limited research indicates that a similar situation may occur in young psittacine birds. For example, work with PBFD virus suggests that maternally derived antibodies decrease in the chick at between 30 and 45 days of age. For a vaccine to be effective in preventing disease in young companion or aviary birds, it is likely that both the hen and the chick (several weeks after hatch) would need to be vaccinated.

Because the hen is responsible for transferring temporary protection to her eggs, some have suggested that there is no need to vaccinate males for viral diseases that do not cause them specific problems. This theory is flawed. It should be noted that the immunity that a chick receives from a vaccinated hen is only temporary, and thus the chick at some point will be susceptible to infection. By vaccinating both the males and females, flock resistance to a viral infection is increased. This in turn reduces the likelihood that a chick will be exposed to a virus before its immune system is properly developed or until it can be effectively vaccinated.

TYPES OF VACCINES

There are three main types of vaccines that are used to prevent viral diseases: attenuated-live (also called modified live), inactivated (also called killed) or synthetic (also called recombinant or subunit). In the laboratory, attenuated-live virus vaccines are produced by growing a strain of virus that is altered so that it infects an animal but does not cause severe disease. The animal's immunologic response is able to fight off the weak strain of virus contained in the attenuated-live virus vaccine and the vaccinated animal is then protected against future infections by the same virus. Inactivated vaccines are produced by using chemicals to destroy the infectivity of a large quantity of virus present in a solution. Synthetic vaccines are produced by laboratory manipulations which separate the specific portion of the virus that is responsible for inducing an immunogenic response, and concentrating this portion of the virus in a vaccine.

Another method of vaccinating an animal against disease is to expose the animal to a virus by an unnatural route which stimulates an immune response rather than disease. This technique is effective only with certain types of viruses in certain species of birds. For example, if infectious laryngotracheitis virus is given to a gallinaceous bird via the trachea (a natural route of infection), the bird develops a severe upper respiratory disease. If the same virus is introduced through the cloaca (an unnatural route of infection), an immunologic response occurs that protects the bird from infection if the virus is later introduced to the trachea.

A final type of vaccination takes advantage of the fact that some types of viruses that cause disease in young animals cause subclinical infections in adults. With these types of viruses, adults can be exposed to the live virus which will induce an immune response. The antibodies that develop in the infected, but unaffected, hen may then be passed into her developing eggs. The chicks from this hen would then be protected from infection until the maternally derived antibodies begin to decrease.

Attenuated-live Virus Vaccines

The advantage of attenuated-live virus vaccines is their ability to stimulate a stronger, more complete, natural immune response than that induced by an inactivated vaccine. This superior immune response occurs because the vaccinated bird is actually being infected by a virus that is able to replicate in infected cells. Most attenuated-live virus vaccines are produced by growing a virus repeatedly in cell culture until mutants of the virus are identified that are less virulent than their wild-type ancestors. In some cases, the strain of virus found in an attenuated-live virus vaccine may be modified to such a point that it does not protect the vaccinate from a field strain of virus.

Many of the attenuated-live virus vaccines are easy to administer because they can be given orally, in eye drops or by inhalation of a mist. Attenuated-live virus vaccines administered by these natural routes of exposure are designed to stimulate a local immune response in the mucosa of the respiratory or gastrointestinal tract and thus improve a bird's ability to prevent infectious agents from entering the body through these routes. By replicating in the vaccinate, a single dose of an attenuated-live virus vaccine commonly produces a high antibody titer of long duration.

While there are advantages to attenuated-live virus vaccines, there are potential dangers. The function of an attenuated-live virus vaccine is to produce an infection (thus inciting an immune response) without producing disease. Because the virus present in the vaccine is able to infect cells, there is a possibility the vaccine strain of virus can revert to a virulent form, causing severe rather than mild disease. Additionally, attenuated-live virus vaccines may be virulent in animals that are immunosuppressed. The vaccine may itself be immunosuppressive, may cause a low level of morbidity that effects a bird's reproduction and must be handled with care to prevent inactivation. Attenuated-live virus vaccines can also be contaminated with other infectious agents that cannot be detected.

When given to an animal species other than the one they were developed to protect, attenuated-live virus vaccines may induce severe rather than the intended mild disease following vaccination. For example, the live-attenuated canine distemper vaccine that protects dogs from disease may cause an infection that leads to death if administered to ferrets or foxes. A similar problem occurs in psittacine birds vaccinated with some of the attenuated-live virus vaccines that protect poultry from Newcastle disease virus. Poultry are protected when vaccinated with these vaccines, but psittacine birds may develop a severe disease. With the large number of different genera of companion and aviary birds, and the potential for each genus to react differently to the same attenuated-live virus vaccine, it is unlikely that any attenuated-live virus vaccine could be considered safe for widespread use in this group of birds.

INACTIVATED VACCINES

Inactivated vaccines are produced by taking a sample of virus that has been grown in the laboratory and destroying its ability to replicate (by damaging the DNA or RNA software of the virus) without changing the virus' ability to elicit an immune response (ie, minimal damage to the virus' protein coat). These vaccines are generally safe if all the virus particles truly have been inactivated.

Attenuated-live virus vaccines are potent stimulants of the immune system. By comparison, inactivated vaccines poorly stimulate the immune system. To compensate for their reduced immunogenicity, inactivated vaccines must be manipulated in order to induce a protective immune response that lasts for any significant period of time. To be sufficiently immunogenic, inactivated vaccines must contain large quantities (frequently over one million particles per ml) of the inactivated virus, and many require the addition of compounds called adjuvants, which enhance the body's reaction to the vaccine.

Inactivated vaccines are used in place of attenuated-live virus vaccines because of their stability and safety. The first one to two times that a naive bird is vaccinated with an inactivated or synthetic vaccine, it generally develops a low antibody titer that decreases rapidly. When the bird receives a booster vaccination, the antibody titer that develops

is generally higher and of longer duration *Figure 5.10.*

SYNTHETIC VACCINES

The crude inactivated or attenuated-live virus vaccines of the 20th century are being replaced with more effective, safer vaccines that contain only specific parts of a virus that stimulate an immune response. In the near future, most vaccines used to protect animals and humans from disease will contain only the portion of a virus' protein covering that is responsible for producing a protective immunologic response.

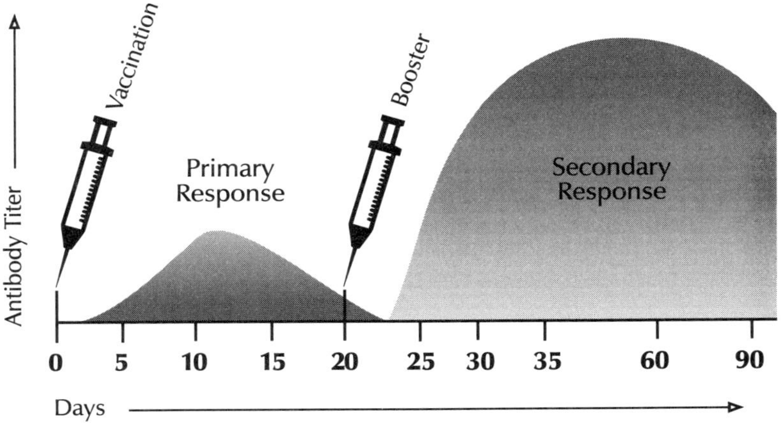

These subunit vaccines are the safest vaccines available because they contain only a portion of the virus, and the DNA or RNA software that is necessary for the virus to replicate is missing *Figure 5.11.* With the DNA or RNA absent, there is no potential for the virus contained in these vaccines to cause disease, which can occur if the virus in an attenuated-live virus vaccine converts to a virulent strain. Subunit vaccines also eliminate the possibility that virus contained in an inactivated vaccine has not been completely inactivated.

FIG 5.10
The first one to two times that a naive bird is vaccinated with an inactivated or synthetic vaccine, it generally develops a low antibody titer that decreases rapidly. When the bird receives a booster vaccination, the antibody titer that develops is generally higher and of longer duration.

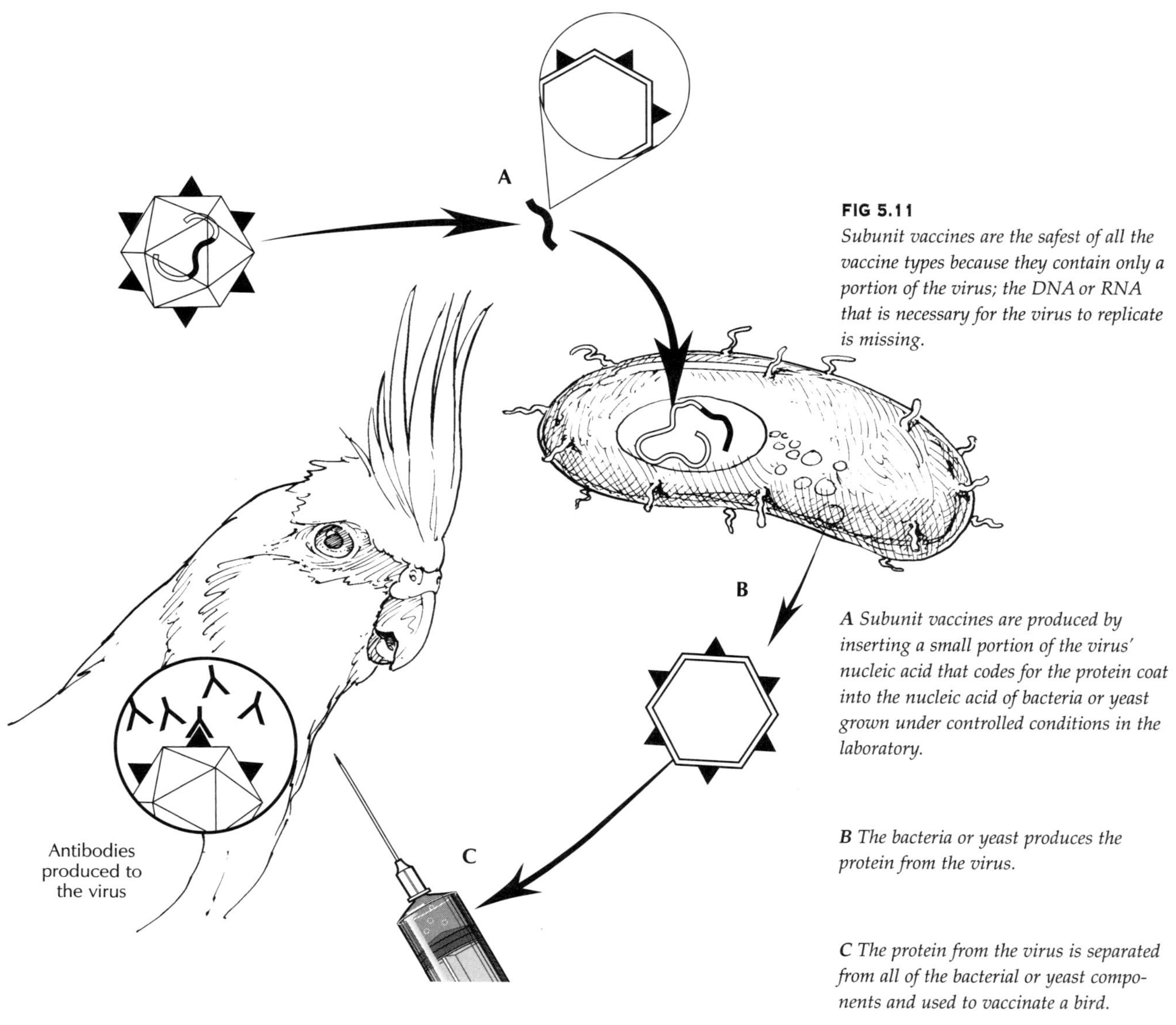

FIG 5.11
Subunit vaccines are the safest of all the vaccine types because they contain only a portion of the virus; the DNA or RNA that is necessary for the virus to replicate is missing.

A Subunit vaccines are produced by inserting a small portion of the virus' nucleic acid that codes for the protein coat into the nucleic acid of bacteria or yeast grown under controlled conditions in the laboratory.

B The bacteria or yeast produces the protein from the virus.

C The protein from the virus is separated from all of the bacterial or yeast components and used to vaccinate a bird.

Other advantages of a subunit vaccine are that it: 1) is a pure vaccine that is free of extraneous proteins or other contaminating infectious agents; 2) has a high degree of stability; and 3) can be a consistently similar product among batches. Several subunit proteins from the same virus can be combined in a single injection to broaden the immunologic response (as is seen with a natural infection) without the risk of inducing disease. Subunit proteins from several different viruses also could be incorporated into the same vaccine to protect a vaccinate from varying serotypes of the same virus or from completely different viruses. The key to successful subunit vaccines is that they must be produced from a portion of the virus that is stable to mutations and they must be produced from a portion of the virus that induces a protective immunologic response.

VACCINE FAILURES

A certain percentage of the birds in any population will not respond to vaccination. A vaccine that induces approximately 80% protection in a group of vaccinates is considered sufficiently efficacious. The failure of a vaccine to elicit an appropriate immunologic response in a larger percentage of vaccinates can be caused by factors associated with the vaccine or the bird. The most common causes of vaccine failures are improper administration of the vaccine or improper handling (exposure to excessive heat or moisture) of the vaccine prior to administration.

Vaccines can also fail to stimulate a protective immune response if a vaccinate is immunocompromised; if a bird is infected by an antigenic variant of the virus that is not contained in the vaccine, if maternally derived antibodies are present (as occurs in recently hatched chicks), if the target virus is already infecting the bird at the time of vaccination or if the vaccinate is receiving immunosuppressive therapy (steroids, some antibiotics). Additionally, certain populations of birds may exhibit a genetic variability in their response to vaccination against certain viruses. Other factors that have been shown to affect the post-vaccination immunologic response include poor nutrition, increased levels of progesterone, increased levels of testosterone, heat stress, aging and decreased levels of thyroid hormone, growth hormone or thyroid stimulating hormone. It should be noted that any vaccine or drug, no matter how safe it may appear, should be expected to cause a reaction in a certain percentage of animals.

ADJUVANTS

Adjuvants are compounds that are added to inactivated vaccines to increase the level and duration of the immune response they induce. Some types of animals are sensitive to certain types of adjuvants. Many companion and aviary birds, particularly cockatoos and pigeons, appear to be very sensitive to vaccines that use an oil-emulsion as an adjuvant.

Reactions that may occur in birds vaccinated with oil-containing adjuvants include mild discoloration at the injection site or open, draining abscesses with large areas of necrosis. In some cases, granulomatous or necrotizing lesions caused by oil-adjuvants may not be noted until several months after vaccination. In other birds, lesions may be noted within several weeks of administration of a booster vaccine.[5,6,8-10]

The granuloma formation following the use of oil adjuvants in some companion and aviary birds is not unlike that described with the use of similar vaccines in other avian species. Chickens vaccinated intramuscularly at 13 to 20 weeks of age with oil-containing vaccines had residual lesions at slaughter 48 to 92 weeks post-vaccination.[7] These lesions were characterized by the formation of discolored yellowish cysts in the tissues adjacent to the vaccination site. The high oil content of some emulsions also has been associated with abscess formation and discomfort at the injection site in mammals.[2]

Several adjuvants have been identified that are safe for use in companion birds.[13] Of specific adjuvants tested in companion birds, Acemannan, aluminum hydroxide, Pemulem and Equimune have been shown to be safe.

REFERENCES

1. Aieta EM, Berg JD, Roberts PV, et al: Comparison of chlorine dioxide and chlorine in waste water disinfection. J Water Pollut Control Fed 52:810-822, 1980.

2. Amyx HL: Control of animal pain and distress in antibody production and infectious disease studies. J Am Vet Med Assoc 191:1287-1289, 1987.

3. Bates RC, Shaffer PTB, Sutherland SM: Development of poliovirus having increased resistance to chlorine inactivation. Appl Environ Microbiol 34:849, 1977.

4. Clipsham R: Environmental preventive medicine. Proc Midwest Avian Res Expo, 1992, pp 48-58.

5. Curtis-Velasco M: Vaccination reaction in umbrella cockatoos. J Assoc Avian Vet 4:206, 1990.

6. Curtis-Velasco M: Further vaccine reactions. J Assoc Avian Vet 5:10, 1991.

7. Droual R, et al: Investigations of problems associated with intramuscular breast injection of oil-adjuvanted killed vaccines in chickens. Avian Dis 34:473-478, 1990.

8. Fudge AM: High risk birds still benefit. J Assoc Avian Vet 5:10-11, 1991.

9. Gaskin JM: Adverse reactions to current pet bird vaccines. J Assoc Avian Vet 5:12-14, 1991.

10. Harrison GJ: The vaccine dilemma. J Assoc Avian Vet 5:7, 1991.

11. Llabres CM, Ahearn PG: Antimicrobial activities of N-chloramines and diazolidinyl urea. Appl Environ Microbiol 49:370-373, 1985.

12. Noss CI, Olivieri VP: Disinfecting capabilities of oxychlorine compounds. Appl Environ Microbiol 50:1162-1164, 1985.

13. Ritchie BW, Niagro FD, Latimer KS, et al: Antibody response and local reactions to adjuvanted avian polyomavirus vaccines in psittacine birds. J Assoc Avian Vet 8:21-26, 1994.

14. Sittig M: Handbook of Toxic and Hazardous Substances and Carcinogens. Park Ridge, NJ, Noyes Publications, 1985.

15. Springthorpe VS, Sattar SA: Chemical disinfection of virus-contaminated surfaces. Crit Rev Environ Control 20:169-229, 1990.

16. Thurman RB, Gerba CP: Molecular mechanisms of viral inactivation by water disinfectants. Adv Applied Microbiol 33:75-105, 1988.

Papovaviridae

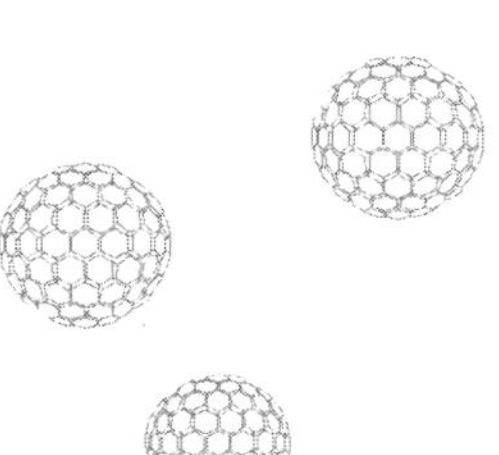

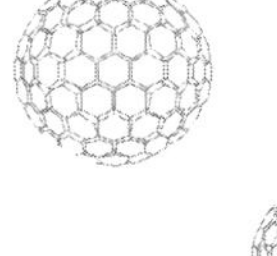

The Papovaviridae family of viruses consists of two subfamilies, Papillomavirinae and Polyomavirinae, which vary in virion size and biological features, as well as in genome size and organization.[22,61]

Viruses in the Papillomavirinae subfamily are generally associated with the formation of benign skin tumors commonly referred to as papillomas. Papillomaviruses have been identified in a wide variety of animals, including birds and humans. These viruses tend to be highly host-specific. The papillomavirus strains that infect humans and other mammals probably do not infect birds, and the strains that infect birds would not be expected to infect mammals. Only a few of the human papillomaviruses have been grown in cell culture, but other advanced techniques have been used to study these viruses. The characteristics and pathobiology of papillomaviruses in mammals and birds have been determined principally using molecular biology evaluation techniques (sequencing, cloning and hybridization) with viral nucleic acid derived from infected tissues.

Papillomas that occur on the skin of finches and African Grey Parrots have been shown to be caused by a papillomavirus. The etiology of papilloma-like lesions in the intestinal tract, oral cavity and cloaca of psittacine birds remains undetermined. It should be noted that all attempts to prove that papillomatous masses in the cloaca of psittacine birds are caused by a virus have been inconclusive.

Members of the Polyomavirinae subfamily are called polyomaviruses. Various strains of these viruses are known to infect rodents, rabbits, birds and primates, including humans. These viruses are relatively specific in host range, suggesting that virus derived from rodents would not be expected to infect primates or birds, while virus derived from birds would not be expected to infect rodents or primates. In mammals, polyomaviruses are generally associated with persistent infections that result in the formation of tumors when they are introduced to an unnatural host.[110] Interestingly, polyomavirus infections generally are not associated with clinical disease in a natural host whose immune system is functioning normally.

Polyomavirus infections were thought to be restricted to subclinical infections in mammals until it was determined that an avian polyomavirus, called Budgerigar fledgling disease, was responsible for an acute generalized infection in nestling Budgerigars. Avian polyomaviruses are unique among the subfamily Polyomavirinae because they can cause an acute infection ending in the death of the affected bird. Studies evaluating the antigenic and genomic relationship of virus recovered from various genera of psittacine birds suggest that the virus that infects a wide variety of psittacine birds is similar.[45,100]

PAPILLOMAVIRUSES AND PAPILLOMAS

CLINICAL FEATURES

Papillomas occur commonly in mammals. Most clinical changes are localized to epithelial tissues, particularly skin. Papillomaviruses replicate in the outer layers of epithelium where they

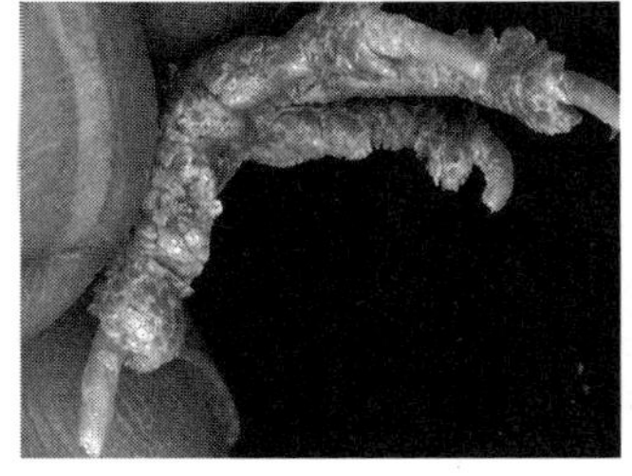

FIG 6.1
Proliferative skin masses on the feet and legs of passerine birds are suggestive of Knemidokoptes mites or papillomavirus infections.

right *Microscopic examination of skin scrapings (for mites) or biopsies (for papillomavirus) can be used to confirm the etiology of these masses.*
photograph courtesy of Greg Harrison

cause thickening of the skin and neoplastic changes in the infected cells. A papillomavirus spreads from its initial site of infection to adjacent cells in the skin. In some mammals, including humans, viral-induced papillomas have been shown to occur on the epithelium of the oral mucosa, upper gastrointestinal tract or genitourinary tract. Infections caused by certain strains of mammalian papillomavirus have been associated with malignant cancers in the reproductive tract of some women. In humans and mammals, the process of developing a clinical sign of a papillomavirus infection (a visible wart) can take weeks to months from the time of initial infection.

The first demonstration of a virus in the Papovaviridae family in a non-mammalian species involved the recovery and characterization of a papillomavirus

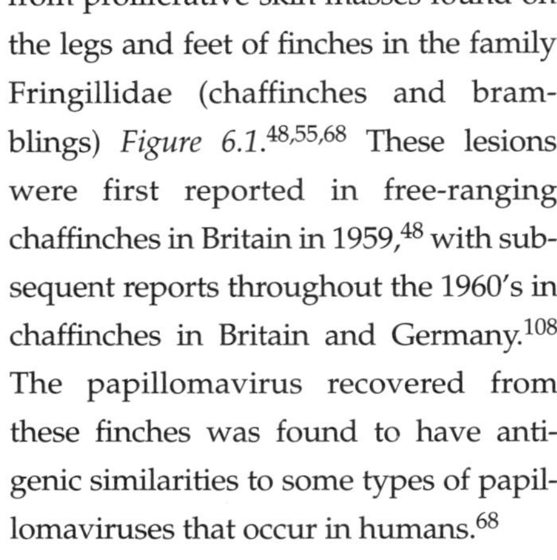

from proliferative skin masses found on the legs and feet of finches in the family Fringillidae (chaffinches and bramblings) *Figure 6.1*.[48,55,68] These lesions were first reported in free-ranging chaffinches in Britain in 1959,[48] with subsequent reports throughout the 1960's in chaffinches in Britain and Germany.[108] The papillomavirus recovered from these finches was found to have antigenic similarities to some types of papillomaviruses that occur in humans.[68]

Papillomavirus infections occur commonly in certain species of finches. However, birds housed in direct contact with affected greenfinches, bullfinches and chaffinches remain clinically normal, suggesting that the virus is highly host-specific.[16,96] In one study, proliferative lesions suggestive of papillomavirus were identified on the legs and feet of 330 of the 25,000 (1.3%) chaffinches examined.[55] In another study, lesions were evident on 4 of 242 (1.6%) chaffinches.[108] The two most common Passeriformes in which papillomas have been reported, chaffinches and bramblings, are closely related. In contrast, lesions appear to be extremely rare in other avian species.[55,68]

Papillomaviruses are considered enzootic in some populations of Anseriformes, Gruiformes and Ciconiiformes.[113] A papilloma-like virus has been demonstrated by electron microscopy in proliferative lesions at the commissure of the beak and on the head of young canaries *Figure 6.2*.[16] This differs from the distribution described in other Passeriformes, where changes are noted primarily on the legs and feet.

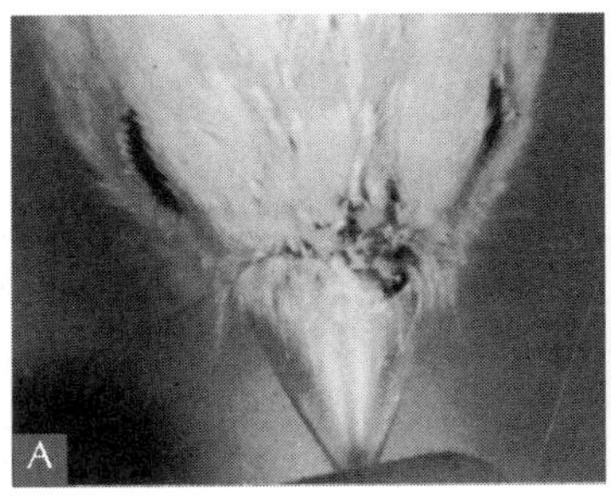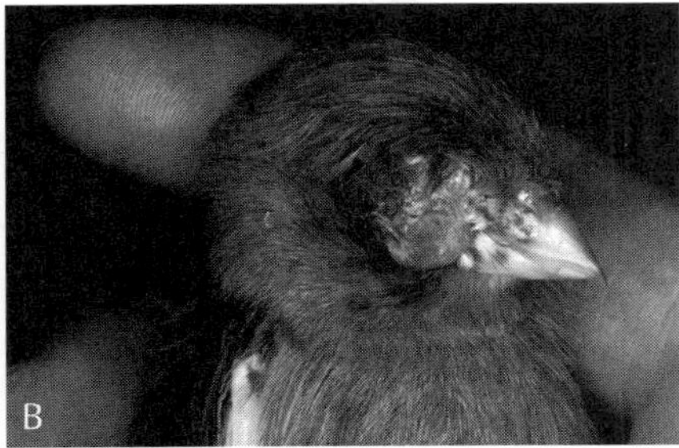

FIG 6.2

A Papillomavirus has been shown to cause proliferative skin masses around the beak of young canaries. These masses can interfere with mastication or, if damaged, become a site for secondary bacterial and fungal infections. B Poxvirus can cause similar-appearing lesions. Microscopic examination may be necessary to determine which of these viruses is causing a suspicious mass.
photographs courtesy of Peter Dom and Louise Bauck, reprinted with permission[16]

A papillomavirus has also been shown to be the cause of persistent cutaneous masses on the head and face of African Grey Parrots *Figure 6.3*. In one study, only one African Grey Parrot in over 5,000 that were examined had a clinically detectable lesion. The virus present in infected cells from an African Grey Parrot cross-reacted with antibody against a bovine papillomavirus (type I), suggesting these two viruses are antigenically related.[46]

Diagnosing that a suspicious lesion on the skin is a papilloma requires the histologic evaluation of biopsy samples collected from the affected area. The histopathology suggestive of a papilloma includes the appearance of long thin folds of hyperkeratotic epidermis accompanied by acanthotic parakeratosis. Confirmation that these suspicious lesions are caused by a papillomavirus requires electron microscopic examination of affected tissues to demonstrate the 45- to 50-nm virus particles that are commonly contained in intranuclear inclusion bodies *Figure 6.4*.[46] Because the incubation period in birds is unknown, it is usually difficult to determine when, where and how the rarely affected bird may have been exposed to a papillomavirus.

Although a virus has yet to be demonstrated, papilloma-like lesions have been diagnosed microscopically in association with proliferative growths originating from skin overlying the phalanges, uropygial gland, mandible, neck, wing, eyelids and beak commissure from various psittacine species, including Amazon parrots, macaws, cockatoos, Quaker parakeets, Cockatiels and Budgerigars *Figure 6.5*.[4,33,74,106] The wart-like growths, in which herpesvirus-like particles have been identified, on the feet of cockatoos and the tumorous form of avian pox (skin or oral mucosa) can appear similar to the lesions induced by papillomavirus *see Figure 7.13*.[29]

The low prevalence of skin lesions suggestive of papillomas in companion birds may indicate any of these possibilities: 1) many suspicious lesions are not caused by a virus; 2) lesions are viral-induced but caused by a virus with a low level of transmissibility; or 3) lesions are viral-induced, but the majority of exposed birds develop subclinical infections and immunity to disease.

Generally, papillomas on the skin of birds need no treatment unless they are causing specific problems. Some lesions can be debilitating if they are damaged and allow secondary infections to

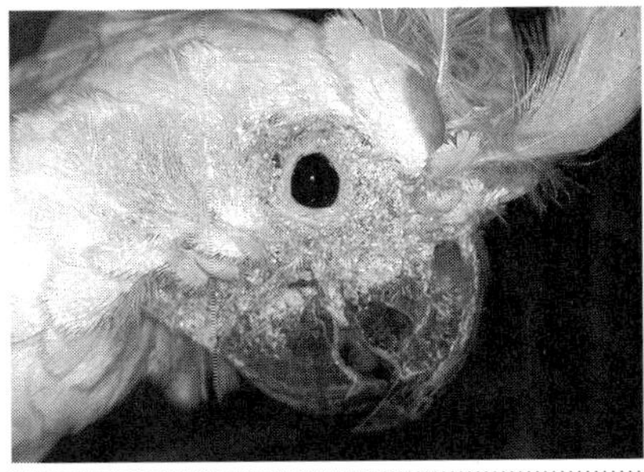

FIG 6.3

A Although rare, papillomavirus infections have been associated with progressive skin masses on the face of African Grey Parrots. These lesions can interfere with mastication, vision or breathing depending on their location and severity.
B A mite, Knemidokoptes, has been shown to cause similar lesions on the face of psittacine birds.

photographs courtesy of Scott McDonald

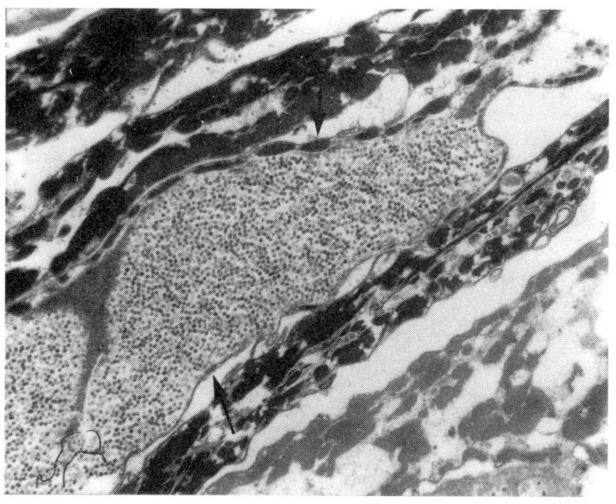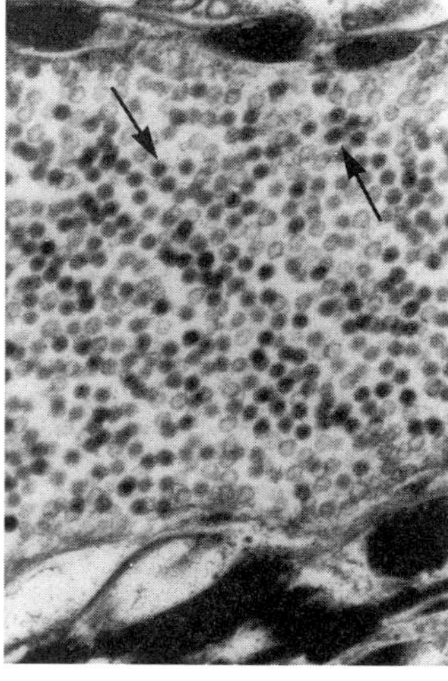

occur, if they inhibit a bird's ability to move or if they interfere with grasping or chewing food. Mild lesions can be observed for changes that would necessitate their removal. Severe lesions should be radiosurgically removed to make a bird more comfortable and to prevent secondary infections. In some mammals, the use of autogenous vaccines have been shown to be effective in stimulating an immune response that results in cessation of clinical lesions. It has not been determined if an autogenous vaccine would be effective in treating papillomavirus-induced skin lesions in birds.

PAPILLOMATOSIS

A papillomavirus has been shown to cause wart-like lesions in the mouth, pharynx and esophagus of rabbits, dogs, humans and cattle, indicating that many different epithelial tissues may be infected by mammalian papillomaviruses *Figure 6.6*.[39,47,63,90,101,109] Similarly, histologic lesions that resemble those induced by papillomaviruses have been described at many locations along the avian alimentary tract. They are particularly common in the oral cav-

ity or cloaca in some species of companion birds. The cause of these lesions in the oral cavity, intestinal tract and cloaca of psittacine birds remains undetermined. The term "papillomatosis"[35] has been suggested to describe papilloma-like growths of unknown etiology that occur on the mucosa of the alimentary tract. This term is used to differentiate these lesions from papillomas on the skin, some of which have been shown to be caused by a papillomavirus.

Some clinicians believe the occurrence and progression of papillomatosis in an aviary setting is characteristic of an infectious agent.[62,106] Other evidence suggests that these lesions may not be caused by an infectious organism. In papillomatous growths collected from the cloaca of psittacine birds, attempts have failed to demonstrate papillomavirus by electron microscopy, by nucleic acid detection techniques (low stringency southern blotting) or by staining affected tissues with viral-specific antibodies (immunocytochemical procedures). Attempts to experimentally induce papillomatous growths in the cloaca of Amazon parrots, macaws and cockatoos using homogenized tissues from affected birds have also been unsuccessful.[57,102] In one case, herpesvirus-like particles were described in a cloacal papilloma from a conure; however, the presence of these herpesvirus-like particles was considered an incidental finding.[31] Because repeated attempts to demonstrate a virus in

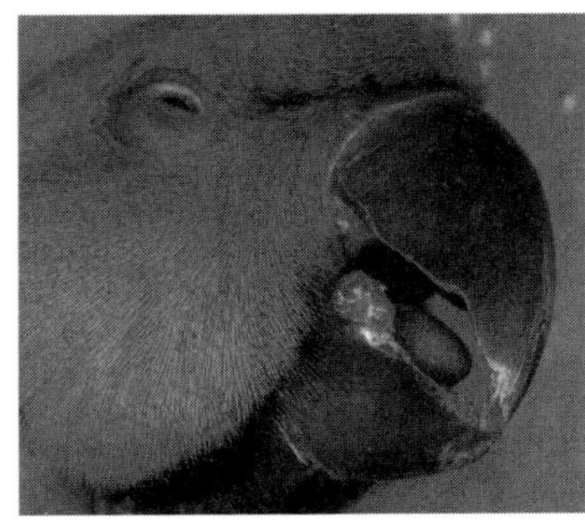

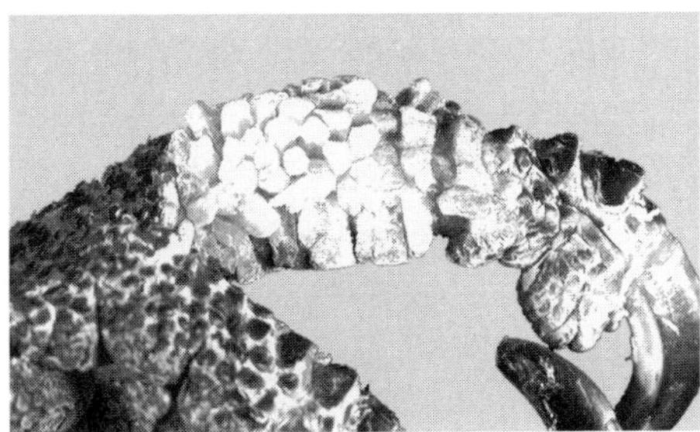

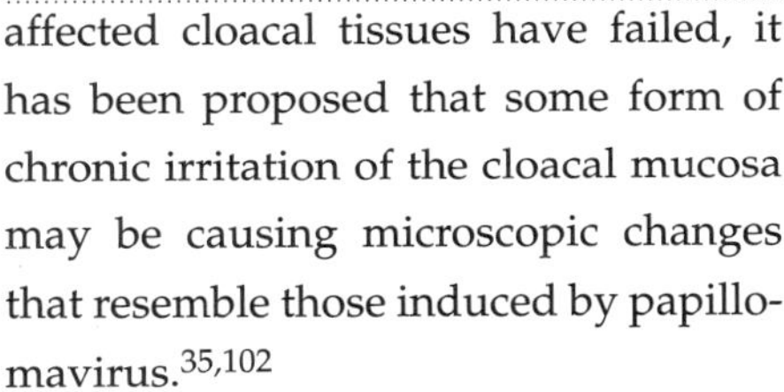

FIG 6.5

In psittacine birds, lesions suggestive of, but not proven to be caused by, papillomavirus are commonly seen on the (A) face, (B) body and (C) feet. A herpesvirus has been demonstrated by electron microscopy in proliferative lesions on the feet of some cockatoos and macaws (see Figure 7.13).

affected cloacal tissues have failed, it has been proposed that some form of chronic irritation of the cloacal mucosa may be causing microscopic changes that resemble those induced by papillomavirus.[35,102]

CLINICAL FEATURES

In one study involving 19 species of New World psittacines, papillomatous masses were documented in decreasing frequency in the cloaca, glottis, choanal slit, esophagus, oropharynx, ventriculus and proventriculus *Table 6.1*.[106] Papillomatous changes have also been reported in the crop, liver, pancreas, nasal mucosa, conjunctiva and nasolacrimal duct.[35] Grossly, papillomatosis of the cloaca may be characterized by the appearance of large, raised, distinct masses or small, coalescing bumps that cover much of the cloacal mucosa *Figure 6.7*.[62] These friable growths may appear as red, pink or white areas that have a tendency to bleed profusely when manipulated or damaged. Papil-

lomatous changes may occur in the oral cavity or esophagus in birds that do or do not have detectable lesions in the cloaca.

Most birds with papillomatous lesions in the gastrointestinal tract are asymptomatic with no changes in the complete blood count (CBC) or blood chemistries. However, the white blood cell count may be elevated in

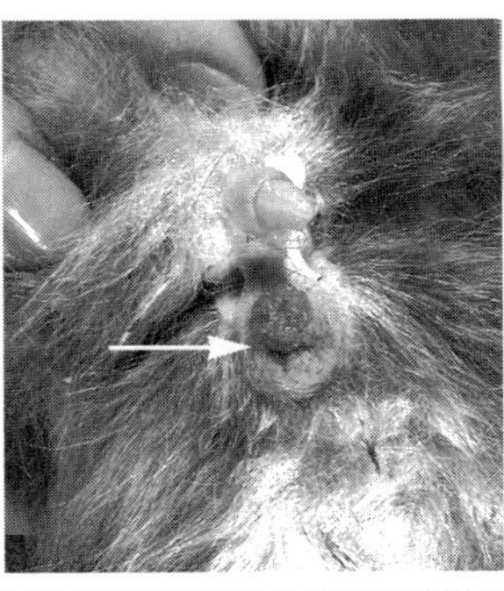

FIG 6.6

In mammals, host-specific papillomaviruses can cause masses in numerous tissues including the mouth, esophagus, vagina and around the rectum. Papillomavirus was the cause of this wart (arrow) in the rectum of a rabbit. photograph courtesy of Scott McDonald

TABLE 6.1

Frequency of papillomatosis in psittacine birds by anatomic location[35]

Cloaca	85 of 103 birds (82%)
Esophagus	11 of 15 birds (73%)
Oropharynx	30 of 46 birds (65%)
Crop	7 of 15 birds (47%)
Ventriculus	6 of 19 birds (32%)
Conjunctiva	3 of 14 birds (21%)
Proventriculus	3 of 19 birds (16%)
Nasal mucosa	1 of 11 birds (9%)
Nasolacrimal duct	1 of 11 birds (9%)

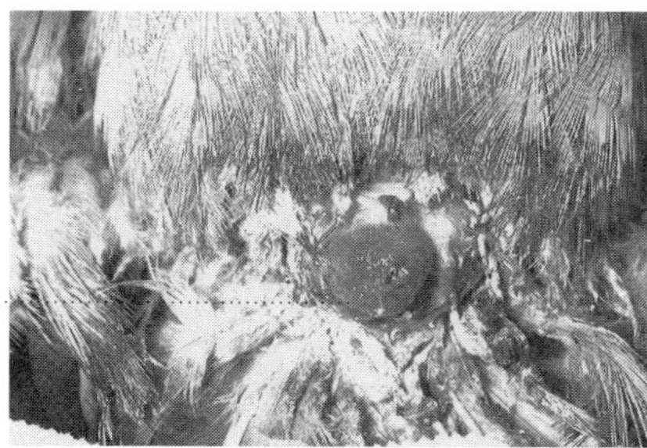

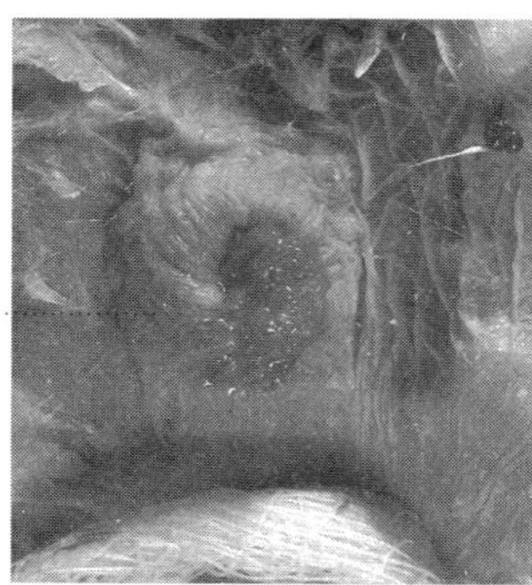

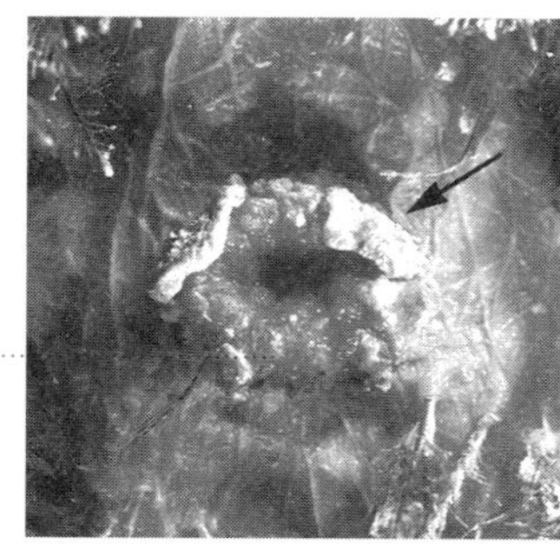

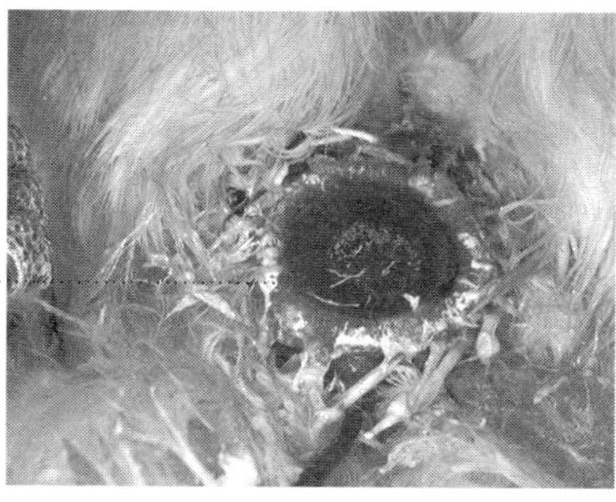

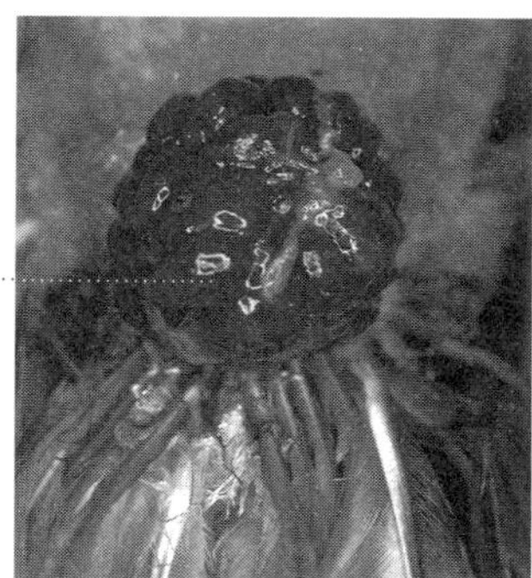

birds with internal lesions that become necrotic, that trap ingesta inciting a local inflammatory response or that are infected with secondary bacteria or fungi.

Clinical signs in birds with papillomatosis generally occur when a proliferative mass mechanically interferes with the normal physiologic activities of swallowing, digestion or defecation. Papillomatous lesions located in the cloaca may be associated with tenesmus, putrid smelling feces, infertility, recurrent enteritis or hematochezia. Those in the oral cavity or esophagus may cause halitosis, dysphagia, dyspnea or wheezing *Figure 6.8*. Cloacal papillomatosis may cause or mimic recurrent prolapses, or droppings may be loose, causing feathers around the vent to be stained or covered with fecobezoars *see Figure 6.7*. Birds with chronic, malodorous feces or bad breath should be carefully evaluated for the presence of papillomatous growths.

Papillomatosis of the upper intestinal tract may obstruct mechanical movement of food, causing anorexia, chronic weight loss or vomiting.[62,106] Birds with papillomatous growths in the ventriculus, proventriculus or crop may have clinical signs that resemble those associated with proventricular dilatation disease, such as regurgitation, poor digestion and weight loss *Figure 6.9*.

It has been speculated that birds with papillomatosis of the cloaca may experience decreased reproductive capabilities. Clinical evidence suggests that some birds with cloacal papillomatosis are able to breed normally, while others may be incapable of successful copulation. The effect of cloacal papillomatosis on reproductive capacity probably depends on the location and severity of the lesions. Severe lesions that block the movement of sperm out of a male and into the opening of the oviduct would be expected to decrease reproductive performance, but mild lesions located in other areas of the cloaca may have no effect. In one study involving a group of 10 breeding pairs of Blue and Gold Macaws with cloacal papillomatosis, the reproductive performance of the affected pairs was at least equal to, if not slightly better than, that of a conspecific group of birds that did not have cloacal papillomatosis.[6a] Theoretically, severe lesions near the opening of the oviduct into the cloaca could cause a hen to have problems delivering an egg.

It has been reported that liver or pancreatic cancers are more common in psittacine birds with cloacal papillomatosis.[35,42] In one study, cancer of the pancreatic duct was present in 8 of 10 psittacine birds with internal papillomatosis.[35] Two of 10 Amazon parrots that died within 15 weeks of being diagnosed with cloacal papillomatosis were diagnosed with bile duct carcinomas.[42] Cancer of the bile ducts was confirmed in the liver of 15 of 24 psittacine birds that had papillomatous masses somewhere in the alimentary tract.[35] A 3.5-year-old Yellow-faced Amazon Parrot and an adult Orange-winged Amazon Parrot with histories of cloacal prolapse were diagnosed with cholangiocarcinoma, a type of liver cancer.[9,82] This report occurred in 1983 before cloacal papillomatosis was frequently discussed in avian literature, and it was

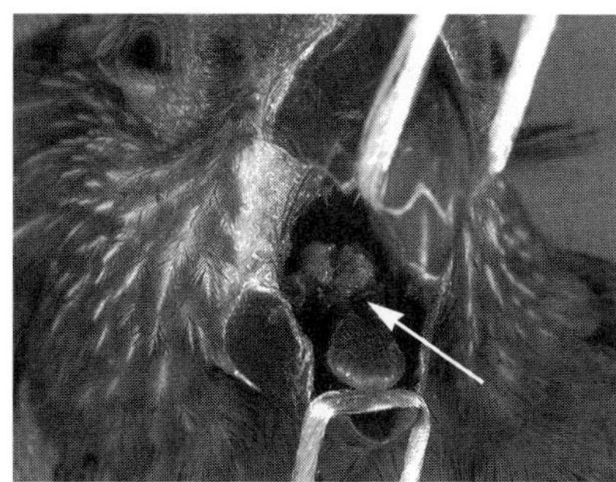

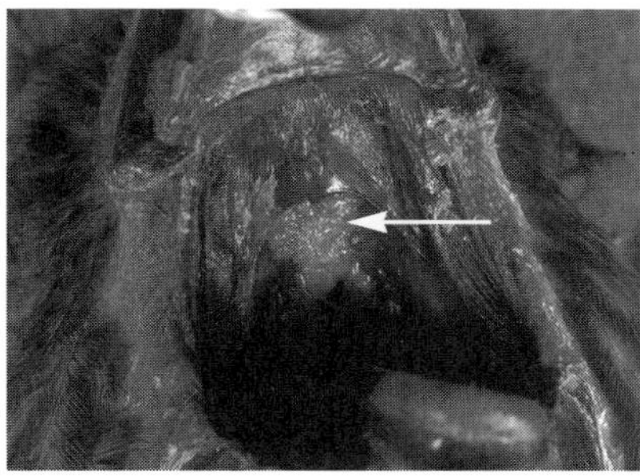

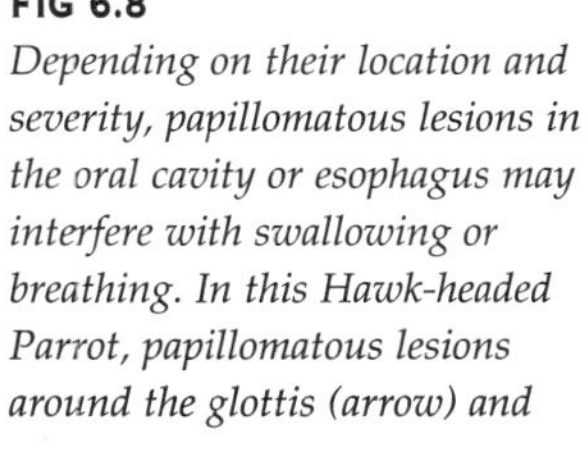

FIG 6.8

Depending on their location and severity, papillomatous lesions in the oral cavity or esophagus may interfere with swallowing or breathing. In this Hawk-headed Parrot, papillomatous lesions around the glottis (arrow) and

choanal slit (arrow) make it difficult for this bird to breathe.

photographs courtesy of Scott McDonald

not mentioned whether the cloacal mucosa in this bird was normal or had papilloma-like growths. Cholangiosarcoma was diagnosed in a 3-year-old Blue-fronted Amazon Parrot in which the integrity of the cloacal mucosa was not reported.[21]

EPIZOOTIOLOGY

Papillomatous lesions have been described in macaws, Hawk-headed Parrots, African Grey Parrots, Amazon parrots, caiques, conures, Budgerigars and Pionus parrots.[10,33,35,62,106] The species of psittacines most commonly reported with oral or cloacal papillomatosis vary with the investigator. In one study, the prevalence of lesions was highest in Hawk-headed Parrots (58%) and Green-winged Macaws (28%).[106] In another study involving 141 psittacine birds with papillomatous lesions in the digestive tract, 60% were macaws, 33% were Amazon parrots and 6 were other species (Cockatiels, Hawk-headed Par-

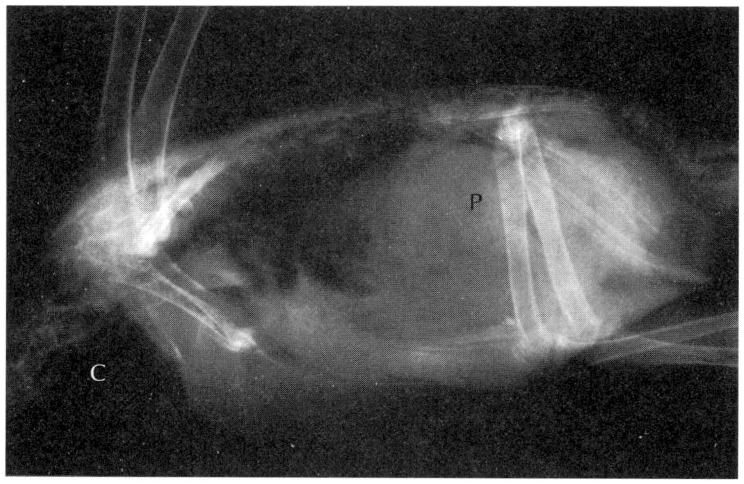

FIG 6.9

Papillomatous lesions located on the mucosa of the proventriculus or ventriculus can interfere with the normal passage of ingesta. Because the ingested food cannot move freely through the digestive tract, it can accumulate in, and cause distention of, the proventriculus (P) and crop (C). B Affected birds may regurgitate. These clinical changes are similar to those noted with proventricular dilatation disease.

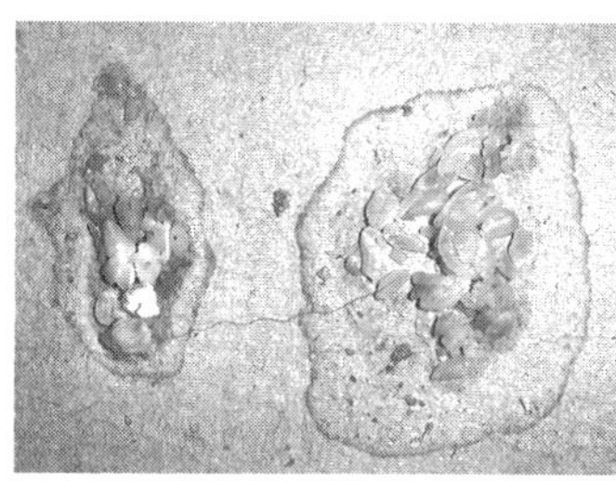

rots, Orange-fronted Conures, African Grey Parrot). In the latter study, affected birds ranged in age from 3.5 weeks to 36 years old (mean 5.5 years). Sixty-one percent of the birds were female, and 39% were male.[35] Papillomatosis has been described in wild-caught and domestically raised birds, but has not been reported in free-ranging psittacines.[62]

TRANSMISSION — Because the etiology of papillomatosis in psittacine birds remains undetermined, it is currently not possible to confirm if transmission of some infectious agent is necessary for lesions to develop. Epizootiologic evidence has been used to suggest that papillomatous lesions in the alimentary tract of psittacine birds are caused by an infectious agent. Several large psittacine aviaries have had epornitic outbreaks of the disease following the introduction of a clinically positive bird.[62,106] However, in other aviaries,

papillomatosis appears to be restricted to individual birds; their exposed mates frequently remain unaffected. Clinical findings would suggest that if papillomatosis is caused by an infectious agent it is of low transmissibility.[6a]

In one study, eggs from Blue and Gold Macaws with cloacal papillomatosis were artificially incubated and the chicks were hand-fed. None of the chicks from these positive parents developed visibly detectable papillomatosis. This suggests that if an infectious agent is involved in this syndrome, it is probably not transmitted to neonates in the egg.[6a]

PATHOLOGY

The gross and histologic lesions associated with papillomatosis are usually proliferative and have been rarely reported to be necrotizing.[29] Proliferation of epithelial cells on a fibrovascular stalk suggests papillomatosis. The classic histologic description involves acanthosis and hyperkeratosis. The tendency of birds with papillomatosis to develop neoplasias of the pancreas or liver[9,35,42] would suggest that cloacal or oral lesions may be caused by a virus that can transform infected cells *Figure 6.10.*

DIAGNOSIS

To be properly examined for the presence of papillomatous growths, the cloacal tissue should be gently everted using a gloved finger, moistened cotton-tipped applicator or speculum *Figure 6.11.* Using a soft paint brush, the surface of the cloacal mucosa can be covered with an acetic acid solution (5%) to help identify papillomatous

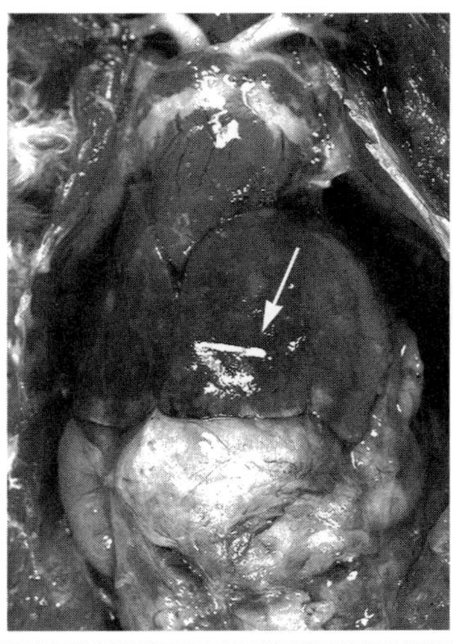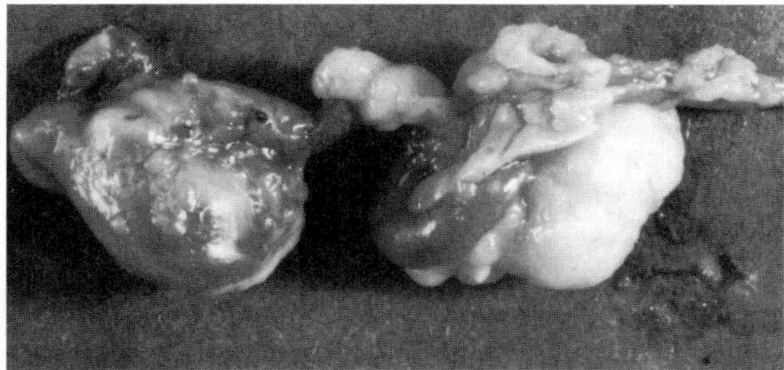

FIG 6.10

*Liver (arrow) and **above** pancreatic cancer appear to occur more commonly in birds with papillomatosis than in birds without these lesions. This would suggest that papillomatosis is being caused by an agent that is capable of transforming cells in multiple tissues.*
photograph courtesy of Cheryl Greenacre

changes. The abnormal tissue will turn white while the normal cloacal mucosa remains pink. Histologic examination of biopsy samples is necessary to confirm a diagnosis.[106]

Suspicious lesions in the oral or cloacal cavity can be viewed directly and appear as smooth, raised pale areas *see Figure 6.8*.[106] Many internal papillomatous lesions are not recognized until necropsy; however, filling defects indicating that a mass is present in the lumen of the alimentary tract may be detected by contrast radiography in birds with suspicious clinical signs such as chronic regurgitation or weight loss. Endoscopy is necessary to identify and obtain diagnostic biopsies of suspect papillomatous lesions in the esophagus, proventriculus or high in the cloaca. A liver biopsy may be helpful in confirming neoplastic changes in birds that have a history of cloacal papillomatosis and clinical changes suggestive of liver disease.

CONTROL AND TREATMENT

Until further information on the etiology of this disease is available, it would be prudent to isolate birds with lesions from the remainder of a flock. Malnutri-

tion and vitamin A deficiencies have been suggested to potentiate papillomatosis.[29] If true, then providing supplemental vitamin A would be indicated.

Oral papillomas are frequently localized, easy to remove and may not recur after removal. In contrast, cloacal papillomas are typically diffuse, difficult to

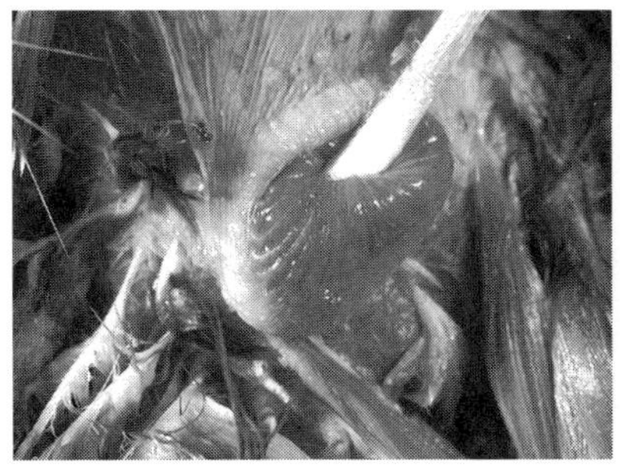

resolve and frequently recur after treatment.[106] None of the proposed therapies is consistently effective in all cases, and papillomatous lesions often recur following what may appear to be a successful treatment regime. Additionally, papillomatous lesions may spontaneously regress, remain grossly undetectable for periods of 2 to 18 months, and then reappear, making the evaluation of therapeutic measures difficult at best.[34,35,62,106]

FIG 6.11

The normal mucosa of the everted cloaca is pinkish-red and smooth.

***right** Papillomatosis should be suspected when the mucosa is irregular in texture. Areas with papillomatous lesions turn white when they are coated with 5% acetic acid (apple cider vinegar).*
photographs courtesy of Scott McDonald

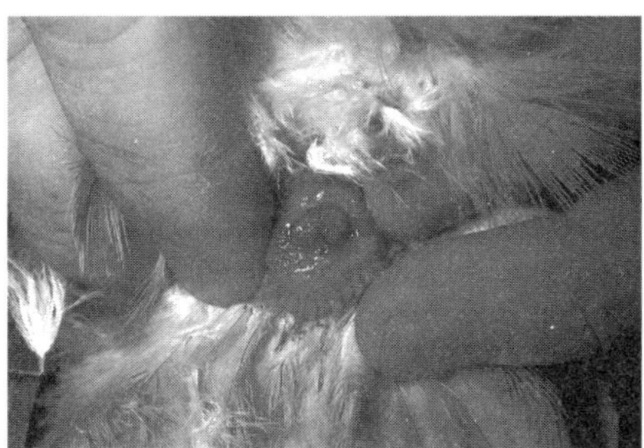

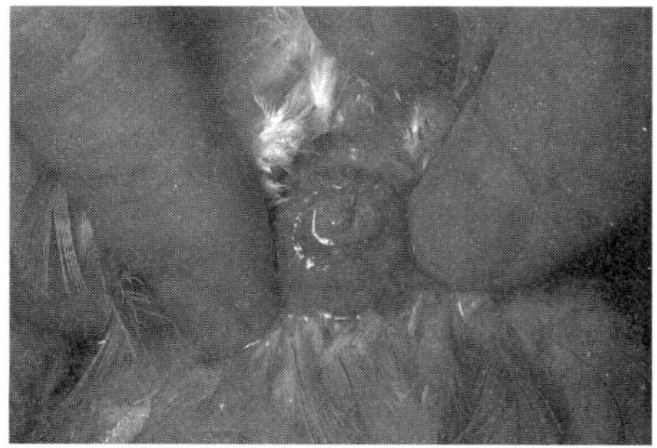

Papillomatous lesions on the mucosa of the cloaca can be removed using silver nitrate sticks. This is a delicate procedure and birds should be anesthetized during treatment.

The affected tissue is everted by placing gentle pressure on opposite sides of the cloaca.

The silver nitrate is applied to the surface of the affected tissue and immediately inactivated by flushing with copious fluids to prevent the liquefied material from burning unaffected mucosal tissues. The cauterized tissue will turn grayish-white.

Suggested therapeutic measures for papillomatous growths in the oral cavity or cloaca have been based on the physical removal of the masses through cryotherapy, radiosurgery or surgical excision using a laser. These removal procedures have been performed alone or in combination with the use of autogenous vaccines. When any technique is used to mechanically remove growths from the cloaca, care should be exercised to prevent excessive tissue damage that may result in severe scarring and reduction in the size of the cloacal lumen. Scarring can result in incontinence, reproductive failure or blockage of the ureters or colon.

One study indicated that surgical use of a laser to remove affected tissues was particularly effective.[37] Another report detailed the use of silver nitrate in the chemical cauterization of cloacal papillomas. In the latter study, only a small number of lesions had been treated, but all of the patients remained papillo-ma-free and clinically normal for months after therapy.[50] In three years of clinical use, this technique has proven to be an easy, safe and readily available way to remove papillomatous lesions from the cloaca *Figure 6.12*. Lesions should be exteriorized by inserting a moistened cotton swab followed by carefully rubbing a small area of the lesion with a silver nitrate stick. The silver nitrate should be immediately inactivated by flushing with copious fluids to prevent the liquified material from burning unaffected mucosal tissues. The procedure should be repeated at two-week intervals until the lesions are resolved. This removal technique is best performed with a bird under isoflurane anesthesia, because the bird must be kept motionless during application of the silver nitrate to prevent the liquified compound from burning normal mucosa. When silver nitrate cauterization is to be used on large masses (greater than 0.25 cm in diameter), it should be done in stages. Severe damage to the cloaca could result in permanent constrictive scarring.

POLYOMAVIRUSES

The first acute, generalized infection associated with a polyomavirus in any species of animal was described in young Budgerigars and was called Budgerigar fledgling disease (BFD).[5,7,14,18,54] A similar avian polyomavirus has been associated with high morbidity and mortality in finches and various other genera of psittacine birds. Some evidence exist that an avian polyomavirus is capable of infecting several species of birds other than psittacines and passerines (see Epizootiology).

The acute nature of some avian polyomavirus infections is most unusual for the Papoviridae family. Polyomavirus infections in mammals are classically associated with apathogenic, subclinical infections in the natural, immunocompetent host. However, some polyomaviruses can cause severe disease in mammals that are immunocompromised or can cause tumor formation when they are introduced to an unnatural mammalian host.[89,110]

Polyomaviruses that infect Budgerigars, other psittacine birds and finches appear to be morphologically and, in some cases, antigenically similar. However, evidence suggests that the genome of the virus that infects various psittacine birds and finches partially differs. Additionally, the clinical presentation, distribution of lesions and epizootiology of these viruses differ dramatically among susceptible species.[7,8,12,19,33,54,85] Because of the differences in the way avian polyomavirus affects Budgerigars versus larger psittacine birds, the extensive information derived from studies in Budgerigars may not be completely applicable to other birds.

BUDGERIGAR FLEDGLING DISEASE

CLINICAL FEATURES

Budgerigar fledgling disease was first reported in the United States and Canada in 1981. One of the initial outbreaks in 1976 in Quebec involved the acute death of 1- to 15-day-old Budgerigars.[5] A 1979 report of hepatitis in Budgerigars in which adenovirus-like particles were seen in the liver may actually have been an early description of an avian polyomavirus outbreak.[44]

Avian polyomavirus appears to be distributed worldwide, but there are some apparent regional differences in the clinical changes associated with the virus. For example, in Europe a more chronic form of the disease is common in Budgerigars, while in the United States and Canada an acute form of disease with high mortality is typical. The type of clinical disease induced by polyomavirus in Budgerigars also may be influenced by the age and condition of a bird when it is exposed to the virus.

Budgerigar neonates from infected flocks may develop normally for 10 to 15 days and then die suddenly without premonitory signs.[60] Other infected hatchlings may develop clinical signs, including abdominal distention, hemorrhage under the skin and reduced formation of down and contour feathers. Some infected Budgerigars have been reported to develop neurologic signs characterized by ataxia and tremors of the head and neck several days before dying *Figure 6.13*.[5-7,12,43,60] Similar neurologic signs have been reported in some Budgerigars that have survived an infection.[6]

Although mortality rates vary with the age of the exposed birds, polyomavirus infections in young Budgerigars usually are rapidly fatal once clinical signs develop. In most aviary outbreaks, the

TABLE 6.2

Summary of polyomavirus-induced feather abnormalities in Budgerigars[6]*

1-15 days old	lack of down on head and neck
15-25 days old	lack of flight feathers or poorly developed feathers
> 25 days old	slow growth of feathers, absence of some flight feathers

Feather lesions are rare in larger psittacine birds.

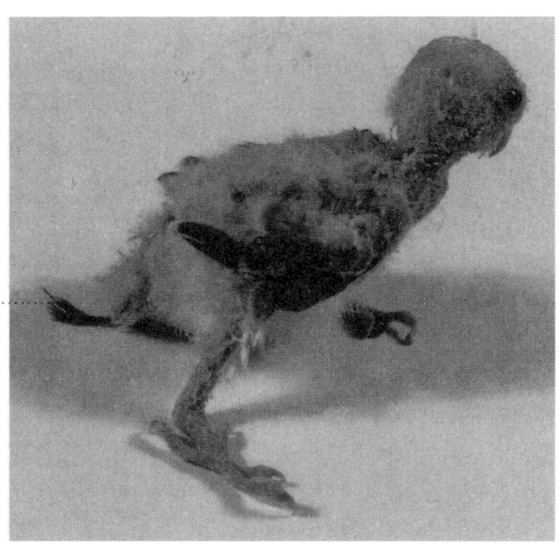

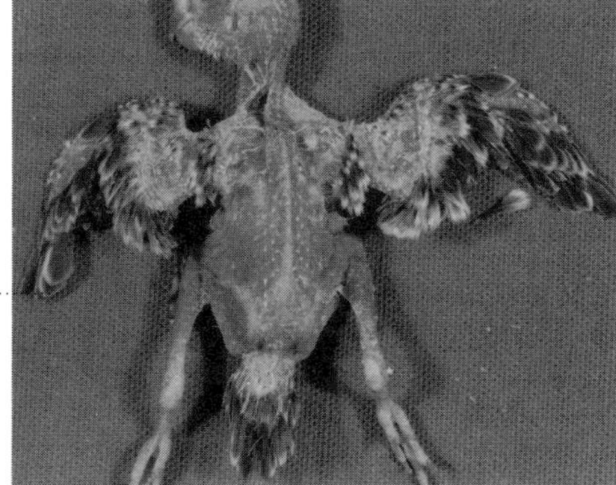

FIG 6.13

Budgerigar neonates infected with avian polyomavirus can die without premonitory signs or may develop abdominal distention, hemorrhage under the skin and

reduced formation of down and contour feathers.

Some affected birds may develop abnormalities of the central nervous system that are noted clinically as depression, ataxia or tremors of the head and neck.

photographs courtesy of Gilles Berneir, reprinted with permission[5]

incidence of clinically recognized disease will progressively increase over a period of several months, with peak virus activity occurring during the most active portion of the breeding season.[6] In one particularly aggressive outbreak of avian polyomavirus, 90% of young Budgerigars were affected within a two- to three-week period.[32]

Budgerigars that develop clinical signs of disease after 15 days of age may survive and exhibit feather abnormalities that occur equally on both sides of the body. These include dystrophic primary and tail feathers, lack of down feathers on the back and abdomen, and lack of filoplumes on the head and neck *Figure 6.14*.[5-7,14,32,43,64] Budgerigars with dystrophic feathers may be unable to fly and have been referred to as "runners." Those with severe feather pathology, often referred to as "French moult," may be featherless.[5,6,43,51,104] However, even Budgerigars with severe feather abnormalities may have no other clinical changes or readily visible internal organ damage that would indicate a polyomavirus infection.[43]

Reports vary on the effect of avian polyomavirus on egg and embryonic development. In one Budgerigar aviary, polyomavirus was thought to have caused an 80-90% decrease in egg hatchability at the same time that hatchling mortality increased from 10% to 60%.[32] In another study, there was no apparent increase in embryonic mortality in eggs from Budgerigar hens that were considered latently infected with avian polyomavirus. Additionally, many of the chicks from these hens were found to be infected, but remained asymptomatic.[80] Findings from the first report would suggest that avian polyomavirus can cause embryonic deaths and decreased hatchability, while findings from the second study would suggest that avian polyomavirus infections do not cause problems in embryonic development. It is possible that both findings are correct, and the effect of the virus may depend on as yet unidentified factors.

"FRENCH MOULT" — is a term that has been used to describe a slow, debilitating disease in Budgerigars characterized by the progressive development of abnormal feathers. Initial observations indicated that this syndrome occurred most commonly in young Budgerigars produced late in the breeding season. In the 1980's, it was hypothesized that the capillaries that supply blood to the developing feathers are extremely delicate, and any minor degree of damage to these capillaries could cause a restriction of the normal blood flow; this would result in sludging and clotting of the blood in the capillaries followed by avascular necrosis of the affected feather.[104] This physiologic explanation for the production of dystrophic feathers suggests that any factor or infectious agent that causes damage to the blood supply of developing feathers could cause visible changes suggestive of French moult.

It has been proposed that French moult may represent a non-fatal form of BFD.[5,6,43,51,104] However, some Budgerigars with classic French moult lesions have polyomavirus antibodies, while others do not.[13,51] In North America and Europe, it has been determined that lesions attributable to French moult can be caused by either avian polyomavirus or psittacine beak and feather disease (PBFD) virus infections.[5,6,43,51] Investigations in Budgerigars in Australia demonstrated clinical signs of French moult associated with the PBFD virus but not with polyomavirus.[71,73,112] In general, feather lesions in Budgerigars caused by polyomavirus resolve after several months,

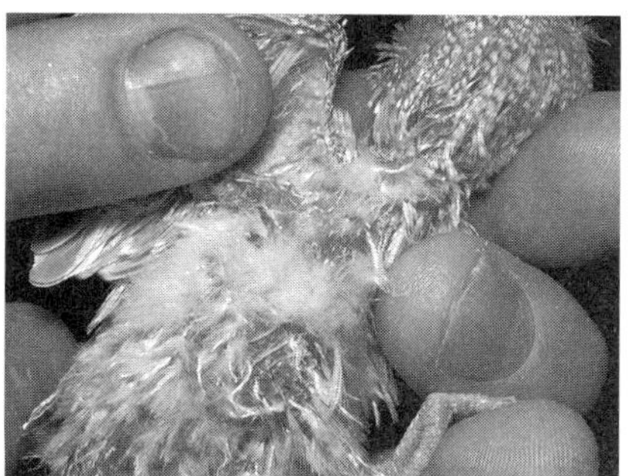

FIG 6.14

Budgerigar neonates that survive an acute polyomavirus infection typically develop dystrophic primary, tail and body feathers. The feather changes caused by avian polyomavirus in budgerigars are grossly indistinguishable from those caused by psittacine beak and feather disease virus.

while those induced by PBFD virus may persist and become progressively worse.[29] Viral-specific DNA probes or viral-specific antibodies can be used to differentiate feather abnormalities caused by polyomavirus from those caused by PBFD virus.[53]

POLYOMAVIRUSES IN NONBUDGERIGAR PSITTACINE BIRDS

CLINICAL FEATURES

The polyomaviruses that infect various nonbudgerigar psittacine birds appear to be morphologically and antigenically similar. However, the clinical presentation, distribution of lesions and epizootiology of polyomavirus infections vary dramatically between Budgerigars and larger psittacine birds *Table 6.3*.[7,8,12,19,33,54,85] Antibody surveys indicate that most infections in nonbudgerigar psittacine birds are subclinical, and the immune response that occurs in these birds prevents the acute form of the disease.

Peracute death with no premonitory signs is the most common clinical finding in young, nonbudgerigar psittacine birds affected by avian polyomavirus. Acute infections are characterized by death following a 12- to 48-hour period of clinical changes that may include depression, anorexia, weight loss, delayed crop emptying, regurgitation, diarrhea, dehydration, bleeding under the skin, difficulty in breathing and polyuria *Figure 6.15*.[8,33,84,85,107] An infected Yellow-headed Amazon Parrot exhibited posterior paresis and paralysis within 18 hours of developing the initial clinical signs suggestive of a polyomavirus infection.[8]

Neonates with polyomavirus infections may bleed profusely, or for a prolonged period, from intramuscular injection sites or from follicles where feathers have been removed. Subcutaneous hemorrhage over the crop and across

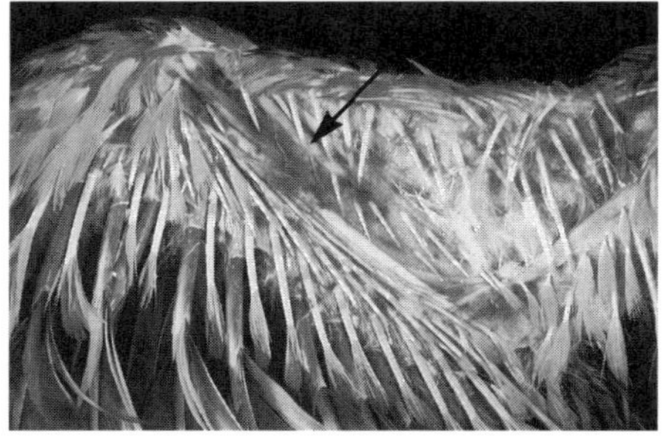

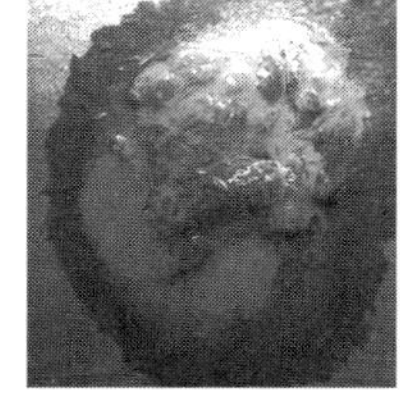

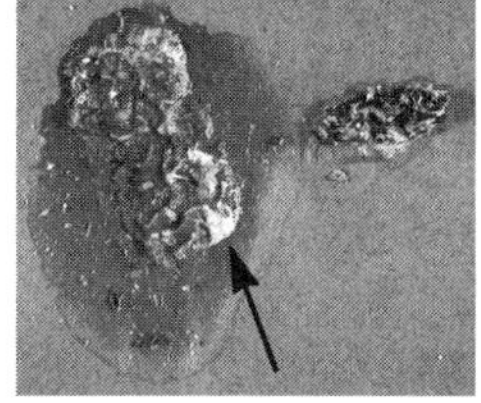

FIG 6.15
Acute polyomavirus infections in nonbudgerigar psittacine neonates can cause:
above *delayed crop emptying, regurgitation,*
above right *bleeding under the skin (arrow) and*
below left *yellowish or greenish discoloration of the urates indicative of liver disease.*
right *Transient diarrhea was the only clinical change noted in nonbudgerigar psittacine birds infected with avian polyomavirus. Excrement from an infected bird (arrow) is seen next to that of an uninfected bird.*

the cranium is common in larger psittacine birds with polyomavirus infections.[45] While a propensity to bleed abnormally is a common finding with birds infected with polyomavirus, this clinical finding is not diagnostic for polyomavirus. Any disease involving vasculitis, clotting disorders or damage to the liver can cause bleeding disorders.

Birds that survive for several days after the onset of clinical signs can develop yellowish urates similar to those described in birds infected with Pacheco's disease virus or chlamydia. Passage of yellowish-colored urates is indicative of the liver damage caused by each of these infectious organisms *Figure 6.15*. Some birds with acute polyomavirus infections may have increased activities of lactate dehydrogenase (LDH), aspartate aminotransferase (AST) and alkaline phosphatase (AP), suggesting the damage that avian polyomavirus causes to the liver and other tissues.[8]

Eclectus Parrots appear to be particularly susceptible to polyomavirus infections, and birds ranging from three weeks to four years of age have been shown to die with characteristic clinical signs. In addition, affected Eclectus Parrots may exhibit a gastrointestinal stasis that mimics proventricular dilatation disease, melena, hematuria and abdominal pain. The clinical detection of hematuria has been suggested as an indication of an active polyomavirus infection in Eclectus Parrots.[99]

In addition to the peracute and acute forms of avian polyomavirus, a chronic disease progression has also been described. It is typified by weight loss, intermittent anorexia, polyuria, poor feather formation and recurrent bacterial or fungal infections.[8,72] Birds that recover from the chronic form of the disease may appear normal, although some birds have been found to die months later from kidney failure.[8] The feather abnormalities that are commonly observed with a polyomavirus infection in Budgerigars are described less frequently in other psittacine birds.[8,36,45,70,72]

It is common for an affected group of birds to exhibit a mixture of clinical changes that include peracute, acute and chronic forms of the disease. For example, in one outbreak, some exposed birds died acutely, a couple died with renal disease 28 and 29 days after the infection was first noted and others recovered but with signs of stunting, maldigestion, polyuria and slowed gastric emptying times.[8] Reported mortality rates during an outbreak have varied from 27% to 41% of the at-risk young birds.[8,12] In one outbreak, 14 of 45 (31%) at risk fledglings ranging in age from 28 to 45 days (mean 38 days) died over a six-week period.[45]

EXPERIMENTAL INFECTIONS

Both Budgerigar and nonbudgerigar psittacine birds have been shown to be susceptible to experimental polyomavirus infections; however, the characteristic disease has been induced only in some Budgerigars. Experimentally infected three- to ten-day-old Budgerigar neonates died 11 days after being given BFD virus intramuscularly.[52] When 25-day-old Budgerigars were exposed intranasally to virus collected from the skin of diseased birds, they developed microscopic lesions characteristic of a polyomavirus infection but remained clinically normal. Budgerigars of a similar age that were given the same virus preparation subcutaneously also remained clinically normal and did not develop any gross or microscopic changes suggestive of a polyomavirus infection.[5]

TABLE 6.3

Clinical signs associated with polyomavirus infections

- Lethargy/depression
- Slow emptying crop
- Regurgitation
- Lack of feeding response

Budgerigars
- Acute deaths in nestlings
- Feather dystrophy in fledglings

Large Psittacines
- Acute death neonates (within 12 to 24 hours)
- Hemorrhage under the skin and throughout the body
- Neurologic disease (adults)
- Acute death (adults)

Finches
- Acute deaths (fledglings, adults)
- Beak abnormalities (fledglings)

In one trial, young seronegative Budgerigars seroconverted within 16 days after being placed in the same enclosure with seropositive birds. Seronegative Budgerigars also seroconverted when they were placed in enclosures adjacent to those containing seropositive birds. These findings suggest that direct and indirect transmission can occur in Budgerigars. An attempt to transmit polyomavirus from a seropositive Sun Conure to seronegative Budgerigars through direct contact was unsuccessful, perhaps because the Sun Conure was not shedding virus during the study period.[12]

Avian polyomavirus isolated in cell culture from Budgerigars and administered orally or intramuscularly to Blue and Gold Macaw chicks induced infections, but the birds remained asymptomatic.[86] When Blue-crowned Conure chicks, 40 to 50 days old were exposed to avian polyomavirus by intramuscular inoculation, they seroconverted, shed virus intermittently for 7 to 14 days and remained clinically normal.[88a] Mature Amazon parrots, African Grey

Parrots and cockatoos exposed to avian polyomavirus by the intramuscular or intravenous routes will seroconvert and shed virus intermittently in the feces. Birds infected by the intravenous route develop transient diarrhea five to ten days after inoculation, then recover *see Figure 6.15*.[88a] Birds experimentally infected with liver homogenates containing avian polyomavirus have been found to respond in a similar fashion to birds that are experimentally infected with cell-culture derived virus.[88a]

The effect of avian polyomavirus (BFDV) on experimentally infected chickens varies dramatically with the age of exposure. Chicken embryos infected at ten days of age died ten days later, and had gross and histologic lesions characteristic of the disease. In contrast, chicken embryos infected at 11 and 12 days of age remained normal, developed precipitating antibodies that could be detected two weeks after hatching, and did not develop gross or microscopic changes suggestive of an infection.[58]

Two-week to four-month-old broilers and specific-pathogen-free (SPF) chickens injected with avian polyomavirus by the intramuscular or intravenous routes developed virus-neutralizing antibodies, suggesting that they had been infected. Some experimentally infected chickens developed transient diarrhea but otherwise remained clinically normal. None of the experimentally infected birds developed gross or histologic changes suggestive of a polyomavirus infection.[88a] In two infectivity trials, several chickens used as uninfect-ed, contact controls seroconverted, suggesting that transmission of the virus had occurred between experimentally infected and seronegative birds. However, chickens administered avian polyomavirus by the oral route did not develop virus-neutralizing antibodies, suggesting that they had not been infected.[88a]

Avian polyomaviruses have a worldwide distribution. Characteristic lesions associated with infections have been demonstrated in the United States,[8,36,45,65,85] Canada,[5,32] Japan,[43] Italy,[69] Hungary,[103] Germany,[51,100] South Africa[1] and Australia.[59,70,72] Unlike most members of the Papovaviridae family which have a restricted host range, avian polyomaviruses appear to infect a wide variety of Psittaciformes, Passeriformes and probably gallinaceous birds. A list of birds considered naturally susceptible to an avian polyomavirus (based on the demonstration of characteristic lesions or antibodies to the virus) is provided in *Table 6.4.*

All species of psittacine birds or finches should be considered susceptible to the virus. It has been suggested that polyomavirus-induced disease occurs most commonly in young Budgerigars, macaws, conures, Eclectus Parrots, lovebirds, Ring-necked Parakeets and caiques.[8,36,70,84] Both of the initial reports documenting avian polyomavirus infections in nonbudgerigar psittacine birds in the United States involved conures.[8,45] In Australia, polyomavirus infections are considered particularly common in lovebirds.[70] In comparison to the species listed above, polyomavirus-induced disease is considered less common in young cockatoos, Grey-cheeked Parakeets, lories, African Grey Parrots, Hawk-headed Parrots and Amazon parrots.[8,36,70,84] It should be cautioned, however, that reported variances in susceptibility may represent a skew in the population of exposed birds and not an actual difference in susceptibility.

Gallinaceous birds also appear to be susceptible to polyomavirus. A virus that morphologically resembles a polyomavirus was recovered from the intestinal contents of asymptomatic turkeys. The recovered virus did not cause a discernible disease in experimentally infected birds.[15] A polyomavirus-like agent was identified in the feces of an ostrich located in the southeastern United States. A polyomavirus with similarities to BFD virus was recovered from the drinking water and feces associated with a chicken layer replacement farm in Germany. It was not determined if the virus in this poultry house originated from chickens or was a contaminate from another source,[100] but serologic studies suggest that chickens are susceptible to polyomavirus infections. Polyomavirus-specific antibodies have been demonstrated in broiler chickens from central Europe and the United States.[88a,100] During a polyomavirus epornitic in a mixed species aviary, virus-neutralizing antibodies were detected in two Golden Pheasants and a Lady Amhurst Pheasant that had been naturally exposed to affected psittacine birds, while a potentially exposed Bantam Chicken and two Toco Toucans

TABLE 6.4

Birds considered naturally susceptible to avian polyomavirus

Psittaciformes

African Grey Parrots	Amazon parrots	Bourke's Parrots
Budgerigars	Cockatiels	Cockatoos
Conures	Grey-cheeked Parakeets	Eclectus Parrots
Hawk-headed Parrots	Kakariki	Lories
Lorikeets	Lovebirds	Macaws
Meyer's Parrot	Parrolets	Pionus parrots
Quaker Parakeets	Rose-ringed Parakeets	Scarlet-chested Parrot
Senegal Parrots	Splendid Parakeets	

Passeriformes

Canaries	Blue bills	Finches
Seedcrackers		

Others

Chickens	Brown Pigeon	Golden Pheasants
Lady Amhurst Pheasant	Turkeys	Ostriches
Peaceful Dove		

remained seronegative.[66a] Broilers and specific-pathogen-free chickens have been shown to develop virus-neutralizing antibodies following experimental infection with avian polyomavirus (see Experimental infections).[88a] Inclusion bodies suggestive of polyomavirus have been described from Australia in a Kakariki, a Peaceful Dove, a Brown Pigeon and a canary.[83a]

There are no reports confirming polyomavirus-induced disease in free-ranging psittacine birds; however, polyomavirus-like particles were identified in the droppings of free-ranging galahs from Western Australia.[58a] Additionally, polyomavirus-neutralizing antibodies have been detected in Sun Conures as they entered the United States from Guyana, and before they were exposed to birds from other areas.[8] This finding suggests that the birds were infected with polyomavirus before leaving Guyana. Antibodies to avian polyomavirus were detected in a group of free-ranging Dusky-headed Parakeets from southeast Peru. Three of 19 birds (19%) had virus-neutralizing antibodies and 5 of 10 birds (50%) had complement-fixing antibodies, suggesting that polyomavirus infections occur in free-ranging birds from this region.[30]

<u>RELATIONSHIP OF VIRUS STRAINS</u> — Polyomaviruses from some infected nonbudgerigar psittacines have been confirmed to be antigenically related to the polyomavirus isolated from Budgerigars and some finches.[36,66,107,111] The genome of the polyomaviruses isolated from Budgerigars, a Blue and Gold Macaw and a chicken house have been shown to be closely related. Because of these similarities, it has been suggested that the avian strains of polyomavirus be placed into the subgenus *Avipolyomavirus* in the Polyomavirinae subfamily *Table 6.5*.[100]

DNA probes have been developed to detect specific segments of the nucleic acid found in the avian polyomavirus. These DNA probes, designed from polyomavirus recovered from Budgerigars, will detect the virus in excretions, secretions and infected tissues of non-budgerigar psittacine birds.[11,65,76] DNA probes designed to detect avian polyomavirus in psittacine birds did not detect polyomavirus nucleic acid in infected seedcrackers, suggesting that these passerines are infected with a polyomavirus that has a nucleic acid sequence that is partially different from the nucleic acid sequence of the virus that infects psittacine birds.[26]

Comparison of nucleic acid recovered from avian and mammalian polyomaviruses indicates that there are some similarities, but also substantial differences between the two.[54] DNA

TABLE 6.5

Suggested classification of avian polyomaviruses[100]

FAMILY	Papovaviridae
SUBFAMILY	Polyomavirinae
GENUS	Polyomavirus
SUBGENUS	Avipolyomavirus

probes designed to detect BFD virus did not cross-react with either of two mammalian polyomaviruses (SV-40 or JC virus).[76]

<u>AGE SUSCEPTIBILITY</u> — The mortality rate associated with naturally acquired avian polyomavirus infections in young Budgerigars may range from 25% to 100%, depending on the age of the birds. Older birds are considered relatively resistant to disease, while at the peak of viral activity, up to 100% of exposed Budgerigars younger than 15 days of age may die. The mortality rate reported for older (over 3 weeks of age) juveniles ranges from 30% to 80% (see Experimental Infections).[5-7,12,14, 32,43,60,64]

Nonbudgerigar psittacine neonates are considered to be highly susceptible to polyomavirus infection and the diseases it can cause. Infections may occur in either parent- or hand-raised neonates;[8,45,84] however, disease may be more common in hand-raised chicks.[81] In nonbudgerigar psittacines, clinical signs are most common at the time of weaning; however, neonates from 14 to 150 days of age have been reported to be susceptible to naturally induced avian polyomavirus infections and disease.[8,36,45, 100] Reported mortality rates vary from 31% to 41% of the at-risk young birds.[8,45]

In most cases, older psittacine birds (over one month in Budgerigars and over five months in nonbudgerigar psittacine birds) exposed to avian polyomavirus seroconvert and remain clinically normal. Occasionally, adult psittacine birds may die acutely with lesions suggestive of a polyomavirus infection. Polyomavirus infections have been documented in an eight-month-old Splendid Parakeet and in sporadic, acute deaths in fully-fledged lovebirds less than one year old. Clinical signs in the lovebirds were related to hepatitis, splenic necrosis or kidney damage. They included depression, chronic wasting, paresis, ataxia, dyspnea and passage of dark red-brown feces for several days before death.[70,72] An adult Moluccan Cockatoo with neurologic signs was diagnosed as having polyomavirus based on the demonstration of suggestive inclusion bodies in the brain.[91]

In addition, polyomavirus has been associated with neurologic signs in a number of adult psittacine birds, particularly cockatoos *Figure 6.16*.[53a] An outbreak of polyomavirus in an aviary with numerous psittacine species resulted in the death of an adult Eclectus Parrot, a Painted Conure and 3 of 11 adult White-bellied Caiques in the collection. The affected birds were 2 to 2.5 years old, and had lesions similar to those described with polyomavirus infections in psittacine fledglings.[85] In another report, polyomavirus was suspected as the cause of death in Eclectus Parrots up to four years of age.[99] Microscopic changes suggestive of a polyomavirus infection were noted in an adult Sun Conure that died following a three-week illness with non-specific clinical signs.[8]

No one has determined why most adult birds exposed to polyomavirus seroconvert following infection by this virus, while some develop clinical abnormalities and die. Factors that govern the susceptibility of young and adult birds to avian polyomavirus-induced disease might include the route of virus exposure, the quantity of virus to which the bird is exposed, a preexisting resistance to disease based on previous exposure to the virus, a selective immunosuppression in the affected birds or the occurrence of strains of polyomavirus with increased virulence for certain species.

<u>PERSISTENT INFECTIONS</u> — In mammals, polyomaviruses typically induce subclinical, persistent infections with periodic virus shedding during periods of excessive stress. Latently infected mammals have been shown to maintain high antibody titers throughout their lifetime, indicating that the virus is repeatedly stimulating the immune system. In mammals, the kidney, brain and salivary glands are common sites where the virus persists.[89,110]

Like mammals, Budgerigars that recover from polyomavirus infections are thought to develop latent infections and maintain a sustained high antibody titer.[27,80,81] Evidence suggests that these latently infected birds are responsible for the persistence, transmission and

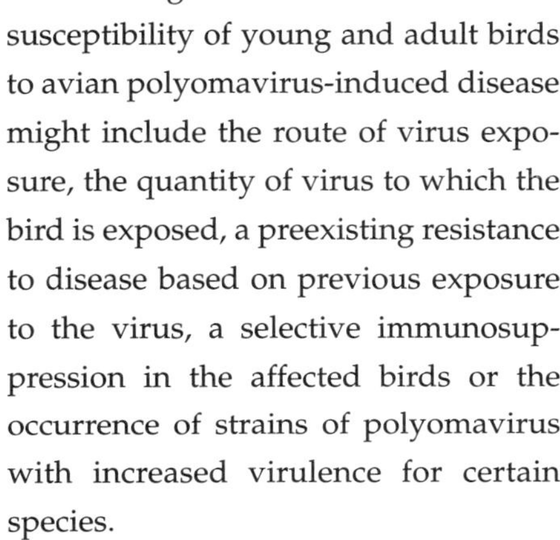

FIG 6.16
Neurologic signs that include ataxia, torticollis (shown here) and tremors of the head and neck have been reported in some adult psittacine birds, particularly cockatoos, with polyomavirus-induced inclusion bodies in the brain.

spread of the virus through various Budgerigar flocks.[5,6,28,79] Stress associated with changes in weather, diet, breeding or concomitant disease may cause latently infected Budgerigars to shed virus resulting in outbreaks of disease. In polyomavirus outbreaks at 23 different Budgerigar aviaries, the onset of disease could be traced to the addition of new, clinically normal breeders.[5,6] This finding suggests that latent infections may become active during the breeding season, resulting in an increased dissemination of the virus when a larger number of young susceptible Budgerigars are present in the flock.

Polyomavirus inclusion bodies frequently can be detected in feathers, feather follicles and renal tissue from persistently infected, clinically normal adult Budgerigars.[6,7] Viral inclusion bodies are common in the renal tubular epithelium. It has been proposed that virus localized to this anatomic area would be protected from the neutralizing effects of antibodies found circulating in the blood. Persistence of virus in the kidneys of Budgerigars, with subsequent excretion in the urine, has been proposed as a method of virus transmission.[5,7,19,28,64]

The incidence of polyomavirus activity in Budgerigars is high. In one study, virus was demonstrated by histopathology and virus isolation from all of ten Budgerigars with clinical signs of avian polyomavirus and from five of ten (50%) clinically healthy Budgerigars *Figure 6.17*.[7] In another study, four of seven (57%) nestling Budgerigars had microscopic changes suggestive of avian polyomavirus, while all of seven nestlings had viral nucleic acid in their tissues.[80] Polyomavirus nucleic acid was demonstrated in the serum of nine of twelve (75%) 9- to 13-day-old Budgerigars from a flock in which virus activity was considered enzootic.[79]

The possibility that up to 100% of the Budgerigars in some flocks could be infected with avian polyomavirus was first discussed in 1984.[6] Subsequent studies using DNA probes have confirmed it. In one study involving a flock of Budgerigars with a history of repeated polyomavirus-induced disease, all of 40 clinically normal Budgerigars had polyomavirus nucleic acid in their tissues. Virus-neutralizing antibodies to avian polyomavirus were detected in all 144 of another group of Budgerigars, and viral nucleic acid was detected in most tissues collected from these Budgerigars.[80] However, between 36% and 47% of the tested Budgerigars had viral nucleic acid in the serum, which could be present in any tissue, making the detection of nucleic acid in specific organs of limited value. Detecting polyomavirus throughout the body of persistently infected Budgerigars is considered an unusual finding for this group of viruses, because mammalian polyomaviruses have been shown to persist in only a few, select tissues.[3,17,105]

Based on findings in Budgerigars, it has been proposed that persistent infections may also occur in larger psittacine birds. However, supporting evidence of this hypothesis is scarce and circumstantial. It should be noted that the progression of polyomavirus-induced disease differs dramatically between Budgerigars and other psittacine birds. There have been no consistent laboratory findings to confirm that nonbudgerigar psittacine birds that recover from an infection become latently infected. However, several findings from natural outbreaks seem to support this theory. A polyomavirus outbreak in an aviary with large psittacine birds occurred 14 months after the birds had been stabilized and released from quarantine; this has been used as supporting evidence for a carrier state.[8] The persistence of antibody titers in some larger psittacine birds for over a year also has been suggested as an indication of latent infections.[8,28] However, these birds with sustained high antibody titers were not maintained in isolation and thus could have been repeatedly exposed to virus in a contaminated environment. Persistent, inapparent infections have been suggested to occur in Cockatiels.[78]

<u>INCUBATION</u> — The incubation period of avian polyomavirus has not been confirmed in nonbudgerigar psittacine birds because experimentally infected individuals do not develop the clinical signs of disease that are characteristic in naturally acquired infections. Based on clinical observation, the incubation period of polyomavirus in nonbudgerigar psittacine birds has been estimated to be as long as 14 days but may be as short as 2 days.[8,28,36] Budgerigar fledglings with naturally acquired infections show peak mortality rates between the 15th and 19th day of life, suggesting that the incubation in this species may be less than 15 days. Budgerigar neo-

FIG 6.17

It has been repeatedly demonstrated that many budgerigars are carriers of and intermittently shed avian polyomavirus. Also it has been shown that the virus recovered from budgerigars is infectious for larger psittacine birds.

Based on these findings, it is not recommended that budgerigars be maintained in the same airspace with other unvaccinated psittacine birds, particularly neonates or recently weaned birds.

photographs courtesy of Greg Harrison

nates experimentally infected by intramuscular inoculation died 11 days after being exposed to the virus.[52]

Infection and subsequent antibody response appear to occur relatively rapidly after exposure. In one study, polyomavirus-neutralizing antibodies were first demonstrated 21 days after a Blue and Gold Macaw chick was exposed to live virus by the oral and intracloacal routes. Because ingestion is probably one of the routes by which a natural infection occurs, this finding suggests that the virus is capable of infecting a bird and inducing a detectable antibody response in less than 21 days following ingestion of the virus. In another Blue and Gold Macaw chick exposed to live virus by the intramuscular route, antibodies were detected when a blood sample was first collected seven days after exposure.[86] Chickens experimentally infected by the intramuscular or oral route developed virus-neutralizing antibodies by three days after inoculation.[88a]

<u>TRANSMISSION</u> — Experimental data and observations with the natural disease suggest that polyomavirus transmission in Budgerigars may occur by both horizontal and vertical routes — that is, both among members of a flock and between generations from parent to offspring.[5,12,14,45,80]

In Budgerigars, exposure to virus-contaminated feces, feather dust, urates and respiratory secretions is thought to play the principal role in the natural horizontal transmission of polyomavirus *Figure 6.18*. The virus also is shed in crop secretions providing an additional method for infected parents to transmit the virus to their young. [6,19,78] In young Budgerigars, polyomavirus is frequently demonstrated in the skin and cells that line the feather follicles, resulting in the presence of virus in "feather dust."[6] Virus that is shed with exfoliated skin and feather debris may contaminate any area where the feather dust is transported, and could enter a susceptible bird through ingestion or inhalation of contaminated materials. Nonbudgerigar psittacine birds with polyomavirus infections usually do not develop skin or feather lesions, and in these birds feather dust may play a less important role in dissemination of the virus in the aviary or nursery environment.

Seronegative young adult psittacines will seroconvert when housed adjacent to seropositive breeding adults, suggesting that indirect transfer of the virus can occur among birds that are in different enclosures but share an air-space.[8,12,13,45,107] Experimental infections can be induced in Budgerigars by inoculating them intranasally with virus collected from the skin of naturally infected birds.[6] These findings suggest that inhalation of aerosolized virus could be a natural route through which polyomavirus enters a susceptible bird.

It has been suggested that infected adult Budgerigars may serve as a source of virus for susceptible chicks. Viral nucleic acid has been detected in the excrement, feathers and oral mucosa of clinically normal Budgerigars.[78] In one study, 33% of 11 Budgerigars with polyomavirus-neutralizing antibodies were found to be shedding virus in the feces when tested with a nucleic acid detection system.[80]

Vertical transmission of polyomavirus has been confirmed in Budgerigars, but has not been confirmed to occur in other psittacine birds. Findings that support vertical transmission include the identification of intranuclear inclusion bodies in one-day-old Budgerigars, the occurrence of polyomavirus infection when eggs from parents that consistently produced diseased neonates were cross-fostered to parents producing normal young and the detection of polyomavirus nucleic acid in Budgerigar eggs.[5,6,32,80] The detection of polyomavirus nucleic acid in the testes, ovary or oviduct of adult Budgerigars provided further evidence that vertical transmission could occur in this species.[80] Eggs produced by birds in flocks with a history of polyomavirus problems can be screened for the presence of

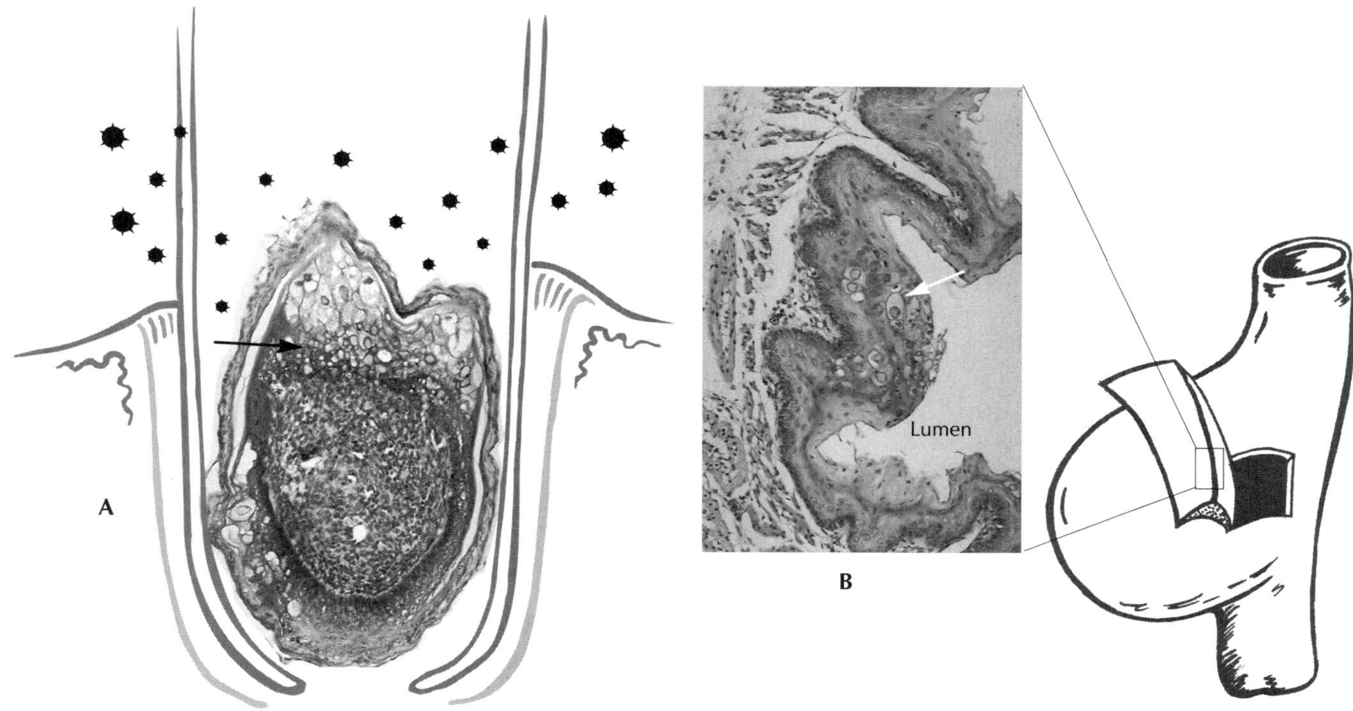

polyomavirus nucleic acid using viral-specific DNA probes.

Some investigators believe that vertical transmission of avian polyomavirus is responsible for infertility and early embryonic death *Figure 6.19*.[32] Others suggest that infected eggs may not be adversely affected. The concentration of viral nucleic acid in the eggs of persistently infected adult Budgerigars reportedly is low. This finding has been used to suggest that the virus is deposited in the egg during development in the hen and that active viral replication is not occurring.[80]

As occurs with Budgerigars, horizontal transmission through contact with contaminated excretions and secretions probably plays a major role in dissemination of the virus through and among flocks. In one outbreak in nonbudgerigar psittacine neonates, disease occurred simultaneously in two nurseries at different geographic locations, which previously had contact with one another through the transfer of birds between the facilities.[8] Polyomavirus nucleic acid can be detected in cloacal swabs taken from nonbudgerigar psittacine birds during a flock outbreak.[11,65,66] During one outbreak in a group of mixed Psittaciformes, 41 of over 200 (20%) birds of 35 different species were found to be shedding polyomavirus in their excrement during the peak of the epornitic. However, only 3 of these 41 (7%) birds were still shedding detectable quantities of virus when they were retested 60 days later, indicating that viral shedding is transient. In adult parrots naturally infected with avian polyomavirus, virus was detected in the excrement in 26% of the birds when they were tested twice at a four- to six-month interval.[81] Nucleic

FIG 6.18

Viruses like polyomavirus, which replicate in the outer layers of developing feathers, are easily shed into the environment with exfoliating debris.

A Inclusion bodies (arrow) containing large concentrations of avian polyomavirus are seen in the outer layers of a developing feather. The virus particles that are released from the feather can be transmitted throughout an aviary or home in association with feather dust.

B Polyomavirus also replicates in the cells lining the crop (arrow) and can be passed to chicks with regurgitated food.

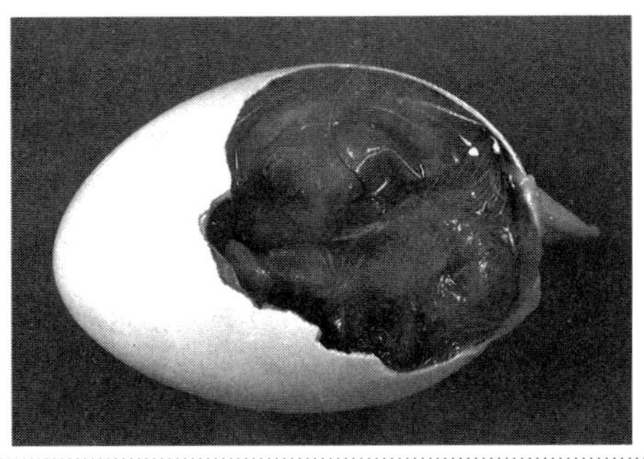

acid also has been detected in the excrement of experimentally infected psittacine chicks starting from two to seven days following intramuscular inoculation.[86] The recovery of viral DNA from the cloaca suggests that the virus could be shed from gastrointestinal, renal or reproductive tissues.[65,66,86]

PATHOLOGY IN BUDGERIGARS

The gross lesions associated with polyomavirus infections in Budgerigars include hydropericardium, cardiomegaly, hepatomegaly with multifocal yellow-white foci, ascites, splenomegaly, hemorrhage under the skin and hemorrhage of the intestines and heart.[5-7,14] Affected Budgerigars develop gross lesions that suggest virus-induced lysis of infected cells. These changes are recognized microscopically as necrosis and inflammation. The development of ascites, pericardial effusion and hemorrhage suggests that the virus damages the cells lining the blood vessels causing the vessels to leak fluids.

Microscopically, the nuclei of infected cells are typically swollen. Histologic changes have been described in most tissues of the body and include necrosis in the heart, liver, kidneys and bone marrow, and atrophy of the lymphoid tissue in the spleen and cloacal bursa. Polyomavirus has been demonstrated microscopically in skin, feathers, feather follicles, kidney, uropygial gland, crop, lung, liver, heart, bone marrow, spleen and brain of infected Budgerigars. Intranuclear inclusion bodies can be found in most of these tissues, as well as in the pancreas, adrenal glands, intestines, testes and ovaries.[5]

PATHOLOGY IN NONBUDGERIGAR PSITTACINE BIRDS

In polyomavirus-infected lovebirds, gross lesions include clear fluid in the abdominal cavity, strands of fibrin on the abdominal organs and a small pale spleen. The liver may appear normal or be pale, congested and hemorrhagic. Microscopic lesions suggestive of polyomavirus infection were noted in the spleen, kidney and liver of all 19 of a group of affected lovebirds.[70]

In larger psittacine birds, neonates that die from avian polyomavirus are usually in excellent overall condition and may have full crops and alimentary tracts, indicating a rapid death secondary to viral-induced tissue damage. Gross lesions may include hepatomegaly with irregular red and yellow mottling, splenomegaly, pale swollen kidneys, pale cardiac and skeletal muscles, feather dystrophy, ascites, hemorrhage under the skin and hemorrhage of the liver, intestines, heart and many serosal surfaces *Figure 6.20*.[8,36,45,85] The frequency of characteristic gross changes in a group of affected psittacine birds is summarized in *Table 6.6*. The pallor that is noted in the major muscles of many affected birds is indicative of the massive hemorrhage that occurs throughout the body. As in Budgerigars, hemorrhage and ascites occur because

AVIAN VIRUSES: FUNCTION AND CONTROL

of damage to the cells that line the blood vessels causing the vessels to leak fluids. Feather lesions associated with polyomavirus infection are rare in nonbudgerigar psittacine birds.

TABLE 6.6

Common gross findings in a group of 39 nonbudgerigar psittacine birds with avian polyomavirus[36]

Age ranged from 2 to 16 weeks
All affected birds were being hand-raised
Pallor described in the muscles of 33 birds (84.6%)
Hepatomegaly in 29 birds (74.3%)
Hemorrhage under the skin or on the serosal surface of organs in 24 birds (61.5%)
Splenomegaly in 16 birds (41%)

Most nonbudgerigar psittacine birds that die from polyomavirus infections will have hepatic necrosis, with karyomegaly in the liver and spleen.[36] In one study, all 45 affected psittacine neonates had microscopic changes suggestive of hepatic necrosis.[45] In another study, karyomegaly was documented in the liver from ten polyomavirus-infected birds (five with and five without inclusion bodies).[24] However, it should be noted that the detection of karyomegaly is neither essential to positive identification of an avian polyomavirus infection, nor exclusive to it. This microscopic change also can occur with adenovirus infections, in healthy individuals during early growth, after mycotoxin ingestion or when the liver is recovering from injury.[36,24]

Other microscopic changes that have been reported in affected nonbudgerigar psittacine birds include depletion and death of lymphocytes in the bursa and membranous glomerulopathy. In one study, damage to the kidneys in the form of membranous glomerulopathy was demonstrated in 29% to 41% of nonbudgerigar psittacine neonates that died from a polyomavirus infection.[36,75] In nonbudgerigar psittacine birds, virus is demonstrated most commonly in the monocytes and macrophages that are responsible for initiating the immune response and helping to control an infection.[78]

Like karyomegaly, characteristic intranuclear inclusion bodies may or may not be present.[36,53] When inclusion bodies are identified, they are most common in the spleen and liver; however, they also have been documented in the walls of blood vessels, in growing

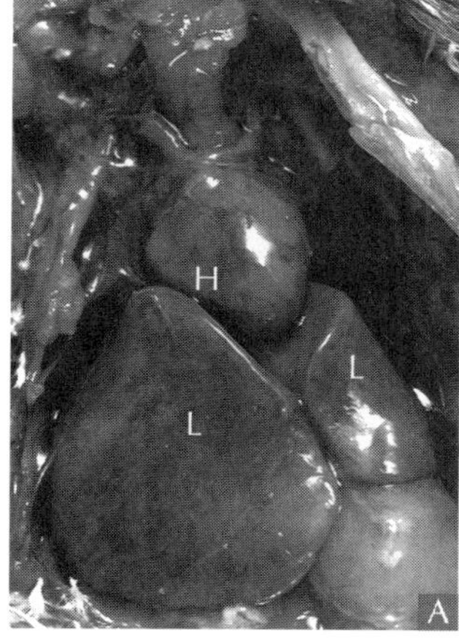
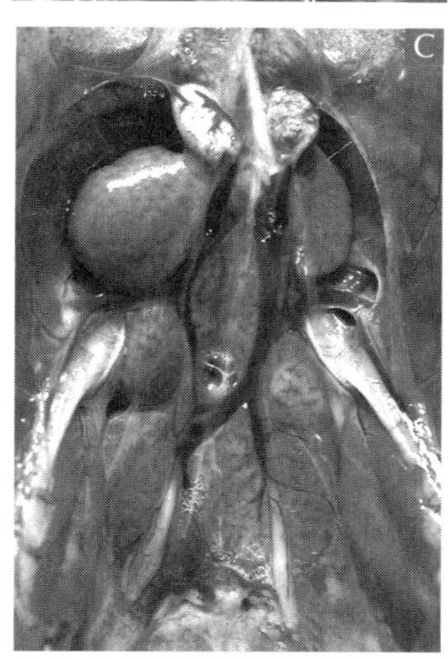
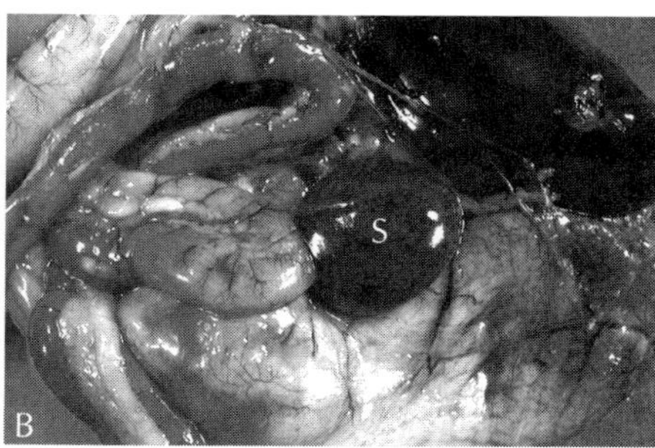
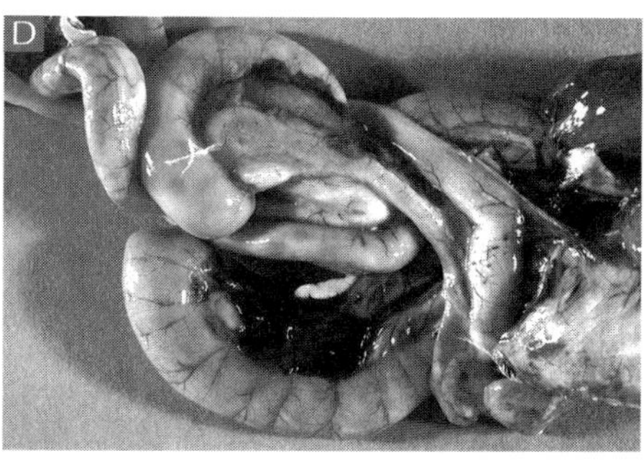

FIG 6.20

Gross lesions associated with avian polyomavirus infections include:
A hemorrhage of the heart (H), and enlargement, hemorrhage and irregular red and yellow mottling of the liver (L) and
B spleen (S).
C The kidneys may be pale, swollen and irregularly hemorrhagic.
D Hemorrhage is common under the skin and throughout the abdomen.

feathers, in the crop, in the kidney, in the cells lining the intestines, in bone marrow, and in a variety of organs including the lung, heart, pancreas, esophagus, myocardium, proventriculus and adrenal gland.[8,36,45,53] In a group of 14 nonbudgerigar psittacine birds that died from an avian polyomavirus infection, inclusion bodies were demonstrated in the spleen of 6 (43 %) birds and in the liver of 5 (36 %) birds.[45] In a group of 32 psittacine birds with liver necrosis, avian polyomavirus was confirmed in 19 (59%) birds, bacterial hepatitis was confirmed in 5 (16%) birds and chlamydiosis was confirmed in 3 (9%) birds. Intranuclear inclusion bodies suggestive of polyomavirus were identified in only 5 of the 32 (16%) birds. The other 14 birds that were infected with avian polyomavirus were diagnosed by using DNA probes to detect viral nucleic acid in the liver, demonstrating that DNA probe testing of tissues is a more sensitive method for diagnosing polyomavirus infections than is the detection of microscopic changes.[24]

PATHOGENESIS

Many aspects of the progression and outcome of an avian polyomavirus infection remain unresolved. A satisfactory explanation for why some psittacines are fatally infected by avian polyomavirus while most develop subclinical infections has not been set forth. No one knows yet why some asymptomatic adult Budgerigars consistently produce infected young, while others have neonates that intermittently may develop clinical signs and die.[5-7,12,32]

It is theorized that the age of a bird at the time of viral exposure may be a principal factor in the progression of a polyomavirus infection. Clinical signs of disease are most common in 1- to 3-week-old Budgerigars and 4- to 16-week-old nonbudgerigar psittacine birds.[5,12,36] Infected Budgerigars that die shortly after hatching routinely develop more severe and widespread lesions than do birds in which the morbid state is more prolonged. Budgerigars infected at greater than 25 days of age develop infections with no signs of disease, suggesting that resistance to disease (but not necessarily infection) occurs with increasing age.[5] When 11- to 12-day-old chicken embryos are experimentally infected with polyomavirus, the hatched chicks remain normal and produce detectable antibodies by two weeks of age. In contrast, chicken embryos infected at ten days of age are susceptible to infection and develop lesions throughout their bodies.[58]

Polyomavirus-infected Budgerigars develop a viremia, which accounts for the presence of the virus in most tissues and probably accounts for the high level of mortality in neonates.[5,18] Viremia has been demonstrated in some nonbudgerigar psittacine birds, but appears to be an inconsistent finding.[65,66,79] In one study, viral nucleic acid was not detected by DNA probe analysis of blood collected from experimentally infected Blue and Gold Macaw chicks. This could indicate any of several possibilities. Perhaps viremia was absent. Alternatively, a transient period of viremia may have occurred during an interval when blood samples were not being collected. Still another possibility is that there was too little circulating virus to be detected by the DNA amplification and detection procedures used.[86]

Birds may be infected by more than one virus at the same time. Polyomavirus infections have been documented in conjunction with poxvirus and reovirus in Amazon parrots and with an adeno-like virus and reovirus in fledgling lories.[107] Concurrent avian polyomavirus and PBFD virus infections have been reported in Budgerigars, lovebirds, African Grey Parrots, Eclectus Parrots and several species of cockatoos.[53,77] These viruses typically infect different cells. In birds infected with both PBFD virus and avian polyomavirus, the PBFD virus had a predilection for the nuclei of feather, follicular and intestinal epithelial cells and the cytoplasm of macrophages in numerous tissues. Avian polyomavirus typically infected nuclei of endothelial cells, hepatocytes and renal tubular epithelial cells.[53]

Mammalian polyomaviruses cause persistent, subclinical infections even though an infected host develops viral-specific neutralizing antibodies. It is theorized that the immune system keeps the polyomavirus infection in check because persistently infected mammals that become immunocompromised may die from a polyomavirus-induced disease.[110] A similar deficiency in the immune system might be responsible for the occasional polyomavirus-induced death in adult psittacine birds or the frequent deaths that occur in some young birds.

Budgerigars are known to shed polyomavirus even though they may develop high levels of virus-neutralizing antibodies. It has been theorized that birds that develop such persistent infections may have become infected before their immune system was completely functional.[28] These birds are defined as being immunotolerant. This definition may not be accurate, however, because with some viral infections, immunotolerant individuals do not develop an antibody titer. In some Budgerigar flocks 100% of the tested birds have antibodies to this virus,[80] and in some nonbudgerigar psittacine flocks, up to 63% of the tested birds have such antibodies.[88]

It has been suggested that some of the microscopic changes that occur in the kidneys of nonbudgerigar psittacine birds are caused by the antibodies that the infected bird's body produces in an attempt to destroy the virus.[78] These observations are based on uncontrolled field cases and are not supported by experimental data. After some nonbudgerigar psittacine birds with preexisting polyomavirus-neutralizing antibodies are vaccinated, they will develop an increase in antibody titer. Some vaccinates have been found to develop antibody titers that are extremely high (greater than 1:16,000) yet remain clinically normal, even when followed for over three years after vaccination. If neutralizing antibodies to polyomavirus were involved in the disease process, some of these experimentally vaccinated birds would be expected to be adversely affected, but in fact, they all remain clinically normal.[87,88] Additionally, Blue and Gold Macaw chicks, Blue-crowned Conure chicks, Amazon parrots and chickens that are experimentally infected with BFD virus will seroconvert, and some of these individuals develop high neutralizing antibody titers (greater than 1:640).[86] Experimentally infected Blue and Gold Macaw chicks remained clinically normal three years after infection. Experimentally infected Blue-crowned Conure chicks and adult Amazon and African Grey Parrots remained clinically normal 18 months after infection.[88a]

Additionally, field studies in non-budgerigar psittacine birds have indicated that up to 63% of the birds in some populations may have polyomavirus-neutralizing antibody titers, indicating that a previous infection occurred. This high seroprevalence has been documented in flocks with no history of polyomavirus-induced disease, even when tracked for years.[87] It is unlikely that such a large number of birds could be infected with a virus that induced an immune system-mediated disease without inducing some recognizable problems in a flock with this concentration of previously infected birds.

Some mammalian polyomaviruses persist by incorporating viral nucleic acid into host cell DNA.[41,89] Research is currently being performed to determine whether the polyomavirus nucleic acid integrates into the DNA found within a bird's cell, as well as to determine the cellular site of viral latency in the carrier state.

Polyomaviruses in mammals can cause tumors when the virus infects an unnatural host,[38] but have not been shown to cause tumors in their natural host. Because psittacine birds appear to be a natural host for avian polyomaviruses, it would be unlikely that infected birds would develop tumors, and thus far, there has been no correlation between polyomavirus infections in birds and an increased incidence of tumors.[19,54,65]

IMMUNITY

Polyomavirus-infected mammals develop persistent infections that can be demonstrated through the presence of high virus-neutralizing antibody titers over a long period of time.[110] Budgerigars also can be persistently infected with avian polyomavirus, but the levels of antibodies appear to increase or decrease depending on a bird's reproductive activity. In one flock of Budgerigars in which polyomavirus was considered enzootic, virus-neutralizing antibody titers ranging from 1:32 to 1:2048 were detected in 65% of nestling birds and in some individuals as young as 9 days of age.[79] In another group, polyomavirus-neutralizing antibody titers that ranged from 1:32 to 1:2048 were detected in all 144 adult Budgerigars that were examined. In Budgerigars that were continuously breeding, antibody titers decreased over a 4 to 18 month period. When breeding Budgerigars rested, 65% of the birds developed a significant increase in antibody titer, irrespective of their gender.[79,80] In another study, polyomavirus-neutralizing antibodies were detected in flocks of Budgerigars experiencing disease outbreaks, but were not detected in flocks of

Budgerigars experiencing no clinical problems.[52]

During outbreaks in nonbudgerigar psittacine bird aviaries, polyomavirus-infected survivors, and some asymptomatic birds exposed to them, have been shown to develop polyomavirus-neutralizing antibodies. The demonstration of virus-neutralizing antibodies in normal-appearing nonbudgerigar psittacines after natural infection suggests that many birds exposed to polyomavirus develop subclinical infections.[8,45,66] In fact, seroprevelance studies suggest that most nonbudgerigar psittacine birds infected with avian polyomavirus will develop an antibody response and remain clinically normal.[8,45,66,88,107] It is undetermined what percentage, if any, of the nonbudgerigar psittacine birds that have antibodies to avian polyomavirus are persistently infected.

Neither normal appearance nor the presence of antibodies correlate in a simple manner with the likelihood that a nonbudgerigar psittacine bird will shed virus. Of 106 serum samples collected from a group of breeding nonbudgerigar psittacines, 33% were positive (titer greater than 1:10), 20.7% were suspect (titer of 1:10) and 46.3% were negative (titer less than 1:10). Adults from one flock that were exposed to diseased birds seroconverted and raised seronegative, normal young in two subsequent breeding seasons.[8] In another aviary, antibody titers were detected in 10 of 15 (67%) blood samples taken from birds ranging in age from 6 weeks old to adults.[45] During a polyomavirus outbreak, 32 of 76 (42%) of the birds in the

TABLE 6.7

Seroprevalence of polyomavirus neutralizing antibodies in two study groups indicating frequent virus activity in nonbudgerigar psittacine birds

	Study One[88]	Study Two[107]
Amazon parrots	4 of 5 - 80%	4 of 35 - 11%
African Grey Parrots	10 of 25 - 40%	—
Cockatoos	85 of 131 - 65%	52 of 131 - 40%
Conures	4 of 5 - 80%	14 of 52 - 27%
Eclectus Parrots	3 of 11 - 27%	—
Hawk-headed Parrots	2 of 2 - 100%	—
Pionus parrots	9 of 11 - 81%	—
Macaws	28 of 43 - 65%	76 of 186 - 41%
Blue and Gold Macaws	—	17 of 57 - 30%
Green-winged Macaws	—	10 of 23 - 43%
Hyacinth Macaws	—	5 of 6 - 83%
Military Macaws	—	4 of 11 - 36%
Red-fronted Macaws	—	1 of 7 - 14%
Scarlet Macaws	—	22 of 62 - 35%

aviary had virus-neutralizing antibody titers that ranged from 1:20 to 1:1280. All of the seropositive birds were clinically normal, yet five of the seronegative birds and three of the seropositive birds were found to be excreting polyomavirus in their excrement.[66]

In three aviaries containing cockatoos, macaws, Amazon parrots and conures with a prior history suggestive of polyomavirus infections in neonates, 11%, 25% and 45% of the birds were found to have polyomavirus-neutralizing antibodies *Table 6.7*. Antibody titers decreased in many of the birds over a two- to three-month period.[107] In two aviaries with large psittacine birds, virus-neutralizing antibodies were detected even though neither of these facilities had ever experienced a polyomavirus-induced death. A few of these birds were found to have antibody titers that decreased from as high as 1:320 to as low as 1:5 over a three-month period.[12]

The incidence of polyomavirus infections is clearly much higher than the level of morbidity or mortality. In still another study, 146 of 233 (63%) mixed species psittacine birds were seropositive. Despite polyomavirus activity in this flock as indicated by the high seroprevalence of virus-neutralizing antibodies, the study population had produced 279 chicks over the previous six years with no losses attributable to avian polyomavirus. In addition, none of the deaths of 28 adult birds over the previous nine years were attributable to polyomavirus.[88]

Based on data collected in Budgerigars, it has been theorized that a nonbudgerigar psittacine bird that survives a polyomavirus infection becomes an asymptomatic carrier and maintains a persistently high antibody titer. In some cases, these neutralizing antibodies have been found to persist, which has been interpreted as an indication of an ongoing infection (carrier bird).[27,28] In other surveys, virus-neutralizing antibody titers were found to decrease, suggesting that the birds were not being continuously stimulated antigenically by the virus.[12,107] The demonstration of rapidly decreasing antibody titers in multiple groups of naturally infected nonbudgerigar psittacines raises the possibility that some birds infected with avian polyomavirus are able to mount an effective, transient immune response and clear the infection. It has never been demonstrated whether the sustained polyomavirus-neutralizing antibody titers that occur in some nonbudgerigar psittacines are caused by exposure to virus that remains in the body or repeated exposure of the immune system to virus that enters the bird from a contaminated environment. Virus-neutralizing antibodies have been detected in some birds that die with polyomavirus; however, virus-neutralizing antibodies are also present in birds that survive acute and subclinical infections.[8,45,79]

It has been reported that polyomavirus antibodies are not passed from the Budgerigar hen to her chicks in the egg.[80] The absence of antibodies would have to be controlled by a mechanism that selected against the transfer of avian polyomavirus antibodies, while allowing the transfer of other antibodies. From an evolutionary perspective, this must be considered an interesting finding. This appears problematic, for the passage of antibodies to the chick would be favored from an evolutionary standpoint. Budgerigar hens that did not pass antibodies to their chicks would be expected to have fewer chicks survive than hens that passed antibodies to their chicks. Over time, the genetic material within a flock from the hens that did not pass protective antibodies would be expected to decrease, and the genetic material within a flock from the hens that did pass protective antibodies to their chicks would be expected to increase.

DIAGNOSIS

Documenting avian polyomavirus infections in dead birds is straightforward. Confirming infections in live birds is more difficult. The most accurate way to confirm the presence of avian polyomavirus involves recovery of the virus in cell culture from the tissues of an infected bird;[7] however, this procedure is time-consuming. Other techniques that can be used to document a polyomavirus infection include: the demonstration by electron microscopy of virus particles in affected tissues; demonstration of a four-fold increase in antibody titer in paired samples; specialized staining of suspect lesions using viral-specific antibodies; and the detection of viral nucleic acid using polyomavirus-specific DNA probes *Figure 6.21*.[8,18,24,45,53,65,80,86]

Clinical changes that are suggestive of a polyomavirus infection include acute death with subcutaneous hemorrhage in larger psittacine birds and acute death or development of feather abnormalities in Budgerigar chicks. However, these changes can also be associated with other infectious and noninfectious disease processes. PBFD virus, malnutrition, endocrine abnormalities, other viral infections, trauma, bacterial infections, fungal infections and drug reactions also can cause feather lesions similar to those noted with polyomavirus infections. In general, feather lesions caused by polyomavirus in Budgerigars resolve after several months, while those induced by PBFD virus may persist and become progressively worse.[29]

Gross changes may not be apparent at necropsy in some birds that die acutely with avian polyomavirus infections. In other birds, hemorrhage may be noted under the skin and on the surface of the liver, spleen and heart (see Pathology). However, it is critical that a polyomavirus infection not be diagnosed based only on gross changes, because liver disease, clotting disorders and a variety of infectious agents including gram-negative bacteria and Pacheco's

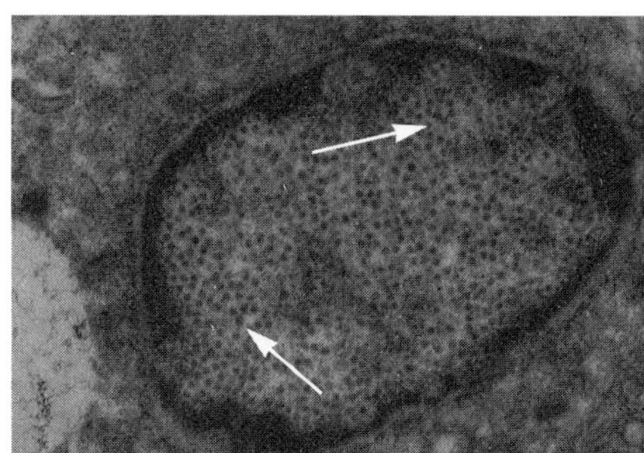

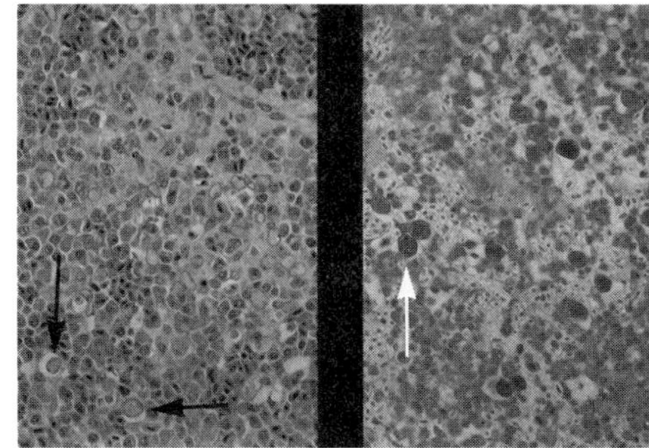

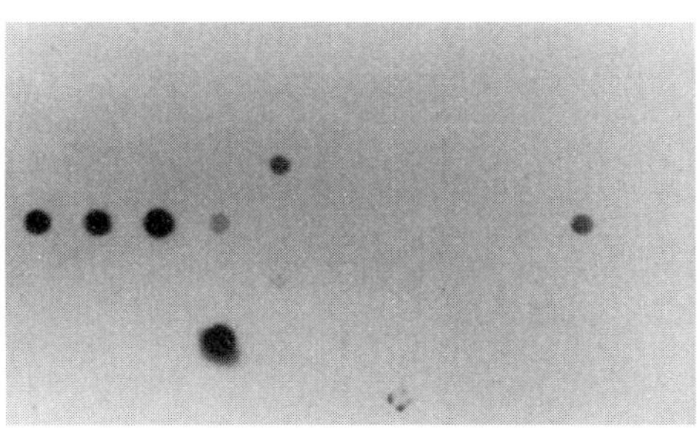

disease virus can cause similar lesions *see Figure 4.2*.

Demonstration of large intranuclear inclusion bodies is considered suggestive of a polyomavirus infection. These intranuclear inclusion bodies from the diseased tissues of numerous species of infected birds have been shown by electron microscopy to contain 40- to 55-nm viral particles *Figure 6.21*.[5,7,18,45,36] In some cases, specialized staining of affected tissues with viral-specific antibodies or viral-specific nucleic acid probes is required to differentiate between intranuclear inclusion bodies

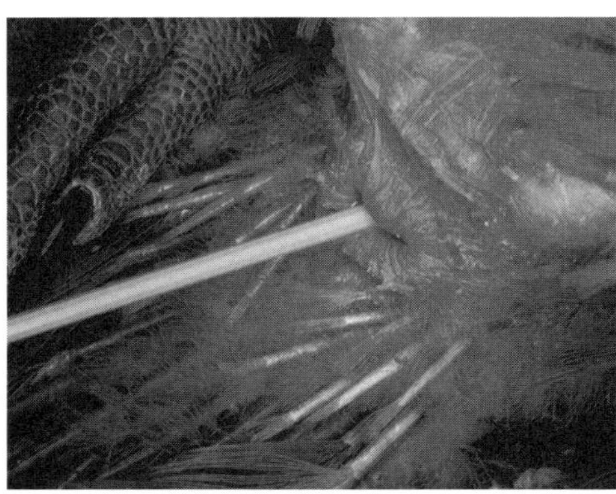

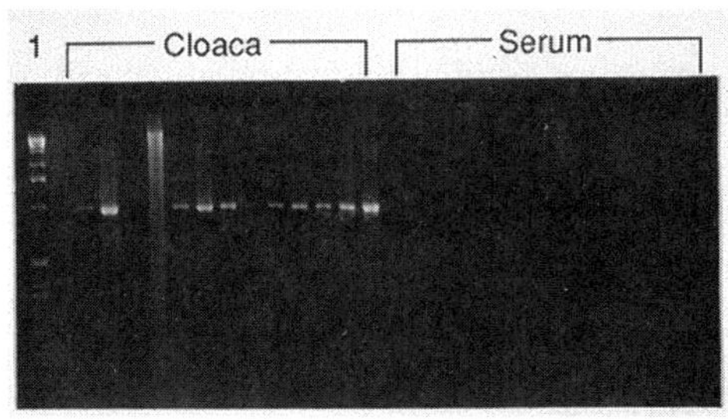

In this study, serum was compared to cloacal swabs for detection of virus from known shedders. The virus was detected in the cloaca of all of the shedders, but was not detected in any the sera from any of the same birds. Lane 1 is a control. The presence of a white line indicates a positive test.

acid can be demonstrated in the serum of some nonbudgerigar psittacine birds by using DNA probes.[65]

Detection of virus-neutralizing antibodies can be used to demonstrate that a bird has been infected previously with avian polyomavirus. The presence of a positive antibody titer indicates that a bird was infected with the virus previously. A lack of an antibody titer may indicate: that the bird has not been infected; that it was infected in the distant past and the antibody titer has decreased below a detectable level; or that it has been infected but has not developed an antibody response (considered rare with avian polyomavirus).

induced by polyomavirus and those caused by PBFD virus, adenovirus or herpesvirus.[53,24,83] In one study, DNA probe testing of suspect tissues was found to be more sensitive than microscopic examination for diagnosing polyomavirus infections.[24] In another study documenting concurrent avian polyomavirus and PBFD virus infections, DNA probing was superior to routine histopathology and in situ hybridization for confirming avian polyomavirus infections in suspect birds.[53]

Polyomavirus particles have been demonstrated by electron microscopy in the serum of some infected Budgerigars. Viral nucleic acid was detected in the serum of 36% to 47% of tested Budgerigars, depending on their breeding status.[6,80] Virus concentrations were considered lower in the tissues of breeding when compared to nonbreeding Budgerigars. Rarely, viral nucleic

In Budgerigars, a sustained high antibody titer is thought to indicate that the bird has a persistent infection.[12,79] A similar antibody response has been theorized to occur in nonbudgerigar psittacine birds, where antibody titers are thought to wane in recently infected birds and stay high in persistently infected birds.[8,19,28,107] However, viral-specific DNA probe testing of nonbudgerigar psittacine birds has indicated that there is no correlation between active shedding of polyomavirus in the excrement and the titer of neutralizing antibodies (see Immunity).[66]

POLYOMAVIRUS-SPECIFIC DNA PROBES — A polyomavirus-specific DNA probe test has been used to demonstrate polyomavirus nucleic acid from various locations, including liver, spleen, kidney, cloacal secretions, intestinal secretions, serum and blood.[66,80,85] A version of this test available through an avian diagnostic laboratory (Avian Research Associates) correctly identified 114 of 116 known positive samples (sensitivity of 98.2%), and correctly identified all of the 56 known negative samples (specificity of 100%).[11] Viral nucleic acid occasionally can be detected in the blood or serum of some nonbudgerigar psittacine birds; however, a cloacal swab is the best sample for detecting birds that are shedding polyomavirus *Figure 6.22*. A positive test indicates that polyomavirus nucleic acid was detected in the submitted sample; a negative test, that it was not. In adult parrots naturally infected with avian polyomavirus, viral nucleic acid was detected in the excrement of 26% of the birds that were tested twice at a four- to six-month interval.[81]

Another use for DNA probes is to detect viral nucleic acid in fresh tissues from birds that may have died from polyomavirus infections. In suspect cases, duplicate tissue samples can be sent for microscopic examination to provide a tentative diagnosis and for DNA probing to confirm the presence of polyomavirus nucleic acid in the submitted sample. The best sample to submit for postmortem confirmation of polyomavirus is a swab of the cut surface of the spleen, liver and kidney (use the same swab for all three tissues) *Figure 6.23*. Eggs that fail to hatch or are infertile can be screened for polyomavirus using viral-specific DNA probes.

Research findings suggest that the strains of polyomavirus that infect Passeriformes have a genome that is slightly different from the genome of the virus that infects Budgerigars.

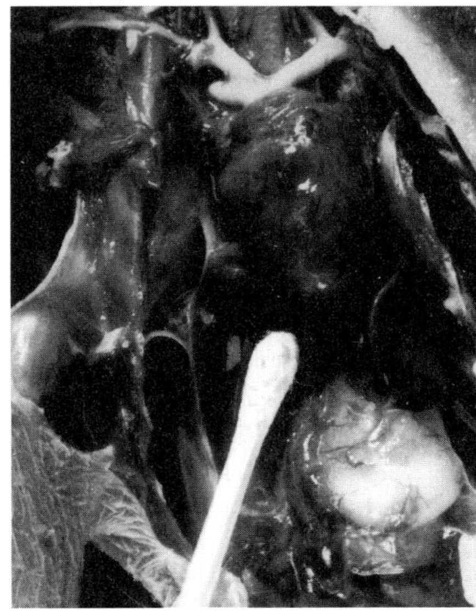

FIG 6.23

The best sample to submit for postmortem confirmation of polyomavirus is a swab of the cut surface of the spleen, liver and kidney.

Because of this difference, DNA probes designed to detect avian polyomavirus in psittacine birds have not been effective in demonstrating virus in some species of Passeriformes.[25,26]

CONTROL

As is the case with many other viral-induced diseases in companion animals, the best method for preventing avian polyomavirus infections is likely to be through the widespread use of an effective vaccine. An inactivated vaccine is currently being evaluated for registration by the USDA.[86-88] While vaccination is the best way to prevent polyomavirus infections, other procedures that will reduce a birds exposure to this virus include, sound hygienic practices, maintaining closed aviaries and preventing visitors from entering avian nurseries. The techniques that help prevent or control an avian polyomavirus outbreak are outlined in *Table 6.8*.

Polyomavirus virions are small, nonenveloped particles that are resistant to severe environmental conditions, to a variety of disinfectants and organic solvents, to freezing and thawing, and to heating to 56°C for two hours.[7,89,110] A polyomavirus that infects primates (SV-40) has been shown to be inactivated by some products containing ethanol and resistant to others containing the same active ingredient.[14,20,92] The environmental stability of avian polyomavirus causes a considerable problem in the aviary or hospital because persistently infected adult Budgerigars have been shown to shed virus in their feather dust or excrement, and clinically affected nonbudgerigar psittacines have been shown to shed virus in their excrement. Nosocomial transmission of the virus can be reduced by strict adherence to hygiene and appropriate use of disinfectants.

TABLE 6.8

Techniques for preventing or controlling polyomavirus outbreaks in the nonbudgerigar psittacine aviary

PREVENTION
    Vaccinate susceptible adults and neonates.
    Do not ship or accept an unvaccinated bird.
    Do not maintain Budgerigars in the same airspace with other unvaccinated psittacine neonates.
    Clean and disinfect the nursery environment regularly.
    Ship only weaned birds.
    Use biosecure shipping containers (Horizon Micro-Environments) to prevent virus exposure during transport.
    Maintain a closed aviary and strictly limit visitations by non-aviary personnel.
    Do not return a neonate to the nursery if it has been exposed to other birds.
    If new birds must be added to the flock, vaccinate and quarantine them for a minimum of 60 to 90 days.
    Do not mix neonates from multiple sources in the same airspace.
    Use separate feeding instruments for each bird.
    Do not use a feeding utensil and place it back in a common food container.
CONTROL
    Isolate clinically affected birds.
    Carefully vaccinate exposed birds making sure that virus transmission is not facilitated by the handling procedure.
    Do not place a clinically ill bird in the same airspace with birds in the nursery.
    Isolate birds in direct contact with clinically ill birds or those that are shedding.
    Completely clean and disinfect the nursery environment.
    Use the DNA probe to test the nursery environment for viral-contamination.

Cleaning of virus-contaminated objects by manual removal of any organic debris followed by the use of appropriate disinfectants is required to remove or inactivate polyomavirus that is likely to be in an aviary environment following an outbreak. *Tables 6.9 and 6.10* illustrate the effectiveness of some disinfectants in inactivating polyomavirus. A commonly used disinfectant, chlorhexidine, was found to be ineffective against avian polyomavirus, which may explain why nurseries that use this product to soak syringes between feedings can still experience polyomavirus outbreaks.

Given the extremely high prevalence of persistent polyomavirus infections, and the frequency with which Budgerigars can shed the virus,[79] one should not maintain young nonbudgerigar psittacines in the same airspace with Budgerigars. The potential for intraspecies transmission of polyomavirus may be a particular problem for pet retailers that maintain both large and small psittacine birds.

A DNA probe-based test is extremely valuable for identifying birds that are shedding virus in their excrement during an outbreak. Birds that are shedding the virus can be separated from others in a nursery to prevent further transmission, while vaccinated birds are developing antibodies to the virus. By testing cloacal swabs of a bird at the time of death one can determine whether it is shedding virus, which in turn will help determine whether its environment may have been contaminated. If the environment is contaminated, there is a potential for viral

Disinfectants found to experimentally inactivate avian polyomavirus and their sources

Agent	Active ingredient	Manufacturer	Dilution	Comments
Avinol-3	Synthetic phenol	Veterinary Products Laboratory PO Box 34820 Phoenix, AZ 85067-4820	1:256	
Clorox	Sodium hypochlorite	Many	1:10	Inexpensive; produces fumes that may be irritating to mucous membranes.
Dent-A-Gene	Stabilized chlorine dioxide	Oxyfresh Independent Distributors 1-800-999-9551 ext 105270	1:400	Considered safe for exposed humans and animals at recommended dilution.
Alcohol	Ethanol 70%	Many	undiluted	

amplification in a susceptible population. If an infected bird dies soon after infection, it may not be shedding virus at the time of death, and thus the bird's environment may not be contaminated with virus. Birds that are clinically ill, are found to be shedding polyomavirus, or are in direct contact with birds that are clinically ill or shedding polyomavirus should be isolated from birds that are clinically normal and not shedding virus.[28,84,85]

Nonbudgerigar psittacine birds that are shedding avian polyomavirus should be isolated, not euthanized. Birds that are shedding avian polyomavirus are likely to be of no further concern when the birds to which they are exposed are protected from infection by vaccination. Nonbudgerigar psittacines that are shedding avian polyomavirus can be best managed by maintaining them in restricted environments in which they do not directly or indirectly expose other nonvaccinated birds, particularly neonates, to the virus.

Testing Budgerigars for the presence of virus-neutralizing antibodies and cul-ling positive birds have been suggested as methods to establish flocks of polyomavirus-free Budgerigars.[28,78] Additionally, depopulation of Budgerigar aviaries experiencing outbreaks, followed by restocking with seronegative birds has been suggested as a method of controlling enzootic infections in this species.[28,32,78]

In nonbudgerigar psittacines, the seroprevalence of avian polyomavirus is so high (up to 63% in some flocks)[88] that culling seropositive birds creates an unacceptable carnage and is neither practical nor recommended for controlling polyomavirus infections. As is the case with many virus-induced diseases, vaccina-

TABLE 6.10

Activity of disinfectants at recommended dilutions against avian polyomavirus

Agent	1 min. exposure	5 min. exposure
Avinol-3	Excellent	Excellent
Clorox	Excellent	Excellent
Dent -A-Gene	Excellent	Excellent
Ethanol	Excellent	Excellent
Mikroklene	Good	Good
Chlorhexidine solution	Poor	Poor
Orange Power	Poor	Poor
Quaternary ammonia	Good	Good

An avian polyomavirus vaccine was first shown to be effective in preventing experimental infections in vaccinated Blue and Gold Macaw chicks. Vaccinated chicks were resistant to experimental challenge with live virus that induced infections in nonvaccinated chicks. A similar vaccine that can be used to prevent infections is now available.

tion will prove to be the best method to prevent avian polyomavirus infections in nonbudgerigar psittacines.[86-88]

Outbreaks of polyomavirus tend to be consistent in Budgerigar aviaries that utilize a constant breeding cycle, while the disease appears to be self-limiting in aviaries raising larger psittacine birds with discontinuous breeding cycles. It has been suggested that polyomavirus disease-free Budgerigar nestlings can be produced by interrupting the breeding cycle, removing all but the older breeding birds and disinfecting the aviary.[32] In one study, older breeding Budgerigars shed fewer virus particles and less frequently than young adults. Additionally, it appeared that older Budgerigars used for continuous breeding may stop shedding virus altogether.[80,81] Breeding cycle manipulation was used to control polyomavirus infections in Budgerigars in an aviary experiencing an avian polyomavirus outbreak. By interrupting the breeding cycle for seven months and thoroughly cleaning the aviary with bleach, return to high hatchability and a cessation of

hatchling deaths was achieved when breeding resumed. However, in other affected aviaries, mortalities and disease have continued when breeding was resumed following a rest period.[32]

TREATMENT — Several therapies, including various immunostimulants and antiviral drugs designed for other viruses, have been suggested for the treatment of birds with avian polyomavirus. Included in the group of frequently mentioned therapies are interferon, acyclovir and AZT. Anecdotal evidence suggests that some of these therapies may be effective in the treatment of birds with clinical signs suggestive of avian polyomavirus. However, none of these treatments has yet been confirmed to be effective in birds with documented active avian polyomavirus infections. Of these speculative therapies, interferon may be the most promising.

VACCINATION — During epornitics in mixed psittacine bird collections, infected survivors and asymptomatic birds exposed to them have been shown to

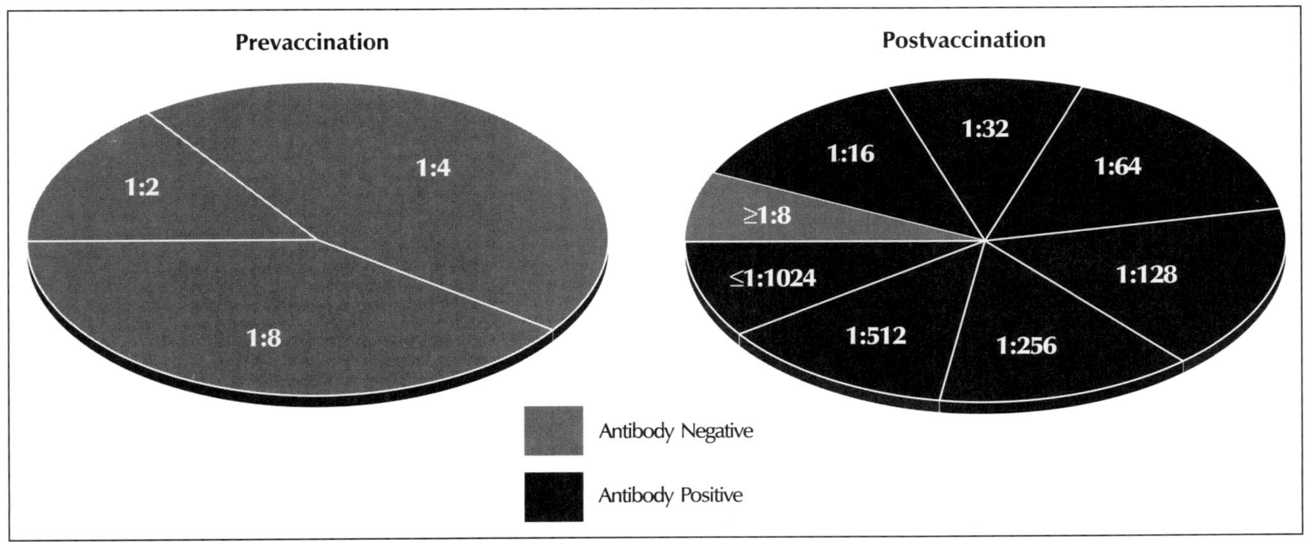

develop polyomavirus-neutralizing antibodies.[8,45,107] Seronegative young adult birds will seroconvert when housed adjacent to seropositive breeding adults, indicating that an antibody response does occur following natural exposure to the virus.[8,12,14,45] The detection of virus-neutralizing antibodies in flocks of birds in which individuals are clinically normal suggests that many infections are subclinical.[8,45,88,107] Collectively, these findings suggest that some exposed birds are able to mount an effective immune response. If a natural immunity to disease occurs, then it should be possible to induce a similar protective immunologic response through vaccination. Experimental studies have indicated, in fact, that an antibody response can be induced through vaccination, and that the resulting immunologic response is protective. In one study using Blue and Gold Macaw chicks, an inactivated vaccine elicited polyomavirus-neutralizing antibodies in all the vaccinates *Figure 6.24*. The induced immunologic response protected the vaccinated chicks from subsequent challenge with live virus.[86] In other studies, an inactivated avian polyomavirus vaccine was shown to protect Amazon parrots, cockatoos, African Grey Parrots and chickens from infection.[88a]

Vaccination studies have indicated that several adjuvants including aluminum hydroxide, Acemannan, Equimune and Pemulem can be used safely in companion birds.[87,88a] The safety and immunogenicity of avian polyomavirus vaccines, administered either intramuscularly or subcutaneously, were evaluated in a group of 233 mixed species Psittaciformes that ranged in age from twelve weeks old to greater than five years old. Vaccination stimulated a marked virus-neutralizing antibody response, particularly in birds that had been seronegative prior to vaccination. The vaccine used in this study elicited an increase in polyomavirus-neutralizing antibodies by two weeks after the second vaccination in 93% of the birds that had been seronegative prior to vaccination, and 63% of the birds that had been seropositive prior to

Virus-neutralizing antibody titers before and after vaccination in a group of psittacine birds that were seronegative prior to vaccination. Chart represents numbers of birds in each category.

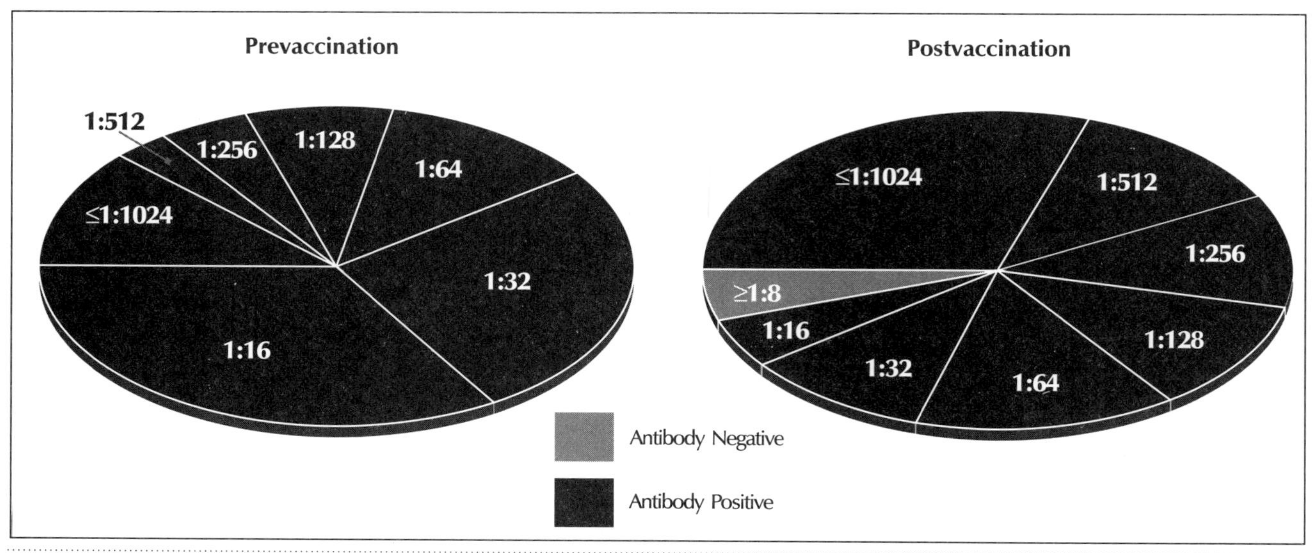

FIG 6.26

Virus-neutralizing antibody titers before and after vaccination in a group of psittacine birds that were seropositive prior to vaccination. Chart represents numbers of birds in each category

vaccination. Seventy-six percent of all the vaccinates had at least a four-fold increase in virus-neutralizing antibody titers at this time. Birds that were seronegative (titer ≤ 8) at the beginning of the study had a greater increase in virus-neutralizing antibody titer than did birds with a high titer (≥ 64) at the beginning of the study. The pre-and post-vaccination virus neutralizing antibody titers in vaccinates that were seronegative on day 0 are presented in *Figure 6.25*. The pre-and post-vaccination virus neutralizing antibody titers in vaccinates that were seropositive on day 0 are presented in *Figure 6.26*.[88]

Serious reactions have not been observed in any vaccinates, and the appetites and attitudes of all vaccinated birds have remained normal. Three types of reactions are expected at the site of subcutaneous vaccination: yellowish discoloration of the skin, thickening of the skin or formation of a knot. Similar reactions undoubtedly occur in many mammals vaccinated with products containing adjuvants, but the reactions are difficult to visualize because of

the fur and thickness of the skin. These reactions indicate that the bird's immune system has responded to the vaccine, and the changes should resolve without treatment three to six weeks post-vaccination *Figure 6.27*. In one field trial, some cockatoos and macaws experienced a heavy molt of up to ten days' duration that started three to five days after the second vaccination. It could not be determined if this molt occurred in response to the vaccination, the stress associated with handling or was induced by climatic or other external factors. The molt was uneventful and appeared to have no adverse affect on the vaccinates.[88]

To establish the safety of a vaccine intended for widespread use, a field trial was performed in a flock of birds in which the virus-neutralizing antibody titers to avian polyomavirus were not determined prior to the study. From previous studies, it was expected that many of the birds in this flock would have preexisting neutralizing antibodies.[66] In fact, 63% of the birds used in this study were considered to have been

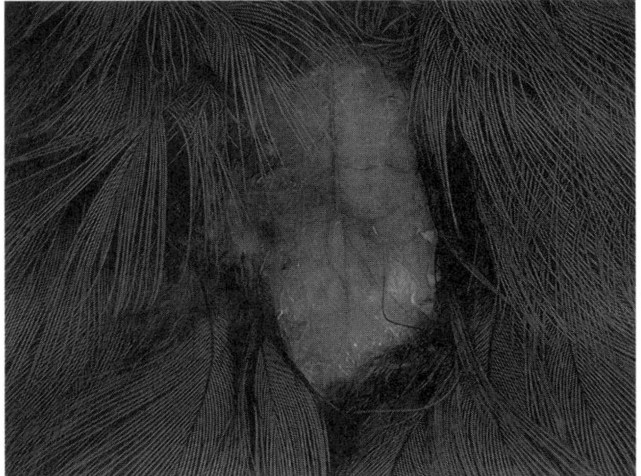

FIG 6.27

When used correctly, the avian polyomavirus vaccine can cause one of three types of reactions at the subcutaneous injection site:
A yellowish discoloration of the skin;
B thickening of the skin with mild scabbing;
C formation of small knots. These reactions indicate that the bird's immune system has responded to the vaccine, and the changes should resolve without treatment three to six weeks after vaccination.
photographs courtesy of Richard Porter

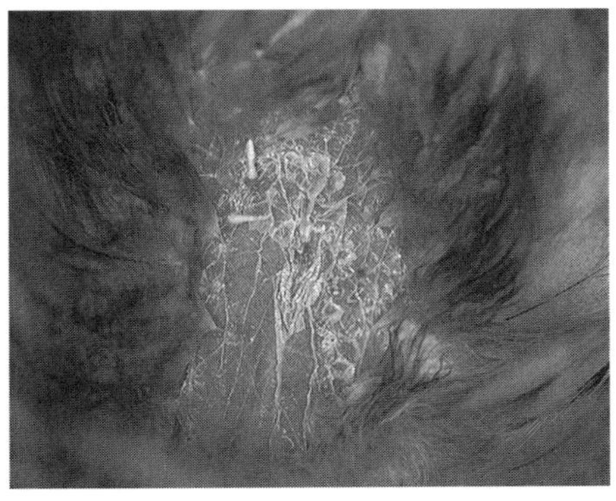

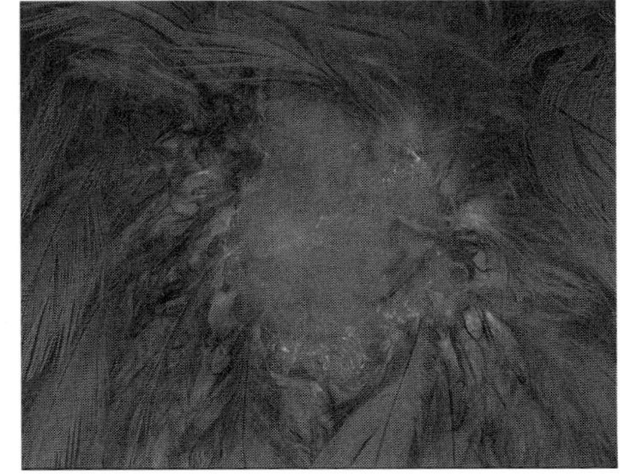

previously exposed to avian polyomavirus because of the detection of virus-neutralizing antibodies prior to vaccination. None of the vaccinates with preexisting neutralizing antibodies developed a severe adverse reaction following vaccination.[88] A similar vaccine was shown to be safe in birds with preexisting neutralizing antibodies even when the birds were vaccinated five times in a 49-day period.[87]

Given the prevalence of polyomavirus infections in companion birds, as indicated by the detection of virus-neutralizing antibodies, it is noteworthy that an inactivated avian polyomavirus vaccine does not cause adverse reactions in vaccinates that were seropositive before vaccination. A safe, effective, inactivated avian polyomavirus vaccine is in the process of being registered by the USDA. Once available, the guidelines provided by the manufacturer should be carefully followed. Any unusual reactions should be reported to the manufacturer.

Aviculturists with large breeding facilities rarely handle or evaluate the overall health of their adults. The infrequent attention provided to these adults allows some problems such as liver disease, kidney disease, heart disease and cancers to slowly progress in what appear to be clinically normal birds. Many of these hidden problems will be detected or exacerbated during the handling procedures necessary for vaccination. In the best managed aviaries in which adult birds are rarely handled, experience suggests that during a vaccination process,

preexisting medical problems will be identified in 2% to 4% of the adults.

POLYOMAVIRUS IN PASSERIFORMES

Lesions suggestive of polyomavirus have been reported in Passeriformes in the United States, Canada, Australia and Italy.[23,25,49,94,95,97] Polyomavirus has been associated with acute mortality in 2- to 3-day old, fledgling, young adult and mature finches. Some affected birds die peracutely, while others exhibit nonspecific signs of illness 24 to 48 hours before death.[23,49,59,111]

A polyomavirus-induced disease was first reported in mature finches in Canada in 1986.[49] Subsequently, polyomavirus infections have been reported in a number of passerine birds. In one outbreak, 36 of 70 (51%) two- to three-day-old Gouldian Finches died. Many of the fledglings that survived had poor feather development and long, tubular, misshapen beaks *Figure 6.28*. Additionally, it was noted that survivors fledged at 31 days rather than the normal 26 days after hatching.[59] Polyomavirus was diagnosed in adult and juvenile Black-bellied Seedcrackers from a zoological collection. The affected birds died acutely or following several days of lethargy.[25,26] Polyomavirus infections have been associated with splenitis and hepatitis in greenfinches[97] and with pneumonia and concurrent poxvirus infection in canaries.[94] Polyomavirus was diagnosed in a six-month-old goldfinch that was one of 35 young birds to die following a two- to three-day history of anorexia, depression and dyspnea.[95]

PATHOLOGY

Polyomavirus-infected finches may die acutely with no detectable pathology or may have gross or histologic changes similar to those described for other birds. Gross pathologic changes may include swollen, pale, mottled livers and an enlarged congested spleen. Other lesions that have been described include perirenal and serosal (or subserosal) intestinal hemorrhage.[23,49,59,95,111] Gross lesions in polyomavirus-infected seedcrackers included hepatomegaly, swollen kidneys, and hemorrhage in the brain and liver and at the base of the skull.[25,26] Histologic changes in affected passerines include inflammation and necrosis of the liver, myocarditis, bone marrow necrosis, hyperplasia of the macrophages in the liver and spleen, lymphoid necrosis and infiltrates of plasma cells in the cells lining the intestinal tract.[23,49,59,95,111]

Intranuclear inclusion bodies have been demonstrated in the spleen, bone marrow, intestines, kidneys, heart and liver.[23,49,59,95,97,111] Inclusion bodies suggestive of those induced by polyomavirus bodies were detected in the spinal cord of a female Purple Grenadier Finch with a two- to three-day history of lethargy.[93] Intranuclear inclusion bodies suggestive of avian polyomavirus have been identified in the liver, spleen, kidney, pancreas and cloacal bursa from various seedcrackers.[26] In one study, intranuclear inclusion bodies were not detected in finches that died at two to three days of age, but were present in birds that died at fledging.[59]

Based on findings in Budgerigars, it has been suggested that surviving finches may develop persistent infections followed by intermittent shedding of the virus. It also has been suggested, but not confirmed, that infected hens may pass virus to their eggs, resulting in early embryonic death.[49,59]

Antibodies to avian polyomavirus derived from psittacine birds were found to cross-react with avian poly-omavirus in the tissues of Painted Finches.[111] This suggests that the avian polyomavirus that infects psittacine birds is antigenically related to the virus that infects finches. However, DNA probes used to detect avian polyomavirus in psittacine birds will not detect the virus in seedcrackers, suggesting that a virus with a varied genome can infect passerines.[26] A DNA probe has been designed that will detect avian polyomavirus in some passerines. In one study using this viral-specific DNA probe, avian polyomavirus nucleic acid was detected in the tissues of 25 of 45 (55%) seedcrackers and Blue-bills in a closed aviary population. The affected populations experienced sporadic but ongoing mortality. Viral nucleic acid was detected in birds ranging in age from two days to four years old. Only 4 of the 25 (16%) positive birds had inclusion bodies suggestive of avian polyomavirus. The birds infected with avian polyomavirus were thought to have a higher incidence of hepatic necrosis, hepatitis, bacterial infections and parasitism than birds that were not infected with avian polyomavirus.[25]

REFERENCES

1. Abrey A: Summary of avian post mortems of Allerton laboratory in the Natal Area of South Africa during the last 18 months. Newsletter of Assoc Avian Vet 5:98-99, 1984.

2. Aieta EM, Berg JD, Roberts PV, et al: Comparison of chlorine dioxide and chlorine in wastewater disinfection. J Water Pollut Control Fed 52:810-822, 1980.

3. Arthur RR, Diagostin S, Shah KV: Detection of BK virus in urine and brain tissue by the polymerase chain reaction. J Clin Microbil 27:1174-1179, 1989.

4. Beach JE: Diseases of budgerigars and other cage birds. A survey of post-mortem findings. Vet Rec 74:134-140, 1962.

5. Bernier G, Morin M, Marsolais G: A generalized inclusion body disease in the budgerigar (*Melopsittacus undulatus*) caused by a papovavirus-like agent. Avian Dis 25:1083-1092, 1981.

6. Bernier G, Morin M, Marsolais G: Papovavirus-induced feather abnormalities and skin lesions in the budgerigar: Clinical and pathological findings. Can Vet J 25:307-310, 1984.

6a. Bond M: Personal communication.

7. Bozeman LH, Davis RB, Gandry D, et al: Characterization of a papovavirus isolated from fledgling budgerigars. Avian Dis 25:972-980, 1981.

8. Clubb SL, Davis RB: Outbreak of papova-like viral infection in a psittacine nursery-a retrospective view. Proc Assoc Avian Vet, 1984, pp 121-129.

9. Coleman CW: Bile duct carcinoma and cloacal prolapse in an orange-winged Amazon parrot (*Amazona amazonica amazonica*). J Assoc Avian Vet 5:87-89, 1991.

10. Cribb PH: Cloacal papilloma in an Amazon parrot. Proc Assoc Avian Vet, 1984, pp 35-37.

11. Dahlhausen B, Radabaugh S: Update on psittacine beak and feather disease and avian polyomavirus testing. Proc Assoc Avian Vet, 1993, pp 5-7.

12. Davis RB, Lukert PD, Avery P: An update on budgerigar fledgling disease (BFD). Proc 33rd West Poult Dis Conf, 1984, pp 96-98.

13. Davis RB: Budgerigar fledgling disease (BFD). Proc 32nd West Poult Dis Conf, 1983, p 104.

14. Davis RB, Bozeman LH, Gaudry OJ, et al: A viral disease of fledgling budgerigars. Avian Dis 25:179-183, 1981.

15. Deshmukh DR, Larsen CT, Dutta SK, et al: Characterization of pathogenic filtrate and viruses isolated from turkeys with bluecomb. Am J Vet Res 30:1019-1025, 1969.

16. Dom P, Ducatelle R, Charlier G, et al: Papillomavirus-like infections in canaries (*Serinus canarius*). Avian Pathol 22:797-803, 1993.

17. Dubensky TW, Murphy EA, Villarreal LP, et al: Detection of DNA and RNA virus genomes in organ systems of whole mice: patterns of mouse organ infection by polyomavirus. J Virol 50:779-783, 1984.

18. Dykstra MJ, Bozeman LH: A light and electron microscopic examination of budgerigar fledgling disease virus in tissue and in cell culture. Avian Pathol 11:11-18, 1982.

19. Dykstra MJ, Dykstra CC, Lukert PD, et al: Investigations of budgerigar fledgling disease virus. Am J Vet Res 45:1883-1887, 1984.

20. Eggers HJ: Experiments on antiviral activity of hand disinfectants. Some theoretical and practical considerations. Zbl Bakt 273:36-51, 1990.

21. Elangham CS, Panciera RJ: Cholangiocarcinoma in a blue-fronted Amazon parrot (*Amazona aestiva*). Avian Dis 32:594-596, 1988.

22. Fenner F, Bachmann PA, Gibbs EPJ, et al: Veterinary Virology. Orlando, Academic Press 1987.

23. Forshaw D, Wylie, SL, Pass, DA: Infection with a virus resembling papovavirus in gouldian finches (*Erythrura gouldiae*). Aust Vet J 65:26-28, 1988.

24. Garcia A, Latimer KS, Niagro FD, et al: Diagnosis of polyomavirus-induced hepatic necrosis in psittacine birds using DNA probes. J Vet Diagn Invest 6:308-314, 1994.

25. Garcia A, Latimer KS, Niagro FD, et al: Diagnosis of polyomavirus infection in seedcrackers (*Pyrenestes* sp.) and blue bills (*Spermophaga haematina*) using DNA in situ hybridization. Avian Pathol 23:525-537, 1994.

26. Garcia AP, Latimer KS, Niagro FD, et al: Avian polyomavirus infection in three black-bellied seed crackers (*Pyrenestes ostrinus*). J Assoc Avian Vet 7:79-82, 1993.

27. Gaskin JM: The serodiagnosis of psittacine viral infections. Proc Assoc Avian Vet, 1988, pp 7-10.

28. Gaskin JM: Psittacine viral disease. A perspective. J Zoo Wild Med 20:249-264, 1989.

29. Gerlach H: Viruses. *In* Ritchie BW, Harrison GJ, Harrison LR (eds):Avian Medicine: Principles and Application. Lake Worth, Wingers Publishing 1994, pp 862-948.

30. Gilardi KVK, Lowenstine LJ, Gilardi JD, et al: A survey for viral, chlamydial and parasitic diseases in wild dusky-headed parakeets (*Aratinga weddellii*) and tui parakeets (*Brotogeris sanctithomae*) in Peru. J Wild Dis In press, 1995.

31. Goodwin MA, McGee ED: Herpes-like virus associated with a cloacal papilloma in an orange-fronted conure (*Aratinga canicularis*). J Assoc Avian Vet 7:23-26, 1993.

32. Gough JF: Outbreaks of budgerigar fledgling disease in three aviaries in Ontario. Can Vet J 30:672-674, 1989.

33. Graham DL: An update on selected pet bird virus infections. Proc Assoc Avian Vet, 1984, pp 267-280.

34. Graham DL: Surprises at necropsy. J Assoc Avian Vet :192-195, 1988.

35. Graham DL: Internal papillomatous disease. A pathologists's view. Proc Assoc Avian Vet, 1991, pp 141-143.

36. Graham DL, Calnek BW: Papovavirus infection in hand-fed parrots: Virus isolation and pathology. Avian Dis 31:398-410, 1987.

37. Greenwood AG, Storm J, Wild DJ: Laser surgery of psittacine internal papilloma. Proc Europ Conf Avian Med Surg, 1993, pp 217-223.

38. Gross L: Oncogenic Viruses. New York, Pergamon Press 1961.

39. Hamada M, Oyamada T, Yashikawa H, et al: Morphological studies of esophageal papilloma naturally occurring in cattle. Jpn J Vet Sci 51:354-351, 1989.

40. Harvey SC: Antiseptics and disinfectants; fungicides; ectoparasiticides. *In* Gilman AG, Goodman LS, Gilman A (eds):The Pharmacological Basis of Therapeutics. New York, McMillian Publishing 1980, 6th ed.

41. Heritage J, Chesters DM, McCance DJ: The persistence of papovavirus BK DNA sequences in normal human renal tissue. J Med Virol 8:143-150, 1981.

42. Hillyer EV, Moroff S, Hoefer H, et al: Bile duct carcinoma in two out of ten Amazon parrots with cloacal papillomas. J Assoc Avian Vet 5:91-95, 1991.

43. Hirai K, Nonaka H, Fukushi H, et al: Isolation of a papova-like agent from young budgerigars with feather abnormalities. Jpn J Vet Sci 46:577-587, 1984.

44. Hunter B, Gagnon A, Onderka D: Viral hepatitis in budgerigars in Southern Ontario. Can Vet J 20:176, 1979.

45. Jacobson ER, Hines SA, Quesenberry K, et al: Epornitic of papova-like virus-associated disease in a psittacine nursery. J Am Vet Med Assoc 185:1337-1341, 1984.

46. Jacobson ER, Mladinich CR, Clubb S: Papilloma-like virus infection in an African Grey Parrot. J Am Vet Med Assoc 183:1307-1308, 1983.

47. Jarrett WFH, et al: Alimentary fibropapilloma in cattle: A spontaneous

tumor, non permissive for papillo-mavirus replication. J Natl Cancer Inst 73:499-504, 1984.

48. Jennings AR: Diseases of wild birds. Bird Study 6:19-22, 1959.

49. Johnston KM, Riddell C: Intranuclear inclusion bodies in finches. Can Vet J 27:432-434, 1986.

50. Kaal: Venereal disease of parrots. Proc 2nd Int Parrot Symposium, 1990.

51. Krautwald ME, Kaleta EF: Relation-ship of French moult and early virus induced mortality in nestling budgeri-gars. Proc 8th Intl Cong World Vet Poult Assoc, 1985, p 115.

52. Krautwald ME, Muller H, Kaleta EF: Polyomavirus infection in budgerigars (*Melopsittacus undulatus*): Clinical and etiological studies. J Vet Med 36:459-467, 1989.

53. Latimer KS, Niagro FD, Campagnoli RP, et al: Diagnosis of concurrent avian polyomavirus and psittacine beak and feather disease virus infections using DNA probes. J Assoc Avian Vet 7:141-146, 1993.

53a. Latimer KS, et al: Unpublished data.

54. Lehn H, Muller H: Cloning and char-acterization of budgerigar fledgling disease virus (BFDV), an avian poly-omavirus. Virology 151:362-370, 1986.

55. Lina P, Van Noord M, deGroot F: Detection of virus in squamous papil-lomas of the wild bird species *Fringilla coelebs*. J Natl Cancer Inst 50:567-571, 1973.

56. Llabres CM, Ahearn PG: Antimicrobial activities of N-chloramines and diazo-lidinyl urea. Appl Environ Microbiol 49:370-373, 1985.

57. Lowenstine LJ: Emerging viral diseases of psittacine birds. *In* Kirk RW (ed):Current Veterinary Therapy IX. Philadelphia, WB Saunders Co. 1986, pp 705-710.

58. Lynch J, Swinton J, Pettit J, et al: Isolation and experimental chicken-embryo-inoculation studies with budgerigar papovavirus. Avian Dis 28:1135-1139, 1984.

58a. MacWhirter P: Personal communication.

59. Marshall R: Papova-like virus in a finch aviary. Proc Assoc Avian Vet, 1989, pp 203-207.

60. Mathey WJ, Cho BR: Tremors of nestling budgerigars with BFD. Proc 33rd West Poult Dis Conf, 1984, p 102.

61. Matthews REF: Classification and nomenclature of viruses. Intervirology 17:1-149, 1982.

62. McDonald SE: Clinical experiences with cloacal papillomas. Proc Assoc Avian Vet, 1988, pp 27-30.

63. Mounts P, Shah KV, Kahima H: Viral etiology of juvenile - and adult-onset squamous papilloma of the larynx.

64. Muller H, Nitschke R: A polyoma-like virus associated with an acute disease of fledgling budgerigars (*Melopsittacus undulatus*). Med Microbiol Immunol 175:1-13, 1986.

65. Niagro FD, Ritchie BW, Latimer KS, et al: Polymerase chain reaction detection of PBFD virus and BFD virus in sus-pect birds. Proc Assoc Avian Vet, 1990, pp 25-37.

66. Niagro FD, Ritchie BW, Lukert PD, et al: Avian polyomavirus: Discordance between neutralizing antibody titers and viral shedding in an aviary. Proc Assoc Avian Vet, 1991, pp 22-26.

66a. Niagro FD, et al: Unpublished data.

67. Noss CI, Olivieri VP: Disinfecting capabilities of oxychlorine compounds. Appl Environ Microbiol 50:1162-1164, 1985.

68. Osterhaus A, Ellens D, Horzinek M: Identification and characterization of a papillomavirus from birds. Intervirolo-gy 8:351-359, 1977.

69. Pascucci S, Maestrini N, Misciattelli M, et al: Malattia da virus papova-simile nel papagallino ondulato (*Melopsittacus undulatus*). Clin Vet (Milan) 106:38-41, 1983.

70. Pass DA: A papova-like virus infection of lovebirds (*Agapornis* sp.). Aust Vet J 82:318-319, 1985.

71. Pass DA, Perry RA: Psittacine beak and feather disease. An update. Aust Vet Pract 15:55-60, 1985.

72. Pass DA, Prus SE, Riddell C: A papova-like virus infection of splendid parakeets (*Neophema splendida*). Avian Dis 31:680-684, 1987.

73. Perry RA, Pass DA: Psittacine beak and feather disease, including French molt, in parrots in Australia. Proc Aust Vet Poult Assoc, 1985, pp 35-38.

74. Petrak ML, Gilmore CE: Neoplasms. *In* Petrak ML (ed):Diseases of Cage and Aviary Birds. Philadelphia, Lea and Febiger Press 1969, pp 461-474.

75. Phalen DN, Ambrus S, Graham DL: The avian urinary system: Form func-tion diseases. Proc Assoc Avian Vet, 1990, pp 44-57.

76. Phalen DN, Wilson VG, Graham DG: Polymerase chain reaction assay for avian polyomavirus. J Clin Micro 29:1030-1037, 1991.

77. Phalen DN, Wilson VG, Graham DL: Epidemiology and diagnosis of avian polyomavirus infection. Proc Assoc Avian Vet, 1991, pp 27-31.

78. Phalen DN, Wilson VG, Graham DL: Avian polyomavirus infection and disease: A complex phenomenon. Proc Assoc Avian Vet, 1992, pp 5-10.

79. Phalen DN, Wilson VG, Graham DL: Avian polyomavirus biology and its

clinical applications. Proc Europ Assoc Avian Vet, 1993, pp 200-216.

80. Phalen DN, Wilson VG, Graham DL: Organ distribution of avian poly-omavirus DNA and virus-neutralizing antibody titers in healthy adult budgerigars. Am J Vet Res 54:2040-2047, 1993.

81. Phalen DN, Wilson VG, Graham DL: A practitioner's guide to avian poly-omavirus testing and disease. Proc Assoc Avian Vet, 1994, pp 251-258.

82. Potter K, Connor T, Gallina AM: Cholangiocarcinoma in a yellow-faced Amazon parrot (*Amazona xanthops*). Avian Dis 27:556-558, 1983.

83. Ramis A, Latimer KS, Niagro FD, et al: Diagnosis of psittacine beak and feather disease (PBFD) viral infection, avian polyomavirus infection, aden-ovirus infection and herpesvirus infection in psittacine tissues using DNA in situ hybridization. Avian Pathol 23:643-657, 1994.

83a. Reece RL: Personal communication.

84. Ritchie BW, Niagro FD, Latimer KS, et al: Avian polyomavirus. An overview. J Assoc Avian Vet 5:147-153, 1991.

85. Ritchie BW, Niagro FD, Latimer KS, et al: Polyomavirus infections in adult psittacine birds. J Assoc Avian Vet 5:202-206, 1991.

86. Ritchie BW, Niagro FD, Latimer KS, et al: Efficacy of an inactivated poly-omavirus vaccine. J Assoc Avian Vet 7:187-192, 1993.

87. Ritchie BW, Niagro FD, Latimer KS, et al: Antibody response and local reac-tions to adjuvanted avian polyomavirus vaccines in psittacine birds. J Assoc Avian Vet 8:21-26, 1994.

88. Ritchie BW, Niagro FD, Latimer KS, et al: An inactivated avian polyomavirus vaccine is safe and immunogenic in various Psittaciformes. Vaccine Sub-mitted for publication, 1995.

88a. Ritchie BW, et al: Unpublished data.

89. Sambrook J: The molecular biology of the papovaviruses. *In* Nayak DP (ed):The Molecular Biology of Animal Viruses. New York, Marcel Dekker 1978, 2nd ed, pp 589-672.

90. Samuel JL, Spradbrow PB, Wood AL: Oral papillomas in cattle. Zentralbl Vet 32:706-714, 1985.

91. Schmidt RE, Goodman GJ, Higgins RJ, et al: Morphologic identification of papovavirus in a Moluccan cockatoo (*Cacatua moluceensis*) with neurologic signs. Assoc Avian Vet Today 1:107-108, 1987.

92. Schurmann W, Eggers HJ: Antiviral activity of an alcoholic hand disinfec-tant. Comparison of the in vitro sus-pension test with in vivo experiments on hands, and on individual fingertips. Antiviral Res 3:25-41, 1983.

93. Shivaprasad HL: Diseases of the nervous system in pet birds. A review

and report of diseases rarely documented. Proc Assoc Avian Vet, 1993, pp 213-222.

94. Sironi G: Dual poxvirus and papova-like virus pneumonia in canary (*Serinus canarius*). La Clinica Veterinaria 111:61-67, 1988.

95. Sironi G: Concurrent papovavirus-like and atoxoplasma infections in a gold finch (*Carduelis carduelis*). Avian Pathol 20:725-729, 1991.

96. Sironi G, Gallazi D: Papillomavirus infection in greenfinches (*Carduelis chloris*). J Vet Med 39:454-458, 1992.

97. Sironi G, Rampin T: Papovavirus - like splenohepatic infection in greenfinches (*Carduelis chloris*). Clinica Veterinaria 110:79-82, 1987.

98. Sittig M: Handbook of Toxic and Hazardous Substances and Carcinogens. Park Ridge, NJ, Noyes Publications 1985.

99. Speer BL: The eclectus parrot, medicine and avicultural aspects. Proc Assoc Avian Vet, 1989, pp 239-247.

100. Stoll R, Luo D, Kouwenhoven B, et al: Molecular and biological characteristics of avian polyomaviruses: Isolates from different species of birds indicate that avian polyomaviruses form a distinct subgenus within the polyomavirus genus. J Gen Virol 74:229-237, 1993.

101. Sunberg JP, Junge RE, El Shazley MO: Oral papillomas in New Zealand white rabbits. Am J Vet Res 46:664-668, 1985.

102. Sunberg JP, Junge RE, O'Banion MK: Cloacal papillomas in psittacines. Am J Vet. Res 47:928-932, 1986.

103. Sztojkov V, Saghy E, Meder M, et al: A hullamous papagaj (*Melopsittacus undulatus*) papovavirus okozta megbetegedesenek hazai megallapitasa. Magy Allatorv Lapja 40:59-63, 1985.

104. Taylor TG: French Moult. *In* Petrak ML (ed):Diseases of Cage and Aviary Birds. Philadelphia, Lea and Febiger 1982. 2nd ed, pp 361-367.

105. Telent A, Aksamit AJ, Proper J: Detection of JC virus DNA by polymerase chain reaction in patients with progressive multifocal leukoencephalopathy. J Infect Dis 162:858-861, 1990.

106. VanDerHeyden N: Psittacine papillomas. Proc Assoc Avian Vet, 1988, pp 23-26.

107. Wainwright PO, Lukert PD, Davis RB, et al: Serological evaluation of some Psittaciformes for budgerigar fledgling disease virus. Avian Dis 31:673-676, 1987.

108. Washington D: Unusual growths on feet of chaffinches. Brit Birds 57:184, 1964.

109. Watrach AM, Hanson LE, Meyer RC: Canine papilloma: The structural characterization of oral papilloma virus. J Natl Cancer Inst 43:453-458, 1969.

110. White DO, Fenner F: Papovaviruses. Orlando, Academic Press 1986.

111. Woods L: Papova-like virus in a painted finch. Proc Assoc Avian Vet, 1989, pp 218-219.

112. Wylie SL, Pass DA: Experimental reproduction of psittacine beak and feather disease/French moult. Avian Pathol 16:269-281, 1987.

113. Zanger N, Muller M: Endemic papillomatosis of viral origin in water fowl of the zoological garden Basle. Schweizer Archiv Tierheilkunde 132:483, 1990.

Herpesviridae

Herpesvirus virions are enveloped and are pleomorphic. Their size ranges from 120 to 200 nm in diameter. Replication occurs in the nucleus, and the envelope is acquired when new virus particles are released from an infected cell by budding through the nuclear membrane. As a group, herpesviruses tend to be well adapted to a particular species and are ubiquitous in a population of these animals. Herpesviruses are one of the most successful of all viruses, and a strain of herpesvirus has been isolated from most species of mammals and birds. Most herpesviruses are host-specific, but some can infect many different species, either naturally or experimentally *Table 7.1*.

Varying types of herpesviruses tend to cause life-long, latent infections that are characterized by the periodic recurrence of viral shedding with or without detectable clinical signs. Depending on the strain of virus and the species of infected bird, various herpesviruses have a propensity to infect lymphatic tissue, epithelial cells or nervous tissue and may cause neoplastic, hemorrhagic or necrotic lesions *Table 7.2*. In general, host-adapted strains cause a mild, subclinical, latent infection in their natural host; however, severe, often life-threatening, disease can occur when a non-adapted herpesvirus enters a new host. Most herpesviruses are associated with the formation of intranuclear inclusion bodies. These inclusion bodies represent areas where damaged portions of the cell's nucleus accumulate.

Based on antibody comparison studies designed to separate closely related viruses in the same family, many serotypes of avian herpesvirus have been defined. Some of these serotypes and the orders of birds that they have been shown to infect are listed in *Table 7.3*. The antigenic relationship of these herpesviruses varies. The herpesviruses that infect owls, pigeons, falcons and eagles are serologically related, as are the viruses that infect quail and cranes.[97] Other unclassified serotypes undoubtedly exist.[112,113]

Most herpesviruses are transmitted from one susceptible individual to another through direct contact. Some, however, can be transmitted through the egg or in contaminated food or water supplies. Herpesviruses generally persist in a population of animals by inducing latent infections. Virus shedding stops and starts at irregular intervals, sometimes with no clinical signs of disease. Shedding is particularly likely during stressful periods like those that occur with concomitant disease, malnutrition, temperature changes (high or low), movement of birds or introduction of new birds to the aviary (situations that may induce territorial behavior), dietary changes or breeding activity. In confined areas with dense populations of birds (conditions that

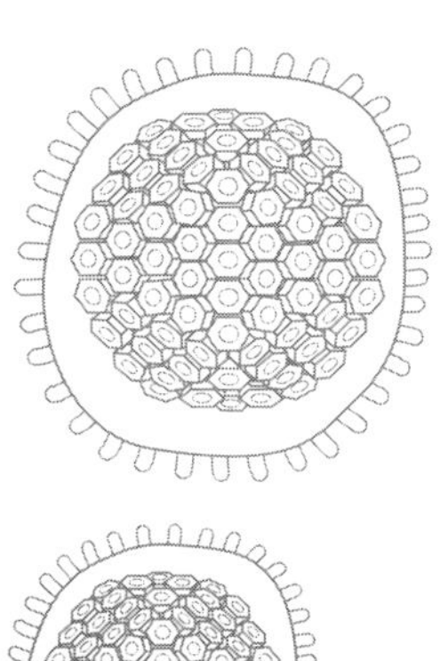

TABLE 7.1

Susceptibility of various orders of birds to experimental or natural herpesvirus infections*

Galliformes
    Infectious laryngotracheitis virus
    Marek's disease virus
    Turkey herpesvirus
    Quail herpesvirus
Anseriformes
    Duck plague virus
    Infectious laryngotracheitis virus (one-day-old ducks susceptible to infection, not disease)
    Marek's disease virus (suggestive lesions in ducks, geese and swans)
    Uncharacterized herpesvirus from ulcerative lesions on the feet of ducks
    Falcon herpesvirus (Muscovy Ducks susceptible)
    Crane herpesvirus (Pekin Ducklings up to 17 days old susceptible)
Psittaciformes
    Pacheco's disease virus (at least three distinct serotypes)
    Amazon tracheitis virus
    Respiratory disease in *Neophema* sp. and *Psittacula* sp.
    Wart-like or flat plaque-like lesions on the skin of psittacine birds
    Budgerigar herpesvirus
    Pigeon herpesvirus (infectious in Budgerigars and Cockatiels)
    Falcon herpesvirus (infectious in Budgerigar and Amazon parrots)
    Marek's disease virus (suggestive lesions in Budgerigars)
Falconiformes
    Falcon herpesvirus
    Owl herpesvirus
    Marek's disease virus (suggestive lesions in a Kestrel)
Strigiformes
    Falcon herpesvirus
    Owl herpesvirus
    Marek's disease virus (suggestive lesions in Great Horned Owls)
Columbiformes
    Pigeon herpesvirus 1
    Pigeons susceptible to a virus isolate from psittacine birds serologically related to pigeon herpesvirus
    Owl herpesvirus infectious in Ring-necked Doves
    Falcon herpesvirus infectious in Ring-necked Doves
Gruiformes
    Falcon herpesvirus (infectious in American coots)
    Crane herpesvirus (infectious in coots)
Ciconiiformes
    Falcon herpesvirus (infectious in Green Herons)
Passeriformes
    Finch herpesvirus
    Weaver herpesvirus
    Infectious laryngotracheitis virus (infectious in canaries)
    Marek's disease virus (suggestive lesions in canaries)

**Not all genera within an order would be expected to be susceptible. Orders of birds not listed have been reported to be susceptible only to their own herpesvirus.*

occur in many aviaries), herpesvirus transmission can occur by exposure of susceptible birds to contaminated aerosols or ingestion of contaminated excrement (fecal/oral transmission).

HERPESVIRUSES IN PSITTACIFORMES

PACHECO'S DISEASE VIRUS
CLINICAL FEATURES

In 1929, a disease investigator named Pacheco encountered an outbreak of acute, fatal hepatitis in psittacine birds maintained in a zoological park in Brazil. This syndrome became known as Pacheco's disease. The clinical progression of Pacheco's disease in affected parrots appeared similar to the clinical changes described with chlamydiosis, but the former disease was shown to be caused by a virus.[142,153] In 1975, Pacheco's disease was confirmed to be caused by an avian herpesvirus, and its presence in psittacine birds in the United States was documented.[166]

Both virulent and avirulent strains of Pacheco's disease virus (PDV) are thought to occur. The clinical signs, progression of disease, and gross and microscopic lesions associated with an infection vary within a flock or an individual bird.[62,125] In addition to the virulence of a virus strain, the varied response to infection may also be governed by the susceptibility of a particular bird or the route through which the virus enters a bird (see Relationship of Virus Strains).[173]

In some PDV outbreaks, the clinical history may indicate that new birds have been added to the aviary, or that aviary residents have been exposed to situations where they might have encountered birds that were shedding the virus (such as bird fairs or sexing clinics). In other cases, PDV will suddenly appear in an otherwise healthy flock and then disappear with the same acuity. Her-

Classic lesions and target cells of persistence for various avian herpesviruses[95]

Virus	Classic Lesions	Target Cells	Site of Persistence
Marek's disease	Lytic/neoplastic	T-lymphocytes, feather follicle	White blood cells
Duck plague	Hemorrhagic	Reticuloendothelial cells	Lymphoid cells, intestines
ILT	Hemorrhagic	Respiratory epithelium	Trachea, trigeminal nerve
Amazon tracheitis virus	Hemorrhagic	Respiratory epithelium	?
Pacheco's disease	Necrotic/hemorrhagic	Hepatocytes, lymphocytes	?
Pigeon herpesvirus	Necrotic	Liver, Pharyngeal epithelium	Pharynx
Owl herpesvirus	Necrotic	Hepatocytes	Pharynx
Falcon herpesvirus	Necrotic	Liver, Pharyngeal epithelium	?
Cormorant herpesvirus	Nonpathogenic	?	?
Crane herpesvirus	Necrotic	Hepatocytes, lymphocytes	?
Stork herpesvirus	Necrotic	Liver, Pharyngeal epithelium	White blood cells
Quail herpesvirus	Necrotic	Hepatocytes	?
Canary herpesvirus	Nonpathogenic	?	?
Finch herpesvirus	Hepatitis	Hepatocytes	?

pesvirus infections are extremely variable: they may involve the death of a single bird in a pair, or massive outbreaks may occur in which numerous birds from throughout the aviary are affected.

The usual clinical history associated with virulent strains of PDV is for a totally normal appearing bird to be found dead in its enclosure. If present, clinical signs may include depression, anorexia, diarrhea (which may be bloody), regurgitation, yellow-green urates, sinusitis, tremors and instability (central nervous system signs), extreme thirst and excessive urination, and conjunctivitis *Figure 7.1*.[125,132,166] Virulent strains of Pacheco's disease virus have been associated with ataxia, hemorrhagic diarrhea and regurgitation of blood-tinged mucus prior to death in affected African Grey Parrots, Amazon parrots and a White-headed Caique.[111-113] As they manifest the final stages of the disease prior to death, many affected birds may develop neurologic signs in the form of tremors, opisthotonos and violent seizures.[65] The greenish-yellow discoloration of the urates and loose stool is an indication of the severe liver damage induced by this virus. Increased activity of some serum enzymes and lipemia may be other indications of viral-induced liver damage. Leukopenia, with or without anemia,

FIG 7.1
Many birds infected with PDV die without showing clinical changes. When they do occur, clinical signs include severe depression;

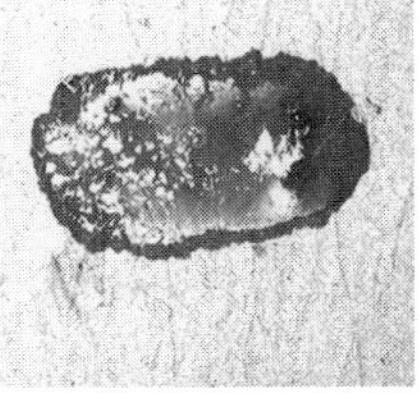
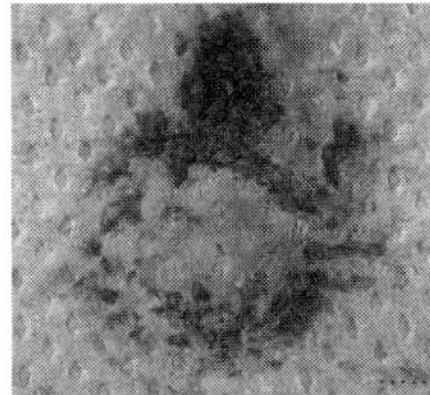

left anorexia (demonstrated here by the excretion of urates with no feces; and
right excretion of yellow-green urates suggestive of the massive liver necrosis caused by the virus.

Central nervous system signs like opisthotonus may be noted as a bird nears death.

has been described in some affected birds.[65,138]

Birds exposed to virulent strains of PDV usually develop clinical signs or die within three to ten days after being exposed to the virus. Most birds die within two days of showing clinical signs. The highest death rates occur when susceptible populations of birds in crowded or stressful aviary conditions are exposed to virulent strains of virus *Figure 7.2*.[30,67,131,166] For example, in a particularly aggressive outbreak at an import station, close to 7,000 birds with representatives of eleven species died.[67] The mortality rates associated with various outbreaks are listed in *Table 7.4*

Less virulent strains of PDV may cause disease outbreaks that differ from those classically described for virulent strains of the virus. Only a few at-risk birds may die, and others may recover following five days to several weeks of mild respiratory distress, diarrhea or polyuria. Any infected bird that survives a PDV infection is thought to become a latently infected carrier of the virus *Figure 7.3*.[64]

EXPERIMENTAL INFECTIONS

Experimental PDV infections can be induced by exposing birds to virus through the oral, intramuscular, intranasal or intraocular routes.[33,67,131,165,166,173,215] The gross and histologic changes that occur in experimentally infected birds are similar to those that occur with natural infections.

In one study, Budgerigars experimentally inoculated with PDV by the intramuscular or intranasal route remained asymptomatic, or died acutely without clinical signs five to seven days after they were exposed to the virus.[33,131,165,166] A Half-moon Conure died five days after

TABLE 7.3

Serologic relationships of avian herpesviruses*

Serotype 1	Gallid HV 2 (Marek's disease virus) Meleagrid HV 1 (turkey herpesvirus)
Serotype 2	Anatid HV 1 (duck plague virus)
Serotype 3	Gallid HV 1 (infectious laryngotracheitis virus) Amazon tracheitis virus
Serotype 4	Psittacid HV 1 (Pacheco's disease virus)
Serotype 5	Psittacid HV 2 (psittacine herpesvirus, Pacheco's-like disease)
Serotype 6	Psittacid HV 3 (psittacine herpesvirus, Pacheco's-like disease)
Serotype 7	Columbid HV 1 (pigeon herpesvirus) Falconid HV 1 (falcon herpesvirus) Strigid HV 1 (owl herpesvirus) - One isolate from a psittacine bird was serologically identical to pigeon herpesvirus. - Budgerigar herpesvirus is serologically related to pigeon herpesvirus.
Serotype 8	Columbid HV 2
Serotype 9	Accipitrid HV 1 (eagle herpesvirus)
Serotype 10	Phalacrocoracid HV 1 (cormorant herpesvirus)
Serotype 11	Perdicid HV 1 (Bobwhite Quail herpesvirus) Gruid HV 1 (crane herpesvirus)
Serotype 12	Ciconiid HV 1 (stork herpesvirus)
Ungrouped herpesvirus isolates	Estrildid HV 1 (finch herpesvirus) Plodeid HV 1 (Weaver Finch herpesvirus) Serinid HV 1 (Canary herpesvirus) Toucan herpesvirus (may be serologically distinct from Pacheco's disease virus) Pigeon herpes encephalitis virus (possible isolate from Spain) Penguin herpesvirus Herpesvirus of *Neophema* sp. Herpesvirus of *Psittacula* sp. Herpesvirus from wart- and plaque-like lesions on the skin of psittacine birds

*The viruses listed within each serotype are antigenically related. The serotype numbers are arbitrarily assigned for purposes of clarity. HV = herpesvirus (adapted from [96,105]).

being exposed to aerosolized virus obtained from an affected cockatoo. A second conure was placed in the same enclosure with the experimentally infected bird, twelve hours after exposure to the virus. The former bird died three days later. A Budgerigar and a Quaker Parakeet experimentally infected by the intraocular route developed clinical signs of disease seven days following exposure to the virus and died two days later.[67] Budgerigars placed in the same airspace with experimentally infected birds were found to be infected and shed virus, but remained asymptomatic.[215] Budgerigars inoculated intramuscularly with PDV died six days later without showing any clinical signs.[131] Another group of Budgerigars died three to four days after being injected intramuscularly with the virus.[80] A group of experimentally infected Quaker Parakeets died from 6 to 14 days (average 8.7) after they were given PDV orally.[65]

The original Pacheco's disease virus isolate was experimentally infectious for parakeets and one- to two-day old chickens, but was not infectious for pigeons, canaries or older chickens.[56,142] Pacheco's disease virus recovered from affected rosellas did not induce disease or a detectable antibody response when given intramuscularly to quail or chicks.[80] Budgerigars and Amazon parrots also can be experimentally infected with falcon herpesvirus, even though this virus is not antigenically related to PDV.[128,130,173,192]

The route of virus entry into a susceptible bird appears to have some effect on the clinical progression of disease. For

FIG 7.3
The single biggest problem in managing PDV is that normal appearing birds can be latently infected and intermittently shed the virus. For the safety of others, any bird that survives a PDV infection should be considered a carrier.

example, a herpesvirus recovered from a racing pigeon was not infectious for Cockatiels exposed by aerosol, but Cockatiels injected with the virus intravenously or intraperitoneally died in four to six days. Budgerigars infected with a pigeon herpesvirus by the intravenous route developed tremors and shallow breathing seven days postexposure and died the following day.[173] Budgerigars exposed to pigeon herpesvirus by the intranasal route were

TABLE 7.4

Mortality rates in various Pacheco's disease virus outbreaks[4,66,100,166]

Species	Mortality Rate	Duration of Outbreak
Macaws and cockatoos	12 of 15 (80 %)	—
Mixed psittacine birds	24 of 75 (32%)	19 days
Amazon parrots, macaws and cockatoos	12 of 29 (41%)	—
Mixed psittacine birds	19 of 43 (44%)	6 days
Mixed psittacine birds	46 of 250 (18%)	11 days

shown to develop systemic infections with severe hepatitis.[192]

EPIZOOTIOLOGY

Pacheco's disease was first reported in birds from Brazil and it is likely that the virus originated in this country.[142] Additionally, many of the initial reports detailing PDV outbreaks in the United States were associated with recently imported birds originating from South America.[67,166] Today, Pacheco's disease virus occurs worldwide. Isolates of the virus have been reported from psittacine birds in the United States,[67,166] Kenya,[107] New Zealand,[53] Argentina, Singapore, England,[131] Brazil,[109] Japan,[186] Spain,[148] Switzerland,[54] Belgium[133] and Germany.[150]

Typically, herpesviruses have a restricted host range, and natural PDV infections have been documented only in Psittaciformes *Table 7.5*. Other species of birds (hornbills, mynah birds, finches, canaries) that have been naturally exposed to PDV-infected Psittaciformes during outbreaks have been shown to remain clinically normal, suggesting that these species are not susceptible to the herpesviruses that cause Pacheco's disease.[56,166]

Herpesviruses can be recovered from clinically affected as well as asymptomatic psittacine birds. Over an eight-year period, PDV was recovered from 25 of 2,274 (1.1%) lots of psittacine birds imported into the United States.[162] Of 500 psittacine birds examined during an 11-month period in a veterinary practice in Europe, a herpesvirus was recovered by culture from the cloaca of 4 of 44 (0.8%) birds and from the trachea of 14 of 44 (2.8%) birds.

Virus was recovered from both body locations in 11 of 44 (2.3%) birds. Many of these birds were asymptomatic.[144] In a study in Japan, a herpesvirus infection (not all isolates were strains of PDV) was diagnosed in 21 of 241 (8.7%) psittacine birds that died within two weeks of being released from quarantine. A herpesvirus was demonstrated in three Amazon parrots, two Cockatiels and two rosellas with changes characteristic for Pacheco's disease, and in fourteen parakeets with respiratory disease.[186] During a five-year study in Austria, herpesvirus inclusion bodies were demonstrated in the liver of 4 of 1,426 (0.3%) of the psittacine birds examined.[60]

SUSCEPTIBILITY — It is postulated that virulent strains of PDV can infect psittacine birds of any age; however, susceptibility to disease has been shown to vary among genera and within the same species. There is currently no precise, discernible pattern to disease susceptibility, but Old World psittacines appear to be more resistant to disease than New World psittacines.[69] Some species of macaws, Amazon parrots and conures are considered highly susceptible, whereas others are relatively resistant.[112,113] Infected lovebirds, Quaker Parakeets, pionus parrots, Amazon parrots and some cockatoos typically die without showing clinical signs; infected macaws and other cockatoos may exhibit several days to several weeks of depression, anorexia, regurgitation or diarrhea before death. Some infected cockatoos have been known to survive following a prolonged (several-week) illness.[62,64]

TABLE 7.5

Psittacine birds reported to be susceptible to Pacheco's disease virus

Ambiona King Parrots	*Aprosmictus* sp.
African Grey Parrots	Amazon parrots
Barnardius sp.	Budgerigars
Caiques	Canary-winged Parakeets
Cockatiels	Cockatoos
Conures	Eclectus Parrots
Lories (genus *Eos*)	Lovebirds
Macaws	*Neophema* sp.
Pionus sp.	*Poicephalus* sp.
Polytelis sp.	*Psittacula* sp.
Quaker Parakeets	Rosellas

Individual species of conures vary widely in their susceptibility to PDV-induced disease. In natural outbreaks, Mitred and White-eyed Conures may be mildly affected and survive; high mortality can occur in Peach-faced, Sun and Orange-fronted Conures.[64] In one aggressive outbreak involving approximately 7,000 birds in a quarantine station, every one of the exposed cockatoos, Canary-winged Parakeets, Quaker Parakeets, Amazon parrots, Hyacinth Macaws and Green-winged Macaws died, while none of the exposed Nanday and Patagonian Conures were affected. In the same outbreak, all of the Half-moon Conures died, but 80% of the White-eyed Conures (which are in the same genus *Aratinga*), and 20% of exposed Blue and Gold Macaws and Cockatiels survived.[67] In another outbreak involving a variety of psittacine birds, none of the exposed "parakeets" developed clinical signs of disease.[183]

RELATIONSHIP OF VIRUS STRAINS — It is likely that numerous serologically distinct, host-adapted herpesviruses occur in specific families or genera of psittacine birds. For example, although at one time, Pacheco's disease was thought to be caused by a single, serologically uniform strain of herpesvirus, herpesviruses recovered in Europe from Amazon parrots and a Mustached Parakeet were serologically unrelated to other psittacine herpesviruses or to each other.[113] It now appears that at least three different serotypes of avian herpesvirus can produce lesions that may be diagnosed as Pacheco's disease.[95,97,100,112,113,185] Likewise, some herpesviruses isolated from psittacine birds indigenous to South America, Africa or South Asia have been shown to be distinct from each other, and each may represent a strain of virus with a preferential host specificity.[84,103]

Because herpesvirus-induced disease is severe in some psittacine birds while others infected with the same virus remain asymptomatic, it is possible that severely affected individuals are being infected with a non-host-adapted herpesvirus. The nucleic acid composition of herpesviruses isolated from a Double-yellow Headed Amazon Parrot and an African Grey Parrot are markedly different, and these differences in the genome of the evaluated viruses could easily account for the variance in their effects.[1]

The evidence of the existence of many different host-adapted herpesvirus strains continues to mount. A virus-neutralization assay showed that an isolate of PDV was not serologically related to either pigeon herpesvirus (PHV) or infectious laryngotracheitis virus.[131] However, another strain of herpesvirus isolated from psittacine birds was found to be related to pigeon herpesvirus,[200] and some Budgerigars and Cockatiels have been shown to be experimentally susceptible to a herpesvirus recovered from a racing pigeon.[173] A herpesvirus isolated from a psittacine bird in Belgium was found to be antigenically identical to pigeon herpesvirus, but different from two other strains of PDV.[200]

Some herpesvirus isolates from psittacine birds are infectious to pigeons, while other strains do not infect pigeons.[142,195,200,208] In one study, an isolate of herpesvirus from a racing pigeon caused experimental infections in Budgerigars, but did not induce disease in pigeons.[173] Collectively, these findings suggest that some strains of pigeon herpesvirus are capable of infecting both pigeons and psittacine birds that are susceptible to PDV. It is theorized that psittacine herpesviruses that are antigenically related to pigeon herpesvirus could spread from species to species in mixed aviaries or zoological collections.[195,200] Although falcon herpesvirus and PDV are not antigenically related, Budgerigars experimentally infected with falcon herpesvirus will develop infections.[128,192]

INCUBATION — Most psittacine birds affected by PDV die within several hours to two days of showing clinical signs, but actual exposure to the virus probably occurred several days earlier.[67] In one natural outbreak, four exposed macaws and eight exposed cockatoos died five to seven days after a group of PDV-infected macaws were introduced to an aviary. In another natural outbreak, susceptible birds died 12 to 14 days after a group of infected birds were added to the aviary.[166]

Experimentally, birds exposed to virulent strains of PDV may develop clinical signs or die within 3 to 14 days after being exposed to the virus.[65,81] The mean time of death in Quaker Parakeets experimentally infected by the intramuscular route was six days.[138] Budgerigars experimentally inoculated with PDV by the intramuscular,

intranasal and aerosol routes either remained asymptomatic or died acutely without clinical signs four to eight days after viral exposure.[33,81,131,166] It should be noted that many of the incubation periods that have been suggested are based on infections with uncharacterized strains of psittacine herpesvirus. The incubation period, or progression of disease, may vary with different viral strains.

TRANSMISSION — Viral spread depends on many factors: the hygiene in the aviary, the species of exposed birds, the distance between enclosures, the strain of the virus and the condition of the flock. Thus, a given PDV outbreak may cause devastating losses or be limited to a single bird in a pair.[62,66,125,132,166] Crowding, poor air circulation, accumulation of excrement and stacking of enclosures increase the likelihood of PDV transmission from an infected to a susceptible bird *Figure 7.4*. Virulent strains of Pacheco's disease virus can spread rapidly through a susceptible flock, with exposed birds becoming clinically ill or dying within two to three days after the index case.[69]

Psittacine birds with clinical signs of Pacheco's disease shed large concentrations of virus in their feces and in pharyngeal secretions, particularly in the last few days of life *Figure 7.5*. Pacheco's disease virus was recovered from the pharynx and cloaca of experimentally infected conures three days after they were exposure to the virus, and for 48 hours prior to death. Virus was recovered from the same anatomic locations in Budgerigars and Quaker Parakeets on post-exposure days 7, 8 and 9. The affected birds died on day 9; thus, shedding was occurring for up to 72 hours prior to death.[67] In another study, virus was shed in the feces of infected Budgerigars in high concentrations (1,000,000 to 10,000,000 virions/gram of feces) within 48 hours of infection; importantly, surviving birds continued to shed virus in the feces for over two weeks.[215]

Using viral-specific antibodies, researchers identified PDV in the liver and intestines of a naturally infected macaw and a cockatoo, suggesting that virus shed in the feces may originate from either of these organs.[148] Inhalation of virus-contaminated aerosols has been implicated in some natural outbreaks and has been shown to occur experimentally.[62,215] Concentrations of virus in the lungs of infected birds can reach 1,000,000 virions/gram of lung. This suggests that virus particles originating in the respiratory tract could be released when a bird coughs, sneezes or exhales.[215]

The frequent detection of PDV antibodies in clinically normal psittacine birds suggests that latent infections are common. In one survey to determine the prevalence of PDV antibodies, 24 of 70 (34%) Amazon and African Grey Parrots maintained in single-bird households had virus-neutralizing antibodies.[113] This would indicate that these solitary birds had been previously infected by PDV. Both imported and domestically raised psittacine birds of several species including Nanday and Patagonian Conures, macaws and Amazon parrots have been implicated as asymptomatic

Pacheco's disease virus is shed primarily in the feces, and birds are most likely to be infected through ingestion of the virus. Thus, birds at greatest risk of infection are those crowded into small areas where feces accumulates or contaminates food or water.

Stacking enclosures also favors the transmission of PDV through a group of birds.

photographs courtesy of Deanna Shafar

carriers.[2,62,63,66,125,169] It is safest to assume that any bird that recovers from a PDV infection is latently infected.

The routes by which latently infected birds intermittently shed virus have not been confirmed.[62] When latently infected birds are involved in an aviary outbreak, the pattern of morbidity and mortality frequently involves birds in direct contact with a suspected carrier *Figure 7.6*. However, once an outbreak has gained momentum, the exact source of PDV is usually difficult to identify. Many aviculturists will incorrectly blame a new group of birds that have recently been added to a flock for introducing PDV to the aviary. This may be an incorrect assumption. It is just as likely that the addition of new birds to an otherwise stable flock

caused sufficient stress in an established aviary resident to initiate shedding of the virus. Stress factors such as other infections, malnutrition, reproductive activity, a change in temperature or a change in location also could cause latently infected birds to shed virus. Virus shedding occurs most commonly in the absence of clinical signs.

While it is common for an outbreak to occur when a carrier in the aviary begins to shed virus, PDV could also be introduced to an aviary following human contact with birds that are shedding the virus, or from returning a bird to the aviary that has been in contact with

symptomatic or asymptomatic viral shedders *Figure 7.7.*[63,132] Practicing sound personal hygiene and avoiding exposure of any bird in the aviary directly or indirectly to other birds will prevent these methods of PDV introduction into a flock or nursery.

It is theorized that some latently infected hens can pass the virus (and antibodies to the virus) to their eggs. The resulting chicks would be latently infected carriers that might not develop detectable levels of antibodies.[64]

More than one serologically distinct psittacine herpesvirus can cause clinical, gross or microscopic changes suggestive of Pacheco's disease.[113] Affected birds may have no gross lesions or may have abnormal changes in the liver, spleen, kidney and intestines. Frequently, birds that die from PDV are in excellent general condition, and the crop and remainder of the gastrointestinal tract may be full of ingesta, indicating how rapidly the virus killed the bird. Specific lesions that have been noted in association with PDV infections include an enlarged, swollen, reddish-to-green or yellowish-brown liver with areas of hemorrhage; splenomegaly with or without areas of hemorrhage; enlarged kidneys; hyperemia or hemorrhage of the interior and exterior surfaces of the intestines; blood in the abdominal cavity; congested blood vessels in the brain; and blood in the lumen of the intestinal tract *Figure 7.8.* Hyperemia or hemorrhage in association with the intestinal tract appears to be a common finding with peracute deaths.[64,113,150,165,166] In a PDV outbreak in a Spanish zoo that killed twelve birds of the genera *Amazona*, *Ara* and *Cacatua*, gross lesions included yellowish discoloration of an enlarged liver, bloodstained material in the lower digestive tract, small areas of hemorrhage on the lining of the lower esophagus and enlargement and dark discoloration of the kidneys and spleen.[100]

Histologic changes associated with PDV infections include necrotizing lesions in many organs as well as hemorrhage and congestion of the liver, spleen and kidney. The necrosis and hemorrhage may be severe, even though a minimal inflammatory response is noted. Other microscopic changes reported in birds that die from PDV include necrosis in the cells that line the bronchi, hemorrhagic enteritis, rhinitis, pneumonia and air sacculitis.[64,165,166] Necrosis of the pancreas was the only microscopic change noted in Quaker Parakeets that were experimentally infected with PDV.[138]

The presence of intranuclear inclusion bodies (Cowdry type A) are considered suggestive of a herpesvirus infection. Inclusion bodies are most commonly found in the liver, but have also been demonstrated in the kidney, spleen, pancreas, small intestines, parathyroid, ovary and bone marrow of some PDV-infected birds.[186] Inclusion bodies may be difficult to find in birds that die soon after being infected with PDV and do not manifest clinical signs of disease.[33,69,166,173] A herpesvirus infection should be considered in any psittacine bird that dies with characteristic clinical and gross changes, irrespective of whether or not inclusion bodies can be detected. It should also be noted that the clinical, gross and microscopic changes associated with PDV can appear similar to those caused by an avian polyomavirus infection (see Diagnosis).

PATHOGENESIS

Host-adapted herpesviruses generally cause a mild, subclinical, latent infec-

Many aviary outbreaks of PDV start when a latently infected bird begins to shed virus; however, the virus could also be introduced to the aviary through the addition of new birds, by visitors, by boarding birds or following any human contact with birds that are shedding the virus. Birds that leave an aviary and are exposed to any other birds should be quarantined before being added back to the flock.

tion that persists for a lifetime. Non-host-adapted strains of herpesvirus can cause severe, life-threatening infections that are commonly associated with lysis of infected cells. The severe liver disease associated with PDV infections in some psittacine birds results from lysis of infected liver cells.

As is the case with herpesvirus infections in mammals, the primary spread of virions within an infected bird's body occurs by direct cell-to-cell contact without exposure of extracellular spaces to the virus see Figure 3.22.[113] This mechanism protects the virus from antibodies that the immune system secrets into the fluids that surround a cell. Herpesviruses are able to use this mechanism to establish persistent infections in a bird, even though it may develop antibodies to the virus.

IMMUNITY

Birds that recover from a PDV infection probably develop latent infections as well as long-lasting immunity that protects them against the severe changes that can be induced by the same strain of virus.[64] Birds that recover from PDV infections may develop low, but detectable, levels of virus-neutralizing antibodies. Precipitating antibodies have been shown to develop early during an infection but decrease in abundance within months.[63,64,113] A detectable increase in antibodies may occur in some

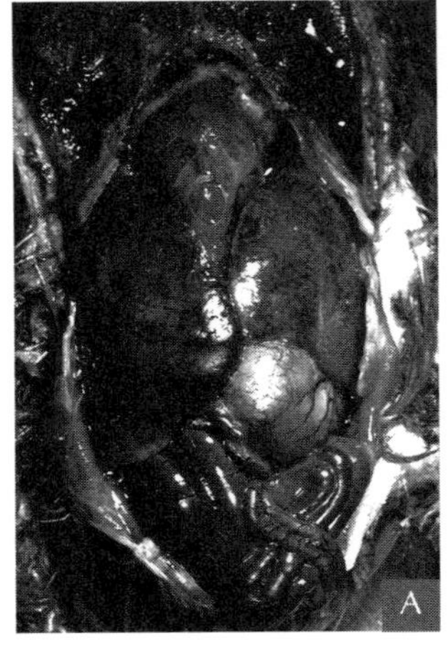
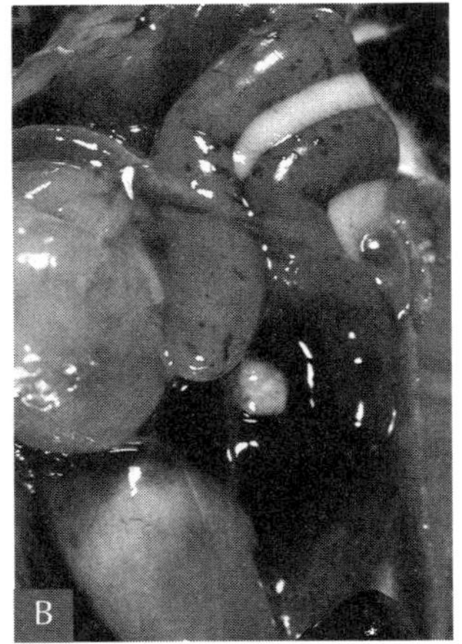
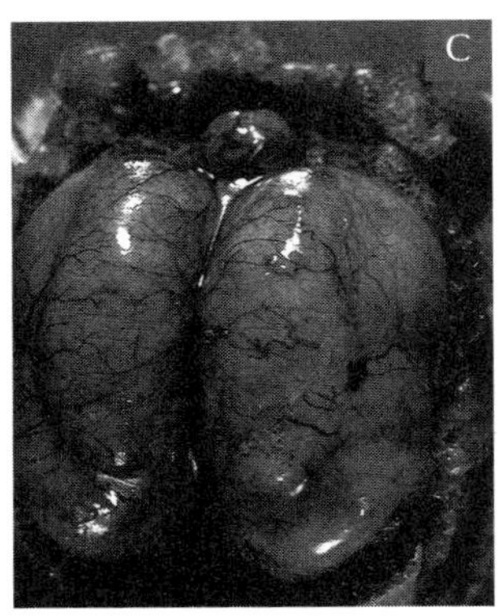
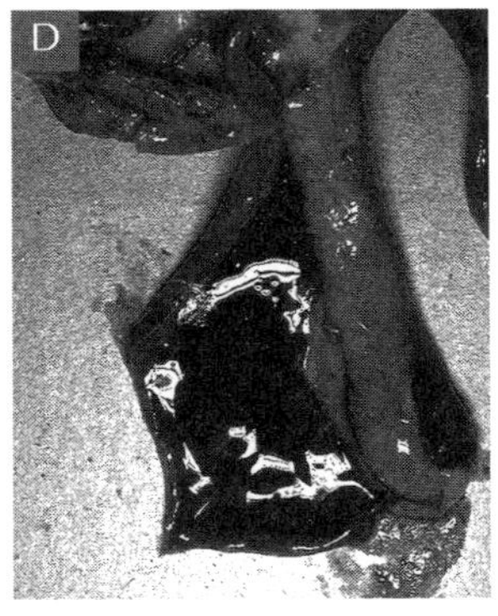

FIG 7.8

The gross lesions associated with PDV infections are generally a result of the severe necrosis caused by the virus.

A The most common change noted at necropsy is an enlarged, swollen reddish to green or yellowish-brown liver with areas of hemorrhage. Other changes that also may occur include:

B accumulation of blood in the abdominal cavity;

C hyperemia of the brain; and

D accumulation of blood in the intestinal tract. Many of the gross changes associated with PDV infections are similar to those caused by avian polyomavirus.

photographs courtesy of Scott McDonald

latently infected birds when they are actively shedding the virus.[64,81] However, a decrease in the antibody level with no detectable increase in viral shedding also has been demonstrated to occur in latently infected birds.[63]

It is theorized that hens with PDV antibodies can pass protective immunity to their chicks. These chicks can remain clinically normal, be latently infected and be neutralizing antibody-negative when maternal antibodies derived from the yolk sac wane.[64]

DIAGNOSIS

A PDV infection should be suspected in any bird that dies suddenly without clinical signs, dies following a brief period of depression associated with passage of yellowish-green urates or has an enlarged, necrotic liver with hemorrhagic lesions in internal organs.

Bacterial hepatitis, chlamydiosis, salmonellosis, lead poisoning, avian polyomavirus, reovirus and adenovirus can cause clinical and gross changes similar to those noted with PDV.

The microscopic demonstration of severe liver necrosis is common in birds that die from PDV. Intranuclear inclusion bodies may be noted in the liver, kidney or spleen, but may be difficult to detect in birds that die soon after they are infected with PDV.[33,69,166,173,186] Additionally, the intranuclear inclusion bodies induced by PDV can appear microscopically similar to those caused by avian polyomavirus or adenovirus. Confirmation that a bird died from PDV requires demonstration of the virus in infected tissues by electron microscopy, cell culture, viral-specific DNA probes or viral-specific antibodies.[148] Sections of the liver, spleen, kidney and brain of infected birds are the best tissues to submit for herpesvirus isolation on cultured cells.[112] Viral-specific DNA probes designed to detect PDV have been developed.[149]

Psittacine birds that survive herpesvirus infections are thought to become latently infected, although the site of viral persistence has not been demonstrated. Some birds that are infected with PDV may survive and develop antibodies to the virus; however, antibodies to PDV develop inconsistently, and many assays may or may not be of value in determining whether a bird is latently infected. Because antibodies to the virus are difficult to detect, finding them is clearly an indication of a previous infection. If the antibody titer is high, it probably indicates that an active infection has recently occurred. For the safety of other birds, those with antibodies should be considered latently infected. Precipitating antibodies have been shown to develop early during an infection but wane within months. Detection of precipitating antibodies has been suggested as a method to document an active or recent (within months) PDV infection.[62,63]

It has been suggested that latently infected birds might be identified by vaccinating a bird with a inactivated vaccine, followed by collection of serum seven days later; birds that had previously been infected would develop a rapid, high titer (> 32), whereas the antibody levels in birds that had not been previously infected is generally not detectable.[74a]

In clinically affected birds, measuring certain enzyme activities in the blood might provide another diagnostic avenue. An increase in serum alanine transaminase (ALT) activity is common when the liver is severely damaged. An increase in ALT activity was detected in most Quaker Parakeets experimentally infected with PDV, probably as a result of severe liver necrosis induced by the virus.[70]

CONTROL

Most birds are infected with PDV after they ingest contaminated excrement. Thus, aviary hygiene is critical in preventing PDV outbreaks. Herpesviruses are enveloped, and thus are generally unstable when outside of a bird's body and can be easily inactivated. Most herpesviruses are sensitive to common disinfectants, high heat (56°C for 5 to 10 min or 37°C for 22 hrs) or acid conditions (pH less than 5). As with most viruses, organic debris such as blood, soil, nesting material or feces would be expected to protect PDV from disinfectants that do not contain detergents *Figure 7.9*.

Because clinically affected birds shed large quantities of virus in their feces and respiratory secretions, any bird suspected of having Pacheco's disease or having been exposed to PDV should be maintained in strict isolation in a separate location from the aviary.

FIG 7.9
Disinfectants would be ineffective in the presence of organic debris.

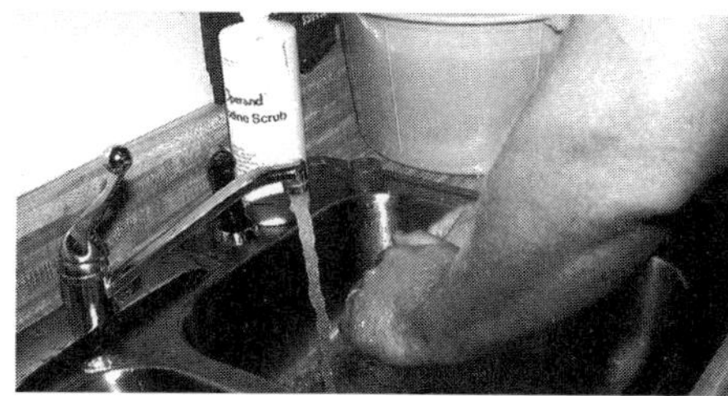

Superior personal and aviary hygiene are critically important to preventing the spread of PDV through an aviary.

Pacheco's disease virus is awkward to manage in an aviary because of the difficulty in identifying latently infected birds.[66] These birds can intermittently shed the virus, and repeated attempts to demonstrate shedders through virus isolation procedures or electron microscopy are time-consuming, expensive and provide inconsistent results. Although viral-specific DNA probes that will detect PDV have been developed, a bird must be shedding virus for the test to be effective.[137a]

Sentinel birds are of limited value in detecting latently infected birds because of varied susceptibility among species, the intermittent nature of virus shedding and the uncertainty that a sentinel bird will be sufficiently exposed to a latently infected bird when the latter is shedding the virus. In one study, Quaker Parakeets that survived infection were placed with birds with no known exposure to PDV. The immunologically naive test birds remained clinically normal for two years.[138] Even after being injected intramuscularly with corticosteroids to suppress their natural immunities, neither Quaker Parakeets that were experimentally exposed to PDV nor a group of psittacine birds that were naturally exposed to PDV were found to be shedding virus in their feces or pharyngeal secretions.[67] This finding suggests that virus shedding from latently infected birds may not occur following all "stressful" events.

Historically, antibody screening has been considered of minimal practical value in controlling infections within an aviary.[63,64,113] However, theories that predict a rapid increase in antibody levels following vaccination in previously infected birds may improve the value of antibody assays for identifying latently infected birds (see Diagnosis).[74a]

VACCINATION — Vaccines developed for the herpesviruses that infect mammals are generally effective in reducing the severity of clinical signs and the degree of viral shedding; however, these vaccines rarely prevent an exposed animal from developing a latent infection. The herpesvirus vaccines designed for use in birds also probably protect a vaccinate from severe disease, but fail to prevent latent infections. Vaccinated Quaker Parakeets, for example, are protected from virus-induced disease, but still will shed virus following challenge.[65] This finding would suggest that vaccinated birds are protected from disease, but not infection. Newer human herpesvirus vaccines that prevent both disease and latent infections are being developed. Application of the same principles to the development of an avian vaccine holds the promise of preventing avian herpesvirus infections.

Herpesvirus vaccines are available for pigeons, psittacine birds and domestic fowl. The vaccines produced for pigeons and domestic fowl are of no value in psittacine birds because the viruses that infect these species are serologically different from the strains that infect psittacine birds. Other herpesvirus vaccines such as those for Marek's disease virus, infectious laryngotracheitis virus and duck plague virus do not protect psittacine birds from PDV.[67,165]

Studies of experimental PDV vaccines have shown some protection. An inactivated PDV vaccine protected a select group of psittacine birds following experimental challenge. The serologic response following vaccination in these birds was minimal, and survivors of viral challenge did not develop a high level of virus-neutralizing antibodies.[81] In another study, vaccinated Quaker Parakeets were protected from virus challenge even though neutralizing antibody titers were low (1:8 to 1:64).[65] A PDV vaccine developed in Europe was found to induce detectable levels of virus-neutralizing antibodies that remained detectable in some birds for over seven months; however, whether these vaccinated birds were protected by virus challenge was not determined.[100]

Based on the protective nature of these experimental vaccines, several inactivated PDV vaccines were registered by the USDA for use in psittacine birds in the late 1980s. An oil emulsion was used as an adjuvant. These vaccines are thought to provide good protection against disease, but the oil adjuvant has been associated with reactions in a small number of vaccinates, particularly cockatoos. Reported adverse reactions include formation of persistent abscesses when the vaccine is injected under the skin, or muscle necrosis when the vaccine is given intramuscularly. In some vaccinates, these lesions were not noted for several months after vaccination; in others, lesions were noted within several weeks of administration of a booster vaccine.[39,40,61,156] Because of this potential for side effects, it is best to use the oil containing PDV vaccine only in high risk situations such as quarantine stations, retail outlets, broker operations and birds exposed during an outbreak. When indicated by the specific needs of a group of birds, use of these vaccines can prevent catastrophic losses. The potential for side effects following subcutaneous injection should be considered negligible in comparison to the damage that can be induced by an outbreak in susceptible birds.

When using any vaccine, the guidelines provided by the manufacturer should be carefully followed. At the time of printing, it was recommended that Biomune's inactivated PDV vaccine be administered to young birds two to three weeks before fledging, followed by a booster vaccination four to eight weeks later, and yearly thereafter. When used according to the manufacturer's recommendations, the vaccine should provide at least seven months of protection.

An outbreak in which 24 of 75 (32%) exposed birds died of PDV over a 19-day period was stopped by administering acyclovir in combination with vaccination.[4] However, vaccinating birds during an outbreak can be dangerous because humans could theoretically serve as mechanical vectors for PDV. The attending clinician must exercise extreme caution to prevent aviary personnel or the vaccination procedure itself from spreading the virus from bird to bird. Concurrent treatment with acyclovir may reduce the chances of virus transmission through the flock.

TREATMENT — In humans, treatment with acyclovir has been shown to shorten the duration of viral shedding, decrease the degree of morbidity and reduce the redundancy of some herpesvirus infections.[13,152] Acyclovir is widely distributed in most tissues, including the brain, and is excreted principally through the kidneys. Unfortunately, acyclovir has not been demonstrated to eliminate latent infections in humans.[140,172]

In psittacine birds, acyclovir has been demonstrated to reduce the levels of sickness and death during PDV outbreaks. The role that acyclovir may play in producing or preventing the development of latent infections in psittacines has not been reported.[4,64,143,169] Therapy is most effective when treatment is initiated before PDV-infected birds develop clinical signs of disease. Acyclovir has also been shown to reduce the death rate when treatment is initiated soon after clinical signs are noted.[64,169] In experimental trials, seven of eight birds infected with PDV and then treated with acyclovir survived, while all of the infected birds that did not receive acyclovir died.[62,138] Because acyclovir has been associated with kidney damage in some species, this drug should be used with caution and only when specifically indicated.

Experimentally, acyclovir powder has been shown to be effective in preventing disease when administered by gavage at a dose of 80 mg/kg three times per day. For large collections of birds where individual therapy is not feasible, powdered acyclovir may be added to the food and appears to be nontoxic even at dosages of 240 mg/kg three times per day. Exposed birds should be treated for seven to ten days.[64,138]

Oral administration is recommended for exposed birds that are clinically normal. Exposed birds that are already exhibiting clinical signs of disease, including the excretion of yellowish-colored urates, can be given injectable acyclovir intravenously or subcutaneously at 40 mg/kg every eight hours. Intramuscular administration is not recommended because it usually results in severe muscle necrosis.[64,138,139]

AMAZON TRACHEITIS VIRUS

A herpesvirus that is serologically distinct from Pacheco's disease virus but similar to the infectious laryngotracheitis (ILT) virus that infects gallinaceous birds has been recovered from the trachea of Amazon parrots in Europe.[69,209] A similar virus may affect Psittaciformes in the United States.[143a] Called Amazon tracheitis virus (ATV), it will produce a mild ILT-like disease (necrotic tracheitis) in experimentally infected chickens and pheasants. Based on this finding, it has been suggested that Amazon tracheitis virus may be a variant of ILT.[69,109]

The progression of ATV infections can be peracute, acute or chronic. Like ILT, the AT virus usually causes a fibrinonecrotic tracheitis. Sheets of necrotic cells that accumulate on the surface of the trachea are referred to as diphtheritic membranes. Some infected birds die rapidly following the development of severe, diphtheritic lesions in the airways. Other naturally infected Amazon parrots may develop an upper respiratory disease that can persist for up to nine months.[69] Conjunctivitis, severe dyspnea, sinusitis, coughing and rales are characteristic clinical changes in affected Amazon parrots *Figure 7.10*. The most severe clinical changes appear to occur when secondary bacteria or fungi invade respiratory tissues that have been damaged by the virus. As occurs with ILT in chickens, death in companion birds usually occurs when necrotic debris accumulates in the trachea, causing asphyxiation *Figure 7.11*.

FIG 7.10

Amazon tracheitis virus has been associated with clinical signs involving the upper respiratory tract. These changes include conjunctivitis, severe dyspnea, coughing and rales.

Amazon parrots experimentally infected with the AT virus died three to four days after infection. Birds in close contact with infected birds were found to develop clinical signs and die within six days, suggesting that virions were being shed from the experimentally infected birds within two days after they were exposed to the virus.[69]

The most common gross lesion associated with an Amazon tracheitis virus infection is the formation of diphtheritic membranes in the trachea. Virus isolation should be attempted in any companion or aviary birds with necrotic tracheitis *Figure 7.11*. Histologic lesions include hemorrhage and necrosis in the tracheal epithelium, pseudomembranous tracheitis, pharyngitis, ingluvitis and air sacculitis. The mucosa of the trachea where the virus is replicating dies and sloughs, making the demonstration of viral-induced intranuclear inclusion bodies more difficult. In severe cases, inclusion bodies may be found only within debris located in the tracheal lumen *Figure 7.12*. Microscopic examination of the trachea from an affected bird may reveal a plethora of secondary bacterial or fungal organisms, making diagnosis of the viral infection difficult.[69,209]

The gross and microscopic changes associated with AT virus appear similar to those caused by many other conditions, including bacterial tracheitis or rhinitis, the diphtheritic form of poxvirus, Newcastle disease virus, chlamydiosis, influenza A virus, aspergillosis, trichomoniasis, *Syngamus* and hypovitaminosis A. Diagnosing an AT virus infection in a live bird may be possible by

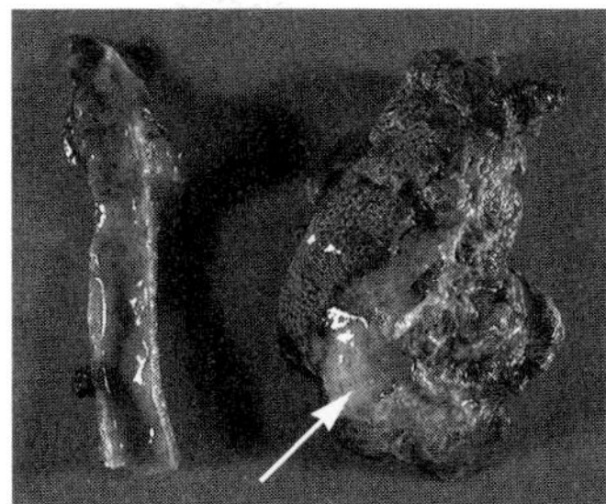

submitting swabs of the trachea or blood-tinged mucus for virus isolation.

Available ILT vaccines have been shown to protect chickens from AT virus, and an experimental AT virus vaccine will partially protect chickens from ILT. No one has yet reported whether an ILT vaccine will protect Amazon parrots or other psittacine birds from AT virus.[69] Vaccination of psittacine birds with an attenuated-live ILT virus vaccine would be considered dangerous.

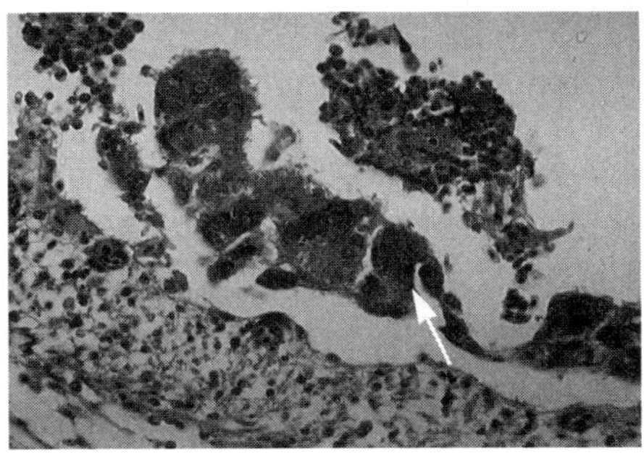

PARAKEET HERPESVIRUSES

A herpesvirus has been shown to be associated with severe respiratory disease and death in a number of parakeets in the genus *Neophema*. Infections have been reported in the United States and Japan. The relationship of the herpesviruses that cause these diseases has not been determined; they are discussed together here because of the similarity in the clinical, gross and microscopic lesions that have been described in affected parakeets.

The first reported case of a herpesvirus-induced disease in the lower respiratory tract of a psittacine bird occurred in a Bourke's Parakeet (*Neophema* sp.) in the United States. The bird was found dead in its enclosure, and intranuclear inclusion bodies containing herpesvirus particles were demonstrated by electron microscopy in the cells lining the lungs.[79] A similar herpesvirus was demonstrated in the lungs and bronchi of other *Neophema* spp. that were found to have proliferative lesions in the mucosa of the trachea and bronchi. Affected birds were found dead or died following a brief period of severe, difficult breathing and central nervous system signs that included "stargazing," stumbling, tremors and torticollis. Accumulations of necrotic debris (pseudomembranes) on the mucosal surface of the nasal passages, pharynx and trachea were characteristic gross findings.

Histologic lesions in affected Bourke's Parakeets included pneumonia, air sacculitis and necrosis of the epithelial cells lining the air passages in the lungs. Herpesvirus-like particles were identified within intranuclear inclusion bodies in the kidneys and in the cells lining the bronchi.[79,124,125] Pancreatitis and inclusion bodies were present in several affected birds. Because it was not possible to recover this virus in cell culture, researchers suggested that the virus was unrelated to infectious laryngotracheitis virus, which is easy to grow in cell culture.[124,125]

In another report, 14 of 67 (21%) parakeets (*Psittacula* sp.) died within two weeks of entering a Japanese quarantine station. The affected birds had been recently imported from India. The cells lining the lungs and air sacs were necrotic, and intranuclear inclusion bodies containing herpesvirus-like particles were identified in the lungs and air sacs of six of the dead birds. Only one of the 14 affected birds had inclusion bodies in the cells lining the trachea. Many affected birds had concomitant infections: 13 of 14 had chlamydiosis, 6 of 14 had aspergillosis and 3 of 14 had candidiasis. The herpesvirus demonstrated in these parakeets was thought to be distinct from ILT virus based upon the tissues affected (ILT virus typically affects the trachea, not the lung or air sacs), the host reaction and the cellular changes induced by the virus.[186]

HERPESVIRUSES ASSOCIATED WITH WART-LIKE SKIN LESIONS

A herpes-like virus was identified by electron microscopy in wart-like lesions on the feet of cockatoos and macaws. Lesions in cockatoos are commonly papillary and horny, whereas lesions in macaws are frequently flat and plaque-like *Figure 7.13*. These lesions can persist for years without causing clinical problems. The fact that some affected birds had lesions at the time they were imported into the United States suggests that the herpesvirus demonstrated in these growths was present in the birds' country of origin. Companions of birds with wart-like lesions have not been reported to develop similar changes, suggesting that this herpesvirus has a low level of infectivity. The relationship of this herpes-like virus to others described in companion birds has not been reported.[124,125]

HERPESVIRUSES IN EUROPEAN BUDGERIGARS

A herpesvirus that is serologically unrelated to those isolated from other psittacine birds has been recovered from Budgerigars in Europe. As yet, there have been no reports of this herpesvirus affecting Budgerigars on other continents. Budgerigar herpesvirus is serologically related to pigeon herpesvirus, but is distinct from Pacheco's disease virus. This virus is commonly described in English show Budgerigars with feather abnormalities (referred to as "feather dusters"). What role, if any, that this herpesvirus plays in the occurrence of these feather abnormalities is unknown. Thought to be egg transmitted, this virus has been demonstrated in dead-in-shell chicks and is considered a major cause of early embryonic death (decreased egg hatchability) in affected flocks. In naturally infected birds, virus can be isolated from abdominal organs, feces and blood, indicating a viremia that results in virus distribution in numerous tissues.[69]

OTHER PSITTACINE HERPESVIRUSES

A herpesvirus was recovered from a 12-year-old Yellow-headed Amazon Parrot that died following a three-day history of nonspecific illness. The bird had been housed in a pet shop, where it was directly exposed only to Budgerigars. Intranuclear inclusion bodies were detected only in the crop and pancreas. The liver and spleen had microscopic changes consistent with chronic inflammation.[124]

An uncharacterized herpesvirus was associated with pancreatitis in an Ama-

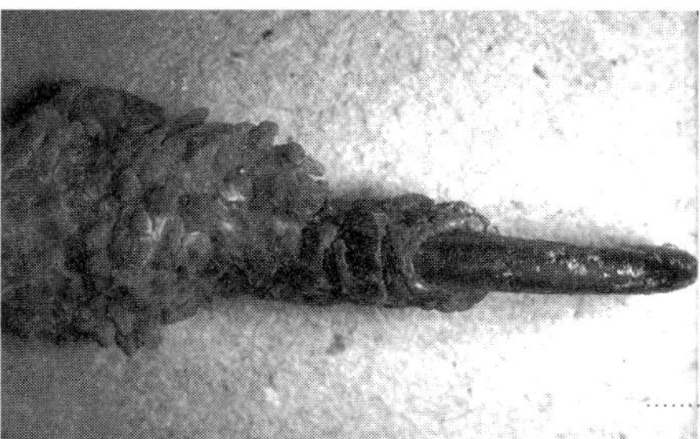

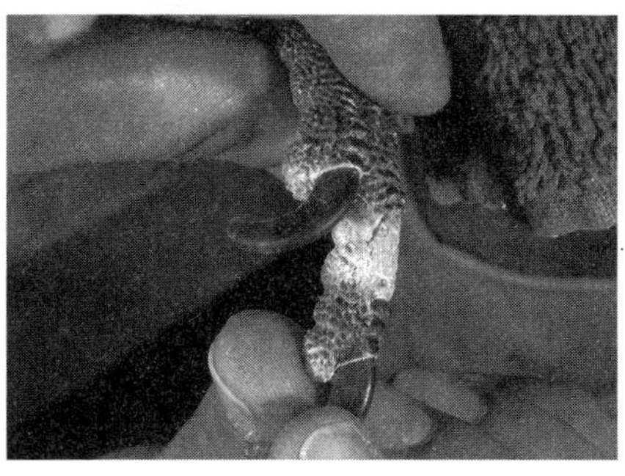

FIG 7.13

Herpesvirus-like particles have been demonstrated by electron microscopy in proliferative lesions on the feet of cockatoos and macaws. The lesions tend to be: papillary in cockatoos; and

flat and depigmented in macaws. Therapy is not necessary for these lesions unless they are secondarily infected or cause problems with normal movement or perching.

photographs courtesy of Scott McDonald

zon parrot.[68] Additional strains of herpesvirus that infect psittacine birds will probably be identified.[64,113,125]

HERPESVIRUSES IN COLUMBIFORMES

PIGEON HERPESVIRUS

Pigeon herpesvirus (PHV-1) is also referred to as "infectious esophagitis" or "inclusion body hepatitis" of pigeons based on the gross or histologic lesions that can be associated with infections. Although all strains of pigeon herpesvirus that have been compared have been shown to be antigenically related, differences in the progression of disease have been associated with varying strains. Some strains can cause severe morbidity and mortality in susceptible pigeons, while other strains are considered apathogenic.[98]

CLINICAL FEATURES

The clinical signs associated with PHV-1 infections vary in severity, with young birds being most commonly affected. Initial clinical changes suggestive of a PHV-1 infection include mild rhinitis and conjunctivitis. These may progress to a moderate to severe upper respiratory disease characterized by dyspnea.[36,37,72] Small ulcerated areas may be noted on the mucus membranes of the larynx, pharynx, cere and at the commissure of the beak of some affected pigeons.[24,37] Young birds may develop other clinical signs that are indicative of a systemic infection, including depression, anorexia, swelling and protrusion of the nictitating membrane, greenish diarrhea, vomiting, polyuria and neurologic signs.[24,36,37,197,204]

In some young birds, death may occur several hours after the first clinical changes are observed.[24] Clinical disease is most severe in young, debilitated pigeons that develop concomitant infections.[52] Secondary bacterial, fungal or chlamydial infections can cause severe mucopurulent rhinitis or conjunctivitis. In an outbreak in young pigeons (ranging in age from six weeks to four months) 8 of 12 birds died over a six month period. Some affected pigeons died with no premonitory signs while others were anorectic for up to one week prior to death.[116]

Older pigeons infected with PHV-1 generally develop mild or subclinical infections that may go unrecognized unless the bird is immunosuppressed by concomitant disease or environmental factors. Most older pigeons infected with the virus recover uneventfully within one to two weeks. Latently infected pigeons can develop severe disease if they become immunosuppressed.[188,195,199] Within an affected loft, levels of sickness may be over 80% during stressful times of the year, espe-cially during the racing and breeding seasons.[190,194] Pigeons from lofts with recurrent *Trichomonas* infections should be evaluated for underlying herpesvirus infections.[69]

EXPERIMENTAL INFECTIONS

The results of experimentally induced PHV-1 infections are inconsistent. Some infected pigeons remain asymptomatic, some develop mild, localized lesions, and still others develop severe systemic infections. The route by which pigeons are experimentally infected and the functional capacity of the immune system have been shown to be important factors in the progression and severity of experimental infections. Systemic infections with hepatitis occur when pigeons are injected with PHV-1 intraperitoneally or intramuscularly.[37,196] By comparison, pigeons exposed to the virus by painting it onto the mucosa of the pharyngeal area generally develop mild, localized lesions.[195,199]

Interestingly, a herpesvirus isolate from a racing pigeon in Australia was not infectious for helmut, tumbler or racing pigeons that were exposed to the virus either by aerosol or by intravenous or intraperitoneal injections. One racing pigeon that was given virus by the combined intravenous and intraperitoneal routes remained asymptomatic, but developed microscopic changes characteristic of an infection one week post-exposure. Cockatiels injected intravenously or intraperitoneally with this same herpesvirus died in four to six days, while Cockatiels exposed to the virus by aerosol remained normal. Budgerigars experimentally injected with the same virus became lethargic, developed tremors and exhibited shallow respiration seven days after infection; the following day, they died. At the time of that study, Pacheco's disease virus had not been reported in Australia.[173]

A group of ten-week-old pigeons remained clinically normal after they were exposed to PHV-1 by coating the surface of the pharynx with virus, or by intravenous or intraperitoneal injection. A group of pigeons injected intramuscularly with corticosteroid (an immunosuppressive agent) at the same time they were exposed to the virus also remained clinically normal.[24] Whether the experimentally infected pigeons had antibodies to the virus from a previous, natural infection was undetermined.

Encephalitis has been shown to occur in experimentally infected pigeons, but not in those that are naturally infected.[37,196] Pigeon herpesvirus has been recovered from the brain of viremic pigeons.[196] However, when a virus is present in the blood, the recovery of the virus from a specific organ in the absence of microscopic changes is of limited importance.

Experimentally infected pigeons were found to be susceptible to a strain of Pacheco's disease virus that was antigenically related to PHV-1, but not to strains of Pacheco's disease virus that were antigenically unrelated to PHV-1.[200] Budgerigars exposed to PHV-1 by the intranasal route were shown to develop systemic infections with severe hepatitis.[192] Doves inoculated with owl herpesvirus died three to nine days after intramuscular injection.[17,129] Some ducklings, goslings, American Kestrels, Great Horned Owls, quail and chickens have been shown to be resistant to PHV-1 infections, following intramuscular infection of the virus.[20,36,129,178,190]

EPIZOOTIOLOGY

The first case of PHV-1 was reported in the United States in 1943.[168] Numerous reports documenting pigeon herpesvirus infections began to appear in the literature after the virus was demonstrated in a group of racing pigeons in the United Kingdom in 1967. Subsequently, PHV-1 infections were documented in racing, homing, show and commercial meat pigeons in North America,[116] Europe,[36,114,190] Australia,[11,24,173] Africa,[145] and New Zealand. While the isolates of PHV that have been studied appear to be serologically related, the viruses from differing geographic regions appear to vary in pathogenicity.

In Europe, some studies indicate that up to 63% of clinically normal carrier pigeons have antibodies to PHV-1, suggesting a high incidence of virus activity in flocks of these birds.[188,190,194] In Belgian lofts with a history of chronic respiratory disease, PHV-1 was recovered from 60% of all pigeons and from 82% of pigeons with active upper respiratory disease.[78,188,190,194] In another study, a herpesvirus was recovered from 4 of 23 (17%) pigeons presented to a diagnostic lab in England during a one-year period.[72] During a five year study in Austria, herpesvirus inclusion bodies were demonstrated in the liver of 3 of 226 (1%) of the pigeons examined.[60]

Infections may occur throughout the year but are most common in the summer months and during the racing season.[24,37,52,116] Pigeons of any age are susceptible to infection, but young birds (less than six months old) are most commonly affected.[52]

Pigeon herpesvirus infections can cause high mortality in some susceptible lofts. In an outbreak in a squab production facility, 93 of 600 (15%) pigeons died over a three-month period. Affected birds ranged in age from 10 days to 18 months, with the majority of affected pigeons ranging from 10 to 16 weeks in age; 45% of the pigeons that hatched during the outbreak died.[24]

Annual outbreaks of PHV-1 are usually associated with moderate levels of clinical illness (50%) with low mortality (15%). In an outbreak in Scotland, clinical disease occurred annually for several years in separated lofts of racing pigeons. Only five- to six-week old pigeons were affected.[37,195]

<u>RELATIONSHIP OF VIRUS STRAINS</u> — Isolates of pigeon herpesviruses appear to be closely related to each other and to the herpesviruses recovered from owls and falcons. A herpesvirus isolated from a pigeon in Scotland was found to be antigenically related to virus strains recovered from pigeons in the United States.[37,168] Pigeon herpesvirus is serologically indistinguishable from owl herpesvirus (OHV) or falcon herpesvirus (FHV), and these viruses may represent varying isolates of an identical virus.[20,24,129] Ring-necked Turtle Doves are susceptible to experimental infections with FHV and OHV, but pigeons have been shown to be resistant to both viruses, even though they are serologically identical to PHV-1. Failures in experimental transmission of FHV and OHV to pigeons may have been caused by the presence of neutralizing antibodies to PHV-1, which would also neutralize FHV or OHV. The genomes of pigeon and falcon herpesviruses have been found to be similar to each other, while the genome of Pacheco's disease virus is different from both PHV-1 and FHV.[1]

The relationships of PHV-1 isolates to those of Pacheco's disease virus vary. A strain of Pacheco's disease virus recovered from a psittacine bird in Belgium was found to be identical to PHV-1, while two strains of Pacheco's disease virus that were antigenically identical (one from Belgium and the other from Switzerland) were antigenically distinct from PHV-1.[200] Pigeon herpesvirus has not been shown to have a serologic relationship with other avian herpesviruses. Some Budgerigars and Cockatiels have been shown to be experimentally susceptible to PHV-1,[173] and Budgerigars have been shown to be naturally susceptible to PHV-1. These findings suggest that PHV may represent a herpesvirus strain with an expanded host range that infects the birds susceptible to Pacheco's disease virus as well as Columbiformes.[192,193,200] It is considered possible that PHV could infect psittacine birds that are maintained in mixed aviaries or zoological collections with latently infected pigeons.[193]

 — As is typical for herpesviruses, PHV-1 persists in a pigeon flock by establishing latent infections in recovered birds. Outbreaks are thought to occur when susceptible birds are exposed to carriers that sporadically excrete virus. Latently infected adults are particularly likely to shed virus during the breeding season, resulting in the exposure of highly susceptible young to the virus.[191] It is possible that feral or free-ranging pigeons can introduce PHV-1 to a flock. An outbreak of PHV-1 in a squab production facility was thought to have originated from a stray racing pigeon that entered the loft.[24]

There appears to be no correlation between the shedding of PHV-1 and the levels of neutralizing antibodies. Birds with high levels of neutralizing antibodies can shed virus, and there is no corresponding increase in viral shedding that can be predicted when the antibody titers decrease to undetectable levels. Many infected pigeons can be induced to shed virus when their immune systems are suppressed.[189,197,202]

Virus excretion occurs principally in the excrement and pharyngeal secretions. Experimentally infected squab begin shedding virus within 24 hours of infection and continued to excrete virus for seven to ten days. The highest concentrations of virus were detected one to three days post-exposure when clinical lesions first appeared.[189] Herpesvirus was recovered from the transportation boxes used to ship pigeons throughout Europe during the racing season. Virus was isolated from 2% of the boxes at the start of the season and 62% of the boxes at the end of the season, and fluctuations in temperature seemed to correlate with an increase in virus shedding.[83]

Virus present in pharyngeal secretions of latently infected hens can be easily transferred to squab when these contaminated secretions are mixed with regurgitated food provided to the neonates.[191] On the other hand, chicks from latently infected hens also can receive antibodies through the yolk sac that will protect the chicks from severe disease unless they are immunosuppressed by concomitant infections or environmental factors.[191,202] As a result, chicks from latently infected hens would be expected to develop latent infections; this maintains virus activity in a new generation of pigeons. Pigeon herpesvirus also can be transmitted from infected adults to immunologically naive squab shortly after hatching.[195]

It is possible that virus could enter the developing egg of a viremic hen during the acute phase of infection; however, PHV has not been shown to be egg-transmitted, probably because the primary infection causes a cessation of breeding activity.[195,199]

PATHOLOGY

The gross and microscopic changes associated with PHV-1 infections vary depending on the virulence of a virus strain and the susceptibility of the host. In some cases, infections cause no gross changes, while other pigeons may develop mild to severe ulcerations on the mucosa of the upper alimentary tract, nasal mucosa and salivary glands.[24,116] Systemic infections are most severe; affected birds may have lesions in the liver, spleen, pancreas and intestinal tract. Gross lesions include hepatomegaly, splenomegaly, congestion of the intestines, ulcerated necrotic areas on the mucosa of the gastrointestinal tract, and ulcers with grey-yellow diphtheritic membranes on the esophageal, pharyngeal and laryngeal mucosa *Figure 7.14*. Hyperemia of the area around the eyes and nostrils and ulcers or small hemorrhagic spots on the oral and pharyngeal mucosa may also be observed. It has been suggested that sialoliths occur secondary to PHV-1 infections, but a direct association between this lesion and a PHV-1 infection has not been confirmed.

Histologic lesions include necrosis of the liver, spleen, pancreas, lungs, kidneys, nasal mucosa and proventriculus. Intranuclear inclusion bodies have been identified in association with necrotic areas in the liver, pancreas, pharynx, small intestines, esophagus, salivary glands, proventricular glands, and nasal and tracheal mucosa.[24,36,37,173] Among eight affected pigeons that died from PHV-1, all had diffuse hepatic necrosis with intranuclear inclusion bodies.[116] It is of interest that hepatitis is frequently the prevailing microscopic change in affected pigeons, even though the principal clinical signs may be associated with the upper respiratory or alimentary tract.[37]

PATHOGENESIS

In some cases, PHV-1 infections are limited to the epithelial cells that line the upper respiratory and alimentary tracts. In other pigeons, the progeny virions that are released from the initially infected epithelial cells are transported throughout the body in the blood, causing subsequent infections in the liver, spleen, pancreas and kidneys.[37] The type of infection that occurs may depend in part on the route of viral exposure and the functional capacity of an infected pigeon's immune system. For example, pigeons experimentally infected with PHV-1 by coating the surface of the pharynx with virus developed infections that were localized to the site of inoculation.[195,199] Experimental infections in immunocompetent pigeons appear to remain localized to the site of infection, whereas experimentally infected pigeons that are immunosuppressed by the injection of cyclophosphamide develop viremia and hepatitis.[196]

As is typical for herpesvirus infections, virus can pass from cell to cell, even though a pigeon develops high levels of neutralizing antibodies. Viremia may occur during the primary infection and following reactivation of a latent infection.[196]

IMMUNITY AND DIAGNOSIS

A pigeon herpesvirus infection can be diagnosed by demonstrating the presence of antibodies to the virus or by detecting the virus in secretions, excretions or affected tissues. Intranuclear inclusion bodies suggestive of an infection may be detected by microscopically examining epithelial cells collected by swabbing ulcerative lesions in the oral cavity or pharynx. The virus can be isolated in cultured cells from swabs of the pharynx or feces in pigeons that are actively shedding the virus, or from the pharynx, liver, spleen or pancreas in birds that die from PHV-1 infections.

The virus is easiest to recover when swabs of the pharynx are collected soon after clinical signs are first noted. The liver is the best tissue to submit for virus isolation in birds with hepatitis. The trachea, lung and larynx are the best tissues to submit for virus isolation in pigeons with lesions in the respiratory tract.[37] Suggestive intranuclear inclusion bodies can usually be detected in association with mucosal ulcerations or in necrotic hepatocytes of dead pigeons. Chlamydiosis, trichomoniasis, salmonellosis, the diphtheritic form of poxvirus, adenovirus or pigeon paramyxovirus can induce clinical or gross changes that are similar to those caused by PHV-1.

Virus-neutralizing antibodies can be detected within a week after a pigeon is infected with PHV-1. Antibody levels that have been reported in naturally infected pigeons range from 1:16 to 1:128. Antibody levels reach a peak around three weeks after infection and then decrease.[189,190] Pigeons that recover from a primary PHV infection would be expected to be resistant to any future disease unless they become severely immunosuppressed. Antibodies may not be present in latently infected birds, and virus shedding can be demonstrated in pigeons even though high levels of neutralizing antibodies may be detected.[36,189,190] Antibody levels may decrease in pigeons even though they are frequently shedding the virus. Squab from infected hens receive maternal antibodies that protect them

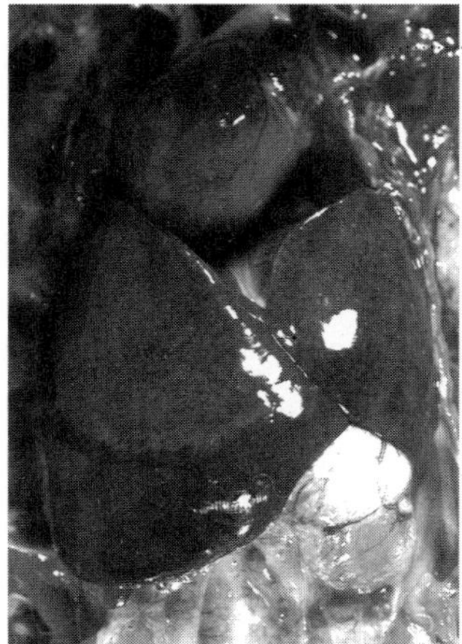
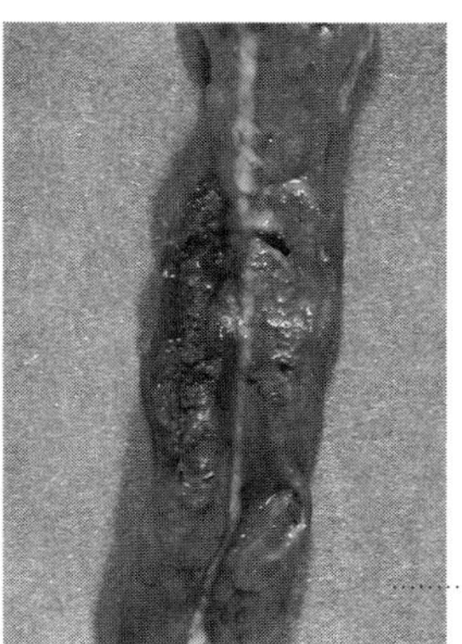

left *Pigeons with herpesvirus infections may have an enlarged, congested liver that is mottled with yellowish, discolored areas suggestive of necrosis.*

right *The intestinal tract may be congested and contain ulcerated areas filled with gray-yellow diphtheritic membranes. Similar ulcerative lesions also may occur in the esophagus or oral cavity.*

from severe disease, but these birds can become latently infected.[195,199] Pigeon herpesvirus-induced disease would be expected to be more common in young pigeons from hens that do not have antibodies to the virus.[52]

CONTROL AND VACCINATION

Pigeon herpesvirus can be inactivated by heating to 56°C for 15 to 30 minutes or by exposure to acid conditions (pH 4), and would be expected to be quickly inactivated by most common disinfectants.[11,37] Herpesvirus infections are enzootic in some flocks because any age pigeon is susceptible to infection, latently infected pigeons intermittently shed the virus and latently infected hens can produce squab that are latently infected. Controlling infections in a flock is arduous because of the difficulty in identifying the latently infected pigeons that intermittently shed virus. To protect captive pigeons from this virus, avian medicine needs either a reliable test to determine whether pigeons are subclinically infected with PHV-1 or a vaccine that will prevent the carrier state.

There is currently no commercially available vaccine to prevent PHV-1. However, pigeons vaccinated experimentally with either an attenuated-live or inactivated virus vaccine were found to have less severe clinical signs and shed fewer virions for a shorter period of time than unvaccinated pigeons exposed to virulent strains of PHV. Seroconversion was superior in pigeons receiving the inactivated vaccine, but neither of these vaccines prevented infected birds from becoming carriers that would shed virus when they were immunosuppressed. The fact that vaccination reduced the degree and duration of viral shedding could be important in controlling infections within a pigeon loft. If fewer virions are shed from pigeons following a primary or latent infection, then fewer virions will be present for dissemination in the flock.[201]

An experimental attenuated-live virus vaccine was mildly pathogenic and caused edema in the pharynx one to two days after vaccination.[201] Because of this side effect and the fact that the inactivated vaccine produced satisfactory protection from disease, it is likely that future research will involve the testing of an inactivated vaccine.

Because most squab are infected shortly after birth, vaccination of neonates would not be expected to be effective in controlling infections. Vaccinated hens may pass maternally derived antibodies to their chicks that would reduce the level of disease in neonates.[201] Repeated vaccination in the pectoral muscles of pigeons that need to fly is not recommended.

Acyclovir will prevent some, but not all, herpesvirus-induced disease. This drug was not found to prevent infections in pigeons exposed to PHV-1, but provided temporary protection to Budgerigars exposed to the same virus. Interestingly, Budgerigars treated with acyclovir and exposed to PHV-1 remained clinically normal while receiving acyclovir, but died from the infection shortly after treatment was stopped.[179] Trisodium phosphonoformate, an anti-herpesvirus drug, will not protect pigeons from clinical disease or prevent them from developing latent infections.[198]

PIGEON HERPES-ENCEPHALITIS VIRUS

In 1977 in Iraq, a herpesvirus was isolated from the brain of pigeons that were experiencing clinical signs of encephalitis (tremors, paralysis, torticollis) and high mortality.[3] Similar infections were subsequently reported in Egypt and South Africa.[145,177] Some confusion exists over whether or not the virus recovered from these pigeons represents a new type of herpesvirus or an antigenic variant of pigeon herpesvirus 1. This confusion arose because the pigeons from which the herpesvirus was initially recovered were also infected with pigeon paramyxovirus, which characteristically causes the clinical signs that were attributed to the herpesvirus. Thus, some researchers believe that the initial reports describing pigeon herpes-encephalitis actually involved pigeons that were infected with paramyxovirus and pigeon herpesvirus 1.[99,203] Other researchers have shown that outbreaks of encephalitis in pigeons have occurred in the absence of detectable pigeon paramyxovirus infections.[29]

The herpesvirus, potentially contaminated with paramyxovirus and recovered from the brain of pigeons with encephalitis, has been shown to be antigenically different from Marek's disease virus, turkey herpesvirus, infectious laryngotracheitis virus and duck plague virus. The herpesvirus recovered from the brain of pigeons does share some antigenic similarities (a precipitating antigen) with pigeon herpesvirus 1, owl herpesvirus and falcon herpesvirus. Thus, the herpesvirus recovered from the brain of pigeons with encephalitis is considered related to pigeon herpesvirus 1.[176,178]

A group of pigeons in Spain experienced high mortality after developing diarrhea and central nervous system signs, including paralysis of the wings and legs, torticollis and head tremors *Figure 7.15*. Clinical signs were particularly common in young pigeons. Gross lesions included congestion and hemorrhage in the brain, liver, pancreas and intestines. Histologic lesions in infected pigeons include perivascular cuffing, degeneration of the nerves and intranuclear inclusion bodies in the granular cell layers of the cerebellum.[29]

The herpesvirus recovered from these pigeons was thought to be free of contamination with pigeon paramyxovirus. Experimentally infected pigeons developed paralysis 17 to 19 days after infection and died within seven days of developing clinical signs. Experimentally infected pigeons had lesions in the

FIG 7.15
Central nervous system signs that include paralysis of the wings and legs, head tremors and torticollis (shown here) have been described in pigeons with paramyxovirus and some strains of herpesvirus.
photograph courtesy of Antonio Ramis

brain that were similar to those described in the index case.[29]

OTHER PIGEON HERPESVIRUSES

A herpesvirus that is antigenically distinct from Marek's disease virus has been associated with neoplastic changes in pigeons.[17]

HERPESVIRUSES THAT PRINCIPALLY AFFECT GALLIFORMES

INFECTIOUS LARYNGOTRACHEITIS VIRUS

A herpesvirus has been shown to cause a severe upper respiratory disease in chickens called infectious laryngotracheitis (ILT). While this virus primarily affects domestic fowl, some evidence exists that a similar virus may affect other avian species. Young turkeys can be experimentally infected, while adults are resistant to disease. Pheasants are susceptible, though less so than chickens, and develop characteristic disease when virus is placed in the eyes. Additionally, ILT caused the death of two pheasants and a peafowl in a zoo in Canada.[38] Quail, guinea fowl, starlings, crows, pigeons, doves, ducks and sparrows are refractory to disease; however, experimentally infected Japanese quail and one-day-old ducklings developed microscopic changes characteristic of an infection, even though they remained clinically normal.[38,94,141,213]

It is uncertain what role, if any, ILT virus plays in causing disease in companion and aviary birds. It should be noted, however, that a herpesvirus serologically similar to ILT virus has been recovered from Amazon parrots and other psittacine birds.[69,143a] Furthermore, canaries may also be susceptible to infectious laryngotracheitis virus.[69]

CLINICAL FEATURES AND EPIZOOTIOLOGY

Infectious laryngotracheitis virus may cause subclinical infections or mild-to-severe upper respiratory disease associated with high mortality. Clinical signs, including anorexia, depression, coughing, sneezing and dyspnea, are common three to twelve days (peak four to six) after infection. Dyspnea can be severe, and an affected bird may gasp for air with its mouth open, head raised and neck extended. A squeaking sound is often audible as a bird struggles to breathe. Swelling of the sinuses and discharge from the eyes, nose or mouth are common in severe cases *Figure 7.16*. Bloody mucus may be expelled when a diseased bird coughs or shakes its head.[184] Many severely affected birds die from asphyxiation as necrotic debris accumulates in, and occludes, the trachea. Mildly affected birds commonly recover two to three weeks after the initial clinical signs are noted. Poor air quality as well as concomitant bacterial, fungal, parasitic or other viral infections will exacerbate the clinical signs of disease and may prolong the recovery period.[93]

In an outbreak of ILT in a zoo, a six-month-old Indian Peafowl and a male Argus Pheasant died peracutely without demonstrating premonitory signs. A female Argus Pheasant expelled a plug of mucus containing blood from the trachea and died the following day. Seven other peafowl in the same building were depressed, coughing and gasping, while exposed turkeys and ducks remained clinically normal. A group of exposed Bantam Chickens began to show clinical signs of disease ten days after the initial outbreak was noted.[38]

<u>TRANSMISSION</u> — Serologically related strains of ILT virus have been recovered on multiple continents; however, there is substantial variance in the pathogenicity of virus strains. Some cause asymptomatic infections, while others are associated with severe disease and high mortality. Birds of any age are considered susceptible to infection, but some natural resistance to disease occurs as chickens mature toward six weeks of age.[55] Infected birds are more likely to develop clinical signs of disease when ambient temperatures are high. Males are more commonly affected than females.[167]

Infected birds shed virus principally in respiratory secretions. The virus can enter a susceptible bird through contact of the respiratory tract or conjunctiva with contaminated aerosols.[184] Following experimental infections, virus can

Difficulty with breathing, manifested clinically as gasping for breath and extension of the head and neck, is common in gallinaceous birds with infectious laryngotracheitis virus infections.

A discharge from the eyes or nostrils may be noted in some birds, particularly when secondary bacterial infections occur.

photographs courtesy of Jean Sander

be recovered from the trachea for six to eight days.[5,86] Because ILT virus is more durable than are other herpesviruses, contaminated fomites also could serve to transport the virus from bird to bird or flock to flock. The source of ILT virus in an isolated group of pheasants, peafowl and chickens in a zoo could not be determined.[38]

Some early reports indicated that very few (2%) exposed chickens developed latent infections.[91] More recent findings indicate that up to 80% of the birds in some flocks are latently infected.[86] Moving previously infected chickens to a new location induced one of five birds to shed virus, while the onset of egg laying was associated with virus shedding in nine of ten chickens. The one relocated chicken that shed virus began five days after the move and continued for five days. Interestingly, administration of corticosteroid or cyclophosphamide (immunosuppressive agents) was not associated with an increase in viral shedding. Clinical signs of disease have not been noted during the short period of time that latently infected chickens shed virus.[85,86]

As is the case with many viruses, birds vaccinated with attenuated-live virus vaccines may become latently infected and can shed virions that may infect susceptible birds.[5] The ILT virus has not been shown to be egg-transmitted.

PATHOLOGY

Gross lesions characteristic of ILT virus infections include thickening and hemorrhage of the tracheal mucosa, accumulation of necrotic debris in the lumen of the trachea and blood-tinged mucus in the larynx, trachea or bronchi *Figure 7.17*. Yellow necrotic plaques may be noted in the oropharynx, particularly near the entrance to the trachea. These lesions appear similar to those caused by the diphtheritic form of poxvirus, vitamin A deficiencies, trichomoniasis and candidiasis. The air sacs are occasionally affected and appear swollen, cloudy and thickened.

Infected epithelial cells that line the trachea are enlarged, and intranuclear inclusion bodies may be detected early in the disease process. These inclusion bodies are more difficult to demonstrate later in the disease process because the infected epithelial cells containing the inclusion bodies die, slough and become part of the debris that accumulates in the lumen of the trachea *see Figure 7.12*. Thus, the absence of inclusion bodies in birds with characteristic lesions does not rule out ILT.

PATHOGENESIS, DIAGNOSIS AND IMMUNITY

The ILT virus principally affects the epithelial cells lining the respiratory tract. If a viremia occurs, it is transient.[82] The virus persists in the tracheal epithelial cells of latently infected birds.[6] In some early cases, impression smears made from exudates collected from the trachea or conjunctiva may contain suggestive intranuclear inclusion bodies. Infectious laryngotracheitis virus is easiest to recover from tracheal swabs collected within the first few days after clinical signs develop. Attempts to isolate a virus from any companion or aviary bird with necrotic tracheitis should be encouraged.

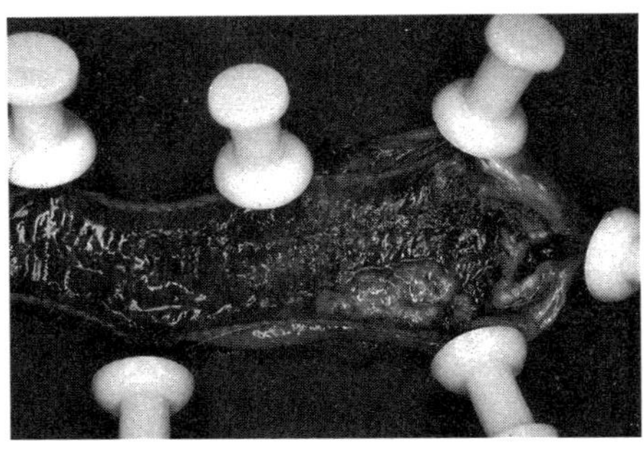

FIG 7.17

Thickening and hemorrhage of the tracheal mucosa, accumulation of necrotic debris in the lumen of the trachea and presence of blood-tinged mucus in the larynx, trachea or bronchi are suggestive of an ILT infection.

photograph courtesy of Jean Sander

Antibodies to ILT virus can be demonstrated by any of several assays, including indirect fluorescence, agar-gel immunodiffusion and virus neutralization. Virus-neutralizing antibodies to ILT virus can be demonstrated as soon as four days after infection, but there has been no correlation demonstrated between the presence or absence of neutralizing antibodies and protection from infection or disease.[214] In chickens, maternally transmitted antibodies decay as quickly as two weeks after chicks hatch and have not been shown to protect them from virus challenge.[77,92]

Other diseases that can cause clinical or gross changes suggestive of ILT in gallinaceous birds include *Hemophilus* (infectious coryza), infectious bronchitis virus, mycoplasmosis, poxvirus (diphtheritic form), aspergillosis, gramnegative bacterial tracheitis, *Syngamus* sp. or tracheal mite infection.

CONTROL AND VACCINATION

Unlike other herpesviruses that are relatively unstable outside the host, the ILT virus can persist for extended periods in a protected environment (three months at room temperature in tracheal exudate, twenty days in deep litter, three days in fresh droppings).[94] Because of its stability, ILT virus could be transmitted from flock to flock on fomites. The ILT virus is susceptible to sunlight, 5% phenol, formalin, sodium hypochlorite and iodophors.[76]

Attenuated-live and inactivated ILT virus vaccines have been developed for use in poultry. Vaccinated birds develop a rapid protective response, and vaccination is generally recommended in the face of an outbreak. Birds vaccinated with attenuated-live virus vaccines may become latently infected and can shed the virus, causing disease in susceptible contact birds.[6] A subunit vaccine developed for ILT virus has been shown to protect experimentally infected chickens from both acute infection and disease, and may prove effective against latent infections as well.[214]

Unfortunately, attenuated-live virus vaccines that are safe and effective in poultry may cause severe disease in other birds. In one study, clinical signs in vaccinated pheasants included conjunctivitis, nasal discharge, occlusion of the nostrils, dyspnea, swelling and crusting of the eyelids, and sometimes death. Golden and Silver Pheasants were mildly affected; Reeve's, Impeyan and Eared Pheasants were severely affected. One vaccinated Reeve's Pheasant died. Other vaccinated pheasants recovered after a two-week period. Young pheasants were more severely affected than adults.[38] Studies to determine the factors that affect susceptibility to infection and the severity of disease in pheasants have not been performed.

MAREK'S DISEASE VIRUS

Marek's disease is caused by a highly contagious herpesvirus with a worldwide distribution. Strains of this virus can be divided into broad groups based on the clinical disease they induce. Varying pathotypes: 1) are apathogenic; 2) transform cells causing cancer; or 3) do not cause cancer. Marek's disease virus (MDV) is antigenically related to an apathogenic turkey herpesvirus.[97]

CLINICAL FEATURES AND EPIZOOTIOLOGY

Marek's disease virus infections are most common in chickens, but lesions resembling those caused by MDV also have been reported in Great Horned Owls, pheasants, partridge, ducks, geese, swans and a Kestrel.[9,32,75] The virus has been demonstrated to infect turkey and quail, and antibodies to MDV have been demonstrated in free-ranging jungle fowl, turkeys and quail.[146] Experimentally infected ducks remain asymptomatic.[163] Lesions that resemble those caused by MDV have been described in Budgerigars, canaries and a toucan, suggesting that a related virus can infect companion birds; however, its prevalence or involvement in disease in these species has not been confirmed.[146]

The classic form of Marek's disease is characterized by lymphocyte accumulation in the nerves resulting in varying degrees of partial or complete paralysis.[9] The mortality rate with this form of the disease is low (10 to 15%). An acute form of the disease is characterized by the formation of lymphoid tumors in numerous organs, including the liver, spleen, kidneys, skin, muscles, bones and gonads. The mortality rate with this form of the disease is typically high (from 30 to 70%).[146] Marek's disease virus has also been associated with a transient encephalitis in young chick-ens that is characterized by acute paralysis followed by complete recovery, as well as decreased egg production and damage to the immune system. Marek's disease virus has been loosely associated with atherosclerosis in chickens.[134] The incubation period varies in chickens from two weeks to several months.

Marek's disease virus primarily infects young birds prior to sexual maturity; 12- to 24-week-old chickens are most commonly affected,[10,146] but chicks as young as 9 days old have been shown to be infected.[87,210] The incidence of infections ranges widely (none to 80%) depending on the geographic location of the flock, the genetic susceptibility of a particular strain of chickens and the virulence of an endemic virus strain.[10] An age-related resistance to disease develops in chickens that are not genetically susceptible to the virus. Infected females develop disease more commonly than males.[25]

Gross and microscopic lesions that are suggestive of MDV were described in a young (15- to 18-month-old), free-ranging Great Horned Owl in the United States. The ataxic, partially paralyzed, emaciated bird was found on the ground and was unable to fly.[75]

In the spinal cord and peripheral nerves of a toucan that exhibited a chronic, slowly progressive ataxia with kidney enlargement, herpesvirus proteins were demonstrated by an agar-gel immunodiffusion test. The microscopic lesions that occurred in this bird were consistent with Marek's disease, but a virus could not be isolated.[152a]

<u>TRANSMISSION</u> — Marek's disease virus has been shown to be transmitted through direct and indirect contact with contaminated aerosols. This virus has not been shown to be vertically transmitted from one generation to another through eggs.[9,10] The virus is shed principally from feather-derived epithelium being passed along as a developing feather differentiates, and fragments of the feather are discarded. Virus is present in the follicular epithelium of chickens within two to three weeks after they are experimentally infected.[26] Indirect routes of transmission are unusual for herpesvirus because most strains are unstable when outside of the host. In contrast, the Marek's disease virus is extremely stable. At room temperature it can remain infectious for four months in feces and up to eight months in feather dust. It is theoretically possible that insects could serve as mechanical vectors for dissemination of the MDV; however, it has not been possible to show that contaminated insects transmit the virus.

Surviving chickens are latently infected and intermittently shed virus. Lymphocytes are considered the principal site where the virus remains latent. Because of the stability of MDV outside of the host, a few latently infected birds, which intermittently shed the virus in their feather debris, can create an environment that is heavily contaminated with infectious virus. It is uncertain what role, if any, infected chickens may play in disseminating MDV to other susceptible species of captive or free-ranging birds.[27]

PATHOLOGY

Marek's disease virus causes neoplastic changes in the lymphoid portion of the immune system. The cancerous lesions caused by this virus are most common in the peripheral nerves, bursa, thymus and visceral organs. The peripheral nerves of affected birds are characteristically enlarged, edematous and grayish in appearance. Small grayish nodules may also be noted on the ovary, kidney, spleen, skin, skeletal muscle or liver *Figure 7.18*.[75,163] Atrophy of the bursa and thymus, which induces an immunodepression, has also been associated with MDV infections. The characteristic microscopic change associated with MDV is the accumulation of neoplastic lymphocytes in the peripheral nerves of various organs and muscles.

In a Great Horned Owl suspected to have MDV, the sciatic nerves were two to three times normal size, gray-white in color and edematous. Gray-white masses were also seen in the kidney, pancreas and enlarged spleen. Lymphoblastic cells were identified microscopically in the liver, kidney, pancreas, spleen and sciatic nerves.[75,163]

DIAGNOSIS

Marek's disease virus can be recovered from many organs (liver, spleen, kidneys) and from viral-induced tumors in affected birds. Virus can be recovered also from the white blood cells of many latently infected chickens. The highest levels of virus are present in the blood about four weeks after infection.[163] The virus is present in the highest concentration in the blood before tumors can be clinically detected, and a sample for virus isolation is not typically available during this highly viremic period. Virus also can be detected in the feather pulp of infected birds. Virus isolation should be attempted routinely in companion birds with tumors to determine whether this virus is causing problems that have yet to be defined.

Antibodies to MDV can be detected using several assays (agar-gel immuno-diffusion, indirect fluorescence, virus-neutralization or ELISA). Infected chickens develop antibodies that can be detected throughout their lifetime.[163] The development of antibodies has been of little value in predicting whether an infected bird will develop disease or remain asymptomatic. Seropositive hens pass antibodies to their chicks in the yolk; these provide some protection from disease for up to three weeks after hatching.[34]

PATHOGENESIS

No one knows for sure how MDV initially replicates in a chicken. Researchers theorize that the virus is inhaled with feather dust and is then consumed by macrophages in the respiratory tract. Virus has been shown to replicate in the bursa, thymus and spleen within three days after experimental infection. The replicating virus causes rupture of infected cells in these organs, resulting in an immunosuppression that makes infected birds more susceptible to other pathogens. Virus is then carried in infected lymphocytes to the feather epithelium where the virus can be detected within two weeks after infection. At this point, the affected birds are highly infectious. Proliferative lesions in the nerves, kidneys, liver and

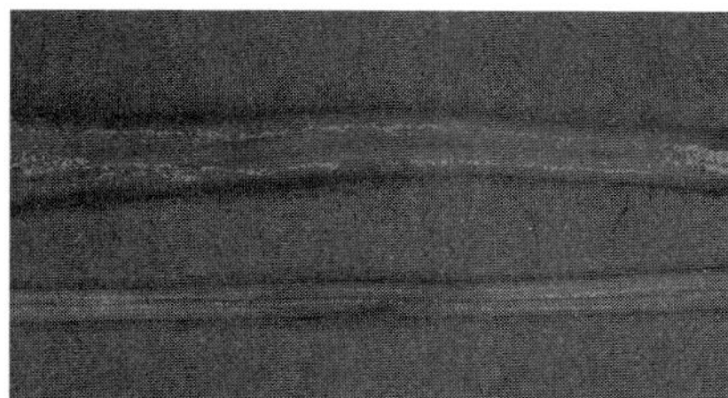
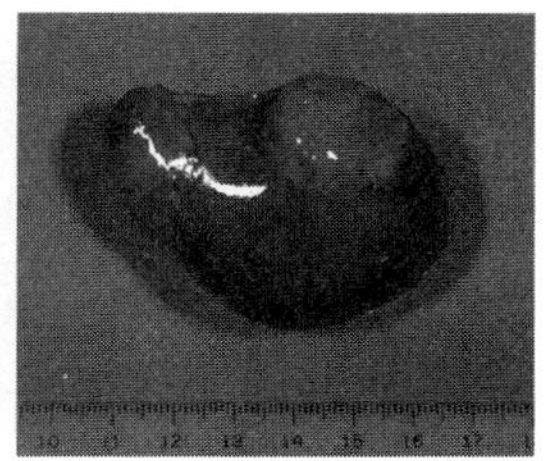
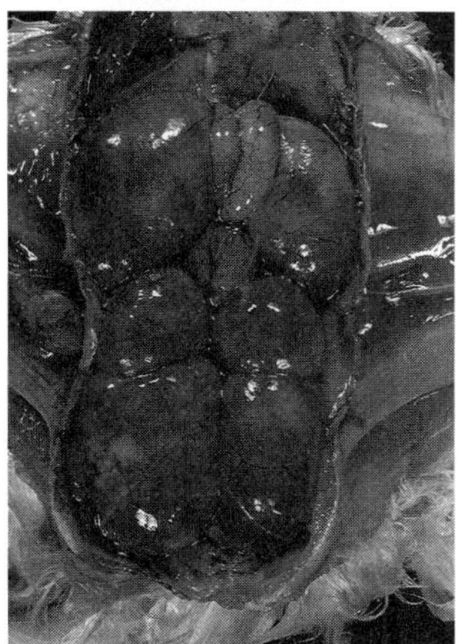

FIG 7.18

In gallinaceous birds with Marek's disease, the peripheral nerves are typically enlarged, edematous and grayish in appearance.
*A nerve **top left** from a bird with Marek's disease is shown above a nerve from a healthy bird. Many affected birds have grayish nodules within multiple tissues including the **top right** spleen and **left** kidneys.*

photographs courtesy of Richard Davis and Jean Sander

gonads are noted three to four weeks after infection.

CONTROL

Marek's disease virus has been shown to be stable at 25°C for 4 days, at 37°C for 18 hours and at 56°C for 30 minutes. If protected in feather debris or feces, the virus can remain infective for 4 to 8 months at cool temperatures.[28] Marek's disease virus would be expected to be rapidly inactivated following contact with most virus-inactivating disinfectants.

Inactivated and attenuated-live virus vaccines have been used in chickens to control Marek's disease. Maternally derived antibodies can interfere with early vaccination of chicks. Vaccinated birds become carriers and intermittently shed the virus. However, they do not develop the severe changes that occur in unvaccinated birds.[163] Acyclovir has been shown to reduce the development of neoplasia in birds experimentally infected with Marek's disease virus, but has not been shown to prevent latent infections.[154]

QUAIL HERPESVIRUS

A herpesvirus that is serologically related to crane herpesvirus was recovered from Bobwhite Quail that were exhibiting clinical signs of depression, anorexia and diarrhea. The virus causes disease in quail less than four weeks of age, and affected birds generally die two to three days after developing clinical signs. Enlargement of the liver and spleen with yellowish, necrotic areas is characteristic.[105]

TURKEY HERPESVIRUS

A virus serologically related to Marek's disease virus has been demonstrated in domestic and free-ranging turkeys. No clinical, gross or microscopic lesions have been associated with natural infections of turkey herpesvirus.[35] While serologically related to Marek's disease virus, turkey herpesvirus does not appear to transform infected cells and is thus not cancer-causing. Latent infections are common in birds with neutralizing antibodies. Turkey herpesvirus replicates in the feather epithelium and virus is shed in the feather dust. Transmission occurs through direct contact and by virus-contaminated aerosols. In some cases, turkey herpesvirus can infect chickens.[146]

HERPESVIRUSES OF FALCONIFORMES AND STRIGIFORMES

FALCON HERPESVIRUS

Herpesvirus infections were first described in raptors from Austria in the early 1900s and have since been documented in free-ranging and domestic raptors in the United States, Europe and Asia.[17,20,73,127,129,155] Frequently referred to as "inclusion body hepatitis" of falcons, the herpesvirus that causes this disease has serologic similarities to the herpesviruses that are found naturally in pigeons and owls.

Infected falcons usually die acutely with no premonitory signs, but may exhibit depression, lethargy and anorexia for 24 to 72 hours before death. Mortality rates in affected birds approach 100%, with death usually occurring one to two days after clinical signs develop.[73] Lymphopenia may occur prior to death.[73,151]

Both young and adult falcons have been found to be susceptible to falcon herpesvirus (FHV) infections.[73,110] Naturally occurring FHV infections have been reported in the Peregrine Falcon, Merlin, Red-headed Falcon, Prairie Falcon, Gyrfalcon and American Kestrel.[73,110,129] Gyrfalcons and Prairie Falcons are considered highly susceptible while Peregrine Falcons appear to be relatively resistant.[151] Experimental infections following intramuscular injection of the virus have been documented in the Prairie Falcon, Merlin, Kestrel, Cooper's Hawk, Sharp-shinned Hawk, Swainson's Hawk, Great Horned Owl, Screech Owl, Long-eared Owl, Snowy Owl, Green Heron, American Coot, Muscovy Duck, Ring-necked Dove, Budgerigar and Amazon parrot.[127-130]

The clinical progression of disease appears to be similar in naturally and experimentally infected falcons. Prairie falcons developed a brief period of depression and anorexia followed by death six days after being injected intramuscularly with FHV. Kestrels died four to six days after receiving an intramuscular injection of the virus. Some experimentally infected Kestrels died following 24 to 48 hours of depression and anorexia, while other Kestrels died with no premonitory signs.[73] Great Horned Owls died 7 to 10 days after being injected intramuscularly with FHV; a Screech Owl died eight days after injection. The experimentally infected Great Horned Owls were anorectic for 24 to 72 hours prior to death.[73,129] A Barred Owl injected intramuscularly with FHV developed clinical signs twelve days after infection and was euthanized two days later.[110] Ring-necked Doves died three days after receiving an intramuscular injection of FHV. Falcon herpesvirus has not been shown to cause experimental infections in pigeons, turkeys or chickens.[73,129] Experimentally, American Kestrels have been shown to be susceptible to owl herpesvirus and resistant to pigeon herpesvirus.[129]

The genomes of pigeon and falcon herpesviruses are similar to each other, while the genome of Pacheco's disease virus is different from both pigeon and falcon herpesviruses.[1] The antigenic and pathogenic relationships of the currently recognized avian herpesvirus are reviewed in *Tables 7.2 and 7.3.*

The gross and microscopic lesions associated with FHV are similar in natural and experimental infections. Gross lesions include hepatomegaly, splenomegaly and hyperemia of the small intestines *Figure 7.19.* Histologic lesions include necrosis in the liver, spleen, intestines and bone marrow. Intranuclear inclusion bodies have been demonstrated in these same tissues.[73,127,129,206]

The route by which FHV is naturally transmitted has not been confirmed; however, it has been suggested that falcons could become infected when they ingest pigeons or other prey species contaminated with FHV.[73] In captive falcons, ingestion of contaminated prey is considered to be a primary route of viral exposure. Domestically raised pigeons that have not been infected with FHV should be used for training purposes and as a food source for captive raptors.[151] To prevent cross-species transmission, it is recommended that raptors not be housed with pigeons.[206]

Falcon herpesvirus has not been shown to be vertically transmitted from parents to offspring. Artificial incubation of eggs and hand-raising of neonates in an area separated from previously exposed adults could be used to produce disease-free young. However, these birds would be expected to be susceptible to infection when exposed later in life. A test to determine whether pigeons are subclinically infected with falcon herpesvirus or a vaccine that will prevent the disease in falcons will be necessary to control the disease in captive raptors.

EAGLE HERPESVIRUS

A herpesvirus serologically distinct from duck plague virus, herpes simplex (a human herpesvirus), falcon herpesvirus and crane herpesvirus was recovered from a dead Bald Eagle chick.[49] Another herpesvirus was isolated from a South

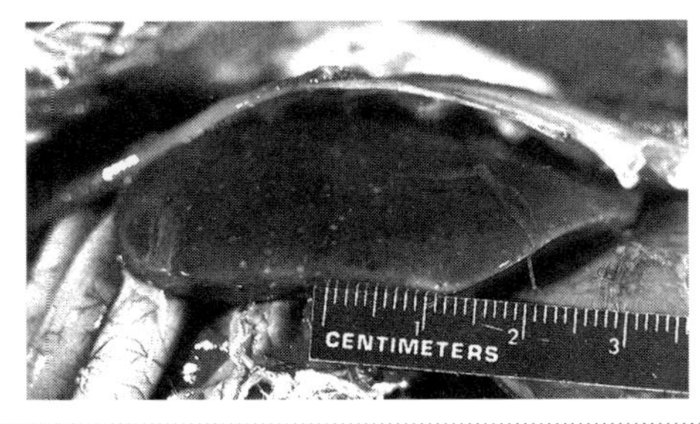

FIG 7.19

Typically, falcons infected with herpesvirus die. The most common change noted at necropsy is an enlarged liver with pinpoint areas of yellowish discoloration suggestive of necrosis. photograph courtesy of National Wildlife Health Center

American Eagle with clinical signs similar to those described in falcons.[97]

OWL HERPESVIRUS

Owl herpesvirus has been reported in captive and free-ranging owls in Asia, Europe, Canada and the United States. Affected birds are typically depressed and anorectic for two to five days prior to death. Occasionally, small yellow nodules may be noted on the pharyngeal mucosa.[17] These lesions may be secondarily infected with bacteria, fungi or trichomonads. Leukopenia may be noted during the acute phase of an infection.[69]

In captive settings, affected owls usually die; however, the demonstration of antibodies in free-ranging owls indicates that some birds survive the initial infection. Antibody-positive owls are considered to be latently infected.[101] The effect of the virus on free-ranging populations of susceptible owls has not been determined.[7]

The natural susceptibility of differing species of Strigiformes to OHV appears to vary. Natural infections have been described in the Great Eagle Owl, Great Horned Owl, Long-eared Owl and Snowy Owl. The Little Owl, Boreal Owl, Spotted Owl, Screech Owl, American Kestrel, Common Kestrel and Ring-necked Dove have been shown to be susceptible to experimental infections. The Tawny Owl, Great Gray Owl, Barn Owl, chicken, Pekin Duck, pigeon, Budgerigar, swift, White Stork, Blackbird, Hooded Crow and sparrow have been demonstrated to be resistant to experimental infection.[20,74,127,164]

Experimentally infected owls will develop characteristic clinical, gross and microscopic lesions following exposure to the virus by the intramuscular or oral routes, as well as following exposure to contaminated aerosols. Kestrels, Great Horned Owls and doves inoculated with owl herpesvirus died three to nine days after they received an intramuscular injection of virus.[17] In one study, owls ranging in age from seven weeks old to adults were susceptible to experimental infections.[20]

Owls have been shown to be experimentally susceptible to falcon herpesvirus. Experimentally infected owls died 7 to 12 days after receiving an intramuscular injection containing an isolate of falcon herpesvirus.[110] Pigeon herpesvirus was not found to be infectious to Great Horned Owls.[20,127] Virus recovered from an Eagle Owl was not infectious to a White Stork.[20]

Owl herpesvirus has been shown to be excreted in the pharyngeal secretions and urine of infected birds.[158] Ingestion is considered the principal method of virus transmission; however, virus can be demonstrated in the epithelium of feather follicles, and owls are susceptible to infections following exposure to contaminated aerosols.[20,69] In owls that were experimentally infected by intramuscular injection, virus activity was considered greatest in the liver, spleen, bone marrow, bursa and thymus. By comparison, the greatest viral activity was demonstrated in the lung, small intestines and trachea of two owls exposed to virus-contaminated aerosols.[20]

Necrotic lesions in the liver, spleen and bone marrow are characteristic of OHV infections *Figure 7.20*.[17] In one study, microscopic lesions consistent with herpesvirus were demonstrated in 67% of the owl tissues examined.[127] Inclusion bodies may or may not be present in association with damaged tissue. Tissues should be submitted for virus isolation in suspect cases.[20,69] In one study, bone marrow was considered the best tissue to submit for virus isolation.[158]

HERPESVIRUSES IN ANSERIFORMES

DUCK PLAGUE VIRUS

Herpesvirus infections have been described in captive and free-ranging populations of a variety of Anseriformes.[135] The disease caused by this virus is commonly referred to as duck enteritis virus or duck plague virus (DPV). The former name is used because the virus can cause severe necrosis of the cells lining the alimentary tract, and the latter name is used because of the high levels of sickness and death associated with outbreaks in susceptible populations of waterfowl. Epornitics of DPV in large flocks of free-ranging Anseriformes can cause death tolls reaching the tens of thousands.[14]

CLINICAL FEATURES

Susceptibility to DPV among waterfowl appears to vary based on the strain of the virus and the species and age of the bird.[44,117,171] Some infected waterfowl remain asymptomatic while other birds infected with the same strain of virus develop overwhelming disease.[211] Mor-

bidity and mortality rates may vary from 0 to 100% of the at-risk population. Mortality rates are highest in flocks with concomitant bacterial infections.[42] Captive or free-ranging populations of Anseriformes that contain latently infected birds experience intermittent deaths when these carriers shed virus.

FIG 7.20
An enlarged liver and spleen mottled with areas of yellowish discoloration are suggestive of an owl herpesvirus infection.

photographs courtesy of Gilles Berneir

Most waterfowl that develop clinical signs of duck plague will die. Conversely, many affected birds die without developing clinical signs of disease and are in excellent overall condition at the time of death. In massive die-offs, deaths of exposed birds may continue for three to five weeks after the first case is noted. In one study, DPV was recovered from ten Muscovy Ducks that died acutely or following a brief period of vomiting and diarrhea, two Mallard Ducks that died acutely with egg-related peritonitis, and a Mute Swan, Canada Goose and Shelduck that had been found dead.[72]

Clinical changes associated with natural and experimental DPV infections include depression, lethargy, anorexia, drooping of the wings, nasal discharge, conjunctivitis, dyspnea, profuse diarrhea (occasionally containing blood), excessive thirst, penile prolapse and central nervous system signs *Figure 7.21*. Usually, death follows in one to ten days, although some older birds develop a more protracted disease in which clinical signs may persist for weeks.[89,211]

Affected birds may be unable to fly and may float on the water with their head and neck in extension or they may swim in circles.[69] The eyelids may be swollen shut late in the disease process, and blood may be noted on the feathers around the cloaca. Erosions at the opening of the sublingual salivary glands in the oral cavity were common in ducks and geese that were documented to be latently infected.[14,89] The diarrhea that occurs in some affected birds may contain mucus and frequently contains bile characteristic of the anorexia and liver damage associated with the infection. Infections in flocks of commercial ducks have been associated with a 20% to 100% decrease in egg production.[137]

EXPERIMENTAL INFECTIONS

When susceptible birds are exposed to DPV by oral, intranasal, intravenous, subcutaneous, intraperitoneal, intramuscular or intracloacal routes, progression of an infection in a particular species is typically similar to that of natural infections.[14,117,119,187]

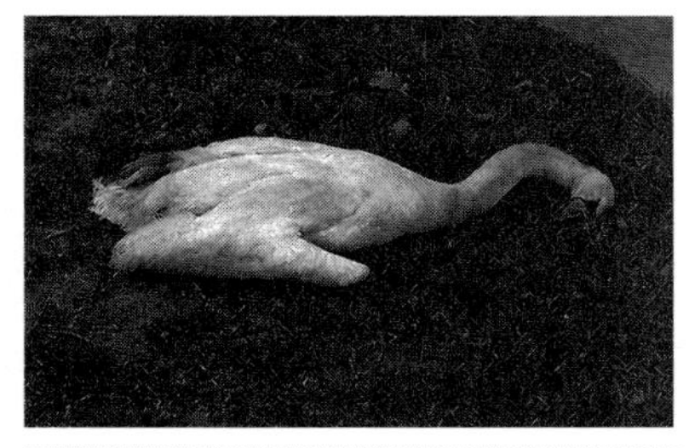

FIG 7.21

Duck plague virus should be considered in waterfowl with depression, anorexia, profuse diarrhea (occasionally containing blood) and central nervous system signs that include difficulty in standing or convulsions. photograph courtesy of Greg Harrison

Experimentally infected adult teals died 63 to 78 hours after being infected intramuscularly with DPV, and 72 to 86 hours after oral inoculation. Some teals died with no premonitory signs, while others died up to 14 hours after the first evidence of lethargy, ataxia and an inability to stand. Contact controls died 161 to 162 hours after exposure, suggesting an incubation period in teals of approximately three days, with infected birds shedding virus two to three days

after being infected. Experimentally infected goslings were depressed at 90 to 133 hours after infection, and three of six goslings died 119 to 133 hours after infection with signs of acute weakness, ataxia, tremors and convulsions.[211]

Experimentally infected Pintails and Pekin Ducks died four to seven days after intramuscular injection.[115] Experimentally infected ducklings developed anorexia, lacrimation and diarrhea prior to death, three to nine days after they were exposed to DPV by oronasal inoculation.[175] Herring Gulls, Black-headed Gulls, pigeons and adult chickens have been shown to be resistant to experimental infection. By comparison, one-day-old to two-week-old chickens inoculated intracerebrally are susceptible.[21,89,118]

EPIZOOTIOLOGY

Duck plague virus has been demonstrated to cause disease in waterfowl that range in age from seven days to at least eight years old. Given this range, any age waterfowl is considered susceptible to DPV.[118,135] The virus is considered to be enzootic in most countries where it occurs, and outbreaks have been reported in Europe,[48,89,126] North America,[59,123] and Asia.[90,136,174] Reports of DPV infections could not be found for waterfowl in South America or Australia.

The earliest report of DPV may have been in The Netherlands in 1923, when 90% to 100% of a group of ducks died acutely with necropsy findings that included enteritis and hemorrhage throughout the body.[8] Suspected outbreaks associated with the virus occurred in Europe throughout the 1940s and 1950s. The first reported outbreak of DPV in the United States occurred in commercial Pekin Ducks on Long Island in 1967.[122] The importation of latently infected waterfowl from Europe is considered the most likely way that the virus was introduced to North America and Asia.[137] Free-ranging waterfowl may have disseminated the virus throughout Europe. However, there have been no reports of DPV in free-ranging waterfowl in Great Britain, and annual outbreaks in that country have been commonly reported since 1972.[72]

Outbreaks of DPV have been associated with massive mortality in captive and free-ranging waterfowl, particularly when large groups of birds congregate during migration. In one outbreak on a wildlife refuge in South Dakota, an estimated 43,000 of 100,000 (42%) ducks and 250 of 9,000 (3%) Canada Geese died. While only a few epornitics have been reported in free-ranging waterfowl, numerous devastating outbreaks have been reported in captive birds.[135,170] An outbreak in Thailand killed 850,000 of the estimated 20 million waterfowl in that country.[174] In an outdoor aviary in California, all of 46 Muscovy Ducks died over a two-month period.[170] In an outbreak in a zoological park, 27 adult ducks that had been in captivity from 1.5 to 8 years, as well as several free-ranging ducks that used the same pond, died acutely without premonitory signs during a three-week period. The mortality rate was low and the duration of the outbreak was short, probably because the exposed birds were vaccinated.[135] In captive waterfowl, infected survivors will become latently infected and the disease will be self-limiting if susceptible birds are not introduced to the flock.

All the DPV isolates that have been studied are serologically related; however, virulent and avirulent strains of the virus have been recovered. Both susceptibility to and progression of DPV infections vary in ducks, geese and swans. Muscovy Ducks are highly susceptible and frequently die. When a mix of varying waterfowl species are naturally exposed to DPV, the Muscovy Ducks frequently die first.[137,170] Waterfowl that have been reported to be affected by DPV following natural or experimental infection are listed in *Table 7.6*.

Mallard Ducks, European Teals and Common Pintails are susceptible to infection, but are considered relatively resistant to disease. Many infected Mallard Ducks will remain asymptomatic, develop virus-neutralizing antibodies

TABLE 7.6

Waterfowl shown to be susceptible to natural or experimental DPV infections[43,90,135,207]

DUCKS

American Widgeon	Australian Grey Teal
Black Duck	Bufflehead Duck
Canvasback Duck	Common Eider
Common Goldeneye	Common Merganser
Common Pochard	Common Shelduck
European Widgeon	Gadwall
Garganey Teal	Goosander
Greater Scaup	Indian Runner Duck
Khaki Campbell Duck	Lesser Scaup
Mandarin Duck	Merganser
Muscovy Duck	Pekin Duck
Pintail	Red-headed Duck
Ring-necked Duck	Shell Shoveler
Teal	Tufted Duck
Wood Duck	

SWANS

Black Swan	Mute Swan

GEESE

Bean Goose	Canada Goose
Egyptian Goose	Grey Goose
White-fronted Goose	

and remain latently infected. Mallard Ducks frequently survive challenge with even the most virulent strains of DPV.[89,118] In one study, Blue-winged Teal were considered relatively susceptible to disease; Canada Geese were considered relatively resistant to the same strain of virus, with some experimentally infected geese developing disease and others remaining asymptomatic.[171] In a natural outbreak in a zoological park, eight species of ducks were affected, but exposed geese and swans remained clinically normal.[135] While Canada Geese are considered relatively resistant to disease, natural outbreaks of DPV have been reported in this species.[122,171] Mallard Ducks experimentally immunosuppressed with cyclophosphamide were found to be more susceptible to DPV-induced disease than non-immunosuppressed ducks.[71]

<u>CARRIERS AND TRANSMISSION</u> — Duck plague virus can be transmitted through both direct and indirect contact, and ingestion of contaminated water is thought to be the major route of natural transmission. Naturally and experimentally infected waterfowl that survive an infection have been shown to be latently infected.[14] These latently infected birds intermittently shed virus in their feces and respiratory secretions. Considered to serve as reservoirs of DPV, they have been shown to intermittently shed virus in their feces for up to five years.[14,89] Virus shed in the feces of latently infected birds can contaminate eggs in the nest and be a source of virus for newly hatched chicks.[120]

Mallard Ducks appear to be particularly resistant to DPV-induced disease and may serve as a primary reservoir for the virus. Free-ranging birds are frequently implicated in the dissemination of DPV, but the virus was not recovered from 5,000 free-ranging waterfowl sampled from across the United States, and the virus has not been documented in free-ranging waterfowl in Great Britain.[12]

Erosions under the tongue and in the oral cavity are common in ducks and geese that are documented carriers of the virus *Figure 7.22*. Virus has been shown to be present in these erosions, although not all latently infected birds that are shedding virus will develop this lesion. The detection of virus in these erosions implicates contaminated oral secretions in the transmission process.[14,89] Virus can be isolated from the cloaca and esophagus in birds that have recovered from active infections, indicating the importance of latently infected birds in the maintenance of flock infections.[14,43,135]

The frequency of virus shedding in individual latently infected ducks was evaluated over a one-year period. It ranged from 42% to 93% for Black Ducks and from 43 to 85% for Mallard Ducks. In another group of individual birds infected with a different strain of DPV, virus shedding was detected in 50% to 100% of the pintails, 66% to 100% of the gadwalls and 33% to 100% of the Mallard Ducks. Virus shedding reportedly occurred simultaneously in groups of latently infected ducks; the factors that initiated this synchronized shedding were not determined.[14]

Outbreaks are common in captive birds housed on ponds that are visited by latently infected, free-ranging waterfowl. The virus has been shown to be stable in cool (22°C) lake water for up to a month.[14,41,135,211] Once DPV has been introduced to a flock or a body of water, a cycle of continuous infection can be maintained when new susceptible waterfowl ingest contaminated food or water *Figure 7.23*. Because DPV is relatively stable outside of the host, it is likely that contaminated fomites are involved in the transmission of the virus from flock to flock. While free-ranging pigeons, starlings, sparrows and seagulls that frequent waterfowl ponds have not been shown to be susceptible to DPV infections, they could serve as mechanical vectors for dissemination of the virus.[137]

Duck plague virus was recovered from a fully formed egg that was in the cloaca of a hen that died.[89] It was subsequently proven that DPV could be vertically transmitted in some species of waterfowl, and that ducklings from infected hens will shed virus in their feces.[14,15,89] However, vertical transmission appears to be of little importance in the epizootiology of DPV because egg production is decreased dramatically in affected hens.

PATHOLOGY

Pathologic changes associated with DPV vary with the aggressiveness of the virus and the age, species and condition of the host. Gross lesions may or may not be present, but are most common

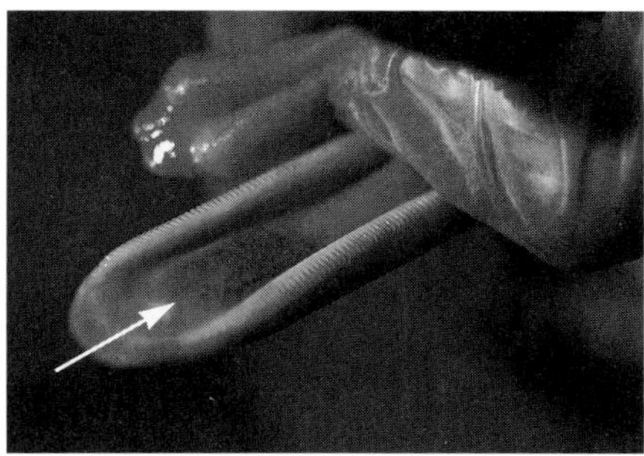

and most severe in waterfowl that survive for several days after the onset of clinical signs. In these birds, diphtheritic membranes and small foci of hemorrhage are common on the mucosa of the oral cavity, pharynx, esophagus, intestinal mucosa and cloaca. Annular bands of hemorrhage are common in the intestines of affected birds. Petechial hemorrhages on the surface of the esophagus and intestines are characteristic *Figure 7.24*. Other gross lesions that may be noted include an orange-colored liver, and hemorrhage of the bursa, liver, pancreas, lungs, heart, spleen, thymus and oviduct, and free blood in the body cavities. In reproductively active hens, the ovaries are frequently hemorrhagic.[43,121]

Gross lesions vary in experimentally infected Blue-winged Teals and geese. In the teals, lesions were limited to a small, dark spleen and occasional erosions in the esophagus and cloaca. Because lesions are uncommon in Blue-winged Teals and Wood Ducks, infections are usually diagnosed by histopathology or virus isolation. Lesions in infected goslings are more conspicuous and include an enlarged friable liver, and erosions and hemorrhage in the intestinal tract.[211] In other experimentally infected teal, gross lesions included petechiae on the heart, small dark spleens, hemorrhage in the

colonic mucosa and hemorrhage at the junction of the esophagus and proventriculus.[171]

Histologic lesions associated with DPV may include enteritis, proventriculitis, and necrosis of the liver, spleen, pancreas and epithelial cells lining the esophagus and intestines. Intranuclear inclusion bodies have been demonstrated in the liver and esophageal, proventricular and intestinal mucosa of affected birds.[43,121,135] In one report, intranuclear and intracytoplasmic inclusion bodies were identified in the cells lining the alimentary tract of experimentally infected ducks.[175]

PATHOGENESIS

As DPV damages the cells that line blood vessels, it causes bleeding into body cavities and the alimentary tract. The damage to blood vessels reduces the flow of blood to certain areas of an organ, causing the cells to die because they do not receive sufficient nutrients and oxygen. This virus also damages the cells that line the digestive tract, causing ulcers, and replicates in and damages the tissues that form the immune system, causing at least temporary immunosuppression.[147] It persists in the lymphoid tissues and intestines, and some surviving ducks and geese were found to be shedding DPV for years after the initial infection.[14,89]

IMMUNITY AND DIAGNOSIS

Detection of high levels of virus-neutralizing antibodies is suggestive of recent exposure to the virus. Demonstration of a rising titer in paired serum samples collected several weeks apart indicates an active infection.[42] In one group of waterfowl, 33% to 60% of the exposed birds had detectable levels of virus-neutralizing antibodies 19 to 35 days after clinical signs of disease were first noted in the flock. The levels of virus-neutralizing antibodies increased in experimentally infected ducks and geese by 21 days after infection, but were highest from 38 to 42 days after infection.[44,45]

The demonstration of antibodies against DPV in a flock would indicate previous exposure to the virus. In one study, experimentally infected Black Ducks had relatively low virus-neutralizing antibody titers (1:4 to 1:32) 17 months after they had been exposed to the virus.[14] In another study, neutralizing antibodies were not detected in 22 clinically normal birds two weeks after an outbreak.[135] Antibody assays are of little value in detecting latently infected waterfowl, because birds with and without detectable levels of neutralizing antibodies have been shown to shed the virus. Seropositive hens pass protective immunity to their chicks: however, these chicks are susceptible to infection by 13 days of age. Vaccinated hens were not found to pass maternally derived immunity to their chicks.[182]

At one time, waterfowl that recovered from DPV infections were considered to be latently infected and solidly immune against DPV-induced disease. In one study, birds that developed neutralizing antibodies following experimental infection were shown to be resistant to disease when challenged with live virus.[14] However, some latently infected waterfowl and chicks that are infected in the uterus will die when injected intramuscularly with virulent strains of DPV.[16] Nonetheless, waterfowl that recover from an infection are probably resistant to disease when they are exposed to virus through natural routes.

Duck plague should be considered in any waterfowl that dies acutely with or without clinical signs of gastrointestinal or neurologic disease. Influenza A virus, duck hepatitis virus, pasteurellosis and bacterial infections may cause similar-appearing changes. The virus is most likely to be isolated on cultured cells from the esophagus, liver and spleen of affected birds. Virus may be intermittently recovered from the feces or respiratory secretions of waterfowl with latent infections. However, DPV was not detected in cloacal swabs of any of 104 ducks, geese or swans tested two weeks after an initial outbreak on a pond in a zoological garden.[14,89,135] Electron microscopy may be used to detect viral particles in feces or swabs of oral lesions in samples that contain sufficient quantities (more than 1,000,000 particles per ml) of virus.

CONTROL

The likelihood of a DPV epornitic in captive waterfowl can be reduced through judicious quarantine procedures, good hygiene, rapid removal of clinically affected birds from a pond, and reduced exposure of captive birds to free-ranging Anseriformes. The cycle of infection within a flock tends to be self-limiting in closed, captive popula-

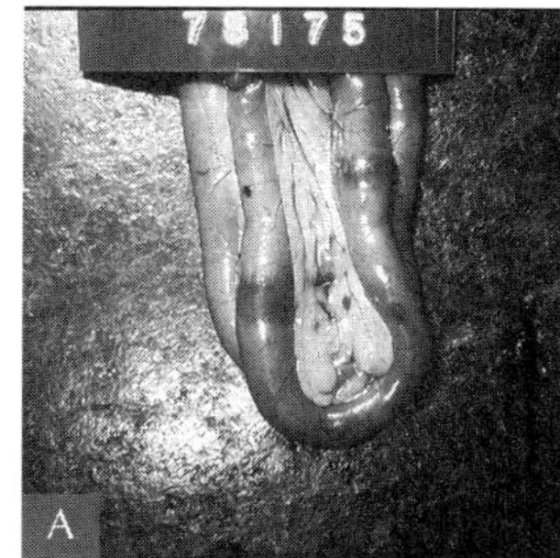
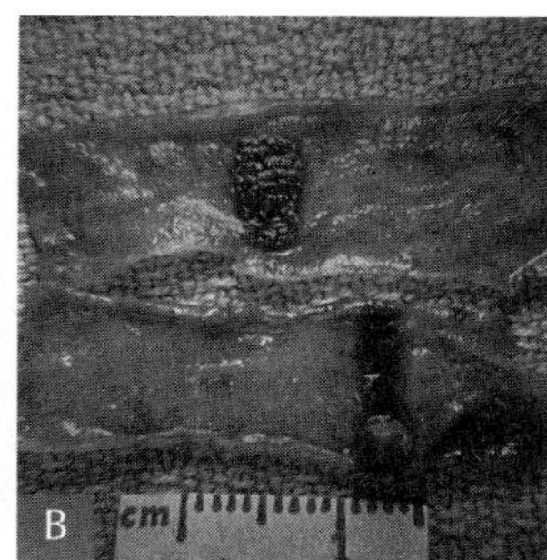
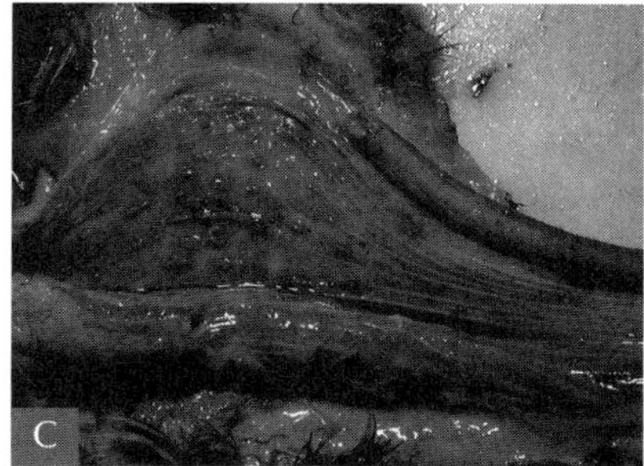
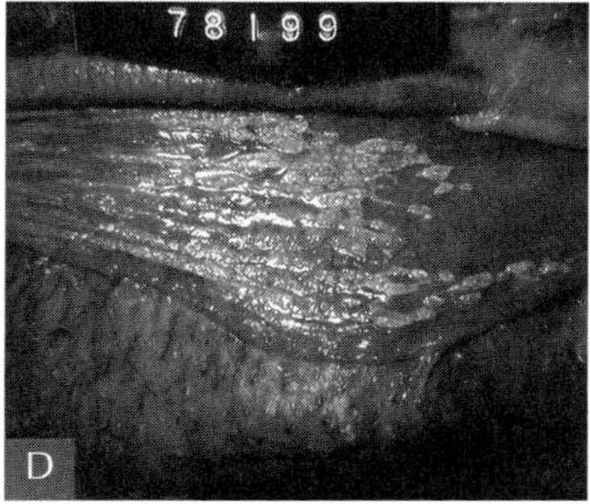
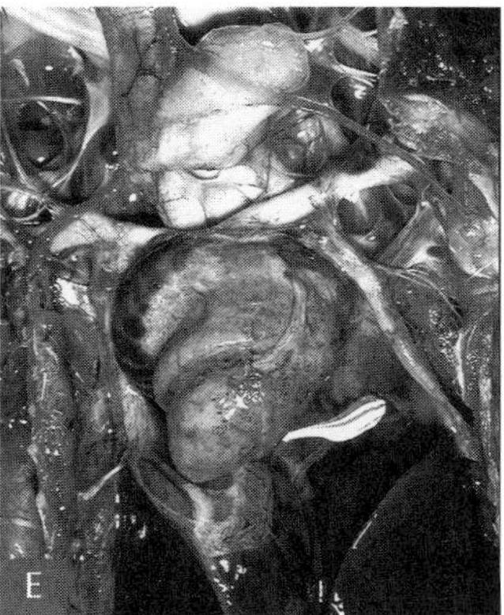

FIG 7.24
A, B Annular bands of hemorrhage are common in the intestines of birds infected with duck plague virus.
C, D Areas of hemorrhage and accumulation of sheets of necrotic debris are common on the mucosa of the esophagus and intestines.
E Petechial hemorrhages may occur in most organs but are particularly common in the heart , spleen and ovary.

photographs courtesy of Roy Montgomery and John Olsen

tions of waterfowl to which no new susceptible birds are added. Outbreaks are most likely to occur in crowded ponds or lakes that are frequently visited by free-ranging waterfowl, particularly Mallard Ducks.

The sensitivity of DPV to disinfectants is typical for a herpesvirus. The virus can be inactivated with most disinfectants and by heating to 60°C for 10 minutes or 50°C for two hours. The virus is stable at pH 5 to 9, but is inactivated at pH 3 and 11. Unlike other herpesviruses that are unstable when outside of the host, DPV has been shown to remain infectious in 4°C water for two months and in 2°C water for one month.

Attenuated-live virus vaccines have been shown to protect ducks and geese from DPV-induced disease and should be used during an outbreak to reduce the level of mortality.[89,135] In one study, the mortality rate in 200 unvaccinated ducklings challenged by intramuscular injection with live virus was 88%, while the mortality rate in vaccinated ducklings was 0.77%.[180,181] Vaccinating during an outbreak stopped mortality within two weeks.[137]

An attenuated-live virus vaccine is available in The Netherlands (Intervet International). The vaccine can be given by intramuscular or subcutaneous injection, and should be administered annually in areas where the virus is endemic. Because the isolates of DPV recovered from around the world appear to be similar, it is likely that the vaccine available from The Netherlands would be protective for most susceptible waterfowl.

Vaccinated birds develop rapid resistance to disease. In some studies, most of the ducklings infected with DPV and vaccinated four hours later remained clinically normal, while the unvaccinated birds died. In another study, some birds vaccinated from one to three days after exposure to the virus were shown to be resistant to disease. The immunity induced by vaccination in these birds is probably mediated by interference (the virus in the vaccine blocks cellular receptors) or interferon mediated resistance because neutralizing antibodies would not be expected to be produced so rapidly.[88,115] Other researchers have found that effective immunity does not occur until two to four days after vaccination.[180] The presence of virus-neutralizing antibodies does not correlate with protection from disease because vaccinated birds with no detectable antibodies were resistant to disease following experimental exposure.[89] Hyperimmune serum administered intramuscularly also was effective in preventing disease in experimentally challenged ducklings.[115]

An attenuated-live virus vaccine designed for oral administration has been shown to be effective in protecting waterfowl from disease following experimental infection. An oral vaccine would be superior to an injectable product for protecting large populations of captive and free-ranging birds.[22] Some research has shown that inactivated DPV vaccines are effective, while in other studies inactivated vaccines have performed poorly.[22,23,180,181] In either case, inactivated vaccines would not be expected to be as effective as the attenuated-live virus vaccines.

Hens with antibodies derived from natural infection will pass some protection to their chicks, but the resistance lasts less than two weeks. Maternally derived antibodies generated by vaccination have not proven to protect chicks from infection.[181]

HERPESVIRUS IN THE FEET OF DUCKS

An uncharacterized herpesvirus was recovered from the skin and plantar surface of the feet of free-flying ducks with ulcerative lesions.[212] The relationship of this herpesvirus to those recovered from waterfowl or other avian species remains unreported.

HERPESVIRUSES IN GRUIFORMES AND CICONIIFORMES

CRANE HERPESVIRUS

In the winter of 1973 at a zoological park in Austria, 12 Grey-crowned Cranes and 7 Demoiselle Cranes died over a 15-day period.[18,19] In 1978 in a breeding facility in the United States, 17 of 51 (33%) captive adult and immature Sandhill, Japanese, Stanley, Red-crowned and Hooded Cranes died over a 15-day period.[50] In 1982, a similar disease was described in a group of cranes from a zoological park in France.[57] A herpesvirus was implicated in all of these outbreaks. The disease induced by this virus is commonly referred to as "inclusion body disease" of cranes. Affected cranes died within two days of developing clinical signs that included depression, anorexia, diarrhea and an unwillingness to stand. Mortality ranged from 35% to 100% of the at-risk cranes.[19] Infections have been confirmed in captive Lesser Sandhill, Greater Sandhill, Manchurian, Sarus, Blue, Hooded, Demoiselle, Japanese, Stanley and Crowned Cranes. There is no evidence that infections occur in free-ranging cranes.[50,51,160,161]

Differences in the susceptibility of varying species of cranes to disease has been documented in natural and experimental infections. All five of a group of Common, Red-crowned and Sandhill Cranes exposed to the virus by intramuscular, intracloacal, intranasal or oral routes died within six to twelve days. A White-necked Crane experimentally exposed to the same virus preparation remained clinically normal and developed neutralizing antibody titers of 1:16 by the 14th day after exposure to the virus by the combined intracloacal and intranasal routes. Antibodies were no longer detectable in this bird by the 40th day after infection.[161] In another experimental trial, infected Sandhill and Blue Cranes were most susceptible to disease, while Sarus Cranes exposed to the same virus remained clinically normal but seroconverted, indicating that they had been infected.[50]

In a seroprevalance study, antibodies to crane herpesvirus were demonstrated in banked serum in a group of captive cranes collected three years before an outbreak of disease occurred. During a two-year test period, antibodies to the virus could not be demonstrated in the serum of 95 free-ranging Sandhill Cranes. In comparison, 9 of 11 captive cranes in an area where an outbreak occurred had antibodies.[50,51] In another seroprevalence study, 2 of 16 (12%) Red-crowned Cranes, 2 of 37 (5%) White-naped Cranes, 2 of 20 (10%) Hooded Cranes, 2 of 49 (4%) Black-crowned Cranes and 1 of 114 (0.9%) Grey-crowned Cranes had antibodies to the virus. Antibodies also were detected in 2 Blacked-necked Cranes. Even though the seroprevalence was considered low, it appeared from this study that cranes of Asian origin were more likely to be infected than cranes originating from other geographic regions.[46]

The crane herpesvirus is serologically related to the herpesvirus isolated from Bobwhite Quail.[57] Crane herpesvirus is serologically distinct from duck plague virus. However, a herpesvirus isolated from cranes did produce experimental disease in Pekin Ducklings up to 17 days old and in adult coots. Ducklings died 2 to 11 days after being injected intramuscularly with the virus. Ducklings 3 days of age were found to be more susceptible than those that were 17 days old suggesting an age-related predisposition. Muscovy Ducks and 16-day-old chickens were not susceptible to infection.[50]

Crane herpesvirus could be isolated from the oropharynx and cloaca in cranes as early as day two after infection. The virus was recovered in the highest concentrations from the feces just prior to death; however, many affected birds were not found to be shedding detectable concentrations of virus. Virus was isolated from the liver and spleen of experimentally affected cranes. The demonstration of virus in the trigeminal nerve of a previously infected crane confirmed that latent infections occur. However, shedding could not be induced in seropositive birds following injection of cyclophosphamide or dexamethasone (immunosuppresive agents). In an infected group of cranes, neutralizing antibodies to crane herpesvirus decreased over a three-year period, suggesting that the birds were not continuously exposed to endogenous or exogenous virus. Shedding in previously infected cranes was rare. Virus was not recovered from 1,300 cloacal or oropharyngeal swabs collected over a 22-month period. Crane herpesvirus has not been demonstrated in eggs from previously infected birds.[159,161]

Crane herpesvirus appears to have high affinity for parenchymal and lymphoid tissue. Gross lesions include hepatomegaly, splenomegaly, hemorrhage of the thymus, and hemorrhage and ulceration with the formation of diphtheritic membranes on the mucosa of the oral cavity, esophagus, small intestines and colon.[50] The most common changes appear to be diphtheritic lesions throughout the alimentary tract and swelling and tan-to-white discoloration of the liver.[160]

Microscopic changes include necrosis in the liver, spleen, bone marrow, mucosa of the alimentary tract, thymus and bursa. In one study, necrosis of the liver and spleen was present in all of the affected cranes examined.[160] Hemorrhage and congestion have been documented in the thymus, proventriculus, ventriculus, small intestines, liver, spleen and kidneys. Intranuclear inclusion bodies are most common in the liver, but may also occur in other tissues including the spleen, bursa, thymus and intestines.[50,159,160]

STORK HERPESVIRUS

A herpesvirus that is serologically distinct from Marek's disease virus, turkey herpesvirus, Lake Victoria cormorant virus, Pacheco's disease virus, owl herpesvirus and pigeon herpesvirus was isolated from two Black Storks that died acutely with no clinical signs. Because of the formation of inclusion bodies in the liver of infected birds, the disease has been referred to as "inclusion body disease" of storks. Infections have been confirmed in White and Black Storks in Germany.[106] Each affected stork had an enlarged, congested liver and yellowish, necrotic areas in the liver, spleen, esophagus and bone marrow. Ulcerated areas with diphtheritic membranes were described in the choana, esophagus, pharynx and larynx.[97]

Infected storks developed a persistent viremia even though neutralizing antibodies were detected in the serum. This herpesvirus circulates within the blood stream in association with leukocytes and can be recovered from the white blood cells of infected storks. Infected birds developed antibodies inconsistently. Thus, antibody assays are of limited value in predicting the status of disease or infection.[102,104] Experimentally inoculated young chickens, ducklings, a pigeon and an owl remained clinically normal.[106]

HERPESVIRUSES IN PASSERIFORMES

Uncharacterized herpesviruses that cause necrosis in the liver, spleen and bone marrow have been reported in finches (Estrildidae), weavers (Ploceidae) and canaries (Corduclidae).[96,97] As with other herpesviruses, infections in Passeriformes are probably characterized by latent infections with periodic shedding during periods of stress. Virus may be shed from the cloaca or oropharynx, and swabs collected from these areas can be used for virus isolation. Shedding appears to be particularly common during the breeding season.[96,97] An uncharacterized herpesvirus was found to selectively kill Gouldian Finches housed in mixed-species aviaries in Austria and Switzerland.[157]

FINCH CYTOMEGALOVIRUS

A herpesvirus that was placed in the cytomegalovirus subgroup of Herpesviridae was recovered from several groups of captive finches in Europe.[47,205] Cytomegaloviruses differ from other herpesviruses in being limited in host range and having a long replication cycle in which infected cells become enlarged.

In one outbreak caused by this herpesvirus, 15 species of captive finches, including Gouldian Finches, Crimson Finches, Red-faced Waxbills and Zebra Finches, developed clinical signs of disease, with mortality rates in Gouldian Finches reaching 70% of the at-risk population. All affected birds were indigenous to Australia. Clinical signs, including depression, anorexia, chemosis, conjunctivitis and dyspnea, lasted for 10 to 14 days in most affected finches. Gouldian Finches were particularly susceptible, with most affected birds dying over a five-day to two-week period after the first clinical signs were noted. The outbreak lasted for seven weeks.[47,205]

Gross changes in these finches included swollen, hyperemic eyelids that were sealed with crusts, melena in the small intestines, air sacculitis, hyperemia of the liver and the serosal surface of the jejunum, and esophageal ulcers. Histologic lesions included conjunctivitis, cytomegaly and karyomegaly of degenerating epithelial cells, hemorrhage in the lung, syrinx and bronchi, and diphtheritic lesions in the esophagus and choana. Intranuclear inclusion bodies were demonstrated within epithelial cells of the conjunctiva, esophagus and respiratory tract.[47,205]

Other infectious organisms that have been associated with conjunctivitis in Passeriformes include Newcastle disease virus, paramyxovirus-2, poxvirus, chlamydia, mycoplasma, candida, oxyspirura, thelazia, mycobacteria and many other bacteria.[47]

OTHER HERPESVIRUSES

LAKE VICTORIA CORMORANT VIRUS

Lake Victoria cormorant virus was isolated from the blood of a three-week-old Little Pied Cormorant in New South Wales, Australia. This herpesvirus isolate could not be connected with any specific lesions and was serologically distinct from infectious laryngotracheitis virus, Marek's disease virus, owl herpesvirus, Pacheco's disease virus and pigeon herpesvirus. A similar virus could not be recovered in the blood of 112 other birds, including other cormorants, Black Swan, Darter, Plover and English Sparrow. The virus was not experimentally infectious for chickens, Budgerigars or pigeons.[58]

PENGUIN HERPESVIRUS

A herpes-like virus was associated with loss of condition and respiratory distress in two adult Black-footed Penguins. The only abnormal changes noted at necropsy were the presence of thick plaques and exudate on the air sacs with hyperemia of the tracheal mucosa. Microscopic lesions, similar to those described for infectious laryngotracheitis virus, included syncytial cells containing intranuclear inclusion bodies in the sinuses, trachea and bronchi. The method by which the captive birds were exposed to this virus could not be determined.[108]

TOUCAN HERPESVIRUS

A herpesvirus that was not serologically related to Pacheco's disease virus was recovered from a toucan that died the day after developing clinical signs of depression and anorexia. The toucan had been shipped with two macaws that died five days earlier and were diagnosed with herpesvirus-induced hepatitis (Pacheco's disease virus) based on microscopic changes. The liver and spleen of the toucan were enlarged, and the liver was yellowish and friable. Necrosis was evident microscopically in the liver and spleen, and intranuclear inclusion bodies were seen in both tissues.[31]

REFERENCES

1. Aini I, Shih LM, Castro AE, et al: Comparison of herpesvirus isolates from falcons, pigeons and psittacines by restriction endonuclease analysis. J Wildl Dis 29:196-202, 1993.

2. Akrae M: Die Identifizierung und Charaktersierung eines Falkenpocken-und eines Psittaciden-pockenvirus aufgrand ihres Verhaltens in der Eikultur, souie serologischer Reaktionen und ihres Wirtsspektrums in Tierversuch, München, 1980.

3. Al Falluji S, Sheikhly FA, Tantawi HH: Viral encephalomyelitis of pigeons: Pathology and virus isolation. Avian Dis 23:777-784, 1979.

4. Arnold ID: An outbreak of psittacine herpesvirus in rosellas. Proc Assoc Avian Vet, 1990, pp 292-300.

5. Bagust TJ: Laryngotracheitis (Gallid-1) herpesvirus infection in the chicken 4. Latency establishment by wild and vaccine strains of ILT virus. Avian Pathol 15:581-595, 1986.

6. Bagust TJ, Calnek BW, Fahey KJ: Gallid-1 herpesvirus infection in the chicken. 3. Reinvestigation of the pathogenesis of infectious laryngotracheitis in acute and early post-acute respiratory disease. Avian Dis 30:179-190, 1986.

7. Barkoff M: Die Krankheiten des Uhus (*Bubo bubo*) und ihre Bedeutung fur die Wiederein-burgerung in der Bundersrepublik Deutshcland. University of Gießen, 1987.

8. Baudet AERF: Een sterfte onder eend-edn in Nederland, veroorzaakt door een filtreerbaar virus (vogelpest). Tijdschr Diergeneesk 50:455-459, 1923.

9. Biggs PM: Marek's disease. Vet Rec 81:583-588, 1967.

10. Biggs PM: The epidemiology of avian herpesviruses in veterinary medicine. Dev Biol Standard 52:3-11, 1982.

11. Boyle DB, Binnington JA: Isolation of a herpesvirus from a pigeon. Aust Vet J 49:54, 1973.

12. Brand GJ, Docherty DE: A survey of North American migratory waterfowl for duck plague (duck virus enteritis) virus. J Wildl Dis 20:261-266, 1984.

13. Bryson EJ, et al: Treatment of first episodes of genital herpes simplex virus infection with oral acyclovir. N Engl J Med 308:916-921, 1983.

14. Burgess EC, Ossa J, Yuill TM: Duck plague: A carrier state in waterfowl. Avian Dis 24:940-949, 1979.

15. Burgess EC, Yuill TH: Vertical transmission of duck plague virus (DPV) by apparently healthy DPV carrier waterfowl. Avian Dis 25:795-800, 1981.

16. Burgess EC, Yuill TM: Superinfection in ducks persistently infected with duck plague virus. Avian Dis 26:40-45, 1982.

17. Burtscher H:. Die virusbedingte Hepatosplenitis infectiosa strigum. 1. Mitteilung: Morphologische Untersuchungen. Path Vet 2:227-255, 1965.

18. Burtscher H, Grunberg W: Epizootic virus hepatitis in cranes (*Balearica pavonina* and *Anthropoidea virgo*). Proc Int Symp Ertankungen Zootiere, 1975, pp 277-279.

19. Burtscher H, Grunberg W: Herpesvirus - hepatitis bei Kranichen (Aves: Gruidae). 1. Pathomorphologische Befunde. Zbl Vet Med 26:561-569, 1979.

20. Burtscher H, Sibalin M: Herpesvirus strigis: Host spectrum and distribution in infected owls. J Wildl Dis 11:164-169, 1975.

21. Butterfield WK, Ata FA, Dardiri AH: Duck plague virus distribution in embryonating chicken and duck eggs. Avian Dis 13:198-202, 1969.

22. Butterfield WK, Dardiri AH: Serologic and immunologic response of wild waterfowl vaccinated with attenuated duck plague virus. Bull Wildl Dis Assoc 5:99-102, 1969.

23. Butterfield WK, Dardiri H: Serological and immunological response of ducks to inactivated and attenuated duck plaque virus. Bact Proc 100, 1968.

24. Callinan RB, Kefford B, Borland R, et al: An outbreak of disease in pigeons associated with a herpesvirus. Aust Vet J 55:339-341, 1979.

25. Calnek BW: Influence of age at exposure on the pathogenesis of Marek's disease. J Natl Can Instit 51:929-939, 1973.

26. Calnek BW, Addinger HK, Kahn DE: Feather follicle epithelium: a source of enveloped and infectious cell-free herpesvirus from Marek's disease. Avian Dis 14:219-233, 1970.

27. Calnek BW, Hitchner SB: Survival and disinfection of Marek's disease virus and the effectiveness of filters in preventing airborne dissemination. Poultry Sci 52:35-43, 1973.

28. Calnek BW, Witter RL: Marek's disease. *In* Calnek BW (ed): Diseases of Poultry 9th ed. Ames, Iowa, Iowa State Univ Press, 1991, pp 342-385.

29. Carranza J, Poveda JB, Fernandez A: An outbreak of encephalitis in pigeons (*Columba livia*)in the Canary Islands (Spain). Avian Dis 30:416-420, 1986.

30. Cartwright M, Spraker TR, McCluggage D: Psittacine inclusion body hepatitis in an aviary. J Am Vet Med Assoc 187:1045-1046, 1985.

31. Charlton BR, Barr BC, Castro AE, et al: Herpes viral hepatitis in a toucan. Avian Dis 34:787-790, 1990.

32. Cho BR, Kenzy SG: Marek's disease virus infection in zoo birds. Proc Am Assoc Zoo Vet, 1975, pp 3-4.

33. Cho BR, McDonald TL: Isolation and characterization of a herpesvirus of Pacheco's parrot disease. Avian Dis 24:268-277, 1980.

34. Chubb RC, Churchill AE: Effect of maternal antibody on Marek's disease. Vet Rec 85:303-305, 1969.

35. Colwell WM, Simpson CF, Williams LE, et al: Isolation of a herpesvirus from wild turkeys in Florida. Avian Dis 17:1-11, 1973.

36. Cornwell HJC, Weir AR, Follett EAC: A herpesvirus infection of pigeons. Vet Rec 81:267-268, 1967.

37. Cornwell HJC, Wright NG: Herpesvirus infection of pigeons. 1. Pathology and virus isolation. J Comp Path 80:221-227, 1970.

38. Crawshaw GJ, Boycott BR: Infectious laryngotracheitis in peafowl and pheasants. Avian Dis 26:397-401, 1982.

39. Curtis-Velasco M: Vaccination reaction in umbrella cockatoos. J Assoc Avian Vet 4:206, 1990.

40. Curtis-Velasco M Further vaccine reactions. J Assoc Avian Vet 5:10, 1991.

41. Dardiri AH: Transmission and certain disease features of duck plague. Proc 14th World Poult Congress, 1971, pp 55-64.

42. Dardiri AH: Duck viral enteritis (duck plague) characteristics and immune response of the host. Am J Vet Res 36:535-538, 1975.

43. Dardiri AH: Duck plague. *In* Rohner H Jena (ed): Hanbuch der Viruskrankheiten bei Tieren. Fischer, 1978, pp 1284-1286.

44. Dardiri AH, Gailiunas P: Response of Pekin and mallard ducks and Canada geese to experimental infection with duck plague virus. Bull Wildl Dis Assoc 5:235-247, 1969.

45. Dardiri AH, Hess WR: The incidence of neutralizing antibodies to duck plague virus in serums from domestic ducks and wild waterfowl in the United States of America. Proc 71st US Livestock San As, 1067, pp 225-237.

46. Dein FJ, Docherty DE: Preliminary findings from serosurvey for inclusion body disease of cranes. Proc Assoc Zoo Vet, 1992, pp 178.

47. Desmidt M, Duchatelle R, Uyttebroeck E, et al: Cytomegalovirus like conjunctivitis in Australian finches. J Assoc Avian Vet :132-136, 1991.

48. Devos A, Viaene N, Staelens M: Eendenpest in Belgie. Vlaams diergeneesk Tijdschr 33:260-266, 1964.

49. Docherty DE, et al: Isolation of a herpesvirus from a bald eagle nestling. Avian Dis 27:1162-1165, 1983.

50. Docherty DE, Henning DJ: The isolation of a herpesvirus from captive cranes with an inclusion body disease. Avian Dis 23:278-283, 1980.

51. Docherty DE, Romaine RI: Inclusion body disease of cranes: serological followup to the 1978 die-off. Avian Dis 27:830-835, 1982.

52. Dorrestein GM: Viral infections in racing pigeons. Proc Assoc Avian Vet, 244-257, 1992.

53. Durham PJK, Gumbrell RC, Clark RG: Herpesvirus hepatitis resembling Pacheco's disease in New Zealand parrots. New Zealand Vet J 25:168, 1977.

54. Ehrsam H, Hauser B, Metzler A: An acute outbreak of Pacheco's parrot disease in a domestic flock of parrots. Schweizer Archiv fur Tierheilkunde 120:23-28, 1978.

55. Fabey KJ, Bagust TJ, York JJ: Laryngotracheitis herpesvirus infection in the chicken. Avian Pathol 12:505-514, 1983.

56. Findlay GM: Pacheco's parrot disease. Vet J 89:12, 1933.

57. Foerster S, Chastel C, Kaleta EF: Crane hepatitis herpesviruses. J Vet Med 36:433-441, 1989.

58. French EL, Purchase HG, Nazerian K: A new herpesvirus isolated from a nestling cormorant. Avian Pathol 2:3-15, 1973.

59. Friend M, Pearson G: Duck plague: The present status. Proc 53rd Conf West Assoc State Fisheries Soc, 1973.

60. Fuchs A, Weissenbock H: Inclusion body hepatitis in psittacine birds and pigeons. Comparative histological and ultrastructural findings. Proc Europ Conf Avian Med Surg, 1993, pp 552-557.

61. Fudge AM: High risk birds still benefit. J Assoc Avian Vet 5:10-11, 1991.

62. Gaskin JM: Considerations in the diagnosis and control of psittacine viral infections. Proc Assoc Avian Vet, 1987, pp 1-14.

63. Gaskin JM: The serodiagnosis of psittacine viral infections. Proc Assoc Avian Vet, 1988, pp 7-10.

64. Gaskin JM: Psittacine viral disease. A perspective. J Zoo Wild Med 20:249-264, 1989.

65. Gaskin JM, Arnold BS, Robbins BS: An inactivated vaccine for psittacine herpesvirus infection (Pacheco's disease). Proc Am Assoc Zoo Vet, 1980, pp 102-105.

66. Gaskin JM, Raphael B, Major A, et al: Pacheco's disease: The search for the elusive carrier bird. Proc Am Assoc Zoo Vet, 1981, pp 24-28.

67. Gaskin JM, Robbins CM, Jacobson ER: An explosive outbreak of Pacheco's parrot disease and preliminary experimental findings. Proc Amer Assoc Zoo Vet, 1978, pp 241-253.

68. Gerlach H: Viral diseases. *In* Harrison GJ, Harrison LR (eds): Clinical Avian Medicine and Surgery. Philadelphia, WB Saunders Co, 1986, pp 408-433.

69. Gerlach H: Viruses. *In* Ritchie BW, Harrison GJ, Harrison LR (eds): Avian Medicine: Principles and Application. Lake Worth, FL, Wingers Publishing, 1994, pp 862-948.

70. Godwin JS, Jacobson ER, Gaskin JM: Effects of Pacheco's parrot disease virus on hematology and blood chemistry values of Quaker parrots. J Zoo An Med 13:127-132, 1982.

71. Goldberg DR, Yuill TM, Burgess EC: Mortality from duck plague virus in immunosuppressed adult mallard ducks. J Wildl Dis 26:299-306, 1990.

72. Gough RE, Alexander DJ, Collins MS, et al: Routine virus isolation or detection in the diagnosis of diseases in birds. Avian Pathol 17:893-907, 1988.

73. Graham DL, Mare CJ, Ward FP, et al: Inclusion body disease (herpesvirus infection) of falcons (IBDF). J Wildl Dis 11:83-91, 1975.

74. Green RG, Shillinger JA: A virus disease of owls. Amer J Path 12:405-410, 1936.

74a. Grimes JE: Personal communication.

75. Hallilwell WH. Lesions of Marek's disease in a great horned owl. Avian Dis 15:49-55, 1971.

76. Hanson LE, Bagust TJ: Laryngotracheitis. *In* Hofstad MS, et al (eds): Diseases of Poultry. Ames, Iowa State University Press, 1991, pp 485-495.

77. Hayles LB, Hamilton D, Newby WC: Transfer of parental immunity to infectious laryngotracheitis of chicks. Can J Comp Med 40:218-219, 1976.

78. Heffels U, Fritzsche K, Kaleta EF, et al: Serological examination for viral infections in pigeons in Germany. Dtsch Tierartzl Wschr 88:97-102, 1981.

79. Helfer DH, Schmitz JA, Seefeldt SL, et al: A new viral respiratory infection in parakeets. Avian Dis 24:781-783, 1980.

80. Hirai K, Hitchner SB, Calnek BW: Characterization of paramyxo- herpes- and orbiviruses isolated from psittacine birds. Avian Dis 23:148-163, 1979.

81. Hitchner SB, Calnek BW: Inactivated vaccine for parrot herpesvirus infection (Pacheco's disease). Am J Vet Res 41:1280-1282, 1980.

82. Hitchner SB, Fabricant J, Bagust TJ: A fluorescent-antibody study of the pathogenesis of infections laryngotracheitis. Avian Dis 21:185-194, 1977.

83. Holz B: Virus isolations from fecal samples collected from pigeon cabin cars. Proc VIII Symp Avian Dis, 1992, pp 1-6.

84. Horner RF, Parker M, Abrey ANS, et al: Isolation and identification of psittacid herpes-virus 1 from imported psittacines in South Africa. J South Afr Vet Assoc 63:59-62, 1992.

85. Hughes CS, Gaskell RM, Jones JC, et al: Effects of certain stress factors on the re-excretion of infectious laryngotracheitis virus from latently infected carrier birds. Res Vet Sci 46:274-276, 1989.

86. Hughes CS, Jones RC, Gaskell RM, et al: Demonstration in live chickens of the carrier state in infectious laryngotracheitis. Res Vet Sci 42:407-410, 1987.

87. Hughes WF: Marek's disease and avian encephalomyelitis. Proc 18th West Poulty Dis Conf, 1969.

88. Jansen J: The interference phenomenon in the development of resistance against duck plague. J Comp Path 74:327, 1964.

89. Jansen J: Duck plague. J Am Vet Med Assoc 152:1009-1016, 1968.

90. Jansen J, Kunst H: The reported incidence of duck plague in Europe and Asia. Tijdschr Diergeneesk 89:765-769, 1964.

91. Jordan FTW: A review of the literature on infectious laryngotracheitis. Avian Dis 10:1-26, 1966.

92. Jordan FTW: Immunity to infections laryngotracheitis. *In* Rose ME, et al (eds): Avian Immunology. British Poultry Science, 1981, pp 245-254.

93. Jordan FTW: Respiratory conditions of the fowl. Practice 4:64-73, 1982.

94. Jordan FTW, Evanson HM, Bennet JM: The survival of the virus of infectious laryngotracheitis. Zbl Vet Med B14:135-150, 1967.

95. Kaleta EF: Herpesvirus-induzierte infectionen und krankheiten des vogels. Tierarrztliche Praxis 11:67-75, 1983.

96. Kaleta EF: Herpesviruses of free-living and pet birds. *In* Purchase H, et al (eds): A Laboratory Manual for the Isolation and Identification of Avain Pathogens. Dubuque, Kendall/Hunt Publishing Co, 1989, pp 97-102.

97. Kaleta EF: Herpesviruses of birds - A review. Avian Pathol 19:193-211, 1990.

98. Kaleta EF: Taubenherpesvirus (Plagues grob-klein-Impfstoff). Briettaube, 1991.

99. Kaleta EF, Alexander DJ, Russell PH: The first isolation of the avian PMV-1 virus responsible for the current panzootic in pigeons? Avian Pathol 14:553-558, 1985.

100. Kaleta EF, Brinkmann MB: An outbreak of Pacheco's parrot disease in a psittacine bird collection and an attempt to control it by vaccination. Avian Pathol 22:785-789, 1993.

101. Kaleta EF, Druner K: Hepatosplenitis infectiosa strigum und andere Krankheiten der Greifvogel und Eulen. Ztbl Vet Med Suppl 25:173-180, 1976.

102. Kaleta EF, et al: Herpesvirus and Newcastle disease viruses in white storks (*Ciconia ciconia*). Avian Pathol 12:347-352, 1983.

103. Kaleta EF, Heffels U, Neumann N, et al: Serological differentiation of 14 avain herpesviruses by plaque reduction tests in cell cultures. Proc 2nd Intl Symp Vet Lab Diag, 1980.

104. Kaleta EF, Kummerfeld N: Persistent viremia of a cell-associated herpesvirus in white storks (*Ciconia ciconia*). Avian Pathol 15:447-453, 1986.

105. Kaleta EF, Marschall HJ, Gluender G, et al: Isolation and serological differentiation of a herpesvirus from Bobwhite Quail (*Colinus virginianus*). Arch Virol 66:359-364, 1980.

106. Kaleta EF, Mikami T, Marschall HJ, et al: A new herpesvirus isolated from black storks (*Ciconia nigra*). Avian Pathol 9:301-310, 1980.

107. Kaliner G: Intranuclear hepatic inclusion bodies in an African Grey Parrot. Avian Dis 19:640-642, 1975.

108. Kincaid AL, Bunton TE, Cranfield M: Herpesvirus-like infection in black-footed penguins. J Wild Dis 24:173-175, 1988.

109. Kitzing D: Zur Charakterisierning einies mit dem ILT-virus serologisch verwandten Herpesvirus ans Amazonen. Diss Med Vet München, 1981.

110. Kocan AA, Potgieter LND, Kocan KM: Inclusion body disease of falcons (herpesvirus infection) in an American kestrel. J Wildl Dis 13:199-201, 1977.

111. Kraft V, Teufel P: Nachweis eines Pockenvirus bei Zwergpapageien (*Agapornis personata* und *Agapornis roseicollis*). Berl Muench Tieraerztl Wochenschr 84:83-87, 1971.

112. Krautwald ME, et al: Nachweis eines neuen Herpesvirus bei einem ungewohnlichen Fall von Pacheco'scher Kranheit bei Amazonen and Graupapageien. J Vet Med 35:415-420, 1988.

113. Krautwald ME, Kaleta EF, Forester S: Heterogenicity of Pacheco's disease and its causative agents. Proc Assoc Avian Vet, 1988, pp 11-20.

114. Krupicka V, Smid B, Valicek L, et al: Isolation of a herpesvirus from pigeon on the chorioallantoic membrane of embryonated eggs. Vet Med (Praha) 15:609-612, 1970.

115. Lam KM, Lin W: Antibody-mediated resistance against duck enteritis virus infection. Can J Vet Res 50:380-383, 1986.

116. Lehner NDM, Bullock BC, Clarkston TB: Intranuclear inclusion disease of pigeons. J Am Vet Med Assoc 151:939-941, 1967.

117. Leibovitz L: Progress Report: Duck plague surveillance of American Anseriformes. Bull Wild Dis Assoc 4:87-91, 1968.

118. Leibovitz L: Gross and histopathologic changes of duck plague (duck virus enteritis). Am J Vet Res 32:275-290, 1971.

119. Leibovitz L: Duck plague (duck viral enteritis). *In* Hofstad ME (ed): Diseases of Poultry. Ames, Iowa State University Press, 1978, pp 621-632.

120. Leibovitz L: Duck virus enteritis. *In* Purchase HG, et al: A Laboratory Manual for the Isolation and Identification of Avian Pathogens 3rd ed. Dubuque, Kendall/Hunt Publishing Co, 1989, pp 95-96.

121. Leibovitz L: Duck virus enteritis (duck plague). *In* Calnek BW, et al (eds): Diseases of Poultry 9th ed. Ames, Iowa State University Press, 1991 pp 609-618.

122. Leibovitz L, Huang J: Duck plague in American Anseriformes. Bull Wildl Dis Assoc 4:13-14, 1968.

123. Leibovitz L, Hwang J: Duck plague on the American continent. Avian Dis 12:361-378, 1988.

124. Lowenstine LJ: Diseases of psittacines differing morphologically from Pacheco's disease, but associated with herpes-like particles. Proc 31st West Poult Dis Conf, 1982, pp 141-142.

125. Lowenstine LJ: Emerging viral diseases of psittacine birds. *In* Kirk RW (ed): Current Veterinary Therapy IX. Philadelphia, WB Saunders Co, 1986, pp 705-710.

126. Lucam F: La peste aviare en France. Proc XIVth Intern Vet Cong, 1949, p 380.

127. Mare CJ: Herpesviruses of birds of prey. J Zoo Anim Med 6:6-11, 1975.

128. Mare CJ, Graham DL: The isolation and characterization of herpesvirus of birds of prey. Proc US Anim Health Assoc, 1972, pp 444-451.

129. Mare CJ, Graham DL: Falcon herpesvirus, the etiological agent of inclusion body disease of falcons. Infect Immun 8:118-126, 1973.

130. Mare CJ, Graham DL: Pathogenicity and host range of the falcon herpesvirus. *In* Page LA (ed): Wildlife Diseases. New York, Publishing Corporation, 1976, pp 471-482.

131. Martin HT, Early JL: The isolation of herpesvirus from psittacine birds. Vet Rec 105:256-258, 1979.

132. McCluggage D: Pacheco's parrot disease in a psittacine breeding aviary. Proc Assoc Avian Vet, 1988, pp 115-120.

133. Meulemans G, Dekegel D, Peeters J, et al: Isolation and characterization of a herpesvirus from parrots. Vlaams Diergeneeskundig Tijdschrift 6:455-461, 1978.

134. Minick CR, Fabricant CG, Fabricant J, et al: Atheroarteriosclerosis induced by infection with a herpesvirus. Am J Path 96:673-700, 1979.

135. Montali RJ, Bush M, Greenwell GA: An epornitic of duck viral enteritis in a zoological park. J Am Vet Med Assoc 169:954-958, 1976.

136. Mukerjii AMS, Gosh BB, Ganguly JL: Duck plague in West Bengal. Indian Vet J 40:457, 1963.

137. Newcomb SS: Duck virus enteritis (duck plague). Epizootiology and related investigations. J Am Vet Med Assoc 153:1897-1902, 1968.

137a. Niagro FD: Personal communication.

138. Norton TM, Gaskin JM, Kollias GV, et al: Efficacy of acyclovir against herpesvirus infection in Quaker Parakeets. Am J Vet Res 52:2007-2009, 1991.

139. Norton TM, Kollias GV, Gaskin JM, et al: Acyclovir pharmacokinetics and the efficacy of acyclovir against Pacheco's parrot disease in Quaker Parakeets. Proc Assoc Avian Vet, 1989, pp 3-5.

140. O'Brien JJ, Campolin-Richards DM: Acyclovir: An updated review of its antiviral activity, pharmacokinetic properties, and therapeutic efficacy. Drugs 37:233-309, 1989.

141. Odagiri Y: Experimental pathology on infectious laryngotracheitis of poultry. Bull Univ Osaka Prefect 34:75-131, 1982.

142. Pacheco G, Bier O: Epizootic chez les perroquets du Bresil. Relations avec la psittacose. CR Soc Biol [Paris] 105:109-111, 1930.

143. Parrot T: New clinical trials using acyclovir. Proc Assoc Avian Vet, 1990, pp 237-238.

144. Pieper K, Kaleta EF: Virus isolations from psittacine birds. Proc Europ Chap Assoc Avian Vet, 1991, pp 199-201.

145. Pollard B, Harais ES: Pigeon herpesvirus confirmed in South Africa. J S Afr Vet Assoc 54:247-248, 1983.

146. Powell PC, Payne LN: Marek's disease. *In* McFerran JB, McNulty MS (eds): Virus Infection of Birds. New York, Elsevier Science Publishers, 1993, pp 37-75.

147. Procter SJ: Pathogenesis of duck plague in the bursa of Fabricius, thymus and spleen. Am J Vet Res 37:427-431, 1976.

148. Ramis A, Fondevila D, Tarres J, et al: Immunocytochemical diagnosis of Pacheco's disease. Avian Pathol 21:523-527, 1992.

149. Ramis A, Latimer KS, Niagro FD, et al: Diagnosis of psittacine beak and feather disease (PBFD) viral infection, avian polyomavirus infection, adenovirus infection and herpesvirus infection in psittacine tissues using DNA in situ hybridization. Avian Pathol 23:643-657, 1994.

150. Randall CJ, Dagless MD, Jones HGR, et al: Herpesvirus infection resembling Pacheco's disease in Amazon parrots. Avian Pathol 8:229-238, 1979.

151. Redig PT: Health management of raptors trained for falconry. Proc Assoc Avian Vet, 1992, pp 258-264.

152. Reichman RC, et al: Treatment of recurrent genital herpes simplex infections with oral acyclovir. J Am Med Assoc 251:2103-2107, 1984.

153. Rivers TM: A recently described disease of parrots and parakeets differing from psittacosis. J Exp Med 29:155-156, 1931.

154. Samorek-Salamonomicz E, Cakala A, Wijaszka TL: Effect of acyclovir on the replication of turkey herpesvirus and Marek's disease virus. Res Vet Sci 42:334-338, 1987.

155. Schettler CH: Eine infektiose Leberentzundung (Hepatitis) der Eulen. Tieraertzl Umschau 24:163-167, 1969.

156. Schmidt RE: From a pathologist's perspective. J Assoc Avian Vet 5:10, 1991.

157. Schonbauer M, Kohler H: Uber eine Virusinfektion bei Prachfinken (Estrildidae). Kleintier Praxis 27:149-152, 1982.

158. Schroder D. Untersuchungen am Herpesvirus der Hepatosplenitis infectiosa strigum. VII DVG-Tagung Vogelkrht, 1990, pp 298-304.

159. Schuh J, Yuill T: Inclusion body disease of crane virus: A herpesvirus of captive cranes. Proc Am Assoc Zoo Vet, 1982, pp 19-20.

160. Schuh JCL, Sileo L, Siegfried LM, et al: Inclusion body disease of cranes: Comparison of pathologic findings in cranes with acquired versus experimentally induced disease. J Am Vet Med Assoc 189:993-996, 1986.

161. Schuh JCL, Yuill TM: Persistence of inclusion body disease of cranes virus. J Wildl Dis 21:111-119, 1985.

162. Seene DA, Pearson JE, Miller LD, et al: Virus isolations from pet birds submitted for importation into the United States. Avian Dis 27:731-744, 1983.

163. Sharma JM: Marek's disease. In Purchase HG, et al (eds) A Laboratory Manual for the Isolation and Identification of Avian Pathogens. Dubuque, Kendall/Hunt Publishing Co, 1989, pp 89-94.

164. Sileo LH, Carlson HC, Crumley SC: Inclusion body disease in a great horned owl. J Wild Dis 11:92-96, 1975.

165. Simpson CF, Hanley JE: Pacheco's parrot disease of psittacine birds. Avian Dis 21:209-219, 1977.

166. Simpson CF, Hanley JE, Gaskin JM: Psittacine herpesvirus infection resembling Pacheo's parrot disease. J Inf Dis 131:390-396, 1975.

167. Sinkovic B: The influence of environmental temperatures, age, sex and breed on the mortality from infections laryngotracheitis. 14th World Poultry Congress, 1970, pp 484-486.

168. Smadel JE, Jackson EB, Harman JW: A new virus disease of pigeons. Recovery of the virus. J Exptl Med 81:385-399, 1945.

169. Smith CG: Use of acyclovir in an outbreak of Pacheo's parrot disease. J Assoc Avian Vet 1:55-57, 1987.

170. Snyder SB, Fox JG, Campbell LH, et al: An epornitic of duck virus enteritis (duck plague) in California. J Am Vet Med Assoc 163:647-652, 1973.

171. Spieker JO: Virulence assay and other studies of six North American strains of duck plague virus tested in wild and domestic waterfowl. University of Wisconsin, 1978.

172. Strauss SE, Croen KD, Sawyer MH, et al: Acyclovir suppression of frequently reocurring genital herpes: Efficacy and diminishing need during successive years of treatment. J Am Vet Med Assoc 260:2227-2230, 1988.

173. Surman PG, Purcell DA, Tham VL, et al: The isolation of a herpesvirus from a pigeon and experimental infection in psittacine birds. Aust Vet J 51:537-538, 1975.

174. Suwatanaviorj V, Tantaswasdi U, Kuhawanta S, et al: Duck plague or duck virus enteritis. Bull Off Int Epizoot 88:619-624, 1977.

175. Tantaswasdi U, Wattanavijarn W, Methiyapun S, et al: Light, immunofluorescent and electron microscopy of duck virus enteritis (duck plague). Jpn J Vet Sci 50:1150-1160, 1988.

176. Tantawi HH, Al-Abdulla JM, Abdul-Mohaimen N, et al: Viral encephalomyelitis of pigeons. VI. Some physico-chemical properties of the virus and extracted viral DNA. Avian Dis 25:272-278, 1981.

177. Tantawi HH, Hassan FK: Pigeon herpes encephalomyelitis virus in Egypt. Trop Anim Health Prod 14:20-22, 1982.

178. Tantawi HH, Iman ZI, Mare J, et al: Antigenic relatedness of pigeon herpes encephalomyelitis virus to other avian herpesviruses. Avian Dis 27:563-568, 1983.

179. Thiry E, Vindevogel H, Leroy P, et al: In vivo and in vitro effect of acyclovir on pseudorabies virus, infectious bovine rhinotracheitis virus, and pigeon herpesvirus. Ann Rech Vet 14:239-245, 1983.

180. Toth TE: Active immunization of white pekin ducks against duck viral enteritis with modified live vaccine. Immunization of ducklings. Am J Vet Res 31:1275-1281, 1970.

181. Toth TE: Active immunization of white pekin ducks against duck enteritis (duck plague) with modified-live-virus vaccine: Serologic and immunologic response of breeder ducks. Am J Vet Res 32:75-81, 1971.

182. Toth TE: Two aspects of duck virus enteritis: Parental immunity and persistence/excretion of virulent virus. Proc 74th Ann US Animal Hlth Assoc, 1971, pp 304-314.

183. Trapp AL, Lowrie PM, Roberts AW: Inclusion body hepatitis resembling Pacheco's parrot disease in psittacine birds in Michigan. Proc Am Assoc Vet Lab Diag AAVLD, th Ann Proc, 1977, pp 365-374.

184. Tripathy DN, Hanson LE: Laryngotracheitis. In Purchase HG, et al (eds): A Laboratory Manual for the Isolation and Identification of Avian Pathogens 3rd ed. Dubuque, Kendall/Hunt Publishing Co, 1989, pp 85-88.

185. Tritt S, Spenkoch-Piper H: Serotyping of psittacine herpesvirus. Proc European Conf Avian Med Surg, 1993, pp 241-252.

186. Tsai SS, Park JH, Hirai K, et al: Herpesvirus infections in psittacine birds in Japan. Avian Pathol 22:141-156, 1993.

187. Van Dorssen CA, Kunst H: Over de gevoeligheid van eenden en diverse andere waterrogels voor eendenpest. Tijdscher diergeneesk 80:1286-1295, 1955.

188. Vindevogel H: Le coryza infectieux du pigeon. University of Liege Faculty of Veterinary Medicine.

189. Vindevogel H, Aguilar-Setien A, Dagenais L, et al: Diagnostic de l'infection herpetique du pigeon. Ann Med Vet 124:407-418, 1980.

190. Vindevogel H, Dagenais L, Lansival B, et al: Incidence of rotavirus, adenovirus, and herpesvirus infection in pigeons. Vet Rec 109:285-286, 1981.

191. Vindevogel H, Debruyne H, Pastoret PP: Observation of pigeon herpesvirus 1 reexcretion during the reproduction period in conventionally reared homing pigeons. J Comp Path 95:105-112, 1985.

192. Vindevogel H, Duchatel HP: Receptivite de la perruche au virus herpes du pigeon. Ann Med Vet 121:193-195, 1977.

193. Vindevogel H, Duchatel JP, Burtonboy G: Infection herpetique de psittacides. Ann Med Vet 122:167-169, 1978.

194. Vindevogel H, Kaeckenbeeck A, Pastoret PP: Frequence de l'ornithose-psittacose et de l'infection herpetique chez la pigeon voyageur et les psittacides en Belgique. Rev Med Liege 36:693-696, 1981.

195. Vindevogel H, Pastoret PP: Pigeon herpetic infection: Natural transmission of disease. J Comp Path 90:409-413, 1980.

196. Vindevogel H, Pastoret PP: Pathogenesis of pigeon herpesvirus infection. J Comp Path 91:415-426, 1981.

197. Vindevogel H, Pastoret PP: Pathogenesis and latency of pigeon herpesvirus 1 (PHV-1). Proc III Conf Avian Dis, 1983, pp 78-85.

198. Vindevogel H, Pastoret PP, Aguilar-Setien A: Assessment of phosphonoformate-treatment of pigeon herpesvirus infection in pigeons and Budgerigars, and Aujesky's disease in rabbits. J Comp Path 92:177-180, 1982.

199. Vindevogel H, Pastoret PP, Burtondoy G: Pigeon herpes infection: Excretion and re-excretion of virus after experimental infection. J Comp Path 90:401-408, 1980.

200. Vindevogel H, Pastoret PP, Leroy P, et al: Comparison of three strains of herpesvirus isolated from psittacine birds with the pigeon herpesvirus. Avian Pathol 9:385-394, 1980.

201. Vindevogel H, Pastoret PP: Vaccination trials against pigeon herpesvirus infection (Pigeon herpesvirus 1). J Comp Path 92:483-494, 1982.

202. Vindevogel H, Pastoret PP, Thiry E: Latency of pigeon herpesvirus 1. Latent Herpesvirus Infections in Veterinary Medicine. Martinus Nijhof Publ, 1984, pp 489-499.

203. Vindevogel H, Pastoret PP, Thiry E, et al: Reapparition de formes graves de la maladie de Newcastle chez le pigeon. Ann Med Vet 126:5-7, 1982.

204. Vogel K, et al: Herpesvirus-1-Infektion der Tauben. In Heider, et al (eds): Kranheiten des Wirtschaffgeflugels. Ein Handbuch fur Wissenschaff und Praxis. Gusfav Fischer Verlag, Stuggart, Jena, 1992, pp 439-450.

205. Von Rotz A, et al: Lethal verlauferide Herpesvirus infektion bei Gouldsamandinen (Cholebia gouldiae). Schweiz Arch Tierheilk 126:651-658, 1984.

206. Ward FP, Fairchild DG, Vuicich JV: Inclusion body hepatitis in a prairie falcon. J Wildl Dis 7:120-124, 1971.

207. Weingarten M. Entenpest: Klinik, diagnose, bekamfung. VI DVG-Tagung Vogelkuht, 1988, pp 197-203.

208. Winterroll G: Herpesvirus infection of psittacines. Praktische Ticrarzt 58:321-322, 1977.

209. Winterroll G, Gylstorff I Schwere durch Herpesvirus Verursachte Erkankung des Respirationsapparates bei Amazonen. Berl Munch Tierar Wochen 92:277-288, 1979.

210. Witter RL, Moulthrop JL, Burgoyne GH, et al: Studies on the epidemiology of Marek's disease herpesvirus in broiler flocks. Avian Dis 14:255-267, 1970.

211. Wobeser G: Experimental duck plaque in blue-winged teal and Canada geese. J Wild Dis 23:368-375, 1987.

212. Wojcinski ZW, Wojcinski HSJ, Barker IK, et al: Cutaneous herpesvirus infection in a mallard duck (Anas platyrhynchos). J Wild Dis 27:129-134, 1991.

213. Yamada S, Matsuo K, Fukuda T, et al: Susceptibility of ducks to the virus of infectious laryngotracheitis. Avian Dis 24:930-938, 1980.

214. York JJ, Fahey KJ: Vaccination with affinity-purified glycoproteins protects chickens against infectious laryngotracheitis herpesvirus. Avian Pathol 20:693-704, 1991.

215. York SM, York CJ: Pacheco's virus vaccine studies. Proc 32nd West Poult Dis Conf, 1983, pp 101-103.

Circoviridae

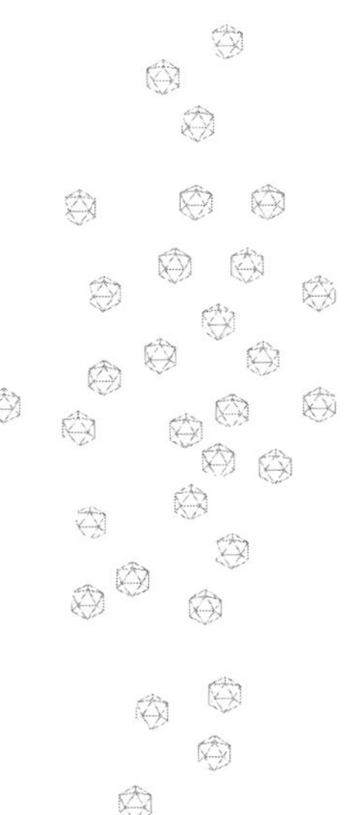

PSITTACINE BEAK AND FEATHER DISEASE

A chronic disease — characterized by symmetric feather dystrophy and loss, development of beak deformities and eventual death — was first described in various species of Australian cockatoos in the early 1970s.[29] Subsequently named psittacine beak and feather disease (PBFD), the syndrome has been diagnosed in many other psittacine species. It is possible that the disease had been noticed as early as 1887 by Australian explorers who described characteristic feather changes in free-ranging Red-rumped Parrots (*Psephotus* sp.) in South Australia.[1]

Throughout the 1970s and early 1980s, both infectious and noninfectious etiologies were proposed for PBFD. Two suspected endocrine abnormalities were disproved by showing normal pre- and post-adrenocorticotropic hormone (ACTH) plasma corticosterone concentrations and normal thyrotropin concentrations after administration of TSH.[9] A group of infectious agents that commonly infects domestic fowl was ruled out by showing that serum samples from diseased birds were negative for *Mycoplasma gallisepticum, M. synoviae*, Newcastle disease virus, reovirus, adenovirus, *Salmonella pullorum*, infectious bursal disease virus, reticuloendotheliosis virus and infectious bronchitis virus.[29] Some early literature describing PBFD suggested that avian polyomavirus might be the etiologic agent. However, polyomavirus intranuclear inclusion bodies that contain 40 to 50 nm viral particles were not consistently demonstrated in PBFD-affected birds, and PBFD-positive birds did not consistently have polyomavirus-neutralizing antibodies.[5,8,9,17] In North America and parts of Europe, an avian polyomavirus was isolated from Budgerigars with feather lesions similar to those described for French moult and PBFD (see Chapter 6).[11,26,27,31]

Within the past seven years, the etiology of PBFD has been confirmed by identifying a virus that produces the disease in susceptible psittacine birds.[41,48] The disease has been experimentally reproduced in neonatal Budgerigars and Galahs by using crude feather homogenates containing 19 to 22 nm viral particles,[48] in young Sulphur-crested Cockatoos and Galahs by using a partially purified virus preparation produced from the feathers of infected birds[34] and in neonatal Budgerigars, African Grey Parrots and Umbrella Cockatoos by using purified concentrated PBFD virions.[40,41]

VIRUS CHARACTERISTICS

The PBFD virus is a 14 to 16 nm nonenveloped virion containing a single-stranded circular DNA molecule. Virus with similar morphologic characteristics, protein composition, antigenic properties and nucleic acid sequence can be consistently demonstrated in many species of psittacine birds with clinical or histologic lesions suggestive of PBFD.[16,37]

Based on the virion dimension, protein composition and nucleic acid size and conformation, the etiologic agent of PBFD has been shown to be a member of a new family of pathogenic animal viruses.[41] The virion size and nucleic acid characteristics described for the PBFD virus are similar to those found for the chicken anemia virus (CAV), pigeon circovirus and porcine circovirus (PCV).[45,46] Because of their unique characteristics, these viruses have been tentatively placed in a new family of animal viruses named Circoviridae. These are currently the smallest pathogenic animal viruses that have been described. Each of these viruses appears to be specific in the host it infects, and they have distinct protein and nucleic acid characteristics. Thus, PBFD virus would not be expected to infect pigeons or chickens, and pigeon circovirus would not be expected to infect psittacine birds or chickens.

CLINICAL FEATURES

The first clinically detectable sign of PBFD is the appearance of necrotic, abnormally formed feathers. Feathers are normally arranged in distinct tracts called pterylae. Both the normal pattern of molt and the appearance of dystrophic feathers in birds with PBFD occur separately in each specific feather tract. The type of feathers that are initially involved and the distribution of feather loss depend on the stage of molt when the clinical signs of disease are manifest. In young birds (less than two months old), all of the feather tracts may be affected during a one-week period, whereas in older birds, the disease is more prolonged with progressive feather changes during ensuing molts *Figure 8.1*.[26,27,31] A temporary cessation in the appearance of new, abnormally formed feathers may be noted in the periods between molts. Progression of beak abnormalities may occur during the inter-molt period.[5]

Peracute, acute and chronic disease patterns that differ markedly in clinical presentation have been described in association with PBFD virus infections. The type of clinical disease is controlled principally by the age of the bird when feather abnormalities first occur, but also may be influenced by the route of viral exposure, the titer of the infecting virus and the condition of the bird when viral exposure occurs.

Peracute infections should be suspected in neonatal psittacines that show signs of septicemia accompanied by pneumonia, enteritis, rapid weight loss and death.[26] Peracute disease appears to be particularly common in young cockatoos and African Grey Parrots, and these birds may die before feather abnormalities are easily recognized. Because feather lesions may be difficult to detect or are absent in birds that die with a peracute PBFD virus infection, a correct diagnosis may be missed if a complete necropsy and thorough histologic examination are not performed *Figure 8.2*. Histologic changes associated with peracute PBFD virus infections may be limited to edema in the cells that line the feather follicle and severe

FIG 8.1
Psittacine beak and feather disease virus can cause differing clinical changes depending on the stage of molt when a bird begins to develop feather abnormalities.

Chicks that develop clinical lesions while the majority of feathers are still in a developmental stage exhibit the most severe feather pathology. This Umbrella Cockatoo appeared normal up to 34 days after being exposed to the virus when hemorrhage was noted in a single feather (arrow).

Within a week, all of this cockatoo's feathers were dying and falling out of their follicles. The affected bird died soon after this photograph was taken.

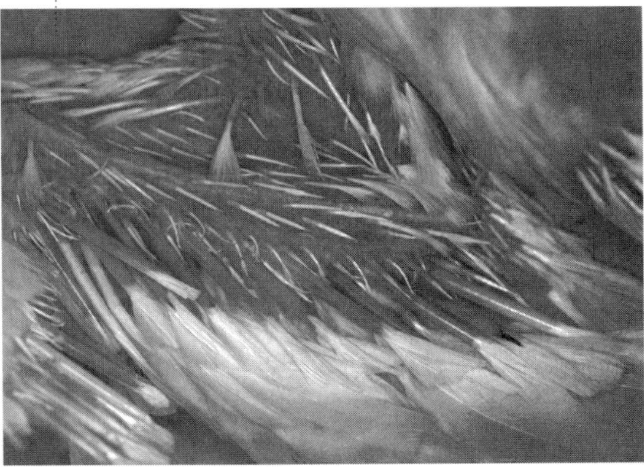

Little Corellas from an infected hen were completely normal ten days before this photograph was taken. Note that all of the developing body, flight and crest feathers have been damaged by the virus.

This Little Corella's body feathers were already developed before clinical signs of disease occurred. Thus, the bird's overall feather condition appears good; however,

when the still developing flight feathers were removed from the bird, constrictions and hemorrhage suggestive of PBFD were evident.

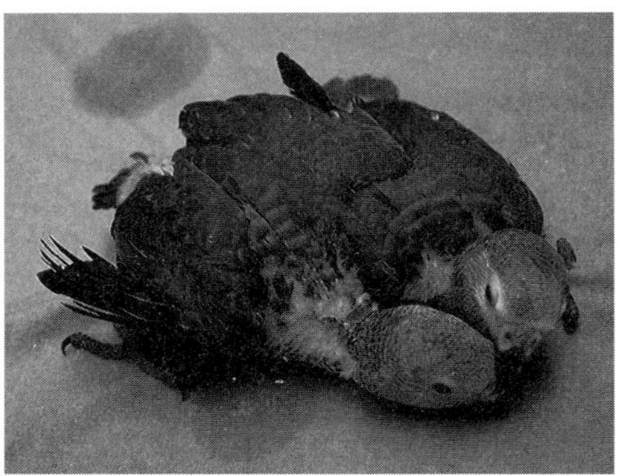

necrosis of the bursa and thymus. It has been suggested that the disease may also be a cause of embryonic mortality.[29] Collecting blood for DNA probe testing is the easiest way to diagnosis the presence of PBFD virus in a suspect bird (see Diagnosis).

The acute form of PBFD, commonly called "French moult" in Australia, is most frequently reported in young birds during their first feather formation after replacement of the neonatal down.[30] Chicks as young as 28 to 32 days of age have been described with classic lesions. Acute infections are characterized by several days of depression followed by sudden changes in developing feathers including necrosis, fractures, bending, hemorrhage or premature shedding of diseased feathers.[40,48] In some acute cases of PBFD, birds with minimal feather changes may be depressed and develop crop stasis and diarrhea followed by death in one to two weeks.[26,27,29,31]

In the acute form of the disease, gross feather lesions can be quite subtle, with only a few feathers showing dystrophic changes. This clinical picture is described as being particularly common in young Sulphur-crested Cocka-

toos and lovebirds in Australia.[27,31] Experimentally infected Galah, Umbrella Cockatoo and African Grey Parrot chicks were found to develop acute infections characterized by a rapid onset of depression, followed in 24 to 48 hours by the appearance of abnormal feathers (see Experimental Infections).[40,48] Chicks that develop clinical lesions while the majority of feathers are still in a developmental stage exhibit the most severe feather pathology. These birds may appear totally normal one day and exhibit 80% to 100% feather dystrophy within a week.[40] The clinical progression of disease is less dramatic in neonates that develop clinical signs after body contour feathers are mature. In these birds, feather changes may be limited to the still-developing flight and tail feathers *see Figure 8.1*.[27,31]

A non-regenerative anemia (packed cell volume = 14% to 25%) has been reported in African Grey Parrots with an acute presentation of the disease. These birds may die suddenly, and a diagnosis is confirmed by demonstrating virus-containing inclusion bodies in the bone marrow, thymus or bursa. Research has not determined whether the anemia is caused by the PBFD virus or is induced by some secondary pathogen that takes advantage of a weakened immune system.

Chronic PBFD is characterized by the symmetric, progressive appearance of abnormally developed feathers during each successive molt. This clinical presentation occurs in birds that survive the acute phase of the disease. Gross changes include retention of feather sheaths, hemorrhage within the pulp cavity and fractures of the feather shaft. Short clubbed feathers, deformed curled feathers, stress lines within vanes and circumferential constrictions may also be present in PBFD-positive birds *Figure 8.3*.[9,16,26,27,29,31] Many of the gross feather lesions are induced by the retention of hyperkeratotic feather sheaths.[9,27] Replacement feathers become increasingly abnormal.

The distribution of dystrophic feathers within individual pterylae is variable and depends upon the stage of molt when the bird begins to develop clinical signs. In older birds the first sign of PBFD is the replacement of normal powder down and contour feathers with dystrophic, necrotic feathers that stop growing shortly after emerging from the follicle. It has been assumed that powder down feathers are frequently the first to be affected because of their consistent molt pattern compared to the seasonal loss found in other feather tracts. The disease then progresses to involve the contour feathers in most tracts, followed by dystrophic changes in the primary, secondary, tail and crest feathers. If birds live long enough, they will eventually develop baldness as the feather follicles become inactive *Figure 8.4*.[9,27,42] In most birds, primary feathers are the last to manifest abnormalities. In contrast to the classic presentation just described, some birds may have substantial involvement of the flight, tail and crest feathers with only minimal changes in the powder down feathers.[9,27,42]

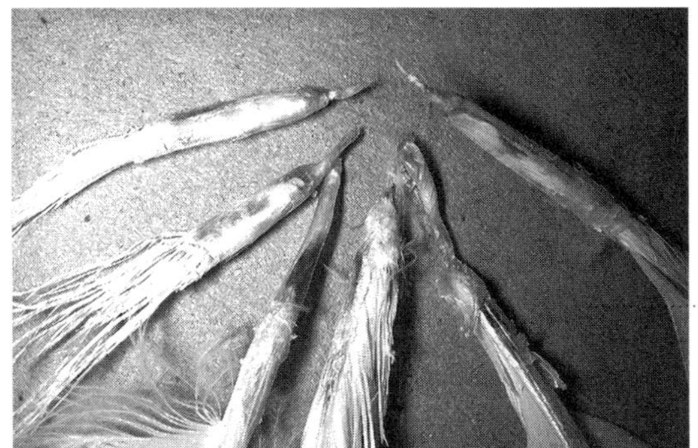

Feather abnormalities that can occur in PBFD-infected birds include retention of feather sheaths, hemorrhage within the pulp cavity, fractures of the feather shaft and circumferential constrictions.

Several suspicious feather abnormalities can be seen while the feathers are still in their follicles, including a failure of the feathers to grow to normal length, hemorrhage within the pulp cavity and constriction of the developing feathers (arrow).

Free-ranging birds with severe feather changes may have a brownish discoloration of the skin that is thought to occur from exposure of normally sheltered skin to the sun.[22,26] In some Psittaciformes with pigmented feathers, abnormal coloration has been associated with histologic lesions consistent with PBFD virus infection. This is particularly true in African Grey Parrots, where affected feathers maybe red instead of grey *Figure 8.5*.[4] However, the appearance of abnormally colored feathers is not pathognomonic for PBFD. Any abnormality that causes a feather to develop improperly may cause it to appear discolored. Black feathers on the head or body have been frequently discussed by aviculturists as an indication of PBFD virus; however, the appearance of black, normally developed feathers in white cockatoos is not a characteristic clinical sign of

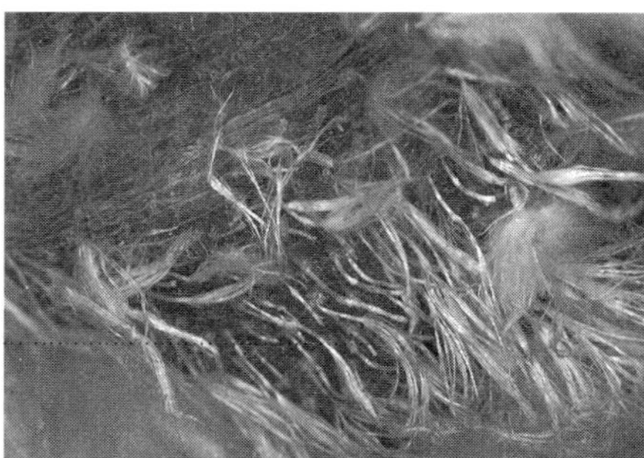

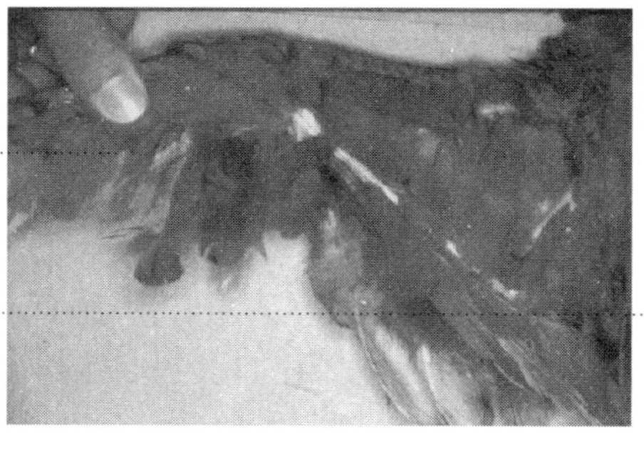

FIG 8.4

In older birds, the first sign of PBFD is the replacement of normal powder down and contour feathers with dystrophic, necrotic feathers that stop growing shortly after emerging from the follicle.

The disease then progresses to involve contour feathers in most tracts, followed by dystrophic changes in the primary, secondary, tail and crest feathers, as in this affected Scarlet Macaw.

If birds live long enough, like this Vasa Parrot, they eventually develop baldness as the feather follicles become inactive.

photographs courtesy of Joey Rodgers

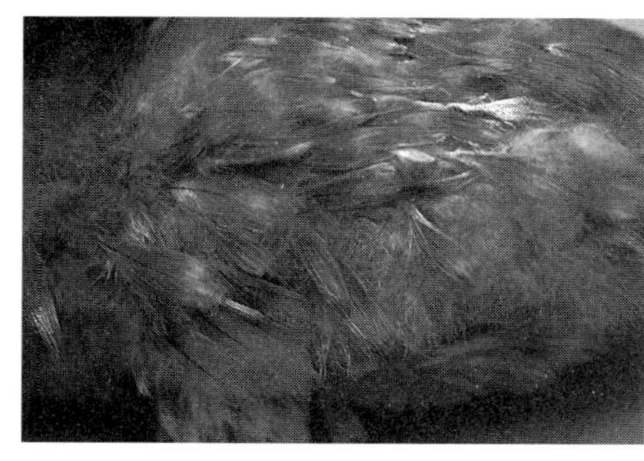

FIG 8.5

Many birds infected with the PBFD virus will develop abnormally colored feathers. PBFD virus-positive African Grey Parrots may show red body feathers.

PBFD. In one study, 75% of a group of normal, young Sulfur-crested Cockatoos had black feathers at some time during their development.[18]

<u>BEAK AND NAIL LESIONS</u> — Depending on the species involved and other factors that remain unresolved, beak changes may or may not be present in birds with PBFD. In one study involving 22 cockatoos (Sulphur-crested, Moluccan, Umbrella, Citron-crested, Triton and Goffin's), infected birds over one year of age had a lower incidence of beak lesions than did birds that were under one year of age.[9] Beak pathology does not routinely occur with some affected species, but in others, such as the Sulphur-crested Cockatoos, Galahs, Little Corellas and Moluccan Cockatoos, beak lesions are relatively common.[9,26,42]

If present, changes in the beak and oral mucosa of clinically affected birds may include progressive beak elongation with transverse or longitudinal fractures, palatine necrosis and ulcerations in the mouth *Figure 8.6*.[9,26,42] The upper beak is generally more severely affected than is the lower beak.[9] Necrosis of the upper beak progresses from the tip toward the base. In normal cockatoos and some other birds, the powder down feathers produce a fine granular material that is spread over the feathers and beak, giving the beak a soft, greyish tone. In birds with diseased powder down feathers, the beak may appear to be semi-gloss or gloss black instead of its normal greyish color.[27]

Typically, beak deformities develop in birds following a protracted course of PBFD where substantial feather changes have occurred. However, some individuals develop severe beak lesions with relatively minor feather pathology, and cracking of the hard corneum at the tip of the beak may be the initial complaint requiring veterinary attention. Rarely, abnormal elongation and fractures may also be recognized in the nails.[9,22,27,29]

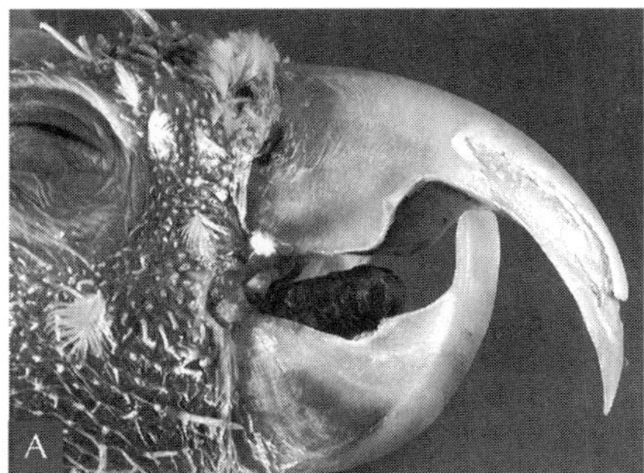

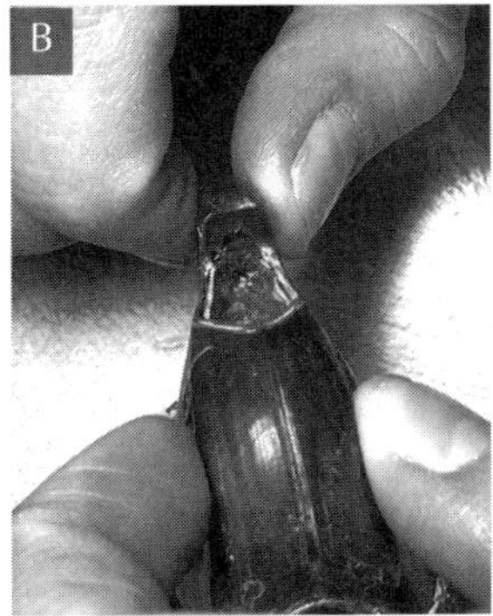

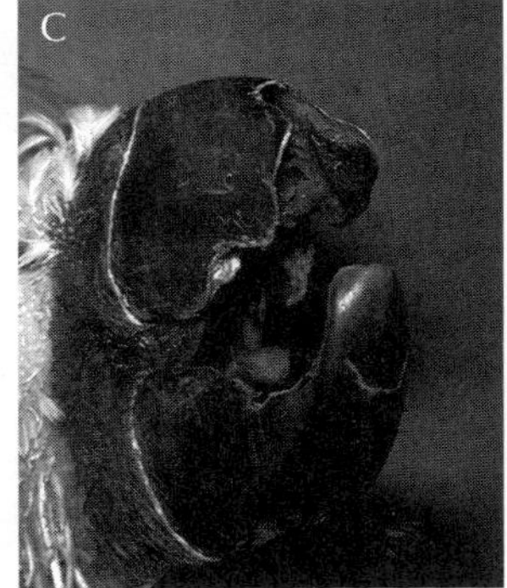

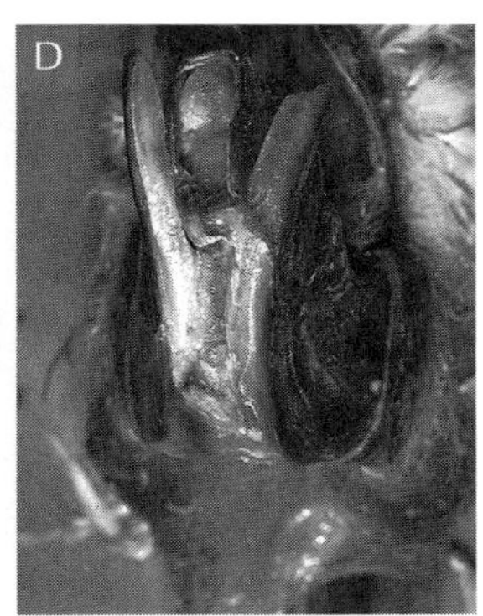

FIG 8.6

If present, changes in the beak and oral mucosa of birds infected with PBFD virus may include:
A progressive elongation of the beak with
B necrosis of the palate area and
C transverse or longitudinal fractures.
D In most cases, the upper beak is more severely affected than the lower beak; however, the lower beak may be abnormal in some birds.

EXPERIMENTAL INFECTIONS

Susceptible psittacine birds can be experimentally infected with the PBFD virus through oral, intracloacal, subcutaneous, intraocular or intranasal routes.[40,48] Characteristic gross and histologic lesions of PBFD in the feathers, thymus and cloacal bursa have been induced experimentally in neonatal Budgerigars and Galahs, using virus-containing feather homogenates from diseased birds. Neonatal Budgerigars were exposed to preparations containing virus at one, three, five, seven, ten and fourteen days of age, while Galahs were inoculated at one, three and five days of age. All birds inoculated with this homogenate at seven days of age or younger developed PBFD. Some of the Budgerigars infected at ten to fourteen days of age developed PBFD, while others remained asymptomatic. Some uninoculated control Budgerigars in direct contact with infected birds also developed PBFD.[48]

Experimentally infected Galah chicks became acutely anorectic and depressed approximately four weeks after infection. Twenty-four hours later, the feathers appeared to lose their luster and became pale and brittle. Subsequently, dystrophic feathers began to appear as the neonates developed their adult plumage. Gross lesions in experimentally infected Budgerigar and Galah chicks included slowed maturation, retention of feather sheaths, twisting of the shafts and the presence of stress lines in the remiges and rectrices. Histologic lesions included necrosis of the cells lining the developing feather and the presence of large purple intracytoplasmic inclusion bodies containing an accumulation of virus particles. The gross and histologic lesions were considered most severe in the birds exposed to virus soon after hatching. These findings were used to suggest that the age of a bird at the time of infection influenced the progression of the induced disease. Attempts to

FIG 8.7

Birds exposed to the PBFD virus, such as this African Grey Parrot chick, may develop normally for three to four weeks after infection. They then become depressed, develop diarrhea and regurgitate intermittently.

Within several days, the birds progressively begin to lose abnormal-appearing body, flight and tail feathers. Many severely affected young birds will die.

infect one-day-old specific-pathogen-free chickens with homogenates that were infectious to psittacine neonates were unsuccessful.[48] Some attempts to infect Budgerigars with feather homogenates containing virus also have been unsuccessful. These transmission failures may be attributed to the routes of viral exposure, titer of infecting virus or presence of maternal antibodies.[27]

Psittacine beak and feather disease was experimentally induced in Budgerigar, African Grey Parrot and Umbrella Cockatoo chicks administered purified, concentrated PBFD virus by the combined oral, subcutaneous and intracloacal routes.[40,41] In one study, African Grey Parrot and Umbrella Cockatoo chicks were administered virus at three to five and six to eight days of age. An infected African Grey Parrot chick developed normally until it was 30 days old, at which time the bird seemed

depressed and began to regurgitate intermittently. Three days later, the bird started to lose powder down and contour feathers. Several of the primary and secondary feather shafts appeared dark red-brown and were loose in their follicles. Over the next five days, the bird continued to lose contour, flight, tail and crest feathers that appeared progressively dystrophic *Figure 8.7*. An infected Umbrella Cockatoo chick developed normally until it was 40 days old, at which time the bird seemed depressed and anorexic. Two days later, several primary and secondary feather shafts appeared dark red-brown and were loose in their follicles. Over the next five days the bird continued to lose contour, flight, tail and crest feathers that appeared progressively dystrophic. This bird died when it was 47 days old. Histologic evaluation of tissues from both affected chicks revealed basophilic intranuclear and intracytoplasmic inclusion bodies in feather epithelium, thymus and cloacal bursa. Intranuclear inclusion bodies were located within epithelial cells, and intracytoplasmic inclusion bodies were in macrophages.[40]

Sulphur-crested Cockatoo chicks administered PBFD virus obtained from the affected feathers of a Sulphur-crested Cockatoo developed inappetence, lethargy, crop stasis and feather abnormalities; they died three to four weeks after exposure to the virus by the combined oral and intramuscular routes. Galah chicks infected with the same virus preparation developed feather abnormalities three to four weeks after infection but remained otherwise clinically normal. Necropsy findings included hepatomegaly, small kidneys and atrophy of the thymus and bursa.[34]

Adult birds experimentally infected with PBFD virus have been found to remain clinically normal and develop antibodies to the virus.[42a] In another study, 3 of 18 adult Galahs exposed to the virus by the oral route had a mild increase in antibody titer. One of these birds developed diarrhea and died four days after administration of the virus. It was not reported whether this bird had microscopic changes suggestive of a PBFD virus infection. All the adult Galahs administered virus by the intramuscular route remained clinically normal and developed an increase in antibody titer.[34]

EPIZOOTIOLOGY

Psittacine beak and feather disease has been reported in Australia, North America, Europe and Asia. It is likely that PBFD virus originated in psittacine birds indigenous to Australia and that it has been introduced to other continents through the worldwide movement of birds to meet demands of the companion bird industry.[22] Continued intercontinental movement of birds could easily result in the introduction of PBFD virus or other more devastating viruses into free-ranging populations of the world's more endangered psittacine species.

Historically, PBFD virus was thought to affect only Old World and South Pacific psittacine birds, with white and pink cockatoos being particularly susceptible. However, the disease has also been documented in several genera of black

TABLE 8.1

Birds currently considered susceptible to PBFD virus*

Cacatuidae
    Black Palm Cockatoo (*Probosciger aterrimus*)
    Citron Cockatoo (*Cacatua citrinocristata*)
    Galah (Rose-breasted Cockatoo) (*Eolophus roseicapillus*)
    Gang-gang Cockatoo (*Callocephalon fimbriatum*)
    Goffin's Cockatoo (*Cacatua goffini*)
    Long-billed Corella (*Cacatua tenuirostris*)
    Major Mitchell's Cockatoo (*Cacatua leadbeateri*)
    Moluccan Cockatoo (*Cacatua moluccensis*)
    Red-vented Cockatoo (*Cacatua haematuropygia*)
    Little Corella (Bare-eyed Cockatoo) (*Cacatua sanguinea*)
    Sulphur-crested Cockatoo (*Cacatua galerita*)
    Triton Cockatoo (*Cacatua triton*)
    Umbrella Cockatoo (*Cacatua alba*)
Psittacidae
    African Grey Parrot (*Psittacus erithacus*)
    Black Parrot (*Coracopsis nigra*)
    Blue and Gold Macaw (*Ara ararauna*)
    Blue-fronted Amazon Parrot (*Amazona aestiva*)
    Bourke's Parrot (*Neopsephotus bourkii*)
    Budgerigar (*Melopsittacus undulatus*)
    Cockatiel (*Nymphicus hollandicus*)
    Eastern Rosella (*Platycercus eximius*)
    Eclectus Parrot (*Eclectus roratus*)
    Golden-shouldered Parrot (*Psephotus chrysopterygius*)
    Green-winged Macaw (*Ara chloroptera*)
    Hooded Parrot (*Psephotus dissimilis*)
    Indian Ring-necked Parakeet (*Psittacula manillensis*)
    Jandaya Conure (*Aratinga auricapilla*)
    King Parrot (*Alisterus scapularis*)
    Mallee Ring-necked Parakeet (*Barnardius barnardi*)
    Meyer's Parrot (*Poicephalus meyeri*)
    Northern Rosella (*Platycercus venustus*)
    Orange-bellied Parrots (*Neophema chysogaster*)
    Pale-headed Rosella (*Platycercus adscitus*)
    Pennant's Rosella (*Platycercus elegans*)
    Port Lincoln Parrot (*Barnardius zonarius*)
    Princess of Wales Parakeet (*Polytelis alexandrae*)
    Red-bellied Parrot (*Poicephalus rufiventris*)
    Red-lored Amazon (*Amazona autumnalis*)
    Red-rumped Parrot (*Psephotus haematonotus*)
    Rose-ringed Parakeet (*Psittacula krameri*)
    Scaley-headed Parrot (*Pionus maximiliani*)
    Scarlet Macaw (*Ara macao*)
    Senegal Parrot (*Poicephalus senegalus*)
    Swift Parrot (*Lathamus discolor*)
    Vasa Parrot (*Coracopsis vasa*)
    Western Rosella (*Platycercus icterotis*)
Agapornis sp.
    Fisher's Lovebird (*Agapornis fischeri*)
    Masked Lovebird (*Agapornis personatus*)
    Nyassa Lovebird (*Agapornis lilianae*)
    Peach-faced Lovebird (*Agapornis roseicollis*)
Loridae
    Blue-streaked Lory (*Eos reticulata*)
    Mitchell's Lory (*Trichoglossus haematodus mitchellii*)
    Rainbow Lory (*Trichoglossus haematodus*)
Others
    Laughing Turtle Doves (*Streptopelia senegalensis*)

**This list includes only species in which suggestive lesions have been identified, and probably does not represent all the species of birds that may be infected.*

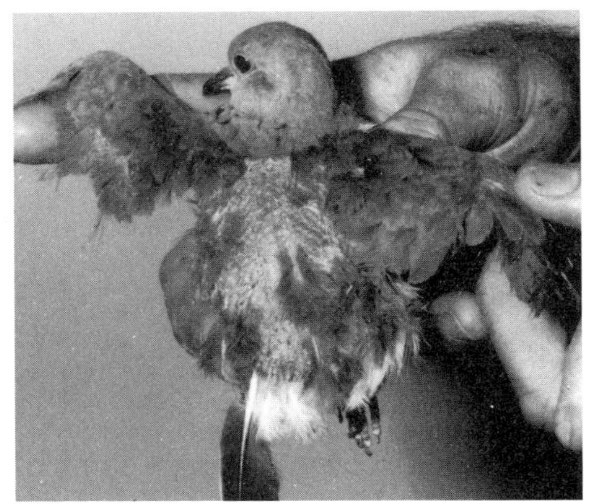

cockatoos and in New World psittacines including Amazon parrots, macaws and a species of *Pionus*. Histologic or clinically suggestive lesions of PBFD have now been described in over 40 species of psittacine birds, and new species are being defined as susceptible every year *Table 8.1*. Until 1993, PBFD had been documented only in birds within the Psittaciformes order, but a similar virus has been recovered from free-ranging doves in Australia with feather abnormalities similar to those described in affected psittacines *Figure 8.8*.[25] The actual host range of the PBFD virus remains largely unknown.

It appears that the species of birds listed in *Table 8.1* are of varied susceptibility, and the clinical and pathological signs that these birds may develop is markedly different. DNA probe testing indicates that Old World Psittaciformes are most likely to be infected. This finding would be expected because this group of birds appears to be more susceptible and is more likely than New World Psittaciformes to develop feather abnormalities. Of specific psittacine birds tested with a DNA probe, 30% of lovebirds, 10.2% of Eclectus Parrots, 8.7% of cockatoos and 8% of African Grey Parrots were positive. The inci-

dence of virus detection in macaws and Amazon parrots (New World Psittaciformes) was lower (4%).[3]

Epizootiologic studies in one import station in the United States suggested that 0.5% of the imported Lesser Sulphur-crested Cockatoos, Umbrella Cockatoos, Citron Cockatoos and Moluccan Cockatoos had gross lesions consistent with PBFD. Finding clinically affected birds when first placed in quarantine suggests that these birds had been infected in their country of origin.[9] In Australia, 75% of the captive Sulphur-crested Cockatoos examined in one veterinary hospital had clinical signs consistent with PBFD, and it was estimated that up to 90% of all Sulfur-crested Cockatoos sold in the retail bird trade developed PBFD.[18,29] The incidence of the disease in other commonly maintained captive psittacine birds in Australia, including Galahs and Budgerigars, is believed to be much lower.[29]

Psittacine beak and feather disease is reportedly endemic in free-ranging populations of Sulphur-crested Cockatoos, Galahs, Gang-gang Cockatoos, Little Corellas, Major Mitchell's Cockatoos, Crimson Rosellas, Orange-bellied Parrots, Budgerigars and Rainbow Lorikeets.[21,27,31,35] In 1887 in South Australia, explorers recorded a decrease in the population of free-ranging Red-rumped Parrots that was caused by abnormally formed feathers that prevented flight. This finding may have been the earliest report of PBFD virus activity, and if correct, would support the theory that this virus originated in Australia.[1]

It has been reported that up to 20% of some free-ranging Sulphur-crested Cockatoos in Victoria, Australia, may have clinical signs of PBFD in any one year.[22] In another study, a flock of 120 Sulfur-crested Cockatoos decreased to 20 individuals over a nine-month period. Many of the dead or dying members of this flock were diagnosed with PBFD. Half of the 20 birds in a free-ranging flock of Crimson Rosellas had PBFD.[21] Although not formally documented, there is frequent discussion of clinical changes consistent with PBFD occurring in free-ranging populations of Moluccan Cockatoos, Philippine Red-vented Cockatoos, lovebirds, Umbrella Cockatoos and Citron Cockatoos. Free-ranging Laughing Turtle Doves in Australia have been shown to develop feather abnormalities characteristic of those described for psittacine birds infected with the PBFD virus.[25]

Antibodies to PBFD virus were documented in clinically-normal, free-ranging Rainbow Lorikeets in Australia. From 41% to 94% of individual birds in some flocks of free-ranging Sulphur-crested Cockatoos, Galahs, Little Corellas and Eastern Long-billed Corellas were found to have PBFD virus antibodies. Two adult free-ranging Mallee Ring-necked Parrots were also found to have PBFD virus antibodies. In one group of 135 Sulphur-crested Cockatoos from Camden, six birds (4.4%) had clinical lesions suggestive of PBFD and 95 (73%) had PBFD virus antibodies. In another group of 17 Sulphur-crested Cockatoos, one bird had lesions suggestive of PBFD and 15 birds had PBFD virus antibodies.[35] These findings indicate that PBFD virus activity is widespread in New South Wales, that free-ranging birds are frequently infected and that many infected birds mount an effective immune response that protects them from disease.

SUSCEPTIBILITY — While many species of psittacine birds have been shown to be susceptible to disease *Table 8.1*, the clinical progression and outcome of an infection vary widely based on the age of the bird when clinical signs first occur and presumably based on the species of the bird. In one study, experimentally infected Galah and Sulphur-crested Cockatoo chicks developed characteristic feather changes approximately three to four weeks after infection. The Sulphur-crested Cockatoos died, while the Galahs developed only the characteristic feather lesions.[34]

Generally, PBFD is a disease of young birds (up to three years), but older individuals (up to twenty years) that had been clinically normal through much of their life have been diagnosed with PBFD *Figure 8.9*.[42] In a study of 176 cockatoos with PBFD, 92% were less than three years of age and only 8% were more than three years old.[6] Older birds that are experimentally infected with the PBFD virus generally produce antibodies against the virus and remain clinically normal.[34] Because attempts to experimentally induce disease in older birds have failed, it is possible that older birds that develop clinical signs of PBFD were infected at a younger age,

remained latently infected, and developed abnormal feathers at some later date.[40]

An age-related susceptibility to the PBFD virus was suggested by one experimental transmission study. Neonatal Budgerigars experimentally infected at less than 7 days of age were found to develop severe disease, while birds infected at 10 to 14 days were reported to experience lower levels of morbidity, and some remained asymptomatic.[48] It should be noted that the incubation period is shorter in young birds than in older birds. Adult birds experimentally infected with PBFD virus develop a transient viremia that can be detected using DNA probes, then seroconvert and remain clinically normal.[34,42a]

Early reports suggested that males were more commonly diagnosed with PBFD than females; however, more recent investigations indicate there is no gender predisposition.[9,22,29] There does seem to be an increased occurrence of PBFD late in the breeding season in both captive and free-ranging cockatoo populations.[29,31] This observation may reflect an accumulation of young, susceptible birds as the breeding season progresses or increased stresses late in the breeding season that increase a bird's susceptibility to infection.

<u>RELATIONSHIP OF VIRUS STRAINS</u> — Currently, only one strain of PBFD virus has been identified. The demonstrated similarity of the PBFD virus affecting varying genera has been important in the use of effective diagnostic tests (DNA probes) and will be important in the development of an effective and safe recombinant vaccine.

Comparison of PBFD virus from a Sulphur-crested Cockatoo, a Black Palm Cockatoo, a Red-lored Amazon Parrot and a Peach-faced Lovebird demonstrated that recovered viruses were morphologically and antigenically similar, and that the microscopic changes induced by the virus in each of these birds was similar.[38] Long-term use of DNA probes designed to detect the PBFD virus has confirmed the presence of virus nucleic acid in the blood or tissues of numerous infected psittacine birds, again suggesting that the detected virus is similar.

Further support for the similarity of the PBFD virus among various genera has been generated in experimental transmission studies. Virus purified from the infected feathers of a Moluccan Cockatoo was used to develop an experimental vaccine that stimulated a protective immunologic response when injected into Umbrella Cockatoos and African Grey Parrots. Virus purified from an

Umbrella Cockatoo caused gross and microscopic changes characteristic of PBFD in the feathers of experimentally infected Umbrella Cockatoo and African Grey Parrot chicks.[40] Galah and Sulphur-crested Cockatoo chicks experimentally infected with virus obtained from the affected feathers of a Sulphur-crested Cockatoo developed characteristic gross and microscopic lesions.[34] Collectively, these findings strengthen the theory that an antigenically related virus causes PBFD in various Psittaciformes.

INCUBATION — Neonates and fledglings infected with the PBFD virus may develop the first recognizable clinical changes during their initial feather development after replacement of the neonatal down.[40] Galah chicks experimentally infected with PBFD virus have been reported to develop clinical signs of PBFD approximately four weeks after infection.[48] African Grey Parrot chicks experimentally infected at 3 to 8 days of age became depressed by 30 days of age and developed progressive feather dystrophy by 33 to 44 days old. Umbrella Cockatoo chicks infected at 3 to 8 days of age became depressed by 40 days of age and developed progressive feather dystrophy from 42 to 47 days of age.[40] Sulphur-crested Cockatoo chicks approximately six weeks old were given PBFD virus obtained from the affected feathers of a Sulphur-crested Cockatoo; they developed inappetence, lethargy, crop stasis, feather abnormalities and died three to four weeks after exposure to the virus by the combined oral and intramuscular routes. Galah chicks developed feather abnormalities

three to four weeks after infection but remained otherwise clinically normal.[34] The time variance in developing clinical signs associated with PBFD among different psittacine chicks may be attributed to concentrations of maternally transmitted antibodies, titer of virus in the inoculum or host responses to the virus.[40]

These experimental infection studies suggest that the minimum incubation period for the appearance of dystrophic feathers is 21 to 25 days.[34,40,48] Virus can be detected in the blood using DNA probes as soon as two days after natural exposure to the virus, and long before an infected bird may develop clinical signs of disease. Most adult birds infected with the PBFD virus will develop a transient viremia that can be detected using DNA probes, then seroconvert and remain clinically normal.

Both clinical experience and DNA probe testing of blood for the presence of PBFD virus nucleic acid suggest that the maximum incubation period can be years. Viral nucleic acid could be detected in the blood of one clinically normal Ducorp's Cockatoo for 18 months before feather abnormalities developed. The bird was DNA probe-positive when tested intermittently during the 18 months before clinical changes occurred. In another study, DNA probes detected viral nucleic acid in the blood of three artificially-incubated Little Corella chicks from an affected hen when they were first tested at 20 days of age. Two of these Little Corella chicks appeared clinically normal until 32 days of age, then became depressed

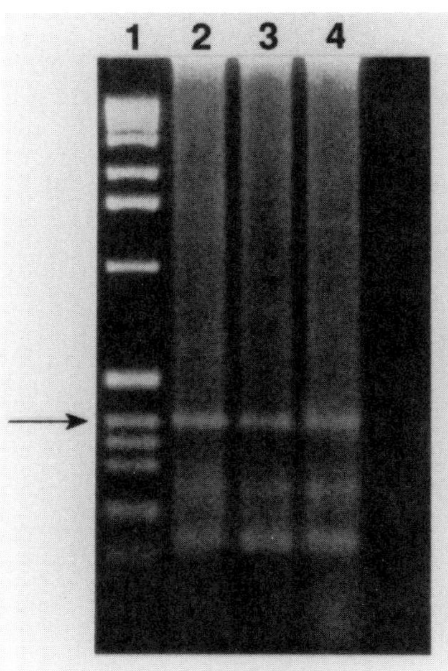

FIG 8.10

Three eggs from a Little Corella Cockatoo hen with PBFD virus were artificially incubated. PBFD virus nucleic acid was detected in the blood of all three of the chicks using a DNA probe at 20 days of age (lanes 2, 3, 4; lane 1 is a control).

The chicks whose blood samples are in lanes 2 and 4 developed clinical signs of PBFD starting at 32 days of age. Shown here are the affected chick 2 and its still clinically normal sibling, chick 3. Chicks 2 and 4 developed progressive feather changes over a two- to three-week period and died. Chick 3 remained clinically normal until it was 80 days of age. Because all three of these chicks were presumably infected by the hen, and all three of the chicks were blood-positive at 20 days of age, these findings suggest that the time from infection to the development of clinical signs can vary.

and began to exhibit abnormal feathers. Within ten days, most of the developing contour, flight, tail and crest feathers on these two birds were abnormal. The third chick remained clinically normal until 80 days of age, at which time abnormally formed flight feathers were noted *Figure 8.10*. Because these three birds were offspring of the same infected hen, the eggs had been removed and artificially incubated, and virus nucleic acid had been detected from all three birds at 20 days of age, it was assumed that all of these chicks were infected in the egg. These findings suggest that the natural incubation period can vary even in birds that presumably are infected at the same point in time, by the same route and with a similar quantity of virus.

<u>TRANSMISSION</u> — One of the critical pieces of information that veterinarians and bird enthusiasts should understand is how the PBFD virus can be transmitted. Armed with this information and the DNA probe test, one can implement testing and hygiene practices which will substantially decrease, if not eliminate, the possibility of a PBFD virus outbreak in the United States and Europe, where the virus does not infect free-ranging birds. Testing and hygiene will be less effective in controlling PBFD in Australia where infected free-ranging birds can easily introduce the virus to susceptible captive birds.

The collection and congregation of large numbers of neonates prior to distribution in the retail market is thought to favor the spread of PBFD virus among highly susceptible young birds.[18] The flocking nature of many of the species of birds that are susceptible to PBFD virus favors its spread through direct contact or through virus contamination of water or feeding areas shared by numerous free-ranging birds. In captivity, PBFD virus was detected in the nest box of a pair of Princess Parrots that were producing infected young, and neonates from the nest continued to develop PBFD even after the nesting container was thoroughly cleaned.[33]

Psittacine chicks can be experimentally infected with the PBFD virus through the oral, intracloacal, subcutaneous, intramuscular or intranasal routes. Three of 18 (17%) adult Galahs administered virus by the oral route seroconverted, indicating that they were infect-

AVIAN VIRUSES: FUNCTION AND CONTROL

ed. Additionally, some Budgerigar neonates in direct contact with affected birds developed clinically apparent infections. These findings suggest that inspiration of aerosolized viral particles or ingestion of virus-contaminated materials may account for natural routes of viral transmission.[40,48]

With the large number of inclusion bodies that can be demonstrated in the feathers of PBFD-positive birds, it has been postulated that feather dust may serve as a major source for PBFD virus excretion and transmission. High concentrations of the virus were demonstrated in feather dust collected from a room where birds with active cases of PBFD were being housed, implicating contaminated dust from any source (feather dander or dried, aerosolized excrement) as a major vehicle for the natural transmission of this virus *Figure 8.11*.[39] Given the ease with which feather dust can be dispersed both through natural air flow and through contact with clothing, nets, bird carriers, food dishes or insects, it is likely that contaminated feather dust is a major method for transmission and environmental persistence of the PBFD virus *see Figure 3.1*. One can determine whether PBFD virus is present in a home or aviary by using viral-specific DNA probes to test environmental swabs for the presence of viral nucleic acid (see Diagnosis).

Psittacine beak and feather disease virus was demonstrated in the feces and crop washings from various species of psittacine birds diagnosed with PBFD.[33,39] The recovery of PBFD virus in the feces and crops of diseased

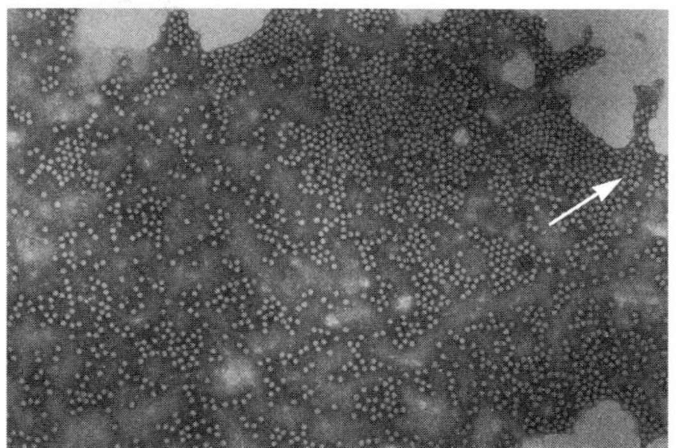

birds suggests that contaminated excretions and secretions from infected birds may be involved in disease transmission. It has been postulated that the demonstration of inclusion bodies containing PBFD virus in the cells lining the lumen of the palate, esophagus, crop, intestines and bursa could easily account for the virus shed in the feces.[15]

Because viral nucleic acid can be demonstrated in the blood of infected birds, vertical transmission would be suspected. It has been determined that artificially incubated chicks from PBFD-infected hens will consistently develop PBFD, suggesting that vertical transmission of the virus does occur *see Figure 8.10*. Thus, attempts to prevent PBFD virus infections through artificial incubation are futile. For comparative purposes, the chicken anemia virus (CAV), which is in the same family with the PBFD virus, can be transmitted to the eggs of hens that are infected during the laying period.[7,19,51] Chickens infected with CAV shed virus in the feces for up to five weeks after infection. Recovered hens that are reinfected can shed virus in their feces for up to four days. Neither stress factors nor administration of betamethasone induced previously infected, recovered hens to shed CAV, suggesting that a latent infection

FIG 8.11
High concentrations of the PBFD virus (arrow) can be recovered from feather dust collected from areas where affected birds are housed. It was estimated that the preparation of virus used for this photograph contained over a trillion virus particles per milliliter of sample.

does not occur. Cessation of virus shedding in the feces coincided with the detection of circulating antibodies.[7]

Several reports indicate that asymptomatic adult psittacines may produce progeny with clinical signs of PBFD in successive breeding seasons. These findings suggest a carrier state may exist with vertical or horizontal transmission of PBFD virus from parent to offspring.[29,31,44] However, in most suspected cases of parent to offspring transmission, epizootiologic investigations indicate probable exposure to the PBFD virus occurring through sources other than the parents.

PATHOLOGY

Gross feather and beak changes associated with PBFD are described under clinical features. Gross changes in internal organs are rarely noted during necropsy. When they are present, they are variable and differ with the age of the bird and the types of secondary infections that are present.[9,15,22,27,48] In young birds, the cloacal bursa may be small with poorly developed folds, and the thymus may be small with pale necrotic areas. In mature birds, the spleen is frequently small and depleted of lymphocytes; occasionally necrosis of the reticular cells can be seen microscopically.[4]

The primary microscopic changes associated with PBFD have been described in the feather shaft where necrosis and ballooning degeneration of epithelial cells lining the developing feather are common.[9,16,27,37] The follicular epithelium may also be necrotic, but this lesion is less commonly reported. The PBFD virus appears to target feather epithelial cells more frequently than follicular epithelial cells.[16] Disease-induced thickening of the developing feather sheath prevents the normal process that releases the developed feather. The virus also induces changes that cause midshaft constrictions and clubbing at the base of the developing feather.[8,9] Inflammatory cells (including plasma cells, lymphocytes, macrophages and heterophils) accumulate in the affected feather pulp.[8,16,27,37] Granulomatous dermatitis with vesicle formation were described in a group of lovebirds infected with PBFD virus.[28]

In peracute cases, suggestive microscopic changes in the bursa and thymus include atrophy and accumulation of necrotic cells. In these cases, microscopic changes may or may not be present in the feathers. If present, the feathering changes may be limited to mild swelling in the cells lining the feather follicle rather than the characteristic necrosis of severely affected cells.[9,16,27]

Microscopic changes in the beak of PBFD birds are similar to those described in their feathers, including necrosis of epithelial cells. Hyperkeratosis and separation of the outer layer of the beak from the underlying tissues and bone may also be evident, often accompanied by secondary necrosis and osteomyelitis of associated tissues.[9,16,27] In birds with beak pathology, necrosis and inflammation of the cells lining the tongue, mouth and crop have also been reported.[9,15,27] Secondary gram-negative bacteria and fungi are commonly isolated from beak lesions, and may be associated with acute or chronic inflammatory reactions.[8,9,27,29,31]

Within internal organs, consistent PBFD lesions including atrophy, have been described only in the thymus and bursa, with focal aggregates of necrotic cells in these lymphoid tissues.[9,15,26,31] In experimental infections, microscopic changes in internal organs were limited to the thymus and bursa of Fabricius and included atrophy of bursal follicles and depletion of thymic lymphocytes.[48]

INCLUSION BODIES — Psittacine beak and feather disease virus may cause intranuclear or intracytoplasmic inclusion bodies that are most consistently demonstrated in the feathers, thymus and bursa of affected birds.[9,16,27] Additional sites where inclusion bodies have been detected include the beak and hard palate, tongue, parathyroid gland, crop, esophagus, spleen, intestines, bone marrow, liver, thyroid, testes, ovary and adrenal glands.[15] Particularly in early cases, intracytoplasmic inclusion bodies have been reported to be more consistently demonstrated than are intranuclear inclusion bodies.[9,16] Both intranuclear and intracytoplasmic inclusion bodies were identified in 23 of 32 birds examined in one study. In this group, intranuclear inclusion bodies were restricted to epithelial cells and intracytoplasmic inclusion bodies were found only within macrophages.[16]

Intracytoplasmic inclusions are thought to originate in epidermal cells and attain their greatest size within macrophages that engulf these infected cells.[9,16,27] However, the occurrence of

viral antigen within macrophages in the bone marrow and within circulating monocytes suggests that these cells may be infected directly.[16]

PATHOGENESIS

Many PBFD virus-infected birds that develop feather abnormalities die from six months to one year after the onset of clinical signs. Some birds have been known to survive over 10 to 15 years in a featherless state. Death usually occurs either from changes induced by secondary bacterial, chlamydial, parasitic, fungal or other viral agents or from terminal changes that necessitate euthanasia.[9,27,29,49] Cockatoos with PBFD have been diagnosed with severe cryptosporidial infections, which generally occur only in patients with immunodeficiencies *Figure 8.12*.[12]

The deaths of PBFD virus-affected birds from secondary or opportunistic patho-

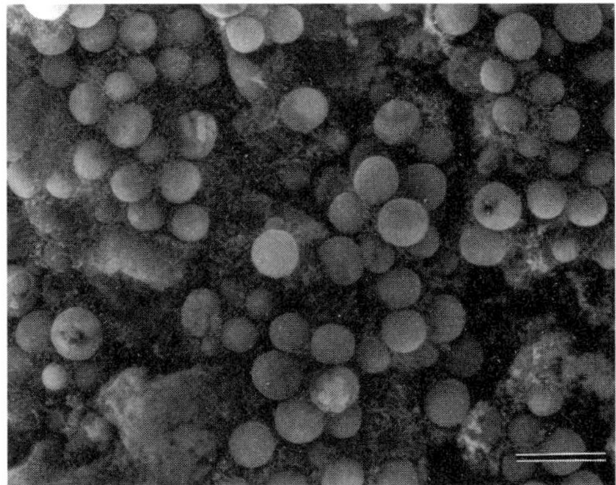

FIG 8.12
Severe cryptosporidial infections have been documented in cockatoos with PBFD, suggesting that cockatoos infected with the PBFD virus may have damaged immune systems that are unable to resolve the cryptosporidial infection.

photograph courtesy of Kenneth Latimer, reprinted with permission[12]

gens and their susceptibility to these agents have been used to suggest that infected birds are immunosuppressed; however, only limited work has been performed to document this theory. The virus has been shown to cause necrosis and atrophy of the bursa and thymus, which would be expected to correlate with at least a temporary reduction in the efficiency of the immune system.[10,15,26] In one study, affected birds were found to have low concentrations of albumin and gammaglobulin (vital components of the immune system) as indicated by serum electrophoresis.[9] Other investigations with serum electrophoresis in birds with PBFD virus have indicated that some birds had hypogammaglobulinemia (usually birds with severe beak necrosis or other clinical abnormalities) and some birds had hypergammaglobulinemia.[42a]

The peracute and acute forms of the disease, which are most common in neonatal psittacines, are associated with high levels of mortality and seem to be associated with the most severe pathologic changes in the bursa and bone marrow. Young birds with these forms of the disease probably have immature immune systems when exposure to the virus occurs. If they do not have maternal antibodies against the PBFD virus, the birds would not be expected to be capable of mounting an effective immune response.

Virus-infected birds with inclusion bodies located only within the nucleus of epithelial cells have been found to recover spontaneously. On the other hand, psittacine birds with intracytoplasmic inclusion bodies located in macrophages usually succumb to the disease. Because the macrophage is critical for the initial processing and pre-sentation of viral antigen to the immune system, *see Figure 3.18*, the determining factor in whether an infected bird develops a chronic fatal PBFD virus infection or develops a protective immune response may be based on how the body processes the virus before it begins to persist in the cytoplasm of macrophages.[40]

Most adult birds infected with the PBFD virus develop a transient viremia that can be detected using DNA probes. They then mount an effective immune response that eliminates the virus from the body and remain asymptomatic. Most infected birds that do develop feather abnormalities eventually succumb to the disease, although Budgerigars, lorikeets and lovebirds occasionally have been reported to recover from suspected PBFD virus-induced feather abnormalities over a two- to twelve-month-period.[27,31] In Old World psittacines such as cockatoos, Eclectus Parrots and African Grey Parrots, recoveries from PBFD virus-induced feather abnormalities have not been satisfactorily confirmed. In contrast, New World psittacine birds, including a pionus parrot and several macaw neonates with PBFD virus-induced feather abnormalities and intracytoplasmic inclusion bodies, have been demonstrated to recover *Figure 8.13*.[42a]

In documenting recoveries in a pionus parrot and several Scarlet Macaw chicks, it was found that the affected birds developed characteristic feather lesions and their blood was DNA probe-positive for PBFD virus nucleic acid. Over a period of several weeks

Scarlet Macaws that had severe PBFD virus-induced feather abnormalities are now clinically normal and DNA probe-negative for the virus. It is undetermined why some New World Psittaciformes are able to recover, while affected Old World Psittaciformes eventually die. photograph courtesy of Joey Rodgers

(macaws) to months (pionus parrot), the blood of these infected birds became DNA probe-negative, and new feather growth was normal. In the new (normal) feathers, PBFD virus-induced inclusion bodies could not be identified, but they continued to be present in the already differentiated and clinically abnormal feathers. As the diseased feathers were replaced with new healthy feathers, the birds returned to a normal appearance, and PBFD virus could not be detected in the blood or feathers.

IMMUNITY

Psittacine birds have been shown to develop precipitating and hemagglutination-inhibition (HI) antibodies following recovery from natural infections or through vaccination.[38] It is apparent from data collected in captive and free-ranging birds that most individuals exposed to the PBFD virus develop subclinical infections that result in the development of a protective immunologic response *Figure 8.14*. The factors that determine whether a susceptible bird mounts an immune response or is fatally infected could depend on the age of the bird when virus exposure occurs, the presence and levels of maternal antibodies, the route of viral exposure and the titer of the infecting virus.

Birds with active PBFD virus infections have been found to have low-to-undetectable levels of HI antibodies, whereas birds that have been exposed to the virus but remain clinically normal have high levels of HI antibodies *Table 8.2*.[35,38] In one study, HI antibodies could not be demonstrated in 42 birds with PBFD

TABLE 8.2

Hemagglutination-inhibition (HI) titers of clinically normal birds exposed to PBFD-affected (+) birds[38]

Species	PBFD Virus Exposure	HI titer
Moluccan Cockatoo	Housed with PBFD+ mate	2560
Moluccan Cockatoo	PBFD+ bird in collection	2560
Moluccan Cockatoo	PBFD+ bird in collection	1280
Umbrella Cockatoo	Housed with PBFD+ mate	1280
Umbrella Cockatoo	PBFD+ bird in collection	80
Umbrella Cockatoo	PBFD+ bird in collection	640
Umbrella Cockatoo	Housed with PBFD+ mate	>5120
Goffin's Cockatoo	Housed with PBFD+ mate	2560
Goffin Cockatoo	Housed with PBFD+ mate	2560
Cockatoo	PBFD+ bird in collection	160
Cockatoo	PBFD+ bird in collection	1280
Cockatoo	PBFD+ bird in collection	1280
Cockatoo	PBFD+ bird in collection	2560
Cockatoo	PBFD+ bird in collection	>5120
Cockatoo	PBFD+ bird in collection	40

virus, and antibody titers in normal birds ranged from 1:20 to 1:2560.[35] The factors that determine why some PBFD-positive birds develop low HI antibody titers whereas others do not have not been determined *Table 8.3*.[38] It is possible that birds infected with the PBFD virus under certain conditions may become immunotolerant, whereas birds exposed under different conditions are able to develop an effective antibody response. Additionally, the low titers demonstrated in infected birds may be explained by the severe damage that occurs to the bursa and thymus and/or by the persistent infections that appear to occur in macrophages.

FIG 8.14
Most mature birds that are exposed to the PBFD virus develop a transient infection followed by an appropriate immune response that destroys the virus; the bird remains clinically normal.

In the United States, antibody surveys suggest that most birds of a susceptible species are infected by the PBFD virus

HI titers of PBFD-positive birds[38]

Species	PBFD virus exposure	HI titer
African Grey Parrots	Clinical disease at 12 wks of age, housed with PBFD+ birds	640
	Clinical disease at 12 wks of age, sibling of affected bird	<10
Black Palm Cockatoo	Unknown exposure	<10
Eclectus Parrot	Housed with PBFD+ birds	40
	Housed with PBFD+ birds	640
	Housed with PBFD+ birds	320
	Housed with PBFD+ birds	<10
	Housed with PBFD+ birds	640
	Housed alone	40
	Housed with PBFD+ birds	1280
Moluccan Cockatoos	Housed with PBFD+ birds	640
	Housed with PBFD+ birds	160
	Housed with PBFD+ birds	320
	Housed with PBFD+ birds	320
	Housed with PBFD+ birds	320
	Single bird in household	20
	Housed with PBFD+ birds	160
	Housed with PBFD+ birds	160
Red-vented Cockatoos	Housed with PBFD+ birds	80
	Mate of clinically normal bird	40
	Housed with PBFD+ birds	40
Sulphur-crested Cockatoos	Housed with PBFD+ birds	80
	Housed with PBFD+ birds	320
	Clinical disease at 24 wks of age, housed with PBFD+ birds	80
	Housed with PBFD+ birds	320
	Housed with PBFD+ birds	320
	Clinical disease at 16 wks of age, housed with PBFD+ birds	40
Umbrella Cockatoos	Housed with PBFD+ birds	80
	Housed with PBFD+ birds	640
	Housed with PBFD+ birds	320
	Housed with PBFD+ birds	1280
	Housed with PBFD+ birds	1280
	Clinical disease at 16 wks of age, housed with PBFD+ birds	320

at some time in their life and develop an antibody response.[38] The seroprevalence of infection in some populations of free-ranging psittacines in Australia is also high. In one study, 41% to 94% of clinically normal free-ranging cockatoos tested were found to have antibody titers to PBFD virus that ranged from 1:20 to 1:10,240. In one group of 135 Sulphur-crested Cockatoos from Camden, Australia, 95 (73%) of the birds had PBFD virus HI antibodies. In another group of 17 Sulphur-crested Cockatoos, 15 (88%) had detectable levels of antibodies. Two clinically normal adult Mallee Ring-necked Parrots were also found to have HI antibodies.[35] These findings suggest that free-ranging birds are frequently exposed to the virus, as are their captive conspecifics in aviaries in which PBFD virus-affected birds are housed.

In adult birds and in 30- to 45-day-old African Grey Parrot, Umbrella Cockatoo and Sulphur-crested Cockatoo chicks, antibodies have been induced by experimental vaccination with PBFD virus treated with β-propiolactone. Neonates from hens with HI antibody titers greater than 1:1280 were shown to be protected from challenge with live virus, whereas neonates from hens with HI antibody titers less than 320 were shown to be susceptible.[40] The fact that chicks from vaccinated hens remain clinically normal when exposed to live virus suggests that vaccinated hens transfer protective factors (probably antibodies) that provide at least temporary resistance to their chicks. The level of antibodies that is needed to protect a bird from infection and the specific period when these levels are most critical has not been determined.

For comparative purposes, chicken hens pass protective antibodies to CAV to their chicks that persist for up to 21 days after hatching.[23] Maternally derived antibodies decline to an undetectable level by three weeks of age, and most young chickens develop their own antibodies to the virus by eight to nine weeks of age.[20] Antibodies to CAV can be demonstrated within two weeks after experimental infection, and these antibodies have been demonstrated to persist for at least six months. Once antibody titers decrease sufficiently, hens reexposed to the virus will develop an increase in antibody titer. It has also been demonstrated that the cessation of CAV shedding in the feces coincides with the detection of circulating antibodies.[7]

DIAGNOSIS

Psittacine beak and feather disease should be suspected in any bird with progressive feather loss involving malformed feathers *see Figure 8.3*. However, one cannot determine if a bird is infected with the PBFD virus strictly by examining the feathers. Visible feather changes grossly similar to those caused by PBFD virus can be induced by any factor that disrupts the blood supply to the developing feather, including avian polyomavirus, adenovirus, trauma, bacterial folliculitis, fungal folliculitis, septicemias, malnutrition, endocrine abnormalities and some drug reactions, particularly to penicillins and cephalosporins. Feather lesions identical to those caused by PBFD virus also can be produced by pinching developing feathers at or near the pulp cap *Figure 8.15*. Conversely, birds with normal-appearing feathers can have the PBFD virus nucleic acid in their blood stream. The only effective method available for determining if a bird is infected with PBFD virus before feather lesions are present is DNA probe testing of blood *Figure 8.16*.

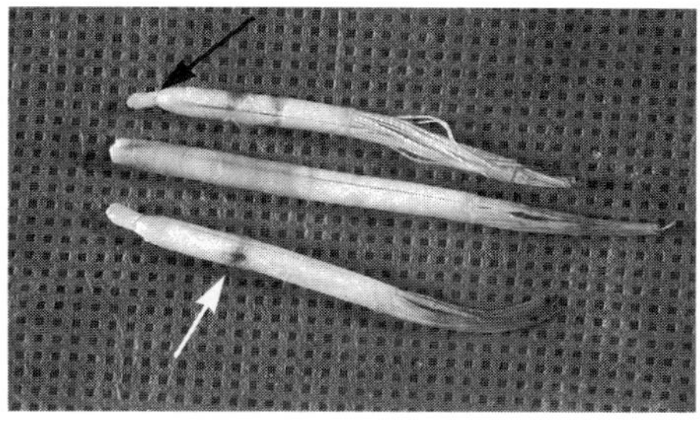

In birds with feather abnormalities, PBFD virus infections can be confirmed by microscopic evaluation of affected feathers for the presence of characteristic intracytoplasmic inclusion bodies or by demonstrating viral nucleic acid in the blood using PBFD virus-specific DNA probes. Identifying basophilic intracytoplasmic inclusion bodies in birds with characteristic feather lesions is considered diagnostic. However, adenovirus and polyomavirus can cause intranuclear inclusion bodies that appear similar to those caused by the PBFD virus.[24] Therefore, a confirmatory diagnosis of PBFD requires the use of viral-specific antibodies to demonstrate PBFD virus antigen or the use of DNA probes to detect PBFD virus nucleic acid.[14,16,36]

Psittacine beak and feather disease virus does not infect already formed feathers and does not appear to affect all newly developing feathers at the same time. Because of the progression of this disease, it would be anticipated that feather biopsies of birds with normal feathers or biopsies of clinically normal feathers from diseased birds would be

FIG 8.15
Visible feather changes similar to those caused by PBFD virus can be induced by other factors that disrupt the blood supply to the developing feather. The clubbing (arrow) and necrosis (white arrow) of the feathers in this photograph were created by "pinching" the feathers at the level of the pulp cap.

FIG 8.16
Unlike its PBFD virus affected siblings (above) this African Grey Parrot had minimal feather abnormalities. However, it was determined by DNA probe testing to be infected with the PBFD virus and subsequently developed severe feather abnormalities.

FIG 8.17

When feather biopsies are used for diagnosing PBFD, it is important that a recently formed abnormal feather be collected. If an abnormal feather (arrow) were examined microscopically, it is likely this bird would be correctly diagnosed with PBFD, however, if a normal feather (white arrow) were examined, the bird may be incorrectly diagnosed as normal.

ineffective in determining which birds have or could develop PBFD virus infections *Figure 8.17*. In one study, feather biopsies were taken from 50 clinically normal cockatoos prior to sale, and these birds were revaluated at three-month intervals for nine months. Three of the birds (6%) that were originally negative by feather biopsy developed PBFD over a six-month period, and eight birds (16%) had clinical lesions suggestive of PBFD that could not be confirmed by microscopic evaluation of feather biopsies.[18] In a study comparing routine histopathology, immunoperoxidase staining and DNA probe detection of PBFD virus in tissues from suspect patients, DNA testing was found to be the most sensitive (97.7%) and specific (100%). Routine microscopic evaluation of tissues was found to be the least sensitive (72.4%) and specific (93.8%).[14]

Based on histologic changes, concurrent infections of avian polyomavirus and PBFD virus have been reported in Budgerigars, lovebirds, African Grey Parrots, Eclectus Parrots and several species of cockatoos.[13,32] While either avian polyomavirus or PBFD virus can cause similar-appearing feather abnormalities, it has been noted that feather lesions seen with polyomavirus typically resolve after one or two molts,

whereas PBFD lesions as a rule progress from molt to molt.[4] In birds infected with both PBFD virus and avian polyomavirus, the PBFD virus had a predilection for the nuclei of feather, follicular and intestinal epithelial cells, and for the cytoplasm of macrophages in numerous tissues. Avian polyomavirus typically infected nuclei of endothelial cells, hepatocytes and renal tubular epithelial cells.[13]

Changes in the complete blood count (CBC) or serum chemistry are not typical in PBFD virus-infected birds with the chronic form of the disease. In several studies, there were no differences in the white blood cell count, red blood cell count or serum chemistries of PBFD virus-affected birds when compared to normal birds of the same species.[9,42a] Thus, many of the reported changes in the blood cell counts or serum chemistries in birds with PBFD probably are caused by concomitant infections rather than being directly related to alterations in the body caused by the PBFD virus. In general, changes in the complete blood count or serum chemistries should be considered of little to no value in diagnosing a PBFD virus infection. The notable exception may be anemia in young African Grey Parrots with acute infections.

<u>DNA PROBE TESTING</u> — Use of viral-specific DNA probes is the most sensitive and specific test for detecting PBFD virus. In validating the DNA probe test designed to detect the PBFD virus in circulating white blood cells, it was found that the test correctly identified 377 of 378 known positive samples,

resulting in a test sensitivity of 99.7%. One hundred known negative samples were correctly identified for a specificity of 100%.[3] These probes can be used on biopsy samples of suspect feathers to confirm an infection or on a blood sample to demonstrate viral nucleic acid in circulating white blood cells of infected birds before clinical changes in the feathers are apparent *Figure 8.18*. This test is available in the United States and in several European countries see Appendix.

For DNA probe detection of active (feather lesions present) or subclinical (birds showing no feather abnormalities) infections, the recommended sample to submit is whole anticoagulated blood (0.2 to 1.0 ml in heparin). In birds that have feather abnormalities, whole anticoagulated blood should be submitted for DNA probe testing and biopsy samples of diseased feathers should be placed in 10% formalin and held for further diagnostic testing, should any be needed *Figure 8.19*.

A positive DNA probe test in a bird that has feather abnormalities suggests that the bird has an active PBFD virus infection. A positive blood test in a bird that does not have feather abnormalities may indicate that the bird is latently infected or that it recently has been exposed to the PBFD virus and is viremic. A bird that tests positive and has no feather abnormalities should be retested in 90 days. If the bird is still positive, then it should be considered to be latently infected or is continuously being exposed to the virus. A negative test 90 days later would indicate that the viral nucleic acid is no longer

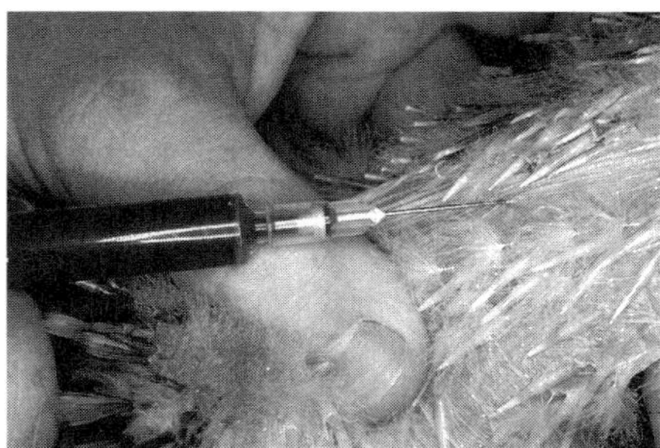

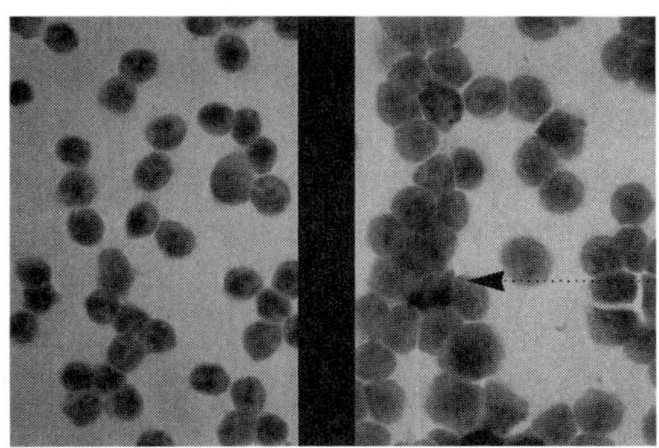

detectable in the blood and that the bird has eliminated the virus. Virus contamination of a sample from a toe-nail clip can produce a positive test in a bird that does not have the PBFD virus in its blood *see Figure 4.18*.

A negative DNA probe test for PBFD virus indicates that viral nucleic acid was not detected in the submitted sample. Some birds that have PBFD virus infections may be DNA probe-negative. These birds generally have severe infections with a majority of their feathers being abnormal. These birds may be negative because the high concentration of nucleic acid in the sample overwhelms the test reagents, or the white blood cell count may be decreased because of concomitant infections, resulting in an insufficient number of virus-positive cells to be detected.[3]

It is common for birds that have recently been exposed to the PBFD virus to be DNA probe-positive, even though the birds have normal-appearing feathers *see Figure 8.14*. The majority of these

Test blood for PBFD virus using DNA probes

A positive test in a bird with no feather abnormalities indicates that the bird has been exposed to PBFD virus and that viral nucleic acid is present in the blood. This bird must be retested in 90 days.

A negative test indicates that PBFD virus nucleic acid was not detected in the blood.

If the bird is still positive when retested at 90 days, this indicates that the bird is either subclinically infected or that the bird is being repeatedly exposed to the virus. Subclinically infected birds can develop feather lesions at some future date.

If the bird is negative when retested, this indicates that the bird was transiently infected and that the bird's immune system was able to clear the virus from the blood. Birds with normal feathers that have cleared an infection should be considered resistant to PBFD. Most birds that are exposed to the PBFD virus will have virus present in their blood for a brief period.

Test blood for PBFD virus using DNA probes

A positive DNA probe test in a bird with abnormally developing feathers suggest that the bird has an active PBFD virus infection.

If negative, a feather biopsy (including the feather follicle) should be submitted for microscopic examination. In some cases, birds infected with the PBFD virus may have a negative blood test if the concentration of virus in the blood stream is extremely high or if the bird's white blood cell count is extremely low. These birds usually have severe feather lesions and can be diagnosed by microscopic examination of affected feathers. It should be noted that some PBFD-infected psittacines of South American descent have spontaneously recovered from the disease.

FIG 8.19
Diagnostic flow chart for PBFD virus

MANAGEMENT OF A POSITIVE BIRD:

If a bird from a breeding aviary with feather abnormalities is found to be positive, remove the bird from the area as quickly as possible. Virus-infected birds with feather abnormalities shed large concentrations of virus in their feather dust, which can be easily carried to other birds by the wind or on clothes, skin or hair. All areas, supplies and equipment that could be contaminated with feather dust from the infected bird should be repeatedly cleaned and disinfected. One can determine whether cleaning efforts following an outbreak have been sufficient by DNA probe testing of air ducts, carpets, enclosures or any dusty area.

If a companion bird with feather abnormalities is found to be positive, the bird must never be exposed, directly or indirectly, to other birds outside of the household. Infected companion birds can live a long life when provided a stress-free environment; however, anyone who maintains a PBFD-positive bird must be aware that the virus can be transported to other locations on clothing or in hair. Caregivers should not expose other birds to this virus by entering aviaries, pet shops or bird shows.

birds will be transiently positive and will eliminate the virus. These birds develop disease only if their immune system is unable to clear the infection. In 10,000 blood samples tested over a period of one year, DNA probes indicated that approximately 5% were positive for PBFD virus. Many of these birds were clinically normal and were either subclinically infected or were transiently positive and were negative when retested 90 days later.[3] The DNA probe test can also be used to screen walls, enclosures, air-circulating ducts and equipment or any dusty area in the home or hospital to determine whether PBFD virus is contaminating these surfaces *Figure 8.20*. The appropriate sample for testing for environmental contamination is a swab collected from the premises in question.

<u>DETECTING HI ANTIBODIES</u> — The hemagglutination-inhibition (HI) test provides a rapid, specific technique to assess the immunologic response of psittacine birds to the PBFD virus. However, the clinical value of information provided by detecting antibodies to the virus is limited. Many birds of a susceptible species have some detectable anti-PBFD virus antibodies, indicating previous exposure to the virus. Antibody titer surveys in the United States would suggest that most birds of a susceptible species are exposed to the virus at some time in their lives but are able to mount an effective immune response. Both clinically affected birds and those that have not been exposed to the virus recently can have low HI antibody titers.[35,38] Because these low HI antibody titers

can suggest that a bird is infected or susceptible, HI antibody titers would not be expected to provide clinically relevant information in a field situation.[40] In one experimental study, affected birds had HI antibody titers that were similar to those induced by natural infections in which the adults recovered or in birds that were vaccinated with inactivated virus.[34,35,38] These findings further support the judgement that HI antibody titers provide little information that is relevant to the clinical management of PBFD virus infections. Additionally, HI antibody levels provide no information as to whether a bird is subclinically infected.[35,38]

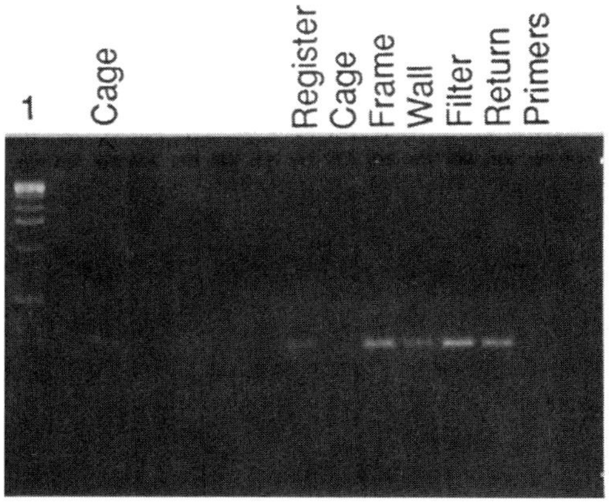

CONTROL

The environmental stability of the PBFD virus is unknown. It would be prudent to consider its stability to be similar to that described for CAV, which is similar in ultrastructure and DNA composition to the PBFD virus. CAV has been found to be environmentally stable and remarkably resistant to inactivation. In liver tissue, CAV remained infectious when treated with amphoteric soap (10%), orthodichlorobenzene (10%), iodine (1%), sodium hypochlorite (1%, bleach), methyl alcohol, ethyl alcohol,

1　Cage　Register　Cage　Frame　Wall　Filter　Return　Primers

FIG 8.20

Lane 1 is a control. In the other lanes, the presence of a white band indicates that a respective environmental sample contained PBFD virus nucleic acid. In this case, the home of an aviculturist with a PBFD-positive bird was screened for viral contamination. Before cleaning, the enclosure was positive for PBFD virus nucleic acid (lane 3). After cleaning, the enclosure was negative for PBFD virus nucleic acid (lane 9). However, the registers, returns and air filter associated with the heating system, as well as the wall and a picture frame in the room where the bird was housed were positive for PBFD virus nucleic acid.

chloroform and heating to 80ºC for one hour, and even when boiled for five minutes. Dried material containing CAV remained infectious after treatment with ethylene oxide for two hours. Fumigation with formaldehyde for 24 hours only partially inactivated CAV. A 10% solution of sodium hypochlorite or iodine was necessary to inactivate the virus in liver tissue. Cell culture-derived CAV was susceptible to iodine (1%), sodium hypochlorite, beta-propiolactone (0.4%), glutaraldehyde (1%) and heating to 80ºC for one hour, but was resistant to phenol (5%), sodium azide (0.1%), thimerosal (0.1%) and urea.[50]

Until a safe subunit or cell culture-derived vaccine to prevent PBFD virus infections is available, DNA probe tests in conjunction with sound hygiene are the best ways to prevent PBFD. This disease can virtually be eliminated from captive birds in the United States and Europe by making certain that only DNA probe-negative birds without feather abnormalities are added to an established group or aviary. In an effort to reduce the number of new cases of PBFD, it is advisable that all birds of a susceptible species be tested to determine whether they are latently infected with the PBFD virus. This is particularly true with respect to breeding birds, birds being added to established breeding aviaries, birds being sent to pet retailers and birds being evaluated during post-purchase examinations. Recommendations for preventing PBFD virus infections are summarized in *Table 8.4.*

A companion bird that is diagnosed with PBFD virus or that is latently infected with the PBFD virus can live a long life when provided a stress-free environment and supportive medical care. However, the PBFD virus is extremely contagious, particularly to young birds, and PBFD-positive birds should definitely not be in direct or indirect contact with psittacine neonates or endangered species. Additionally psittacine neonates must not be exposed to areas that may have been contaminated by feces or feather dust from a PBFD-positive bird.[33,39,48]

THERAPY — Numerous therapeutic trials using a variety of anti-viral drugs, immune system stimulants and herbal extracts have been attempted for PBFD virus-infected birds. Many of these therapies may improve the attitude or feather condition of an infected bird; however, none have been shown to resolve an infection by removing PBFD virus nucleic acid from the blood stream. Infected birds of South American descent have been shown to spontaneously recover from advanced PBFD. Secondary bacterial, fungal or parasitic infections should be treated with appropriate medications.

VACCINATION — Because the PBFD virus is currently limited in host range and most of the susceptible birds are restricted to enclosures, a widespread testing program can be used to control this disease in companion birds in the United States and Europe. Prevention of the disease through testing will be more difficult in Australia, because free-ranging psittacine birds, and at least one species of dove, can serve as reservoirs for the virus. In all captive groups of susceptible birds, the use of an effective, safe vaccine would be the simplest and most economical way to prevent infections.

A number of factors suggest that vaccination will be successful in preventing PBFD: 1) investigations indicating that many birds naturally exposed to the PBFD virus remain clinically normal and develop a protective immune response;[35,38] 2) comparative studies of PBFD virus isolates recovered from various genera of psittacine birds indicating that an antigenically similar virus infects a wide range of susceptible birds;[37] and 3) results of experimental vaccination studies demonstrating that vaccinated birds are protected from disease[34,40] and that vaccinated hens pass at least temporary protection to their chicks.[40]

Historically, the first vaccines that were used to protect mammals from virus infections were produced by recovering a virus from the tissues of a diseased animal, treating the recovered virus with chemicals to destroy its ability to replicate and then mixing it with an adjuvant which serves as an irritant to stimulate the immune system. For at least three reasons, these vaccines were inherently dangerous. First, it was frequently difficult to determine whether the virus in the vaccine remained infectious. Second, other viruses and infectious agents that were not detectable could be in the diseased tissues from which the vaccine was produced. Finally, some of the tissue proteins derived from one species of animal were found to be toxic to others. The next generation of vaccines that was developed for

TABLE 8.4

Recommendations for preventing PBFD

For companion bird enthusiasts:
    Use DNA probes to test all companion birds.
    Use DNA probes to test any new birds that are added to an avian family.
    If the supplier of a bird does not provide you with a three-week guarantee to have the bird tested, change suppliers.
    If an outbreak occurs, use DNA probes to evaluate the success of cleaning efforts.
For breeders:
    Use DNA probes to test all breeding birds.
    Use DNA probes to test any adult or neonate before it is shipped from a facility.
    Use DNA probes to test any new birds during the post-purchase quarantine period.
    Ship birds in containers (Horizon Micro-Environments) that prevent them from being exposed to PBFD virus during transport.
    If an outbreak occurs, use DNA probes to evaluate the success of cleaning efforts.
For pet shop owners:
    Use DNA probes to test all birds before they are sold.
    Deal only with breeders who test their birds before shipping.
    Request birds be shipped in containers that prevent them from being exposed to PBFD virus during transport.
    Use DNA probes to test the shop twice a month to detect any virus contamination.

mammals was derived from virus that had been propagated in cell culture. These vaccines were safer than their predecessors because the infectivity of the virus could be easily determined, and well defined cell cultures known to be uncontaminated with other infectious agents could be used to replicate the virus.

Difficulties were encountered in propagating the PBFD virus in cell culture. Thus, in 1989, a vaccine developed from the tissues of infected birds was tested for its ability to prevent PBFD virus infections. This vaccine was shown to be effective in preventing the disease. A duplicate of this tissue-derived vaccine is now available in Australia. Use of this vaccine may be warranted in Australia where psittacine birds are plentiful, many are destroyed

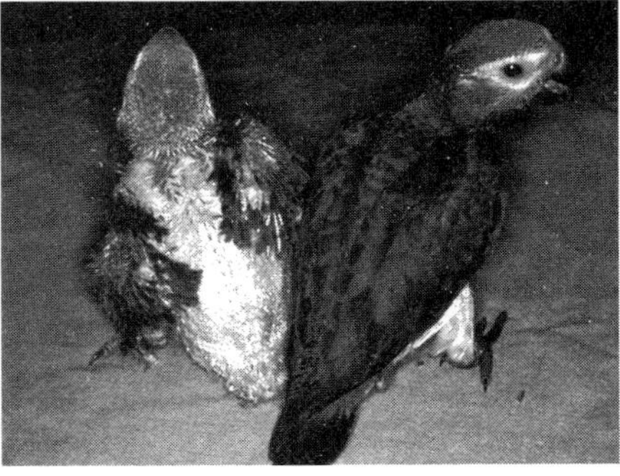

FIG 8.21

An experimentally vaccinated, protected bird stands next to an unvaccinated infected bird.

Hemagglutination-inhibition and precipitating antibody titers at intervals after vaccination with β-propiolactone-treated PBFD virus

Species Age and sex	Post-vaccination days			
	0	7	14	21*
African Grey Parrot Adult female	160	5120	5120	5120
African Grey Parrot Adult male	320	640	5120	5120
Amazon Parrot Adult male	320	160	2560	2560
Moluccan Cockatoo Adult male	640	640	2560	5120
Moluccan Cockatoo Adult female	160	320	640	1280
Umbrella Cockatoo Adult male	160	1280	2560	5120
Umbrella Cockatoo Adult male	80	1280	1280	1280
Umbrella Cockatoo Adult female	320	1280	2560	2560
Umbrella Cockatoo Adult female	80	160	2560	2560
African Grey Parrot 45 days (sex unknown)	160	2560	2560	5120
African Grey Parrot 45 days (sex unknown)	80	320	640	2560
African Grey Parrot 30 days (sex unknown)	640	2560	2560	5120
African Grey Parrot 30 days (sex unknown)	640	640	2560	2560
African Grey Parrot 30 days (sex unknown)	640	640	——	5120
Umbrella Cockatoo 45 days (sex unknown)	<40	160	640	128
Umbrella Cockatoo 45 days (sex unknown)	<40	320	640	1280
Sulphur-crested Cockatoo 45 days (sex unknown)	80	2560	2560	2560

gerous option where the disease can be prevented through testing *Figure 8.21*. Development of a subunit or cell culture-derived vaccine will be necessary to ensure that a vaccine intended for use in companion birds is adequately safe.

While tissue-derived vaccines are dangerous for use in a field setting, they have provided valuable research data. In one study, adult Umbrella Cockatoos, Moluccan Cockatoos, African Grey Parrots and a Yellow-headed Amazon Parrot were inoculated intramuscularly or subcutaneously with β-propiolactone-treated PBFD virus. Thirty- to 45-day-old chicks of African Grey Parrots, Umbrella Cockatoos and Sulphur-crested Cockatoos were vaccinated with the same inoculum. All the adult and neonate vaccinates had increased concentrations of HI antibodies by day 21 after inoculation *Table 8.5*. To demonstrate that chicks from vaccinated hens are protected from PBFD virus challenge, three African Grey Parrot chicks and two Umbrella Cockatoo chicks from vaccinated hens and one African Grey Parrot chick and one Umbrella Cockatoo chick from unvaccinated hens were exposed to purified PBFD virus. Chicks from the vaccinated hens remained clinically normal during the 50-day test period. Chicks from the unvaccinated hens developed clinical and histologic lesions of PBFD. These findings indicate that adult and 30- to 45-day-old psittacine birds will seroconvert following vaccination with β-propiolactone-treated PBFD virus. Also, hens inoculated with β-propiolactone-treated PBFD virus produce chicks

as pests and the disease is endemic. However, because of the resilience of the virus and its unique methods of replicating, use of this vaccine in the United States and Europe is an extremely dan-

that are at least temporarily resistant to virus challenge.[40]

In another study, adult Gang-gang and Galah Cockatoos injected with an experimental vaccine derived from the affected feathers of a cockatoo developed antibodies to the virus within two weeks of vaccination. Nestling Galahs, Short-billed Corellas, Sulphur-crested Cockatoos, Major Mitchell's Cockatoos and Gang-gang Cockatoos vaccinated with the same virus preparation also develop antibodies to the virus. Galah and Sulphur-crested Cockatoo nestlings vaccinated as early as 14 to 21 days after hatching were found to develop antibodies to the virus; however, the titers that occurred in chicks were lower than the titers that developed in adults. Vaccinated chicks challenged with live virus by the combined oral and intramuscular routes at six to eight weeks of age were found to be protected from disease, while unvaccinated chicks exposed to live virus developed gross and microscopic changes suggestive of PBFD. Antibody titers as high as 1:160 were found to persist in some vaccinates for six months. In this study, an HI antibody titer as low as 1:80 was considered protective.[34]

CIRCOVIRUS IN PIGEONS

A virus that morphologically resembles PBFD virus was first described in pigeons from Canada in 1986, followed by reports in Australia in 1989 and in Northern California in the autumn of 1990.[46] Clinically affected birds ranged in age from six weeks to twelve months and developed clinical signs of lethargy, anorexia, diarrhea and poor weight gains. Infected birds were diagnosed with varying fungal, bacterial, viral or parasitic diseases, suggesting that they were immunosuppressed. A group of affected birds in Australia had concomitant chlamydiosis, mycoplasmosis, trichomoniasis, coccidiosis, ascaridiasis and bacterial infections.[2] Many of the young birds (two to five months of age) in an affected flock had clinical signs that included weight loss and polyuria. Mortality was highest in weanlings (seven to eight weeks old).[2] Reported mortality rates range from less than 1% to 100% of young affected birds, with death occurring three to four days after clinical signs are first noted. However, it has not been determined whether death in young pigeons is caused by the circovirus or concomitant infections. A circovirus has been demonstrated in cytoplasmic inclusions in macrophages within the spleen, gut-associated lymphoid tissue and cloacal bursa of various infected pigeons.[43,47] The circovirus that infects pigeons appears to differ antigenically from the PBFD virus.[46]

REFERENCES

1. Ashby E: Emu 6:193, 1907.

2. Crowley A, Anapath PO, Marshall R, et al: Circovirus-like infection in pigeons. Proc Xth World Vet Poult Assoc Congress, 1993, p 251.

3. Dahlhausen R, Radabaugh S: Update on psittacine beak and feather disease and avian polyomavirus testing. Proc Assoc Avian Vet, 1993, pp 5-7.

4. Gerlach H: Viruses. In Ritchie BW, Harrison GJ, Harrison LR (eds): Avian Medicine: Principles and Application. Lake Worth, Wingers Publishing, 1994, pp 862-948.

5. Graham DL: Parrot reovirus and papovavirus infections and feather and beak syndrome. Proc 34th West Poult Dis Conf, 1985, pp 118-120.

6. Graham DL: Feather and beak disease: Its biology, management, and an experiment in its eradication from a breeding aviary. Proc Assoc Avian Vet, 1990, pp 8-11.

7. Hoop RK: Persistence and vertical transmission of chicken anaemia agent in experimentally infected laying hens. Avian Pathol 21:493-501, 1992.

8. Jacobson ER: Cockatoo beak and feather disease syndrome. In Kirk RW (ed): Current Veterinary Therapy IX. Philadelphia, WB Saunders Co, 1986, pp 710-713.

9. Jacobson ER, Clubb S, Simpson C, et al: Feather and beak dystrophy and necrosis in cockatoos. Clinicopathologic evaluations. J Am Vet Med Assoc 189:999-1005, 1986.

10. Jacobson ER, Hines SA, Quesenberry K, et al: Epornitic of papova-like virus-associated disease in a psittacine nursery. J Am Vet Med Assoc 185:1337-1341, 1984.

11. Krautwald M-E, Kaleta EF: Relationship of French moult and early virus induced mortality in nestling Budgerigars. Proc 8th Intl Cong World Vet Poult Assoc, 1985, p 115.

12. Latimer KS, et al: Cryptosporidiosis in four cockatoos with psittacine beak and feather disease. J Am Vet Med Assoc 200:707-710, 1992.

13. Latimer KS, Niagro FD, Campagnoli RP, et al: Diagnosis of concurrent avian polyomavirus and psittacine beak and feather disease virus infections using DNA probes. J Assoc Avian Vet 7:141-146, 1993.

14. Latimer KS, Niagro FD, Rakich PN, et al: Comparison of DNA dot-blot hybridization, immunoperoxidase staining and routine histopathology in the diagnosis of psittacine beak and feather disease in paraffin-embedded cutaneous tissues. J Assoc Avian Vet 6:165-168, 1992.

15. Latimer KS, Rakich PM, Kircher IM, et al: Extracutaneous viral inclusions in psittacine beak and feather disease. J Vet Diag Invest 2:204-207, 1990.

16. Latimer KS, Rakich PM, Steffens WL, et al: A novel DNA virus associated with feather inclusions in psittacine beak and feather disease. Vet Pathol 28:300-304, 1991.

17. Lowenstine LJ: Emerging viral diseases of psittacine birds. In Kirk RW (ed): Current Veterinary Therapy IX. Philadelphia, WB Saunders Co, 1986, pp 705-710.

18. Marshall R, Crowley A: A field study for the control of PBFD virus in wild-caught cockatoos. Proc Assoc Avian Vet, 1992, pp 37-47.

19. McNulty MS, Connor TJ, McNeilly F: Influence of virus dose on experimental anaemia due to chicken anaemia agent. Avian Pathol 19:167-171, 1990.

20. McNulty MS, Connor TJ, McNeilly F, et al: A serological survey of domestic poultry in the United Kingdom for antibody to chicken anaemia agent. Avian Pathol 17:315-324, 1988.

21. McOrist S: Some diseases of free-living Australian birds. ICBP Technical Publication 16:13-68, 1989.

22. McOrist S, Black DG, Pass DA, et al: Beak and feather dystrophy in wild sulphur-crested cockatoos (*Cacatua galerita*). J Wildl Dis 20:120-124, 1984.

23. Otaki Y, Saito K, Tajima M, et al: Persistence of maternal antibody to chicken anemia agent and its effect on the susceptibility of young chickens. Avian Pathol 21:147-151, 1992.

24. Pass DA: Inclusion bodies and hepatopathies in psittacines. Avian Pathol 16:581-597, 1987.

25. Pass DA: Natural infection of wild doves (*Streptopelia senegalensis*) with the virus of psittacine beak and feather disease. Proc Xth World Vet Poult Assoc Congress, 1993, p 165.

26. Pass DA, Perry RA: The pathology of psittacine beak and feather disease. Aust Vet J 61: 69-74, 1984.

27. Pass DA, Perry RA: Psittacine beak and feather disease: An update. Aust Vet Pract 15:55-60, 1985.

28. Pass DA, Perry RA: Granulomatous dermatitis in peach-faced lovebirds. Aust Vet J 64:285-287, 1987.

29. Perry RA: A psittacine combined beak and feather disease syndrome. Proc Courses Veterinarians. Australia, 1981, pp 81-108.

30. Perry RA: Some feather characteristics of acute French moult in fledgling budgerigars (*Melopsittacus undulatus*). Aust Vet Pract 13:128, 1983.

31. Perry RA, Pass DA: Psittacine beak and feather disease, including French moult, in parrots in Australia. Proc Aust Vet Poult Assoc, 1985, pp 35-38.

32. Phalen DN, Wilson VG, Graham DL: Epidemiology and diagnosis of avian polyomavirus infection. Proc Assoc Avian Vet, 1991, pp 27-31.

33. Raidal S, Sabine M, Cross GM: Laboratory diagnosis of psittacine beak and feather disease by haemagglutination and haemagglutination inhibition. Aust Vet J 70:133-137, 1993.

34. Raidal SR, Firth GA, Cross GM: Vaccination and challenge studies with psittacine beak and feather disease virus. Aust Vet J 70:437-441, 1993.

35. Raidal SR, McElnea CL, Cross GM: Seroprevalence of psittacine beak and feather disease in wild psittacine birds in New South Wales. Aust Vet J 70:137-139, 1993.

36. Ramis A, Latimer KS, Niagro FD, et al: Diagnosis of psittacine beak and feather disease (PBFD) viral infection, avian polyomavirus infection, adenovirus infection and herpesvirus infection in psittacine tissues using DNA in situ hybridization. Avian Pathol 23:643-657, 1994.

37. Ritchie BW, Niagro FD, Latimer KS, et al: Ultrastructural, protein composition and antigenic comparison of psittacine beak and feather disease virus recovered from four genera of psittacine birds. J Wildl Dis 26:196-203, 1990.

38. Ritchie BW, Niagro FD, Latimer KS, et al: Hemagglutination by psittacine beak and feather disease virus and use of hemagglutination-inhibition for detection of antibodies against the virus. Am J Vet Res 52:1810-1815, 1991.

39. Ritchie BW, Niagro FD, Latimer KS, et al: Routes and prevalence of shedding of psittacine beak and feather disease virus. Am J Vet Res 52:1804-1809, 1991.

40. Ritchie BW, Niagro FD, Latimer KS, et al: Antibody response to and maternal immunity from an experimental PBFD virus vaccine. Am J Vet Res 53:1512-1518, 1992.

41. Ritchie BW, Niagro FD, Lukert PD, et al: Characterization of a new virus from cockatoos with psittacine beak and feather disease. Virology 171:83-88, 1989.

42. Ritchie BW, Niagro FD, Lukert PD, et al: A review of psittacine beak and feather disease. J Assoc Avian Vet 3:143-148, 1989.

42a. Ritchie BW, et al: Unpublished data.

43. Shivaprasad HL, Chin RP, Jeffrey JS, et al: Particles resembling circovirus in the bursa of Fabricius of pigeons. Avian Dis 38:635-641, 1994.

44. Smith R: Psittacine beak and feather disease: A cluster of cases in a cockatoo breeding facility. Proc Assoc Avian Vet, 1986, pp 17-20.

45. Todd D, Niagro FD, Ritchie BW, et al: Comparison of three animal viruses with circular single-stranded DNA. Archives Virol 117:129-135, 1991.

46. Woods LW, Latimer KS, Barr BC, et al: Circovirus-like infection in a pigeon. J Vet Diag Invest 5:609-612, 1993.

47. Woods LW, Latimer KS, Niagro FD, et al: A retrospective study of circovirus infection in pigeons: nine cases (1986-1993). J Vet Diagn Invest 6:156-164, 1994.

48. Wylie SL, Pass DA: Experimental reproduction of psittacine beak and feather disease/French moult. Avian Pathol 16:269-281, 1987.

49. Wylie SL, Pass DA: Investigations of an enteric infection of cockatoos caused by an enterovirus-like agent. Aust Vet J 66:321-324, 1989.

50. Yuasa N: Effect of chemicals on the infectivity of chicken anemia virus. Avian Pathol 21:315-319, 1992.

51. Yuasa N, Yoshida I: Experimental egg transmission of chicken anemia agent. Natl Inst Avian Health Q 23:99-100, 1983.

Paramyxoviridae

Viruses in the family Paramyxoviridae are enveloped and can vary in shape and size from 150 to 300 nm. These viruses replicate in the cytoplasm of an infected cell, but virus-induced inclusion bodies can be found in the nucleus and cytoplasm. The genera included within the Paramyxoviridae — *Paramyxovirus, Pneumovirus* and *Morbillivirus* — are thought to be closely related to the rhabdoviruses and the orthomyxoviruses. Paramyxoviruses have been associated with influenza-like diseases in birds and mammals, including humans. Pneumoviruses have been associated with upper respiratory disease in turkeys and in mammals, including humans. Morbilliviruses have been described only in mammals; they cause measles in humans and canine distemper in dogs. With respect to birds, the most important member of the Paramyxoviridae is Newcastle disease virus (paramyxovirus-1).

The paramyxoviruses that infect birds differ in host range. Some infect only specific groups of birds. Others infect a wide range of avian and mammalian hosts. Currently, these viruses have been grouped into nine distinct serotypes, with several additional isolates from birds that have yet to be characterized *Table 9.1*.[3] By convention, these viruses are identified according to the type of bird from which the isolate was originally recovered, geo-graphic location of initial isolation and reference number or name/year. For example, Goose/Delaware/1053/76 indicates that this isolate was originally recovered from a goose in Delaware, that its reference number is 1053 and that the isolate was recovered in 1976.

TABLE 9.1

The 9 classified serotypes of avian paramyxovirus[3]

PMV 1 Newcastle disease virus
PMV 2 Chicken/California/Yucaipa/56
PMV 3 Turkey/Wisconsin/68
PMV 4 Duck/Hong Kong/D3/75
PMV 5 Budgerigar/Japan/Kunitachi/75
PMV 6 Duck/Hong Kong/199/77
PMV 7 Dove/Tennessee/4/75
PMV 8 Goose/Delaware/1053/76
PMV 9 Duck/New York/22/78

Paramyxovirus serotype 1 (PMV-1), commonly referred to as Newcastle disease virus, has the broadest host range. It has been shown to infect most species of birds and many species of mammals, including humans. Other serotypes of PMV are thought to be more restricted in host range. An overview of the avian host range and the clinical changes associated with each serotype of paramyxovirus is provided in *Table 9.2*.

PARAMYXOVIRUS 1

The most important paramyxovirus that infects birds is PMV-1 or Newcastle disease virus (NDV). This disease was first described in Java, Indonesia in 1926. It was linked to a fatal disease in domestic poultry near Newcastle, England in 1927. The virus subsequently spread to

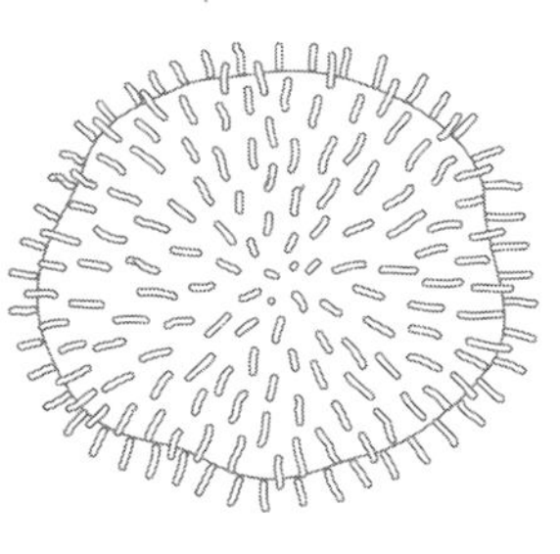

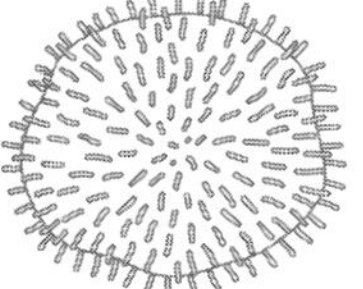

Overview of clinical changes, species susceptibility and geographic distribution for serotypes of avian PMV[3]

SEROTYPE	HOST RANGE	GEOGRAPHIC DISTRIBUTION	CLINICAL DISEASE
PMV-1	All species of birds considered susceptible	Globally distributed	Varies with species, age, condition of host and virulence of the virus strain. Infected birds may remain asymptomatic, develop disease and recover, die acutely with no premonitory signs or die following a protracted illness. Mild-to-severe disease involving the respiratory, gastrointestinal or neural system.
PMV-2	Passeriformes, Psittaciformes, domestic fowl, ducks, coots, egrets	Europe, Africa, Asia, South America, North America	Passeriformes — asymptomatic or develop mild respiratory disease. Psittaciformes — asymptomatic or develop severe pneumonia, mucoid tracheitis, diarrhea and high mortality.
PMV-3	Psittaciformes, Passeriformes, domestic fowl, ducks, geese, coots, pigeons	Europe, North America, Japan	Psittaciformes — acute death or die following a brief period of neurologic signs and dyspnea. Passeriformes — asymptomatic or develop conjunctivitis followed by anorexia, yellowish diarrhea, dyspnea and neurologic signs.
PMV-4	Ducks, geese, rails, domestic fowl and pheasants	North America, Europe, Japan, China, New Zealand and Australia	Asymptomatic.
PMV-5	Confirmed in Budgerigars, suspected in lories	Japan, Europe and possibly Australia	Acute onset of depression, diarrhea, torticollis and high mortality (90%).
PMV-6	Ducks, geese and turkeys	Canada, Europe, Japan, China and Hong Kong	Asymptomatic in waterfowl; mild respiratory disease and decreased egg production in turkeys.
PMV-7	Pigeons and doves	United States, Japan and England	Asymptomatic.
PMV-8	Ducks and geese	United States and Japan	Asymptomatic.
PMV-9	Domestic ducks	United States	Asymptomatic.

all continents, probably through the shipment of contaminated poultry and poultry products. Most species of domestic, aviary and free-ranging birds have been found to be susceptible to some strain of PMV-1;[2,3] however, PMV-2 and PMV-3 are more commonly recovered from companion and aviary birds than is PMV-1.

The strains of NDV that infect birds have been divided into four large groups based on their virulence and the type of disease they induce in chickens: lentogenic, mesogenic, velogenic neurotropic and velogenic viscerotropic *Table 9.3*. Because the same virus isolate can vary in virulence and the type of disease induced in differing avian hosts, this classification scheme cannot be generalized beyond chickens.

The lentogenic and mesogenic groups of NDV are common in domestic fowl in the United States; the velogenic strains are foreign pathogens that do

Strains of Newcastle disease virus grouped by their virulence in chickens*

LENTOGENIC - mild or inapparent infections.

MESOGENIC - mild to severe disease.

VELOGENIC NEUROTROPIC (VNND) - severe disease with high mortality. These strains do not cause hemorrhagic lesions in the gastrointestinal tract.

VELOGENIC VISCEROTROPIC (VVND) - severe disease with high mortality. These strains cause hemorrhagic lesions in the gastrointestinal tract.

This grouping is not consistent for other birds.

not naturally occur in this country. The fear of introducing virulent strains of NDV and avian influenza was the principal reason the USDA established guidelines in the early 1970s to monitor and restrict the importation of companion and aviary birds.[95,96] In the United States, psittacine birds were linked to epornitics of the most pathogenic form of Newcastle disease (VVND) in 1929, 1962, 1970 and 1974.[42] The 1970 outbreak in poultry cost approximately 19 million dollars to control. An outbreak linked to parrots from 1971 to 1973 in California cost 56 million dollars to control.[127] Of concern to the veterinary community and professional aviculturists is the fact that since 1971, VVND has been recovered from companion birds in the United States every year except 1978 and 1990.[92] These isolates were from birds suspected of being smuggled into the United States, particularly young Amazon parrots. Because of its importance to the poultry industry, PMV-1 is the best studied of the avian paramyxoviruses. Much of the information on other PMV strains is derived by inference from these data.

CLINICAL FEATURES

The response of birds to a PMV-1 infection varies with the strain of virus and the age, species and condition of the host. Birds infected with PMV-1 can remain clinically normal, develop disease and recover, die acutely with no premonitory signs or die following a protracted illness. Clinically affected birds can develop almost any combination of mild to severe disease involving the respiratory, gastrointestinal or nevous system. Severe disease, particularly of the nervous system, frequently ends in death. However, a strain of paramyxovirus that causes severe disease in one group of birds may cause subclinical infections in others.[4] For example, in a paramyxovirus outbreak in a mixed species aviary, all exposed *Neophema* sp. developed clinical signs. Although in direct contact with the infected birds, Rosella Parakeets and Cockatiels remained asymptomatic.[111] A list of the potential variations in the progression of PMV-1 induced disease is provided in *Table 9.4*.

TABLE 9.4

Variations in clinical progression of PMV-1 infections[49]

- Subclinical infection.

- Peracute death following several hours of depression in response to viremia.

- Acute gastrointestinal disease; voluminous greenish diarrhea accompanied by anorexia and lethargy.

- Acute respiratory disease; exudates from the upper respiratory tract, rales, conjunctivitis, dyspnea.

- Acute gastrointestinal and respiratory disease.

- Acute gastrointestinal or respiratory disease followed by development of central nervous system signs.

- Chronic central nervous system signs that persist for months.

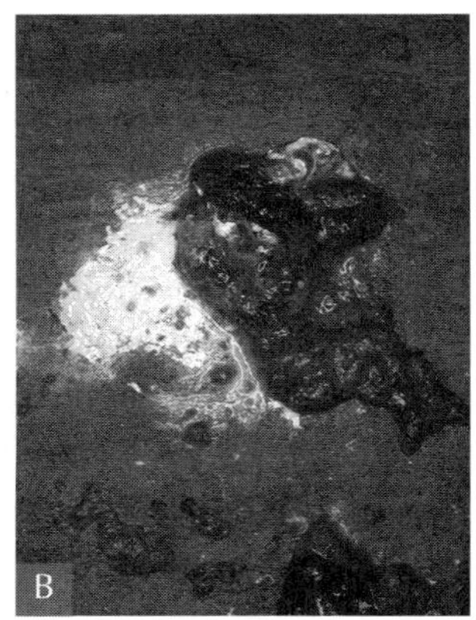

FIG 9.1

*Clinical signs of Newcastle disease infection include: **A** depression, ocular and nasal discharge, conjunctivitis, rhinitis, sneezing, coughing and dyspnea; **B** diarrhea; **C** ataxia, torticollis; **D** opisthotonos, convulsions, circling, tremors; and **E** paralysis of the wings and legs.*

photographs courtesy of Greg Harrison, Barbara Oglesbee and S.S. Tsai

Particularly common in psittacine birds that are smuggled into the United States, clinical changes suggestive of PMV-1 infection include depression, anorexia, ocular and nasal discharge, conjunctivitis, rhinitis, sneezing, coughing, dyspnea, bluish discoloration of facial appendages, diarrhea, ataxia, torticollis, opisthotonos, convulsions, circling, tremors and paralysis of the wings and legs *Figure 9.1*.[2,29,43,57,71,83,86] Nervous system abnormalities usually occur after other clinical signs are noted, and frequently end in death. Affected birds may die within four to five hours of becoming paralyzed. Neurologic signs may intensify when infected birds are excited or disturbed.[30] Birds that survive the virus-induced damage to the nervous system can manifest intermittent episodes of convulsions and tremors that persist for months following apparent recovery.[49,74]

Newcastle disease virus infections in lovebirds have been associated with ataxia, heart failure and moderate mortality. Amazon parrots, Psittaculidae, Plum-headed Parakeets and Eclectus Parrots are considered highly susceptible to NDV and may develop diarrhea, dyspnea and central nervous system signs. Some Rose-winged Parakeets are considered highly susceptible; others are relatively resistant. Some species of cockatoos appear to be more susceptible to infection than others. Cockatiels are moderately susceptible to infection. Budgerigars appear to be relatively resistant to natural infections; however, they are highly susceptible to experimental infections and develop disease similar to that described for macaws. Infected rosellas frequently remain asymptomatic and are thought to develop persistent infections. Lories appear to be refractory to infection.[49]

Collectively, an infected group of psittacine birds in a quarantine station developed diarrhea, head tremors, incoordination, torticollis and respiratory distress; however, the clinical changes in individual species of psittacine birds tended to vary. Respiratory signs and conjunctivitis were most common in cockatoos, Amazon parrots and Cockatiels. Diarrhea occurred in Cockatiels and in some cockatoos and macaws. Neurologic signs including head-bobbing, opisthotonos, torticollis, ataxia, dilated pupils and unilateral or bilateral wing paralysis were most common in Goffin's Cockatoos, Yellow-crowned Amazon Parrots and conures. Moluccan Cockatoos were likely to die acutely with few clinical signs or die following a brief period of depression and diarrhea. Goffin's Cockatoos experienced a more prolonged disease that lasted for several days and included high incidence of neurologic signs. It was of interest that exposed cockatoos had a higher level of morbidity than macaws maintained under identical conditions. Clinical changes in exposed African Grey Parrots were considered mild even though 61% of the exposed birds were serologically positive, indicating that they had been infected. Exposed finches, Secretary Birds, swans, mynahs and flamingos remained asymptomatic.[30]

Velogenic viscerotropic Newcastle disease (VVND) virus is the most virulent form of PMV-1 affecting poultry. A review of an outbreak in a group of Amazon parrots provides an example of the type of disease progression that can occur in companion birds infected with some strains of VVND virus. A group of three six-week-old Double Yellow-headed Amazon Parrots developed an acute onset of depression and crop stasis. It was considered likely that these young parrots had entered the United States without passing through a USDA quarantine station. One chick was recumbent on presentation and entered a coma-like state the following day. At this point, the other chicks became depressed and developed crop stasis. Radiographs indicated gaseous distention of the proventriculus and intestines suggestive

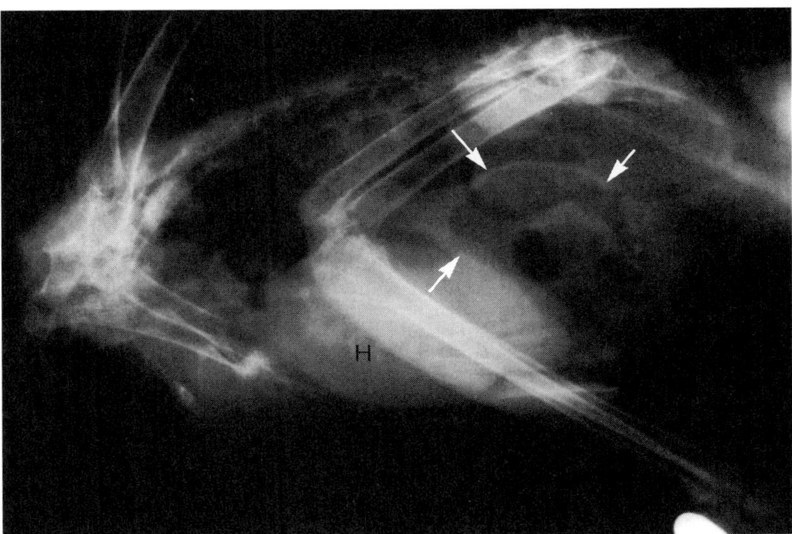

FIG 9.2

Crop stasis and gaseous distention (arrows) of the proventriculus and intestines, all suggestive of ileus may occur in birds infected with PMV-1.

of ileus *Figure 9.2*. Less than 12 hours after becoming depressed, one of the remaining chicks developed a mild head tremor. Meanwhile, other chicks had been in contact with these chicks; three of these were dead and a fourth chick was ill.[120] Changes in blood parameters from these three chicks are provided in *Table 9.5*.

In another group of Amazon parrots infected with VVND, depression, greenish diarrhea, severe dyspnea, tremors and paralysis were noted prior to death.[92] During an outbreak in a quarantine station, paramyxovirus was recovered from birds with enteritis.[98]

A paramyxovirus serologically related to, but distinct from, NDV caused the death of over 200 of 300 (67%) Peach-faced Lovebirds. Affected birds died suddenly or following a brief period

TABLE 9.5

Changes in blood parameters in three Amazon parrots with PMV-1[120]

Parameter	Affected birds	Normal values
Total solids	low, 2.3-2.5 gm/dl	3.2-5.2 gm/dl
Total proteins	low, 1.9-2.2 gm/dl	3.5-4.7 gm/dl
Albumin	low, 0.1-0.7 gm/dl	1-1.7 gm/dl
CPK activity	high, 6060-10,509 U/L	193-1261 U/L

Morbidity and mortality in a group of birds exposed to VVND virus[42,43]

| | Aerosol Exposure | | | Contact Exposure | | | Overall Mortality | |
	#	Affected	Died	#	Affected	Died	by day 28	by day 203
Budgerigars	70	100%	16%	35	100%	6%	12%	22%
Amazon parrots	28	100%	29%	14	100%	0%	19%	29%
Canaries	88	19%	6%	44	11%	2%	4%	25%
Half-moon Conures	35	100%	40%	17	100%	35%	38%	55%
Lesser Hill Mynahs	32	84%	22%	16	50%	6%	17%	21%
Black-headed Nuns	53	98%	17%	18	83%	11%	13%	21%

of lethargy, and gross lesions included swollen spleens and kidneys. Other psittacine and passerine birds in the same air space remained clinically normal.[57] A non-hemagglutinating paramyxovirus was recovered from a group of pionus parrots with accumulations of lymphocytes and plasma cells in the brain and pancreas. These pionus parrots exhibited tremors, circling, ataxia, depression and torticollis. An unclassified paramyxovirus was recovered from lories with diarrhea, edema, hemorrhage of the gastrointestinal tract and enlarged spleens.[74]

EXPERIMENTAL PMV-1 INFECTIONS

The progression and outcome of experimentally induced PMV-1 infections differ in various species of companion and aviary birds. Chickens may remain asymptomatic or die with severe disease when experimentally infected with strains of virus that cause VVN-like disease in parrots. A strain of NDV that is 100% fatal to young chickens may cause few-to-no deaths in pigeons or psittacine birds.[25] Less virulent strains of NDV can cause high morbidity and mortality in young chickens, while older birds are relatively resistant to disease. Virulent strains are likely to infect and cause disease in birds of any age.

In one study involving Amazon parrots, Budgerigars, conures, nuns, canaries and mynah birds, some birds exposed to VVND virus by nebulization developed ruffled plumage, conjunctivitis and signs of central nervous system damage from three days to two weeks after infection. Others remained asymptomatic. Still others showed inappetence and ruffled feathers from three days to two weeks after infection.[43] Many affected birds recovered. Birds that developed mild neurologic signs commonly survived, whereas those with severe signs of paresis or paralysis commonly died one to three weeks after infection. Both clinical signs of disease and mortality rates were higher in directly exposed birds than in those exposed by contact to infected birds.[42] The mortality rates were highest in conures (55%) and lowest in Black-headed Nuns and mynahs (21%) *Table 9.6*. Clinical signs of disease were most severe in experimentally infected Budgerigars; however, mortality rates were considered low (22%). Mynah birds and canaries exposed to PMV-1 exhibited no clinical signs of disease, but mortality rates in these birds were 21% and 25%, respectively. None of the experimentally infected Amazon parrots, Budgerigars, canaries, mynahs or nuns developed gross lesions characteristic of the disease in chickens. Conures that died up to several weeks after infection had hemorrhagic lesions in the gastrointestinal tract, particularly at the junction of the ventriculus and proventriculus *Figure 9.3*.[43]

To determine the host susceptibility of a paramyxovirus recovered from affected *Neophema* sp., the virus was injected intramuscularly, intracerebrally or intranasally into one-day-old chicks, 10- to 16-week-old chickens, one-day-old ducklings, Japanese Quail, pigeons, Budgerigars, Red-rumped Parakeets and Cockatiels. Budgerigars and Cockatiels were refractory to infection following intramuscular inoculation. All four Red-rumped Parakeets died within a week of intramuscular inoculation but did not develop the torticollis that was characteristic in the originally affected *Neophema*. Experimentally infected Japanese Quail, one-day-old ducklings and pigeons did not develop clinical signs of disease, but seroconverted. The one-day-old chicks injected intracerebrally developed weakness and anorexia three to four days after inoculation and died within 48 hours. Older chickens developed mild respiratory disease and high antibody titers by two weeks after infection.[111]

A paramyxovirus recovered from a lovebird was infectious for young chickens, Budgerigars and quail but caused only pathogenic changes and was recovered from the feces of experimentally infected Japanese Quail. Infected quail developed greenish, watery diarrhea 4 to 7 days after infection followed by a cessation of clinical signs. All three groups of birds developed HI antibodies to the virus by four weeks after infection.[57]

EPIZOOTIOLOGY

Although the exact distribution of each is unknown, the 9 avian serotypes of paramyxovirus have been reported throughout the world *see Table 9.2*. VVND is considered a nonindigenous disease by the USDA. This virulent strain of the virus is most common in Asia, but occurs also in Europe and Africa. The strains of NDV that are commonly recovered from native and domestic birds in North America are of low virulence. Smuggling of birds is considered the only route by which VVND virus could enter the United States.

Most species of captive and free-ranging birds are considered susceptible to NDV, but the clinical progression of disease varies widely among infected birds. Paramyxovirus-1 can infect susceptible birds of any age. Infections have been documented in Anseriformes, Columbiformes, Psittaciformes, Passeriformes, Falconiformes, Cuculiformes, Strigiformes, Sphenisciformes, Gruiformes, Piciformes, Phasianidae, Struthioniformes and Pelicaniformes.[5,29,41,49,89,93,95,105,106] Persistent infections have been described in free-ranging Anseriformes, Psittaciformes, Strigiformes and Passeriformes, implicating these groups in the distribution of the

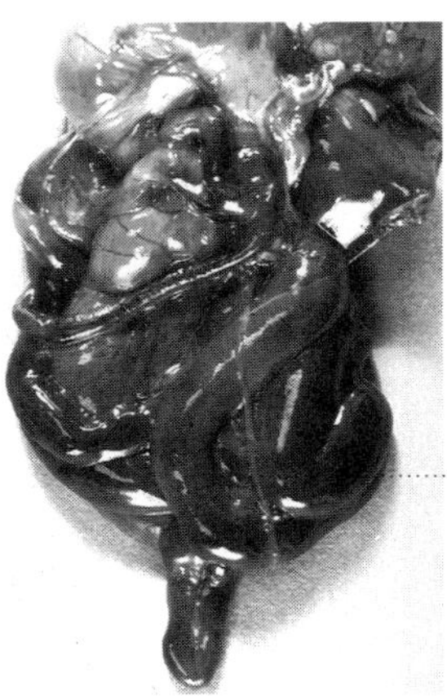

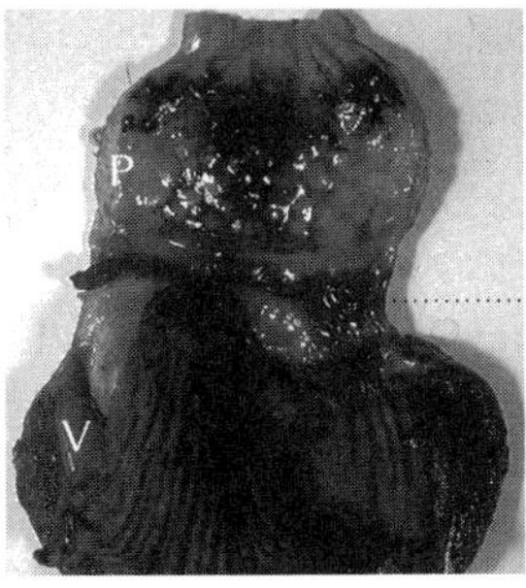

FIG 9.3
Some psittacine birds infected with PMV-1 develop lesions that are similar to those described in chickens including: hemorrhage throughout the gastrointestinal tract;

particularly between the proventriculus (P) and ventriculus (V).

photographs courtesy of Jean Sander

virus.[43,49,55,112] Epornitic outbreaks of PMV-1 appear to occur every 10 to 12 years.[2]

During a 1980 outbreak of VVND in a group of companion birds in a quarantine station, 8,223 birds of 89 species died or were destroyed over a 17-day period. Because of this outbreak, a total of 30,307 birds from 583 premises throughout the United States was killed.[30] In 1991, a group of 10 Amazon parrots was purchased at a "swap meet" by the owner of a Texas aviary with over 606 birds. These parrots, which were subsequently shown to be infected with VVND virus, were shipped throughout the country, exposing 400 birds in Illinois, 115 birds in Indiana and 400 birds in Michigan to the virus. Some exposed Amazon parrots, Cockatiels and conures developed respiratory or central nervous system signs. Birds in all the collections that were in direct contact with the affected Amazon parrots were killed. The remaining birds were screened for NDV by taking cloacal swabs from each bird three to four times at seven- to ten-day-intervals. In another VVND outbreak, 60 of 70 (86%) twelve-week-old Yellow-naped Amazon Parrots purchased from a street vendor in California developed central nervous system signs and died.[92]

RELATIONSHIP OF VIRUS STRAINS —
In gallinaceous birds PMV-1 viral isolates are divided into groups based on their virulence *see Table 9.3*, but this classification scheme is of minimal value in other avian species because clinical signs and pathologic changes vary widely among avian hosts.[6,25,41,43]

Some antigenic cross-reactivity has been described between PMV-1 and PMV-3 antiserum, particularly in companion bird species.[16,32] This cross-reaction may cause companion birds with antibodies to PMV-3 to appear as though they are infected with PMV-1. A PMV-3 isolate from Budgerigars protected chickens from experimental infection with NDV.[7]

INCUBATION PERIOD — The incubation period can vary from 3 to 28 days, depending on the virulence of the virus strain, quantity of virus to which the bird is exposed and susceptibility of the host. The experimental incubation period in a group of psittacine birds infected with VVND virus ranged from 3 to 14 days, whereas the incubation period in mynahs infected with the same virus was up to 28 days. The incubation period in a group of naturally infected psittacine birds ranged from 5 to 16 days. The incubation period in chickens is 2 to 15 days (average, 5 days). [42,49,120]

TRANSMISSION — Paramyxoviruses can be shed from an infected bird in all secretions (but primarily respiratory) and excretions (but primarily feces) for varying lengths of time. Ingestion of contaminated materials or inhalation of contaminated aerosols is the most common method by which a bird is exposed to the virus. Aerosolized fecal dust and contaminated bedding are considered potential sources for indirect exposure to paramyxoviruses. Newcastle disease virus is stable outside the host. Thus, insects, rodents and humans should be considered potential mechanical vectors for the dissemination of PMV

among susceptible flocks. Chicken-to-chicken transmission by the feather mite has been reported.[99]

Persistently infected migratory birds, particularly waterfowl and passerines, have been suggested as potential vectors for PMV.[2,4] During an outbreak of VVND in chickens in southern California, virus was recovered from sparrows and a crow that had been in direct contact with infected poultry, as well as from free-ranging doves and ducks.[93] This finding indicates that infected chickens can be a source of virus for free-ranging birds, which then could disseminate the virus to other flocks. The low incidence of virus recovery from free-ranging birds suggests that they are of minimal importance in virus dissemination.[93] Additionally, no NDV outbreaks in captive birds have been linked to free-ranging birds. Free-ranging birds are probably less important in disseminating the virus than are contaminated humans or transcontinental shipping of infected birds, particularly pigeons and Galliformes. Vertical transmission is possible, but is considered unlikely because infected hens will generally stop laying eggs when they are viremic. Eggs covered with virus-laden feces could contaminate an incubator and serve as a source of virus for recently hatched chicks.

In chickens, high concentrations of NDV are present in the upper respiratory tract for up to three days after infection, and virus still may be present twelve days after infection.[19] In a study involving psittacine and passerine birds, all experimentally infected par-

TABLE 9.7

Duration of VVND virus shedding in the feces or oral secretions in experimentally infected birds[42,43]

CANARIES - rapidly eliminate the virus, stop shedding by 28 days after infection.

MYNAHS- shedding up to 119 days after infection.

BUDGERIGARS - shedding up to 137 days after infection.

CONURES - shedding up to 84 days after infection.

NUNS - shedding up to 84 days after infection.

AMAZON PARROTS - shedding for over one year after infection.

rots were found to be excreting PMV-1 by the oral or cloacal route by twelve days after infection. Some birds were found to be shedding virus as soon as three days after infection. Many infected birds were found to remain asymptomatic, yet still shed virus in their feces. The duration of shedding by various species following experimental VVND virus infections ranged from 28 days to over a year *Table 9.7*. These findings implicate chronically infected birds in the transmission of PMV-1 within a flock.[25,42,43]

Newcastle disease virus has been shown to cause a transient conjunctivitis and generalized malaise in humans. Most cases of NDV-induced conjunctivitis in humans occur in individuals involved in poultry production. It should be noted that antibodies to the mumps virus can cross-react in the assay to detect NDV antibodies, causing a false-positive serologic test.[28]

PATHOLOGY

The pathology associated with PMV-1 infections varies with the isolate, the species of infected bird and the virulence of the infecting strain of virus. Affected birds may develop no gross lesions, or necropsy findings may include cardiomegaly, pericardial effusion, splenomegaly, mucus in the upper respiratory tract, hemorrhage in the trachea, hemorrhage of the ovaries, hyperemia of the brain and hemorrhage and edema in the respiratory and gastrointestinal tracts *Figure 9.4*.[25,29,43,49,83,120]

In gallinaceous species, inflammation and hemorrhage of lymphoid tissue in the jejunum is characteristic of virulent PMV-1 infections. This lesion is an inconsistent finding in other avian species, but when present should warrant suspicion.[43] Many affected psittacine birds do not develop gross lesions, while others may have hemorrhagic lesions in the trachea, lungs, heart, proventriculus, intestines, cerebellum and cerebrum, and necrotic plaques on the pharynx and glottis.[30] In one study, none of the experimentally infected Amazon parrots, Budgerigars, nuns, canaries or mynahs developed gross lesions characteristic of the disease in chickens. However, conures that died within two weeks of inoculation had hyperemia, hemorrhage and necrosis of the gastrointestinal tract, particularly at the junction of the proventriculus and ven-

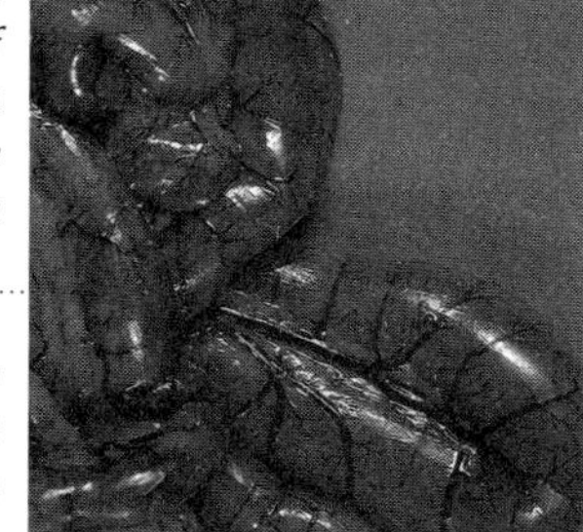

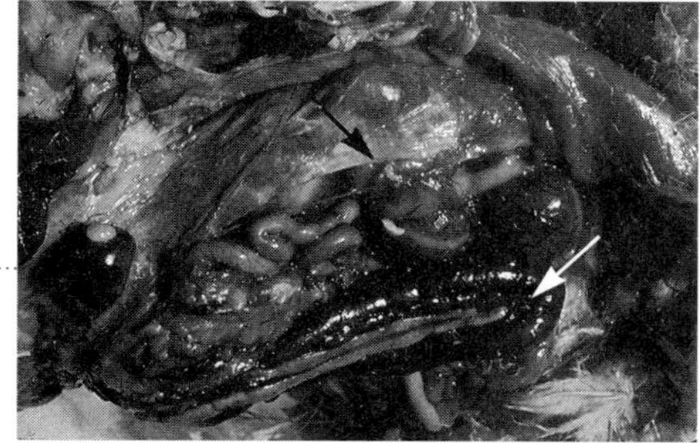

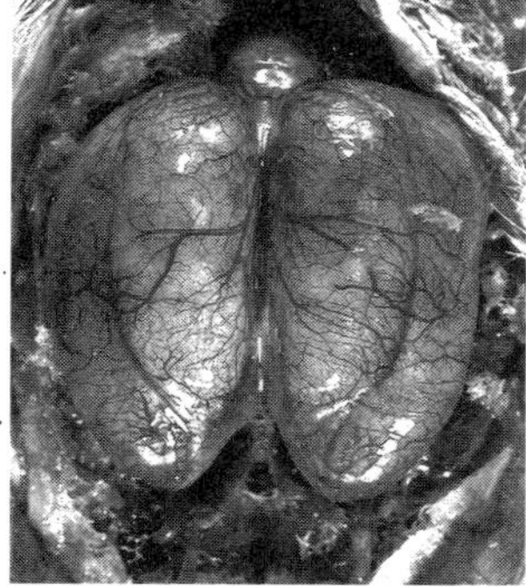

triculus *see Figure 9.3*. Other lesions noted in the conures included serous to fibrinous peritonitis, hepatomegaly and splenomegaly.[43] An enlarged and discolored liver was noted in quail experimentally infected with a PMV recovered from lovebirds.[57]

Microscopic lesions associated with PMV infections include evidence of hemorrhage, necrosis and perivascular cuffing in the brain. Rarely, intranuclear or intracytoplasmic inclusion bodies may be noted in the brain.[30,49]

PATHOGENESIS

The progression of PMV-1 infections varies among avian families as well as between different species within the same family.[43,49,97] Virulent strains spread rapidly through the body by binding to erythrocytes; in chickens the virus can be found in most tissues within 24 hours of exposure. This virus damages the cells that line the blood vessels, resulting in the hemorrhage that is common in affected birds.

Some strains of NDV have been shown to infect the intestinal tract preferentially. Other strains infect mucosal cells that line the respiratory tract from the nasal passages to the lungs.[80] Newcastle disease virus replicates primarily in the respiratory and gastrointestinal tracts, and invades the central nervous system late in the disease process. Thus, neurologic signs usually occur in birds after the development of respiratory and gastrointestinal signs.

DIAGNOSIS AND IMMUNITY

Diagnosis of PMV-1 infections can be accomplished: 1) by using paired serum samples to demonstrate an increase in antibody titer; 2) by the electron microscopic identification of paramyxovirus particles in the feces; or 3) by culturing the virus from diseased tissues, feces or trachea. Because of the many different PMV serotypes and the variations in the antigenicity of each strain, serology is less effective than virus isolation for diagnosing infections. Multiple samples of feces or respiratory secretions may be necessary for successful virus isolation.

In some species, antibodies can be detected from four to ten days after infection by several assays including hemagglutination-inhibition (HI), agar-gel immunodiffusion (AGID) or enzyme-linked immunosorbent assay (ELISA). These antibody assays can be used in appropriate species to screen a population of birds for previous NDV activity. During one outbreak in a quarantine station, 96% of Blue and Gold Macaws, 100% of Yellow-crowned Amazons, 93% of Nanday Conures, 61% of African Grey Parrots, 80% of Moluccan Cockatoos, 57% of Goffin's Cockatoos and 75% of plovers, but no lories, Secretary Birds, swans, flamingos or mynahs, were seropositive.[30] A paramyxovirus recovered from a lovebird induced HI antibodies by four weeks after infection in young chickens, Budgerigars and Japanese Quail.[57] Japanese Quail, one-day-old ducklings and pigeons experimentally infected with a paramyxovirus isolated from *Neophema* sp. remained asymptomatic but seroconverted. Older chickens developed mild respiratory disease and high antibody titers that were

detectable by two weeks after infection.[111] Antibodies protect chickens from disease caused by the same virus strain, and antibodies that are transferred from a hen to her chicks will provide protection for up to seven days after hatching.

As a diagnostic test in living birds, demonstration of HI antibodies is most reliable in chickens with prolonged infections, because these antibodies may not be detectable for up to one week after infection. Infected Budgerigars, raptors and some Columbiformes have been shown to develop low or undetectable levels of HI antibodies. Infected Amazon parrots develop moderate (up to 1:64) HI antibody titers. Cockatoos may develop higher titers (up to 1:320).[49] Monoclonal antibodies that can differentiate between types of NDV and pigeon paramyxovirus have been developed.[69]

For virus isolation, feces or swabs of the pharyngeal area containing secretions from the respiratory tract should be placed in appropriate transport media, packed on ice (4°C) and shipped immediately to the diagnostic laboratory. Postmortem samples for virus isolation should include trachea, lung, spleen, liver and brain. Frozen sections of the nasal or tracheal mucosa may be processed for staining with fluorescent antibodies, but cross-reactions can occur, causing a false-positive result. Fluid from the aqueous humor can be collected to detect virus (using hemagglutination assay), and may be the fastest way to diagnose an infection (hours to days).[49] Serology results (HI or AGID) generally require two days, whereas culture results may take from three to five days to several weeks.

Formalin-fixed tissues from the brain and trachea can be used for microscopic examination. As a rule, histopathologic lesions are most difficult to identify in long-term infections in which central nervous system signs are noted.

Infectious and noninfectious causes of gastrointestinal or respiratory tract disease should be considered in the rule-out list. Comparable clinical signs may be noted with chlamydiosis, salmonellosis, encephalomalacia, lead toxicity and calcium deficiencies. Many infectious agents that affect the respiratory tract cause sinusitis, while many uncomplicated cases of NDV are not associated with sinusitis.

CONTROL

Although paramyxoviruses are enveloped and would be expected to be rapidly inactivated when outside of the host, they are in fact relatively stable in the environment and are resistant to many commonly used disinfectants. Newcastle disease virus is stable at 50°C for 134 days, 40°C for 30 days and 27°C for four weeks.[54,100] The virus has been found to remain active in moist soil for 22 days, on feathers at 20°C for 123 days and in lake water for 19 days.[22,82,90] The virus can be inactivated by extremes of pH (< 2 and > 11), high temperatures (56°C), sunlight, detergents, chloramine (1%), sodium hypochlorite, lysol, phenol and 2% formalin.[49,57,87]

Because NDV is stable outside the host, it is possible that contaminated insects, rodents and humans can serve as mechanical vectors for the dissemination of the virus among susceptible flocks. Sound hygiene coupled with insect and rodent control and exclusion will help block these methods of virus exposure. Preventing secondary infections and practicing thorough hygiene can reduce the level of mortality and restrict spread of the virus through a flock. It has been suggested that intramuscular administration of 2 ml/kg of body weight of hyperimmune serum may protect exposed birds if they are treated prior to the development of clinical signs.[49]

All birds legally presented for importation into the United States are placed in a USDA-approved quarantine station for 30 days. During this quarantine period, samples are taken to detect viruses, predominantly PMV and influenza virus. If a hemagglutinating virus that is pathogenic to chickens is recovered, then the infected lot of birds is refused entry into this country. Birds in which HA viruses are not detected are released from quarantine.[74] This system provides an excellent screening method to prevent birds with PMV-1 infections from entering the United States. Thus, VVND virus should not be a concern for professional aviculturists who avoid birds that have entered this country illegally.

Newcastle disease virus occurs worldwide, and many free-ranging birds have been shown to be susceptible to infection. There has not been an outbreak of NDV in captive birds that was confirmed to have been caused by exposure to free-ranging birds. However, it is best for the health of aviary residents to prevent their exposure to free-ranging birds, particularly waterfowl, Passeriformes, doves or pigeons.

<u>VACCINATION</u> — In domestic fowl, Newcastle disease is controlled principally through vaccination with avirulent, attenuated-live virus vaccines. Dosing with a live-virus vaccine followed by a booster in three weeks has been shown to provide three to four months of immunity. Use of an inactivated NDV vaccine has been shown to provide five to seven months of immunity in poultry.[119] Inoculation with a live-virus vaccine, followed in two to three weeks by an inactivated vaccine might provide nine to twelve months of protection.[49]

Unfortunately, the apathogenic strains of NDV that have been used to vaccinate and protect poultry from virulent strains of the virus can induce fatal infections when administered to companion birds. In a study in Europe, an inactivated PMV-1 vaccine designed for use in pigeons was used successfully to vaccinate a group of mixed species birds, including ratites, swans, ducks, passerines, toucans, quail and psittacines.[33] In Germany, a combination of inactivated and live-virus vaccines has been used to successfully immunize aviary birds.[49,74] In the United States, it is not recommended that attenuated-live virus vaccines intended for use in chickens be administered in non-gallinaceous birds. The potential infectivity

of the vaccine strain of virus in a non-adapted host has not been adequately studied.

Effective vaccination regimes would be helpful in controlling infections in aviaries, breeding farms and zoo collections. However, NDV is a notifiable disease in many countries, and governmental regulations may restrict vaccination of avian species other than commercial fowl. Inactivated vaccines produced for chickens may be safe in other species provided that there are no governmental regulations that restrict their use.

PARAMYXOVIRUS 2

Paramyxovirus 2 (PMV-2) has been isolated from captive or free-ranging Passeriformes, Hanging Parrots, mynahs, *Neophema* sp., lovebirds, African Grey Parrots, chickens, turkeys, ducks, rails and Budgerigars from numerous geographic locations.[2,7,17,79,88,106,117,121] Epizootiologic evidence suggests that PMV-2 infections are most common in Passeriformes, and this virus may be endemic in some populations of free-ranging birds including Senegal Parrots and Passeriformes *Figure 9.5*.[5,50,51] Most passeriform isolates have been from asymptomatic birds; however, some isolates will cause mild upper respiratory disease in experimentally infected passerines. An isolate of PMV-2 was thought to have caused the death of a Blue Waxbill, but the isolated virus caused only mild respiratory disease in experimentally infected birds.[79] Amaduvade Finches experimentally infected with PMV-2 remained clinically normal.[77]

FIG 9.5

Most PMV-2 infected passerines remain asymptomatic; others may develop respiratory disease, depression and reduced activity. photograph courtesy of Louise Bauck

In an elaborate trial to evaluate the effects of PMV-2 on behavior, African Cut-throat Finches were infected with PMV-2 by the combined nasal, oral, eye drop and intraperitoneal routes. Over a three-week period, they showed significantly reduced activity when compared to controls and to their own behavior following recovery. Activity was decreased 24% during the first week, 33% during the second and 30% during the third week after infection.

In another study, it was found that activity increased exponentially prior to death. These finches seroconverted following experimental infection; however, virus was not detected in the cloacal swabs from any of the finches, and sentinel birds did not seroconvert. The

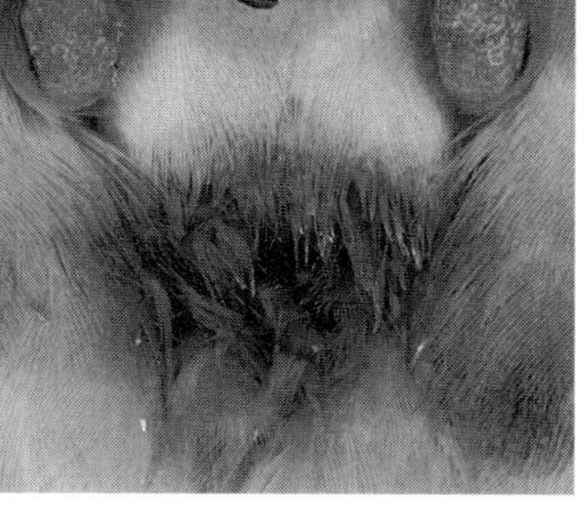

FIG 9.6

*While PMV-2 infections in Passeriformes appear to be mild and self-limiting, infections in psittacines, particularly African Grey Parrots, can cause difficulty with breathing; **right** pasty droppings and high mortality.*

antibody titers ranged from 1:32 to 1:128 by day 26 after infection. This study clearly demonstrates that although infections in which easily observable clinical changes are not detected may be characterized as subclinical, in fact subtle changes can occur that may substantially affect avian health, particularly that of a free-ranging bird.[51]

While PMV-2 infections in Passeriformes appear to be mild and self-limiting, infections in psittacines can vary from asymptomatic to severe. Signs include pneumonia, mucoid tracheitis, diarrhea and high mortality. This clinical presentation is particularly common in African Grey Parrots. In one report, an infected African Grey Parrot was listless and dyspneic, had pasty droppings and died within eight days of showing clinical signs *Figure 9.6*.[31]

The original PMV-2 isolate was recovered from chickens with laryngotracheitis. Infections in turkeys and chickens may be subclinical or can be associated with mild respiratory disease, decreased egg production or up to 90% mortality. Secondary bacterial or other viral infections may contribute to the disease process because experimentally infected chicks generally remain asymptomatic.[5] In several studies, PMV-2 isolates from finches were not pathogenic in chicks, and experimentally infected pigeons remained clinically normal.[45,64] Turkey poults experimentally infected with virus isolated from free-ranging passerines in Costa Rica seroconverted, indicating that they had been infected; however, the infected turkeys remained clinically normal.[50] Both turkeys and chickens are susceptible to infection by PMV-2 and PMV-3, but turkeys are more commonly infected with these viruses than are chickens. The increased incidence of infections in turkeys may be related to their production in open fields, where they are easily exposed to infected, free-ranging birds.

Isolates of PMV-2 have been recovered from free-ranging or indigenous captive birds in Europe, Indonesia, Senegal, Kenya, Israel and Costa Rica. Virus has been recovered from imported passerines or psittacines in Japan, Northern Ireland, England, Canada and the United States.[50,106] In Africa, PMV-2 was isolated from Passeriformes indigenous to Senegal prior to export and from 23 of 613 (3.7%) free-ranging birds in the same area.[44,45] In Germany, PMV-2 was recovered from the trachea of 34 of 108 (31%) free-ranging Passeriformes, and in Czechoslovakia, from the cloaca of two wrens (out of 477 small birds sampled).[88,117] During a PMV-2 outbreak in turkeys in Israel, the virus was recovered from free-ranging Cattle Egrets that frequented the turkey lots.[72] During a one-year period in Great Britain, 38 of 61 (62%) paramyxovirus isolates were characterized as PMV-2. These isolates were recovered from Passeriformes, a lovebird and African Grey Parrots.[5]

Monoclonal antibodies have been used to separate 53 isolates of PMV-2 into 4 groups *Table 9.8*. All of the group 3 and group 4 isolates were from passerine birds. Most of the chicken isolates were from group 1, as were isolates from Passeriformes, ducks and Psittaciformes that had been recently imported into Europe.[91] Of the hemagglutinating (HA) viruses isolated at a United States quarantine station over an eight-year period, 46% were PMV-2, and this serotype was recovered eight times more frequently from Passeriformes than from Psittaciformes.[106] Over a

five-year period, 102 isolates of PMV were made from cloacal swabs of mynahs originating from Southeast Asia. Of these isolates, 57 (56%) were characterized as PMV-2 and 42 (41%) as PMV-3.[92]

TABLE 9.8

Grouping of PMV-2 isolates based on antigenic variance determined using monoclonal antibodies[91]

GROUP 1 - isolates from Psittaciformes, a gadwall and some Passeriformes, a Mallard Duck, coot and turkeys

GROUP 2 - chickens

GROUP 3 - two strains from Passeriformes

GROUP 4 - most strains from Passeriformes

Antibody surveys indicate that the prevalence of PMV-2 may be greater than has been demonstrated by virus isolation or disease. Antibodies to PMV-2 have been demonstrated in asymptomatic free-ranging raptors, Passeriformes, ducks, geese and homing pigeons *Table 9.9*.[50,64,76] The incidence of infection in free-ranging birds was highest in the summer (July to September) when young birds were most common.[76] An inactivated PMV-2 vaccine was shown to induce antibodies in vaccinated turkeys.[102]

PARAMYXOVIRUS 3

Paramyxovirus 3 (PMV-3) has been isolated from Passeriformes, Galliformes, waterfowl and Psittaciformes including lovebirds, African Grey Parrots, Cockatiels, Grey-cheeked Parakeets, conures, *Neophema* sp., Moluccan Cockatoos, *Psittacula* sp., Senegal Parrots, Amazon parrots, Budgerigars, macaws, finches, mynahs, ducks, rails, flamingos, chickens and turkeys.[2,5,17,]

TABLE 9.9

Prevalence of PMV-2 antibodies in various avian species[76]

SPECIES	# SAMPLED	PMV-1	PMV-2	PMV-3
Domestic pigeons	51	—	13.7%	2%
Wood Pigeon	3	—	66.7%	66.7%
Partridges	123	—	23.6%	2.4%
Quail	68	—	4.4%	—
Japanese Quail	10	—	20%	—
Pheasants	34	—	5.9%	—
Ducks	247	1.2%	8.5%	6%
House Sparrows	35	—	68.6%	8.6%
Greylag Goose	9	—	11.1%	—
Aquatic birds	371	1.6%	7.9%	4.7%
Non-aquatic birds	44	—	61.4%	11.4%

[71,104,106,110,121] Most isolates of PMV-3 have been from dead or clinically normal birds in quarantine. Infections with PMV-3 appear to be most common in Psittaciformes, but surveys to determine the prevalence of PMV-3 in free-ranging Psittaciformes have not been reported *Figure 9.7.*

A paramyxovirus that probably represents a strain of PMV-3 was recovered in The Netherlands from a flock of *Neophema* sp. experiencing high flock morbidity and low mortality. Affected birds had central nervous system signs that included torticollis and walking in circles *Figure 9.8.* The disease could be experimentally reproduced in *Neophema* sp. and Red-rumped Parakeets, but not in Budgerigars or Cockatiels.[111] In one outbreak in *Neophema* sp., birds of all ages were susceptible to infection, but mortality rates (up to 40%) were most severe in the neonates. Necropsy findings in these birds included pulmonary edema and congestion with enlargement of the liver.[34] During another study involving *Neophema* sp., the only gross change noted at necropsy was pancreatic atrophy.[118]

Experimentally infected Bourke's Parakeets developed clinical signs and died within a week of intramuscular inoculation.[57a] In nestling Cockatiels, PMV-3 was associated with opisthotonos, tremors, leg paralysis, dyspnea and high mortality. The creatine kinase, lactate dehydrogenase and aspartate aminotransferase activities were elevated in affected birds. Necropsy findings included cardiomegaly and pericardial effusion. Virus was recovered from the brain and heart of affected birds. During this outbreak, exposed adult Cockatiels remained clinically normal.[121]

A paramyxovirus suspected to be PMV-3 based on microscopic changes in the brain was demonstrated in a Moluccan Cockatoo with ataxia that progressed rapidly to an inability to stand, twitching and death. Intracytoplasmic (and occasionally intranuclear) inclusion bodies suggestive of PMV were identified in the brain. Several Grey-cheeked Parakeets that had been added to the household without a quarantine period had died several weeks prior to the cockatoo, and the rest of these parakeets subsequently died. Paramyxovirus-like particles were demonstrated by electron microscopy within inclusion bodies in the brain. Amazon parrots, conures, Cockatiels, Budgerigars and finches that were in the same air space remained clinically normal.[71]

FIG 9.7

Of the paramyxovirus strains, PMV-3 appears to be most common in psittacine birds.

Finches with clinically apparent PMV-3 infections develop conjunctivitis initially, followed by anorexia, yellowish diarrhea and dyspnea *Figure 9.9*. Some affected birds die within a few days of developing clinical signs. Others recover. This presentation has been described in the Gouldian Finch, Blue Waxbill, common canary, White-rumped Canary, Orange-cheeked Waxbill, Black-throated Grassfinch, Double-barred Finch and Avadavat. When five finches were experimentally infected with this virus, three remained normal, one died five days after infection with central nervous system signs that had been noted on day four, and the last bird died without clinical signs 42 days after inoculation. Virus was recovered from brain of the latter bird, indicating that the incubation period can be up to six weeks. The virus recovered from the finches caused mortality from one to nine days after infection in one-day-old chicks inoculated intracerebrally, but intravenously exposed chicks seroconverted and remained clinically normal.[104]

The PMV-3 virus has been recovered from asymptomatic chickens and turkeys as well as from birds with respiratory disease and decreased egg production. Secondary bacteria or other viruses may be important in creating clinical disease, because experimentally infected chicks generally remain asymptomatic.[8]

Epizootiology

Over the years, PMV-3 has been documented in psittacine or passerine birds from the United States, Canada, Germany, The Netherlands, England, Spain, France, Belgium and Japan (*see Table 9.9*). During a one-year period in Great Britain, 9 of 61 (15%) paramyxovirus isolates were characterized as PMV-3. The virus was recovered mainly from psittacine birds, although a few infected Passeriformes were also detected.[5]

It appears that PMV-3 is the predominant paramyxovirus recovered from psittacine birds in the United States. Of the hemagglutinating viruses isolated from a United States quarantine station over an eight-year period, 34% were PMV-3, and 47% of these isolates were recovered from Psittaciformes.[106] This virus was isolated eight times more frequently from Psittaciformes than from Passeriformes.[2,71,106,121] Over a five-year period, 102 isolates of PMV were made from cloacal swabs of mynahs originating from southeast Asia, of which 42 (41%) were PMV-3.[92]

The PMV-3 isolates from psittacines are more closely related to each other than to turkey isolates.[5] Using monoclonal antibodies, turkey strains can be differentiated from psittacine strains.[16] Some cross-reactivity occurs between PMV-1 and PMV-3 antiserum, particularly in companion bird species.[32] This cross-

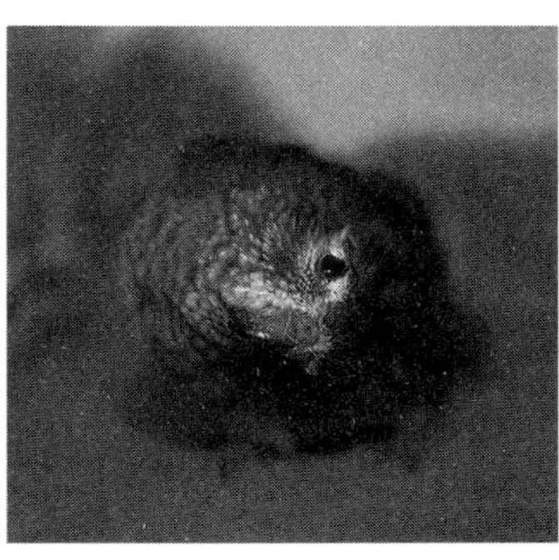

FIG 9.8

PMV-3 affected Neophema sp. typically develop central nervous system signs.
photograph courtesy of Doug Mader

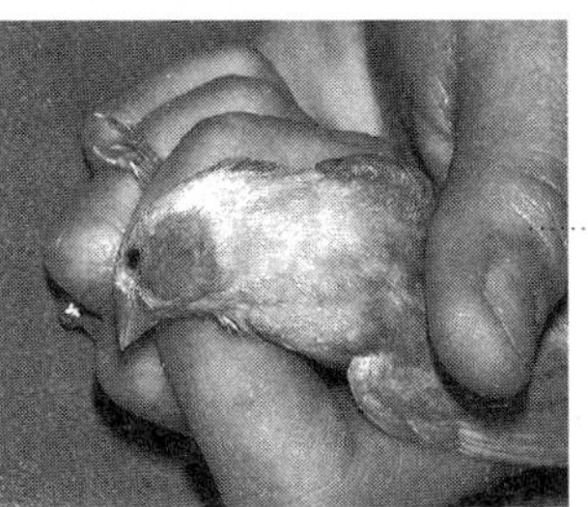

FIG 9.9

Paramyxovirus 3 should be considered in finches with depression, conjunctivitis, yellowish diarrhea, dyspnea

and central nervous system signs.

photographs courtesy of Louise Bauck and Greg Harrison

reaction may cause companion birds infected with PMV-3 to appear as though they were infected with PMV-1. A PMV-3 isolate from Budgerigars was found to protect chickens from NDV.[2] Some Passeriformes, Japanese Quail and domestic pigeons have been implicated as latently infected carriers.[49,104]

The diagnostic techniques used for PMV-3 infections are similar to those described from PMV-1. An inactivated PMV-3 vaccine designed for use in turkeys in Great Britain was shown to induce HI antibodies in canaries and Budgerigars. Following challenge, unvaccinated birds developed clinical signs and shed virus in their excrement while vaccinated birds remained clinically normal.[20] Commercially available PMV-1 vaccines designed for use in chickens or pigeons and PMV-3 vaccines designed for use in turkeys were evaluated for efficacy in a group of Bourke's Parakeets. The parakeets were vaccinated with one of these vaccines, followed by a booster in six weeks. Two weeks after the last vaccination, the birds were injected intramuscularly with live virus. Both vaccinated and unvaccinated birds developed clinical signs and died within one week after challenge. By comparison, birds vaccinated with inactivated PMV-3 virus recovered from a Bourke's Parakeet remained asymptomatic following virus challenge, suggesting that this disease may be preventable in companion birds through vaccination.[57a] An inactivated PMV-3 vaccine designed for use in turkeys is available (Appendix).

FIG 9.10

Paramyxovirus 4 has been demonstrated in asymptomatic, free-ranging waterfowl.

PARAMYXOVIRUS 4

Paramyxovirus 4 (PMV-4) has been demonstrated in asymptomatic, free-ranging waterfowl (ducks, geese and rails) in North America, Europe, Japan, China, New Zealand and Australia *Figure 9.10*. This virus also has been recovered from free-ranging pheasants in Czechoslovakia.[3,109,129] In a study in the United States, five of eight (62.5%) paramyxoviruses recovered from free-ranging ducks were characterized as PMV-4; the other three isolates were PMV-1.[2,129] This virus has been shown to be apathogenic in chickens.[4]

PARAMYXOVIRUS 5

The first isolation of Paramyxovirus 5 (PMV-5) occurred in the mid-1970s in Budgerigars from Kunitachi, Japan that were experiencing an acute infection characterized by depression, diarrhea and high mortality (90%).[132] Following experimental infection with PMV-5 by the intranasal, oral or intra-abdominal route, 18 of 23 (78%) Budgerigars developed diarrhea, torticollis and dyspnea, and died within two weeks after infection *Figure 9.11*.[2,85,87,132] Affected Budgerigars were viremic at the time of death, and virus could be recovered from the brain, liver, lung, spleen and blood.[87] Survivors were found to develop virus-neutralizing antibodies. Although it has not been studied in other avian species, PMV-5 has not been shown to be pathogenic for chickens or pigeons.[83,85,87] Unlike other paramyxoviruses, PMV-5 does not hemagglutinate red blood cells, so the hemagglutination-inhibition assay cannot be used to demonstrate antibodies to this virus.[87] A virus serologically related to PMV-5 was recovered from the liver and spleen of Budgerigars that died in a European aviary. In this outbreak, 20% of the Budgerigars in the aviary died over a two-year period. Clinical changes in affected birds included diarrhea and vomiting.[53]

A paramyxovirus that was serologically unrelated to PMV-1 or PMV-2 was recovered from a group of captive Budgerigars in Australia that were experiencing enteritis and high mortality. A definitive report on the serologic classification of this isolate could not be located; however, it has been suggested that this isolate was PMV-5. Approximately 40 of 80 (50%) birds died during a two-week period. Surviving Budgerigars developed antibodies to the virus. Two months prior to this outbreak in Budgerigars, a group of 70 free-ranging Rainbow Lories in the same area had been depressed for three to four days before developing diarrhea and dying. Affected Budgerigars had hyperemia of the abdominal organs. In the Rainbow Lories, gross lesions included enlargement of the liver and spleen, ulcerative-hemorrhagic enteritis and edema of the intestinal wall.[83,84] The gross and clinical lesions in the affected lories were also similar to those described with enteritis caused by *E. coli*, *Salmonella* and other gram-negative bacteria.

Young Budgerigars experimentally infected with the virus by the intranasal, oral or intra-abdominal route developed acute, fatal enteritis, while older birds were ill for four to seven days and recovered. Gross lesions in

these birds included hemorrhage in the proventriculus and intestines, discoloration of the liver, and splenomegaly. This PMV isolate was not pathogenic for chick embryos, young or mature chickens or pigeons; however, infected birds seroconverted within two weeks of infection.[83]

Budgerigars infected with PMV-5 develop diarrhea, torticollis, dyspnea and die within two weeks.

OTHER PARAMYXOVIRUSES

Paramyxovirus-6 (PMV-6) has been recovered from free-ranging ducks and geese in Canada, Europe, Japan and China, and from domestic ducks and contaminated pond water in Hong Kong.[108] In turkeys, PMV-6 may cause mild respiratory disease and decreased egg production.[4] Paramyxovirus-7 (PMV-7) has been recovered only from Columbiformes in the United States, Japan and Great Britain.[2,3,5,67] The fact that the virus has been isolated from free-ranging doves in Tennessee, pigeons in Japan and pigeons and doves in Britain suggests that the virus is widespread; however, the exact relationship of the virus recovered from various regions has not been reported.[2,4] In the United States and Japan, PMV-8 has been isolated from ducks and geese. In the United States, PMV-9 has been recovered from domestic ducks.[112]

PARAMYXOVIRUSES IN COLUMBIFORMES

Pigeons and doves are susceptible to strains of Newcastle disease virus that commonly infect chickens, as well as to a strain of PMV-1 that causes severe neurologic disease in pigeons but is only mildly pathogenic for chickens. This latter virus appears to be a variant of the mesogenic form of NDV and is commonly referred to as pigeon PMV-1 to differentiate it from other related strains of virus.[12,13,103,126] It has been suggested that pigeon PMV-1 is a mutant derived from a velogenic strain of paramyxovirus that infects chickens.[81]

Pigeon PMV-1 differs from other strains of Newcastle disease virus serologically, biochemically and in the disease it produces. It has been associated with high mortality, particularly in young pigeons.[21] Antibodies to pigeon PMV-1 and NDV cross-react in the HI test but can be distinguished using monoclonal antibodies.[48,69] It is of interest that paramyxovirus isolates from a duck, a sparrow and a Kestrel were found to be serologically identical to pigeon PMV-1.[3] Pigeons infected with PMV-2 remain clinically normal.[46]

NEWCASTLE DISEASE VIRUS IN COLUMBIFORMES

Pigeons and doves are susceptible to velogenic and lentogenic strains of NDV *see Table 9.3*, but appear to be relatively resistant to disease. These birds are most commonly infected with NDV during epornitics in domestic poultry. Because Columbiformes are susceptible to strains of virus that infect poultry, free-ranging pigeons may serve as mechanical vectors that disseminate the virus among domestic flocks of susceptible species.[125] Additionally, poultry are considered the primary source for virus exposure in free-ranging Columbiformes.[40] In 1971-73, an outbreak of a highly pathogenic strain of PMV-1 in racing pigeons coincided with an epornitic in European poultry.[125] In one study, VVND virus was recovered from the cloacal swab of 1 of 9 (11%) asymptomatic free-ranging pigeons in Africa.[39]

The clinical changes that occur in Columbiformes infected with NDV are governed by the virulence of the infecting virus and the susceptibility of the individual host. Lentogenic strains of NDV generally cause a mild respiratory disease characterized by conjunctivitis, rales and dyspnea. Experimentally infected birds shed virus from the larynx three to seven days after infection, and develop a mild respiratory disease and conjunctivitis within six days.[126]

Young pigeons and doves infected with VVND can develop the classic respiratory, digestive and central nervous system signs described in other species. Most infected birds develop a conjunctivitis followed by watery (occasionally hemorrhagic) diarrhea and central nervous system signs. Some birds may exhibit central nervous system signs in the absence of other clinical problems.[125] Up to 70% of exposed pigeons may develop clinical signs, with mortality rates approaching 40%. Death is most common in birds that develop central nervous system signs. The incubation period of NDV in pigeons and doves is 6 to 16 days.[49,125]

Gross lesions associated with NDV in pigeons include hemorrhage throughout the gastrointestinal tract and tracheitis. Edema and congestion of pharyngeal and laryngeal mucosae are frequent. Inflammation of the kidneys and brain is a common microscopic finding.

Virus is shed in the feces early in the disease process but ceases as a bird begins to recover. Virus can be detected in the lung and trachea for up to four weeks, and in the brain for up to five weeks after infection. Unlike some other species of birds, recovered pigeons do not appear to develop chronic NDV infections.[125] Newcastle disease virus can be prevented in pigeons with vaccination, but the low incidence of disease makes vaccination impractical.

Pigeon Paramyxovirus 1

Pigeon PMV-1 is distributed worldwide and is a particularly important pathogen in racing, homing and show pigeons. It tends to be mesogenic in poultry but causes severe neurologic disease in pigeons.[48,126] The predominant clinical changes in pigeons with PMV-1 include central nervous system signs that are frequently accompanied by polydipsia, polyuria and diarrhea, which may be watery or hemorrhagic *Figure 9.12*. Clinical signs may occur in 20% to 80% of exposed birds. Mortality, which may approach 90%, is highest in young birds and in individuals with concomitant disease. Older birds typically recover in three to four weeks. Some birds with severe neurologic changes will recover over a two- to six-month period if provided supportive care that includes electrolyte replacement, control of secondary infections and supportive alimentation.[1,35,60,116]

Affected free-ranging pigeons either were found dead or exhibited neurologic disease. Other possible causes of the neurologic problems in these free-ranging pigeons included organophosphate

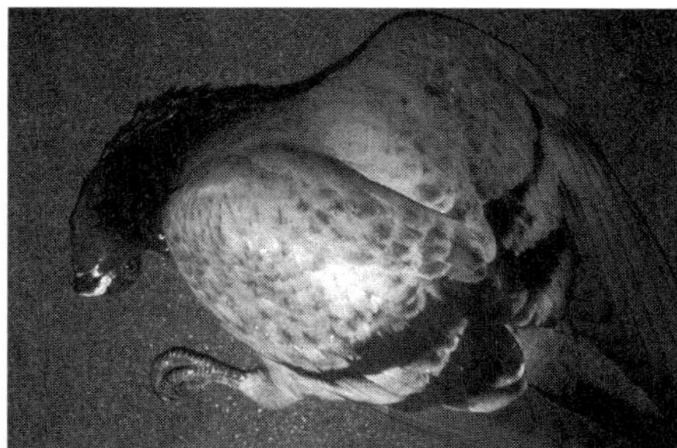

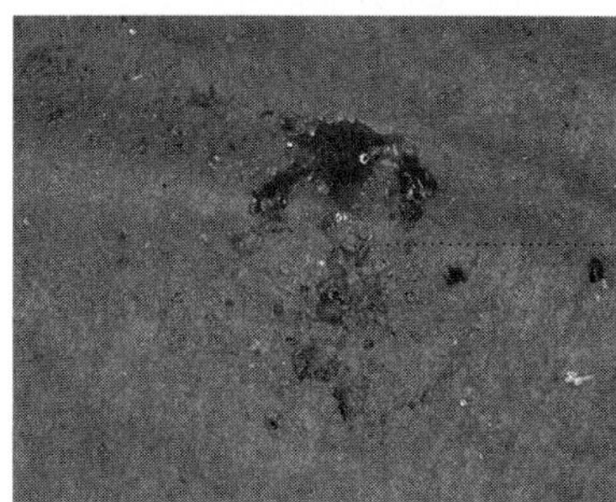

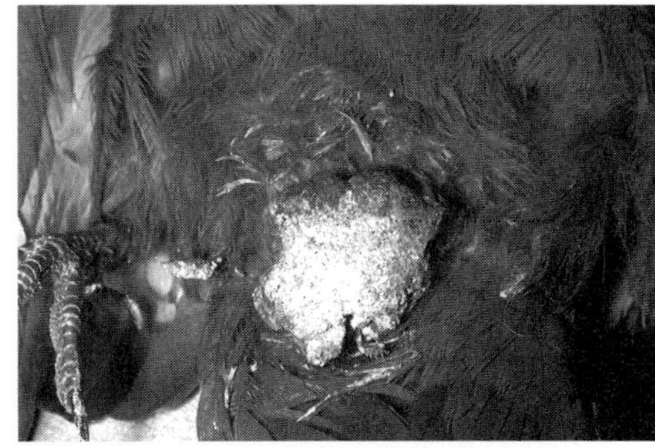

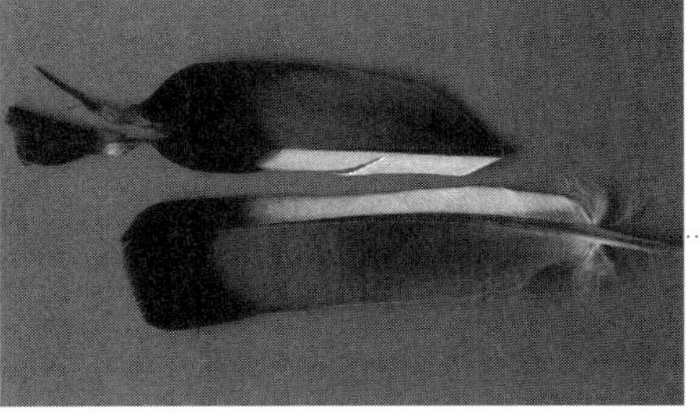

FIG 9.12

Clinical changes in pigeons with PMV-1 include central nervous system signs frequently accompanied by:

diarrhea and

polyuria.

Deformities in the developing feathers have been described in experimentally infected pigeons.

photographs courtesy of E.F. Kaleta and Jean Sander

toxicity, lead poisoning, bacterial encephalitis and trauma. Respiratory changes, which are common in pigeons infected with NDV, are rare in birds infected with pigeon PMV-1.[49,63]

In a PMV-1 outbreak in a loft, 45 of 90 (50%) pigeons developed clinical signs that included ataxia, torticollis, diarrhea and death within ten days after the first case was noted. It was suspected that the virus was introduced to the loft by pigeons returning from a race. The aviculturist chose not to vaccinate the remaining birds in the flock, and only 25% to 30% of the exposed birds survived.[131]

The clinical changes associated with pigeon PMV-1 have decreased in severity since the disease was first recognized in the 1980s. Currently, many infected pigeons may develop a brief period of lethargy and loose droppings followed by a complete recovery. Other infected pigeons remain asymptomatic and develop antibodies to the virus.[114] Pigeon PMV-1 is experimentally infectious to chickens.[73,97]

EPIZOOTIOLOGY

A neurologic disease that caused clinical changes resembling Newcastle disease was described in racing pigeons in the Mediterranean in 1981. Over the following two years, the virus spread rapidly through racing pigeons in Europe, and 800,000 poultry in Great Britain were slaughtered in 1983 due to a paramyxovirus epornitic.[14,15,126] These outbreaks were traced to poultry food contaminated with PMV-1 originating from the feces of infected pigeons.[73] Pigeon PMV-1 has been documented in Europe, Asia, Africa and North America *Table 9.10*.[14,15] Pigeon PMV-1 was isolated in 1985 from racing pigeons in Ontario, but was not detected in feral pigeons from this area until 1989.[63] This finding would suggest that the virus was introduced to Canada by infected racing pigeons, and subsequently spread to free-ranging pigeons. Serologic surveys indicated that from 7% to 19% of racing pigeons in France had PMV-1 antibodies.[126]

TABLE 9.10

Chronological and geographic appearance of pigeon PMV-1[14,48,63,65,106,126]

Newcastle disease virus in Columbiformes
    Prior to 1971 - sporadic reports worldwide
    1971-1973 - increased incidence in Europe
      in conjunction with epornitic in poultry
Pigeon PMV-1
    1978 - First report likely in Iraq
    1981 - racing pigeons in Italy
    1981 - pigeons throughout Mediterranean
    1983 - reports in most of Europe
    1984 - pigeons in New York City
    1985 - racing pigeons in Canada
    After 1985 - globally distributed
    1989 - feral pigeons in Canada

Pigeon PMV-1 infections can occur year-round, but are particularly common in unvaccinated pigeon flocks during the racing season (August-October in Europe; April-June or September-October in the United States).[35] All ages of Columbiformes are susceptible to infection, but young birds are most commonly affected. Other birds that have been shown to be susceptible to pigeon PMV-1 include Wood Pigeons, Cracidae, Pavoninae, Phasianinae, Blackbirds, House Sparrows, Barn Swallows, European Kestrels, common buzzards, Amazon parrots and Eastern Rosellas.[49]

It has been suggested that passage of PMV-1 strains through pigeons could result in new strains with increased pathogenicity for domestic poultry. The virulence of strains recovered from pigeons increased for chickens between 1985 and 1989 in Canada, whereas the virulence of strains for chickens decreased between 1983 and 1985 in Britain.[63]

Observations from field cases suggest that the incubation period can range from five days to over a month, and new cases may continue for up to five weeks after the initial onset in a flock. The experimental incubation period is five to eighteen days.[11,35,123] Virus transmission can occur through direct or indirect contact with infected birds, or by ingestion or inhalation of contaminated respiratory secretions or feces. Infected pigeons excrete virus with laryngeal secretions from two to nine days and with the feces two to fourteen days after infection.[37,123] Shedding may continue for up to a month, and virus can remain infectious in contaminated feces for months. Domestic poultry have been shown to be infected following the ingestion of food contaminated with virus-laden pigeon feces.[11,14,15] Biologically secure shipping containers could be used to reduce the spread of the virus during transport (Horizon Micro-Environments). The greatest risk for an outbreak occurs when infected domestic or free-ranging pigeons come in direct contact with members of an unvaccinated flock.[103]

The gross changes associated with pigeon PMV-1 infections include hyperemia of the brain and abdominal organs, catarrhal enteritis, enlarged kidneys and hemorrhage of the pancreas; however, these changes may not be present in some affected birds. Enteritis is most common during the acute phase of an infection. Microscopically, the accumulation of lymphocytes and plasma cells in the kidneys was the most consistent finding in free-ranging pigeons; hepatitis was present in nine of ten infected, free-ranging pigeons. Other microscopic changes that have been described include non-suppurative encephalitis, hepatitis and pancreatitis.[63]

Methods of diagnosing pigeon PMV-1 are similar to those described for other paramyxovirus infections and include virus isolation or demonstration of a rising antibody titer in paired serum samples. Virus can be recovered most easily from the feces or respiratory secretions during the acute phase of an infection. Newcastle disease virus, pigeon herpesvirus, heavy metal toxicity, organophosphate poisoning and salmonella can cause similar-appearing clinical changes.[35]

CONTROL

Many birds infected with pigeon PMV-1 will recover if provided fluids, supportive feeding and antibiotics to prevent secondary infections. Any birds with clinical signs, and those birds to which they are directly exposed, should be removed from the flock to prevent them from infecting other birds. These birds should not be returned to the flock until six to eight weeks after they are clinically normal because the virus can be shed in the feces of recovering pigeons for up to a month. Free-ranging pigeons can serve as a source of virus for captive birds and should be restricted from the loft.[73] Pigeons returning from a show or race should be quarantined for six weeks to reduce the chances that they may introduce the virus to the loft. In the United Kingdom, pigeon PMV-1 is a reportable disease.

In general, vaccination should be expected to prevent disease induced by pigeon PMV-1, but it does not prevent infections or shedding of the virus.[66]

Both attenuated live-virus and inactivated vaccines containing various adjuvants have been tested for their ability to protect pigeons from disease.

An inactivated aqueous phase vaccine developed in Europe appears to provide the best level of protection and causes the fewest post-vaccination side effects. This vaccine has been shown to induce antibodies that can be detected as soon as one week after a single (0.2 ml) subcutaneous vaccination.[36,38,124] By ten days post-vaccination, birds developed a protective immune response that conferred resistance to disease when vaccinates were challenged twelve months later with live pigeon PMV-1 by the intramuscular, ocular or intranasal route. Pigeons as young as three weeks could be vaccinated successfully irrespective of the presence of maternal antibodies.[36,56] Because of the

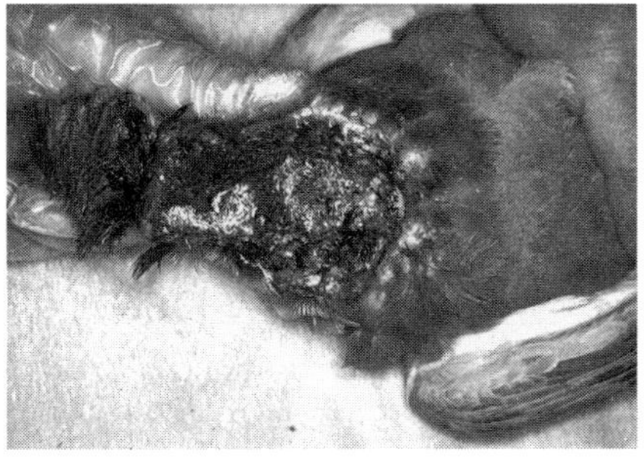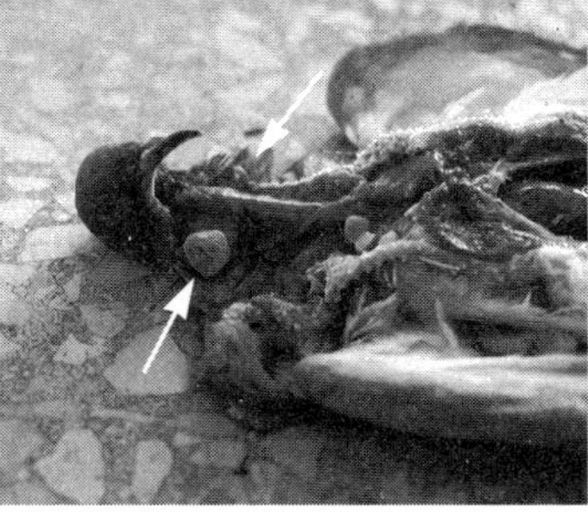

*Vaccines that use oil as an adjuvant can cause diffuse necrosis (particularly when secondary bacterial or fungal infections occur) or **right** formation of large granulomas (arrows).*
photographs courtesy of E.F. Kaleta

rapid immune response induced by this vaccine, it is possible to vaccinate in the face of an outbreak and safely reduce the level of morbidity and mortality in a flock. In one study, it was demonstrated that hens vaccinated as early as three months prior to laying may not pass protective antibodies to their chicks.[68]

Inactivated vaccines containing an oil adjuvant have also been shown to pro-

tect vaccinates from disease.[68] However, oil-containing vaccines have been associated with granuloma formation at the site of intramuscular inoculation in up to 2% of vaccinates. Additionally, these vaccines have been associated with reactivation of latent herpesvirus infections.[65,122,128] Inactivated vaccines containing oil produced larger masses of longer duration than vaccines containing aluminum hydroxide *Figure 9.13*.[66]

All new pigeons that are being added to a flock should be vaccinated during the quarantine period. Adults are best vaccinated four months prior to the breeding, racing or show season. Squab should be vaccinated at three to four weeks of age.[12,113] Subcutaneous inoculations are recommended in pigeons because intramuscular inoculations carry the potential for inflammation that may impair flight performance. Pigeons can be safely vaccinated in the lower third of the neck near the midline, thus avoiding an extensive vascular area found higher in the neck which, if severely damaged, can cause immediate death *Figure 9.14*.[66] The recommendations provided by the manufacturer should be followed carefully.

OTHER SPECIES-SPECIFIC PARAMYXOVIRUS DISEASES

PARAMYXOVIRUSES IN WATERFOWL

All paramyxovirus serotypes except PMV-5 have been demonstrated in captive waterfowl in North America, Europe or Asia *see Table 9.2*. All PMV serotypes except 5 and 9 have been reported in free-ranging ducks and geese.[4,24,129,130] Waterfowl are considered relatively resistant to paramyxoviruses and generally develop subclinical infections. Some ducks and geese infected with virulent strains of PMV-1 may develop mild respiratory signs, watery diarrhea and ataxia. A few epornitics of paramyxovirus have been associated with high mortality in captive and free-ranging waterfowl.[70] During a PMV outbreak in Canada, more than 5,000 cormorants and 140 pelicans died. Young birds were most likely to die, while affected adults exhibited clinical signs of wing and leg paralysis. A Common Teal from Iran was thought to have died from a PMV-1 infection.[23] One isolate of paramyxovirus from a duck was found to be serologically identical to pigeon PMV-1.[10]

It is theorized that some waterfowl may develop chronic infections and serve as reservoirs for the virus. An isolate of VVND virus was recovered from the cloaca of 2 of 21 (9%) asymptomatic free-ranging ducks in Africa.[39] Canada Geese infected with PMV-1 remained asymptomatic, shed virus in the feces from three to six days after infection, and seroconverted.[49] Paramyxovirus serotypes 1, 4, 6 and 8 were isolated from a group of free-ranging waterfowl in Louisiana. Isolates were made from 605 Blue-winged Teal, 75 Mottled Ducks, 375 Gadwalls, 334 Green-winged Teal and 20 Mallard Ducks. Paramyxoviruses were recovered more commonly in the fall (September) than in the winter (December, January), and juveniles were more commonly infected than adults. Virus was recovered from 4% of juveniles and from less than 1% of adults.[112] In another study,

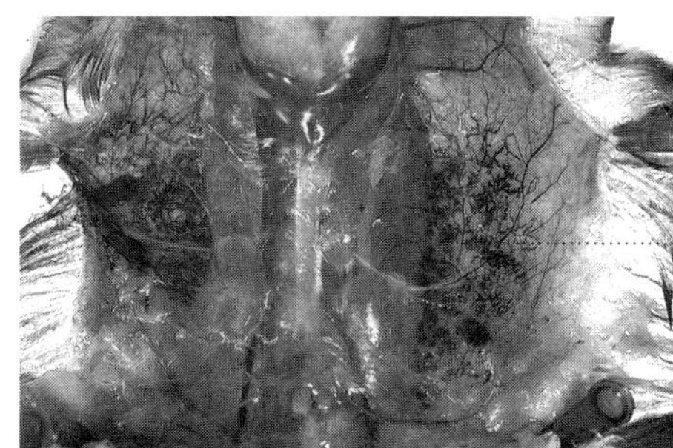

FIG 9.14

Pigeons have an extensive network of blood vessels in the upper portion of the neck.

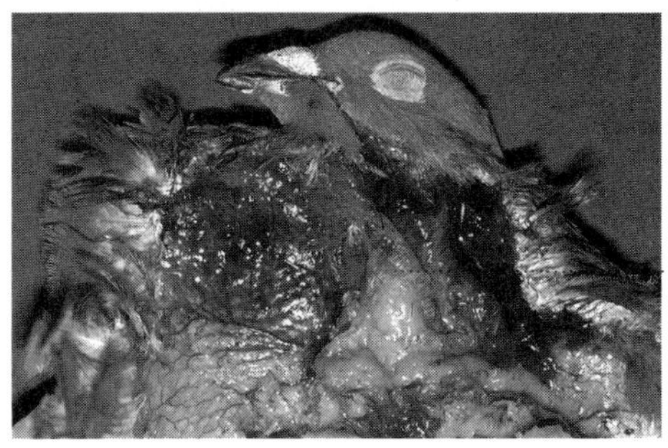

Vaccine-induced damage to the vascular plexus can cause fatal hemorrhage. Subcutaneous injections should be given to pigeons in the lower third of the neck

photographs courtesy of E.F. Kaleta, reprinted with permission[66]

paramyxovirus was recovered most commonly in the winter months from domestic ducks.[108]

PARAMYXOVIRUS IN PASSERIFORMES

Paramyxoviruses are commonly recovered from free-ranging Passeriformes where they are thought to cause primarily asymptomatic infections. However, it should be noted that free-ranging birds that become clinically ill may be killed quickly by predators, and thus may not be available for evaluation. Passeriformes are susceptible to NDV and could serve as reservoirs for virus transmission in endemic areas, but the importance of free-ranging Passeriformes in transmitting PMV to captive populations of susceptible birds has not been documented. Paramyxovirus infections should be considered in captive or free-ranging Passeriformes that either die following a brief period of neurologic signs, die acutely without clinical signs or die following a brief period of upper respiratory disease.[49]

Paramyxovirus in Raptors

Some raptors are highly susceptible to NDV. Others are considered relatively resistant to disease. Affected falcons, hawks and eagles can remain asymptomatic, die after a brief period of lethargy, or exhibit one to two weeks of anorexia and diarrhea followed by death soon after the occurrence of central nervous system signs. Birds with clinical signs may recover if provided appropriate supportive care. Owls appear to be relatively resistant to disease, but infected birds may shed virus in their feces.[49]

In England, a VVND virus outbreak in a group of captive raptors coincided with an epornitic in chickens, and use of exposed chickens as a food source for the affected raptors. During this four-year epornitic, VVND was recovered

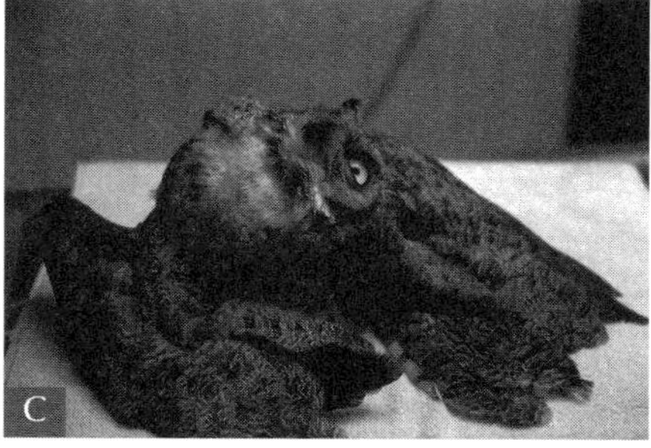

FIG 9.15

Raptors with paramyxovirus infections show: A torticollis; B incoordination; C convulsions and tremors of the head.

photographs courtesy of Stuart Porter

from 11 of 44 (25%) captive birds, including Red-headed Falcon, European Kestrels, Secretary Birds, Barn Owls and a Little Owl. Affected birds

developed neurologic signs of varying intensity that included torticollis, incoordination, convulsions and head-twitching *Figure 9.15*. Gross lesions were inconsistent, but included reddish discoloration of the gastrointestinal tract and congestion of the lung. The unaffected raptors in the collection were experimentally inoculated with a Hitchner B-1 attenuated-live virus vaccine, which successfully stopped deaths in the collection. Virus was not recovered from raptors in two subsequent years after the epornitic in chickens subsided.[29]

In a study in Nigeria, 21 of 37 (57%) Lanner Falcons were found to have NDV antibodies ranging from 0 to 1:640. This indicates that natural infections occur in raptors in Nigeria.[89] A VVND virus caused the death of seven of eight (87%) Shaneen Hawks at a quarantine station.[93] Paramyxoviruses have also been recovered from a captive White-tailed Eagle, and from a free-ranging osprey and Great Horned Owl.[61,105,133] In an uncontrolled study, LaSota vaccine was not reported to cause clinical signs of disease when administered intranasally or by eye drop to two Red-tailed Hawks.[59]

Paramyxovirus in Galliformes

Isolates of NDV are serologically similar, but variations occur with respect to the clinical and gross changes that occur in chickens. In poultry, strains vary from those that cause asymptomatic infections to those that kill birds in less than a week *see Table 9.3*.[80] The strains that kill fowl quickly are less likely to spread than are the less virulent strains of the virus that cause prolonged disease of the respiratory or gastrointestinal tract. In addition to PMV-1, domestic fowl are also susceptible to PMV-2 and PMV-3.

Newcastle disease virus has been reported in free-ranging pheasants in Europe.[78] Pheasants generally are considered highly susceptible to velogenic strains, but free-ranging birds are rarely infected. In captive pheasants, infections cause clinical signs similar to those described in chickens. Crested Firebacks and Silver Pheasants may be anorectic and unable to stand, and may develop dyspnea or hemorrhagic diarrhea. Partridges naturally infected with virulent strains may die with no clinical signs, or survivors may develop neurologic signs. In one outbreak in a zoological collection, NDV killed 35 Indian Partridges. Pheasants can be experimentally infected with NDV by subcutaneous inoculation or through direct contact with an infected bird.[115]

Guinea fowl are considered moderately to highly susceptible to NDV, and develop a clinical disease that is similar to that described in pigeons.[39] Grouse may be latently infected with moderately virulent strains which can cause conjunctivitis and weight loss.

PARAMYXOVIRUS IN RATITES

Ostriches and cassowaries are considered moderately susceptible to NDV.[101] Affected birds develop enteritis followed by a sudden onset of ataxia. Their long necks may droop, be draped over the back or be held in a cork-screw like position *Figure 9.16*. Other birds may develop respiratory disease and watery diarrhea. Affected ratites that are provided supportive care may recover in several weeks. Hemorrhagic lesions in the intestinal or respiratory tract are suggestive. In a study in Zimbabwe, NDV antibodies were detected in the serum of 23% of 149 ostriches with an ELISA. The percentage of seropositive ostriches from different farms ranged from 5% to 60%.[27] Paramyxovirus type-1 was recovered from the intestines of 2 of 30 (6.6%) ostriches and PMV-2 was recovered from the intestines of 1 of 30 (3.3%) ostriches.[107]

FIG 9.16
Ratites affected with Newcastle disease virus develop enteritis followed by a sudden onset of ataxia. Their long necks may droop, be draped over the back or

be held in a corkscrew-like position.

photographs courtesy of B. Perelman

PARAMYXOVIRUS IN PENGUINS

Viruses that belong to the family Paramyxoviridae have been recovered from free-ranging penguins in Antarctica and Sub-Antarctica. Isolates were recovered from four King Penguins, one Royal Penguin and three Adelie Penguins. One isolate from a King Penguin was found to be a lentogenic strain of NDV. The other isolates were not antigenically related to the PMV-1 through PMV-9 serotypes that have been identified. Antibodies to paramyxovirus were detected in free-ranging Adelie Penguins and Skuas.[9,18] VNND was recovered from a group of Adelie Penguins in quarantine. The birds had been collected from Antarctica and were thought to have been exposed to the virus prior to or during shipping. In this group, eight of twelve (67%) penguins died, and concomitant aspergillosis and the stresses associated with captivity were considered important factors in the high level of mortality.[93] The repeated demonstration of paramyxovirus or antibodies to the virus from populations of penguins at various locations in Antarctica suggests that this virus is widespread and persistent in this region.[18]

PNEUMOVIRUSES - TURKEY RHINOTRACHEITIS VIRUS

Pneumovirus infections have been associated with severe rhinotracheitis in turkeys, and with disease of the respiratory tract (swollen head syndrome) and decreased egg production in chickens. Infections have been documented in Europe and South Africa.[26] Antibodies to turkey rhinotracheitis virus were detected in 99% of 149 asymptomatic ostriches in Zimbabwe, suggesting that this virus commonly infects these birds.[27] Pheasants and guinea fowl are experimentally susceptible to pneumovirus infection, but ducks, geese and pigeons are refractory.[52]

Infections in turkeys and chickens are associated with sneezing, swelling of the infraorbital sinuses and conjunctivitis. Morbidity and mortality can be high, particularly when secondary infections occur. Some infected turkeys have been shown to develop antibodies to the virus, indicating they are able to recover from an infection. Chickens that develop clinical signs typically die. Other paramyxoviruses (PMV-1 and PMV-3, for example) and influenza virus can cause similar-appearing clinical signs.

REFERENCES

1. Al Falluji S, Sheikhly FA, Tantawi HH: Viral encephalomyelitis of pigeons: Pathology and virus isolation. Avian Dis 23:777-784, 1979.

2. Alexander DJ: Avian paramyxoviruses. Vet Bull 50:737-752, 1980.

3. Alexander DJ: Taxonomy and nomenclature, of avian paramyxoviruses. Avian Pathol 16:547-552, 1987.

4. Alexander DJ: Newcastle Disease and other paramyxovirus infections. In Calnek BW, et al (eds):Diseases of Poultry. Ames, Iowa State University Press 1991, 9th ed, pp 496-519.

5. Alexander DJ, Allan WH, Parsons G, et al: Identification of paramyxoviruses isolated from birds dying in quarantine in Great Britain during 1980 to 1981. Vet Rec 111:571-574, 1982.

6. Alexander DJ, Allen WH: Newcastle disease virus pathotypes. Avian Pathol 3:269-278, 1974.

7. Alexander DJ, Chettle NJ: Relationship of parakeet/Netherlands 449/75 virus to other paramyxoviruses. Res Vet Sci 29:105-106, 1978.

8. Alexander DJ, Collins MS: Pathogenicity of PMV-3/parakeet/Netherlands/449/75 for chickens. Avian Pathol 11:179-185, 1982.

9. Alexander DJ, Manvell RJ, Collins MS, et al: Characterization of paramyxoviruses isolated from penguins in Antarctica and sub-Antarctica during 1976-1979. Arch Virol 109:135-143, 1989.

10. Alexander DJ, Manvell RJ, Kemp PA, et al: Use of monoclonal antibodies in the characterization of avian paramyxovirus Type I isolates submitted to an international reference laboratory. Avian Pathol 16:553-565, 1987.

11. Alexander DJ, Parsons G: Avian paramyxovirus type 1 infection of racing pigeons. 2. Pathogenicity experiments in pigeons and chickens. Vet Rec 114:466-469, 1984.

12. Alexander DJ, Parsons G, Marshall R: Avian paramyxovirus type 1 infections of racing pigeons. 4. Laboratory assessment of vaccination. Vet Rec 118:262-266, 1986.

13. Alexander DJ, Russell PH, Collins MS: Paramyxovirus type 1 infections of racing pigeons. 1. Characterization of isolated viruses. Vet Rec 114:444-446, 1984.

14. Alexander DJ, Russell PH, Parsons G, et al: Antigenic and biological characterization of avian paramyxovirus Type I isolates from pigeons-an international collaborative study. Avian Pathol 14:365-376, 1985.

15. Alexander DJ, Wilson GWL, Thain JA, et al: Avian paramyxovirus type 1 infection of racing pigeons. 3. Epizootiological considerations. Vet Rec 114:213-216, 1984.

16. Anderson C, Kearsley R, Alexander DJ, et al: Antigenic variation in avian paramyxovirus type 3 isolates detected

by mouse monoclonal antibodies. Avian Pathol 16:691-698, 1987.

17. Ashton WLG, Alexander DJ: A two-year survey on the control of the importation of captive birds into Great Britain. Vet Rec 106:80-83, 1980.

18. Austin FJ, Webster RG: Evidence of ortho and paramyxoviruses in fauna from Antarctica. J Wildl Dis 29:568-571, 1993.

19. Bang FB, Foard M, Bang BG: Acute Newcastle disease viral infection of the upper respiratory tract of the chicken. I. A model for the study of environmental factors on upper respiratory tract infections. Am J Pathol 76:333-348, 1974.

20. Bennewitz D: On the immunoprophylaxis of paramyxovirus-3 infections in parakeets and passeriformes. Proc Europ Conf Avian Dis, 1988, pp 86-103.

21. Biancifiori F, et al: An occurrence of Newcastle Disease in pigeons. Virological and serological studies on the isolates. Comp Immunol Microbiol Infect Dis 6:247-252, 1983.

22. Boyd RJ, Hanson RP: Survival of Newcastle disease virus in nature. Avian Dis 2:82, 1958.

23. Bozorgmehri-Fard MH, Keyranfar H: Isolation of Newcastle disease virus from teals (*Anas crecca*) in Iran. J Wild Dis 15:335-337, 1979.

24. Bradshaw JE, Trainer DO: Some infectious diseases of waterfowl in the Mississippi flyway. J Wildl Mgmt 30:570, 1966.

25. Brugh M, Beard CW: Atypical disease produced in chickens by Newcastle disease virus isolated from exotic birds. Avian Dis 28:482-488, 1984.

26. Buys SB, du Preez JH, Els HJ: The isolation and attenuation of a virus causing rhinotracheitis in turkeys in South Africa Onderstepoort. J Vet Res 56:87-98, 1989.

27. Cadman HF, Kelly PJ, Zhou R, et al: A serosurvey using enzyme-linked immunosorbent assay for antibodies against poultry pathogens in ostriches (*Struthio camelus*) from Zimbabwe. Avian Dis 38:621-625, 1994.

28. Chang PN: Viral zoonoses. *In* Steele JH (ed):Series in Zoonoses. Boca Raton, CRC Press 1981, 2nd ed, pp 261-274.

29. Chu HP, Trow EW, Greenwood AG, et al: Isolation of Newcastle disease virus from birds of prey. Avian Pathol 5:227-233, 1976.

30. Clubb SL, Levine BM, Graham DL: An outbreak of viscerotropic velogenic Newcastle disease in pet birds. Proc Assoc Zoo Vet, 1980, pp 105-109.

31. Collings DF, Fitton J, Alexander DJ, et al: Preliminary characterization of a paramyxovirus isolated from a parrot. Res Vet Sci 19:219-221, 1975.

32. Collins MS, Alexander DJ, Brockman S, et al: Evaluation of mouse monoclonal antibodies raised against an isolate of the variant avian paramyxovirus type I responsible for the current panzootic in pigeons. Arch Virol 104:53-61, 1989.

33. Cornelissen H: Vaccination of over 200 bird species against Newcastle disease: Methods and vaccination reactions. Proc Europ Conf Avian Med Surg, 1993, pp 275-287.

34. Crosta L, Rampin T, Sironi G, et al: Paramyxovirus serotype 3 infection in Neophema parakeets. Proc Europ Conf Avian Med Surg, 1993, pp 269-273.

35. Dorrestein GM: Viral infections in racing pigeons. Proc Assoc Avian Vet, 1992, pp 244-257.

36. Duchatel JP, Flore PH, Hermann W, et al: Efficacy of an inactivated aqueous-suspension Newcastle disease virus vaccine against paramyxovirus Type I infection in young pigeons with varying amounts of maternal antibodies. Avian Pathol 14:257-267, 1992.

37. Duchatel JP, Leroy P, Coignoul F, et al: Essais de vaccination de pigeons contre la paramyxovirose par injection sous-cutanee de vaccins inactives. Ann Med Vet 129:39-50, 1985.

38. Duchatel JP, Vindevogel H: Effects of vaccination with a live or an inactivated aqueous-suspension NDV vaccine in pigeons previously infected with paramyxovirus type I. Diergeneesk T 56:135-140, 1987.

39. Echenonwu GON, Iroegbu CU, Emeruwa AC: Recovery of velogenic Newcastle disease virus from dead and healthy free-roaming birds in Nigeria. Avian Pathol 22:383-387, 1993.

40. Erickson GA, Brugh M, Beard CW: Viscerotropic velogenic Newcastle disease in pigeons: Clinical disease and immunization. Avian Dis 14:257-267, 1980.

41. Erickson GA, Gustafson GA, Pearson JE, et al: Velogenic viscerotropic Newcastle disease in selected captive avian species. Proc Am Assoc Zoo Vet, 1975, pp 133-135.

42. Erickson GA, Mare CJ, Beran GW, et al: Epizootiologic aspects of viscerotropic velogenic Newcastle disease in six pet bird species. Am J Vet Res 39:105-107, 1978.

43. Erickson GA, Mare CJ, Gustafson GA, et al: Interactions between viscerotropic velogenic Newcastle disease virus and pet birds of six species. 1. Clinical and serological responses, and viral excretion. Avian Dis 21:642-654, 1977.

44. Fleury HJA, Alexander DJ: Paramyxovirus Yucaipa. Bull Intl Inst Pasteur 76:175-186, 1978.

45. Fleury HJA, Alexander DJ: Isolation of 23 Yucaipa-like viruses from 616 wild birds in Senegal, West Africa. Avian Dis 23:742-744, 1979.

46. Fleury HJA, Faivre R: Experimental infection of *Columba livia* with paramyxovirus Yucaipa. Res Vet Sci 34:376-377, 1983.

47. Garnett S, Flanagan M: Survey for Newcastle disease virus in northern Queensland birds. Aust Vet J 66:129-134, 1989.

48. Gelb J, Fries PA, Peterson FS: Pathogenicity and cross-protection of pigeon paramyxovirus -1 and Newcastle disease virus in young chickens. Avian Dis 31:601-606, 1987.

49. Gerlach H: Viruses. *In* Ritchie BW, Harrison GJ, Harrison LR (eds):Avian Medicine: Principles and Application. Lake Worth, Wingers Publishing 1994, pp 862-948.

50. Goodman BB, Hanson RP: Isolation of avian paramyxovirus-2 from domestic and wild birds in Costa Rica. Avian Dis 32:713-717, 1988.

51. Goodman BB, Hanson RP, Moermond TC, et al: Experimental avian PMV-2 infection in a domesticated wild host: Daily behavior and effect in activity levels. J Wild Dis 26:22-27, 1990.

52. Gough RE, Alexander DJ, Collins MS, et al: Routine virus isolation or detection in the diagnosis of diseases in birds. Avian Pathol 17:893-907, 1988.

53. Gough RE, Manvell RJ, Drury SEN, et al: Deaths in Budgerigars associated with a paramyxovirus-like agent. Vet Rec July:123, 1993.

54. Gratzel E, Kohler H: Spezielle Pathologie and Therapie de Geflugelkrankheiten. Ferdinand Euke Verlag, 202-255, 1968.

55. Hanson RP: Issues related to control of velogenic viscerotopic Newcastle disease (VVND) and the importation of pet birds. Proc 31st West Poultry Dis Conf, 1982, pp 165-166.

56. Hermann-Dekkers, et al: Vaccination of pigeons against paramyxovirus infection in infected areas, a comparative trail. Euro Symp Bird Dis, 32-38, 1987.

57. Hirai K, Hitchner SB, Calnek BW: Characterization of paramyxo-herpes- and orbiviruses isolated from psittacine birds. Avian Dis 23:148-163, 1979.

57a. Hooimeijer J: Personal communication.

58. Hore DE, Campbell J, Turner AJ: Aust Vet J 49:238, 1973.

59. Hornbuckle JD: The recent outbreak of exotic Newcastle disease in California. Hawk chalk 9:48-49, 1972.

60. Ide PR: Virological studies of paramyxovirus type 1 infection of pigeons. Can Vet J 28:601-603, 1987.

61. Ingalls WL, Vesper RW, Mahoney A: Isolation of Newcastle virus from the great-horned owl. J Am Vet Med Assoc 119:71, 1951.

62. Johnson DC, Couvillion CE, Pearson JE: Failure to demonstrate viscerotropic velogenic Newcastle disease in psittacine birds in the Republic of the

Philippines. Avian Dis 30:813-815, 1986.

63. Johnston KM, Key DW: Paramyxovirus-1 in feral pigeons (*Columba livia*) in Ontario. Can Vet J 33:796-800, 1992.

64. Kaleta EF: Detection of antibodies against paramyxovirus-1,2,3 in sera from pigeons. Deutsch Tierartl Wschr 89:31-32, 1982.

65. Kaleta EF, Alexander DJ, Russell PH: The first isolation of the avian PMV-1 virus responsible for the current panzootic in pigeons? Avian Pathol 14:553-558, 1985.

66. Kaleta EF, Bruckner D, Goller H: Acute fatalities following subcutaneous injection of paramyxovirus type 1 vaccines in pigeons. Avian Pathol 18:203-210, 1989.

67. Kida H, Yanagana R: Isolation of a new avian paramyxovirus from rock pigeon (*Columba livea*). Zentralblatt fur Baketeriologie Originale 245:421, 1978.

68. Knoll M: Immunitatsdauer nach Iimpfung gegan die Paramyxovirose der Tauben mit einer homologen Olemulsionsvakzine-Ergebnisse eines Langzeitverslches unter Laborbedingungen. V. DVG-Taugng Vogelkrth, 168-177, 1986.

69. Lana DP, Snyder DB, King DJ, et al: Characterization of a battery of monoclonal antibodies for differentiation of Newcastle disease virus and pigeon paramyxovirus-1 strains. Avian Dis 32:273-281, 1988.

70. Lancaster JE: Newcastle disease. A review 1926-1964. Can Dept Agr 1966.

71. Leach MW, Higgins RJ, Lowenstine LJ, et al: Paramyxovirus infection in a Moluccan cockatoo (*Cacatua moluccensis*) with neurologic signs. J Assoc Avian Vet 2:87-90, 1988.

72. Lipkind M, Shihmanter E, Weisman Y, et al: Characterization of Yucaipa-like avian paramyxoviruses isolated in Israel from domestic and wild birds. Ann Virol 133:157-161, 1982.

73. Lister SA, Alexander DJ, Hogg RA: Evidence for the presence of avian paramyxovirus type 1 in feral pigeons in England and Wales. Vet Rec 118:476-479, 1986.

74. Lowenstine LJ: Emerging viral diseases of psittacine birds. *In* Kirk RW (ed):Current Veterinary Therapy IX. Philadelphia, WB Saunders 1986, pp 705-710.

75. Mackenzie JS: Proc Aust Poultry Stock Feed Conv, 1981, p 129.

76. Maldonado A, Arenas A, Tarradas MC, et al: Prevalence of antibodies to avian paramyxoviruses 1, 2 and 3 in wild and domestic birds in Southern Spain. Avian Pathol 23:145-152, 1994.

77. Matsuoka Y, Kida H, Yanagawa R: A new paramyxovirus isolated from an amaduvade finch (*Estrilda amandava*). Jpn J Vet Sci 42:161-167, 1980.

78. McDiarmid A: Modern trends in animal health and husbandry. Some infectious diseases of free-living wildlife. Brit Vet J 121:245, 1965.

79. McFerran JB, Connor TJ, Allen GM, et al: Studies on a paramyxovirus isolated from a finch. Archi fur die gesamte Virusforschung 46:281-290, 1974.

80. McFerran JB, Nelson R: Some properties of an avirulent Newcastle disease virus. Arch ges Virustorsch 4:64-74, 1971.

81. Meulemans G, Carlier MC, Petit P, et al: Antigenic and biological characterization of avian paramyxovirus type 1 isolates from pigeons. Arch Virol 87:151-161, 1986.

82. Moses HE, Brandly CA, Jones EE: The pH stability of the viruses of Newcastle disease and fowl plague. Science 105:477, 1947.

83. Mustaffa-Babjee A, Spradbrow PB, Samuel JL: A pathogenic paramyxovirus from a Budgerigar (*Melopsittacus undulatus*). Avian Dis 18:226-230, 1974.

84. Mustaffa-Babjee A, Spreadborrow PB: Acute enteritis in rainbow lorikeets. Kajian Vet 5:16-19, 1973.

85. Nakayama M, et al: Characterization of virus isolated from Budgerigars. Med Biol 93:449-454, 1976.

86. Narayan O, Lang G, Rouse ST: A new influenza. A virus infection in turkeys. IV Experimental susceptibility of domestic birds to virus strain ty/ontario/7732/1966. Arch ges Virusforsch 26:149-165, 1969.

87. Nerome K, Nakayama M, Ishida M, et al: Isolation of a new avian paramyxovirus from a Budgerigar (*Melopsittacus undulatus*). J Gen Virol 38:293-301, 1978.

88. Nymadawa P, Konstantinow-Siebelist I, Schulze P, et al: Isolation of paramyxoviruses from free-flying birds of the order Passeriformes in the German Democratic Republic. Acta Virol 21:443, 1977.

89. Okoh AEJ: Newcastle disease in falcons. J Wildl Dis 15:479-480, 1979.

90. Olesink OM: Influence of environmental factors on viability of Newcastle disease virus. Am J Vet Res 12:152, 1951.

91. Ozdemir I, Russell PH, Collier J, et al: Monoclonal antibodies to avian paramyxovirus type 2. Avian Pathol 19:395-400, 1990.

92. Panigrahy B, Senne DA, Pearson JE, et al: Occurrence of velogenic viscerotropic Newcastle disease in pet and exotic birds in 1991. Avian Dis 37:254-258, 1993.

93. Pearson GL, McCann MK: The role of indigenous wild, semi-domestic and exotic birds in the epizootiology of velogenic viscerotropic Newcastle disease in Southern California. J Am Vet Med Assoc 167:610-614, 1975.

94. Pieper K, Kaleta EF: Virus isolations from psittacine birds. Proc Europ Chap Assoc Avian Vet, 1991, pp 199-201.

95. Pierson GP, Pfow CJ: Newcastle disease surveillance in the United States. J Am Vet Med Assoc 167:801-803, 1975.

96. Pilchard EL: Exotic Newcastle disease in pet birds. Foreign Animal Disease Report 19:1-2, 1991.

97. Richter R: Utersuchungen zur Steigerung der immunantwort durch den Paramunitaetsinducer PIND ORF beim Gefluegel. Muenchen. 1983.

98. Rigby CE, Pettit JR, Papp-Vid G, et al: The isolation of Salmonellae, Newcastle disease virus and other infectious agents from quarantined imported birds in Canada. Can J Comp Med 45:366-370, 1981.

99. Rotov VI, Litvrishko NT, Dunaev GV, et al: Poultry tick Dermanyssus gallinae as vector of the Newcastle disease virus. Vopr Vet Virusol 1:397, 1964.

100. Saber MS, Alfalluji M, Siam MA, et al: Survival of AG68 strain of Newcastle disease virus under certain local environmental conditions in Iraq. J Egypt Vet Med Assoc 28:73-82, 1968.

101. Samberg Y, Hadash DU, Perelman B, et al: Newcastle disease in ostriches (*Struthio camelus*): field case and experimental infection. Avian Pathol 18:221-226, 1989.

102. Samberg Y, Horenstein K, Peleg B, et al: Use of an oil adjuvant paramyxo-Yucaipa vaccine in poultry and serological response. Develop Biol Standard 51:75-78, 1982.

103. Sanford SE, Hampson RJ: Pigeon paramyxovirus. Can Vet J 30:523, 1989.

104. Schemera B, Toro H, Kaleta EF, et al: A paramyxovirus of serotype 3 isolated from African and Australian finches. Avian Dis 31:921-925, 1987.

105. Schoop G, Siegert R, Galassi D, et al: Newcastle-Infektionen beim Steinkaug (*Athene noctua*), Hornraben (*Bucorvus* sp.), Seeadler (*Haliaetus albicilla*) and Rieseneisvogel (*Dacelo gigus*). Monatshefte fur Tierheilkunde Stuggart 7:223-235, 1955.

106. Seene DA, Pearson JE, Miller LD, et al: Virus isolations from pet birds submitted for importation into the United States. Avian Dis 27:731-744, 1983.

107. Shivaprasad HL, Woolcock PR, Chin RP, et al: Identification of viruses from the intestines of ostriches. Proc Assoc Avian Vet, 1994, p 442.

108. Shortridge KF: Isolation of ortho and paramyxoviruses from domestic poultry in Hong Kong between November 1977 and October 1978 and comparison with isolations made in the preceding two years. Res Vet Sci 28:296-301, 1980.

109. Shortridge KF, Alexandar DJ: Incidence and preliminary characterization of a hitherto unreported, serologically distinct, avian paramyxovirus isolated in Hong Kong. Res Vet Sci 25:128-130, 1978.

110. Shortridge KF, Burrows D, Herday J: Potential danger of avian paramyxovirus type 3 to ornithological collections. Vet Rec 192:363-364, 1991.

111. Smit T, Rundhuis PR: Studies on a virus isolated from the brain of a parakeet (*Neophema* sp.). Avian Pathol 5:21-30, 1976.

112. Stallkneckt DE, Seene DA, Zwank PJ, et al: Avian paramyxoviruses from migrating and resident ducks in coastal Louisiana. J Wild Dis 27:123-128, 1991.

113. Stone HD: Efficacy of oil-emulsion vaccines prepared with pigeon paramyxovirus -1, Ulster, and LaSota Newcastle disease viruses. Avian Dis 33:157-162, 1989.

114. Tangredi BP: Avian paramyxovirus type 1 infection in pigeons: recent changes in clinical observations. Avian Dis 32:839-841, 1988.

115. Thompson CH: Virulent foreign Newcastle disease in partridges. Vet Med 50:399, 1955.

116. Tudor DC: Pigeon health and disease. Ames, Iowa State University Press 1991.

117. Tumova B, Stumpa A, Janout V, et al: A further member of the Yucaipa group isolated from the common wren (*Troglodytes troglodytes*). Acta Virol 23:504-507, 1979.

118. Uyttebroek E, Ducatelle R, Alexander DJ: Steatorrhea and pancreatic lesions in Neophoma parrots with paramyxovirus serotype 3 infections. Vlaams Diergen Tijdschr 60:55-58, 1991.

119. Van Eck JHH: Immunity to Newcastle disease in fowl of different breeds, primarily vaccinated with commercial inactivated oil-emulsion vaccines; a laboratory experiment. Vet Quart 9:296-303, 1987.

120. VanDerHeyden N: Velogenic viscerotropic Newcastle disease in three Amazon chicks. Proc Assoc Avian Vet, 1992, 158-161.

121. VanDerHeyden N, Reed WM: Paramyxovirus group 3 infection of cockatiels. J Assoc Avian Vet 1:53-54, 1987.

122. Vindevogel H, Duchatel JP: Reactions postvaccinates apres injection de vaccin inactive huileux contre la paramyxovirose chez la pigeon. Ann Med Vet 129:471-473, 1985.

123. Vindevogel H, Duchatel JP: Paramyxovirus Type I infection in pigeons. *In* McFerran JB, et al (eds):Acute Virus Infections of Poultry. Martinus Nijhoff Publishers for the Commission of the European Communities 1986, pp 67-77.

124. Vindevogel H, Duchatel JP, Coignoul F, et al: Essais de vaccination de pigeons contre la paramyxovirose au moyen de vaccins inactives. Ann Med Vet 128:543-649, 1984.

125. Vindevogel H, Meulemans G, Halen P, et al: Sensibilite du pigeon voyageur adulte au virus de la maladie de Newcastle. Ann Rech Vet 3:519-532, 1972.

126. Vindevogel H, Pastoret PP, Thiry E, et al: Reapparition de formes graves de la maladie de Newcastle chez le pigeon. Ann Med Vet 126:5-7, 1982.

127. Walker JW, Heron BR, Mixson MA: Exotic Newcastle disease evidication program in the United States. Avian Dis 17:486-503, 1973.

128. Wallis SA: Mortality associated with vaccination of pigeons against paramyxovirus Type I. Vet Rec 114:51-52, 1984.

129. Webster RG, Morita M, Pridgen C, et al: Ortho-and paramyxoviruses from migrating feral ducks. Characterization of a new group of Influenza A viruses. J Gen Virol 32:217-225, 1976.

130. Wobeser G, Leighton FA, Norman R, et al: Newcastle disease in wild water birds in western Canada. Can Vet J 34:353-359, 1990.

131. Yeisley CL: Paramyxovirus in pigeons. Proc Assoc Avian Vet, 1994, pp 219-223.

132. Yoshida N, et al: Properties of paramyxovirus isolated from Budgerigars with an acute fatal disease. J Jap Vet Med Assoc 30:599-603, 1977.

133. Zuydam DM: Isolation of Newcastle disease virus from the osprey and the parakeet. J Am Vet Med Assoc 120:88-89, 1952.

Poxviridae

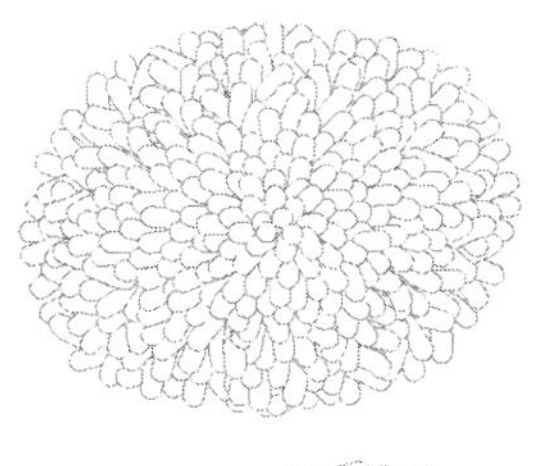

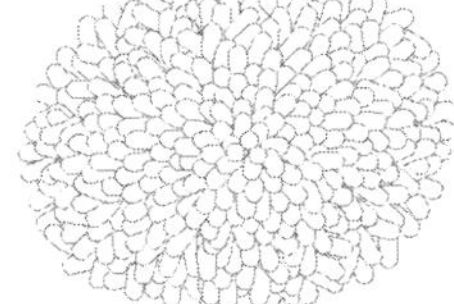

Poxviruses are among the largest and most complex of all animal viruses. With a brick-shaped virion that measures up to 400 nm in diameter, these viruses are only slightly smaller than many common bacteria *see Figure 1.6*. Unlike most DNA viruses which replicate in the nucleus, poxviruses replicate in the cytoplasm of infected cells.

Because of its large size and ability to replicate in cultures of living cells, the vaccinia virus (a strain of mammalian poxvirus) was the first animal virus to be visualized microscopically and to be grown in a laboratory. There are currently seven genera of poxvirus that infect vertebrates. There is one report in the literature describing the recovery of an avian-like poxvirus from characteristic lesions in a rhinoceros.[58] Other than this case, avian poxviruses are not known to infect mammals. The poxviruses that infect birds have been placed in the genus *Avipoxvirus*. All avian poxviruses are morphologically similar, but they exhibit varied host specificity, and thus have been placed into specific groups based largely on the type of birds they infect. A summary of the cross-reactivity of some of the more common strains of avian poxviruses is provided *Table 10.1*.

All species of birds, regardless of age, sex or species, are considered susceptible to some strain of poxvirus *Table 10.2*; however, many companion and aviary birds are rarely exposed to a strain of virus that will induce disease.[16,84] In contrast, poxvirus-induced disease is common in unvaccinated flocks of canaries and pheasants, and infections occur sporadically in pigeons, ratites and raptors. Fowl poxvirus is the best studied of the avian poxviruses, and much of the information concerning this group of viruses is derived from these studies. Some of the avian poxviruses have a limited host range, while others are capable of infecting a wide range of birds.

CLINICAL FEATURES

GENERAL CONSIDERATIONS

Poxvirus infections may cause cutaneous, diphtheritic or systemic changes. The type of disease that develops is thought to be influenced by the strain of infecting virus, the route of viral exposure, and the species, age and condition of the infected bird.[28] The varied clinical changes in differing birds can create problems in evaluating poxvirus-induced diseases. In general, the clinical signs that occur are controlled by the location of poxvirus-induced damage. Lesions around the mouth or eyes may prevent a bird from eating or seeing, respectively. Ulcerations in the oral cavity may make an affected bird reluctant to drink or eat, causing dehydration or starvation. Changes in the pharynx, larynx, trachea or lungs may cause dyspnea, wheezing or gasping for air.

Relationships and demonstrated susceptibility of birds to various poxviruses

- Psittacine poxvirus antigenically distinct from fowl, pigeon and quail poxviruses.
- *Agapornis* poxvirus antigenically distinct.
- Amazon poxvirus experimentally infectious for conures, Amazon parrots and young chickens.

- Canary poxvirus antigenically related to fowl poxvirus.
- Canaries resistant to turkey, pigeon and some strains of fowl poxvirus.
- One isolate of canary poxvirus infectious for chickens, quail and turkeys, but not for house sparrows and pigeons.
- One strain of canary poxvirus infectious for chickens, pigeons and sparrows.[46]

- Pigeon poxvirus antigenically related to fowl poxvirus.
- Some pigeon poxviruses cause mild disease in chickens and turkeys.
- Pigeons experimentally susceptible to turkey, fowl and canary poxviruses.

- Fowl poxvirus antigenically related to turkey, quail, waterfowl and falcon poxviruses.
- Chickens susceptible to some strains of pigeon poxvirus.
- Turkey poxvirus is more closely related to fowl than pigeon poxvirus.
- Quail poxvirus antigenically distinct from fowl, pigeon and psittacine poxviruses.
- Pheasant poxvirus isolate infectious for chickens and pigeons.
- Peafowl poxvirus highly infectious for chickens.
- Some grouse poxvirus isolates antigenically related to fowl poxvirus.

- Magpie poxvirus more closely related to pigeon poxvirus than fowl poxvirus.
- Magpie poxvirus causes mild lesions in chickens.
- Some Starling poxvirus strains infectious for mynahs and chickens.
- Sparrow poxvirus infectious for canaries, sparrows, pigeons, turkeys and chickens.

- Falcon poxvirus antigenically related to pigeon, fowl and turkey poxviruses.
- Accipiter poxvirus not infectious to chickens or pigeons.

- Waterfowl poxvirus antigenically related to fowl poxvirus.
- Ducks susceptible to turkey poxvirus and some strains of fowl poxvirus.
- Geese susceptible to some strains of fowl poxvirus.
- Canada goose poxvirus infectious for geese, but not chickens or domestic ducks.

Typically, poxviruses cause either discrete nodules on the unfeathered skin (cutaneous form, or "dry pox") or fibrinonecrotic lesions on mucous membranes (diphtheritic form, or "wet pox") *Figure 10.1*.[46,60,84] Poxvirus-induced lesions in the mouth, trachea and esophagus are characterized by the formation of sheets of necrotic epithelial cells called diphtheritic membranes. Mortality rates with the cutaneous form of poxvirus are low. By comparison, birds with diphtheritic lesions in the trachea or esophagus commonly die, particularly when the lesions are infected with bacteria or fungi that invade the damaged epithelial cells *Figure 10.2*. Depending on the species of bird and as yet undefined factors, cutaneous and diphtheritic forms may occur independently, together, or combined with a septicemic form that is associated with acute depression, anorexia, dyspnea and death.[27,32,46,60,84] The highly fatal septicemic form of pox is most common in canaries and finches.

Psittaciformes, Falconiformes, Passeriformes and waterfowl frequently develop the cutaneous form of poxvirus. Cutaneous poxvirus lesions are characterized by the formation of papules followed by the development of vesicles that open to form an erosion that quickly scabs. These lesions are common on the unfeathered skin around the eyes, beak margins, nares, cere, lower legs or feet *Figure 10.3*.

TABLE 10.2

Defined strains of poxvirus and some susceptible birds

Defined poxvirus species

fowl pox	turkey pox
canary pox	pigeon pox
quail pox	sparrow pox
starling pox	junco pox
psittacine pox	

Probable poxvirus species

peafowl pox	penguin pox
mynah pox	albatross pox

Some families of birds known to be susceptible to poxvirus

Accipitridae	Alaudidae
Alcidae	Anatidae
Apodidae	Certhiidae
Corvidae	Falconidae
Fringillidae	Gruidae
Icteridae	Laridae
Meleagrididae	Mimidae
Parulidae	Pelecanidae
Phaethontidae	Phasianidae
Picidae	Procellariidae
Psittacidae	Sphenicidae
Struthionidae	Sturnidae
Tetraonidae	Turdidae

Once opened, poxvirus-induced vesicles are susceptible to secondary infections, and the severity and duration of lesions are determined by the presence or absence of bacterial or fungal agents. Cutaneous changes caused by poxvirus can resolve within a month or persist for over a year.[46] Papules usually resolve without scarring if no secondary invaders are involved. If the papule is damaged, or fungal or bacterial agents are present, then severe scarring with loss of pigmentation may occur as the erosions heal *Figure 10.4*. In some species, cutaneous lesions may appear only as dark, discolored areas of the skin that do not form into papules.[32]

Diphtheritic lesions frequently occur in infected Psittaciformes, pheasants,

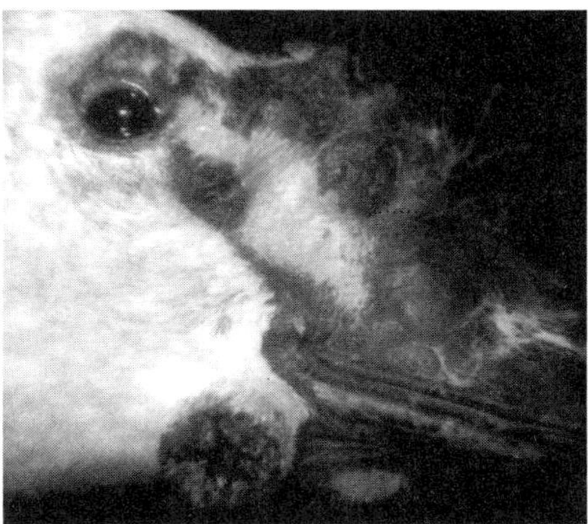

FIG 10.1
Discrete nodules on the unfeathered skin of this swan are characteristic of "dry pox."

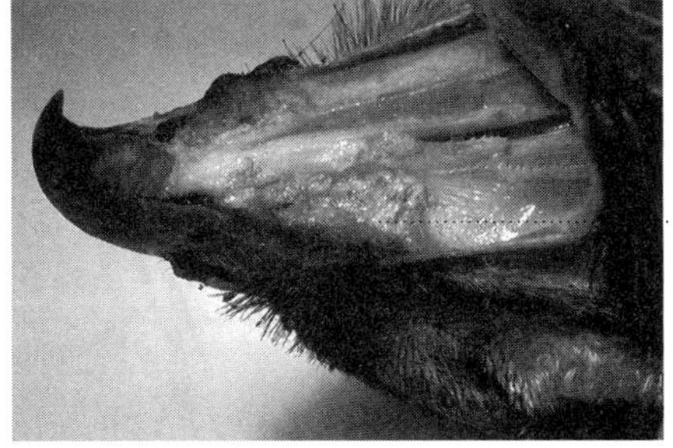

A hawk exhibits necrotic lesions on mucous membranes ("wet pox").

photographs courtesy of Roy Montgomery and National Wildlife Health Center

quail and Columbiformes.[32] The erosions associated with "wet pox" appear as grayish-yellow or brown plaques filled with necrotic tissue, and are common on the mucosa of the beak, tongue,

FIG 10.2
The diphtheritic form of pox can cause high mortality,

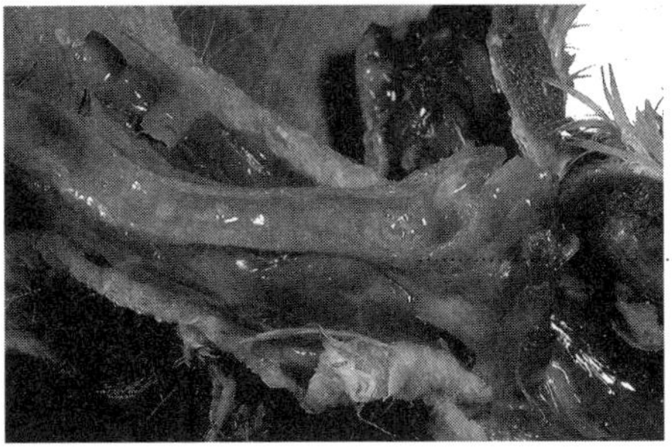

particularly when virus-induced necrotic lesions in the trachea or esophagus are secondarily infected with bacteria or fungi.

photographs courtesy of Scott McDonald

hard palate, larynx, pharynx, trachea, bronchi, esophagus, crop or conjunctiva *Figure 10.5*.[7] These diphtheritic membranes are "cheesy" in appearance and

FIG 10.3

*Steps in development of cuta-neous poxvirus lesions include: **A** the formation of papules; **B** erosion of vesicles; **C** scabs. **D** These lesions are common on the unfeathered skin.*
photographs courtesy of S.S. Tsai and the National Wildlife Health Center

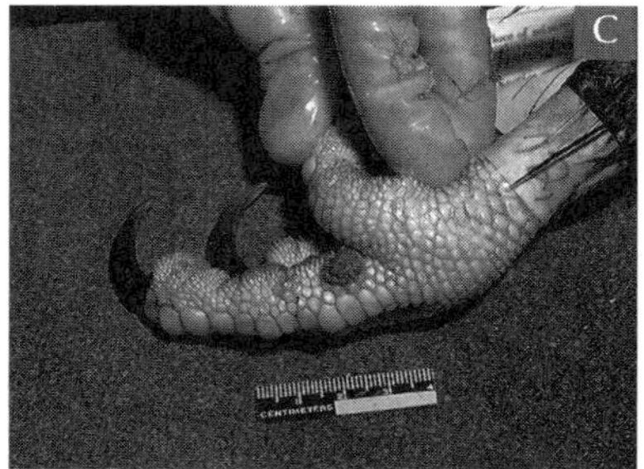
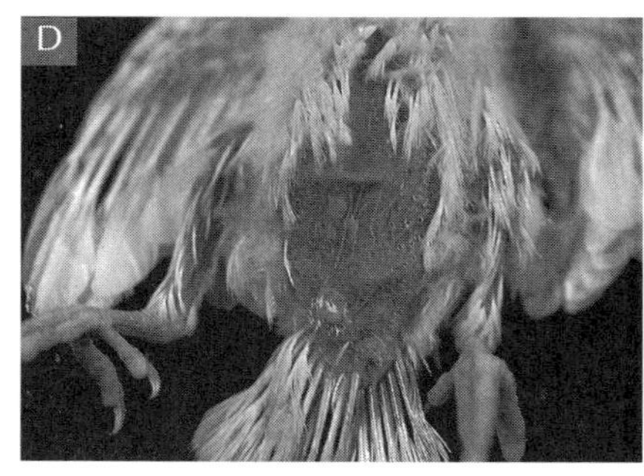

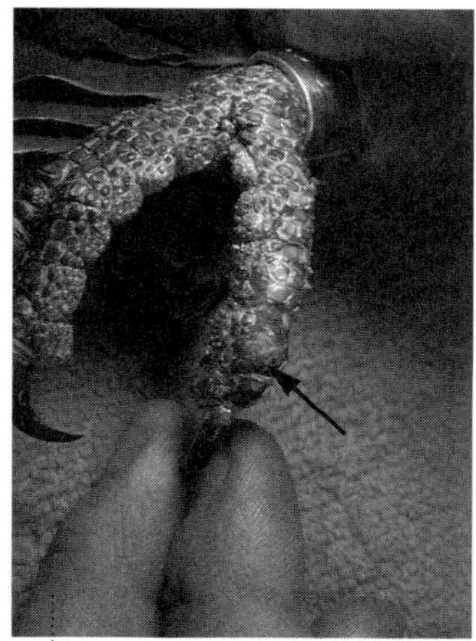

FIG 10.4

Severe scarring with loss of pigmentation (arrow) may occur if virus-induced ulcers are secondarily infected with bacteria or fungi.
photograph courtesy of Scott McDonald

FIG 10.5

The erosions associated with "wet pox" appear as grayish-yellow or brown plaques filled with necrotic tissue.

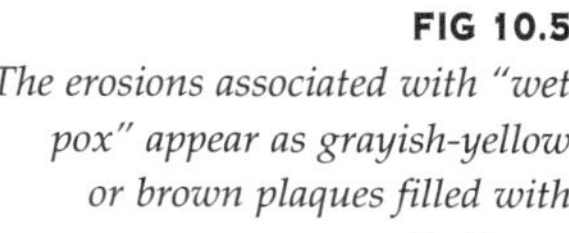
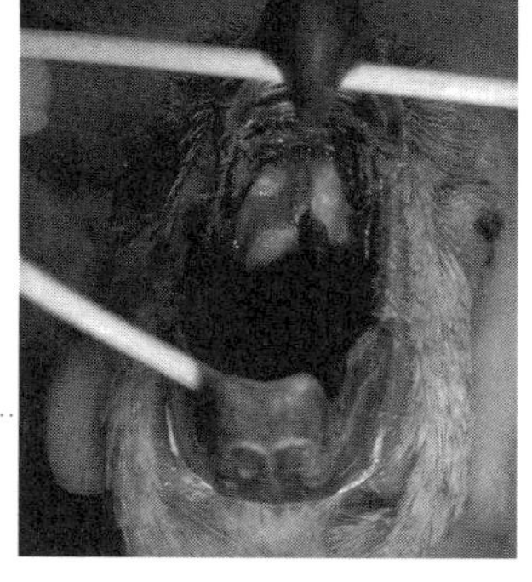

These diphtheritic membranes may bleed profusely if damaged or removed.

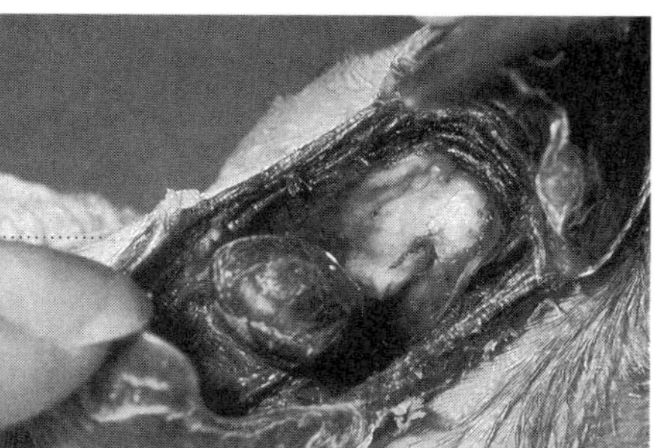

Infections around the eyes are associated with "wet pox."

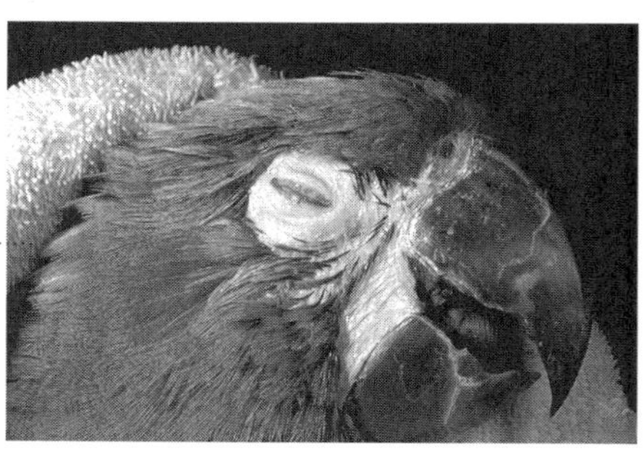

photographs courtesy of Scott McDonald and Barbara Oglesbee

may bleed profusely if damaged or removed. The oral and nasal passages may be occluded with opaque, white, friable plaques in birds with severe damage on the pharyngeal or laryngeal mucosa. These birds may have rhinitis, dysphagia or dyspnea. The pain associated with severe lesions in the mouth, esophagus or crop may cause a bird to stop swallowing. Starvation or suffocation are common causes of death in birds with severe poxvirus-induced lesions in the upper respiratory or alimentary tract. Infections around the eyes are associated with edema, ulceration of the margins of the eyelids, partial to complete closure of the eyelids and inflammation of the conjunctivae *Figure 10.5*.

All members of the Poxviridae family have oncogenic properties, and it has been suggested that poxviruses can induce neoplastic changes in the lung or skin of some infected birds.[40] Skin tumors following a poxvirus infection are particularly common in Columbiformes, and poxvirus-induced tumors are common in the lungs of affected canaries. These tumors are unique and appear as wart-like growths that bleed profusely if disturbed. Poxvirus-associated tumors of the skin can be removed surgically.

CLINICAL FEATURES IN PSITTACIFORMES

Poxvirus infections are reported most commonly in quarantine stations housing recently imported Amazon parrots, macaws and pionus parrots. In many cases, the most severe clinical lesions that occur resolve before these birds are released from the quarantine station.

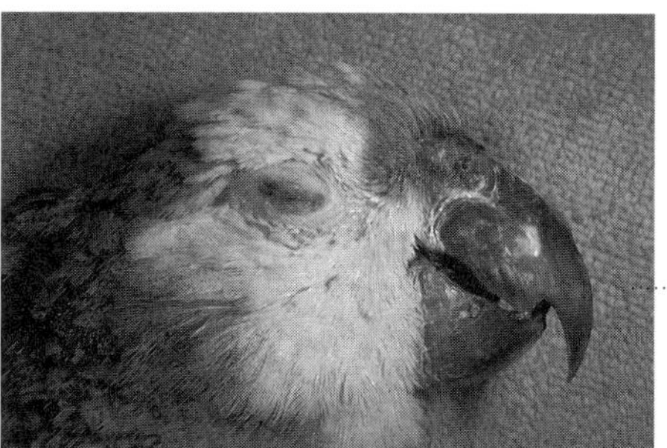

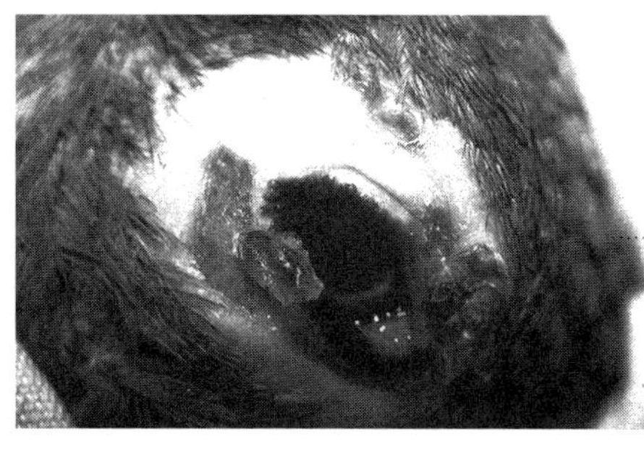

FIG 10.6

The earliest poxvirus-induced changes in Amazon parrots are characterized by a serous ocular discharge, blepharitis, rhinitis and conjunctivitis,

followed by the appearance of ulcerations on the lid margins, and

dry crusty lesions and scabs on the lid margins.

photographs courtesy of Scott McDonald and David Williams

Psittacine poxvirus strains have been shown to vary in virulence in different hosts.[10] Of the various Psittaciformes, Amazon parrots and particularly the Blue-fronted Amazon Parrots appear to be most susceptible to poxvirus infections. The earliest clinical changes in affected Amazon parrots occur 10 to 14 days after infection and are characterized by a serous ocular discharge, rhinitis and conjunctivitis, followed by the appearance of ulcerations on the eyelid margins. By 12 to 18 days after infection, dry crusty lesions are noted on the lid margins and at the lateral and medial canthi of the eyes *Figure 10.6*. In otherwise healthy birds, these lesions generally resolve in two to six weeks if secondary infectious agents are not pre-

Ocular lesions commonly associated with poxvirus infections[44,45]

	Blue-fronted Amazons	Blue-headed Pionus	Mynahs
	(n = 97)	(n = 87)	(n = 90)
Lid deformities	35%	19%	32%
Loss of eye lashes	33%	—	—
Lid depigmentation	—	—	23%
Corneal neovascularization	16%	14%	—
Corneal crystals	15%	—	4%
Conjunctivitis	9%	10%	4%
Corneal ulcers/keratitis	9%	—	12%
Sunken globe	6%	—	—
Ocular discharge	6%	—	—
Ptosis	3%	—	1%
Uveitis	—	4%	—
Cataracts	—	4%	16%

sent. If bacterial or fungal pathogens invade the damaged tissues, the lesions may require months to resolve or can be life-threatening.[7,44,60]

Either permanent damage to the eyes and surrounding skin or chronic sinusitis may occur in some psittacines that survive the initial infection.[11] The occur-

FIG 10.7

In Amazon parrots, the diphtheritic form of pox can cause high levels of mortality,

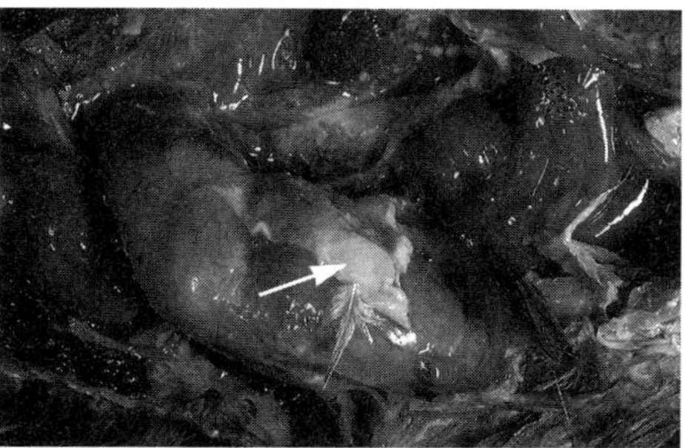

particularly in birds that develop necrotic lesions in the lungs and air sacs (arrow).

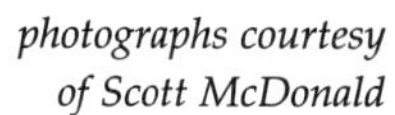

photographs courtesy of Scott McDonald

rence of unilateral or bilateral distortions in the margins of the eyelids, or loss of pigmentation or filoplumes around the eyes may indicate a previous

poxvirus infection.[44,45] In rare cases, poxvirus may cause lesions to the eye that include ulcers or crystallization of the cornea and anterior uveitis.[7,44] In a study involving Blue-fronted Amazon Parrots and Blue-headed Pionus Parrots, 85 of 184 (46%) birds that recovered from a poxvirus infection had lesions involving the eyes or surrounding tissues. The types of ocular lesions that were reported are listed in *Table 10.3*.

Infected Amazon parrots may also develop a severe upper respiratory tract disease in conjunction with diphtheritic lesions on the oral, pharyngeal, esophageal or crop mucosa. Anorexia and respiratory distress are common in birds with oral and pharyngeal lesions.[7,10,36,44,60] In one poxvirus outbreak in a quarantine station, half of the Blue-fronted Amazon Parrots examined had lesions in the oral cavity.[60] Depression, anorexia, diarrhea and hematochezia have been described in Amazon parrots and macaws with poxvirus-induced diphtheritic lesions in the gastrointestinal tract.[32]

Mortality rates are highest when diphtheritic lesions cause defects in the mucosal barrier of the alimentary and respiratory tracts, allowing secondary bacterial, fungal or chlamydial organisms unrestricted access to the affected bird.[7,36,39,60] Most infected psittacines that develop pneumonia or air sacculitis can be expected to die *Figure 10.7*; however, the mortality rate in birds with other poxvirus-induced lesions varies widely.[60] In one outbreak, 70 of 250 (28%) Double Yellow-headed Amazon Parrots and Lilac-crowned Amazon

Parrots died, and most of the survivors were anorectic, listless and had swollen eyelids.[7] In a quarantine station, 208 of 801 (26%) Amazon parrots died within the first 17 days of quarantine. A total of 651 of the 801 (81%) birds died or were euthanized before release from quarantine. Many of the dying birds had lesions characteristic of poxvirus.[11] In another poxvirus epornitic in a quarantine station, several Blue-fronted Amazon Parrots with mild swelling of the eyelids and conjunctiva were placed in a room with 466 Blue-fronted Amazon Parrots, 77 Blue-crowned Conures, 24 Golden-capped Conures, 15 Red-fronted Macaws, 7 Yellow-collared Macaws and 5 Blue-headed Pionus. Amazon parrots begin to die by the third week of quarantine. By the fourth week of quarantine, 40 affected Blue-fronted Amazon Parrots were dead, and 33 other birds, including 3 Red-fronted Macaws and 3 Blue-headed Pionus, were ill. Other exposed birds remained clinically normal.[60] In a holding facility in Argentina, 270 of 511 (53%) Blue-fronted Amazon Parrots had lesions suggestive of poxvirus.[11]

Lovebirds can develop either a cutaneous or diphtheritic form of poxvirus. Cutaneous lesions may simply become areas of dry, darkened skin rather than papules that erupt. Periocular infections may be noted initially when a lovebird develops conjunctivitis that becomes increasingly severe, as secondary pathogens invade the damaged epithelial cells.[1,53] Poxvirus infections can cause up to 75% morbidity and mortality in highly susceptible populations of lovebirds *Figure 10.8*.[32]

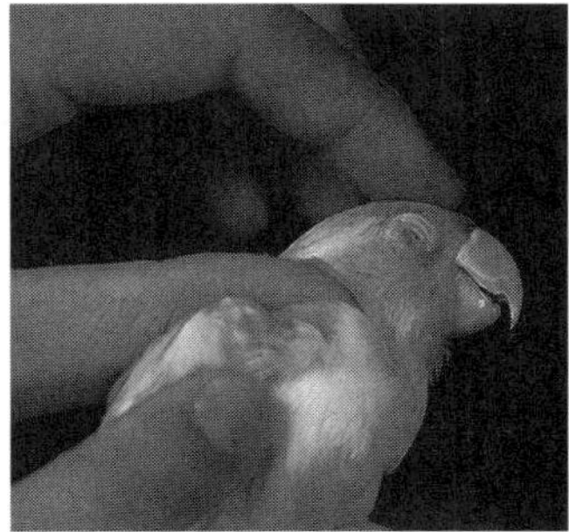

Experimental Infections in Psittaciformes

Psittacine poxvirus strains have been shown to vary in virulence with different hosts. For example, virus isolated from psittacine birds produced mild lesions and conferred protection against subsequent infection in experimentally inoculated Peach-faced Lovebirds and Red-lored Amazon Parrots, but the same virus preparation caused severe disease in Blue-fronted Amazon Parrots.[10] A poxvirus recovered from Amazon parrots was shown to be more pathogenic for psittacine birds than for chickens.[7]

Amazon parrots and conures experimentally inoculated by rubbing poxvirus onto the conjunctiva, oral mucosa and defeathered follicles developed characteristic lesions on the skin around the eyes within seven days after exposure. Interestingly, only one experimentally infected parrot developed lesions in the oral cavity, suggesting that other as yet undefined factors may be involved in the development of the diphtheritic form of this disease. A poxvirus was isolated from Budgerigars with abnormal feathers ("feather dusters"); however, Budgerigars experimentally infected with the isolated virus developed dyspnea and not feather abnormalities.[32]

FIG 10.8
Lovebirds may develop poxvirus-induced skin lesions that appear as areas of dry, darkened skin on the feet, legs or around the eyes.
right *Mortality rates are highest in susceptible populations of lovebirds maintained in crowded conditions.*
photographs courtesy of Louise Bauck and S.S. Tsai

A poxvirus recovered from an Amazon parrot caused typical skin lesions in young specific-pathogen-free (SPF) chickens experimentally inoculated by wing-web injection and when virus was painted into defeathered follicles. By comparison, young SPF chickens remained clinically normal when the same virus preparation was placed in the trachea or esophagus.[7] In another study using poxvirus recovered from a Blue-fronted Amazon Parrot, one-day-old and six-week-old chickens inoculated by the feather follicle route developed characteristic skin lesions, but those infected by intramuscular, ocular or wing web injection remained clinically normal.[74]

EPIZOOTIOLOGY

GENERAL CONSIDERATIONS

Poxvirus infections have been identified worldwide in over 60 species of birds in over 20 different families. Orders of infected birds include Psittaciformes, Galliformes, Columbiformes, Falconiformes, Charadriiformes, Struthioniformes, Anseriformes and Passeriformes. Most species of birds, regardless of age, sex, or species, are considered susceptible to some strain of poxvirus. The results of several studies documenting the prevalence of poxvirus lesions in various captive birds are listed in *Table 10.4*.

Infections have been documented in captive and free-ranging populations of Passeriformes, Falconiformes, Columbiformes and Anseriformes.[41,87] The prevalence of poxvirus lesions in free-ranging birds appears to vary with the species of bird and the time of year. Lesions occur most commonly during periods with extensive rainfall when mosquito activity is at its highest level.

Poxvirus lesions were described in 5% of free-ranging Amakihi and 2% of free-ranging Iiwi in Hawaii, and in up to 50% of House Finches trapped at some banding stations in the United States.[68,86] It has been suggested that the introduction of poxvirus may have been responsible for the rapid decline in populations of several species of birds indigenous to Hawaii and New Zealand.[5,86] In a group of mynah birds in a United States quarantine station, 57 of 434 (13%) birds had detectable poxvirus lesions. These birds had lesions at the time they were shipped from Malaysia, suggesting that the birds were exposed to the virus prior to or during the initial days of captivity in their native land.[45] Poxvirus lesions are common in some flocks of free-ranging quail and turkeys and are rarely seen in others.[6,15,19]

PSITTACIFORMES — Poxvirus infections are common in recently imported Amazon and pionus parrots, but are infrequently seen in established collections of other psittacine birds. Epornitics are occasionally reported in Psittaciformes housed in outdoor

TABLE 10.4

Prevalence of poxvirus in various captive birds in Europe and Australia[35,75]

Europe (one-year study, virus isolated by cell culture)

psittacines 2 of 38 (5%)	pigeons 5 of 23 (22%)
albatross 1	passerines 0 of 11
ducks 0 of 45	geese 0 of 10
swans 0 of 4	pheasants 0 of 72
chickens 0 of 89	turkeys 0 of 41
partridges 0 of 8	

Australia (characteristic lesions)

canaries 39 of 282 (14%)	raptors 6 of 70 (9%)
pigeons 14 of 541 (3%)	

enclosures in southern Florida and other coastal areas of the United States with high densities of companion birds and mosquitoes. Poxvirus infections are rare in companion birds maintained in a home environment. Psittacine birds considered most susceptible to poxvirus include Amazon and pionus parrots from South and Central America, as well as macaws, lovebirds, Quaker Parakeets, conures and parakeets from South America. Poxvirus has been reported also in Budgerigars and lories. Cockatoos and Cockatiels appear to be more resistant to infections than other Psittaciformes.[7,10,27,60]

<u>RELATIONSHIP OF VIRUS STRAINS</u> — The various *Avipoxvirus* strains are thought to be variants from the same original poxvirus, some of which remained antigenically similar, while others mutated to become immunologically distinct. Thus, the relationships of the various avian poxviruses vary, with some strains being more closely related than others. The genomes of fowl, pigeon and junco poxviruses have been shown to be similar, while the genome of quail, canary and mynah poxviruses are different from fowl poxvirus.[78] While all avian poxviruses are morphologically similar, they do exhibit varied host specificity, which is important in controlling cross-species infections. A summary of the cross reactivity of some of the more common avian poxviruses is provided in *Table 10.1*.

Avian poxviruses have been placed into specific groups (eg, canary poxvirus, pigeon poxvirus, magpie poxvirus, *Agapornis* poxvirus, psittacine pox-

virus) based largely on the type of birds they infect. Some strains infect only one species of bird, and other strains can infect several species. For example, the poxvirus that infects canaries is not known to infect psittacine birds, and the poxvirus that infects psittacine birds is not known to infect canaries. The antigenic relationship of the poxviruses isolated from various companion birds remains largely undetermined; however, it is known that the poxvirus commonly recovered from psittacine birds is antigenically distinct from fowl, pigeon and quail poxviruses.[7,9,10,60,90,91] Chickens vaccinated with fowl or pigeon poxvirus vaccine remained susceptible to psittacine poxvirus, and chickens experimentally infected with psittacine poxvirus remained susceptible to fowl or pigeon poxviruses, indicating that these three strains of virus are antigenically distinct. Psittacine poxvirus will cause cutaneous lesions when experimentally introduced into one- or ten-day old SPF chickens, even though psittacine poxvirus is not antigenically related to fowl poxvirus.[7]

It is common for poxviruses to infect heterologous avian species, but disease is typically most severe in the natural host. A psittacine poxvirus isolate was found to be highly virulent for Amazon parrots, yet caused only mild disease in experimentally infected chickens.[7] In some cases, birds infected with one strain of virus will be protected from infection by another strain. For example, chickens infected with pigeon poxvirus develop a mild disease, and may become resistant to infection by fowl poxvirus.

In addition to broad differences in host range among genera of *Avipoxvirus*, there are also differences in host range among individual poxvirus isolates *see Table 10.1*. A similar difference in the pathogenicity of varying strains of psittacine poxviruses is also suspected.

INCUBATION — The incubation period of poxvirus varies from four days to over a month, depending on the strain of the virus and the individual host.[46] Following natural exposure, most susceptible birds develop lesions in 7 to 14 days, and outbreaks in large flocks of susceptible birds may continue for several months.[7,87] In one outbreak involving captive raptors, new cases of poxvirus occurred from six days to three months after the initial infection was diagnosed.[87] The incubation period in flickers may be up to one month and appears to be longer than in other species.[48]

CARRIERS — It has been suggested that some birds that recover from poxvirus infections may develop persistent infections and intermittently shed virions from the gastrointestinal system, skin or feathers.[32,51,56] Persistent infections have been suggested to occur in chickens, pigeons, Rice Finches, raptors, lovebirds and a swan. Persistent infections of up to 13 months in duration have been reported in chickens.[82] Stress factors are thought to be associated with activation of latent infections.[25] A swan developed a poxvirus infection six weeks after being relocated to a different enclosure. It was

suggested that the bird in this case was latently infected with poxvirus, and the stress associated with movement induced a disease process; however, it was undetermined whether the enclosure in which the bird was placed was contaminated with poxvirus.[65] Stress was thought to have activated a latent infection in a lovebird, but it could not be determined if the bird could have been exposed to an exogenous source of virus.[67]

TRANSMISSION — As a group, poxviruses survive by being extremely durable outside of the host. This durability increases the likelihood that viable virus particles will come in contact with a susceptible host. Poxvirus transmission can occur through direct contact with an infected bird or through indirect contact with a contaminated object or insect. However, poxviruses are not capable of penetrating intact epithelium, and must enter the body through abraded skin or mucous membranes. Any type of trauma to the skin, including abrasions caused by cannibalism, territorial aggression, feather picking or aggressive preening of one another, can provide a route for poxvirus to enter the body. Young birds may be infected through lesions that are created in the oral cavity during exuberant feeding behavior or through sibling aggression in the nest. In a quarantine setting, hand-feeding was considered to be a route of virus transmission from infected to susceptible birds.[11]

In most areas, mosquitoes, mites and other blood-sucking insects serve as the primary mechanical vectors for poxviruses. One report suggested that contaminated blowflies may serve also as mechanical vectors of the virus. Poxvirus is found in the blood of infected birds; the mouthparts of a mosquito or other insects that consume blood can be contaminated with the virus. The virus is then injected into a new bird with subsequent feedings. Mosquitoes have been shown to harbor the virus from several weeks to several months after feeding on an infected bird *Figure 10.9*. Because of the importance of mosquitoes in transporting the virus from bird to bird, poxvirus epornitics are common in the spring and fall when the density of mosquitoes increases during periods of heavy rains.[17]

It has been theorized that poxvirus infections in the oral cavity and upper respiratory tract may occur after a bird inhales aerosolized viral particles found in contaminated droppings or soil. The demonstration of virus replication in the lungs of experimentally infected chickens suggests that virus may be excreted in respiratory secretions.[62] In turkeys, poxvirus lesions may occur on the mucosa of the cloaca, and virus transmission can occur through artificial insemination.[3] It has been suggested that latently infected chickens can transmit the virus to their chicks through the egg.[38]

Birds with poxvirus infections should be considered most infectious while lesions or scabs are present. During these periods, poxvirus may be transmitted if an infected bird comes in contact with a susceptible bird. The virus shed from an infected bird also may

FIG 10.9

Poxviruses cannot pass intact skin and must enter the body through abrasions or cuts. In most cases, mosquitoes serve as mechanical vectors for poxvirus, passing it from bird to bird during each successive feeding.

contaminate soil, water containers, food containers, perches, gloves or enclosures *Figure 10.10*.[30,87] In captive raptors, poxvirus transmission is thought to occur when a susceptible bird contacts a contaminated glove or perch. Contaminated shipping containers may also be a source of virus exposure. Feeders and water baths that serve to congregate free-ranging birds must be kept clean and should be disinfected frequently to prevent them from serving as a source of poxvirus.

Free-ranging birds, particularly Passeriformes and pigeons, can be a source of virus for susceptible species of captive

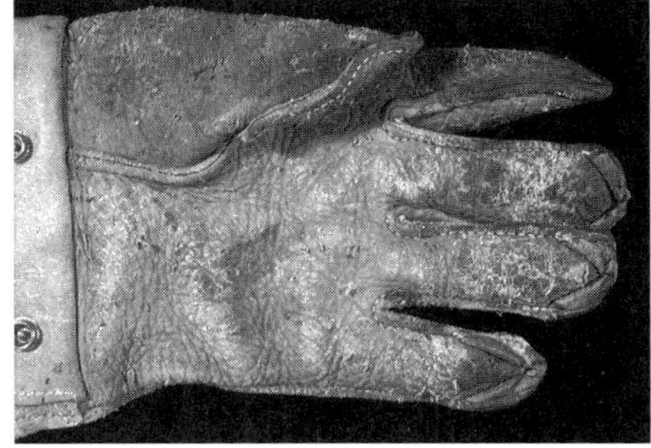

FIG 10.10

Poxviruses are extremely stable when outside of a host. Because of their stability, they can be transmitted from bird to bird through contact with contaminated gloves, soil, perches, or enclosures. Hospital and aviary equipment and supplies that come in direct contact with an infected bird should be autoclaved prior to subsequent use.

birds.[21,22,28,41,54] Infected, free-ranging Starlings were thought to have introduced the virus to a group of captive Rothchild's Mynahs.[54] Free-ranging Jungle Fowl were considered the reservoir for a poxvirus that caused an outbreak in a group of captive pheasants.[28]

PATHOLOGY

Many of the gross lesions associated with poxvirus infections are readily visible and have been described under clinical features. Gross lesions that may not be apparent in the conscious bird include fibrinonecrotic lesions in the gastrointestinal tract; ulceration of the nasal, tracheal and bronchial mucosa; and nodules and ulcerations at the junction between the esophagus and proventriculus.[7,36,56]

Histologic lesions that have been associated with poxvirus infections include necrosis of the heart and liver, as well as air sacculitis, pneumonia, peritonitis and accumulation of necrotic debris on the surface of the alimentary tract. Demonstration of intracytoplasmic inclusion bodies (Bollinger bodies) in suspect lesions is considered pathognomonic *Figure 10.11*. Characteristic inclusion bodies may be noted in association with lesions in the skin or mucosa of the sinuses, trachea, crop, esophagus or throat. Most affected birds develop only intracytoplasmic inclusion bodies; however, both intranuclear and intracytoplasmic inclusion bodies have been described in juncos and flickers.[34] Inclusion bodies may be hard to locate because of necrosis or inflammatory reactions induced by secondary bacteria or fungi. Birds with the septicemic form of the disease often die acutely, and microscopic demonstration of the virus may be difficult.

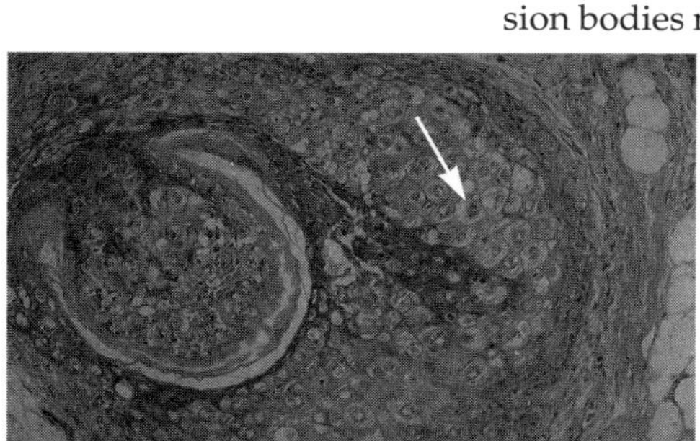

FIG 10.11

Demonstration of Bollinger bodies (arrow), in suspect lesions is considered pathognomonic for a poxvirus infection. photograph courtesy of S.S. Tsai

PATHOGENESIS

A poxvirus infection is initiated when a viable virus particle comes in contact with damaged epithelium on the skin or mucous membrane. The type of poxvirus infection that develops in a particular bird is thought to be dependent on the route of viral entry into the host. Cutaneous poxvirus lesions most likely occur when virus enters the body through damaged skin or insect bites. The pathogenesis of the diphtheritic form of poxvirus is unknown, but may be caused by inhalation or ingestion of virus-contaminated materials. The diphtheritic form of pox has been difficult to reproduce experimentally, and may require a prolonged incubation period or other contributing factors such as malnutrition, stress, debilitation or damaged respiratory mucosa to facilitate the development of the commonly recognized lesions.[7]

Some poxvirus infections are limited to the cells adjacent to the site of viral entry. Others are characterized by viremia with severe organ damage. Generalized infections are characterized by replication of virus at the initial site of virus entry, release of new virus particles to the blood stream (primary viremia), and transport of virus to the liver and bone marrow, where the virus again replicates. As the virus replicates in these target organs, a new generation of virus particles are released into the blood (secondary viremia), causing more substantial lesions throughout the body. In chickens, poxvirus can be detected in the blood stream three to seven days after infection. Virus replication was demonstrated in the lungs of chickens experimentally infected by aerosol or wing-web injection, suggesting that the lung may be an important site of viral replication in chickens and that virus may be excreted in respiratory secretions.[62]

The severity of a poxvirus-induced disease appears to depend on whether or not a secondary viremia occurs following the primary infection.[7,28] Lesions are most severe in birds that develop a secondary viremia. Less pathogenic strains appear to cause infections that are limited to the inoculation site.

A particular strain of poxvirus may be narrow in host range or may produce disease of varying virulence in several avian species. Virus strains that are not adapted to a particular bird may cause more, or less, severe lesions than are characteristic in the common host. For example, a strain of avian poxvirus that causes severe disease in lovebirds might not cause any signs of disease in other psittacine birds.[7] Immunosuppression is considered to occur in birds infected with poxvirus because they commonly develop secondary bacterial, fungal or chlamydial infections. However, little work has been done to confirm the effects of poxvirus on an infected bird's immune system.

IMMUNITY AND DIAGNOSIS

Poxvirus infections generally stimulate the production of antibodies that protect a bird from infection by the same strain of virus for six months to one year.[7,51] Chickens that recover are considered immune to disease following natural infections.[82] Cell-mediated immunity is more important in resolv-

ing poxvirus infections than is the production of antibodies. While birds that recover from infection are thought to develop immunity, it has been suggested that some birds are latently infected. In chickens and pigeons, characteristic poxvirus lesions can be induced by giving previously infected birds an immunosuppressive drug.[25,30,84]

The clinical changes associated with the cutaneous form of poxvirus are often suggestive. However, trauma, *Trichophyton*, *Knemidokoptes* mites, papillomavirus and bacterial infection can cause similar lesions. Diphtheritic lesions caused by poxvirus may be confused with those associated with candidiasis, hypovitaminosis A, aspergillosis, trichomoniasis or herpesvirus *Figure 10.12*. The lesions caused by infectious laryngotracheitis virus can appear particularly similar to those caused by poxvirus.

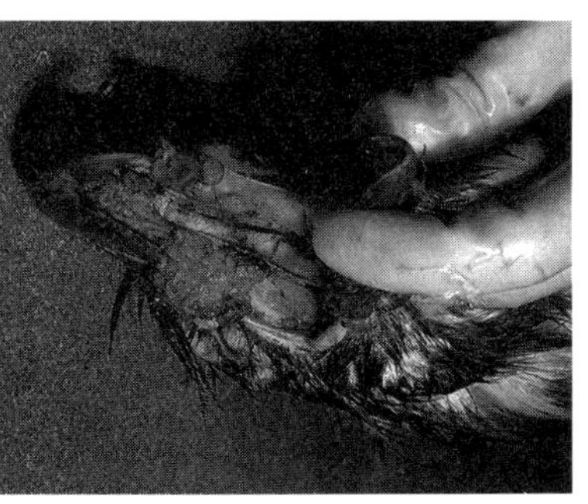

FIG 10.12
Depending on the species, diphtheritic poxvirus lesions can appear similar to those associated with candidiasis, hypovitaminosis A, aspergillosis, trichomoniasis or herpesvirus.
photograph courtesy of National Wildlife Health Center

The presence of poxvirus in papules, vesicles or scabs can be confirmed by virus isolation, demonstration of virus particles by electron microscopy or microscopic identification of Bollinger bodies in a biopsy sample.[84] Poxvirus particles are large enough to be detected through a light microscope, and impression smears of suspect lesions can be stained with Gimenez stain.[83] Poxvirus can be easily propagated in various cell cultures, making laboratory isolation a viable option for diagnosis.

Poxvirus may be shed intermittently in the feces of some birds and can be identified by culture or electron microscopy. DNA probes have been developed that can be used to detect fowl poxvirus nucleic acid in fluid collected from lesions or in biopsy samples.[85] Because the genomes of fowl, pigeon and junco poxviruses are similar,[78] DNA probes designed for use in chickens might also work in pigeons and juncos.

Antibodies to poxvirus can be detected using an agar-gel diffusion, virus neutralization, ELISA or hemagglutination-inhibition test. In chickens, hemagglutination-inhibition antibodies may be detected as early as one week after infection and may persist for more than 15 weeks. Virus-neutralizing antibodies can be detected by one to two weeks after infection.[9,66] In some birds, poxvirus antibodies are detectable only in the early convalescent period; however, the demonstration of a rising titer can be used to confirm the presence of a poxvirus infection in birds with chronic ocular or respiratory disease that may be associated with some poxvirus infections.[31]

CONTROL AND TREATMENT: GENERAL CONSIDERATIONS

Poxviruses are environmentally stable, can survive for years in dried organic debris such as feces, blood, soil or scabs, and are resistant to many commonly used disinfectants. Fowl poxvirus remains infectious when placed in 1% phenol or a 1:1000 dilution of formalin for up to nine days, but can be inactivated with 1% potassium hydroxide, heating to 50°C for 30 minutes, heating to 60°C for 8 minutes, or application of steam, 2% NaOH or 5% phenol.[4,32]

To prevent the spread of the virus through a flock, birds with active lesions and birds exposed to them should be kept separate from the remainder of the group. The occurrence of new cases can be reduced by using general disease prevention principles.

Poxvirus-induced lesions that are infected secondarily with bacteria or fungi can be treated by gently removing devitalized tissue, thoroughly cleansing the wound, and applying topical antimicrobial agents. Systemic antibiotics are indicated in birds with lesions in the respiratory or gastrointestinal tract. Lesions in the oral cavity are commonly infected with *Candida* sp. or *Trichomonas* sp. Treatment of these secondary invaders will improve recovery. The administration of supplemental vitamin A (10,000-25,000 IU/300 g body weight IM) and vitamin C may enhance the healing of damaged epithelial surfaces.[32] The administration of vitamin A was considered to be particularly beneficial when treatment was initiated early in the disease process.[11] Assisted feeding may be necessary in birds that are not willing to eat. Lesions around the eyes are best treated by cleansing the wounds and applying safe antimicrobial agents. Disfiguring scars are less likely to occur if scabs can be allowed to resolve naturally.[44]

<u>VACCINATION</u> — Vaccination is the best way to prevent poxvirus infections in the species for which an effective vaccine is available. Birds in high-risk areas (those with high densities of birds and

mosquitoes) should be vaccinated. Poxvirus vaccines are available for use in pigeons, chickens, turkeys, canaries, quail, waterfowl, falcons and psittacine birds (other than lovebird pox). Poxviruses that infect pigeons, domestic fowl, canaries, waterfowl and falcons share some similarities, and fowl pox vaccines can be used with some success in these species. Poxviruses that infect Psittaciformes, Charadriiformes and most Passeriformes are serologically unrelated to other poxviruses. Thus, prevention of poxvirus induced-disease in these latter groups of birds requires species-specific vaccines.

In most cases, birds vaccinated with an attenuated-live virus vaccine will be immune to infection within 10 to 14 days after vaccination. Some inactivated vaccines stimulate only partial immunity, which will protect a bird from severe disease but not from infection. It is advisable to vaccinate only healthy, disease-free birds. Vaccinating in the face of an outbreak can cause humans to serve as mechanical vectors. Additionally, strains of avian poxvirus have been shown to have the capacity to combine with other avipoxviruses to create different strains of a new virus. This feature of poxvirus replication can be a particular problem if an attenuated-live virus vaccine strain recombines with a wild-type virus that is circulating in a susceptible group of birds.

Vaccination in Psittacines —
Fowl and pigeon poxvirus vaccines will not protect psittacine birds from psittacine poxvirus. However, Blue-fronted Amazon Parrots experimental-ly vaccinated with inactivated psittacine poxvirus were found to have less severe lesions, to recover faster and to have less mortality than non-vaccinated birds.[7,10] Based on this finding, an inactivated vaccine derived from virus isolated from Amazon parrots was developed for use in psittacine birds. The protection provided by inactivated poxvirus vaccines may be inferior to that provided by attenuated-live virus vaccines, but the former are safer than the latter for use in psittacine birds. Vaccination should be considered in high-risk birds including those in quarantine stations, in pet shops that handle imported birds, and in birds raised in areas with high densities of mosquitoes. In general, high-risk birds should be vaccinated at the beginning of the quarantine period with a booster two to eight weeks later. Adult breeding birds should be vaccinated in the non-breeding season. Chicks can be vaccinated one to three weeks prior to fledging, with administration of a booster four to eight weeks later.[10]

Species-specific Poxvirus Diseases

Pox in Columbiformes
Columbiformes may develop either the cutaneous or diphtheritic form of poxvirus. The dry form is characterized by the development of large wart-like proliferative growths on the skin, eyelids, beak, feet and legs. Unfeathered young pigeons may develop lesions anywhere on the body *Figure 10.13*. Initially, lesions are characterized as small, blister-like areas that progressively enlarge and develop a yellow discoloration. These small yellowish lesions

Unfeathered young pigeons may develop lesions anywhere on the body, particularly around the cloaca and umbilicus.

photographs courtesy of Scott McDonald

infection are particularly common in Columbiformes.

Pigeon poxvirus has been shown to have a worldwide distribution. The incubation period in pigeons is seven to nine days, and all ages of birds are susceptible. Both captive and free-ranging pigeons and Mourning Doves have been shown to be susceptible to poxvirus.[41,79] Infections can occur at any time of year but are most common in the late spring and fall when mosquito densities are the highest. Because the virus is primarily transmitted by mosquitoes, birds that are housed near stagnant bodies of water are at greatest risk.

occur in groups that merge, ulcerate, and form large, brown, scabbed areas. Scabbed lesions may bleed profusely if damaged. Typically, uncomplicated lesions resolve in three to four weeks; however, cutaneous and diphtheritic lesions have been reported to persist in Mourning Doves for several months.[52,79]

The diphtheritic form of pigeon pox is characterized by the appearance of thickened, yellowish, necrotic plaques that may be present in the mouth, throat, esophagus, crop or trachea *Figure 10.14*. The necrotic lesions appear as yellowish-gray, roughened areas. Necrotic debris may accumulate in the trachea of severely infected pigeons, causing death by asphyxiation. Infections that involve both the skin and mucous membranes are common. Central nervous system abnormalities have been described in some Mourning Doves with poxvirus infections.[47] Skin tumors after poxvirus

The virulence of an infecting strain of virus appears to determine the number of birds within a flock that become ill during an outbreak. In most cases, mortality is highest in young pigeons and may approach 50%. Mortality is lower in older birds, but a virulent strain of virus introduced to a highly susceptible flock can cause up to 90% morbidity.

The poxvirus isolated from pigeons is most pathogenic for pigeons and doves, but some strains can cause mild disease in chickens and turkeys. Pigeons are experimentally susceptible to turkey, fowl, pheasant, sparrow and some strains of canary poxviruses.[20,33,46] A virus isolated from an accipiter was not infectious for pigeons.[80]

Sound hygiene and isolation of pigeons with poxvirus lesions will help reduce the spread of this virus through a flock. Ensuring that birds are not overcrowded and instigating thorough cleaning and disinfection in the loft at least every two

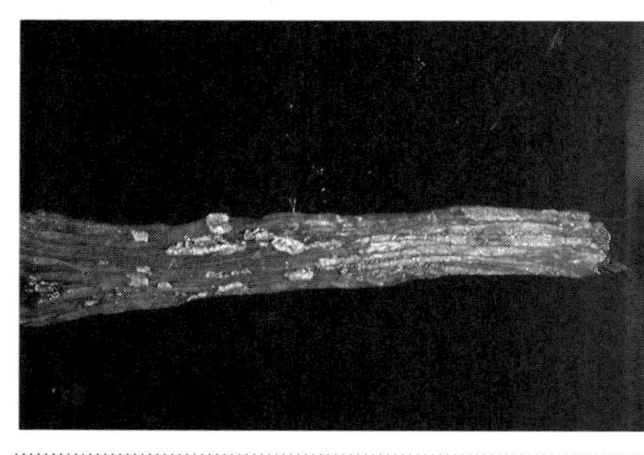

weeks will also help decrease transmission. Pigeons with poxvirus lesions and those that are exposed to them should not be flown for at least four weeks to prevent these birds from transmitting the virus to birds from other lofts.

Poxvirus-induced disease can be prevented in pigeons through vaccination. In high-risk areas, all pigeons over four to six weeks old should be vaccinated. It is best to vaccinate an entire flock at once and then vaccinate any new additions during their quarantine period. Pigeons should be revaccinated a month before the race or show season when they may be exposed to other infected birds. Healthy adults should be vaccinated eight weeks prior to the breeding season.[57]

An attenuated-live virus vaccine is administered by removing feathers from the legs and brushing the vaccine into the open feather follicles. This vaccine is recommended for use in racing pigeons but not in show pigeons, because the vaccine can cause blemishes in the skin or the appearance of abnormal feathers. A properly vaccinated bird will develop swelling and yellowish-brown discoloration at the inoculation site five to seven days post-vaccination. If this lesion does not develop, the bird should be revaccinated. Decreased food consumption and lethargy may occur five to seven days after vaccination. A protective immune response generally develops 10 to 14 days after vaccination. A wing-web injection technique, which is used for canaries and pheasants, can also be used for pigeons, but may cause irritation that can briefly hinder flight performance.

POXVIRUS IN PASSERIFORMES

Nodules caused by a poxvirus have been documented on the eyelids, beak, legs, toes and skin of the wings, neck and abdomen in a number of domestic and free-ranging Passeriformes, including finches, canaries, mynahs and grackles.[21,22,41,46,54,59] Poxvirus lesions that are not secondarily infected with bacteria or fungi should heal in three to four weeks; however, lesions associated with the head may impair vision, breathing or masticating, making it more difficult for a free-ranging bird to survive *Figure 10.15*. Lesions on the wings may impair flight.

Poxvirus infections are particularly common in free-ranging House Finches. At a banding station in California, 17% of the house finches that were captured died from virus-induced lesions. In one report, close to 50% of House Finches trapped at a Hawaiian banding station had lesions suggestive of poxvirus associated with the joints in the wing, on the tarsal joint or on the face. Free-ranging birds can be a source of virus for susceptible captive genera, particularly among Passeriformes.[21,22,41,46,54,59] Virus-contaminated feeders and water baths may play an important role in poxvirus transmission among free-ranging birds. The antigenic relationships among the poxviruses that infect various Passeriformes remains largely undetermined. Magpie poxvirus was found to be more closely related to pigeon poxvirus than

FIG 10.15
Poxvirus lesions on the head interfere with the normal functions of this passerine.

photograph courtesy of National Wildlife Health Center

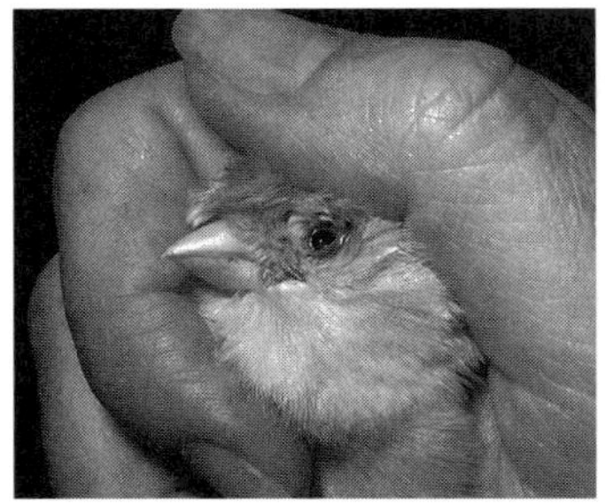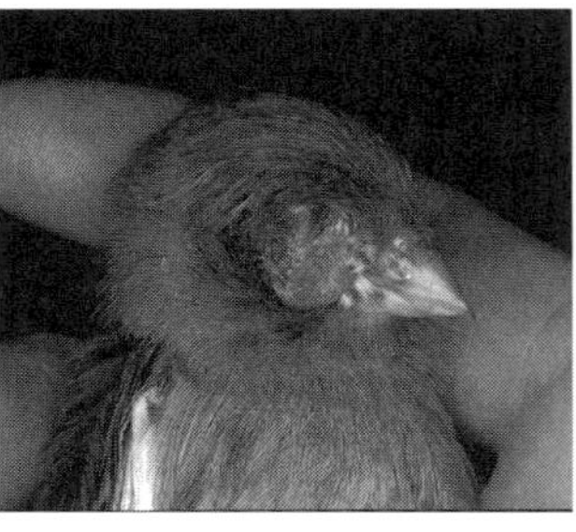

to fowl poxvirus, but four-day-old chickens experimentally infected with magpie poxvirus develop mild lesions.[9]

CANARY POXVIRUS

Canary poxvirus infections were first described in the early 1930s.[8] This virus should be considered enzootic in most flocks of canaries, and birds of any age are susceptible. Canaries can develop the cutaneous, diphtheritic or septicemic forms of the disease. The septicemic form is most common in canaries, and in the finches with which they cross-breed. Infected birds are usually presented with an acute onset of dyspnea, followed by death caused by pneumonia. Some affected canaries can survive for months, but most birds (>70%) die within a few days of developing clinical signs. Canaries may also develop a chronic form of the disease characterized by the appearance of wart-like lesions on the eyelids and beak, or a diphtheritic form characterized by lesions in the trachea, larynx and oral cavity *Figure 10.16*. In

one case, *Helicobacter* (*Campylobacter*) sp. was recovered from poxvirus lesions in affected canaries.[92] Post-poxvirus tumors are common in the lungs of canaries with dyspnea *Figure 10.17*. Similar tumors have been documented on the head, inside the oral cavity and on the skin of the wing or back of Masked Bull Finches.[23]

The poxvirus that infects canaries is the most lethal member of the family, and most birds with the septicemic form of the disease die. Mortality rates in a flock vary from 20% to 100%, depending on the virus strain, the types of lesions that develop and the susceptibility of individual birds as determined by previous exposure to the virus.[32] The most severe losses can be expected in unvaccinated flocks of canaries. The incubation period in canaries can range from four days to three weeks, and infections rapidly spread through a susceptible population.

A poxvirus epornitic involving a flock of 200 canaries and 50 Zebra Finches provides an example of the typical poxvirus outbreak in canaries. Of the 200 canaries in this group, 165 developed clinical disease and 145 affected birds died. The outbreak started with the sudden death of 19 six-month-old canaries. Affected canaries developed swollen eyes and open-mouthed breathing, and died within two to three days of appearing abnormal. Cutaneous lesions were not observed in the initial group of canaries that died with pneumonia, but did occur in other canaries four weeks after birds with respiratory disease first developed clinical signs. Skin lesions

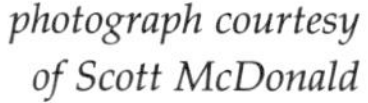

photograph courtesy of Scott McDonald

were common at the margins of the eye lids, around the nostrils and at the commissures of the beak. All of the Zebra Finches exposed to the affected canaries remained clinically normal.[43]

Canary poxvirus infections have been reported to be most prominent in the autumn and winter, suggesting that direct transmission of the virus may be occurring without an insect vector. Alternatively, some as yet undefined mite or other biting insect that increases in numbers during this time of the year may be responsible for the increased incidence of infections. Infections in canaries can be induced by contact with damaged skin or by intramuscular inoculation. Infections can also be experimentally induced by injecting susceptible canaries with blood from infected birds.[8]

Cross-neutralization studies indicate that canary poxvirus is antigenically related to fowl poxvirus; however, canary poxvirus causes severe disease in canaries, but only mild skin lesions on experimentally infected chickens and turkeys. Some strains of fowl poxvirus will cause experimental lesions in canaries. Other studies suggest that canaries are resistant to infection by turkey, fowl and pigeon poxviruses.[8] One isolate of canary poxvirus was found to be infectious for chickens, quail and turkeys, but not for house sparrows and pigeons. Another strain of canary poxvirus was infectious for chickens, pigeons and sparrows.[8] An isolate of poxvirus from a sparrow was infectious for canaries and sparrows.[33] Free-ranging Passeriformes have been suggested as a source of virus for captive canaries.[59]

Microscopic lesions in the septicemic from of disease in canaries may be limited to inflammation of the bronchi, and inclusion bodies may not be present in birds that die soon after infection. In other canaries, Bollinger bodies may be noted in the skin and lungs.[43]

Canary pox infections can be prevented by using a commercially available, attenuated-live vaccine that provides temporary immunity. Because of the severe nature of the disease, all canaries and finches with which they cross-breed should be vaccinated, and vaccination of all susceptible birds in a flock should occur at the same time. Maximum immunity develops three to four weeks after vaccination. Healthy canaries should be vaccinated every six to twelve months, preferably at the time of fledgling, and annually one month prior to the occurrence of increasing mosquito populations in the spring. Vaccination can be performed in the face of an outbreak, and has been shown to reduce mortality in canaries not showing clinical signs of disease at the time of vaccination. Extreme caution should be exercised when vaccinating in the face of an outbreak to prevent the transmission of the virus during the handling and vaccination procedures.

The wing-web method of administration is used for vaccinating canaries. If the vaccination is successful, an area of swelling or a scab will be present seven to ten days after vaccination *Figure 10.18*. If this reaction is not detected in a bird seven to ten days following vaccination, the bird should be revaccinated. Canaries should not be vaccinated dur-

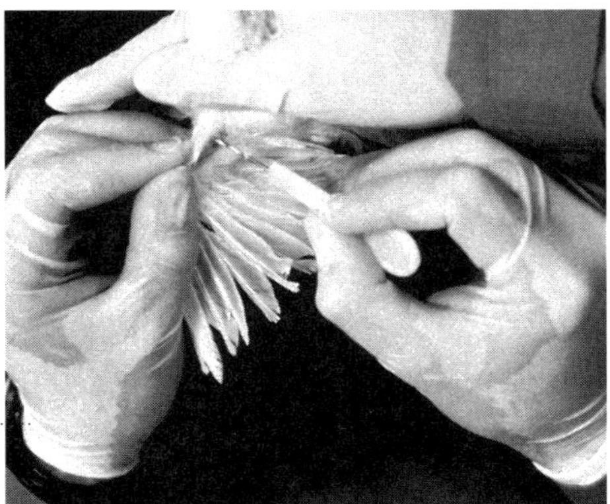

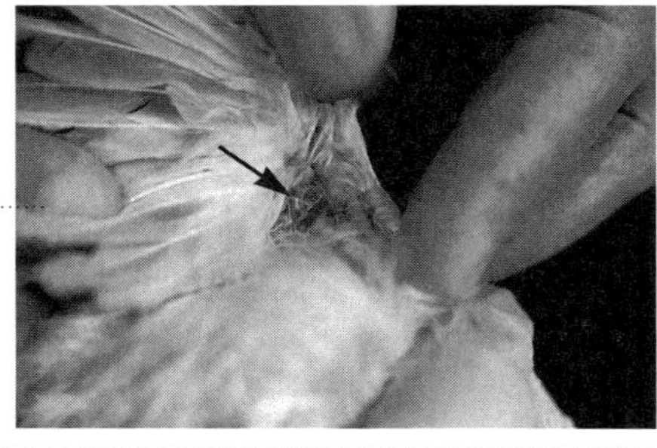

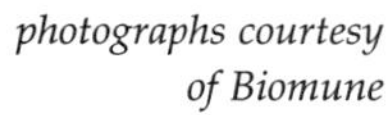

ing or within four weeks of egg-laying. It is important that a fresh needle be used in each vaccinate so that blood-borne diseases (including poxvirus) are not transmitted between birds by the vaccination process.

Poxvirus in Mynahs

Poxvirus infections associated with conjunctivitis, keratitis, corneal ulcers, eyelid deformities and ulceration of the skin on the head and beak have been reported in Hill, Bali and Rothchild's Mynahs. An ocular discharge with edema of the eyelids was the first clinical changes to occur in affected Hill Mynahs.[45] Poxvirus-induced morbidity and mortality appears to be lower in affected mynahs than in psittacines.[45] In a United States quarantine station, 57 of 434 (14%) mynah birds had detectable poxvirus lesions on the eyelid, head and around the beak at the time they were released from quarantine. Many of these mynahs had lesions at the time they

were shipped from Malaysia, suggesting that infections occurred prior to or during the initial days of captivity in their native land. Birds that recovered had distortion and depigmentation of the eyelids *see Table 10.3*.[45]

Poxvirus lesions were noted in 6 of 15 (40%) Rothchild's Mynahs in a flight enclosure in a zoological park. Free-ranging Starlings were thought to have been the source of virus for these captive birds.[54] Mynahs and Starlings belong to the same family, Sturnidae. The poxvirus that infects free-ranging Starlings is considered infectious for mynah birds. Some strains of mynah poxvirus have been shown to be highly infectious to domestic poultry, while others may not affect poultry.[54,70,76] Fowl poxvirus vaccines will not protect poultry from the mynah poxvirus.[76]

Poxvirus in Waterfowl and Murres

Poxvirus lesions have been reported occasionally in waterfowl, including Whistling Swans, Mute Swans, teals, geese, shearwaters, gulls, cormorants, tropicbirds and a Royal Tern.[42,50,65,89] Waterfowl most commonly develop the cutaneous form of poxvirus. Lesions have been documented on the feet, legs, face, nares, carpus, forehead, back, eyelids, commissure of the beak and the mucosa under the tongue.[39,50,89] In waterfowl, lesions are most common on the feet, while in other species of birds, lesions are more common around the eyes, beaks and nares.

Cutaneous lesions start as raised reddened bumps that turn into vesicles and then erupt. Once open, these ulcerative lesions are susceptible to secondary bacterial and fungal invaders. In uncomplicated cases, cutaneous lesions usually heal in several weeks; however, affected areas of skin are usually devoid of pigment following recovery. If secondarily infected, lesions can become severe and life-threatening.

Combined cutaneous and diphtheritic lesions have been documented in White-tailed Tropicbirds, and contributed to the deaths of 6 of 81 (7 %) fledglings in one flock. In another study, the prevalence of poxvirus in White-tailed Tropicbirds was determined to be less than 0.5%, with only fledgling birds affected. One White-tailed Tropicbird had lesions in the cells lining the trachea and bronchi.[89] The prevalence of poxvirus lesions was higher in Red-tailed Tropicbirds, with 19 of 115 (16%) living chicks, 2 of 15 (13%) adults and 4 of 9 (44%) dead chicks exhibiting lesions.[55]

The antigenic relationships of poxviruses recovered from waterfowl have been shown to vary. A poxvirus recovered from a Canada Goose was infectious for geese, but did not produce infection in chickens or domestic ducks. In other studies, poxvirus infections have been experimentally induced in geese and ducks.[14,49] A group of one-week-old ducklings and goslings developed characteristic lesions on the feet following experimental inoculation with fowl poxvirus by the intravenous or intradermal route. The incubation period in these birds was four to eight days. Uninoculated ducks and geese in direct contact with the infected birds also developed characteristic lesions. Experimentally infected gulls remained clinically normal.[49] Some strains of poxvirus recovered from waterfowl have been shown to be antigenically related to fowl poxvirus; while other strains are antigenically unique. Fowl poxvirus vaccine should protect waterfowl from an antigenically related virus.

Poxvirus in Raptors

Poxvirus infections have been documented in free-ranging and captive populations of raptorial birds, including falcons, hawks and eagles.[12,37,41,63,71,87] There are no reports of poxvirus in owls, and 42 owls of 6 species that were exposed in an outbreak in a rehabilitation center remained normal. Additionally, poxvirus infections could not be induced when owls were experimentally inoculated with virus derived from hawks.[87] A poxvirus isolated from an accipiter was not infectious to chickens or pigeons.[80]

In raptors, poxvirus lesions usually occur on the face, feet or legs and may include superficial vesicles, erosions or nodules *Figure 10.19*. In one aviary outbreak, lesions on the feet of affected raptors healed from 16 to 62 days after debridement and application of topical antimicrobial agents. Most lesions resolved within a month after the initiation of therapy. A poxvirus outbreak in a raptor rehabilitation center involved 8 of 87 raptors of four different species.[87] It has been suggested that poxvirus may play a role in some cases of bumblefoot.[30] The poxviruses could cause the initial damage in the skin of the foot, allowing secondary bacterial pathogens to cause chronic infections. Self-inflict-

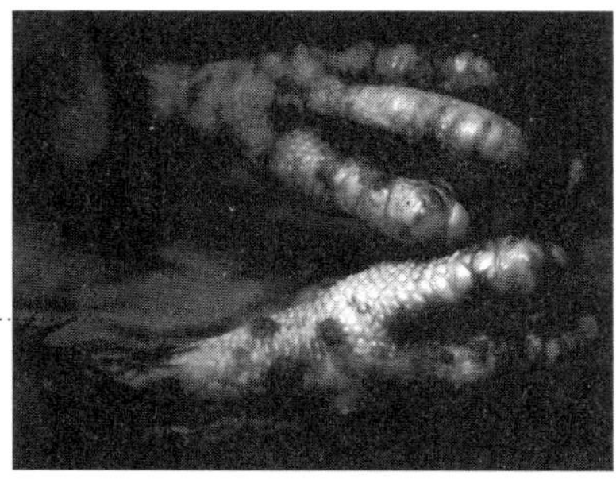

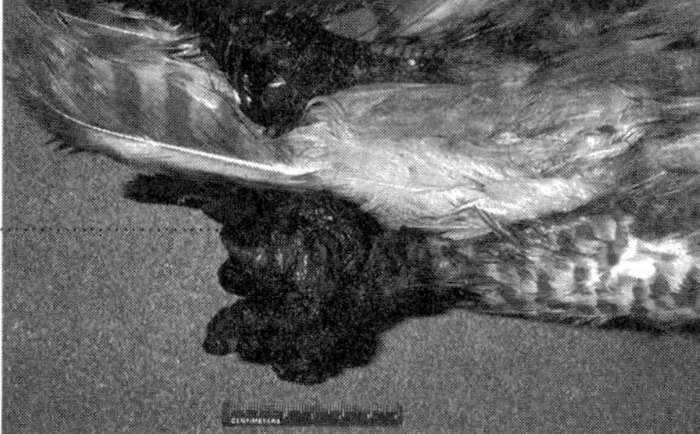

ed puncture wounds and those induced by prey species may cause the necessary damage to the skin on the feet and legs to allow poxvirus a portal of entry.

Poxvirus infections are mild and self-limiting in most healthy raptors; however, a strain of virus that infected birds in the Persian Gulf was associated with severe lesions that inhibited a bird's ability to use its feet and caused central nervous system signs.[47] Recurrence of poxvirus lesions have been reported following stressful events in some raptors that recover from the initial lesions.[30] This observation suggests that some raptors that recover from poxvirus-induced disease may remain latently infected or develop a transient immunity that does not protect the bird from subsequent infections. Some poxviruses that infect falcons are antigenically similar to fowl poxvirus, and fowl poxvirus vaccines may be effective in preventing disease in some raptors.

POXVIRUS IN GALLIFORMES

In some gallinaceous birds, fowl poxvirus may cause a mild disease characterized by small, focal, nodular lesions on the comb and wattles. In other birds, infections may be severe, with characteristic lesions occurring on the unfeathered skin at the corners of the mouth or around the eyes or cloaca.[84] Birds with lesions in the trachea frequently suffocate. Decreased egg production and retarded growth have also been reported in affected chickens. Conjunctivitis and respiratory signs were the initial clinical changes in chickens seven days after infection by placing virus in the eyes or by tracheal inoculation, respectively. Vesicles were typically noted in chickens between the fourth and seventh day after infection by the cutaneous route.[61]

A six-week poxvirus outbreak in a group of pheasants in a zoological park involved nine birds representing five species. Most of the affected pheasants recovered, although two died. Affected free-ranging Jungle Fowl were considered the source of the virus. The outbreak started in two White-eared Pheasants, followed in ten days by the appearance of lesions in Impeyan Pheasants maintained in enclosures several feet away. Lesions resolved in three to four weeks in most of the affected pheasants.[28]

Poxvirus recovered from captive peafowl was highly infectious for chickens, and strains isolated from some grouse appear to be antigenically related to fowl poxvirus.[2,24] Quail poxvirus is immunologically distinct from fowl, pigeon and psittacine poxviruses. Quail, chickens and turkeys inoculated with pigeon and fowl poxvirus were not protected from challenge with quail poxvirus. Quail and chickens vaccinated with quail poxvirus were not protected from challenge with pigeon or fowl poxviruses. Quail and chickens vaccinated with psittacine poxvirus were not protected from challenge by quail poxvirus. Quail poxvirus is experimentally infectious for chickens and turkeys, suggesting the possibility of cross-species transmission.[90] Pigeon poxvirus will partially protect some chickens from naturally induced fowl poxvirus infections.[82] Commercial quail poxvirus vaccine does not protect against fowl poxvirus.[69]

Poxvirus recovered from an Amazon parrot caused typical but mild skin lesions in young SPF chickens experimentally inoculated by wing-web injection and by painting the virus into defeathered follicles. By comparison, young SPF chickens inoculated by injecting the same virus preparation into the trachea or esophagus remained clinically normal.[7] In another study, one-day-old and six-week-old chickens inoculated by the feather follicle route developed characteristic skin lesions, while those infected by intramuscular, ocular or wing-web injection remained clinically normal.[74]

Magpie poxvirus was found to be more closely related to pigeon poxvirus than to fowl poxvirus, but experimentally infected chickens developed mild lesions. Turkey poxvirus is more closely related to fowl poxvirus than to pigeon poxvirus.[19] Cross-neutralization studies indicate that canary poxvirus is antigenically related to fowl poxvirus; however, canary poxvirus causes severe disease in canaries but only mild skin lesions in experimentally infected chickens and turkeys.[8,46] An isolate of poxvirus from a sparrow was infectious for turkeys and chickens.[33] Some reports list Starling poxvirus as being restricted to Starlings and mynahs, both in the family Sturnidae, but others suggest that some strains of this virus are infectious in chickens.[54,76]

Chickens are routinely vaccinated for poxvirus at four weeks of age and again several months before egg production. Gallinaceous birds can be vaccinated using the wing-web or defeathered follicle techniques. Properly vaccinated birds should develop a reaction seven to ten days after vaccination by wing-web injection or four to six days after the virus is painted into defeathered follicles *Figure 10.20*. Successfully vaccinated birds should be considered immune ten days after vaccination. In chickens, vaccination by either aerosol or wing-web injection was found to provide better immunity than when the vaccine was administered orally or in the drinking water. Aerosol vaccination caused a marked bronchopneumonia.[62]

QUAIL POX

Poxvirus lesions, including proliferative masses on the eyelids, commissures of the beak, hard palate, tongue and around the nostrils, have been described in captive and free-ranging quail. Swelling of the sinuses caused by

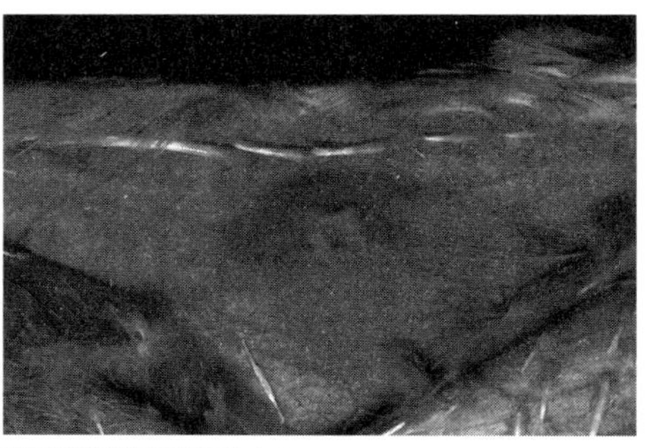

FIG 10.20

Gallinaceous birds can be vaccinated using the wing-web or defeathered follicle techniques.

photograph courtesy of Jean Sander

accumulation of yellow, thick exudate has also been reported *Figure 10.21*.[73]

Quail poxvirus has been associated with substantial losses on some game farms. In one group of captive Bobwhite Quail, 200 of 1000 (20%) birds died; in another outbreak, 30% of the birds were clinically affected.[73,77] Morbidity reached 40% and mortality reached 10% in a group of 500 pen-raised quail.[73] Decreased food consumption and reduced egg production are common in affected flocks. In one outbreak, only egg-laying females were affected even though birds ranging in age from 16 weeks old to adults were exposed. This finding suggest that the stress of breeding may have increased susceptibility to disease.

FIG 10.21

Lesions associated with the head and eyes of quail may alter food-gathering activity, flock behavior or make a bird more susceptible for predation.
photograph courtesy of National Wildlife Health Center

In free-ranging birds, 75% of affected Bobwhite or California Quail had lesions on the legs and feet, and up to 25% had lesions on the head.[15,18] All of a group of Gambel's Quail had lesions on the head.[6] In another study, 79% of Bobwhite Quail and 86% of Scaled Quail had lesions on the wing near the alula, probably secondary to damage caused to the leading edge of the wing as the quail moves through dense, abrasive vegetation.[88] Changes in free-ranging birds that affect activity level may alter food gathering-activity, flock behavior or make a bird more susceptible for predation. In this manner, an infectious agent that itself may be nonlethal may still contribute to a bird's death. For example, free-ranging quail with lesions on their feet had difficulty walking, which put them at increased risk for predation.[15]

The prevalence of poxvirus lesions in free-ranging quail appears to vary with the species and the time of year. In one study, 67 of 256 (26%) quail evaluated over a three-year period had suggestive lesions.[15] In several studies, 11 of 146 (7%) Gambel's Quail from Arizona and 97 of 428 (23%) California Quail from Oregon were found to be infected.[6,15] Poxvirus lesions were noted in 312 of 2,568 (12%) Bobwhite Quail from Florida, Georgia, North Carolina and Tennessee.[18]

Some studies suggest that there is no significant difference in the prevalence of lesions based on the gender of the quail, or based on whether it is mature or immature.[18,88] Other studies suggest that immature quail are more commonly affected. In a group of 89 free-ranging quail in California, 19 of the birds had poxvirus lesions. Of these affected birds, 6% were immature females and 41% were immature males.[15] Poxvirus lesions were not detected in any of 367 free-ranging Bobwhite Quail from Texas, but were present in 54% of Scaled Quail from the same area. In another year, 17% of Bobwhite quail and 34% of Scaled Quail had lesions. In this study, lesions were most common in the late spring and early summer, with a peak

AVIAN VIRUSES: FUNCTION AND CONTROL

prevalence of 55% in June for Bobwhite Quail and 86% in May for Scaled Quail.[88] In another study, the prevalence of poxvirus lesions in Scaled Quail was lowest in the summer and fall and highest in the winter and spring, with the highest prevalence of lesions occurring in January.

Fowl and pigeon poxvirus will not protect quail from quail poxvirus infections.[90] A highly effective quail poxvirus vaccine is commercially available.

POXVIRUS IN TURKEYS

Poxvirus has been diagnosed in captive and free-ranging turkeys, particularly in the southeastern United States. Lesions in free-ranging turkeys in Oregon were present on the head, neck and leading edge of the wing near the alula. Most studies indicate a low prevalence of lesions; however, poxvirus has been associated with debilitating lesions in up to 50% of free-ranging turkeys in some areas.[19] Poxvirus lesions were found in 3 of 113 (2.6%) free-ranging turkeys in Oregon and 1 of 330 (0.3%) free-ranging Rio Grande turkeys in Texas.[81] No gross pox lesions were noted in a group of 1,023 free-ranging turkeys from central Mississippi.[13] Lesions are most common from August to October in the southeastern United States when biting insect populations are highest.[29] In turkeys, poxvirus lesions may occur on the cloacal mucosa, and transmission can occur through artificial insemination.[3]

POXVIRUS IN OSTRICHES

A poxvirus was isolated from young (10- to 60-day-old) ostrich chicks in Israel. The first reported outbreak involved 20 of 100 (20%) at-risk ostrich chicks, with all affected young developing lesions within a ten-day period. Most of the affected birds recovered and the mortality rate was low (15%). Small vesicles containing yellowish fluid were evident on the eyelids, face, feet and around the opening to the external ear. The vesicles ruptured and turned to dry scabs six to ten days after first appearing. Exudate and scabs sealed the eyes closed in severely affected ostrich chicks *Figure 10.22*. Diphtheritic lesions were present on the mucosa of the pharynx, larynx, oral cavity and at the base of the tongue in some affected ostriches. The virus isolated from ostriches caused disease in experimentally infected turkeys.[72]

FIG 10.22
Ostriches can develop poxvirus-induced lesions on the eyelids, face, feet and around the opening to the external ear.

photograph courtesy of B. Perleman, reprinted with permission[72]

Disease was prevented in ostriches by vaccinating them by wing-web injection with a fowl poxvirus vaccine. New cases of pox ceased six days after the at-risk ostrich chicks were vaccinated. Vaccination of ostrich chicks at 10 to 14 days of age is recommended in areas with high mosquito populations.[72] Both facts that the virus recovered from ostriches is infectious to turkeys and that infections in ostriches can be prevented with a fowl poxvirus vaccine suggest that ostriches are susceptible to a poxvirus that is related to fowl poxvirus.

REFERENCES

1. Akrae M: Die Identifizierung und Charaktersierung eines Falkenpocken- und eines Psittacidenpockenvirus aufgrand ihres Verhaltens in der Eikultur, souie serologischer Reaktionen und ihres Wirtsspektrums in Tierversuch. München. 1980.

2. Al Falluji S, Tantawi HH, Albana A, et al: Pox infection among captive peacocks. J Wildl Dis 15:597-600, 1979.

3. Allen B, et al: An unusual presentation of avian poxvirus in breeder farms. Proc 39th West Poultry Dis Conf, 1990, pp 41-42.

4. Andrews C, Pereira HG, Wildy P: Viruses of Vertebrates. London, Bailliere Tindall, 1978.

5. Austin FJ, Bull PC, Chandry MA: A poxvirus isolated from silvereyes from lower Hutt, New Zealand. J Wildl Dis 9:111-114, 1973.

6. Blankenship LH, Reed RE, Irby HD: Pox in mourning doves and Gambel's quail in Southern Arizona. J Wildl Mgmt 30:253-257, 1966.

7. Boosinger TR, Winterfield RW, Feldman DS, et al: Psittacine pox virus: Virus isolation and identification, transmission, and cross-challenge studies in parrots and chickens. Avian Dis 26:437-444, 1982.

8. Burnet FM, Losh D: The immunological relationship between Kikuth's canary virus and fowlpox. British J Exp Pathol 17:302-307, 1936.

9. Chung YS, Spradbrow PB: Studies on poxvirus isolated from a magpie in Queensland. Aust Vet J 53:334-336, 1977.

10. Clubb SL, Eskelund KH: Field trials with a killed psittacine pox vaccine. Proc Assoc Avian Vet, 1988, pp 145-152.

11. Clubb SL, Winterfield RW, Cramm D: Laboratory and field trials with a parrot pox vaccine. Proc Assoc Avian Vet, 1985, pp 71-82.

12. Cooper JE: Two cases of pox in recently imported peregrine falcons (*Falco peregrinus*). Vet Rec 85:683-685, 1969.

13. Couvillion CE, Stacey LM, Hurst GA: Absence of avian pox in wild turkeys in central Mississippi. J Wildl Dis 27:467-469, 1991.

14. Cox WR: Avian pox infection in a Canada goose (*Branta canadensis*). J Wildl Dis 16:623-626, 1979.

15. Crawford JA: Differential prevalence of avian pox in adult and immature California quail. J Wildl Dis 22:564-566, 1986.

16. Cunningham CH: Avian pox. *In* Hofstad MS, et al (eds): Diseases of Poultry. Ames, Iowa State University Press, 1978, pp 597-609.

17. DaMassa AJ: The role of *Culex tarsalis* in the transmission of fowl pox virus. Avian Dis 10:57-66, 1966.

18. Davidson WR, Kellogg FE, Doster GL: An epornitic of avian pox in wild bobwhite quail. J Wildl Dis 16:293-298, 1980.

19. Davidson WR, Nettles VF, Couvillion CE, et al: Diseases diagnosed in wild turkeys (*Meleagris gallopavo*) of the southeastern United States. J Wildl Dis 21:386-390, 1985.

20. Dobson N: Pox in pheasants. J Comp Pathol 50:401, 1937.

21. Docherty DE, Long RIR: Isolation of a poxvirus from a house finch (*Carpodacus mexicannus*). J Wild Dis 22:420-422, 1986.

22. Docherty DE, Long RIR, Flinckinger EL, et al: Isolation of poxvirus from debilitating cutaneous lesions on four immature grackels (*Quiscalus* sp). Avian Dis 35:244-247, 1991.

23. Dorrestein GM, van der Hage MH, Grinwis G: A tumor-like pox lesion in masked bull finches (*Pyrrhula erythaca*). Europ Conf Avian Med Surg, 1993, pp 232-240.

24. DuBose RT: Pox in the sage grouse. Bull Wildl Dis Assoc 1:6, 1965.

25. Duran-Reynolds F, Bryan E: Studies on the combined effects of fowlpox virus and methylcholanthrene in chickens. Ann NY Acad Sci 54:977-991, 1952.

26. Eleazer TN, Harrel JS, Blalock HG: Transmission studies involving a wet fowl pox isolate. Avian Dis 27:542-544, 1983.

27. Emanuelson S, Carney J, Saito J: Avian pox in two black-masked conures. J Am Vet Med Assoc 173:1249-1250, 1978.

28. Ensley PK, Anderson MP, Costello ML, et al: Epornitic of avian pox in a zoo. J Am Vet Med Assoc 173:1111-1114, 1978.

29. Forrester DJ: The ecology and epizootiology of avian pox and malaria in wild turkeys. Bull Soc Vector Ecology, 1991.

30. Garner MM: Bumblefoot associated with poxvirus in a wild golden eagle (*Aquila chrysaetos*). Comp Anim Prac 19:17-20, 1989.

31. Gaskin JM: The serodiagnosis of psittacine viral infections. Proc Assoc Avian Vet, 1988, pp 7-10.

32. Gerlach H: Viruses. *In* Ritchie BW, Harrison GJ, Harrison LR (eds): Avian Medicine: Principles and Application. Lake Worth, Wingers Publishing, 1994, pp 862-948.

33. Giddens WE, Swango LJ, Handerson JD, et al: Canary pox in sparrows and canaries (Fringillidae) and in weavers (Ploceidae): Pathology and host specificity of the virus. Vet Path 8:260-280, 1971.

34. Goodpasture EW, Anderson K: Isolation of a wild avian pox virus inducing both cytoplasmic and nuclear inclusions. Amer J Path 40:437-453, 1962.

35. Gough RE, Alexander DJ, Collins MS, et al: Routine virus isolation or detection in the diagnosis of diseases in birds. Avian Pathol 17:893-907, 1988.

36. Graham CLG: Poxvirus infection in a spectacled Amazon parrot (*Amazona albifrons*). Avian Dis 22:340-343, 1978.

37. Halliwell WH: Avian pox in an immature red-tailed hawk. J Wildl Dis 8:104-105, 1972.

38. Hanson LE, et al: Fowlpox and latency. Proc 24th West Poult Dis Conf, 1975, pp 43-45.

39. Harris JM, Williams AS, Dutra FR: Avian pox in a group of common murres (*Uria aalge*). Vet Med Small Anim Clin July:918-919, 1978.

40. Hartig F, et al: Beitrag zur Pathologie der Lungenform der Kaviarienpocken. Berlin München Tierärztl Nschr 85:352-355, 1972.

41. Hill JR: Epornitic of pox in a wild bird population. J Am Vet Med Assoc 171:993-994, 1977.

42. Jacobson ER, Raphael BL, Nguyen HT, et al: Avian pox infection, aspergillosis and renal trematodiasis in a royal tern. J Wildl Dis 16:627-631, 1980.

43. Johnson BJ, Castro AE: Canary pox causing high mortality in an aviary. J Am Vet Med Assoc 189:1345-1347, 1986.

44. Karpinski LG, Clubb SL: Post pox ocular problems in blue-fronted Amazon and blue-headed pionus parrots. Proc Assoc Avian Vet, 1985, pp 91-100.

45. Karpinski LG, Clubb SL: An outbreak of pox in imported mynahs. Proc Assoc Avian Vet, 1986, pp 35-37.

46. Karstad L: Pox. *In* Davis JW, et al (eds): Infectious and Parasitic Diseases of Wild Birds. Ames, Iowa State University Press, 1971, pp 34-41.

47. Kiel H: Pockeninfektion bei Jadgfalken-Klinik, pathomorphologische Ergebnisse, Parphylaxe and Therapie. DVG-Tagung Vogelkrankheiten München, 1985, pp 202-206.

48. Kirmse P: Host specificity and long persistence of pox infection in the flicker (*Colaptes auratus*). Bull Wildl Dis Assoc 3:14, 1966.

49. Kirmse P: Experimental pox infection in waterfowl. Avian Dis 11:209-216, 1967.

50. Kirmse P: Pox in wild birds. An annotated bibliography. J Wild Dis 49:1-10, 1967.

51. Kirmse P: Host specificity and pathogenicity of pox viruses from wild birds. Bull Wildlife Dis Assoc 5:376-386, 1969.

52. Kossack CW, Hanson HC: Fowl pox in the mourning dove. J Am Vet Med Assoc 124:199, 1954.

53. Kraft V, Teufel P: Nachweis eines Pockenvirus bei Zwergpapageien (*Agapornis personata* und *Agapornis roseicollis*). Berlin München Tierärztl Wochenschr 84:83-87, 1971.

54. Landolt M, Kocan RM: Transmission of avian pox from starlings to Rothchild's mynahs. J Wildl Dis 12:353-356, 1976.

55. Locke LN, Wirtx WO, Brown EE: Pox infection and secondary cutaneous mycosis in a red-tailed tropicbird (*Phaethon rebricauda*). Bull Wildl Dis Assoc 1:60-61, 1965.

56. Lowenstine LJ: Emerging viral diseases of psittacine birds. *In* Kirk RW (ed): Current Veterinary Therapy IX. Philadelphia, WB Saunders Co, 1986, pp 705-710.

57. Marshall R: Management of pigeon diseases. Proc Assoc Avian Vet, 1990, pp 122-135.

58. Mayr A, Mahnel H: Characterization of a fowl pox virus isolated from a rhinoceros. Arch Gesamte Virusforsch 31:51-60, 1970.

59. McCaughey CA, Burnet FM: Avian pox in wild sparrows - I. A note on a spontaneous outbreak. II. A note on the activity of sparrow pox virus in the canary. J Comp Pathol 55:201-205, 1945.

60. McDonald SE, Lowenstine LJ, Ardans AA: Avian pox in blue-fronted Amazon parrots. J Am Vet Med Assoc 179:1218-1222, 1981.

61. Minbay A, Kreier JP: An experimental study of the pathogenesis of fowlpox infection in chickens. Avian Dis 17:532-539, 1973.

62. Mockett APA, Deuter A, Southee DJ: Fowlpox vaccination: Routes of inoculation and pathological effects. Avian Pathol 19:613-625, 1990.

63. Moffatt RE: Natural pox infection in a golden eagle. J Wildl Dis 8:161-162, 1972.

64. Montali RJ, Bush M, Greenwell GA: An epornitic of duck viral enteritis in a zoological park. J Am Vet Med Assoc 169:954-958, 1976.

65. Montgomery RD, Chowdhury KA: Avian pox in a whistling swan. J Am Vet Med Assoc 177:930-931, 1980.

66. Morita C: Role of humoral- and cell-mediated immunity on the recovery of chickens from fowlpox infection. J Immunol 111:1495-1501, 1973.

67. Olsen DE, Dolphin RE: Avian pox. Vet Med Small Anim Clin 73:1295-1297, 1978.

68. Olsen GH: Introduced avian disease and its effects on the Hawaiian ecosystem. Proc Assoc Avian Vet, 1992, pp 279-289.

69. Olsen GH: Common infectious and parasitic diseases of quail and pheasants. Proc Assoc Avian Vet, 1993, pp 146-150.

70. Panigrahy B, Seene DA: Diseases of mynah birds. J Am Vet Med Assoc 199:378-381, 1991.

71. Pearson GL, Pass DA, et al: Fatal pox infection in a rough-legged hawk. J Wildl Dis 11:224-228, 1975.

72. Perelman B, Gur-Lavie A, Samberg Y: Pox in ostriches. Avian Pathol 17:735-739, 1988.

73. Poonacha KB, Wilson M: Avian pox in pen-raised bobwhite quail. J Am Vet Med Assoc 11:1264-1265, 1981.

74. Ramos P: Studies on parrot pox. Proc 33rd West Poult Dis Conf, 1984, pp 90-91.

75. Reece RL: Observations on naturally occurring neoplasms in birds in the state of Victoria, Australia. Avian Pathol 21:3-32, 1992.

76. Reed WM: Immunogenicity and pathogenicity of mynah pox virus. Poult Sci 68:631-638, 1989.

77. Reed WM, Dhillon AS, Winterfield RW, et al: Avian pox outbreak in two flocks of bobwhite quail. Proc 32nd North Central Avian Dis Conf, 1985.

78. Schnitzlein WM, Ghildyal N, Tripathy DN: Genomic and antigenic characterization of avipoxviruses. Virus Res 10:65-76, 1988.

79. Tangredi BP: Avian pox in a mourning dove. Vet Med Small Anim Clin June:700-701, 1974.

80. Tantawi HH, Sheikhly AS, Hassan FK: Avian pox in a buzzard (*Accipiter nisus*) in Iraq. J Wildl Dis 23:249-252, 1981.

81. Thomas JW: Diagnosed diseases and parasitism in Rio Grande wild turkeys. Wilson Bull 76:292, 1964.

82. Tripathy DN, Hanson LE: Immunity to fowlpox. Am J Vet Res 36:541-544, 1975.

83. Tripathy DN, Hanson LE: Smear technique for staining elementary bodies of fowlpox. Avian Dis 20:609-610, 1976.

84. Tripathy DN, Hanson LE: Avian pox. *In* Hitchner SB, et al (eds): Isolation and Identification of Avian Pathogens. Endwell, Creative Publishing Co, 1980, pp 109-111.

85. Tripathy DN, Radzevicius J: Evaluation of cloned DNA fragments of fowlpox virus as diagnostic probes. Abst 42nd North Central Avian Dis Conf, 1991, p 61.

86. Warner RE: The role of introduced diseases in the extinction of the endemic Hawaiian avifauna. Condor 70:101-120, 1968.

87. Wheeldon EB, Sedgwick CJ, Schultz TA: Epornitic of avian pox in a raptor rehabilitation center. J Am Vet Med Assoc 187:1202-1204, 1985.

88. Wilson MH, Crawford JA: Poxvirus in scaled quail and prevalence of poxvirus-like lesions in northern bobwhites and scaled quail from Texas. J Wildl Dis 24:360-363, 1988.

89. Wingate DB, Barker IK, King NW: Poxvirus infection of the white-tailed tropicbird (*Phaethon lepturas*) in Bermuda. J Wild Dis 16:619-622, 1980.

90. Winterfield RW, Clubb SL, Schrader D: Immunization against psittacine pox. Avian Dis 29:886-890, 1985.

91. Winterfield RW, Reed W: Avian pox: Infection and immunity with quail, psittacine, fowl and pigeon pox viruses. Poult Sci 64:65-70, 1985.

92. Woods LW: *Campylobacter* isolation from pox lesions and tissues of canaries. J Assoc Avian Vet 1:174, 1987.

Adenoviridae

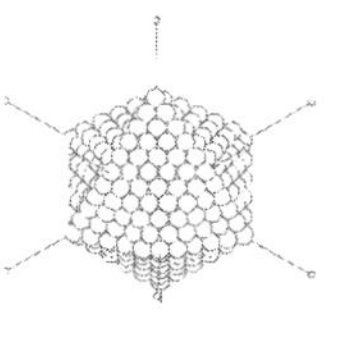

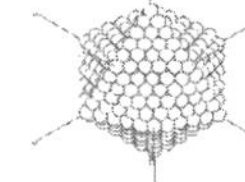

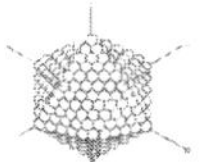

Adenoviruses are nonenveloped and 70 to 90 nm in diameter. Replication occurs in the cell nucleus, where viral particles accumulate to form intranuclear inclusion bodies. Various strains of adenovirus infect mammals and birds. The strains that infect mammals have been placed in the genus *Mastadenovirus*, and the strains that affect birds have been placed in the genus *Aviadenovirus*. The classification of aviadenoviruses is based on the occurrence of group-specific characteristics that are not found in the adenoviruses that infect mammals. *Aviadenovirus* isolates are currently divided into serotypes based on the virus-neutralization test *Table 11.1*. Other classification schemes have been described in which avian adenoviruses are divided into three groups based on shared characteristics *Table 11.2*.

In addition to chickens and turkeys, adenoviruses have been isolated from ducks, geese, guinea fowl, pigeons and psittacine birds.[58] Various serotypes of *Aviadenovirus* have been associated with quail bronchitis, inclusion body hepatitis in chickens, turkey viral hepatitis, duck hepatitis, chicken egg drop syndrome, egg drop syndrome in ducks, chicken splenomegaly, marble spleen disease of pheasants and hemorrhagic enteritis in turkeys. Marble spleen disease in pheasants is the most important problem caused by an adenovirus that has been described in companion, aviary or free-ranging birds.

As a family, adenoviruses are commonly recovered from persistently infected, asymptomatic birds. Some strains of avian adenovirus are host-specific; others can infect various species of birds.[59] Those strains with a broad host range are particularly troublesome in mixed species aviaries. While adenovirus infections have been associated with a number of syndromes *Table 11.3*, experimental infections are usually asymptomatic or associated with mild respiratory disease. This suggests that a diseased bird from which an adenovirus is

TABLE 11.1

Serotypes of avian adenoviruses*

Chickens (12 serotypes)
Turkeys (3 serotypes and turkeys are
    susceptible to some chicken strains)
Turkey adenovirus 1
Turkey adenovirus 2
Turkey hemorrhagic enteritis virus
Ducks
    Egg-drop syndrome (EDS-76)
    Duck adenovirus 1
Pheasants (marble spleen disease virus,
    serologically related to turkey
    hemorrhagic enteritis virus)
Geese (3 serotypes)
Pigeons (susceptible to several serotypes that
    infect chickens, as well as at least two
    unique serotypes)
Budgerigars (several different isolates
    serologically related to some strains
    from chickens)
Unclassified Isolates from:
    Psittacines
    Raptors
    Guinea fowl (susceptible at least to EDS-76
    and turkey hemorrhagic enteritis virus)

*Distribution varies by continent[6,20,56,58,96]

TABLE 11.2

Suggested grouping for avian adenoviruses[27,89,95]

ADENOVIRUS I - Contains 12 serologically distinct adenoviruses, including those that cause quail bronchitis, inclusion body hepatitis of chickens, turkey viral hepatitis and strains that infect pigeons, Budgerigars, ducks, geese and guinea fowl. In general, type I adenoviruses are transmitted horizontally and vertically, and cause severe lesions in birds less than 1 month old.

ADENOVIRUS II - Hemorrhagic enteritis of turkeys, splenomegaly of chickens and marble spleen disease of pheasants. These viruses are serologically identical but genetically distinct. In general, type II adenoviruses are transmitted by direct contact and cause lesions in birds older than 1 month.

ADENOVIRUS III - egg drop syndrome virus (EDS-76).

TABLE 11.3

Clinical features associated with adenovirus infections

CHICKENS - Respiratory disease, decreased egg production, hepatitis, enteritis, pancreatitis, bursal atrophy or anemia.

DUCKS - Acute deaths in 35-day-old Muscovy ducklings with signs of upper respiratory disease.

GOSHAWK - Central nervous system signs and death.

GOSLINGS - High mortality without premonitory signs in 4- to 11-day-old captive goslings. Birds may die following a brief period of respiratory distress or diarrhea. Virus recovery most common in one- to two-month-old birds.

GUINEA FOWL - Marble spleen-like disease with diarrhea and dyspnea followed by death. Guinea fowl experimentally infected with turkey hemorrhagic enteritis virus develop hemorrhagic diarrhea and die. Guinea fowl are susceptible to egg drop syndrome virus.

AMERICAN KESTRELS - Hemorrhagic enteritis and anemia prior to death.

OSTRICHES - Affected birds may be dead in shell to three years of age. Greenish diarrhea, stunting, reduced weight gains and death.

PHEASANTS - Death with no clinical signs or death following a brief period of anorexia, depression, diarrhea and dyspnea. Most common in pheasants from three to eight months of age.

PIGEONS - Birds from hatching to five years of age are susceptible. Most common in two- to four-month-old pigeons that develop depression, anorexia, dyspnea, a crouched stance, polydipsia, polyuria and slimy green diarrhea. Young birds may die within 48 hours of developing clinical signs.

PSITTACINES - Depression, anorexia, diarrhea and cloacal hemorrhage followed by enteritis, hepatitis, pancreatitis, encephalitis, splenitis, conjunctivitis and death. Presence of inclusion bodies in clinically normal birds suggests asymptomatic infections.

QUAIL - Birds less than one week of age die without showing clinical signs. Birds two to three weeks old die after a one- to two-day course of anorexia, depression and passage of greenish diarrhea.

TURKEYS - Asymptomatic infections or hemorrhagic enteritis.

---

isolated may be experiencing concomitant problems. It has been theorized that suppression of the immune system is necessary for some strains of adenovirus to cause disease in domestic fowl.[22] There appears to be a difference in virulence among adenovirus strains.

The morbidity and mortality associated with adenovirus infections appear to vary with the host and the strain of infecting virus. Asymptomatic infections that are not detected until a bird becomes immunocompromised are most common. In gallinaceous species, clinical signs induced by the varying adenovirus serotypes are associated with respiratory disease, decreased egg production, hepatitis, enteritis, pancreatitis, bursal atrophy or anemia.[25,29,39,57] The production of thin-shelled, soft-shelled and shell-less eggs from otherwise normal chickens is considered highly suggestive of an adenovirus infection.[56] It has not been determined if an adenovirus may be causing similar egg-related problems in companion and aviary birds.

CLINICAL FEATURES

Adenovirus infections have been associated with depression, anorexia, diarrhea and cloacal hemorrhage, followed by death in Budgerigars, lovebirds, Cockatiels, Moluccan Cockatoos and Redrumped Parakeets.[32,49,62,64,87] Adenovirus infections also have been associated with pancreatitis, encephalitis, hepatitis, splenitis and conjunctivitis *Figure 11.1*.[28,41,62,65,68,87] Enteritis alone or in combination with hepatitis has been the dominate change associated with adenovirus infections in cockatoos.[32]

An adenovirus outbreak in a group of psittacine birds was described as having a clinical progression similar to Pacheco's disease virus, in which normal-appearing birds were found dead in their enclosures *Figure 11.2*. This clinical presentation has been reported in an Amazon parrot, Patagonian Conure, Eastern Rosella, Hyacinth Macaw, Lesser Sulphur-crested Cockatoo and in Budgerigars.[7,59,78] Adenovirus caused fatal hepatitis in a 3.5-year-old Cockatiel initially thought to have died from Pacheco's disease virus. Other birds in the aviary, including three other species of psittacine birds, remained clinically normal.[78]

Adenovirus-like intranuclear inclusion bodies were found in the enterocytes of a pionus parrot and *Neophema* sp. with central nervous system signs including chronic torticollis.[48] An adenovirus was demonstrated in inclusion bodies in an adult Yellow-naped Amazon Parrot that died 48 hours after developing anorexia, yellowish-diarrhea and a nasal exu-

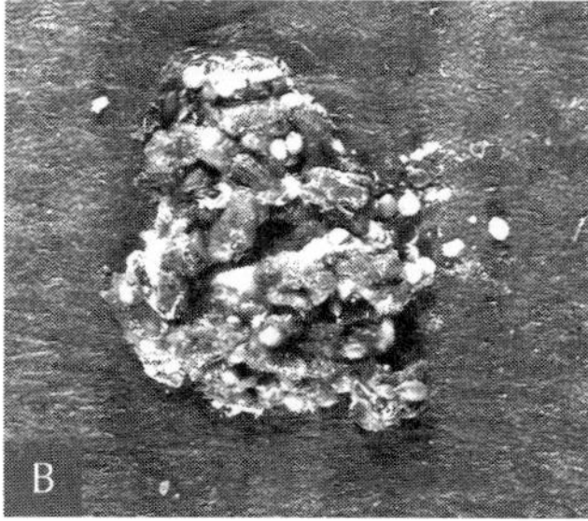
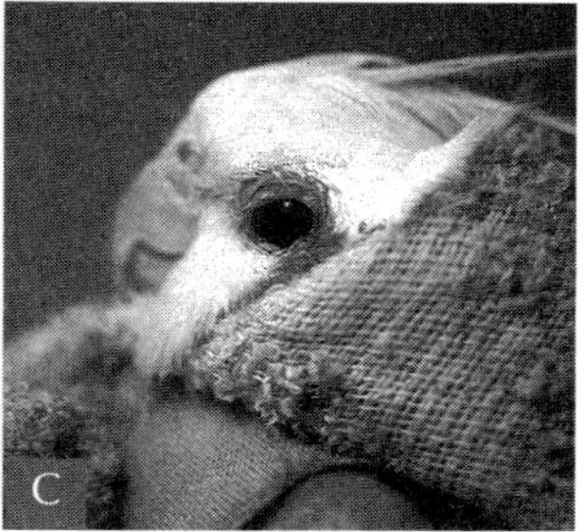

FIG 11.1

In psittacine birds, adenovirus infections have been associated with:
A ataxia or tremors of the head (caused by inflammation of the brain);
B passage of poorly digested food (caused by pancreatitis); and
C conjunctivitis.
photographs courtesy of Louise Bauck

date.[28] An epornitic of adenovirus caused an acute onset of diarrhea and death in all 59 of a group of Eclectus Parrots in Spain. Clinical signs were first noted several days after the birds were moved to a zoological park.[68] This outbreak represents the most severe disease process associated with adenovirus in a group of psittacine birds.

Adenovirus infections have been associated with pancreatitis, renal disease, enteritis and conjunctivitis in lovebirds. However, adenovirus-like inclusions bodies also have been described in asymptomatic lovebirds. In one study, many of the lovebirds with microscopic lesions suggestive of adenovirus had concomitant chlamydiosis, candidiasis, aspergillosis or microsporidiosis.[49] An adenovirus was recovered from the inflamed pancreas of Peach-faced Lovebirds that were depressed and had loose droppings containing poorly digested food.[32] Depression, weight loss and chemosis were the salient features described in an adenovirus out-

In one adenovirus outbreak in a group of psittacine birds, the clinical progression through the flock was described as "Pacheco's disease-like" in which normal-appearing birds were found comatose or dead in their enclosures.

break in White-masked Lovebirds. One-third of the flock of 30 lovebirds died.[41]

Some researchers feel that adenovirus causes a specific, clinical change in Budgerigars, while others feel it is rarely associated with disease.[59,62] Adenovirus has been isolated from Budgerigars that suddenly die following a brief period of enteritis.[59] While enteritis is the most common clinical change in affected Budgerigars, adenovirus was recovered from a flock of Budgerigars in Germany experiencing central nervous system signs including torticollis, opisthotonos, tremors and convulsions *Figure 11.3*. Most of the affected Budgerigars were one to three years of age and some survived. Experimentally infected Budgerigars remained clinically normal, but developed characteristic lesions that could be demonstrated by microscopic examination of tissues.[27]

FIG 11.3

Diarrhea is the most common clinical change associated with an adenovirus infection in Budgerigars, although some infected birds have been reported to develop central nervous system signs.

EPIZOOTIOLOGY

Aviadenoviruses have a worldwide distribution, and outbreaks have been described in companion or aviary birds on most continents.[8,11,24,51,56,92] Because of their widespread distribution and prevalence, it is likely that new isolates of aviadenoviruses will be described.

During a three-year study, 2% of the birds in a Canadian quarantine station were found to have lesions consistent with an adenovirus infection.[73] In another study in Europe, adenovirus was recovered in cell culture from 2 of 38 (5%) psittacines, 14 of 89 (16%) chickens, 12 of 72 (17%) pheasants and 6 of 41 (15%) turkeys. The psittacine birds from which an adenovirus was recovered died acutely. During the same 1-year study, adenovirus was not recovered from 23 pigeons, 11 passerines, 45 ducks, 10 geese, 4 swans or 8 partridges.[31] Inclusion bodies suggestive of adenovirus were detected in 12 of 200 (6%) psittacine birds (eight Budgerigars, three rosellas and one Redrumped Parakeet) that died within two weeks of arriving at a Japanese import station.[62] In one study, 170 of 293 (59%) Budgerigars obtained from a pet dealer had inclusion bodies suggestive of adenovirus.[63] In another study, inclusion bodies containing adenovirus-like particles were detected in the liver of 2 of 1,426 (0.14%) psittacine birds.[26]

In a mixed species aviary containing over 100 birds, five psittacine birds died acutely, while shore birds and waterfowl to which the affected birds were exposed remained clinically normal. The source of the virus in this outbreak could not be determined.[7] In pigeons, adenovirus infections are most common in the late winter and early spring, with only sporadic outbreaks being described during the remainder of the

AVIAN VIRUSES: FUNCTION AND CONTROL

year.[84] A seasonal variation in disease has not been reported in other companion or aviary birds.

<u>RELATIONSHIP OF VIRUS STRAINS</u> — The serologic relationship of the adenoviruses recovered from companion birds remains poorly defined. The adenoviruses that cause hemorrhagic enteritis in turkeys and marble spleen disease in pheasants are serologically related to each other, but not to other aviadenoviruses.[18] Chickens, turkeys, geese and ducks have been shown to be infected with species-specific adenoviruses; however, the adenovirus that infects chickens may also infect pigeons and quail. Some adenoviruses recovered from Budgerigars, pigeons and ducks were found to be serologically similar to those found in gallinaceous birds.[58,82] Nucleic acid analysis indicated that an adenovirus strain isolated from Budgerigars was unique, even though this virus shared antigenic characteristics with strains recovered from chickens.[27]

<u>AGE SUSCEPTIBILITY</u> — The susceptibility of some avian species to adenovirus infections appears to depend on the age of the host. Psittacine birds of all ages appear to be susceptible. Pheasants are most commonly affected from three to eight months of age, with birds less than three months of age experiencing a 20% level of morbidity and mortality. By six months of age, only 12% of exposed pheasants are likely to develop clinical infections.[42,83,84,92] Pigeons from hatching to three years of age were found to be susceptible to disease; infected birds developed clinical signs including depression, dyspnea and diarrhea.[13,30] Adenovirus could be isolated from the feces of goslings from three weeks to one year of age, with the majority of virus recovery occurring in birds that were one to two months old.[8,11,15]

<u>INCUBATION PERIOD</u> — In chickens, the incubation period following a natural infection is considered to be 24 to 48 hours. Even though this virus has a relatively short incubation period, the rate of horizontal spread through a flock can be slow.[57] The incubation period in companion and aviary birds has not been reported.

<u>TRANSMISSION</u> — Some strains of adenovirus are transmitted by both horizontal and vertical routes; others are transmitted only horizontally.[18,89] Experimental studies in chickens indicate that the virus replicates in the intestinal tract following ingestion. Virus is then shed in the feces (up to 20 days) or respiratory secretions (up to 4 weeks).[12,44] Viral particles are commonly identified in hepatocytes, pancreatic cells and enterocytes of infected birds, and virus is shed principally in the feces.[28,35,48,68] Virus shedding in oral or respiratory secretions is considered to be of minimal importance when compared to fecal shedding. Because the virus is relatively durable outside the host, environmental persistence of the virus is common, and transmission can occur through indirect contact with objects contaminated with feces or respiratory secretions from infected birds. Latently infected birds can intermittently shed

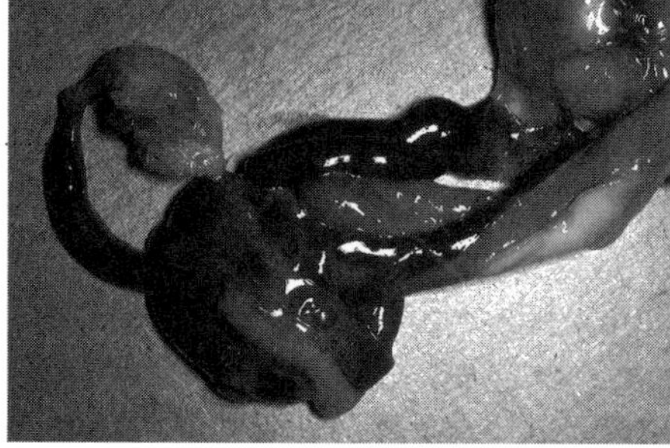

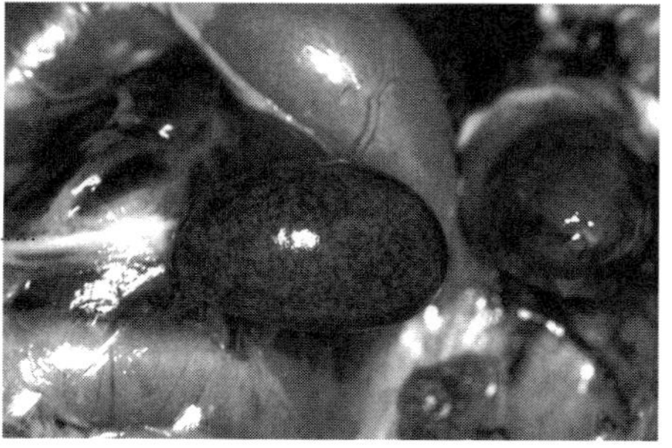

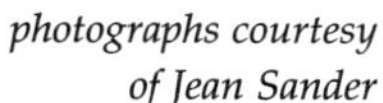

virus, further complicating efforts to reduce virus transmission in a flock.

Vertical transmission has been demonstrated in chickens, but its occurrence appears to depend on the condition of the hen. Reduced hatchability is common in eggs infected while they are developing in the hen.[27] Vertical infections are controlled by maternally derived antibodies. Virus shedding in the feces increases as maternally derived antibodies decrease.[58]

PATHOLOGY

Gross lesions that have been associated with adenovirus in various birds include hepatomegaly, splenomegaly, dilatation of the duodenum and proventriculus, swollen kidneys, and edema, congestion and hemorrhage of the lungs. Focal areas of grayish discoloration in an enlarged, congested spleen are common. Dilated intestinal loops often contain blood *Figure 11.4*. Enlarged friable livers may be hemorrhagic, pale or mottled. Liver lesions are the most consistent gross change in infected psittacine birds.[7,28,67,68,87]

Necropsy findings in an infected Amazon parrot included serous rhinitis and an enlarged, yellowish liver.[28] Necropsy findings in Eclectus Parrots that died following a brief period of diarrhea included severe weight loss and a friable liver with whitish surface mottling and subcapsular hemorrhages. Congestion of the intestines was noted in some affected Eclectus Parrots.[68] Adenovirus-induced hepatitis was described in a Green-cheeked Amazon Parrot, Patagonian Conure, Eastern Rosella, Hyacinth Macaw and Sulphur-crested Cockatoo maintained in a zoological park.[7] Adenovirus-associated pancreatic necrosis was reported in a group of Peach-faced Lovebirds.[87]

Basophilic intranuclear inclusion bodies occur in many affected birds, but may be difficult to detect in birds with severe liver necrosis. Inclusion bodies suggestive of adenovirus infections have been identified from various hosts in hepatocytes, enterocytes, the cells that line the inside of the kidneys and cells in the proventriculus, spleen, lung, liver, intestines, thyroid gland, bursa and bone marrow *Figure 11.5*. Inclusion bodies are routinely seen in association with necrosis in the liver, spleen and pancreas.[7,11,13,24,41,62,63,83,87]

In Budgerigars with adenoviral-induced encephalitis, inclusion bodies have been documented in association with changes in the brain. Inclusion bodies were detected in the cells lining the kidneys in 12 of 200 (6%) psittacine birds that died within two weeks of arriving at a Japanese import station. Inclusion bodies were noted in the liver of only a few birds.[62] Adenovirus-like particles were detected in the pancreas and enlarged nucleus of enterocytes of psittacines.[32]

Adenovirus-like inclusion bodies have been described in asymptomatic love-birds and in birds with necrosis of the cells lining the kidneys.[49] Adenovirus-like particles were demonstrated by electron microscopy in the kidneys and conjunctiva of a group of lovebirds that died with enteritis and conjunctivitis.[41] Adenovirus-like particles were described in the liver and enterocytes of Eclectus Parrots with diarrhea.[68] Adenovirus-like intranuclear inclusion bodies were found in enterocytes of pionus parrots and *Neophema* sp. with central nervous system signs including chronic torticollis.[48] Adenovirus-like particles were demonstrated by electron microscopy in the enlarged necrotic liver of a Cockatiel.[78]

PATHOGENESIS

In mammals, many adenoviruses cause asymptomatic infections that persist even though neutralizing antibodies develop to the virus. Disease is rare in individuals with a functional immune system; however, severe pathologic

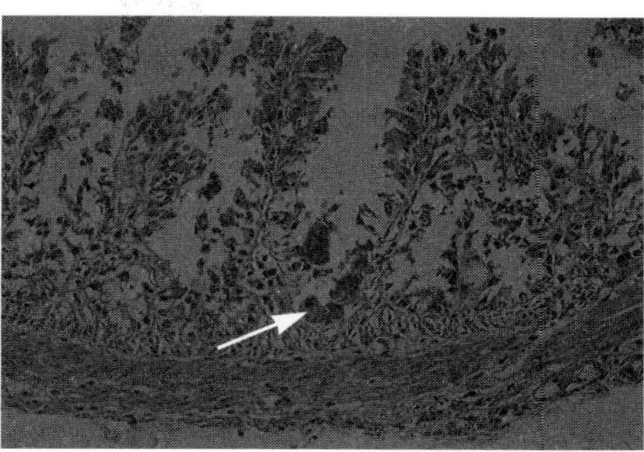

Adenovirus inclusion bodies (arrow) are commonly located in the enterocytes of infected birds. The new virus particles that are produced in these cells are released into the lumen of the intestinal tract and pass out of an infected bird with the feces.
photograph courtesy of S.S. Tsai

changes can occur if the host is immunocompromised. In birds, adenoviruses appear to vary widely in their ability to cause morbidity or mortality. Some strains are mildly pathogenic and primarily cause persistent infections. Others are considered sufficiently virulent to induce disease in an otherwise healthy bird. Generally, adenoviruses are considered to be opportunistic pathogens. Affected birds often have concomitant chlamydial, fungal, bacterial, parasitic or other viral infections, suggesting that immunosuppression is occurring in the affected bird.[49,62,88] In one study, 6 of 170 (3.5%) Budgerigars with inclusion bodies suggestive of adenovirus also had inclusion bodies suggestive of polyomavirus.[63] Because adenoviruses are frequently recovered from birds that die from other causes, the role, if any, that adenoviruses may play in a disease process is difficult to define.

Aviadenoviruses are thought to be more virulent in non-host adapted species than they are in their typical host. In several outbreaks of adenovirus involving psittacine birds, nonpsittacine species in the same facility remained unaffected.[7,27]

Most aviadenoviruses replicate primarily in the gastrointestinal and upper respiratory tracts.[56] Viremia results in

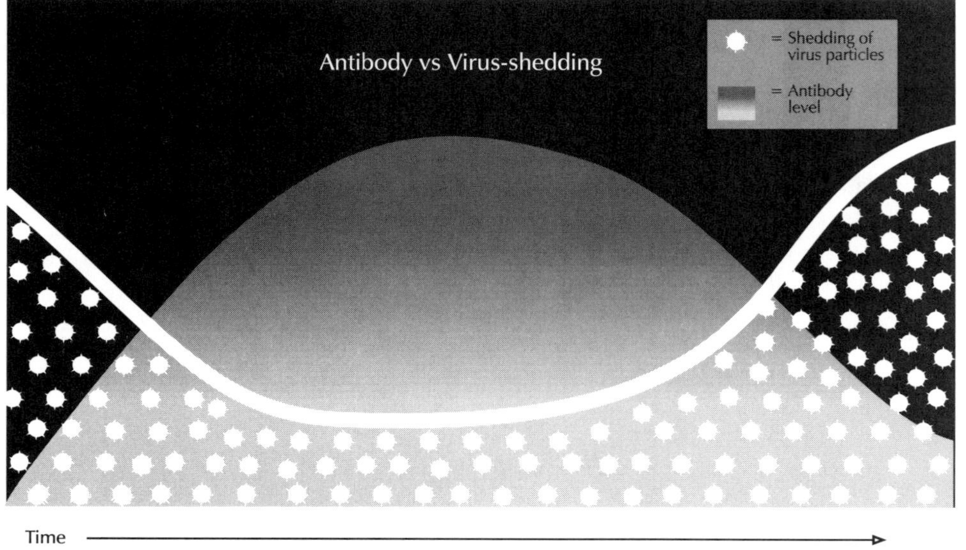

Adenovirus infections appear to be cyclic. High concentrations of antibodies are associated with low levels of virus shedding, and as the antibody levels decrease, there is a subsequent increase in the level of virus shedding.

virus spread to most organs.[12,44] In some cases, adenovirus-associated damage has been noted in the bursa and other lymphoid organs, suggesting that the virus might induce immuno-suppression.[81] Latent infections, with virus shedding occurring during stressful periods, are common in gallinaceous birds.

IMMUNITY

Adenovirus-neutralizing antibodies can be detected in experimentally infected chickens for one to two weeks after infection.[55] In poultry, adenoviruses are commonly recovered from young asymptomatic birds, and surveys for antibodies would suggest that most adult birds have been exposed to numerous strains of adenovirus.[56]

The activation of latent adenovirus infections appears to be cyclic. High concentrations of antibodies are associated with low levels of viral shedding, and as the antibody concentrations decrease, there is a subsequent increase in the level of viral shedding *Figure 11.6.*[27,58] Maternal antibodies may be passed to the developing embryo through the yolk.[27] Chicks from hens with antibodies are refractory to severe disease for up to three weeks after hatching, but become susceptible after four weeks, suggesting a decay in the protective maternally derived antibodies.[23] Chicks with maternally derived antibodies remain susceptible to infection; however, mortality in chicks with maternal antibodies is low compared to chicks without maternal antibodies.[33,55]

DIAGNOSIS

Adenovirus, or antibodies to the virus, can be detected using assays that include agar-gel diffusion, fluorescent antibody or enzyme-linked immunosorbent assay (ELISA).[38,40] As would be expected, the ELISA is more sensitive than an agar-gel diffusion test.[85] Antibodies may also be detected with virus-neutralization assays. Some of the adenoviruses will hemagglutinate various types of red blood cells, and the hemagglutination-inhibition (HI) test provides a rapid method to detect antibodies to these adenoviruses. Some of these tests (agar-gel diffusion, ELISA) can be used to detect any of the adenoviruses; others (virus-neutralization, HI) can be used to detect antibodies to specific strains. It should be noted that adenovirus is frequently demonstrated in asymptomatic birds and the demonstration of an antibody titer to adenovirus may not indicate that this virus is involved in a recognized disease process.

Intranuclear inclusion bodies suggestive of an adenovirus infection are routinely seen in association with necrosis of cells in the liver, spleen or pancreas of infected birds. It should be noted, however, that inclusion bodies may also be demonstrated in asymptomatic birds.[58,62] During one survey of Budgerigars, 170 of 293 (58%) birds examined were found to have suggestive inclusion bodies, but none of these birds showed any specific lesions or disease that could be associated with these inclusion bodies.[63] Adenovirus inclusion bodies were detected in the tissues of 47 of 737 (6%) pigeons.[84]

Depending on the host and the infected tissues, adenovirus can be recovered in cell culture from the feces, pharyngeal secretions, liver (if inclusion body hepatitis is suspected) or kidney. Electron microscopy can be used to detect the virus in feces or pharyngeal secretions, as long as the concentration of virus is sufficient (greater than one million virus particles per ml of sample). Latently infected birds will shed the virus intermittently, making detection of the virus in feces or pharyngeal secretions difficult. Birds with antibodies to adenovirus should be considered to be latently infected.

A viral-specific DNA probe has been developed that can detect adenovirus nucleic acid in the tissues of infected chickens, quail, pigeons and psittacine birds. Using this probe, adenovirus nucleic acid was demonstrated in the liver of pigeons, cockatoos and an African Grey Parrot, and in the small intestines of a cockatoo. In psittacine birds, adenovirus and polyomavirus can cause microscopic changes that appear similar, including enlargement of the nucleus and formation of intranuclear inclusion bodies. A viral-specific DNA probe can be used to distinguish intranuclear inclusion bodies induced by adenovirus from those caused by polyomavirus, herpesvirus or PBFD virus. [67]

Other conditions that can cause problems similar to those induced by adenovirus include chlamydia, paramyxovirus, influenza A virus, bacterial enteritis, bacterial hepatitis, bacterial encephalitis, herpesvirus and polyomavirus.

CONTROL

Adenoviruses are resistant to inactivation when outside a host, and can remain infectious for long periods in litter, food, water or contaminated feces. Adenoviruses are resistant to extremes of heat, organic solvents and extremes in pH (pH 3 and pH 9). The precise heat susceptibility varies with the strain of virus. Some isolates are resistant to 56°C for 20 minutes, while others can withstand this temperature for 22 hours. Some strains are inactivated following 30 minutes of exposure to 60°C or one hour of exposure to 50°C. Many commonly used disinfectants are ineffective.[56] Adenoviruses can be inactivated by treatment for more than one hour with formalin, aldehydes or iodophors.[27] Free-ranging pigeons and waterfowl may serve as a source of virus for birds in an aviary or zoological park and should be prevented from coming in contact with susceptible avian populations.[82]

There is no specific therapy for most adenovirus infections. Supportive care that includes maintaining adequate hydration, assisted feeding and broad-spectrum antibiotics to prevent secondary infections may reduce the level of mortality in affected birds. A vaccine has not been developed for the adenoviruses that infect psittacine birds. The vaccines that are available for other birds are discussed under their respective species.

PHEASANTS - MARBLE SPLEEN DISEASE

CLINICAL FEATURES

Marble spleen disease in pheasants occurs worldwide and is one of the most important adenovirus-associated problems in companion, aviary and free-ranging birds. The adenovirus that causes this disease is serologically related to the turkey hemorrhagic enteritis virus and to the chicken splenomegaly virus, but is serologically unrelated to other aviadenoviruses *Table 11.4*.[18,24]

Marble spleen disease was first reported in Italy in the mid 1960s.[51] Subsequent outbreaks have been documented in the United States, Canada, England, Spain, Germany, Hungary, Bulgaria, Czechoslovakia and Australia.[24,83] Interestingly, marble spleen disease has been reported only in captive pheasants, and neither the virus nor antibodies to the virus were demonstrated in 618 free-ranging pheasants representing 42 species from three areas of the United States.[17] The factors that govern the demonstrated difference in susceptibility between captive and free-ranging pheasants are unknown.

Infected pheasants may die acutely with no clinical signs or following a brief period of anorexia and mild depression that may be accompanied with dyspnea, diarrhea and vomiting.[11,83] The clinical course of the disease is ten days to several weeks. Marble spleen disease is principally associated with acute death secondary to pulmonary edema in three- to eight-month-old pheasants, with birds less than three months of age experiencing a 20% level of morbidity and mortality. By six months of age, only 12% of exposed birds are likely to develop clinical infections.[11,40,83,92] In one outbreak, adenovirus was recovered from 12 of 72 (17%) pheasants found dead in their enclosures. Infected birds ranged in age from ten weeks to adults.[31] In an outbreak in Australia, 20 of 100 pheasants that ranged from three to six months old were ill, and 12% died.[83] Screening for antibodies following outbreaks of adenovirus in pheasant flocks indicates that there is a much higher level of infection than is clinically recognized.[11,42]

Marble spleen disease virus is transmitted principally through the ingestion of contaminated feces, and thus well designed sanitation programs can be used to prevent the spread of the virus through a flock. The virus can remain infectious in contaminated litter for months. This adenovirus has not been demonstrated to be egg-transmitted, and insect vectors are not known to be involved in virus transmission, even though viremia occurs.

The spleen of affected birds is generally three times normal size and is mottled with gray necrotic areas that are high-

Comparison of turkey hemorrhagic enteritis (THE) virus and marble spleen disease (MSD) virus

VIRUS	THE	MSD
GEOGRAPHIC DISTRIBUTION	Europe, North America, Asia and Australia	Europe, North America and Australia
AGE DISTRIBUTION	Most common in 4 to 12 week old birds	Most common in birds 3 to 8 months old
ANTIBODIES	Common in clinically normal turkeys	Common in clinically normal pheasants
TRANSMISSION	Primarily, fecal-oral spread	Primarily, fecal-oral spread
CLINICAL CHANGES	Depression, diarrhea and blood in stool	Acute death following depression or dyspnea
VACCINE	Yes	Yes

lighted by red congested areas *see Figure 11.4*. The lungs are usually edematous. Intranuclear inclusion bodies are commonly found within swollen cells in the spleen, and may be detected also in the liver, lungs, bursa, bone marrow or proventriculus.[11,92]

The adenovirus that causes marble spleen disease replicates in the spleen and can be demonstrated in this tissue using an antigen detection version of an agar-gel diffusion assay. A different version of this assay can be used to demonstrate antibodies to the virus as soon as two weeks after infection.[18,24,40]

Pheasant chicks experimentally inoculated with turkey hemorrhagic enteritis virus developed splenomegaly, seroconverted and remained clinically normal.[19] In pheasants, oral administration of turkey hemorrhagic enteritis virus or an avirulent strain of marble spleen disease virus has been effective in reducing morbidity and mortality.[19,22,24] Based on these findings, a vaccine has been developed.

While the instructions provided by the manufacturer of any vaccine should be carefully followed, in general, the most effective protection from marble spleen disease can be accomplished by adding the attenuated-live virus vaccine to the water of chicks at four to six weeks of age.[19,22] Only clinically healthy pheasants should be vaccinated. Routine vaccination can be used to prevent epornitics.[22] Turkey hemorrhagic enteritis virus antibodies from the serum of previously infected birds will protect susceptible turkeys from disease. However, it has not been determined whether the same antiserum will protect pheasants from the virus that causes marble spleen disease.

ADENOVIRUSES IN TURKEYS

An adenovirus that is related to the marble spleen disease virus has been shown to cause hemorrhagic enteritis in turkeys. The virus can infect turkeys, chickens and pheasants, and has been demonstrated in domestic turkeys from most continents. Reports documenting the presence of this virus in free-ranging turkeys could not be found. Some strains of turkey hemorrhagic enteritis virus are extremely pathogenic and may cause mortality levels that reach 60% of the at-risk population over a five- to ten-day period.[53] Other strains cause asymptomatic infections. This virus is shed principally in the feces and has not been shown to be egg-transmitted. Infections can be prevented through vaccination.[18]

ADENOVIRUSES IN CHICKENS

Chickens are susceptible to a number of strains of adenovirus, some of which may also infect pigeons and quail. Most strains cause subclinical infections, but a few can cause problems that include hepatitis, splenitis and abnormal egg production. Some affected chickens may develop anemia (as low as 10% to 18%) and exhibit bone marrow suppression. Virulent strains of adenovirus have been shown to replicate in the lymphoid tissue, inhibiting the ability of infected chicks to respond to foreign proteins.[76] The production of soft-shelled eggs by otherwise healthy hens is considered highly suggestive of an adenovirus infection.

The strain of adenovirus that causes inclusion body hepatitis in chickens primarily affects chicks that are five to seven weeks old. Suggestive clinical changes include depression, anorexia, white pasty droppings and prostration prior to death *Figure 11.7*. Mortality in these birds can range from 10% to 30% over a brief period of time (less than a week). Mortality rates are highest in birds with concomitant infections. In experimentally infected chicks, mortality peaked on day 3 and remained high through day 5. The

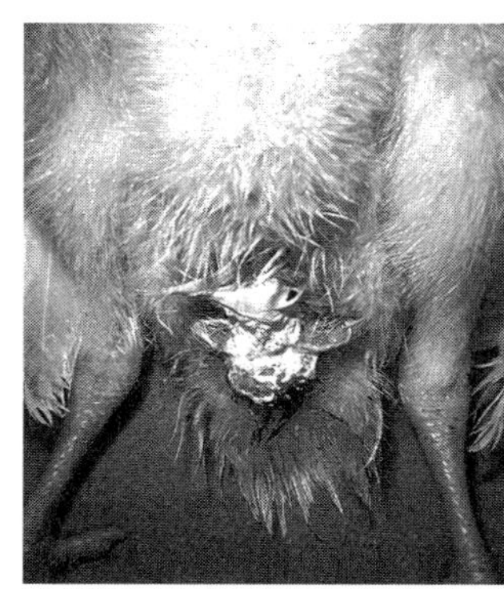

FIG 11.7

Young gallinaceous birds infected with adenovirus may develop white pasty droppings that adhere to the feathers around the cloaca. Bacterial enteritis and other viruses that affect the intestinal tract can cause a similar clinical change.

photograph courtesy of Jean Sander

majority of clinical signs and mortality stopped by day 7 after infection.[76]

Adenoviruses in chickens are transmitted by both vertical and horizontal routes. In experimentally infected chickens, adenovirus was detected in the duodenum, ileum, cecum, cecal tonsils and feces for up to 20 weeks after infection. Virus was detected in these areas even though neutralizing antibodies were present.[75]

In one study using an antigen-capture ELISA, adenovirus was detected in the yolk of 16 of 60 (27%) and in the albumen of 5 of 60 (8%) eggs collected from infected broiler chickens. This finding supports the theory that some adenoviruses can be transmitted to the developing egg. Because this ELISA detects a group-specific antigen, it may be useful for diagnosing the presence of other strains of adenovirus. Adenovirus was detected in the eggs from hens that had neutralizing antibody titers that ranged from 1:320 to 1:1280.[75]

ADENOVIRUSES IN QUAIL

Adenoviruses have been shown to cause bronchitis or hepatitis in captive and free-ranging Bobwhite Quail.[39,46] An adenovirus that was similar to a serotype found in chickens was recovered from Japanese Quail with central nervous system signs.[47] Some adenoviruses that infect chickens also may infect quail.

Quail bronchitis virus is serologically related to an adenovirus (CELO) that infects chickens; however, the nucleic acid composition of each of these viruses appears to be unique.[47] Quail bronchitis virus is highly infectious to young, susceptible quail. Clinical changes suggestive of infection include acute death with no premonitory signs or coughing, sneezing, wheezing, dyspnea, wing droop, conjunctivitis or ocular discharge *Figure 11.8*. The most severe clinical changes occur in the youngest birds (one to three weeks of age). An age-related resistance to disease develops; mortality rates (which may reach 90% to 100%) are highest in birds less than six weeks of age. Many infected adults remain asymptomatic and seroconvert. The incubation period in quail is two to four days, and the disease moves rapidly through a susceptible flock in ten days to two weeks.

The virus is spread through direct and indirect contact with contaminated respiratory aerosols. At necropsy, gross lesions may be absent or can include cloudy, thickened air sacs, fluid in the lungs and air sacs, and accumulation of mucus and debris in the sinuses, trachea and bronchi *Figure 11.9*. Intranuclear inclusion bodies may be noted in the cells lining the trachea and bronchi within two to five days after infection.[91] Virus can be recovered for diagnostic purposes from the lung, trachea, air sacs and fluid in the eye.[90]

Acute necrotizing hepatitis occurred in a group of one- to three-week-old Gambel's Quail, which are indigenous to the Southwestern United States and Northern Mexico. Of 200 orphaned quail, 60 (30%) died. In other outbreaks, 14 of 15 (93%) and 50 of 100 (50%) young quail died. Most quail less than one week of age died without showing clinical

signs, while birds that were two to three weeks old died after a one- to two-day course of anorexia, depression and passage of greenish diarrhea. Affected quail had randomly scattered, pinpoint white foci of necrosis in the liver.[39]

ADENOVIRUSES IN GUINEA FOWL AND GROUSE

An adenovirus has been associated with necrotic pancreatitis and respiratory disease in young guinea fowl. Older guinea fowl appear to be relatively resistant to natural infection, but can be infected by intramuscular inoculation.[69,94] Adenovirus is suspected as the cause of a disease in guinea fowl and Blue Grouse that is similar to marble spleen disease. Affected guinea fowl die from pulmonary congestion and have enlarged, necrotic spleens.

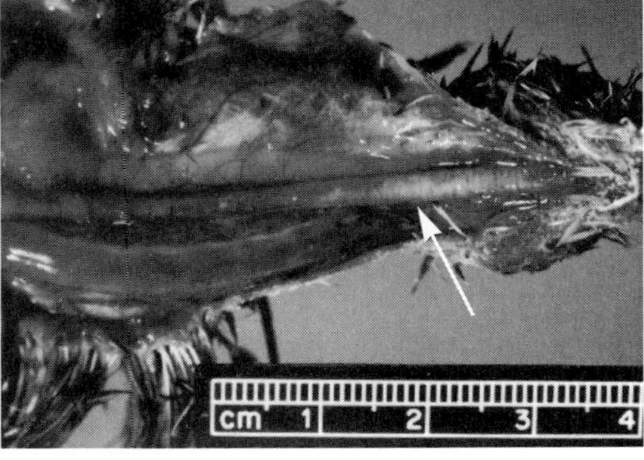

Guinea fowl experimentally infected with turkey hemorrhagic enteritis virus develop clinical and gross lesions similar to those seen with natural infections.[14] Affected grouse develop depression, diarrhea and die. It has not been confirmed that adenovirus is involved in the reported disease

FIG 11.8
Quail bronchitis virus is highly infectious to young, susceptible quail. Many infected chicks die with no premonitory signs; others may develop upper respiratory signs.
photograph courtesy of S.W. Jack

FIG 11.9
Quail that die from an adenovirus infection may have an accumulation of yellowish-discolored mucus and necrotic debris in the trachea (arrow).
photograph courtesy of S.W. Jack

Ostriches naturally or experimentally infected with adenovirus develop severe depression (hunched stance) and mucoid diarrhea.

Pasty excrement may adhere to the feathers around the cloaca.

photographs courtesy of Amy Raines

process.[35] Guinea fowl died following experimental inoculation with an adenovirus recovered from an ostrich.[10] Antibodies to adenovirus were demonstrated in a group of guinea fowl that were producing soft-shelled eggs. It was subsequently demonstrated that guinea fowl are susceptible to experimental infection with the adenovirus that causes egg drop syndrome.[27]

ADENOVIRUS IN OSTRICHES

Adenovirus was recovered from the feces or livers of ostriches ranging from dead-in-shell to three years of age. Affected birds were stunted and had reduced weight gains along with clinical signs of hepatitis and enteritis *Figure 11.10*. Virus was isolated also from the trachea of several young ostriches, and from the yolk sac of dead-in-shell chicks. Some two- to three-month-old ostrich chicks died two to three days after developing diarrhea. Survivors were stunted. In some facilities, up to 95% of ostrich chicks died. *Klebsiella,*

Pseudomonas, E. coli and *Mycoplasma* were also isolated from some affected ostriches, and the role that adenovirus played in the disease process was undetermined.[66] Gross necropsy findings varied with the bird. Common findings included dilatation of the intestinal tract with fluid and gas, enlargement of a mottled liver, impaction of the proventriculus, and pulmonary edema and congestion *Figure 11.11*. No inclusion bodies were found in any ostrich tissues.[66] In other species, adenovirus-induced disease is common as maternal antibodies wane.

Adenovirus was recovered from the pancreas, kidney and lung of a four-month-old ostrich in Italy. The bird was asymptomatic prior to death, except for a brief period of hemorrhage from the mouth. The isolated virus caused the death of 6 of 15 (40%) experimentally infected, young guinea fowl. The affected guinea fowl developed diarrhea as soon as three days after infection, followed by conjunctivitis and pasty eyes on day 5, and death from 7 to 11 days after infection. Gross changes included a congested, enlarged, nodular pancreas, hemorrhage in the abdominal cavity and in the trachea, and edematous lungs. Microscopic changes included hemorrhagic enteritis and pancreatitis. Surviving guinea fowl exhibited microscopic changes suggestive of chronic pancreatitis.[10]

ADENOVIRUS IN PIGEONS

Adenovirus infections have been confirmed in pigeons based on virus isolation and characteristic histologic lesions.[29,30,59] This virus was first reported in pigeons from Northern Ireland in 1976.[59] Subsequently, adenoviruses were demonstrated in pigeons in Japan, Belgium, North America, Hungary and Australia.[13,29,30,45]

Adenovirus has been recovered from diseased pigeons from hatching to five years of age. Infected birds develop clinical signs associated with hepatitis or enteritis. These include depression, anorexia, dyspnea, a crouched stance, polydipsia, polyuria and slimy green diarrhea.[4,13,21,30] The most commonly reported clinical change in young pigeons is diarrhea.[21]

Infections are most common in two- to four-month-old pigeons, although adults may also develop hepatitis. Infected young birds may die within 48 hours of developing clinical signs. Mortality rates can range from 0% to 60%; however, many affected pigeons recover.[13,21,45] Mortality is highest three to four days after infection, but occasional deaths may occur for two to four weeks after the initial case is noted.[21] In some cases, adenovirus has been isolated from pigeons with no other documentable infectious agents. In other cases, the virus is recovered from birds with concomitant infections, including chlamydiosis, pigeon herpesvirus and various bacteria.[13,21,59]

During an adenovirus outbreak on a Japanese commercial pigeon farm, 30 of 200 (15%) birds developed clinical signs that included depression, difficulty breathing and loose, green diarrhea. Of the 30 clinically ill pigeons, 3 died.[30] Over a five-month period in Europe, adenovirus hepatitis was documented in 21 of

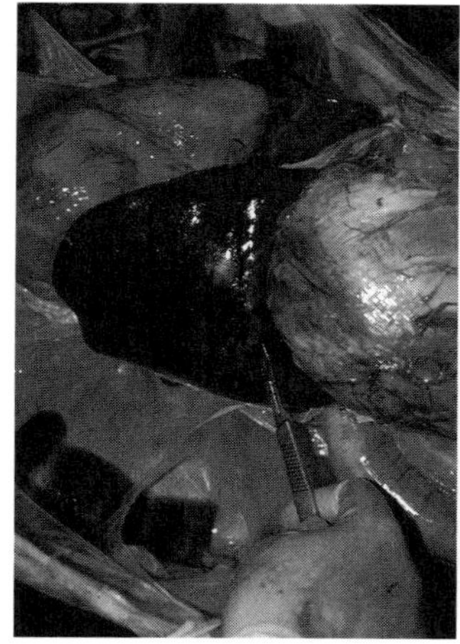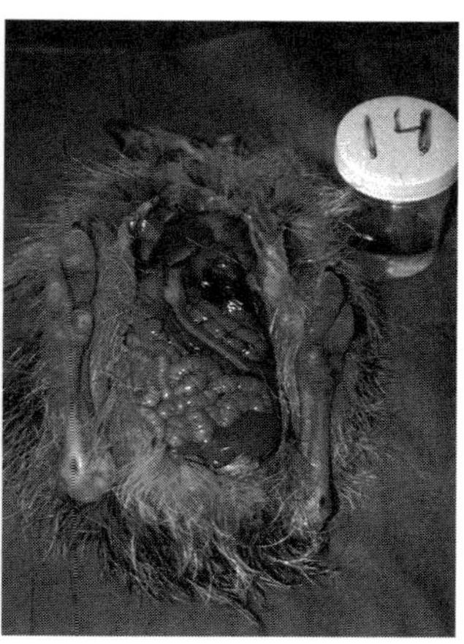

FIG 11.11
Common necropsy findings in ostriches with adenovirus infections include: enlargement and mottling of the liver and **right** *dilatation of the intestinal tract with fluid and gas.*
photographs courtesy of Amy Raines

57 (37%) pigeon flocks from which tissues had been submitted for evaluation. Affected pigeons were anorectic, crouched and had mucoid-greenish stools, polydipsia, polyuria, vomiting and respiratory distress prior to death.[20] In an outbreak of adenovirus in Australia, 10 of 22 (45%) one-year-old racing pigeons began to lose weight and were lethargic 2 weeks after a race. All but three of the affected birds recovered.[45]

In another 23-month study in Europe, adenovirus was detected in 47 of 737 (6%) racing pigeons submitted for necropsy. Disease was most common in the late winter and spring, particularly in June, with only sporadic outbreaks occurring in pigeon flocks during the remainder of the year.[84] Of 75 racing pigeons presented for evaluation, only two (2.7%) had detectable levels of adenovirus antibodies.[86] In a one-year study in Europe, adenovirus was not recovered in cell culture from 23 pigeons.[31] In another study in Europe, inclusion bodies containing adenovirus-like particles were detected in the liver of 5 of 226 (2%) pigeons.[26] Adenovirus was demonstrated in the necrotic liver of a free-ranging pigeon that was found dead in a waterfowl exhibit of a zoological park in Ohio.[74]

Latently infected chickens periodically shed virus in their feces. It is safe to assume that surviving pigeons are also latently infected. Adenovirus particles have been demonstrated by electron microscopy in the feces of infected pigeons.[13] Because adenovirus is shed in high titers in feces, racing and feral pigeons can serve as source of virus for captive groups.[82]

Some adenoviruses isolated from pigeons have been shown to be serologically related to those recovered from chickens; other strains appear to be serologically distinct.[21,58,82] The existence of several strains of adenovirus that can infect pigeons may explain the variance in reported clinical changes. Pigeons are susceptible to some adenoviruses recovered from chickens, and vice versa, raising concerns of cross-species transmission.[52] For example, an adenovirus recovered from a pigeon caused a fatal pancreatitis when it was injected subcutaneously into one-day-old chickens. Affected chickens begin dying five days after infection. Free-ranging pigeons could serve as a source of virus for captive pigeons or susceptible gallinaceous species.[82]

Enlargement of the liver and spleen are the most characteristic necropsy findings. Other gross changes may include yellowish-to-copper-gold, patchy discoloration of the liver and mild reddening of the lining of the intestinal tract. Some affected birds may have greenish fluid in the crop and greenish mucus in an otherwise empty intestinal tract.

One pigeon with unusual lesions exhibited swelling of the mucosa in the respiratory tract, hemorrhagic enteritis and hemorrhage of the liver; this bird had an isolate that could not be serologically classified.[4,13,21,30]

Intranuclear inclusion bodies suggestive of an adenovirus infection are most common in the liver and intestines; however, they have also been demonstrated in the pancreas, kidney, proventriculus and cecum.[21,29,30,45,82] In one group of young pigeons that died following a brief period of diarrhea, suggestive inclusion bodies were detected only in the enterocytes. Adenovirus can cause gross and microscopic lesions that appear similar to those induced by pigeon herpesvirus. In psittacines, a viral-specific DNA probe can be used to differentiate between inclusion bodies caused by adenovirus and herpesvirus.[67]

ADENOVIRUS IN RAPTORS

Adenoviruses have been recovered from several raptors including a Goshawk, Merlin and American Kestrels with varying clinical signs. The virus was recovered from the brain of a free-ranging Goshawk with central nervous system signs. This bird died shortly after being recovered from the wild.[27] Adenovirus-like lesions were reported in a group of nine captive juvenile and adult American Kestrels with hemorrhagic enteritis. Affected kestrels developed melena and anemia prior to death. Gross necropsy findings included splenomegaly and accumulation of blood in the colon and cloaca. Adenovirus-like particles were demonstrated by electron microscopy within intranuclear inclusion bodies in the liver.[79] Adenovirus was also described in association with hemorrhagic enteritis in a free-ranging raptor in Australia.[70]

An adenovirus was demonstrated in intranuclear inclusion bodies within the liver of an adult, free-ranging female Merlin with hepatitis. This bird was recovered from the east coast of the United States.[77]

ADENOVIRUS IN WATERFOWL

Several serologically distinct adenoviruses have been recovered from captive and free-ranging ducks and geese. In addition, some virus isolates from ducks are serologically similar to those found in gallinaceous birds.[58] The adenoviruses that cause egg drop syndrome in waterfowl and chickens are serologically identical. Some isolates of adenovirus from ducks are infectious for chickens, and it is likely that some strains recovered from chickens would be infectious to ducks.[1] During a one-year study in Europe, adenovirus was not recovered in cell culture from 45 ducks, 10 geese or 4 swans.[31]

Three serotypes of adenoviruses recovered from goslings were shown to be serologically distinct from adenoviruses that have been recovered from fowl, turkeys and ducks. Some isolates have been associated with liver disease, tracheitis and high mortality in captive geese; however, goslings experimentally infected with these isolates remained clinically normal.[71,72,96]

Adenovirus was isolated from a group of four- to eleven-day-old captive goslings that died without premonitory signs or following a brief period of respiratory distress. Egg production and fertility remained unchanged in the flock, and naturally exposed chickens and ducks remained clinically normal. Opaque plugs of fibrin and cellular debris were noted in the trachea. The liver of affected goslings was mottled, and hemorrhage was evident in the abdominal cavity and heart of a few birds. Intranuclear inclusion bodies were evident in the cells lining the trachea.[72] The progression of disease in these goslings was similar to that reported with quail bronchitis and in a group of Muscovy Ducks infected with adenovirus.[5]

In Europe, adenovirus-containing inclusion bodies have been reported in the bursa of geese with Derzsy's disease,[3] and an adenovirus of unknown significance has been isolated from dead geese.[96] Adenovirus was isolated from the feces of goslings from three weeks to one year of age, with the majority of virus recovery occurring in birds that were one to two months old.[16]

Adenovirus was described in a group of captive Muscovy ducklings that began to die acutely at 35 days of age. Affected birds were unwilling to move and were emaciated. The outbreak lasted for about ten days, with a combined mortality of 1% to 1.5% of the at-risk population.[6] In another outbreak, adenovirus-like particles were demonstrated in the cells lining the trachea in a group of two- to three-week-old Muscovy ducklings with signs of upper respiratory disease. Ten percent of the at-risk population was affected.

Necropsy findings included accumulation of necrotic debris on the surface of an ulcerated trachea. Some birds also had bronchitis and pneumonia.[5]

EGG DROP SYNDROME

The adenovirus that causes egg drop syndrome (EDS-76) has been demonstrated to infect a variety of captive and free-ranging birds. Isolates of this virus from ducks will induce experimental infections in chickens. Virus activity has been demonstrated in ducks, coots, grebes, geese, cattle egrets, gulls, owls, storks, swans, chickens and turkeys in Europe, Asia, Australia and North America.[2,9,34,43,50,57] Birds of all ages appear to be susceptible to the EDS-76

FIG 11.12

In ducks and chickens, EDS-76 causes a decrease in egg quality and reduced hatchability. The production of thin-shelled, soft-shelled, shell-less or poorly pigmented eggs is considered highly suggestive of an EDS-76 infection.
photograph courtesy of Richard Davis

virus. Serologic surveys suggest that the virus originated in free-ranging waterfowl, which may serve to disseminate the virus among susceptible captive birds.[1,54]

In ducks and chickens, EDS-76 causes a decrease in egg quality and reduced hatchability. The production of thin-shelled, soft-shelled, shell-less or poorly pigmented eggs is considered highly suggestive of an EDS-76 infection *Figure 11.12*.[56] Other clinical changes associated with an infection include anorexia, lethargy, transient diarrhea and abnormally moist albumin. The experimental incubation period in chickens is 7 to 17 days.[80] In chickens, clinical problems usually resolve within a month from the time that they are originally noticed; however, recovered hens are latently infected. Whether an adenovirus may be causing similar egg-related problems in companion and aviary birds has not been determined. Other viruses that have been associated with decreased egg production in gallinaceous birds include infectious bronchitis virus, Newcastle disease virus and influenza A.[56]

This virus can be transmitted vertically or through direct or indirect contact with contaminated droppings or eggs.[80] Infected birds shed the virus in the excrement for up to two weeks after infection.[36,37] Vertically infected chicks can shed virus immediately after hatching. Contaminated droppings are considered the main route of horizontal transmission, but viral-induced lesions have not been described in the cells lining the intestinal tract.[80] This finding suggests that the virus present in the droppings originates in the reproductive tract and is simply expelled from the cloaca with the excrement. Ponds contaminated with virus excreted by infected free-ranging waterfowl may serve as a source of virus for captive birds *Figure 11.13*. Infected birds are viremic for up to five weeks after infection, creating the possibility for insects to serve as vectors for virus dissemination.

In chickens, the EDS-76 virus replicates in the shell gland from seven to twenty days after infection, causing edema and the accumulation of mucoid exudates. Intranuclear inclusion bodies are transiently present in the oviduct and may or may not be detected, depending on when during the infection cycle the tissue samples were collected.[80] The highest concentration of virus is found in the uterus at the time that abnormally developed shells are noted.[93]

In the latently infected hen, the virus persists in the infundibulum. The virus is generally dormant except during the laying season when the infection becomes active and causes the characteristic clinical changes.[56] Antibody titers increase during periods when the virus is active, and these protect a hen from disease.[54,80]

Detecting the EDS-76 virus is difficult because it is present in the shell gland for only a brief period. Virus isolation is

FIG 11.13

egg drop syndrome virus is excreted primarily in the feces and is relatively durable when outside of the host. Ponds contaminated with virus excreted by infected free-ranging waterfowl may serve as a source of virus for captive birds.

best achieved by taking samples from the cloaca or shell gland when clinical changes are first noted. Antibodies to the virus can be detected as soon as five days after infection using a hemagglutination-inhibition or agar-gel diffusion assay.[60] Hemagglutination-inhibition antibodies can persist for up to six months, and detecting these antibodies indicates recent virus activity in the body.[61] Monitoring a flock for changes in antibody titers is the best way to determine whether active infections are occurring. Surveys are needed to determine whether antibodies to this virus are present in companion birds. Inactivated vaccines are effective in protecting chickens from the disease.[37]

REFERENCES

1. Bartha A, Meszaros J: Experimental infection of laying hens with an adenovirus isolated from ducks showing egg drop syndrome. Acta Vet Hung 33:125-127, 1985.

2. Bartha A, Meszaros J, Tanyi J: Antibodies against EDS-76 avian adenovirus in bird species before 1975. Avian Pathol 11:511-513, 1982.

3. Bergmann V: Pathology of Derzsy's disease of goslings (parvovirus infection) and use of electron microscopy for virus detection from tissue. Arch Exp Veterinärmed 41:212-221, 1987.

4. Bergmann V, Kiupel H: EinschluBkorperchen enteritis bei Tauben, hervorgerufen durch Adeno- und Parvovirus. Archiv Exper Vet Med 36:445-453, 1982.

5. Bergmann VV, Heidrich R, Kinder E: Pathomorphologische und elektronenmikroskopische Feststellung einer Adenovirus-Tracheitis bei Moschusenten (Cairina moschata). Monatsh fur Veterinärmed 40:313-315, 1985.

6. Bouquet JF, et al: Isolation and characterization of an adenovirus isolated from Muscovy Ducks. Avian Pathol 11:301-307, 1982.

7. Bryant WM, Montali RJ: An outbreak of a fatal inclusion body hepatitis in zoo psittacines. Proc Assoc Avian Vet, 1987, p 473.

8. Bygrave AC, Pattison M: Marble spleen disease in pheasants (Phasianus colchicus). 92:534-535, 1973.

9. Calnek BW: Hemagglutination-inhibition antibodies against an adenovirus (virus 127) in white Pekin ducks in the United States. Avian Dis 22:798-801, 1978.

10. Capua I, Gough RE, Scaramozzino P, et al: Isolation of an adenovirus from an ostrich (Struthio camelus) causing pancreatitis in experimentally infected guinea fowl (Numida meleagris). Avian Dis 38:642-646, 1994.

11. Carlson HC, Pettit JR, Hemsley RV, et al: Marble spleen disease of pheasants in Ontario. Can J Comp Med 37:281-286, 1973.

12. Cook JKA: Avian adenovirus alone or followed by infections bronchitis virus in laying hens. J Comp Pathol 82:119-128, 1972.

13. Coussement W, Ducatelle R, Lemahieu P, et al: Pathology of adenovirus infection in pigeons. Vlaams Diergeneesk Tijdschr 53:227-283, 1984.

14. Cowen BW, et al: A case of acute pulmonary edema, splenomegaly and ascites in Guinea fowl. Avian Dis 32:151-156, 1988.

15. Csontos L: Isolation of adenoviruses from geese. Acta Veterinaria Academiae Scientairum Hungaricae 17:217-219, 1967.

16. Csontos L, Csatari MMK: Etiologic studies on gosling influenza: Isolation of virus. Acta Vet Acad Sci Hung 17:107-114, 1976.

17. Domermuth CH, et al: Serologic examination of wild birds for hemorrhagic enteritis of turkey and marble spleen disease of pheasants. J Wildl Dis 13:405-408, 1977.

18. Domermuth CH, et al: Hemorrhagic enteritis and related infections. *In* Calnek BW, et al (eds): Diseases of Poultry 9th ed. Ames, Iowa State University Press, 1991, pp 567-572.

19. Domermuth CH, Schwartz LD, Mallinson ET, et al: Vaccination of ring-necked pheasants for marble spleen disease. Avian Dis 23:30-38, 1979.

20. Dorrestein GM: Viral infections in racing pigeons. Proc Assoc Avian Vet, 1992, pp 244-257.

21. Dorrestein GM, Hage MH: Adenovirus inclusion body hepatitis: A new pigeon disease? Proc VII Symp Avian Dis, Univ Munich, 1992, pp 7-14.

22. Fadly AM, Cowen BS, Nazerian K: Some observations on the response of ring-necked pheasants to inoculation with various strains of cell culture-propagated type II avian adenovirus. Avian Dis 32:548-552, 1988.

23. Fadly AM, Winterfield RW: Isolation and some characteristics of an agent associated with inclusion body hepatitis, hemorrhages, and aplastic anemia in chickens. Avian Dis 17:182-93, 1973.

24. Fitzgerald SD, Reed WM: A review of marble spleen disease of ring-necked pheasants. J Wildl Dis 25:455-461, 1989.

25. Fitzgerald SD, Reed WM, Burnstein T: Detection of type II avian adenoviral antigen in tissue sections using immunohistochemical staining. Avian Dis 16:341-347, 1992.

26. Fuchs A, Weissenbock H: Inclusion body hepatitis in psittacine birds and pigeons. Comparative histological and ultrastructural findings. Proc Europ Conf Avian Med Surg, 1993, pp 552-557.

27. Gerlach H: Viruses. *In* Ritchie BW, Harrison GJ, Harrison LR (eds): Avian Medicine: Principles and Application. Lake Worth, Wingers Publishing, 1994, pp 862-948.

28. Gomez-Villamandos JC, Mozos E, Sierra MA, et al: Inclusion bodies containing adenovirus-like particles in the intestine of a psittacine bird affected by inclusion body hepatitis. J Wildl Dis 28:319-322, 1992.

29. Goodwin MA, Davis JF: Adenovirus particles and inclusion body hepatitis in pigeons. J Assoc Avian Vet 6:37-39, 1992.

30. Goryo M, Ueda Y, Umemura T, et al: Inclusion body hepatitis due to adenovirus in pigeons. Avian Pathol 17:391-401, 1988.

31. Gough RE, Alexander DJ, Collins MS, et al: Routine virus isolation or detection in the diagnosis of diseases in birds. Avian Pathol 17:893-907, 1988.

32. Graham DL: An update on selected pet bird virus infections. Proc Assoc Avian Vet, 1984, pp 267-280.

33. Grimes TN, King PJ: Effect of maternal antibody on experimental infections of chickens with a type 8 avian adenovirus. Avian Dis 21:97-112, 1977.

34. Gulka CM, Piela TH, Yates VJ, et al: Evidence of exposure of waterfowl and other aquatic birds to the hemagglutinating duck adenovirus identical to EDS-76 virus. J Wildl Dis 20:1-5, 1984.

35. Gylstorff I: Adenoviridae. *In* Gylstorff I, Grimm F (eds): Vogelkrankheiten. Stuttgart, Eugen Ulmer, 1987, pp 275-278.

36. Heffels U, Fritzsche K, Kaleta EF, et al: Serological examination for viral infections in pigeons in Germany. Dtsch Tierärztl Wschr 88:97-102, 1981.

37. Heffles U, Khalaf SEP, Kaleta EF: Studies on the persistence and excretion of egg drop syndrome 76 virus in chickens. Avian Pathol 11:441-452, 1982.

38. Ianconescu ME, et al: An enzyme linked immunosorbent assay for detection of hemorrhagic enteritis virus and associated antibodies. Avian Dis 28:667-692, 1984.

39. Jack SW, Reed WM: Further characterization of an adenovirus associated with inclusion body hepatitis in bobwhite quail. Avian Dis 34:526-530, 1990.

40. Jackowski RM, Wyand DS: Marble spleen disease in pheasants: Demonstration of agar gel precipitating antibody in pheasants from an infected flock. J Wildl Dis 8:261-263, 1972.

41. Jacobson ER, Gardiner C, Clubb S: Adenovirus-like infection in white-masked lovebirds (*Agapornis personata*). J Assoc Avian Vet 1:32-34, 1989.

42. Jarkowski RM, et al: Marble spleen disease in pheasants: Demonstration of agar gel precipitating antibody in pheasants from an infected flock. J Wildl Dis 8:261-263, 1972.

43. Kaleta EF, Khalaf SED, Siegmann O: Antibodies to egg drop syndrome 76 virus in wild birds in possible conjunction with egg-shell problems. Avian Pathol 9:587-590, 1980.

44. Kawamura H, Sato TT, Tsubahaua H: Isolation of CELO virus from chicken trachea. Natl Inst Avian Health 3:1-10, 1963.

45. Ketterer PJ, Timmins BJ, Prior HC: Inclusion body hepatitis associated with an adenovirus in racing pigeons in Australia. Aust Vet J 69:90-91, 1992.

46. King DJ, Pursglove SR, Davidson WR: Adenovirus isolation and serology from wild bobwhite quail (*Colinus virginianus*). Avian Dis 25:678-682, 1981.

47. Logemann K, et al: Comparative studies for the characterization of avian adenovirus from quail and chickens. Proc VII Symp Avian Dis, Univ Munich, 1990, pp 294-295.

48. Lowenstine LJ: A potpourri of interesting avian cases. Proc Assoc Avian Vet, 1987, pp 105-107.

49. Lowenstine LJ, Fry M: Adenovirus-like particles associated with intranuclear inclusion bodies in the kidney of a common murre (*Uria aalge*). Avian Dis 29:208-213, 1985.

50. Malkinson M, Weisman Y: Serological survey for the prevalence of antibodies to egg drop syndrome 1976 virus in domesticated and wild birds in Israel. Avian Pathol 9:421-426, 1980.

51. Mandelli G, Rinaldi A, Cervio G: A disease involving the spleen and lungs in pheasants: Epidemiology, symptoms and lesions. Clinica Vet 89:129-138, 1966.

52. McCraken, et al: Experimental studies on the etiology of inclusion body hepatitis. Avian Pathol 5:324-329, 1976.

53. McCraken RM, Adair BM: Avian adenoviruses. *In* McFerran JB, McNulty MS (eds): Virus Infections of Birds. New York, Elsevier Science Publishers, 1993, pp 123-144.

54. McFerran JB: Egg drop syndrome, 1976 (EDS-76). Vet Quart 1:176-180, 1979.

55. McFerran JB: Immunity to adenoviruses. *In* Rose ME, et al (eds): Avian Immunology. Edinburgh, British Poultry Science Ltd, 1981, pp 187-203.

56. McFerran JB: Adenoviruses. *In* Purchase HG, et al (eds): A Laboratory Manual for the Isolation and Identification of Avian Pathogens 3rd ed. Dubuque, Kendall/Hunt Publishing Co, 1989, pp 77-81.

57. McFerran JB: Adenovirus infections. *In* Calnek BW, et al (eds): Diseases of Poultry 9th ed. Ames, Iowa State University Press, 1991, pp 552-582.

58. McFerran JB, Adair BM: Avian adenoviruses. A review. Avian Pathol 6:189-217, 1977.

59. McFerran JB, Conner TJ, McCraken RM: Isolation of adenoviruses and reoviruses from avian species other than domestic fowl. Avian Dis 20:519-524, 1976.

60. McFerran JB, Connor TJ, Adair BM: Studies on the antigenic relationship between an isolate from the egg drop syndrome 1976 and a fowl adenovirus. Avian Pathol 7:629-636, 1978.

61. McFerran JB, McNulty MS: Virus Infections of Birds. London, Elsevier Science Publishers, 1993.

62. Mori F, Touchi A, Suwa T, et al: Inclusion bodies containing adenovirus-like particles in the kidneys of psittacine birds. Avian Pathol 18:197-202, 1989.

63. Okita M: Pathology of adenovirus infection in the kidney of Budgerigars. Jpn J Vet Res 37:128, 1989.

64. Pass DA: Inclusion bodies and hepatopathies in psittacines. Avian Pathol 16:581-597, 1987.

65. Phalen DN: Acute pancreatic necrosis in an umbrella cockatoo. Proc Assoc Avian Vet, 1988, pp 203-205.

66. Raines AM: Adenovirus infection in the ostrich (*Struthio camelus*). Proc Assoc Avian Vet, 1993, pp 304-312.

67. Ramis A, Latimer KS, Niagro FD, et al: Diagnosis of psittacine beak and feather disease (PBFD) viral infection, avian polyomavirus infection, adenovirus infection and herpesvirus infection in psittacine tissues using DNA in situ hybridization. Avian Pathol 23:643-657, 1994.

68. Ramis A, Marlasca MJ, Majo N, et al: Inclusion body hepatitis (IBH) in a group of eclectus parrots (*Eclectus roratus*). Avian Pathol 21:165-169, 1992.

69. Reece RL, et al: Inclusion body hepatitis in Guinea fowl (*Numida meleagris*). Aust Vet J 63:26-27, 1986.

70. Reece RL, Pass DA: Inclusion body hepatitis in a tawny frogmouth (*Podargus strigoides*). Aust Vet J 62:426, 1985.

71. Riddell C: Virus hepatitis in domestic geese in Saskatchewan. Avian Dis 28:774-782, 1984.

72. Riddell C, den Hurk JV, Copeland S, et al: Viral tracheitis in goslings in Saskatchewan. Avian Dis 36:158-163, 1992.

73. Rigby CE, Pettit JR, Papp-Vid G, et al: The isolation of salmonellae, Newcastle disease virus and other infectious agents from quarantine imported birds in Canada. Can J Comp Med 45:366-370, 1981.

74. Sagartz JE, Swayne DE: Inclusion body hepatitis in a pigeon associated with an adenovirus. J Zoo Wildl Med 22:485-487, 1991.

75. Saifuddin M, Wilks CR: Vertical transmission of avian adenovirus associated with inclusion body hepatitis. New Zeal Vet J 39:50-52, 1991.

76. Saifuddin M, Wilks CR: Effects of fowl adenovirus infection on the immune system of chickens. J Comp Path 107:285-294, 1992.

77. Schelling SH, Garlick DS, Alroy J: Adenoviral hepatitis in a merlin (*Falco columbarius*). Vet Pathol 26:529-530, 1989.

78. Scott PC, Condron RJ, Reece RL: Inclusion body hepatitis associated with adenovirus-like particles in a cockatiel (*Nymphicus hollandicus*). Aust Vet J 63:337-338, 1986.

79. Sileo L, Franson JC, Graham DL, et al: Hemorrhagic enteritis in captive American kestrels (*Falco sparverius*). J Wildl Dis 19:244-247, 1983.

80. Smyth JA, Adair BM: Lateral transmission of egg drop syndrome-76 virus by the egg. Avian Pathol 17:193-200, 1988.

81. Spradbrow PB, Bains BS: Further isolations of adenovirus from the bursa of Fabricius of domestic chickens. Aust Vet J 50:81-82, 1974.

82. Takase N, Yoshinaga N, Egashira T, et al: Avian adenovirus isolated from pigeons affected with inclusion body hepatitis. Jpn J Vet Sci 52:207-215, 1990.

83. Tham VL, Thies NF: Marble spleen disease of pheasants. Aust Vet J 65:130-131, 1988.

84. Uyttebroek E, Ducatelle R: Epidemiology of adenovirus infections in pigeons. Proc Europ Assoc Avian Vet, 1991, pp 289-292.

85. Van der Hurk JV: Quantitation of hemorrhagic enteritis virus antigen and antibody using enzyme-linked immunosorbent assays. Avian Dis 30:662-671, 1986.

86. Vindevogel H, Dagenais L, Lansival B, et al: Incidence of rotavirus, adenovirus, and herpesvirus infection in pigeons. Vet Rec 109:285-286, 1981.

87. Wallner-Pendleton E, Helfer DH, Schmitz JA, et al: An inclusion-body pancreatitis in *Agapornis*. Proc 32nd West Poult Dis Conf, 1983, pp 99.

88. Ward JM, Young DM: Latent adenoviral infection of rats: Intranuclear inclusions induced by treatment with a cancer chemotherapeutic agent. J Am Vet Med Assoc 169:952-953, 1976.

89. Winterfield RW: Adenovirus infections of chickens. *In* Hofstad MS, et al (eds): Diseases of Poultry. Ames, Iowa State University Press, 1984, pp 497-506.

90. Winterfield RW, Dubose RT: Quail bronchitis. *In* Hofstad MS, et al (eds): Diseases of Poultry 8th ed. Ames, Iowa State University Press, 1984, pp 508-510.

91. Winterfield RW, et al: Quail bronchitis. *In* Calnek BW, et al (eds): Diseases of Poultry 9th ed. Ames, Iowa State University Press, 1991.

92. Wyand DS, Jakoswski RM, Burke CM: Marble spleen disease in ring-necked pheasants: Histology and ultrastructure. Avian Dis 16:319-329, 1972.

93. Yamaguchi S, Imada T, Kawamura H, et al: Pathogenicity and distribution of egg drop syndrome-76 virus in inoculated laying hens. Avian Dis 25:642-649, 1982.

94. Zellen GK, et al: Adenoviral pancreatitis in Guinea fowl (*Numida meleagris*). Avian Dis 33:596-589, 1989.

95. Zhang C, Nagaraja KV: Differentiation of avian adenovirus-type II strains by restriction endonuclease fingerprinting. Am J Vet Res 50:1466-1470, 1989.

96. Zsak L, Kisary J: Characterization of adenoviruses isolated from geese. Avian Pathol 13:253-264, 1984.

Reoviridae

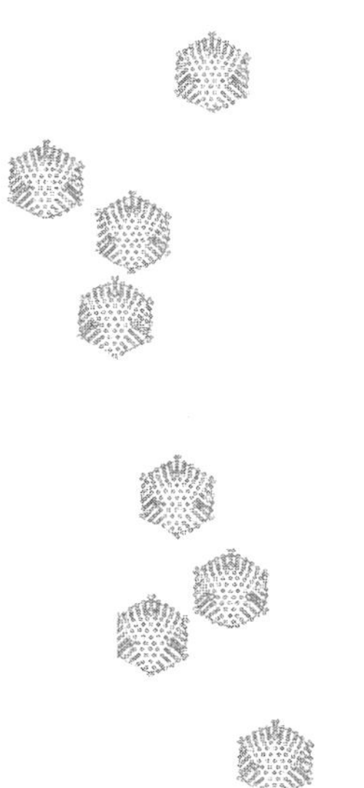

The Respiratory, Enteric, Orphan (REO) viruses were so-named because they were initially isolated from the respiratory or enteric system in animals exhibiting no specific signs of disease. The Reoviridae family consists of three genera that have been shown to infect birds: *Orthoreovirus*, *Orbivirus* and *Rotavirus*. Most viruses in the Reoviridae are considered nonpathogenic; however, a few strains have been associated with high levels of morbidity and mortality in specific populations of susceptible birds and mammals. In many cases, animals infected with these viruses will develop clinical changes that are attributable to secondary infectious agents that take advantage of the immunosuppressed host.

Orthoreoviruses (heretofore called simply reoviruses) are recovered most commonly from the gastrointestinal tract of clinically normal birds. A few strains have been associated with disease in several species of gallinaceous birds, psittacine birds and waterfowl.[10,25] The other two Reoviridae genera, *Orbivirus* and *Rotavirus*, have been discussed infrequently in companion, aviary and free-ranging birds.

ORTHOREOVIRUSES

There are currently 11 serotypes of avian orthoreoviruses (reoviruses) that are antigenically distinct from each other and antigenically distinct from the strains that infect mammals. Reoviruses have been recovered from many avian species, including Psittaciformes, pheasants, pigeons, raptors, geese, ducks, chickens, turkeys and quail.[19,44] Most of the serotypes of reovirus that have been defined were recovered from chickens and turkeys, but it is likely that multiple serotypes also infect various species of companion, aviary and free-ranging birds.[15]

Reoviruses are commonly recovered from asymptomatic birds. The type and degree of pathology that occurs appear to vary with the age of the host, virulence of the virus strain and the route of exposure.[62] Experimentally, some reoviruses have been shown to infect differing avian species. For example, a reovirus recovered from ducks and an eagle was infectious to chickens.[33] Although these experimental findings open the possibility of cross-species transfer of avian reoviruses, a free-ranging avian reservoir for these viruses has never been reported.

CLINICAL FEATURES

Reoviruses have been most thoroughly studied in gallinaceous birds where they have been associated with arthritis, tenosynovitis, stunting of young chicks, bursal atrophy, respiratory disease, gastrointestinal tract abnormalities, hepatic necrosis, myocarditis, pericarditis, malabsorption of nutrients, osteoporosis, feather abnormalities and inflammation of the thymus.[8,60,63,67] Because reoviruses are commonly demonstrated in the

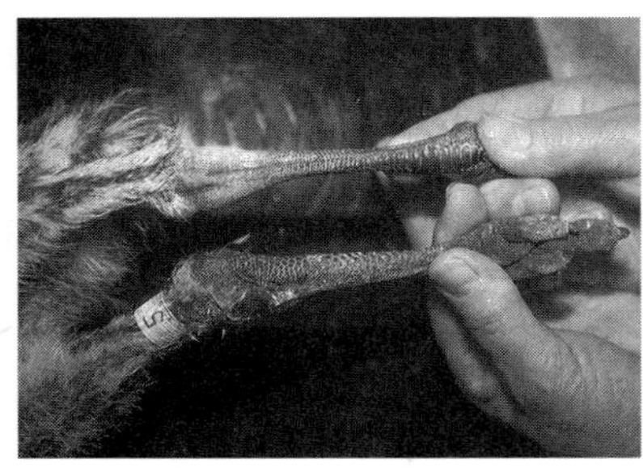

Reovirus infections may cause deformities in the maturing bones of young fowl. Ducks, geese and ratites frequently develop leg abnormalities. In this ratite chick, one leg has rotated backward, while the other is in the correct anatomic position. It is undetermined whether reovirus plays a role in these developmental abnormalities. photograph courtesy of Brett Hopkins

feces of asymptomatic birds, it is uncertain what role these viruses play in causing the clinical problems listed above.[55] It will be of interest to determine if any of the leg deformities that commonly occur in ducks, geese and ratites are being caused by reovirus infections *Figure 12.1*.

The prevalence of reovirus infections or disease in companion, aviary and free-ranging birds has not been reported. Reoviruses have been isolated or identified by electron microscopy from the feces or tissues of asymptomatic psittacine and passerine birds in quarantine stations. Reoviruses have also been recovered from psittacine birds with clinical signs, including anorexia, depression, weight loss, hemorrhage under the skin, diarrhea, abdominal swelling caused by ascites, dyspnea, nasal discharge, ataxia, paralysis, uveitis, hypopyon, and edema in the head and legs *Figure 12.2*.[4,10,25,52,58,64,71-73]

The primary clinical changes noted in affected psittacine birds are associated with hepatitis or enteritis and include

African Grey Parrots and cockatoos with reovirus infections commonly develop clinical signs characterized by depression, anorexia, dyspnea and nasal discharge. photograph courtesy of Louise Bauck

anorexia and diarrhea.[4,25,50,52] Pneumonia was cited as a characteristic problem in some imported psitta-cines from which a reovirus was recovered.[50]

Infections characterized by de-pression, anorexia, dyspnea and a nasal discharge in birds with or without diarrhea are particularly common in African Grey Parrots and cockatoos.[4,9,10,73] In many cases, the clinical progression of reovirus infections in psittacine birds mimics Pacheco's disease virus.

This virus has been suggested as a cause of fledgling mortality in lories and chronic respiratory disease in Amazon parrots.[72] Reovirus should be considered as a cause of sudden death in lories and lorikeets, particularly when hepatomegaly is noted at necropsy.[15] Reovirus was recovered from a Cockatiel with weight loss, ataxia, dyspnea and diarrhea, and has been implicated as a cause of severe anemia in young African Grey Parrots.[17,25,50] Reovirus was recovered from parrots that died with hepatitis, a lovebird with respiratory problems, a toucanette with enteritis and finches with hepatitis and enteritis.[18,19]

Permanent changes in the anterior portion of the eye and posterior synechiae were found to occur in some birds that recovered from reovirus infections.[10,71] Uveitis occurred more frequently in birds with secondary problems than in uncomplicated cases.[10]

Reovirus infections in Old World Psittaciformes, particularly African Grey Parrots and cockatoos, tend to be fatal; New World Psittaciformes appear to be less susceptible. Reported mortality rates in African Grey Parrots vary from 10% to 100% of the exposed population.[4,9,10,50,72] Thus, the prognosis for Old World Psittaciformes with clinical signs of disease is poor, whereas infected New World Psittaciformes that receive appropriate supportive care usually recover.

Severely affected birds typically die 3 to 18 days after developing clinical signs.[4,10,15,73] Morbidity and mortality are highest in birds with concomitant infections.

EXPERIMENTAL INFECTIONS

The experimental incubation period in most birds is two to nine days. The outcome of experimental infections in African Grey Parrots and cockatoos varies, with some infected birds developing severe disease followed by death and other birds remaining asymptomatic.[17,25] In those experimentally infected birds that developed disease, the clinical signs, shedding patterns of the virus and immunologic response were similar irrespective of whether the birds were inoculated by the oral, intratracheal or intramuscular route. However, birds inoculated intramuscularly suffered the highest levels of mortality.[15] Experimentally infected African Grey Parrots died eight to nine days after virus exposure by the oral or intramuscular route. The infected birds developed diarrhea and excreted yellow-orange urates (suggestive of liver disease) prior to death *Figure 12.2.*[25]

Experimental infections can be induced in adult chickens by introducing the virus through the nasal, tracheal or esophageal route.[39] In chickens infected orally, the virus replicates in the intestinal tract lining 24 to 48 hours after infection. Viremia occurs within 30 hours of infection, resulting in widespread distribution of the virus to multiple organs within three to five days after infection. The highest concentration of virus occurs in the intestinal tract.[33,39,60] Virus was recovered from the cloaca of chickens for up to 16 weeks after they were infected by foot pad inoculation, and the virus persisted in the tendons for 13 weeks.[35] This suggests that tendonitis could occur following virus entry through damaged epithelium on the feet *see Figure 12.3.*

EPIZOOTIOLOGY

Reoviruses have a worldwide distribution and have been commonly isolated from companion birds originating from several geographic regions, including

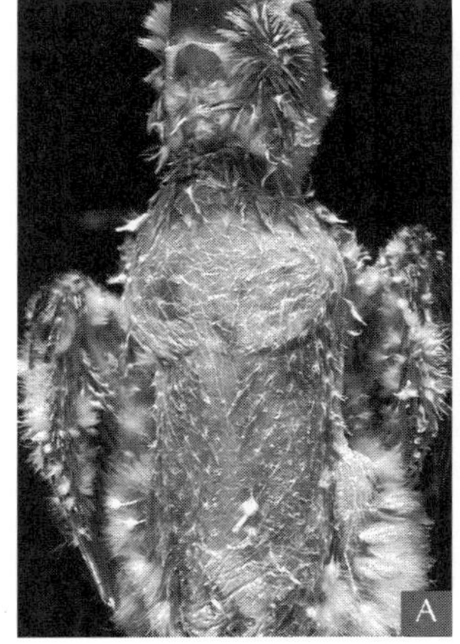
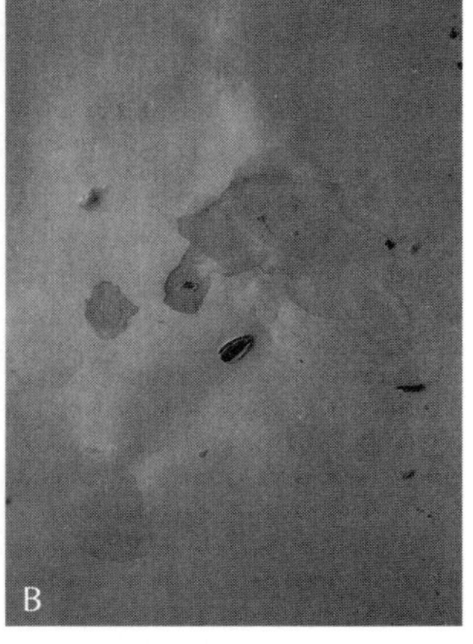
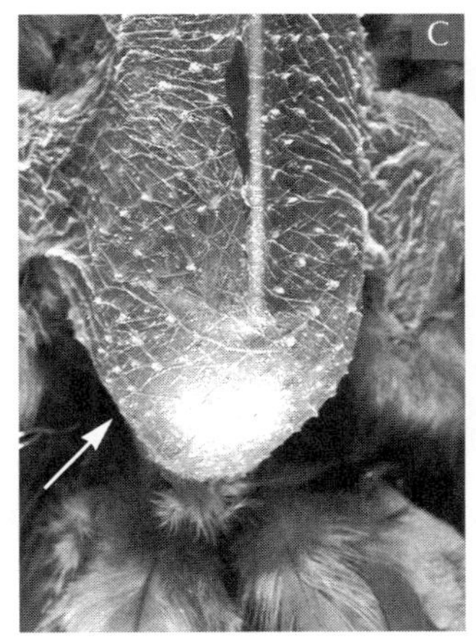

FIG 12.2
In some companion birds, reovirus infections have been associated with:
A hemorrhage under the skin;

B diarrhea with yellow or green urates suggestive of liver disease;

C abdominal swelling (arrow) caused by ascites; and

D edema in the head.

photographs courtesy of Louise Bauck

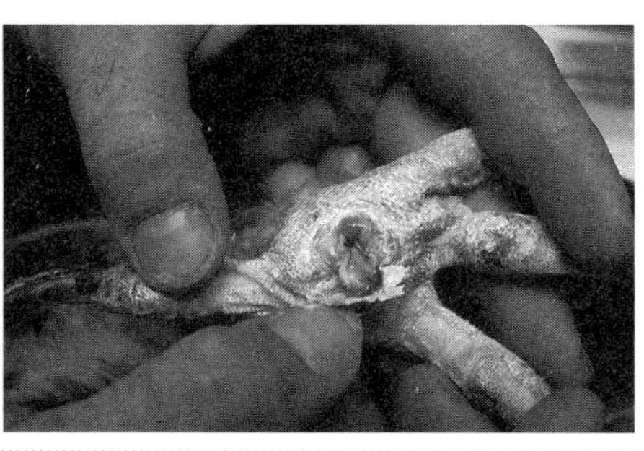

India, Zaire, Senegal, Pakistan, Czechoslovakia, Malaysia, Mali, Ghana, Indonesia and Bolivia.[3,4,10,44,50] Lesions considered consistent with reovirus have also been confirmed in companion and aviary birds within the United States[10,25,52] and Europe.[3,4,44]

Reoviridae-like viruses have been associated with clinical disease in a number of commonly maintained companion and aviary birds, including pigeons, waterfowl, pheasants, finches, Cockatiels, Budgerigars, rosellas, African Grey Parrots, Senegal Parrots, Indian Ring-necked Parakeets, Hawk-headed Parrots, lovebirds, Alexandrine Parakeets, grass parakeets, cockatoos, macaws, Jardine Parrots, Amazon parrots, pionus parrots, Amboina King Parrots, lories, lorikeets and Grey-cheeked Parakeets.[9,10,25,29,44,50,52,70,71]

Reoviruses are commonly recovered from asymptomatic as well as some symptomatic birds in quarantine stations. A reovirus was isolated from 22 of 269 (8%) groups of psittacine and passerine birds in a Canadian import station over a three-year period.[58] Over a seven-year period, reovirus was isolated from 4.1% of the psittacine birds tested during quarantine in the United States.[64] In another study, reovirus was recovered from 24 of 2,274 lots of birds in the United States quarantine system.[64] In Belgium, reovirus was isolated from 15 of 28 (53%) consignments of psittacine birds experiencing enteritis, splenomegaly, and congestion and necrosis of the liver.[50] Reovirus was recovered from the trachea of 3 of 500 (0.6%) companion birds tested in a private practice in Europe. Two of these birds were clinically normal, and the third bird had concomitant aspergillosis.[56]

During a one-year study in Europe, reovirus was recovered from the feces, abdominal organs or lungs of 5 of 89 (6%) chickens, 1 of 41 (2%) turkeys, 2 of 72 (3%) pheasants, 3 of 23 (13%) pigeons, 5 of 38 (13%) psittacines, 2 finches of 11 passerines and 1 of 45 (2%) ducks. Virus was not recovered from 8 partridges, 10 geese or 4 swans.[19] A die-off of 1,000 American Woodcocks in the Northeastern United States was associated with reovirus. Affected birds were severely emaciated.[13] Reovirus antibodies were detected in 19% of 149 ostriches in Zimbabwe. The seroprevalence of the virus on various farms ranged from 0% to 60%.[6]

Reoviruses are frequently recovered from companion birds in conjunction with salmonellosis, *E. coli* and chlamydiosis.[15,50] Reovirus and avian polyomavirus were recovered from an Amazon parrot with hemorrhages throughout the body.[72] Synergistic infections with reovirus and *Cryptosporidium* sp. have been reported in Bobwhite Quail and chickens *see Figure 12.8*.[26,27] The levels of morbidity and mortality appear to be highest in any bird when concomitant bacterial, fungal or other viral infections are present.[10] As an example, a reovirus was implicated in an aviary outbreak in which 250 birds died 20 days after a shipment of African Grey Parrots was added to the flock. Included in the losses were 203 Congo African Grey Parrots, 4 Timneh African Grey Parrots, 13 Jardine's Parrots, 1 Lesser Sulphur-crested Cockatoo, 9 Blue-crowned Amazon Parrots, 16 Red-lored Amazon Parrots, 5 Orange-winged Amazon Parrots, 6 Yellow-naped Amazon Parrots, 4 White-crowned Pionus Parrots, 16 Amboina King Parrots, 2 lories and 6 lorikeets. Concomitant salmonellosis was considered a contributing factor in the deaths of many of the African Grey Parrots and some of the Amazon parrots and lories.[15]

RELATIONSHIP OF VIRUS STRAINS —

Some reoviruses recovered from companion and aviary birds have been shown to be related to strains found in poultry, while others have been shown to be serologically distinct.[10,50] A reovirus isolated from an African Grey Parrot originating from Ghana and another from a cockatoo originating from Indonesia were found to be antigenically unrelated to reoviruses isolated from poultry. Viral isolates from macaws originating in Bolivia were antigenically similar to those recovered from chickens. African Grey Parrots originating from Ghana and Cameroon and cockatoos originating from Indonesia have died with reovirus in quarantine before they were exposed to other birds in the United States;[10] this suggests that these viruses originated in the country of origin, and may naturally infect free-ranging birds in these countries.

A reovirus recovered from American Woodcocks was serologically distinct from the reovirus that causes arthritis/tenosynovitis in chickens.[13] Chickens were shown to be susceptible to experimental infections when inoculated with virus recovered from ducks or a Wedge-tailed Eagle.[33] Isolates of reovirus from ducks in Northern Ireland were shown to be serologically distinct from those that infect poultry.[1]

TRANSMISSION — The exact routes of

reovirus transmission in companion and aviary birds remain speculative. In chickens, reoviruses are primarily transmitted by horizontal routes, principally through direct or indirect contact with contaminated feces.[2,35] Egg transmission has been documented to occur only occasionally in ducks, geese, turkeys and chickens. This route of transmission is considered to be of minimal importance in these species, and has not been confirmed in companion birds.[15,18,39,49]

Reoviruses can enter a host through the gastrointestinal (principal route) or respiratory tract. With the frequent recovery of reovirus from intestinal samples in psittacine birds, pheasants, waterfowl and pigeons, it is likely that contaminated feces serve as a major source of virus in these birds. In psittacine birds experimentally infected by intramuscular inoculation, virus could first be demonstrated in the feces two days after inoculation, and shedding was found to persist for 15 days.[15] Contaminated fecal dust has been suggested as a route for viral dissemination through an aviary. However, during a natural

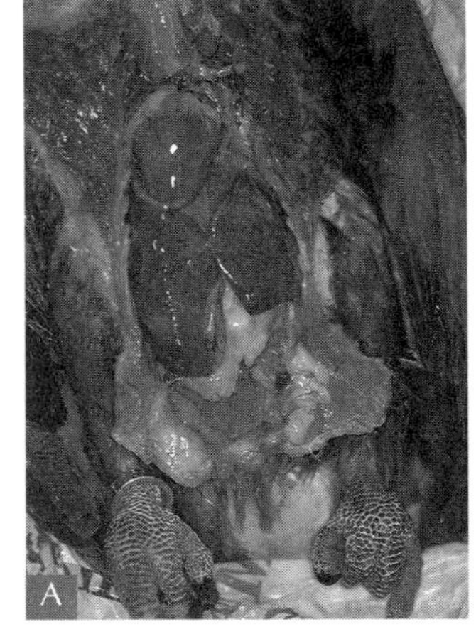
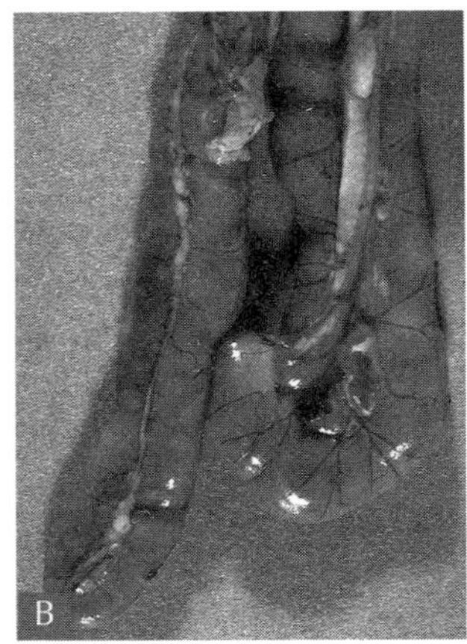
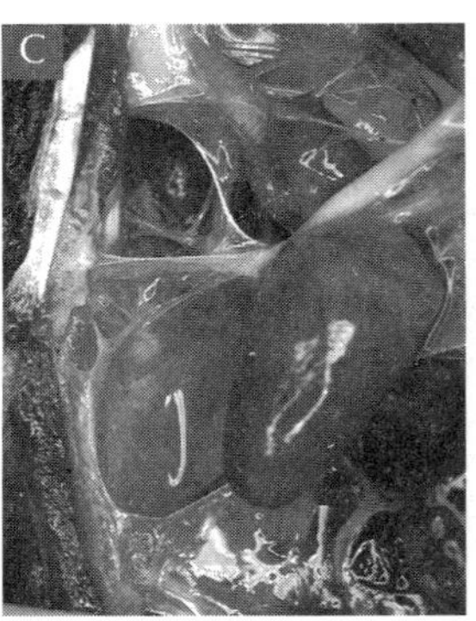
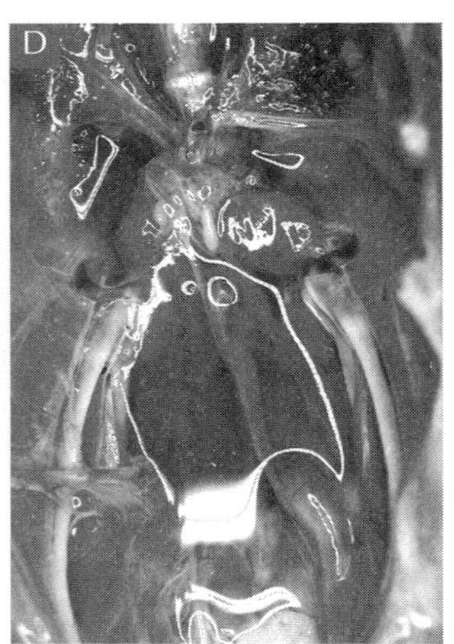
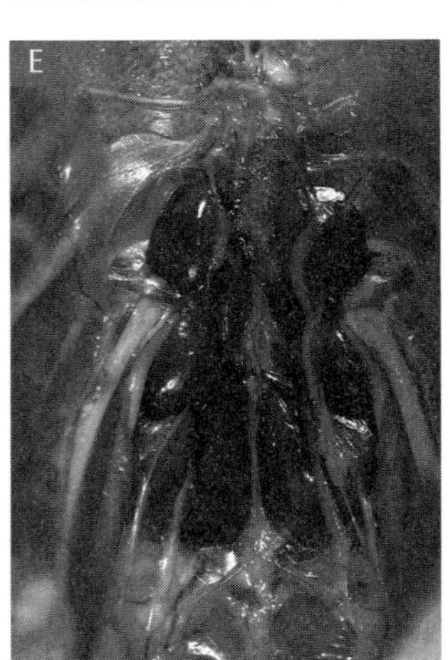

FIG 12.4

The gross changes associated with reovirus infections in psittacine birds are similar to those caused by avian polyomavirus, adenovirus or Pacheco's disease virus. Suggestive necropsy findings include:

A pale yellow mottling of an enlarged liver;
B hyperemia of the intestines; C pericardial effusion;
D ascites; and E swollen congested kidneys.

reovirus outbreak, many African Grey Parrots that shared an enclosure were infected, while other birds that were in the same airspace remained normal.[4,10] Interestingly, the virus was recovered from the trachea, and not the cloaca, of 3 of 500 (0.6%) companion birds tested in a private practice in Europe.[56]

CARRIERS — Some chickens that survive a reovirus infection are thought to develop latent infections. Latently infected hens can transmit the virus to their developing eggs, and birds of either gender can intermittently shed the virus in their feces. The role that persistently infected chickens play in the maintenance of infections in a flock is not clearly defined.

Epizootiologic evidence connecting widely dispersed outbreaks of reovirus to a common source and the deaths of isolated companion birds both suggest that persistent infections may occur.[10,15,52] Further support for the occurrence of subclinical, persistent infections is found in the frequent isolation of reovirus from the feces or from tracheal swabs of clinically normal birds.[10,15,18,56,71] Currently, there is no effective test for detecting persistently infected birds.

PATHOLOGY

The pathology associated with reovirus infections in African Grey Parrots has been found to be similar in both natural and experimentally induced infections.[25] Gross changes include pale yellow mottling of an enlarged liver and spleen, both of which may contain multifocal, grayish-white lesions. An enlarged, discolored liver is the most consistent gross lesion. Other findings may include hyperemia of the intestines, pericardial effusion, ascites, swollen kidneys, congested lungs, thickened air sacs and hemorrhage under the skin and in the heart *Figure 12.4*.[10,24,25,50] Gross lesions may mimic those caused by avian polyomavirus, adenovirus or Pacheco's disease virus.

The predominant histologic lesion is disseminated or focal necrotizing hepatitis. Inflammation in the heart, intestines and air sacs, and necrosis of cells in the spleen, bone marrow, kidney and lining of the gastrointestinal tract may also be present.[10,24,25,50] Experimentally infected psittacine birds were not found to exhibit epicarditis or air sacculitis, suggesting that when present, these lesions are being induced by other etiologic agents.[25] The tenosynovitis that is commonly associated with reovirus infections in domestic fowl has not been described in companion or aviary birds. The hepatocytes of infected birds may or may not contain intracytoplasmic inclusion bodies with viral particles in the 70 to 88 nm size range.[4,25,50,73]

PATHOGENESIS

Reoviruses are commonly recovered from gallinaceous birds that are asymptomatic or are experiencing a variety of clinical changes. The role of reovirus in a disease process if any, is frequently difficult to determine. Experimental infections in chickens are rarely as severe as natural infections. The response of gallinaceous birds infected with reovirus may vary depending on the age of the bird when exposure occurs, the condition of the host, the virulence of the infecting virus, the titer of infecting virus and the route of viral exposure.[38-40,62] Some strains of reovirus clearly are more virulent than others, but the factors that control variances in the virulence and pathogenesis of various strains of reovirus remain poorly defined. Similar parameters probably control the pathogenesis of infections in other susceptible avian species, where infected birds develop specific clinical signs, while others remain asymptomatic.[15,17,25,73]

Reoviruses have a wide tissue distribution, and replication may occur in the liver, lymphoid tissues or mucosal cells lining the digestive and respiratory tracts. In chickens, the cells that line the intestinal tract and cloacal bursa have been shown to be principal sites of virus entry. Primary replication occurs in these same tissues, and progeny virus released from these initial sites of replication are subsequently transported in the blood to other organs throughout the body.[34] In chickens infected by the oral route, reoviruses replicate in the lining of the intestinal tract 24 to 48 hours after infection. Viremia occurs within 30 hours, and the virus can be found replicating in many organs within three to five days.[60]

Some strains of reovirus are thought to induce latent infections that damage the immune system, making the infected bird more susceptible to other pathogens. Chicks infected up to one week of age sustain damage to the bursa and spleen that would be expected to cause immunosuppression.[53,61] Bobwhite Quail infected with reovirus frequently have concurrent cryptosporidial infections, which generally are self-limiting in immunocompetent hosts.[41] Because reoviruses can be long-term residents in an infected host, they should be considered as a complicating factor in infections caused by other microbial organisms.

The liver appears to be the principal target for reovirus in psittacine birds, and infection causes severe necrosis of this organ. Reovirus also may be found in the cells that line the respiratory and digestive tracts.[25,73] In psittacine birds, the affinity of reoviruses to damage a variety of different tissues is thought to be the principal factor in the progression of an infection. Damage to the immune system allows secondary invasion by bacterial, fungal, viral or chlamydial pathogens.[9,15] Reoviruses have been isolated from companion birds in conjunction with paramyxovirus, *Salmonella*, *Aspergillus* and *Mucor* spp. and *E. coli* *Figure 12. 5*.[58] However, some strains of reovirus have been shown to produce clinical and pathologic changes without the influence of secondary pathogens.[9,10,25,50,58]

Reovirus can cause severe liver disease alone or in conjunction with other infectious agents. Concomitant salmonellosis is a contributing factor in many of the deaths associated with reovirus infections in companion birds. photograph courtesy of Gilles Berneir

IMMUNITY

Reovirus antibodies develop in chickens by 17 days after infection and may persist over six months.[30] Reovirus-infected psittacine birds develop precipitating antibodies from seven days to three weeks after viral exposure.[10,15]

These antibodies, which can persist for two to six weeks, are associated with the cessation of viral shedding in the feces *Figure 12.6*.[10,15,17] While disease resistance appears to be associated with the development of neutralizing antibodies, birds with and without titers have been shown to intermittently shed reovirus in their feces.[10,16,38] Maternally transmitted antibodies will protect young chicks from infection.[69]

DIAGNOSIS

Confirming the presence of a reovirus requires its isolation from the feces, gastrointestinal contents, liver, heart, kidney or lung.[50] Samples of joint fluid from gallinaceous birds with tenosynovitis can be submitted for reovirus isolation. Electron microscopy can be used to demonstrate virus particles in the feces, respiratory secretions or joint fluid if a sufficient quantity of virus is present; however, it is common to find reovirus in the feces of normal birds, and demonstration of the virus does not necessarily indicate that a bird will develop disease.

Liver necrosis is a common histologic finding in affected birds. The identification of intracytoplasmic inclusion bodies containing 70 to 88 nm, nonenveloped viral particles is suggestive; however, these inclusion bodies are not always detectable.[25,44,50,73] Because it is frequently difficult to demonstrate reovirus by routine microscopic examination of tissues, this virus may represent an under-diagnosed cause of liver necrosis in companion and aviary birds. In most cases, virus isolation or electron microscopy is necessary to confirm that

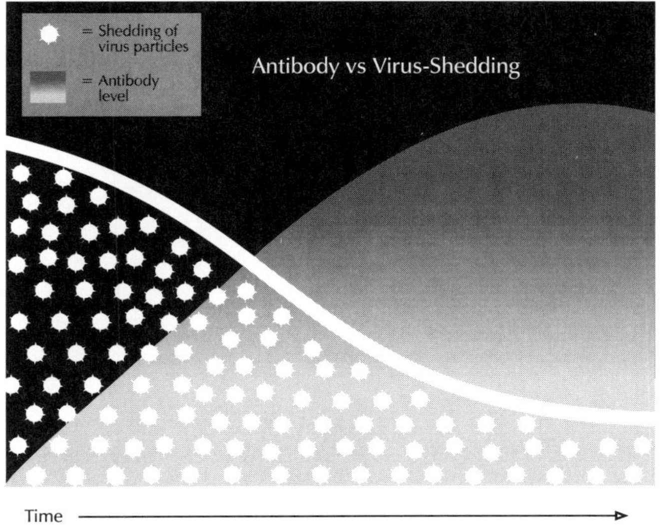

FIG 12.6

Antibody titers increase from seven days to three weeks after infection, causing a cessation of virus shedding in the feces.

liver necrosis is the result of a reovirus infection.

Reovirus antibodies can be detected using agar-gel diffusion, virus-neutralization, enzyme-linked immunosorbent assay (ELISA) or indirect fluorescent antibody assays. Virus-neutralizing assays generally detect strain-specific antibodies, whereas the other tests generally detect group-specific antigens. In chickens, serology is of little value in managing infections because of the prevalence of the disease and the frequency of subclinical infections. In psittacine birds, where reovirus infections are less common, a diagnosis may be possible in live birds by demonstrating precipitating or virus-neutralizing antibodies in birds with suggestive clinical signs.[15,17] Reovirus-specific precipitating antibodies may be detected within ten days of infection; titers decrease to undetectable levels in several months.[15] Because reovirus antibodies decrease rapidly in psittacine birds, demonstrating a high titer suggests that a bird has been recently infected.

Natural and experimental reovirus infections in some psittacine birds have been associated with anemia, hypoproteinemia, hypoalbuminemia and hyperglobulinemia (as detected by serum electrophoresis). Infected psittacine birds developed leukocytosis early in the disease process followed by leukopenia as the infection progressed. When leukopenia occurred, 90% of the white blood cells were lymphocytes. Increases in the aspartate aminotransferase and lactate dehydrogenase activities were considered to be terminal findings suggestive of severe liver necrosis.[10,15,17,52] In experimental infections, the serum protein levels dropped four days after inoculation and returned to normal two to three weeks later. Gamma globulins increased and albumin levels decreased over a ten-day period.[17]

CONTROL

Reoviruses, in general, are stable in the environment and are resistant to pH 3, lipolytic agents (ether and chloroform), 2% lysol, 3% formalin, 3% formaldehyde, 1% hydrogen peroxide, 1% phenol, quaternary ammonium and heating to 56°C for 120 minutes or to 60°C for 10 hours.[12,18,31,50] These viruses are extremely stable in contaminated organic materials and respiratory secretions as long as they remain moist. Infectivity can be reduced by prolonged contact (up to two hours in some cases) with phenols, aldehydes, 70°C, halides, formalin, 70% ethanol, beta-propiolactone and 0.5% iodine.[31] Chlorhexidine at a dose of 20 milliliters per gallon of drinking water for 30 days has been suggested as a way to reduce the infectivity and spread of reovirus in flock outbreaks. However, laboratory studies have shown that reovirus is not completely inactivated by chlorhexidine at normal dilutions.[10,15,60]

Control of reovirus in chickens is difficult because it can be vertically transmitted, is resistant to inactivation, and infected birds may intermittently shed the virus. In gallinaceous birds, infections are best controlled by vaccinating breeding birds, which confers temporary protection to the chicks. Both inactivated and attenuated-live virus vaccines

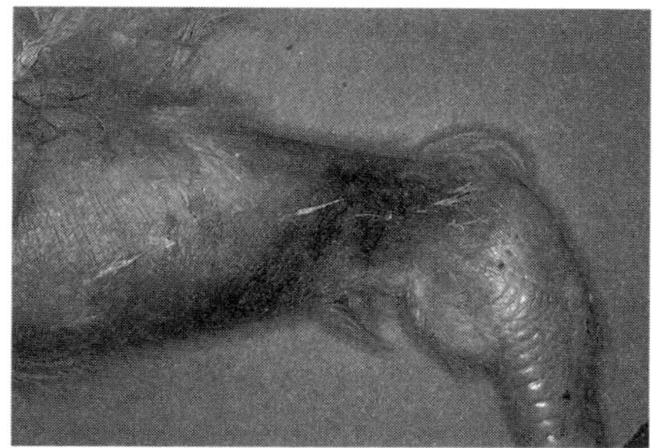

In gallinaceous birds, reovirus can cause inflammation of the tendons and joints of the hock. The affected area may swell and be hemorrhagic. This disease is characterized by enlargement of the area around the hock and progressive lameness. Eventual rupture of the tendon of the gastrocnemius muscle results in an inability to stand or walk. photograph courtesy of Jean Sander

are available for chickens. Unfortunately, reovirus vaccines commercially produced for chickens are of little value in protecting companion and aviary birds from infections because the virus strains commonly found in these species are antigenically unrelated to those found in poultry.[10,17] An experimental inactivated vaccine produced from a psittacine reovirus isolate was found to reduce morbidity and mortality of infected African Grey Parrots and cockatoos.[10] Goslings can be provided temporary protection through subcutaneous administration of hyperimmune serum; however, it has not been determined whether a similar approach would be effective in providing temporary protection for neonates of other species.

Species-specific Reoviral Diseases

Pheasant, Quail, Chickens and Turkeys

Reovirus infections appear to be common in gallinaceous birds. During a one-year study in Europe, reovirus was recovered from chickens, turkeys and pheasants. Affected pheasants died following a brief period of diarrhea.[19]

The most severe reovirus-associated disease in chickens is tenosynovitis-arthritis. This disease is characterized by enlargement of the area around the hock and progressive lameness. Eventual rupture of the tendon of the gastrocnemius muscle results in an inability to stand or ambulate *Figure 12.7*. An age-related resistance to disease has been shown to occur in chickens, with one-day-old chicks being more susceptible than two-week-old chicks.[32] Myocarditis, hepatitis and splenitis in chicks less than seven days of age are thought to be caused by reovirus. A malabsorption syndrome characterized by abnormal feathering, osteoporosis and the passage of poorly digested food has been associated with a reovirus in one- to three-week-old chickens. Mortality is common in affected chicks. Necropsy findings frequently include dilatation of the proventriculus and catarrhal enteritis.[54] It is of interest that reoviruses recovered from ducks and a Wedge-tailed Eagle were infectious for chickens.[33]

A reovirus was recovered from a group of five-day-old to five-week-old Bobwhite Quail experiencing high levels of mortality. These birds were also infected with *Cryptosporidium* sp. Quail experimentally infected with reovirus alone remained clinically normal, while those infected with both reovirus and *Cryptosporidium* sp. developed systemic reovirus infections *Figure 12.8*.[27,59]

PIGEONS

Reoviruses have been recovered from the liver of pigeons with hepatitis and from the feces of pigeons with diarrhea, alone or in conjunction with dyspnea. The virus is shed primarily from the cloaca.[44,68,71] In one study, virus recovered from the liver of an infected bird did not produce disease or stimulate seroconversion in five-week-old pigeons infected orally; however, virus was recovered from the feces two to five days after infection. Virus could not be recovered at any time from the pharynx of these experimentally infected pigeons.[71]

From 8% to 16% of homing pigeons in Europe have been found to have reovirus antibodies.[28,70] Over a one-year period in a European study, reovirus was recovered from 3 of 23 (13%) pigeons with diarrhea, tremors and weight loss.[19]

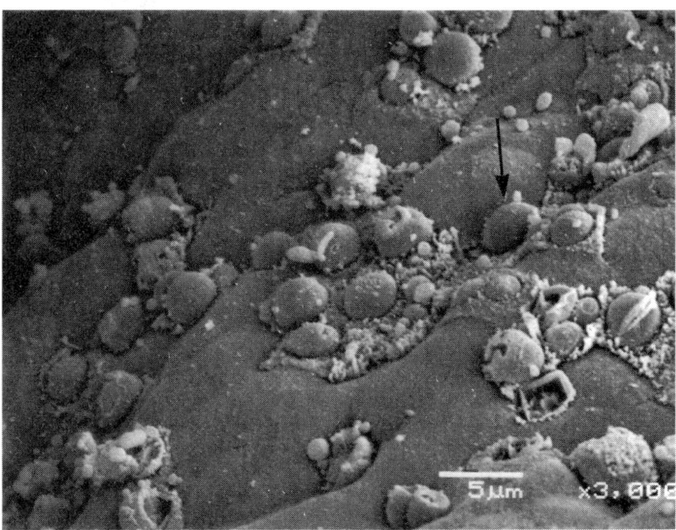

WATERFOWL

Reoviruses have been isolated from Muscovy and Mallard Ducks; some have exhibited clinical signs of disease, while others have been asymptomatic.[44] Clinical problems attributable to reovirus are most common in ducklings between ten days and six weeks of age. Affected birds may exhibit retarded growth and poor feather formation. Commercial crosses between Mallard and Muscovy Ducks appear to be particularly susceptible, whereas pure Muscovy Ducks are relatively resistant. Mortality may vary from 20% to 90% depending on the susceptibility of the flock. Necrosis of an enlarged liver, spleen and kidney is the most common necropsy finding in affected Muscovy ducklings. The disease can be experimentally reproduced by intramuscular inoculation with the virus.[18,42]

Four of 46 (8.7%) serum samples from commercial ducks in Northern Ireland were positive for duck reovirus antibodies, but none of the birds had antibodies to poultry strains of reovirus. In contrast, 19 of 100 (19%) chickens had antibodies against the duck reovirus.[1]

Reoviruses have been recovered from the feces of geese with diarrhea and Derzsy's disease.[11] A reovirus has been shown to cause myocarditis in 5- to 21-day-old goslings. Older geese are resistant to disease. Clinical signs in goslings are first noted three to six days after exposure and include depression, loss of appetite, nasal discharge, ocular discharge, conjunctivitis, dyspnea and occasionally diarrhea. Some affected goslings may develop neurologic signs. Survivors are stunted.[18]

Vertical transmission of reovirus has been documented in ducks and geese.[18] A modified-live vaccine available for chickens was shown to improve survival in vaccinated Muscovy ducklings; 85% of vaccinates survived.[43] Goslings can be provided temporary protection through subcutaneous administration of hyperimmune serum.

ORBIVIRUSES

The second genus of the family Reoviridae is *Orbivirus*. The largest group of orbiviruses that infects birds is transmitted by ticks. These orbiviruses have been recovered most commonly from murres, puffins, penguins and other sea birds.[7] Virus and antibodies to the virus have also been demonstrated in some land birds.[37,76] In general, clinical signs of disease have not been described in birds naturally or experimentally infected with these viral isolates.

A virus with properties similar to an orbivirus was isolated from a Cockatiel that died acutely, and from a Budgerigar that died following a brief period of depression and diarrhea. The Cockatiel had an enlarged liver and spleen, and its air sacs were cloudy. The Budgerigar had catarrhal enteritis and a swollen liver. Unlike reovirus isolates, this virus was found to be sensitive to pH 3.0. The viruses recovered from the Cockatiel and Budgerigar were antigenically related, and neither isolate cross-reacted with other orbiviruses. Experimentally, quail and chickens were found to be resistant to this virus, while Budgerigars developed diarrhea four to eight days after oral inoculation. The virus could be recovered from the feces during episodes of diarrhea. Infected birds recovered and developed neutralizing antibodies to the virus.[29]

ROTAVIRUSES

The third genus of the Reoviridae of importance to birds is *Rotavirus*. Rotaviruses can induce lysis of infected intestinal cells, resulting in severe diarrhea. While rotaviruses are most commonly associated with disease in young animals, they can occasionally cause clinical changes in adults, particularly those that are immunocompromised. This virus generally is excreted in the feces in high concentrations, and can survive in feces for months. Ingestion of virus-contaminated feces and oral contact with fomites are probably the primary routes by which susceptible birds are exposed to the virus. Cross-facility transmission is likely in contaminated aviaries with poor sanitation. Egg transmission of rotavirus has not been documented. Thus far, the rotaviruses that infect birds and mammals have been shown to be serologically distinct, and the transmission of rotavirus from birds to mammals is unlikely.[5]

Rotaviruses are frequently demonstrated in the feces of young domestic fowl with transient diarrhea. Affected young birds may die, while older birds usually recover. Various strains may replicate preferentially at different areas of the intestinal tract from the duodenum to the colon. Infected birds may be asymptomatic or may have dilated intestinal loops filled with watery material and gas *Figure 12.9*. The highest concentration of virus can be found in the feces from three to five days after infection; however, shedding can start as soon as one day after infection. The incubation period in companion and aviary birds is undetermined. Latent infections have not been demonstrated in chickens that recover from the primary infection.[47,75]

Rotaviruses have been recovered from pheasants, ducks, pigeons, lovebirds, guinea fowl and partridges.[22,66] Virus has been recovered in the feces of asymptomatic ducks and free-ranging pigeons, as well as from pigeons with watery diarrhea.[51,66,68] Infected guinea fowl recovered after a five- to eight-day period of enteritis.[18]

Rotaviruses have been reported as a cause of enteritis in pheasants from England and the United States.[22,23,74] Over a one-year period in Europe, rotavirus was recovered from the intestinal contents or feces of 15 of 72 (21%) pheasants and 6 of 28 (21%) partridges. Affected pheasant and partridge chicks ranged in age from six days to eight weeks. Clinical signs in

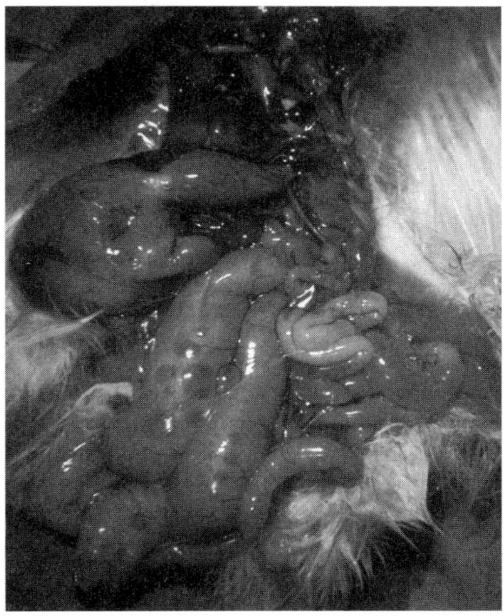

FIG 12.9
Birds with rotavirus infections may have no gross lesions or they may have dilated intestinal loops filled with watery material and gas.

photograph courtesy of Richard Davis

these birds included diarrhea, stunting and death (30% mortality).[19,21] Diarrhea can be induced by oral inoculation.[14,21] A rotavirus that was serologically related to turkey rotavirus was recovered from the intestinal tracts of a group of two- to three-week-old pheasant chicks experiencing diarrhea and high mortality.[57] The fact that rotaviruses isolated from pheasant chicks were serologically related to those recovered from turkeys suggests that turkeys could be a source of infection for pheasants and vice versa.

A seven-month-old male lovebird in England developed weight loss and paralysis of the left leg, then died. This bird was housed with six other birds that remained clinically normal. At necropsy, the gastrointestinal tract (including the crop) contained normal ingesta. The spleen and liver were enlarged and mottled. A virus with characteristics suggestive of a rotavirus was recovered from the liver and gastrointestinal contents of the affected bird. One- to two-day-old chickens experimentally infected with the lovebird isolate remained clinically normal following oral inoculation. It was undetermined if the rotavirus in this lovebird caused the observed clinical problems or was merely an incidental finding.[44] Rotavirus has also been isolated from the feces of an asymptomatic lovebird.[20]

Antibodies to rotaviruses have been demonstrated in ducks and pigeons.[45,48] In Belgium, 8 of 75 (11%) racing pigeons and 6 of 75 (8%) carrier pigeons from the same area had detectable levels of rotavirus antibodies.[70] In one study, up to 32% of free-ranging pigeons in Japan were seropositive.[51]

Birds that survive rotavirus infections develop antibodies that are passed to the chick through the egg; however, antibodies are not protective.[46] Antibodies can be detected in experimentally infected chickens and turkeys within four to six days of infection.[75] Because rotavirus infections are common in chickens and turkeys, detection of antibodies is of limited diagnostic value. Rotavirus may be detected in intestinal contents of suspect birds by electron microscopy or culture. Most infections are subclinical, and demonstration of rotavirus in the feces of a bird with diarrhea does not confirm a rotavirus-induced disease.

Rotaviruses are stable at pH 3, and are not completely inactivated by heating to 56°C for eight hours, or by exposure to 3% sodium hypochlorite even for two hours. Iodophors have been shown to inactive rotavirus in as little as ten seconds.[36,65] Infected birds shed large quantities of rotavirus in their feces, and the stability of the virus outside the bird can result in accumulation of virus in a bird's enclosure. A heavily contaminated environment can be a particular problem in the aviary or zoological facility in which birds have access to contaminated soil or ponds.

REFERENCES

1. Adair BM, Burns K, McKillop ER: Serological studies with reoviruses in chickens, turkeys and ducks. J Comp Path 97:495-501, 1987.

2. Afaleq AIA, Jones RC: Localization of avian reovirus in the hock joints of chicks after entry through broken skin. Res Vet Sci 48:381-382, 1990.

3. Ashton WLG, Alexander DJ: A two-year survey on the control of the importation of captive birds into Great Britain. Vet Rec 106:80-83, 1980.

4. Ashton WLG, Randall CJ, Dagless MD, et al: Suspected reovirus-associated hepatitis in parrots. Vet Rec 114:476-477, 1984.

5. Brussow H, Nakagomi O, Gerna G, et al: Isolation of an avian-like group A rotavirus from a calf with diarrhea. J Clin Microbiol 30:67-73, 1992.

6. Cadman HF, Kelly PJ, Zhou R, et al: A serosurvey using enzyme-linked immunosorbent assay for antibodies against poultry pathogens in ostriches (*Struthio camelus*) from Zimbabwe. Avian Dis 38:621-625, 1994.

7. Chastel C: Tick-borne virus infections of marine birds. Adv Dis Vect Res 5:25-60, 1988.

8. Clark FD, Ni Y, Collisson EW, et al: Characterization of avian reovirus strain-specific polymorphism. Avian Dis 34:304-314, 1990.

9. Clubb SL: A multifocal disease syndrome in African grey parrots (*Psittacus erithacus*) imported from Ghana. Proc Assoc Avian Vet, 1984, pp 135-150.

10. Clubb SL, Gaskin JM, Jacobson ER: Psittacine reoviruses. An update including a clinical description and vaccination. Proc Assoc Avian Vet, 1985, pp 83-90.

11. Csontos L, Csatari MMK: Etiologic studies on gosling influenza. Isolation of virus. Acta Vet Acad Sci Hung 17:107-114, 1976.

12. Deshmukh DR, Larsen CT, Dutta SK, et al: Characterization of pathogenic filtrate and viruses isolated from turkeys with bluecomb. Am J Vet Res 30:1019-1025, 1969.

13. Docherty DE, Converse KA, Hansen WR, et al: American woodcock (*Scolopax minor*) mortality associated with a reovirus. Avian Dis 38:899-904, 1994.

14. Foni E, Gelmetti D, Nigrelli WD, et al: Transmissible enteritis syndrome in pheasants for restocking: Experimental reproduction of the disease. Isolation of rotavirus. Selezione Vet 30:879-888, 1989.

15. Gaskin JM: Considerations in the diagnosis and control of psittacine viral infections. Proc Assoc Avian Vet, 1987, pp 1-14.

16. Gaskin JM: The serodiagnosis of psittacine viral infections. Proc Assoc Avian Vet, 1988, pp 7-10.

17. Gaskin JM: Psittacine viral disease: A perspective. J Zoo Wild Med 20:249-264, 1989.

18. Gerlach H: Viruses. *In* Ritchie BW, Harrison GJ, Harrison LR (eds): Avian Medicine: Principles and Application. Lake Worth, Wingers Publishing, 1994, pp 862-948.

19. Gough RE, Alexander DJ, Collins MS, et al: Routine virus isolation or detection in the diagnosis of diseases in birds. Avian Pathol 17:893-907, 1988.

20. Gough RE, Collins MS, Wood GW, et al: Isolation of a chicken embryo-lethal rotavirus from a lovebird (*Agapornis* sp). Vet Rec 122:363-364, 1988.

21. Gough RE, et al: Viruses and virus-like particles detected in samples from game birds in Great Britain during 1988. Avian Pathol 19:331-342, 1990.

22. Gough RE, Wood GW, Collins MS, et al: Rotavirus infection in pheasant poults. Vet Rec 116:295, 1985.

23. Gough RE, Wood GW, Spackman D: Studies with an atypical avian rotavirus from pheasants. Vet Rec 118:611-612, 1986.

24. Graham DL: An update on selected pet bird virus infections. Proc Assoc Avian Vet, 1984, pp 267-280.

25. Graham DL: Characterization of a reo-like virus and its isolation from and pathogenicity for parrots. Avian Dis 31:411-419, 1987.

26. Guy JS, Levy MG, Ley DH, et al: Interaction of reovirus and *Cryptosporidium bailey* in experimentally infected chickens. Avian Dis 32:381-390, 1988.

27. Guy JS, Levy MG, Ley DH, et al: Experimental reproduction of enteritis in bobwhite quail (*Colinus virginianus*) with cryptosporidium and reovirus. Avian Dis 31:713-722, 1987.

28. Heffels U, Fritzsche K, Kaleta EF, et al: Serological examination for viral infections in pigeons in Germany. Dtsch Tierartzl Wschr 88:97-102, 1981.

29. Hirai K, Hitchner SB, Calnek BW: Characterization of paramyxo- herpes- and orbiviruses isolated from psittacine birds. Avian Dis 23:148-163, 1979.

30. Ide PR, Dewitt W: Serological incidence of avian reovirus infection in broiler breeders and progeny in Nova Scotia. Can Vet J 20:348-353, 1979.

31. Jackson GG, Muldoon RL: Viruses causing common respiratory infection in man. IV. Reoviruses and adenoviruses. J Infect Dis 128:811-866, 1973.

32. Jones RC, Georgiou K: Reovirus-induced tenosynovitis in chickens: The influence of age at infection. Avian Pathol 13:441-457, 1984.

33. Jones RC, Guneratne JRM: The pathogenicity of some avian reoviruses with particular reference to tenosynovitis. Avian Pathol 13:173-189, 1984.

34. Jones RC, Islam MR, Kelly DF: Early pathogenesis of experimental reovirus infection in chickens. Avian Pathol 18:239-253, 1989.

35. Jones RC, Onunkwo O: Studies on experimental tenosynovitis in light-hybrid chickens. Avian Pathol 7:171-181, 1978.

36. Kang SY, Nagaraja KV, Newan JA: Physical, chemical and serological characterization of avian rotaviruses. Avian Dis 32:195-203, 1988.

37. Karabatsos N: International Catalogue of Arboviruses 1985 Including Certain Other Viruses of Vertebrates. American Society of Tropical Medicine Hygiene, 1985.

38. Kibenge FSB, et al: Effects of experimental immunosuppression on reovirus-induced tenosynovitis in light-hybrid chickens. Avian Pathol 16:73-92, 1987.

39. Kibenge FSB, Gwaze GE, Jones RC, et al: Experimental reovirus infection in chickens: Observations on early viremia and virus distribution in bone marrow, liver and enteric tissues. Avian Pathol 14:97-98, 1985.

40. Kibenge FSB, Wilcox GE: Tenosynovitis in chickens. Vet Bull 53:431-444, 1983.

41. Ley DH: Intestinal cryptosporidiosis and reovirus isolation from young pen-raised bobwhite quail with severe diarrhea and high mortality. Proc 34th West Poult Dis Conf, 1985, pp 93-95.

42. Malkinson M, Perk K, Weisman Y: Reovirus infection of young Muscovy ducks. Avian Pathol 10:443-440, 1981.

43. Marius V: Characterization of a reovirus isolated from Muscovy ducks. Study of virulence for different species of fowl and waterfowl. Studies of immunity afforded by reovirus vaccine (S1133). WPSA Europ Waterfowl Dis Symp, 1981.

44. McFerran JB, Conner TJ, McCraken RM: Isolation of adenoviruses and reoviruses from avian species other than domestic fowl. Avian Dis 20:519-524, 1976.

45. McNulty MS: Morphology and chemical composition of rotaviruses. *In* Bricout F, Scherrer R (eds): Viral Enteritis in Humans and Animals. Paris, INSERM, 1979, pp 111-140.

46. McNulty MS: Rotavirus infections. *In* Calnek BW, et al (eds): Diseases of Poultry. Ames, Iowa State University Press, 1991, pp 628-635.

47. McNulty MS, Allan GM, McCracken RM: Experimental infection of chickens with rotavirus: Clinical and virological findings. Avian Pathol 12:45-54, 1983.

48. McNulty MS, Allan GM, Stuart JC: Rotavirus infection in avian species. Vet Rec 103:319-320, 1978.

49. Menendez NA, Calnek BW, Cowen BS: Experimental egg-transmission of avian reovirus. Avian Dis 19:104-111, 1975.

50. Meulemans G, Dekegel D, Charlier R, et al: Isolation of orthoreoviruses from psittacine birds. J Comp Path 93:127-134, 1983.

51. Minamoto N, Oki K, Tomita M, et al: Isolation and characterization of rotavirus from feral pigeons in mammalian cell cultures. Epid Inf 100:481-491, 1988.

52. Mohan R: Clinical and laboratory observations of reovirus infection in a cockatoo and a grey-cheeked parrot. Proc Assoc Avian Vet, 1984, pp 29-34.

53. Montgomery RD, Villegas P, Dawe DL, et al: A comparison between the effect of an avian reovirus and infectious bursal disease virus on selected aspects of the immune system of the chicken. Avian Dis 30:298-308, 1986.

54. Page RK, Fletcher OJ, Rowland GN, et al: Malabsorption syndrome in broiler chickens. Avian Dis 26:618-624, 1982.

55. Page RK, Fletcher OJ, Villegas P: Infectious synovitis in young turkeys. Avian Dis 26:924-927, 1982.

56. Pieper K, Kaleta EF: Virus isolations from psittacine birds. Proc Europ Chap Assoc Avian Vet, 1991, pp 199-201.

57. Reynolds DL, Theil KW, Saif YM: Demonstration of rotavirus and rotavirus-like virus in the intestinal contents of diarrheic pheasant chicks. Avian Dis 31:376-379, 1986.

58. Rigby CE, Pettit JR, Papp-Vid G, et al: The isolation of salmonellae, Newcastle disease virus and other infectious agents from quarantined imported birds in Canada. Can J Comp Med 45:366-370, 1981.

59. Ritter DG, Ley DH, Levy M, et al: Intestinal cryptosporidiosis and reovirus isolation from bobwhite quail (*Colinus virginianus*) with enteritis. Avian Dis 30:603-608, 1986.

60. Robertson MD, Wilcox GE: Avian reoviruses. Vet Bull 56:155-174, 1986.

61. Rosenberger JK: Reovirus interference with Marek's virus vaccination. Proc 32nd West Poultry Dis Conf, 1983, pp 50-51.

62. Rosenberger JK, Olson NO, van der Heide L: Viral arthritis and other reoviruses. *In* Purchase HG, et al (eds): A Laboratory Manual for the Isolation and Identification of Avian Pathogens 3rd ed. Dubuque, Kendall/Hunt Publishing Co, 1989, pp 157-160.

63. Rosenberger JS, Sterner FJ, Butts S, et al: In vitro and in vivo characterization of avian reoviruses. Avian Dis 33:535-544, 1989.

64. Seene DA, Pearson JE, Miller LD, et al: Virus isolations from pet birds submitted for importation into the United States. Avian Dis 27:731-744, 1983.

65. Snodgrass DR, Herring JA: The action of disinfectants on lamb rotavirus. Vet Rec 101:81, 1977.

66. Takase K, Nonaka F, Sakaguchi M, et al: Cytopathic avian rotavirus isolated

from duck feces in chicken kidney cell cultures. Avian Pathol 15:719-730, 1986.

67. Tang KN, Fletcher OJ, Villegas P: The effect on newborn chicks of oral inoculation of reovirus isolated from chickens with tenosynovitis. Avian Dis 31:584-590, 1987.

68. Tudor DC: Pigeon Health and Disease. Ames, Iowa State University Press, 1991.

69. Van der Heide L, Kalbec M, Hall WC: Infectious tenosynovitis (viral arthritis): Influence of maternal antibodies on the development of tenosynovitis lesions after experimental infection by day-old chickens with tenosynovitis virus. Avian Dis 30:641-648, 1976.

70. Vindevogel H, Dagenais L, Lansival B, et al: Incidence of rotavirus, adenovirus, and herpesvirus infection in pigeons. Vet Rec 109:285-286, 1981.

71. Vindevogel H, Meulemans G, Pastoret PP, et al: Reovirus infection in the pigeon. Ann Rech Vet 13:149-152, 1982.

72. Wainright PO, Pritchard NG, Fletcher OJ, et al: Identification of viruses from Amazon parrots with a hemorrhagic disease and a chronic respiratory disease. Proc Assoc Avian Vet, 1987, pp 15-19.

73. Wilson RB, Holscher M, Hodges JR, et al: Necrotizing hepatitis associated with a reo-like virus infection in a parrot. Avian Dis 29:568-571, 1985.

74. Yason CV, Schat KA: Isolation and characterization of avian rotaviruses. Avian Dis 29:499-508, 1985.

75. Yason CV, Schat KA: Experimental infection of specific-pathogen-free chickens with avian rotaviruses. Avian Dis 30:551-556, 1986.

Orthomyxoviridae

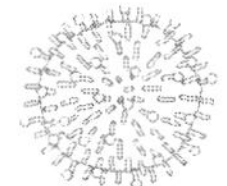

The family Orthomyxoviridae contains the influenza viruses, which are divided into three types (A, B and C). The pleomorphic virions are enveloped and usually measure 60 to 120 nm in diameter. Influenza A viruses are most notable and have been recovered from a wide range of hosts including humans, other mammals and birds. Influenza A virus is the only type that is currently considered of veterinary importance. It has been associated with epornitics of respiratory disease in a number of mammalian and avian species.[35] Influenza B and C viruses have principally been recovered from humans. However, during an influenza C virus outbreak in Hungary, 4% of the captive and free-ranging birds tested had antibodies to the virus. Additionally, common pheasants and Mallard Ducks have been shown to be susceptible to experimental infection.[21]

INFLUENZA A VIRUSES

Two identification schemes are used to describe influenza A virus isolates. The first is based on the variances in the hemagglutinin (H) and neuraminidase (N) proteins found on the surface of the virus. Currently, 14 different hemagglutinin proteins and 9 different neuraminidase proteins have been detected. Individual subtypes of the influenza A viruses found in birds can be composed of any mixture of one of the hemagglutinin and one of the neuraminidase proteins. For example, a subtype of virus designated H5N9 contains hemagglutinin protein number 5 and neuraminidase protein number 9.[28] The second identification scheme has five parameters: the type of virus, host from which the virus was originally recovered, geographic location from which the original isolate was made, reference number and year of isolation. Thus, A/duck/Ireland/113/84 (H5N8) is a type A influenza virus that was recovered from ducks in Ireland in 1984, and the isolate contains the number 5 hemagglutinin and the number 8 neuraminidase protein.

The hundreds of subtypes of influenza A viruses that have been isolated from free-ranging birds, domestic poultry, humans, swine and horses are related. This group of viruses undergoes constant change, resulting in the frequent appearance of new serotypes to which a population of hosts do not have immunity *see Figure 3.29*. This is why "flu" epidemics are common year after year. Large congregations of migratory birds, particularly waterfowl, are thought to serve as reservoirs for the virus.[25] Investigations into the prevalence of influenza A viruses in captive birds are limited; however, there is potential for many avian species to be infected by the virus.[34,70]

CLINICAL FEATURES

In the 1890s, a disease causing high mortality in domestic poultry was first

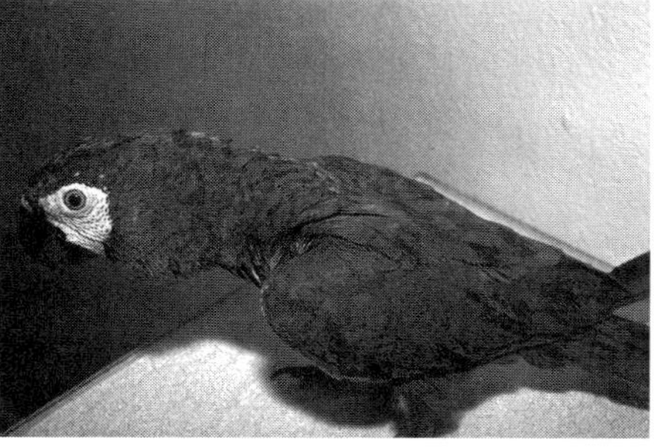

shown to be associated with a filterable agent, and was termed fowl plague.[4] In 1955, the disease was shown to be caused by an influenza virus related to those that infect mammals.[60] Most species of birds of all ages are considered susceptible to influenza A viruses.[11] The morbidity and mortality associated with an infection vary widely with the species of bird and the strain of infecting virus.[7] Birds infected with less virulent strains of the virus usually remain asymptomatic, while highly susceptible birds infected with virulent strains may develop clinical signs and die. Many isolates of influenza A virus are pathogenic in one species and apathogenic in others. Because influenza A viruses are commonly recovered from asymptomatic birds and the clinical changes induced in experimentally infected birds may be less severe than those that occur in natural outbreaks, many of the clinical changes that are attributed to the virus may actually be caused by other organisms or concomitant disease processes.[32] For example, influenza A virus has been recovered from asymptomatic ducks, quail, pheasants and psittacines, as well as from clinically ill individuals of the same species.[43,52]

When clinical changes are present they may include mild-to-severe respiratory signs, depression, anorexia, decreased egg production, diarrhea or edema of the head and neck *Figure 13.1*. Many affected birds recover in two to three weeks. An influenza A virus-induced decrease in egg production has been demonstrated in chickens, turkeys, Japanese Quail and ducks. The virus also should be considered in other species that are having egg-related problems during "flu" epidemics in people. In chickens, egg-related peritonitis has been described as a sequela to influenza A virus infections *Figure 13.2*.[11] Highly virulent strains of influenza A virus cause a viremia that is associated with lymphopenia and damage to endothelial cells, resulting in bleeding disorders.

Influenza A virus has been recovered from asymptomatic psittacine birds as well as from those that die suddenly or following an acute onset of depression, diarrhea and neurologic signs.[41,43] Other infected psittacines may develop a two-week course of lethargy, upper respiratory disease and neurologic signs, including ataxia and torticollis. A strain of influenza A virus related to one recovered from chickens was isolated from an African Grey Parrot that died several days after becoming depressed and passing dark green feces.[21] In a quarantine station, influenza A virus was recovered from the liver, spleen and intestines of a Stanley Parakeet, two Yellow-fronted Kakarikis and a mynah that died one to two days after exhibiting depression. The virus recovered from these birds was not infectious to poultry.[62] Influenza A virus was recovered from a cockatoo that was ataxic and exhibited head-swaying.[5] In companion birds, influenza A virus should be considered a possible cause of respiratory and gastrointestinal disease, particularly when family members are ill or there is an epidemic occurring in humans.

Influenza A virus was recovered from a group of turacos that died following a two-week period of lethargy, anorexia and severe dyspnea. Twenty percent of the affected birds died, and the survivors developed hemagglutination-inhibition (HI) antibodies. The recovered virus was not infectious to Gruiformes, Columbiformes, Psittaciformes or Piciformes.[24] Influenza A virus was recovered from a Siskin with conjunctivitis. The isolate caused conjunctivitis and death in canaries two to four days after inoculation.[21] Mynahs with influenza A virus developed respiratory signs and passed bloody stools.[50]

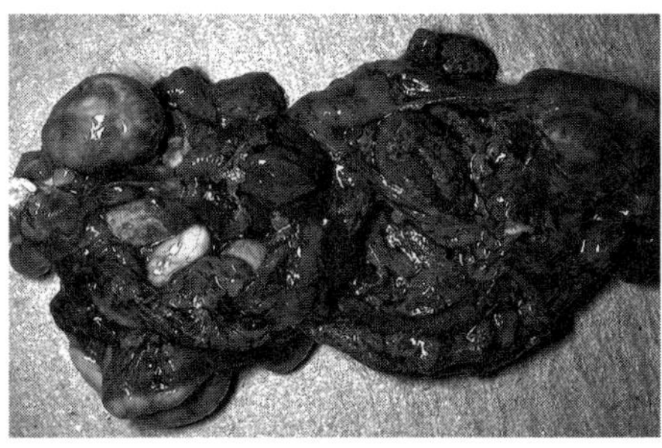

FIG 13.2

In chickens, influenza A virus infections have been associated with egg-related peritonitis. It is undetermined if this virus can cause a similar change in other birds.

In free-ranging waterfowl, influenza A viruses are host-adapted and relatively apathogenic; most infections are asymptomatic. Free-ranging waterfowl are considered refractory to even the most pathogenic subtypes of influenza A virus, whereas domestic ducks appear to be more susceptible.[2,6,56,69] An epornitic caused by influenza A virus in South African Terns resulted in high levels of mortality, which provides sharp contrast to the usually asymptomatic infections in waterfowl. This is the only report of influenza A virus-induced disease in any free-ranging bird.[12]

Influenza A viruses have been recovered from domestic ducks with respiratory and ocular diseases, but these viruses also are recovered frequently from asymptomatic birds.[52] During an influenza A virus outbreak on a mixed species bird farm, the primary clinical change in affected young and adult ducks was sneezing. Older ducks recovered within a two-week period, while 10% of the ten-week-old domestic ducklings died. Turkeys and geese exposed to the ducks remained asymptomatic, but antibodies to the virus were detected in the recovered ducks and in the other birds on the farm.[33] Some studies suggest that influenza A virus can cause growth retardation in ducklings.[56]

Virulent strains of influenza A virus cause an acute onset of clinical disease associated with severe hemorrhagic lesions in domestic fowl. An influenza A virus was isolated from a group of young (two- to eight-week-old) pheasant chicks experiencing high mortality (up to 35%). Adult pheasants remained asymptomatic but developed antibodies to the virus, indicating that they had been infected. The virus recovered from these pheasants did not cause disease in experimentally infected chickens or ducks.[18] During an outbreak in a mixed waterfowl and pheasant farm in the northeastern United States, up to 100 pheasants per day were dying.[23] Some Helmeted Guinea Fowl infected with influenza A virus died without clinical signs; others exhibited decreased egg production, early embryonic mortality, respiratory disease and incoordination. The subtype recovered from these guinea fowl was not infectious for chickens or ducks.[68] Influenza A virus has been associated with 15% to 80% mortality in various flocks of Japanese Quail. The clinical changes in affected quail included sneezing, oculo-nasal discharge, swelling of the sinuses and difficulty breathing. Some birds may develop neurologic signs that include ataxia and convulsions.[24,58] Some subtypes of influenza A virus that are highly pathogenic for chickens and turkeys may cause mild infections in quail, indicating a species variation in the susceptibility to disease.

An influenza virus recovered from clinically affected Budgerigars caused diarrhea in seven of twelve (58%) and death in two of twelve (17%) experimentally infected birds. Budgerigars that were in good condition remained asymptomatic after inoculation, while those in poor condition developed diarrhea and died. The infected birds did not develop HI antibodies, but the virus was recovered from the respiratory tract.[21,36]

In one experimental trial using eight different subtypes of influenza A virus, two-week-old chickens, ducks, quail and turkeys were exposed to virus by the intramuscular or intranasal route, or were placed in direct contact with experimentally inoculated birds. The morbidity and mortality levels with the differing subtypes varied from none to 100% depending on the susceptibility of the various hosts. All of the inoculated ducklings remained asymptomatic. Only two of the subtypes caused consistent seroconversion in the ducklings. However, four subtypes were reisolated from the asymptomatic birds: one from the cloaca only and the other three from the cloaca and trachea. Two subtypes caused clinical signs or death in chickens and turkeys. One of the subtypes was more pathogenic for chickens and the other was more pathogenic for turkeys. Five subtypes caused clinical signs or mortality in quail. The frequency of virus transmission to contact controls varied with the host and the subtype.[7]

Ducks, geese and pigeons remained asymptomatic when infected with an influenza A virus that caused severe disease in chickens and turkeys. The infected birds seroconverted but did not transmit the virus to turkeys with which they were in contact.[45] The same virulent subtype of virus caused asymptomatic infections in pheasants, torticollis and death in 3 of 20 (15%) inoculated quail, and depression and death in 1 of 19 (5%) inoculated pigeons.[64] In another study, experimentally infected adult Mallard Ducks remained asymptomatic but developed microscopic changes consistent with pneumonia.[16] A subtype of virus isolated from a duck caused ocular discharge in one of five (20%) and diarrhea in one of five (20%) experimentally infected four-week-old Pekin ducklings inoculated intratracheally.[73] Other ducks infected with a virulent subtype of virus recovered from chickens remained asymptomatic or developed swelling around the eyes and blindness.[6] Influenza A virus was found to decrease egg production for the week after inoculation in Mallard Ducks. No other reproductive problems were noted, and the ducks returned to normal egg production after one week.[40]

A pathogenic strain of virus recovered from chickens caused 100% mortality in starlings and 30% mortality in house sparrows.[47] Starlings experimentally infected by intratracheal or intravenous or intramuscular inoculation with a turkey isolate did not develop HI antibodies nor could virus be recovered from the cloaca.[59] Another isolate from a chicken caused a mild transient disease, virus excretion and high levels of HI antibodies in pheasants infected by the intratracheal and oral routes.[75]

EPIZOOTIOLOGY

Influenza A virus has a worldwide distribution and has been recovered from many orders of birds, including those that are commonly maintained as companions, in aviaries or in zoological parks *Table 13.1*.

Virus activity has been demonstrated in captive and free-ranging birds, and is

TABLE 13.1

Orders of birds from which influenza A virus has been recovered[9,15,18,19,21,24]

Anseriformes	Galliformes	Podicepediformes
Casuariiformes	Gaviiformes	Procellariiformes
Charadriiformes	Gruiformes	Psittaciformes
Ciconiiformes	Passeriformes	Rheiformes
Columbiformes	Pelecaniformes	Sphenisciformes
Falconiformes	Piciformes	Struthioniformes

particularly common in free-ranging ducks, shorebirds and many passerines, which may serve as an immense reservoir for the virus.[2,71] Influenza A virus is considered enzootic in most flocks of free-ranging and domestic ducks, but infected birds rarely develop clinical signs. In one study of 14,303 free-ranging waterfowl, influenza A viruses were recovered from 15%. In the same study, only 2.2% of the tested Passeriformes were found to be shedding the virus. It is of interest that when the ducks were removed from the study population, the virus shedding rate dropped to 2.2% for the other waterfowl tested.[66] Other studies have indicated that the rate of influenza A virus shedding in migratory waterfowl in North America ranges from 3% to 30%.[65,72]

In a study in Maryland, attempts were made to recover influenza A virus from the feces of 5,013 birds of 16 species during a 33-month period. The virus was recovered from 70 of 3,403 (2%) of the gulls but only from 0.4% of the mute swans. Virus was not recovered from 452 Whistling Swans, 3 Canada Geese, 30 terns, 20 herons, 30 domestic quail, 30 Mallard Ducks, 84 finches, 65 sparrows, 35 cardinals, 27 starlings, 70 Rock Doves nor a Mourning Dove. Even though the virus was not recovered from the Whistling Swans, antibodies to the virus were detected in 43% of these birds. Interestingly, virus was recovered from the gulls more frequently in the warm months than the cold months.[22] This finding is in contrast to influenza epidemics in humans, which are most common during the early winter. A higher incidence of virus activity during the summer has also been reported in flocks of ducks and turkeys.[25,31]

Investigators in India failed to isolate the virus from cloacal swabs of 311 domestic and free-ranging birds.[38] Antibodies against influenza A virus were detected in free-ranging Adelie Penguins at various locations in Antarctica, suggesting that these viruses are widespread and persistent even in this remote region.[9] Antibodies to influenza A virus were not detected in 322 or 442 serum samples of free-ranging turkeys from the southeastern United States or Texas, respectively.[17,49] These findings would suggest that free-ranging turkeys are not a major reservoir for influenza A virus or like many ducks, they do not develop a sustained or detectable level of antibodies following natural infection.

To protect domestic poultry, the companion and aviary birds that are presented for importation into many coun-tries are screened for influenza A virus during quarantine. The virus has been recovered from quarantined birds in the United States, England, Austria and Japan.[3] Influenza A virus was isolated from eight psittacine bird consignments and one finch consignment in 109 groups of birds presented for importation into Canada.[55] Influenza A virus was isolated from 41 of 475 (8.6%) birds in a United States quarantine station, and 20% of the hemagglutinating viruses isolated over an eight-year period from imported canaries, finches and psittacines were influenza A viruses.[61] Over a six-month period, 15 isolates of influenza A virus were made from 12 species of birds, including mynahs, finches and psittacines, that were presented for importation into California. All the birds from which the virus was recovered had originated in Thailand, or had been in contact with birds from Thailand.[13,63] In other studies, the virus was recovered from the tissues of 31 of 475 (6.5%) mynahs in quarantine and from companion bird species that arrived dead at a Tokyo airport.[46,63]

<u>RELATIONSHIP OF VIRUS STRAINS</u> — The outer surface of an influenza A virus contains 1 of 14 different hemagglutinin and 1 of 9 different neuraminidase proteins. The antigenic and pathogenic characteristics of the virus are dependent upon the combination of these two proteins. For example, a virus with the number 5 hemagglutinin protein and the number 7 neuraminidase protein (H5N7) would be antigenically different from a virus that contained the number 4 hemagglutinin and number 9 neuraminidase proteins (H4N9). Irre-

spective of these differences, a particular subtype of virus is likely to be responsible for the majority of disease that is induced by the virus during a specified period in a group of exposed birds. As a hypothetical example, an H5N5 isolate might cause a disease outbreak in chickens, turkeys, psittacine birds, pigs and humans during a particular two- to three-month period. During a subsequent outbreak, an H7N5 isolate may be the virus that is most commonly recovered *see Figure 3.29.*

Isolates of influenza A virus recovered from mynahs, finches, bluebirds and orioles were found to be serologically related to the subtypes recovered from some ducks, indicating the potential for transmission across species.[21] The influenza A viruses isolated during a 1979 epidemic in swine from Belgium were antigenically similar to the strains isolated from ducks in North American and Germany. In this outbreak, it was uncertain whether the influenza virus was the actual cause of disease or if the virus isolated was merely a passenger in the infectious process. This was the first evidence that the influenza A viruses recovered from birds are naturally capable of infecting mammals.[51] Further studies have indicated that strains of influenza A virus isolated from ducks can cause a mild disease in pigs, and subtypes isolated from humans will cause mild disease in chickens.[21]

TRANSMISSION — Various subtypes of the virus have been recovered from the respiratory tract, gastrointestinal tract and ocular secretions of numerous avian species. Some investigators report that virus can be recovered most frequently from the gastrointestinal tract, while others report that the virus is most commonly excreted from the respiratory tract.[10,13,65,72] Observations of natural influenza A virus infections and experimental trials indicate that the duration of virus shedding varies with the virulence of the virus strain, time of the year, and the species, age and condition of the host.[7] Highly virulent subtypes of virus are likely to cause the rapid death of a susceptible host and are less likely to be spread through a flock than less virulent strains that cause a protracted disease and are shed from infected birds over a longer period of time. The incubation period can be several hours to several days depending on the virulence of the subtype and the susceptibility of the host.

Ducks experimentally infected by nebulization remained asymptomatic and shed virus in the trachea from one to three days and in the cloaca from three to ten days after infection. At necropsy, the virus was recovered from the small intestines, cecum, large intestines and bursa.[65] In Pekin ducklings inoculated intratracheally, virus was detected in the trachea and feces from two to ten days after infection in the ducks that developed clinical signs of disease. The virus was not recovered from the trachea or cloaca of asymptomatic ducks.[72] In other birds, virus shedding may start within five days after infection and persist for over four weeks.[7,26] Virus was detected in the cloaca of turkeys by the seventh day after they were infected by the intratracheal or

Captive birds that are directly exposed to migratory waterfowl are at greater risk of developing an influenza A infection than are birds that are in isolated enclosures.

intravenous route with an isolate from a turkey.[59]

Investigations into the incidence of influenza A virus in companion bird species are limited. The precise role that free-ranging birds play in viral transmission to other species remains largely speculative; however, it is suspected that birds serve as a reservoir for influenza A viruses, and that migratory birds (particularly ducks and shore birds) disseminate these viruses worldwide.[6,44,48,74] It has also been suggested that free-ranging pheasants and quail can be asymptomatically infected with and shed influenza A viruses that are pathogenic to turkeys and chickens.[75] Outbreaks in captive birds are frequently associated with direct or indirect exposure to infected free-ranging birds. Captive pheasants and guinea fowl should be considered at risk during epornitics in turkeys.[53]

As evidence for the importance of free-ranging birds in the transmission of influenza A viruses, outbreaks are rare in chickens that are raised in houses that restrict the access of free-ranging birds, while outbreaks are common in turkeys and domestic ducks that are raised in open fields. Additionally, virus activity is more common in domestic ducks than in turkeys, and is more common in turkeys than in chickens. Virus activity is particularly common in turkeys that are raised along migratory flyways used by waterfowl,

especially by free-ranging ducks. A similar situation could be expected in the waterfowl exhibits of zoological parks that are frequented by migratory waterfowl *Figure 13.3*. In one study, an epornitic in turkeys was controlled by placing the birds in areas that restricted the access of free-ranging birds.[39]

Influenza A viruses are principally transmitted through direct contact with feces and aerosols from infected birds. Bodies of water contaminated with virus-laden feces are also considered important sources for virus exposure in susceptible ducks, particularly in small, overcrowded ponds or lakes.[29] Because the virus is frequently recovered from contaminated ponds and young waterfowl, but is rarely recovered from adults, it is theorized that virus activity is maintained in flocks of waterfowl through the continued exposure of young, immunologically naive birds to contaminated ponds.[42] In one study, 60% of young ducks were found to be shedding influenza.[30]

Evidence for vertical transmission is limited. Some researchers believe that vertical transmission does not occur. Others report the recovery of the virus from the yolk, albumen and shell of eggs contaminated with virus-laden feces.[14] The virus also has been recovered from the egg and embryos of infected gulls.[57] It is likely that egg shells contaminated with virus-laden feces could result in transfer of the virus from one flock to another, or could serve as a source of virus for recently hatched chicks in the same incubator.[14]

Direct animal-to-man transmission of influenza has been shown to occur, but is thought to be rare.[37] Nonetheless, evidence suggests that birds and swine may serve as reservoirs for influenza subtypes, which may themselves, or following mutation or recombination with human subtypes of the virus, produce new subtypes of virus to which humans have not developed immunity.[2,28] In this manner, birds can serve as a reservoir for the subtypes of virus that cause pandemics in humans.

A companion bird theoretically could serve as a source of virus exposure for humans, but it is more likely that humans serve as a source of virus exposure for susceptible companion birds. Influenza should be suspected in companion birds with mild upper respiratory disease during "flu" season, particularly if family members in direct contact with the bird currently have or recently have had characteristic clinical signs of the "flu" *Figure 13.4*. Humans should also be considered a mechanical vector for transmission if their shoes or clothing have been contaminated with virus-laden feces.

PATHOLOGY

The gross changes associated with influenza virus infections vary with the host and the virulence of the subtype. Many affected birds that die soon after they are infected with a virulent strain of virus do not develop gross lesions.[11] In others, an accumulation of necrotic debris in the sinuses, air sacs and trachea, hemorrhage in the brain, splenomegaly and congestion or petechial hemorrhage of the lungs or intestinal tract may be noted *Figure 13.5*.[21,62] Congestion and accumulation of fluid in the

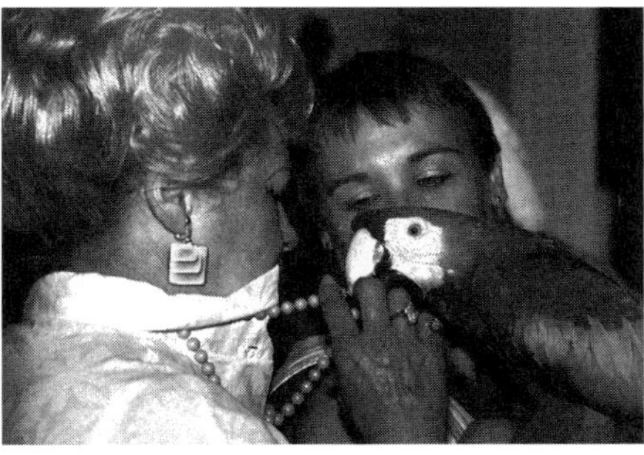

FIG 13.4
People are frequently in close contact with their companion birds. It is possible that a bird could infect a person with influenza A virus; however, it is more likely that an infected person would infect a companion bird. Influenza should be suspected in any companion bird with mild upper respiratory disease during "flu" season.

intestines were the principal gross changes in affected mynahs.[21,50,62] Hemorrhage and edema of the heart have been described in chickens and turkeys.[44] Congestion of the digestive tract was the primary gross finding in an affected Yellow-crowned Amazon Parrot, Plum-headed Parakeet, Ring-necked Parakeet, Singing Parrot and Lesser Sulphur-crested Cockatoo.[61,62] Influenza A virus was recovered from a group of turacos with hyperemic lungs, liver and kidneys.[24] Gross lesions in a group of affected pheasant chicks included an enlarged spleen, cloudy air sacs and accumulation of mucus in the trachea and sinuses.[18]

Histologic lesions associated with influenza A virus infections may include catarrhal or hemorrhagic tracheitis, hemorrhagic enteritis, air sacculitis, sinusitis and pneumonia.[18,33,58] Affected mynahs had hemorrhagic tracheitis, pneumonia, air sacculitis and hemorrhagic enteritis.[50,64]

IMMUNITY AND DIAGNOSIS

During infection, the body's immune system responds to proteins on the surface of the influenza A virus. Many species of mammals and birds infected with influenza A virus will develop antibodies to the virus that assist in clearing the infection and that protect

FIG 13.5

*The gross changes associated with influenza A virus infections vary with the species. Some common findings include: **A** accumulation of necrotic debris in the sinuses; **B** cloudy air sacs (arrow); **C** hemorrhage in the brain; and **D** congestion or petechial hemorrhage of the lungs or **E** intestinal tract.*

photographs courtesy of Gilles Berneir and Jean Sander

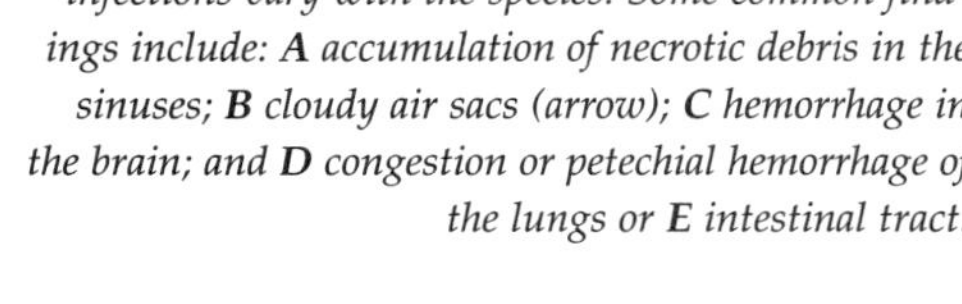

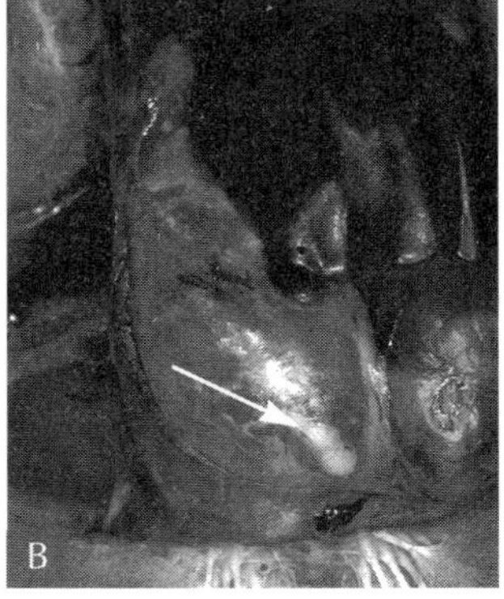
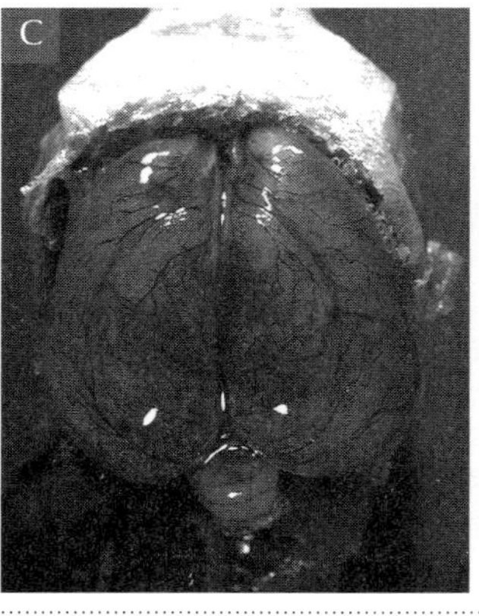
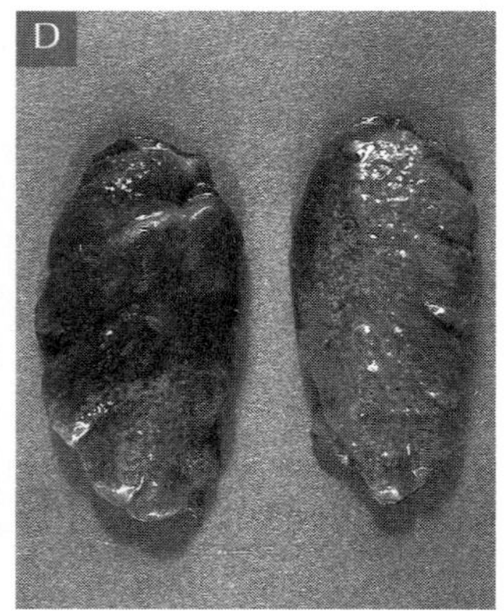
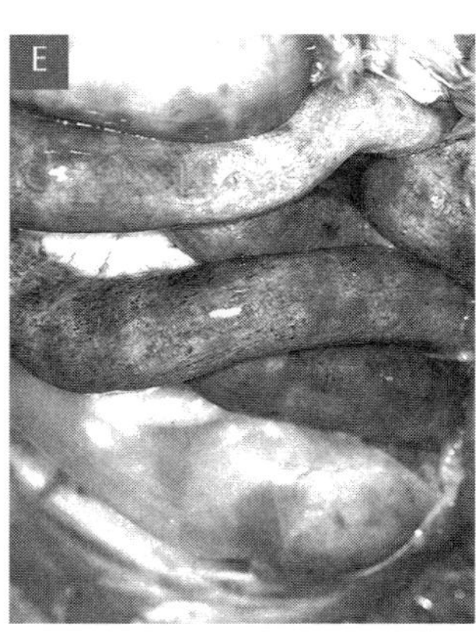

the host from infection by the same subtype. However, annual influenza A virus epornitics in birds and pandemics in humans occur because the changes that constantly occur to the proteins on the surface of the virus result in subtypes to which a host has not developed immunity. Thus, a bird, human or other mammal that has recovered from one subtype of influenza A virus can be infected with different subtypes in subsequent years. Antibodies to the virus have been demonstrated in numerous species of birds including geese, ducks, pheasants, quail, gulls, passerines, swans and flamingos. It has been demonstrated that many species of free-ranging ducks do not develop HI antibodies to the virus following natural or experimental infections.[1,2,26,64] In contrast, exposed domestic ducks developed antibodies to the virus during an outbreak on a commercial farm.[33]

Influenza A virus infections are diagnosed best by isolation of the virus from cloacal swabs, although tracheal swabs are effective in some cases. The best time to collect samples for virus isolation varies widely depending on the strain of virus and the species of bird. In chickens, virus is easiest to recover from several days to 14 days after infection.[11] During epornitics in captive waterfowl, water from ponds or lakes can be cultured to determine whether it is contaminated with the virus. At necropsy, virus isolation can be attempted from the trachea, lung, air sacs, sinuses, spleen, liver, intestines and cloaca.[11,62] Antibodies to the virus can be detected using a group-specific antigen in the agar-gel immunodiffusion assay; however, this assay is relatively insensitive in comparison to others. Virus neutralization and ELISA can also be used to demonstrate antibodies to specific sub-

AVIAN VIRUSES: FUNCTION AND CONTROL

types of the virus. Any respiratory or gastrointestinal pathogen, including bacteria, fungus, *Chlamydia* sp., *Mycoplasma* sp., paramyxovirus, reovirus and some strains of herpesvirus, can cause clinical signs similar to those associated with influenza A virus.

CONTROL

Although influenza A viruses are enveloped, they are considered relatively stable when outside the host, particularly in pond or lake water. One subtype of the virus was shown to be stable in lake water for four days at 22°C or for 30 days at 0°C.[73] Another subtype was even more stable in water and remained infectious in water for 102 days at 28°C and for 207 days at 17°C.[67] In cool areas, virus may remain infectious in feces for over one month. The infectivity of these viruses can be destroyed in a matter of minutes by extremes in pH, heating to 56°C, exposure to sunlight and by most detergents and disinfectants.[11] Infections in captive birds can be prevented by reducing their exposure to infected free-ranging birds (particularly waterfowl and shore birds). The infectivity of virus in a contaminated pond or lake may be reduced by: 1) aerating the lake to bring virus to the surface where it is destroyed by exposure to the sun; 2) draining the lake and allowing the contaminated soil to dry for over a month; or 3) draining, cleaning and disinfecting an artificial pond.

Because many of the clinical problems associated with influenza A virus infections are caused by secondary infections, treatment that includes supportive alimentation, fluids and antibiotics (when needed) would be expected to increase survival. Birds with clinical signs suggestive of an infection should be isolated from other birds while they are ill and for three to four weeks after the clinical signs have resolved. In many countries, influenza A is a reportable disease.

Subtype-specific vaccines have been developed for controlling influenza A viruses in domestic fowl in some areas of the United States. These attenuated-live virus vaccines have limited application because of the speed with which influenza A viruses can change, and because of the capacity of the subtypes of virus found in the vaccines to recombine with field subtypes to create new antigenically distinct viruses. An inactivated vaccine has been shown to reduce the severity of disease in poultry, but does not prevent infection or shedding of the virus. Additionally, these vaccines may favor the natural selection of more virulent strains of the virus.[27,54]

INFLUENZA A VIRUS IN RATITES

Influenza A viruses have been recovered from ostriches, emus and rheas. In one outbreak, a subtype was recovered from emus and rheas in Texas and from a rhea in North Carolina. Many of the rheas in the flock from North Carolina had antibodies to the virus. All of the exposed birds were traced to an animal auction in Texas. The virus isolated from the birds in Texas was not pathogenic for turkeys or chickens. Because of the threat of introducing virulent strains of influenza A virus to poultry, many states now require that ratites be tested for the virus prior to shipping. The specific

FIG 13.6

Influenza A virus can cause depression, diarrhea and death in young ostriches; adults are rarely affected. photograph courtesy of Amy Raines

requirements vary and can be determined by contacting the State Veterinarian's Office in the state of destination.[20]

Influenza A virus was isolated from young ostriches in Africa that died following a period of lethargy, respiratory disease and passage of greenish-colored urates. Clinical signs were most severe in ostriches from five days to fourteen months old, and adult ostriches were rarely affected. Mortality rates were 80% in chicks less than one month of age and these chicks generally died acutely. Mortality rates in chicks up to eight months of age ranged from 15% to 60%. Mortality rates in birds eight to fourteen months of age were below 20%. Most affected birds died within two days of developing clinical signs, but some recovered over a two- to three-week period. Hemagglutination-inhibition antibodies to the virus were demonstrated within three weeks in the ostriches that recovered and in adult birds that remained asymptomatic *Figure 13.6*. Swelling and congestion of the liver were most remarkable in birds that died acutely. The liver was dark green in more chronic cases. Other gross changes included congestion and accumulation of mucus in the small intestines, and swollen, pale kidneys with ureters that were filled with greenish urates.

Historically, young ostriches have been reported to develop clinical signs similar to those described in this outbreak during years with severe droughts. It is suspected that outbreaks occur because free-ranging waterfowl are attracted to the water provided to the ostriches during droughts. Additionally, coprophagia is common in young ostriches. Because influenza A virus is commonly shed in the feces of other avian species, this practice may be responsible for the rapid transmission of the virus through susceptible young ostriches.[8,19]

REFERENCES

1. Abrey A: Summary of avian post mortems of Allerton laboratory in the Natal Area of South Africa during the last 18 months. Newsltr Assoc Avian Vet 5:98-99, 1984.

2. Alexander DJ: Avian influenza: Recent developments. Vet Bull 52:341-359, 1982.

3. Alexander DJ: Isolation of influenza A viruses from exotic birds in Great Britain. Proc Symp Avian Influenza, 1982, pp 79-92.

4. Alexander DJ: Avian influenza: Historical aspects. Proc 2nd Intl Symp Avian Influenza, 1986, pp 4-13.

5. Alexander DJ, Allan WH, Harkness JW: Isolation of influenza virus from psittacines. Res Vet Sci 17:125-127, 1974.

6. Alexander DJ, Allan WH, Parsons D, et al: The pathogenicity of four avian influenza viruses for fowl, turkeys and ducks. Res Vet Sci 24:242-247, 1978.

7. Alexander DJ, Parsons G, Manvell RJ: Experimental assessment of the pathogenicity of eight influenza A viruses of H5 subtype for chickens, turkeys, ducks and quail. Avian Pathol 15:647-662, 1986.

8. Allwright DM, Burger WP, Geyer A: Isolation of an influenza A virus from ostriches (*Struthio camelus*). Avian Pathol 22:59-65, 1993.

9. Austin FJ, Webster RG: Evidence of ortho- and paramyxoviruses in fauna from Antarctica. J Wildl Dis 29:568-571, 1993.

10. Bahl AK, Pomeroy BS, Mangundimedjo S, et al: Isolation of Type A influenza and Newcastle disease viruses from migratory waterfowl in the Mississippi flyway. J Am Vet Med Assoc 171:949-951, 1977.

11. Beard CW: Influenza. *In* Purchase G, et al (eds): A Laboratory Manual for the Isolation and Identification of Avian Pathogens 3rd ed. Kennett Square, Am Assoc Avian Pathol, 1989 pp 110-113.

12. Becker WB: Experimental infection of common terns with tern virus: Influenza virus A/tern/South Africa/1961. J Hyg 65:61-65, 1967.

13. Butterfield WK, Yedloutschnig RJ, Dardiri AH: Isolation and identification of myxoviruses from domestic and imported avian species. Avian Dis 17:155-159, 1973.

14. Cappucci DJ, Johnson DC, Brugh M, et al: Isolation of avian influenza virus (subtype H5N2) from chicken eggs during a natural outbreak. Avian Dis 29:1195-1200, 1985.

15. Castro AE, et al: Isolation and identification of a strain of influenza virus A/quail/California/4794/90 (H4N6) antigenically related to A/duck/Czechoslovakia/56(H4N6) from a quail flock in California. Proc 40th West Poult Dis Conf, 1991, pp 42-43.

16. Cooley AJ, Van Campen H, Philpott MS, et al: Pathological lesions in the lungs of ducks infected with influenza A viruses. Vet Pathol 26:1-5, 1989.

17. Davidson WR, Nettles VF, Couvillion CE, et al: Diseases diagnosed in wild turkeys (*Meleagris gallopavo*) of the southeastern United States. J Wildl Dis 21:386-390, 1985.

18. Dhillon AS, et al: Mortality in young pheasants and avian influenza infection. Proc 35th West Poultry Dis Conf, 1986, pp 28.

19. Easterday BC, et al: Influenza. *In* Calnek BW, et al (eds): Diseases of Poultry 9th ed. Ames, Iowa State University Press, 1991, pp 532-551.

20. Foreign Animal Disease Report. 21:9-10, 1993.

21. Gerlach H: Viruses. *In* Ritchie BW, Harrison GJ, Harrison LR (eds): Avian Medicine: Principles and Application. Lake Worth, Wingers Publishing, 1994, pp 862-948.

22. Graves IL: Influenza viruses in birds of the Atlantic flyway. Avian Dis 36:1-10, 1992.

23. Groocock C: Avian influenza in game-birds in Maryland. Foreign Anim Dis Rep 22:7, 1994.

24. Gylstorff I: Orthomyxoviren. *In* Gylstorff I, Grimm F (eds): Vogelkrankheiten. Stuttgart, Verlag Eugen Ulmer, 1987, pp 249-253.

25. Halverson DK, Kelleher CJ, Seene DA: Epizootiology of avian influenza: Effect of season on incidence in sentinel ducks and domestic turkeys in Minnesota. Appl Environ Microbiol 49:914-919, 1985.

26. Hinshaw VS, Bean WJ, Webster RG, et al: Genetic reassortment of influenza A viruses in the intestinal tract of ducks. Virology 102:412-419, 1980.

27. Hinshaw VS, Sheerar MG, Larsen D: Specific antibody responses and generation of antigenic variants in chickens immunized against a virulent avian influenza virus. Avian Dis 34:80-86, 1990.

28. Hinshaw VS, Webster RG, Rodriguez RJ: Influenza A viruses: Combinations of hemagglutinin and neuraminidase subtypes isolated from animals and other sources. Arch Virol 67:191-206, 1981.

29. Hinshaw VS, Webster RG, Turner B: Water-borne transmission of influenza A viruses? Intervirology 11:66-68, 1979.

30. Hinshaw VS, Webster RG, Turner B: The perpetuation of orthomyxoviruses and paramyxoviruses in Canadian waterfowl. Can J Microbiol 26:622-629, 1980.

31. Hinshaw VS, Wood JM, Webster RG, et al: Circulation of influenza viruses and paramyxoviruses in waterfowl originating from two different areas of North America. Bull World Hlth Org 63:711-719, 1985.

32. Homme PJ, Easterday BC: Avian influenza virus infections. IV Response of pheasants, ducks and geese to influenza A/turkey/Wisconsin/1966 virus. Avian Dis 14:285-290, 1970.

33. Hwang J, Lief FS, Miller CW: An epornitic of type A influenza virus infection in ducks. J Am Vet Med Assoc 157:2106-2108, 1970.

34. Kaplan MM: The epidemiology of influenza as a zoonosis. Vet Rec 23:395-399, 1982.

35. Kaplan MM, Webster RG: Epidemiology of influenza. Sci Am 237:88-106, 1977.

36. Kawano J, Yanagawa R, Kida H: Site of replication of influenza virus A/budgerigar/Hokkaido/1/77 (H4N1) in budgerigars. Zbl Bakt Hyg 1 Abts Org A 245:1-7, 1979.

37. Kawaoka Y, Krauss S, Webster RG: Avian-to-human transmission of the PB 1 gene of influenza A viruses in the 1957 and 1968 pandemics. J Virol 63:4603-4608, 1989.

38. Lalitha Rao B, Khan FU, Bhat HR, et al: Zoonotic studies on influenza in pigs and birds, India, 1980-81. Intl J Zoon 10:40-44, 1983.

39. Lang G: A review of influenza in Canadian domestic and wild birds. Proc Symp Avian Influenza, 21-27, 1982.

40. Laudert EA, Sivanandan V, Halvorson DA: Effect of intravenous inoculation of avian influenza virus in reproduction and growth in mallard ducks. J Wildl Dis 29:523-526, 1993.

41. Lowenstine LJ: Emerging viral diseases of psittacine birds. *In* Kirk RW (ed): Current Veterinary Therapy IX. Philadelphia, WB Saunders Co, 1986, pp 705-710.

42. Markwell DD, Shortridge KF: Possible waterborne transmission and maintenance of influenza viruses in domestic ducks. App Environ Microbiol 43:110-116, 1981.

43. McFerran JB, Connor TJ, Collins DS, et al: Isolation of an avirulent influenza virus from a parrot. Vet Rec 95:466-467, 1974.

44. McNulty MS, Allan GM, McCracken RM: Isolation of a highly pathogenic influenza virus from turkeys. Avian Pathol 14:173-176, 1985.

45. Narayan O, Lang G, Rouse ST: A new influenza A virus infection in turkeys. IV Experimental susceptibility of domestic birds to virus strain turkey/Ontario/7732/1966. Arch ges Virusforsch 26:149-165, 1969.

46. Nerome K, Nakayama M, Ishida M, et al: Isolation and serologic characterization of influenza A viruses from birds that were dead on arrival at Tokyo airport. Arch Virol 57:261-270, 1978.

47. Nestorowicz A, Kawaoka Y, Bean WJ, et al: Molecular analysis of the hemagglutinin genes of Australian H7N7 influenza viruses: Role of passerine birds in maintenance or transmission? Virology 160:411-418, 1987.

48. Nettles VF, Webster RG, Hinshaw VS, et al: The status of wildlife associated with the 1983-84 avian influenza outbreak in Pennsylvania/Virginia. Proc Symp Avian Influenza, 1987, pp 51-60.

49. Nettles VF, Wood JM, Webster RG: Wildlife surveillance associated with an outbreak of lethal H5N2 avian influenza in domestic poultry. Avian Dis 29:733-741, 1985.

50. Panigrahy B, Seene DA: Diseases of mynah birds. J Am Vet Med Assoc 199:378-381, 1991.

51. Pensaert M, Ottis K, Vandeputte J, et al: Evidence for the natural transmission of influenza A virus from wild ducks to swine and its potential importance for man. Bull World Hlth Org 59:75-78, 1981.

52. Pereira HG, Tumova B, Low VC: Avian influenza A viruses. Bull WHO 32:855-860, 1965.

53. Pomeroy BS: Avian influenza in the United States (1964-1980). Proc Symp Avian Influenza, 1982, pp 13-17.

54. Price RJ: Commercial avian influenza vaccines. Proc Symp Avian Influenza, 1981, pp 178-179.

55. Rigby CE, Pettit JR, Papp-Vid G, et al: The isolation of salmonellae, Newcastle disease virus and other infectious agents from quarantine imported birds in Canada. Can J Comp Med 45:366-370, 1981.

56. Ronohardjo P: Growth and economic losses due to influenza A virus infection in local Indonesian ducks. Penyakit-Hewan 18:124-129, 1986.

57. Roslaja LG, Lova DK, Jamnikova SS: Infection rate of black-headed gulls with influenza viruses. Vop Virusol 29:155-157, 1984.

58. Rossi GA, Simonella P, Acciarri C, et al: Ricerche preliminari sa un focolaio di influenza nella quaglia domestica. Atti della Societa Italiana delle. Science Veterinarie 26:565-567, 1972.

59. Samadieh B, Bankowski RA: Transmissibility of avian influenza-A viruses. Am J Vet Res 32:939-945, 1971.

60. Schafer W: Vergleichende sero-immunologische Untersuchungen uber die viren der influenza und klassichen Geflugelpest. Z Naturforsch 10:81-91, 1955.

61. Seene DA, Pearson JE, Miller LD, et al: Virus isolations from pet birds submitted for importation into the United States. Avian Dis 27:731-744, 1983.

62. Shivaprasad HL, Woolcock PR, Sakas PS: Avian influenza in psittacines and a passerine. Proc Assoc Avian Vet, 1994, pp 259-260.

63. Slemons RD, Cooper RS, Orsborn JS: Isolation of type A influenza viruses

from imported exotic birds. Avian Dis 17:746-751, 1973.

64. Slemons RD, Easterday BC: Host response differences among five avian species to an avian influenza virus - A/turkey/Ontario/7732/66(Hav5N?). Bull WHO 47:521-522, 1972.

65. Slemons RD, Johnson DC, Orsborn JS, et al: Type-A influenza viruses in the feces of migratory waterfowl. Am J Vet Res 171:947-948, 1974.

66. Stallknecht DE, Shane SM: Host range of avian influenza virus in free-living birds. Vet Res Commun 12:125-141, 1988.

67. Stallknecht DE, Shane SM, Kearney MT, et al: Persistence of avian influen-za viruses in water. Avian Dis 34:406-411, 1990.

68. Tanyi J: Type A influenza virus infection in guinea fowls. Acta Vet Acad Sci Hung 22:125-131, 1972.

69. Tanyi J, Ronvary J, Derzsy D, et al: Severe sinusitis in ducklings caused by influenza A virus. Acta Veterinaria Academiae Scientiarum Hungaricae 25:261-266, 1975.

70. Webster RG: On the origin of pandemic influenza viruses. Current Topics in Microbiology and Immunology. Berlin, Springer-Verlag, 1972, pp 72-105.

71. Webster RG, Kawaoka Y: Avian influenza. Critic Rev Poult Biol 1:211-246, 1988.

72. Webster RG, Morita M, Pridgen C, et al: Ortho- and paramyxoviruses from migrating feral ducks. Characterization of a new group of influenza A viruses. J Gen Virol 32:217-225, 1976.

73. Webster RG, Yakhno M, Hinshaw VS, et al: Intestinal influenza: Replication and characterization of influenza viruses in ducks. Virology 84:268-276, 1978.

74. Westbury BA, Turner AJ, Amon C: Transmissibility of two avian influenza A viruses (H7N7) between chicks. Avian Pathol 10:481-487, 1981.

75. Wood JM, Webster RG, Nettles VF: Host range of A/chicken/Pennsylvania/83 (H5N2) influenza virus. Avian Dis 19:198-207, 1985.

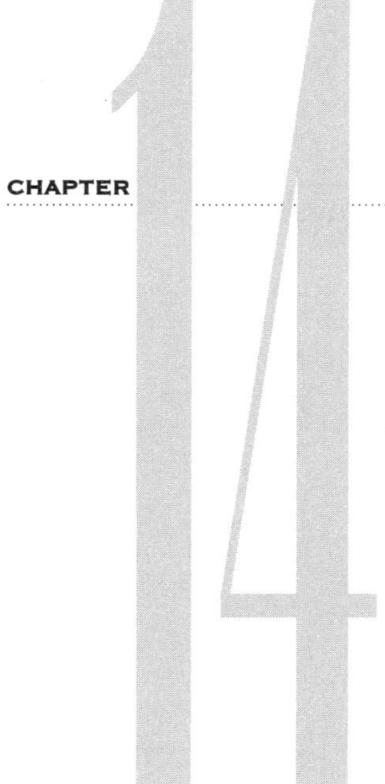

Retroviridae

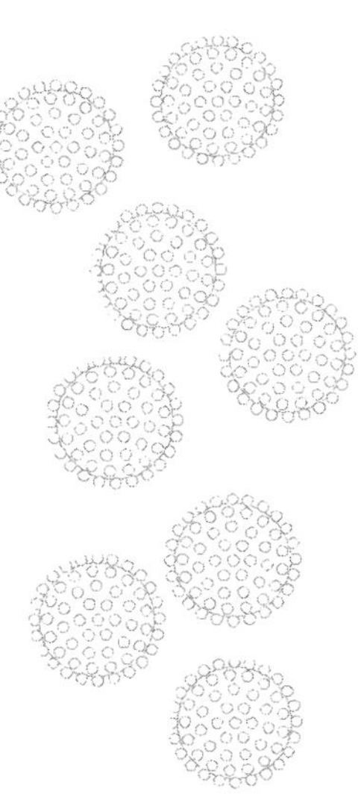

No group of viruses is as diverse as the Retroviridae. This family contains viruses that exhibit markedly different behaviors ranging from asymptomatic infections to rapid formation of tumors and death. Many of these viruses have the ability to integrate into the genome of host cells and cause tumors or otherwise permanently alter the function of infected cells. Retroviruses have been associated with cancers, chronic weight-loss diseases, neurologic disease, damage to the immune system and persistent infections. These viruses are frequently difficult to control because they change rapidly, thus avoiding the host defense systems, or specifically damage important components of the immune system. Host-specific retroviruses have been shown to cause neoplasms (leukemias, lymphomas, sarcomas and carcinomas), immunodeficiencies (AIDS in people) and autoimmune diseases in humans, livestock, companion animals and poultry. Only the avian type C retroviruses have been shown to affect birds.

The avian type C retroviruses of importance to birds are avian leukosis/sarcoma virus (ALSV), which causes neoplasms in chickens and possibly in other species, and reticuloendotheliosis virus, which causes disease in turkeys and ducks. The avian leukosis viruses are divided into subgroups based on differences in the proteins found in the virus envelope. These differences restrict the host range *Table 14.1*. With a few exceptions, neutralizing antibodies to each subgroup of viruses do not cross-react with those of other subgroups.

AVIAN LEUKOSIS/ SARCOMA VIRUSES

Retroviruses were first implicated as a cause of lymphoid tumors in the early 1900s, when the Rous-sarcoma virus was shown to cause neoplastic changes in the B lymphocytes of young chickens. Into the nucleic acid of an infected cell, this virus inserts genes that code for the production of proteins that turn that cell into a tumor-producing factory *Figure 14.1*. Such genes are called oncogenes. Subsequently, avian leukosis/ sarcoma viruses as a group have been associated with a whole range of tumors including lymphoid leukosis (most common), erythroid leukosis (erythroblastosis), myeloid leukosis (myeloblastosis), renal

TABLE 14.1

Host range of avian leukosis virus subgroups[16,35,46]

SUBGROUP A,B,C,D,E,J Chickens
SUBGROUP F Ring-necked Pheasant
SUBGROUP G Golden and Lady Amhurst Pheasants
SUBGROUP H Hungarian Partridge
SUBGROUP I Gambel's Quail
UNCLASSIFIED SUBGROUP Mongolian and Swinhoe Pheasants and Painted Quail

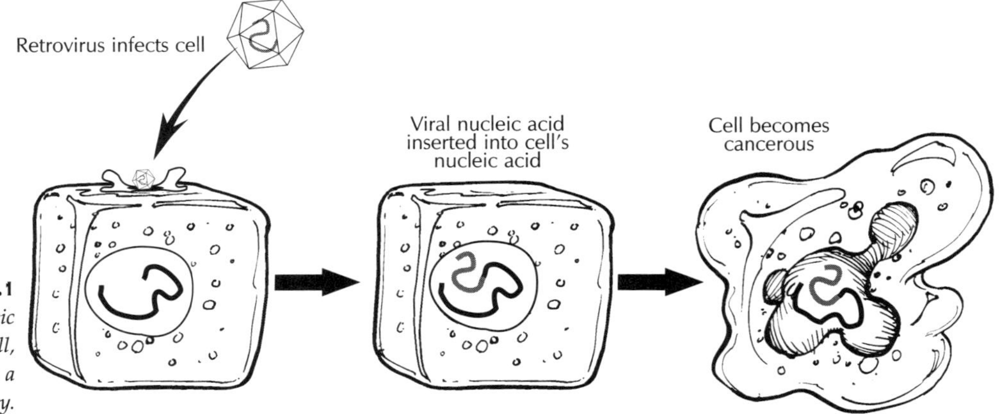

FIG 14.1

Retroviruses insert their nucleic acid into that of the infected cell, which can turn the cell into a tumor-producing factory.

neoplasms (nephroblastomas being most common), hemangiomas and osteopetrosis in chickens.[35]

Some ALSV are endogenous, which means they are spread from infected parent to offspring directly in the transferred genetic material. Others of this group are exogenous, being acquired after hatching through contact with an infected bird. Endogenously derived portions of the genome of a retrovirus have been detected in the cells of most commercial chickens, as well as in some red jungle fowl, pheasants, partridges and grouse, but they have not been demonstrated in other jungle fowl, guinea fowl, peafowl, quail or turkeys.[46]

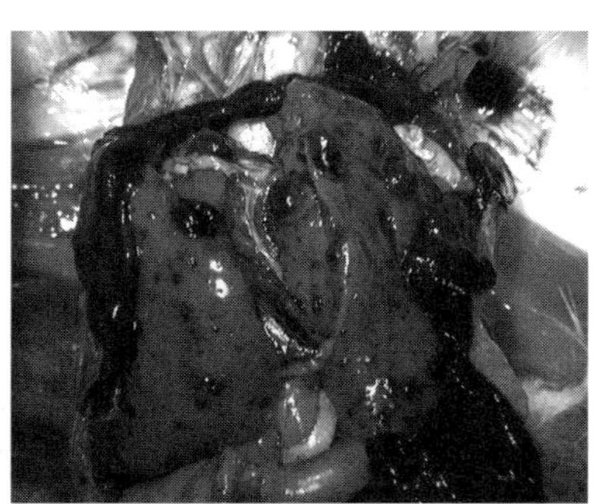

FIG 14.2

Erythroblastosis is characterized by an accumulation of erythroblasts. Affected birds usually have an enlarged, soft, dark red liver and spleen that may contain hemorrhagic areas. photograph courtesy of Jean Sander

The most common neoplasia induced by retrovirus in chickens is lymphoid leukosis; however, the same retrovirus may cause widely varying neoplastic changes depending on the source of the virus, the route of exposure and the age and strain of the chicken. For example, a retrovirus that causes lymphoid leukosis in one strain of chickens may cause erythroblastosis in another. Generally, a particular retrovirus causes one type of neoplasia in most cases, and only occasionally will cause a different neoplasia.[35] Many retroviruses are recovered from asymptomatic hosts. Some groups of chickens have natural resistance to infection or to the development of tumors following infection, and these traits can be enhanced in a flock through selective breeding.[7,30,35]

CLINICAL FEATURES IN CHICKENS

In chickens, retroviruses may induce either neoplasias that affect circulating blood cells or solid tumors that are formed by the accumulation of neoplastic lymphocytes that localize to the bursa, liver, spleen, kidneys or other tissues *Table 14.2*.

The most common neoplastic disease that occurs in chickens is called lymphoid leukosis. Lymphoid leukosis is associated with tumor formation throughout the body but particularly in the liver. Most commercial chickens are exposed to lymphoid leukosis viruses, but commonly fewer than 3% of the infected chickens develop tumors. In

TABLE 14.2

Characteristic tumors and other lesions associated with avian leukosis/sarcoma viruses[17,35]

Erythroblastosis
- Primarily in mature birds but rare.
- Accumulation of erythroblasts.
- Leukemia, anemia and thrombocytopenia.
- Swelling of the liver and spleen, which are usually soft and dark red *Figure 14.2*.
- Petechial hemorrhage in the muscles, under the skin or on abdominal organs.
- Some strains cause lesions in the bone marrow in three days and death within two weeks; other strains have a one- to three-month incubation period.

Myeloblastosis
- Primarily in mature birds but rare.
- Severe leukemia; myeloblasts predominate.
- Anemia and thrombocytopenia are common.
- Grayish tumors in the liver, spleen, kidneys and bone marrow.
- Death as soon as 14 days after infection.

Nephroblastomas
- Enlarged, encapsulated, cystic masses in the kidneys occur two- to six-months after infection *see Figure 14.6*.

Renal adenomas and carcinomas
- Discrete or diffuse nodules in the kidneys occur three weeks after infection *Figure 14.3*.

Osteopetrosis
- Proliferation of osteoblasts.
- Symmetric overgrowth of long bones.
- Proliferation of bone in the marrow *Figure 14.4*.
- Lesions may occur from seven days to over one month after infection.
- Anemia is common.

Other less common neoplasias
- fibromas/fibrosarcomas
- myxoma/myxosarcomas
- osteomas/osteosarcomas
- chondrosarcomas
- mesotheliomas
- endotheliomas
- hemangiomas

rare cases, up to 30% of an infected flock may develop neoplastic lesions.[35] Subclinical infections have been linked to non-neoplastic problems, including decreased egg production, reduced hatchability and decreased growth rates.[14,15] Osteopetrosis can be experimentally induced in young guinea fowl

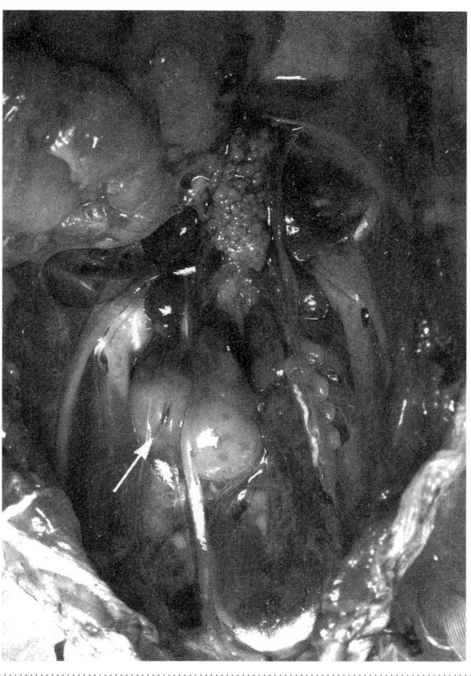

FIG 14.3
Renal adenomas and carcinomas (arrow) can be caused by a retrovirus infection or other cancer-inducing agents.

by infection with a strain of retrovirus isolated from chickens. Affected guinea fowl developed characteristic bony changes, as well as tumors in the pancreas and duodenum.[24]

Lesions caused by lymphoid leukosis are most common in chickens four to nine months of age. This virus is associated with the slow formation of tumors, occurring over a period of more than 16 weeks.[12,35] Clinical changes vary with

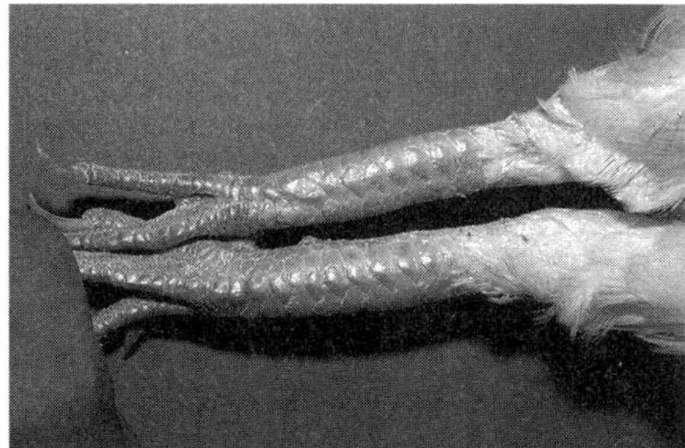

FIG 14.4
Osteopetrosis is characterized by the proliferation of osteoblasts, causing the long bones to be thicker than normal.
photograph courtesy of Jean Sander

the location of tumors. Large masses in the abdomen or enlargement of the liver may cause dyspnea because of pressure on the air sacs *Figure 14.5*. An increased number of heterophils, lymphocytes and monocytes may be noted, but rarely are immature blood cells seen with avian leukosis. Females are more susceptible to disease than males, and young birds develop resistance to the disease as their cloacal bursa matures.

Clinical signs of disease and increased shedding of virus are most common during periods of stress or following damage to, or suppression of the immune system.[35] Marek's disease virus can cause similar clinical and gross changes, but generally affects younger birds and primarily involves the neural system *Table 14.3*.

CLINICAL FEATURES IN BIRDS OTHER THAN CHICKENS

Antigenically distinct subgroups of avian leukosis-sarcoma viruses have been documented in a number of avian species, including pheasants, quail, partridges and chickens.[13] Lesions similar to those described in gallinaceous birds with lymphoid leukosis have been reported in a variety of other birds including passerines, Columbiformes, African Grey Parrot, Amazon parrot, Budgerigars, Red Shining Parrots, conures, canaries, Peregrine Falcon, American Kestrel, eagle, Little Owl and Anseriformes.[2,5,9,23,34,36,41,45]

In one study of 86 birds with neoplastic disease, 16 (18.6%) had proliferative lesions involving the lymphocytes or myelocytes.[18] Occasionally, lesions suggestive of lymphoid leukosis have been detected in captive or free-ranging finches, swans, geese, ducks, pheasants, pigeons, egrets, storks and gulls.[27,33] In another study, 7% of the free-ranging house sparrows in Czechoslovakia had antibodies to lymphoid leukosis virus.[44] Proliferative lesions of the lymphoid cells were described in all the canaries, three Budgerigars, two Cockatiels and a Moluccan Cockatoo in one study. Four of the birds had subcutaneous masses around the face and neck, one bird had a lymphoid mass in the brain and the other eleven birds had generalized lesions with involvement of the abdominal organs.[2]

Clinical signs associated with lymphoid leukosis depend on the location of tumors and the presence of secondary infections. Affected birds may die acutely with no premonitory signs, die follow-

FIG 14.5

Enlargement of the liver or tumors in the abdomen may place pressure on the air sacs making it difficult for an affected bird to breathe. This may be recognized clinically as open-mouthed breathing.

photograph courtesy of Jean Sander

TABLE 14.3

Comparison of Marek's disease virus to lymphoid leukosis virus[3]

	MAREK'S DISEASE VIRUS	LYMPHOID LEUKOSIS VIRUS
AGE	> 6 weeks of age common in chickens 12-24 weeks old	>16 to 24 weeks common in chickens 20-36 weeks old
INCUBATION	incubation period 3 weeks to several months	incubation period more than 14 weeks
CLINICAL AND GROSS CHANGES	paresis and paralysis common enlargement of peripheral nerves tumors in gonads tumors in liver and spleen tumors in skin, muscle, proventriculus	neurologic signs rare no enlargement of peripheral nerves tumors in gonads rare tumors in liver and spleen tumors in the skin and muscle are rare

ing a brief period of depression and diarrhea, or develop chronic problems associated with the formation of proliferative masses in numerous tissues.[5,34,45] Activities of some serum enzymes may be increased if the liver is affected.

The best evidence for a retrovirus-induced neoplasm in companion birds occurs in Budgerigars, which have a high propensity for developing tumors, particularly those of renal origin. In a group of 74 Budgerigars with tumors, 47 (63.5%) had kidney tumors, 22 (30%) had tumors of the gonad and 17 (23%) had liver tumors. The most common clinical changes associated with these tumors were lameness and abdominal enlargement *Figure 14.6*. Within this group of affected Budgerigars, 35 (47%) had evidence of an ALSV infection; however, a direct association between ALSV group-specific antigens and the formation of renal tumors was not confirmed. Tumors were most common in birds three to six years of age, with kidney tumors being most common in the fifth year. If tumors in Budgerigars are induced by ALSV, it is unusual that older birds are affected, because in chickens these viruses most commonly cause tumors in young birds.[32]

Over a five-year period, lymphoid leukosis was diagnosed in 6 of 1,234 (0.5%) captive birds presented for necropsy in London. All affected birds (Cooper Pheasant, Speckled Pigeon, Turtle Dove, Queen Alexandra's Parakeet, Green and Gold Tanager and White-throated Jay) were found dead with no other clinical changes. The most common gross change noted at

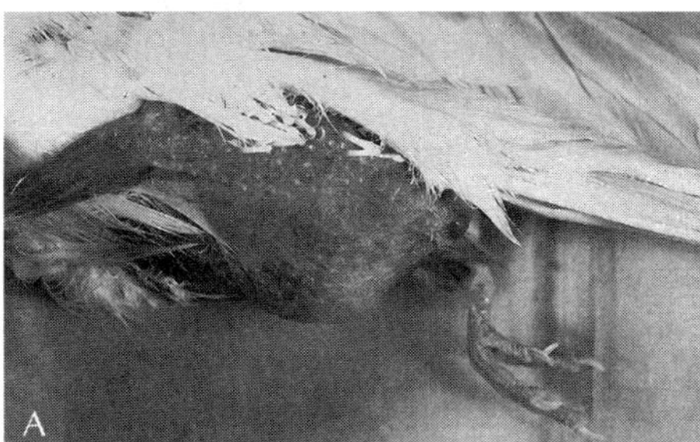

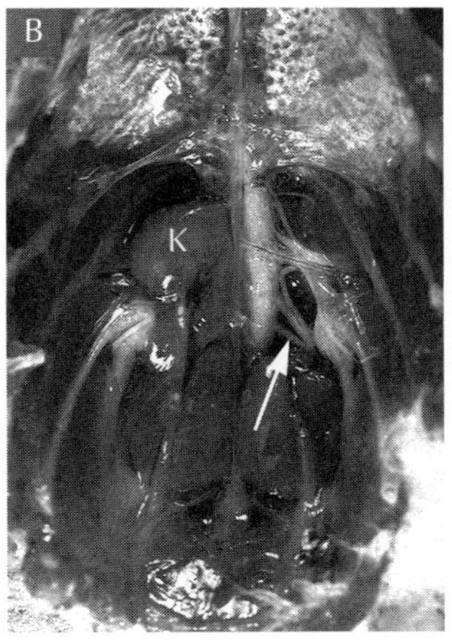

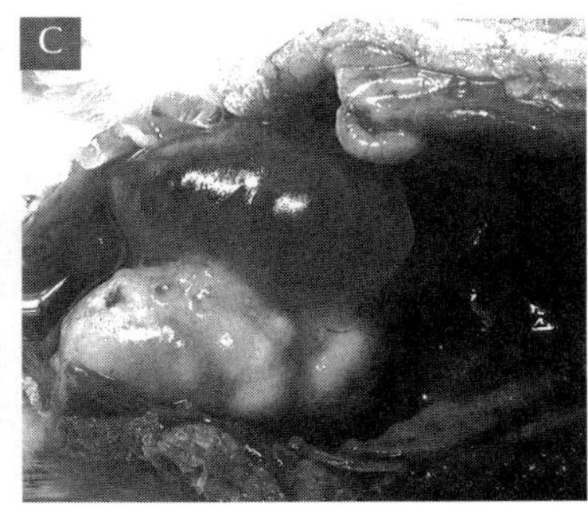

FIG 14.6

A Budgerigars with renal tumors may exhibit abdominal distention with or without rear limb lameness. B Renal tumors cause lameness because the nerves to the rear limbs (arrow) are closely associated with the kidneys (K), and as a renal tumor enlarges, pressure is placed on the nerves.

C Nephroblastomas characterized by the formation of enlarged, encapsulated, cystic masses in the kidneys are commonly caused by retroviruses.

necropsy was an enlarged yellowish-white liver and spleen. Other changes included mesentery tumors and renal enlargement.[45] In another case, a one-year-old African Grey Parrot died following a four-day history of depression and diarrhea. The spleen and liver of this parrot were enlarged and friable, and the liver was mottled. The accumulation of immature lymphocytes in the liver and spleen was considered suggestive of lymphoid leukosis.[34]

Lymphoid leukosis was suspected in a 23-year-old Amazon parrot with a unilateral, periorbital mass that had been slowly increasing in size for 1.5 years. Abnormal blood changes in this bird included heterophilia, lymphocytosis, monocytosis and a partially regenerative anemia. The majority of circulating lymphocytes were mature and exhibited pseudopodia.[5] A similar change in circulating lymphocytes has been reported in chickens experimentally infected with lymphoid leukosis virus.[37] Clinical changes in a Peregrine Falcon diagnosed with lymphoid leukosis included weight loss, ataxia and lymphocytosis characterized by a large proportion of immature lymphocytes, many of which contained mitotic figures.[41]

Lymphoid leukosis is commonly diagnosed microscopically in canaries that either die suddenly, die following a brief period of progressive dyspnea, or develop chronic weight loss and respiratory problems. In one study, the liver and spleen of most affected canaries were enlarged, and some birds had lymphocytosis. An ALSV infection should be considered in any canary with soft, subcutaneous masses of the head and neck, abdominal enlargement or dyspnea.[2]

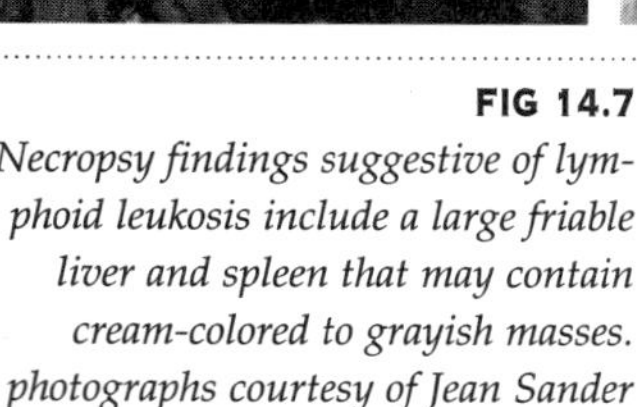

Other neoplasms that are frequently caused by retroviruses also have been reported in canaries, including erythroblastosis, myeloblastosis and stem-cell leukosis. Additionally, erythroblastosis has been reported in cockatoos, a weaver and a grosbeak, and myeloblastosis has been reported in cockatoos, Budgerigars, a Turquoise Parrot and a Pacific Parrotlet.[16,27]

Antibodies to avian leukosis virus were detected in 3% of 149 ostriches in Zimbabwe. Antibodies were detected in only one of the nine flocks evaluated. In this flock, 4 of 10 birds were seropositive.[4]

FIG 14.7

Necropsy findings suggestive of lymphoid leukosis include a large friable liver and spleen that may contain cream-colored to grayish masses. photographs courtesy of Jean Sander

<u>TRANSMISSION</u> — In chickens, ALSV viruses may be transmitted vertically or horizontally, although horizontal transmission is probably more important.[35] The subgroup E viruses *see Table 14.1* are endogenous and are transmitted from parent to offspring through the genome.[8] The potential for endogenous transfer of lymphoid leukosis virus has been suggested for red jungle fowl, some pheasants, partridges and grouse.[13] The other subgroups that occur in chickens — A,B,C,D and J — are transmitted exogenously. These viruses can be transmitted horizontally through direct contact with contaminated blood, saliva, respiratory secretions, semen or feces, or can be transmitted by indirect contact with contaminated insects. The viruses are shed from the oviduct of an infected hen, and can enter the albumen of the developing egg. Infected males can infect hens during copulation.[16,35]

PATHOLOGY

Macroscopic changes associated with lymphoid leukosis include enlarged friable livers and spleens, renal enlargement and cream-colored to grayish masses in the liver, spleen, kidneys, pancreas, gonads, lungs, thymus, bone marrow, mesentery or intestines *Figure 14.7*.[5,34] Microscopic changes are typified by sheets of lymphoblasts with a high mitotic index arranged in diffuse or nodular patterns in parenchymal organs.[5,34,41] The accumulation of lymphoblasts in neural tissue, which commonly occurs with Marek's disease virus, is rare with lymphoid leukosis.

PATHOGENESIS, IMMUNITY AND DIAGNOSIS

Retroviruses can induce neoplasms or cause immunosuppression that allows opportunistic viruses, bacteria, parasites or fungi to induce life-threatening diseases. Retroviruses have an affinity for the immune system and may attack the bone marrow, bursa and thymus. It has been theorized that retrovirus-induced immunosuppression is caused by a cessation of B-lymphocyte maturation and a blockage of T-lymphocyte development. Administration of testosterone has been shown to decrease — and castration has been shown to increase — a chicken's susceptibility to disease.[35]

A complement-fixation and enzyme-linked immunosorbent assay (ELISA) are available for detecting the lymphoid leukosis virus. The virus can be recovered from plasma, serum, feces, vaginal swabs, pharyngeal secretions, feather pulp or albumin of freshly laid eggs. A virus-neutralization assay or ELISA can be used to detect antibodies to the virus in serum or yolks. Because retrovirus infections are common in chickens, antibodies to the virus are frequently detected during seroprevalence studies. Hens that are persistently infected with the virus can be detected by testing multiple eggs for the presence of virus or antibodies to the virus.

Chicks infected in the oviduct or immediately post-hatching become immunotolerant and do not develop antibodies to the virus; however, these chicks do shed virus in their feces and saliva that can infect other birds in the flock. Chicks and adults infected through an exogenous route develop neutralizing antibodies; however, the immunologic response that occurs may not prevent tumor formation. Persistent infections are common.

CONTROL

Retroviruses are relatively unstable outside of a host. Infectivity will be destroyed by most disinfectants and detergents and by temperatures of 37°C (four hours) and 50°C (one minute). Retroviruses, like most viruses, remain stable when frozen. There are no vaccines for ALSV. Control of the virus requires excellent hygiene and preventing exposure to the virus by identifying and removing infected hens from the breeding flock. Selective breeding can be used to produce strains of chickens that are resistant to ALSV. Chicks should be raised separately from adults and not mixed with chicks from varying sources to help reduce transmission through a flock.[47]

Chicks treated with Mibolerone (an androgen analog) from 1 to 49 days post-hatching were found to be resistant to leukosis.[35] The masses in an affected Amazon parrot decreased in size during a 32-week treatment period with oral prednisolone; however, the parrot progressively lost weight and another mass was identified radiographically in the abdomen. In most species of animals with lymphoid leukosis, the expected survival rate following treatment with prednisolone is less than one year.[5]

A group of structurally varied retroviruses can cause a cluster of disease syndromes collectively termed reticuloendotheliosis. These viruses replicate in and cause disease of the reticular and endothelial cells that line capillaries. The reticuloendotheliosis viruses (REV)

FIG 14.8

Stunting and poor feather growth have been reported in chickens with reticuloendotheliosis virus infections. Failure of the feathers to properly exsheathe (arrows) is most common in three- to eight-week-old birds. photograph courtesy of Cs.N. Dren, reprinted with permission[10]

have been associated with the rapid formation of neoplasias, immunodepression, a failure to thrive in young birds (runting) and formation of lymphoid tumors. Ducks, geese, pheasants, quail, turkeys and chickens have been demonstrated to be naturally susceptible, and experimental infections have been reported in these birds as well as in guinea fowl.[10,20,26,48] Subgroups of REV that are related but antigenically distinguishable include duck infectious anemia virus, spleen necrosis virus, chicken syncytial virus and unclassified isolates from Muscovy Ducks, shelducks and racing pigeons.[16,22] Isolates of REV from ducks and pheasants are related to strains that infect turkeys.[10,38]

CLINICAL FEATURES

Most reticuloendotheliosis virus infections in domestic poultry are subclinical, although high mortality has been reported during outbreaks in ducks and pheasants.[10,20,40] The clinical changes in young chickens infected with REV in the oviduct or shortly after hatching include decreased weight gains (stunting), anemia and abnormal feather growth *Figure 14.8*.[19] Failure of feathers to properly exsheathe has been described in three- to eight-week old chickens. When chicks are infected at one day of age, virus can be detected in the feathers from six to nine days later.[48] This virus also has been associated with stunted growth in chickens, a condition which is thought to be secondary to virus-induced damage of the immune system, and inflammation of the digestive tract resulting in the reduced absorption of nutrients. Chronic infections can cause the development of tumors in abdominal organs and nerves. Experimentally infected chicks develop clinical signs of disease two to three weeks after inoculation.[19]

Mortality rates associated with reticuloendotheliosis virus in quail, ducks and pheasants may vary from 2% to 100%. Affected ducklings exhibit poor weight gains and may die. Sporadic deaths are common in ducklings that survive the initial infection, as are recurrent bacterial infections, tumors and poor reproductive performance as the ducks reach sexual maturity. Mortality rates in experimentally infected ducks ranged from 66% to 100%.[26,31]

Tumors of the head and in the mouth were common in pheasants diagnosed with REV *Figure 14.9*. Other clinical changes included severe distention of the infraorbital sinuses and ulcerative lesions in the mouth, similar to those described with poxvirus infections. Many infected pheasants died from six to twelve months of age. At necropsy, grayish-white nodules were noted in the spleen, liver, kidney, lung, air sacs and muscles.[10,31]

Affected quail may die suddenly or develop a wasting disease characterized by the appearance of poorly developed feathers.[42] Some quail may survive for several months with clinical signs of depression and dyspnea, but most eventually die. Japanese Quail develop disease shortly after sexual maturity (six weeks of age).[6]

Reticuloendotheliosis virus has been recovered from domestic and free-ranging turkeys. Affected birds may die without clinical signs or may develop lameness and diarrhea. Generally, less than 30% of a flock will develop clinical changes; however, the mortality in the turkeys that are affected can reach 60%. The incubation period in turkeys is 8 to 11 weeks.[48] Tumors are common in the liver, spleen, kidneys, gonads, thymus, bursa and bone marrow. In one report, a bird was found moribund and another was emaciated and ataxic, and exhibited wing drooping, circling, torticollis and head tremors. Neoplastic lesions were detected in the liver, spleen, lungs, esophagus and oral mucosa of one turkey, and in the distal esophagus and spleen of another.[21,25] The role REV played in the development of neurologic signs noted in a free-ranging turkey was not determined.[21]

In experimentally infected turkeys, virus is present in plasma in four to sev-

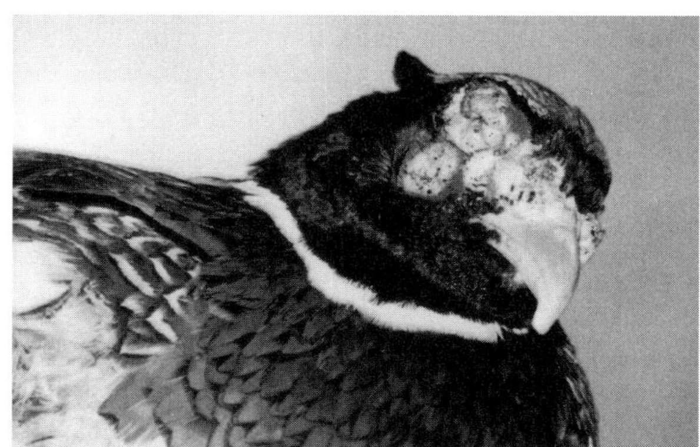

en days, and infected birds become persistently viremic. Young viremic turkeys remained seronegative and shed virus in pharyngeal secretions and feces, while infection in older turkeys was associated with a transient viremia and antibody production. Birds with antibodies can still develop tumors. Poults infected in the oviduct and those infected during the first ten days post-hatching are immunotolerant.[48] The recovery of REV from domestic and free-ranging turkeys, coupled with the fact that virus recovered from turkeys is infectious for chickens, Japanese Quail, ducks, pheasants and guinea fowl, suggests that turkeys can serve as a reservoir for this virus.[21,25]

EPIZOOTIOLOGY

Isolates of REV virus have been recovered from domestic fowl in Australia, North America, South America, Europe and Asia, and from ducks in Australia and the United States.[20,40] Thus far, the virus has been demonstrated in pheasants only in Hungary, but there is no reason to assume that pheasants from other areas are not also susceptible.[10] The demonstration of virus activity in free-ranging geese, ducks and turkeys suggests that these birds could serve as reservoirs for the virus; however, their collective or individual roles in disseminating REV has not been reported.[21,25,39] Antibodies to reticuloendotheliosis virus were detected in 10% of 149 ostriches in Zimbabwe. Within the birds in nine flocks evaluated, the seroprevalence rate varied from 0% to 50%.[4]

Reticuloendotheliosis viruses can be transmitted through horizontal or vertical routes. Transmission from an infected hen to her developing eggs is considered common in ducks,[31] but infrequent in turkeys and chickens.[48] The virus is most commonly transmitted through contact with contaminated pharyngeal secretions, feces, litter or during copulation. Virus transmission was not found to occur when birds were placed in situations that prevented direct contact with one another.[1] Virus infections in domestic fowl have been linked to the use of contaminated vaccines.[48] Mosquitoes and other blood-feeding arthropods that consume the virus during its viremic phase have been suggested as mechanical vectors.[31,39] In Australia, seroconversion in flocks was shown to correlate with increases in the concentration and activity of mosquitoes.[31,43]

Chickens develop an age-related resistance to viremia. Embryos infected in the oviduct or chicks exposed to the virus shortly after hatching develop persistent infections. Chickens infected later in life will seroconvert and do not develop viremia. Ducks of all ages have been shown to be susceptible to REV.[26,31,39] In these birds, the viruses persist in lymphocytes even though the infected birds develop antibodies to them.[26]

PATHOLOGY AND PATHOGENESIS

The lesions that occur with REV are similar in chickens, turkeys, ducks and pheasants. Changes in young birds are characterized by a decrease in size of the bursa, spleen and thymus, abnormal feather growth and reduced weight gain.[47] In older birds, the liver and spleen are typically enlarged and contain multiple, yellowish- to grayish-white nodules. These tumors may also occur in the kidney, bursa, gonads and along the digestive tract. The tumors that may occur include lymphomas, lymphosarcomas, adenomas and carcinomas.[26,31] Peripheral nerve enlargement (similar to the lesion seen with Marek's disease virus) is occasionally noted.

Reticuloendotheliosis virus is considered to be immunosuppressive, predisposing infected birds to secondary bacterial and fungal pathogens. The suppression can be temporary or permanent, depending on the age of the bird when infected, the severity of the disease and the type of tissues damaged.[16] In one study, REV depressed the immune system in experimentally infected one-day-old ducks, but not in those that were more than three weeks old and had a functional bursa. Bacterial infections can cause higher levels of mortality in REV-infected ducks than in noninfected ducks.[26,31] Chickens from which the cloacal bursa has been removed often develop persistent viremia, stressing the importance of the immune system in controlling infections.[48]

IMMUNITY AND DIAGNOSIS

Antibodies to REV can be detected in serum or yolks by a number of assays, including indirect fluorescent antibody, agar-gel immunodiffusion, virus-neutralization and ELISA. In non-immunotolerant birds, antibodies can be detected by two weeks after infection, and they may persist for up to 35 weeks in some individuals.[26,47]

Reticuloendotheliosis virus can be isolated from tumors or the serum or plasma of viremic birds. In persistently infected ducks, the virus can be recovered from the spleen, kidney, ovary or oviduct. In one study, ducks infected after three weeks of age were not viremic, but still developed persistent infections and tumors.[26,31] When chickens are infected at one day of age, the virus can be detected in the feathers six to nine days later.[48]

RETICULOENDOTHELIOSIS VIRUS IN DUCKS AND GEESE

DUCKS

Reticuloendotheliosis virus has been recovered from domestic and free-ranging ducks in Australia, and from domestic ducks in the United States. Ducks of any age are considered susceptible to infection. The incubation period is five to nine weeks. Clinical changes may be limited to depression and decreased weight gain prior to death. Even ducks with high antibody titers have been shown to develop persistent infections associated with lymphocytes. The most common gross findings are an enlarged, friable liver and spleen with multiple yellow to grayish-white nodules. Tumors are common in the liver, spleen, heart, skeletal muscle, pancreas and gastrointestinal tract.[16,26,31,39]

Nucleic acid from reticuloendotheliosis virus was demonstrated in a group of six-month-old Muscovy Ducks that died during a 13-week period. Within the flock, 70 of the 700 ducks died. The remainder of the flock was asymptomatic; however, a second epornitic occurred during the subsequent breeding season. Blast cells judged to be of lymphoid origin were noted in cytologic preparations of the circulating blood. Tumors were present in the thymus, liver, spleen, kidneys, pancreas, lungs and intestines.[29]

Two reticuloendotheliosis viruses that are related to strains recovered from turkeys have been associated with anemia and splenic necrosis in ducks. Duck infectious anemia virus can be transmitted by *Plasmodium lophurae*, the same organism that can transmit malaria. Infections are considered rare, but infected ducks typically develop severe anemia and die. The virus can be detected in the circulating blood.[28]

Ducks infected with the splenic necrosis virus become depressed and anemic, and die within seven to ten days after infection. This virus has been shown to be transmitted through direct contact. Hemorrhage and necrosis of the spleen are the most common necropsy findings.[16]

GEESE

A reticuloendotheliosis virus related to strains isolated from turkeys was recovered from geese.[11] Serologic studies suggest that geese are natural hosts of reticuloendotheliosis virus.[47] Clinical changes in affected goslings include lethargy, lameness, poor feather formation and weight loss; these were first noted in 17-week-old birds. Mortality rates reached 40% by 22 weeks of age.

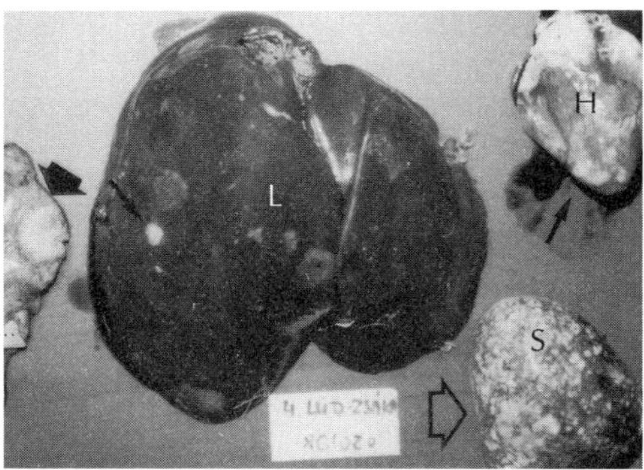

Necropsy findings in affected geese include an enlarged, friable liver (L) and spleen (S) with multiple yellow to grayish-white nodules. Tumors also occur in the heart (H) and other organs.

photographs courtesy of Cs.N. Dren, reprinted with permission[11]

Neoplasms were noted at necropsy in the pancreas, intestines, heart, kidneys or lungs of 14 of 23 (61%) geese. The spleen and liver of affected geese were enlarged and mottled with yellowish-white areas *Figure 14.10*. Osteopetrosis-like lesions were also noted in some affected geese.[11]

Geese and chickens experimentally infected with virus from one of the birds developed neoplasias and died, 11 to 28 days after inoculation. These birds were stunted and lethargic prior to death. Tumors similar to those seen in the original outbreak occurred in the experimentally infected geese. Three other tissue preparations from the originally affected geese caused stunting or chronic lymphoid neoplasia over a 20- to 30-week period. The latter disease presentation was most characteristic of the natural disease. Infiltration of the nerves with lymphocytes, which has been reported in other species, did not occur in the naturally or experimentally infected geese. The origin of the virus in these geese was unknown.[11]

REFERENCES

1. Bagust TJ: Immunodepressive diseases of poultry. Proc Aust Vet Poult Assoc 66:285-293, 1983.

2. Bauck L: Lymphosarcoma/avian leukosis in pet birds: Case reports. Proc Assoc Avian Vet, 1986, pp 241-245.

3. Biggs PM: Criteria for differential diagnosis of lymphoid leukosis and Marek's disease. *In* Payne LN (ed): Differential Diagnosis of Avian Lymphoid Leukosis and Marek's Disease. Luxembourg, Commission of the European Communities, 1976, pp 67-85.

4. Cadman HF, Kelly PJ, Zhou R, et al: A serosurvey using enzyme-linked immunosorbent assay for antibodies against poultry pathogens in ostriches (*Struthio camelus*) from Zimbabwe. Avian Dis 38:621-625, 1994.

5. Campbell TW: Lymphoid leukosis in an Amazon parrot: A case report. Proc Assoc Avian Vet, 1984, pp 229-234.

6. Carlson HC, et al: Reticuloendotheliosis in Japanese quail. Avian Pathol 3:169-175, 1974.

7. Crittenden LB: Two levels of genetic resistance to lymphoid leukosis. Avian Dis 19:281-292, 1975.

8. Crittenden LB, Astrin SM: Genes, viruses and avian leukosis. Bioscience 31:305-310, 1981.

9. Dorrestine GM, van der Sluis J, Zwart P, et al: Diseases of pigeons with emphasis on racing pigeons. Proc Assoc Avian Vet, 1985, pp 181-204.

10. Dren CN, Saghy E, Glavits R, et al: Lymphoreticular tumor in pen-raised pheasants associated with reticuloendotheliosis-like virus infection. Avian Pathol 12:55-71, 1983.

11. Dren CN, Nemeth I, Sari I, et al: Isolation of a reticuloendotheliosis-like virus from naturally occurring lymphoreticular tumors of domestic geese. Avian Pathol 17:259-277, 1988.

12. Fadly AM: Leukosis and sarcomas. *In* Purchase HG, et al (eds): A Laboratory Manual for the Isolation and Identification of Avian Pathogens 3rd ed. Dubuque, Kendall/Hunt Publishing Co, 1989, pp 135-142.

13. Frisby DP, Weiss RA, Roussel M, et al: The distribution of endogenous chicken retrovirus sequences in the DNA of galliform birds does not coincide with avian phylogenetic relationships. Cell 17:623-624, 1979.

14. Gavora JS, Spencer JL, Chambers JR: Performance of meat-type chickens testing positive and negative for lymphoid leukosis virus infection. Avian Pathol 11:29-38, 1982.

15. Gavora JS, Spencer JL, Gowe RS, et al: Lymphoid leukosis virus infection: Effects on production and mortality and consequences in selection for high egg production. Poult Sci 59:2165-2178, 1980.

16. Gerlach H: Viruses. *In* Ritchie BW, Harrison GJ, Harrison LR (eds): Avian Medicine: Principles and Application. Lake Worth, Wingers Publishing, 1994, pp 862-948.

17. Graf T, Beug H: Avian leukemia viruses. Interaction with their target cells in vivo and in vitro. Biochem Biophy Acta 516:269-299, 1978.

18. Graham DL: An update on selected pet bird virus infections. Proc Assoc Avian Vet, 1984, pp 267-280.

19. Grimes TM, Bagust TJ, Dimmock CK: Experimental infection of chickens with an Australian strain of reticuloendotheliosis virus. I. Clinical, pathological and hematological effects. Avian Pathol 8:57-68, 1979.

20. Grimes TM, Purchase HG: Reticuloendotheliosis in a duck. Aust Vet J 49:466-471, 1973.

21. Hayes LE, Langheinrich KA, Witter RL: Reticuloendotheliosis in a wild turkey (*Meleagris gallopavo*) from coastal Georgia. J Wildl Dis 28:154-158, 1992.

22. Heffels U, Fritzsche K, Kaleta EF, et al: Serological examination for viral infections in pigeons in Germany. Dtsch Tierärztl Wschr 88:97-102, 1981.

23. Hughes EP, et al. Hemoproliferative disease in captive and wild birds. Proc 35th West Poult Dis Conf, 1986, pp 83-84.

24. Kirev TT: Neoplastic response of guinea fowl to osteopetrosis virus strain MAV-2I0. Avian Pathol 17:101-112, 1988.

25. Ley DH, Ficken MD, Cobb DT: Histomoniasis and reticuloendotheliosis in a wild turkey (*Meleagris gallopavo*) in North Carolina. J Wildl Dis 25:262-265, 1989.

26. Li J, Calnek BW, Schat HA, et al: Pathogenesis of reticuloendotheliosis in ducks. Avian Dis 27:1090-1106, 1983.

27. Loupal G: Leukosen bei zoo- und wildvogeln. Avian Pathol 13:703-714, 1984.

28. Ludford CG, et al: Duck infectious anemia virus associated with *Plasmodium lophucae*. Exp Parasitol 31:29-38, 1972.

29. Malkinson M: An outbreak of an acute neoplastic syndrome accompanied by undifferentiated leukemia in a flock of Muscovy ducks. Proc 31st West Poult Dis Conf, 1982, p 110.

30. Miles BD, Robinson HL: High frequency of transduction of c-erb B in avian leukosis virus-induced erythroblastosis. J Virol 54:295-303, 1985.

31. Motha MXJ, Egerton JR, Sweeney WA: Some evidence of mechanical transmission of reticuloendothelioses by mosquitoes. Avian Dis 28:858-867, 1984.

32. Neumann U, Kummerfeld N: Neoplasms in budgerigars (*Melopsittacus undulatus*): Clinical pathomorphological and serological findings with special consideration of kidney tumors. Avian Pathol 12:353-362, 1983.

33. Nobel TA: Avian leukosis (lymphoid) in an egret (*Egretta alba*). Avian Pathol 1:75-76, 1972.

34. Palmer GH, Stauber E: Visceral lymphoblastic leukosis in an African grey parrot. Vet Med Small Anim Clin 76:1355, 1981.

35. Payne LN, Purchase HG: Leukosis/sarcoma group. *In* Calnek BW, et al (eds): Diseases of Poultry 9th ed. Ames, Iowa State University Press, 1991, pp 388-439.

36. Petrak ML, Gilmore CE: Neoplasms. *In* Petrak ML (ed): Diseases of Cage and Aviary Birds. Philadelphia, Lea and Febiger, 1969, pp 461-474.

37. Purchase HG, Burmester BR: Leukosis/sarcoma group. *In* Hofstad MS (ed): Diseases of Poultry. Ames, Iowa State University Press, 1978, pp 437-442.

38. Purchase HG, Ludford C, et al: A new group of oncogenic viruses: Reticuloendotheliosis, chick syncytial, duck infectious anemia and spleen necrosis viruses. J Nat Cancer Inst 51:489-499, 1973.

39. Purchase HG, Witter RL: The reticuloendotheliosis viruses. Curr Top Microbiol Immunol 71:103-124, 1975.

40. Ratnamohan N, Spradbrow PB: The reticuloendotheliosis viruses - A review. Pak Vet J 2:101-107, 1982.

41. Turrel JM, McMillan MC, Paul-Murphy J: Diagnosis and treatment of tumors in companion birds II. J Assoc Avian Vet 1:159-165, 1987.

42. Schat KA, Gonzalez J, Solorzano A, et al: A lymphoproliferative disease in Japanese quail. Avian Dis 30:153-161, 1976.

43. Sinkovic B: Studies of the epizootiology of reticuloendotheliosis virus infection in commercial Australian chicken flocks. Proc XXII World Vet Congress, 1983, p 185.

44. Varejka F, Tomsik F: The role of house sparrow (*Passer domesticus*) in the spread of leukosis viruses in poultry. Acta Vet Brno 43:367-370, 1974.

45. Wadsworth PF, Jones DM, Pugsley SL: Some cases of lymphoid leukosis in captive wild birds. Avian Pathol 10:499-504, 1981.

46. Weiss R, Teich N, Varmus H, et al: RNA Tumor Viruses. Cold Spring Harbor Laboratory, 1982.

47. Witter RL: Reticuloendotheliosis. *In* Purchase HG, et al (eds): A Laboratory Manual for the Isolation and Identification of Avian Pathogens 3rd ed. Dubuque, Kendall/Hunt Publishing Co, 1989, pp 143-148.

48. Witter RL: Reticuloendotheliosis. *In* Calnek BW, et al (eds): Diseases of Poultry 9th ed. Ames, Iowa State University Press, 1991, pp 406-417.

Togaviridae

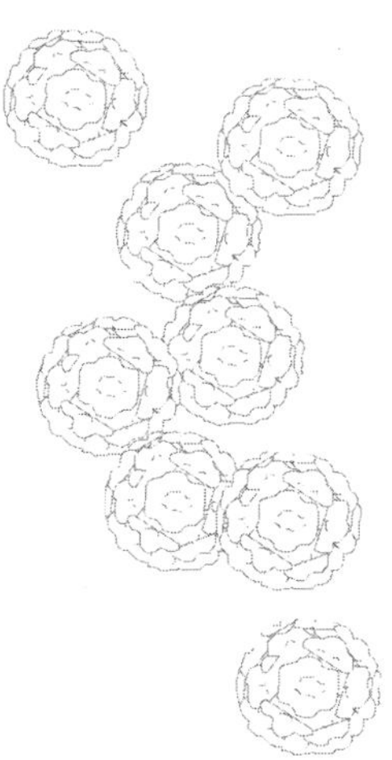

Historically, the term arbovirus has been used — without regard for taxonomic classification — to describe any virus that is transmitted in nature by an arthropod vector. Hundreds of viruses, principally within the families Togaviridae, Flaviviridae, Reoviridae, Rhabdoviridae and Bunyaviridae, have been shown to be transmitted from host to host by arthropods.[52] Many of the viruses within the Togaviridae can cause encephalitis in humans and birds; birds can play various roles in the maintenance, transmission and dissemination of the viruses *Table 15.1*. In their natural host, many arboviruses cause minimal disease, while in unnatural hosts, many of these viruses can cause severe, life-threatening infections. Characteristically, animals involved in the transmission of arboviruses develop a high level of viremia that provides an opportunity for insects that feed on blood to ingest the virus with a blood meal. The ingested virus then replicates inside the insect and is transmitted to a new host when the infected insect bites a susceptible host. Mosquitoes, ticks and biting flies are the most important vectors for the arboviruses, although any insect that feeds on blood could be involved in virus transmission.

Several viruses within the family Togaviridae are clinically important pathogens in companion, aviary and free-ranging birds. These viruses are 60 to 70 nm enveloped particles with a 28 to 35 nm nucleocapsid. Replication occurs in the cytoplasm, and the envelope is obtained when the virus exits the infected cell by budding from the plasma membrane. The family Togaviridae includes three genera: *Alphavirus*, *Rubivirus* and *Pestivirus*. *Alphavirus* contains the encephalitis viruses that can infect and, in some cases, cause disease in mammals and birds. *Rubivirus* contains one member, rubella virus, that has thus far been associated with disease only in humans. The genus *Pestivirus* has several members that have been associated with disease in domestic mammals.

ALPHAVIRUSES AND EASTERN EQUINE ENCEPHALITIS VIRUS

Over 30 alphaviruses have been described. Each occurs in a fairly restricted geographic region that is governed by the natural range of specific insects that transmit the particular virus. Eastern equine encephalitis (EEE) virus, western equine encephalitis (WEE) virus and Highlands J (HJ) virus are the most commonly encountered alphaviruses in North America. Highlands J virus was once thought to be an antigenic variant of WEE virus but now has been shown to be distinct.[10,11] However, Highlands J virus is clearly related serologically to WEE virus, with the former occurring along the eastern coast of North America and

Some characteristics of togaviruses recovered from free-ranging birds[83]

Eastern equine encephalitis virus 　Asymptomatic to death	North, Central and South America
Western equine encephalitis virus 　Asymptomatic to death	North, Central and South America
Highlands J virus 　Isolates from birds 　Asymptomatic to death	Eastern United States
Sindbis virus 　Isolates from birds and mosquitoes 　Asymptomatic to death	Africa
Cabassou 　Isolates from birds and mosquitoes 　Asymptomatic	French Guyana
Chikungunya 　Isolates from birds and mosquitoes 　Asymptomatic	Africa
Fort Morgan 　Isolates from Cliff Swallows 　and other passerines 　Asymptomatic to death	Colorado
Mucambo 　Isolate in a bird and occasional 　detection of antibodies	Brazil
Ross River 　Several isolates from birds 　considered incidental	Australia
Semliki Forest 　Isolates from birds and mosquitoes 　Asymptomatic	Africa
Tonate 　Isolates from free-ranging birds	French Guyana
Venezuelan equine encephalitis virus 　Asymptomatic	Central and South America, southern Texas and Florida
Ockelbo virus 　Asymptomatic	Australia, New Zealand and portions of Europe

the latter occurring principally west of the Mississippi River. Older reports describing WEE virus infections in birds along the east coast of the United States were probably documenting HJ virus infections.

The natural susceptibility of various species of birds to alphaviruses like EEE, WEE and HJ viruses varies widely; however, all avian species are considered susceptible to experimental infections.[88] Response to infection varies with the species of bird and the route of inoculation. Additionally, the response of individual birds to the same strain of virus may be dramatically different. These viruses typically cause subclinical infections in free-ranging birds native to an area where the virus naturally occurs. The apparent resistance to disease exhibited by indigenous birds is thought to be the result of centuries of natural selection in which native birds have co-evolved with these viruses, and now function as reservoirs. In some non-native bird species, these viruses have been associated with encephalitis or enteritis. Some indigenous species of birds infected with these viruses can develop clinical signs of disease following experimental infection, particularly when they are exposed by unnatural routes (such as injection into the brain or abdomen).

While alphavirus infections generally do not cause easily recognizable clinical changes or death, they may cause as-yet-uninvestigated subtle effects on free-ranging birds, including reduced ability to gather food or increased susceptibility to predation.[88]

CLINICAL FEATURES

Depending on the species, birds infected with EEE virus may be totally asymptomatic, die suddenly with no signs of illness or die following a brief period of severe enteritis or neurologic disease.[19,46] When they occur, clinical changes associated with EEE are most common in young birds and may include depression, anorexia, weakness, profuse diarrhea and signs of central nervous system abnormalities. A partial or complete paralysis of one or both legs may be the only sign noted, while other birds may show circling movements with partial paralysis involving one or both wings *Figure 15.1*.[27,51,60] Clinical changes in Lady Gouldian Finches suspected to be infected with EEE virus included partial paralysis and difficulty in breathing.[100] In some species (Chukars and turkeys), EEE virus infections have been associated with reproductive problems and decreased egg production.

While the severity of lesions varies widely among affected species, the types of neurologic changes induced by EEE virus are relatively consistent in any species that develops clinical signs. Some affected birds, even those with neurologic signs, recover completely. Others die within 48 hours of virus exposure, with or without premonitory signs. The anticipated mortality rates vary with the host and are listed in *Table 15.2*.

TABLE 15.2

Reported mortality rates in various birds infected with EEE virus[20,22,27,55,60,73,90]

Species	Mortality in at-risk birds
Most species	Asymptomatic
Turkeys	6%
Exotic pheasants	5% to 80%
Quail	40% to 90%
Chukars	30% to 85%
Whooping Cranes	up to 60%
English Sparrows	up to 50%
Pigeons	up to 25%
Emus	up to 58%

Clinical changes associated with EEE virus infections are most common in avian species that have been imported into North America (non-indigenous

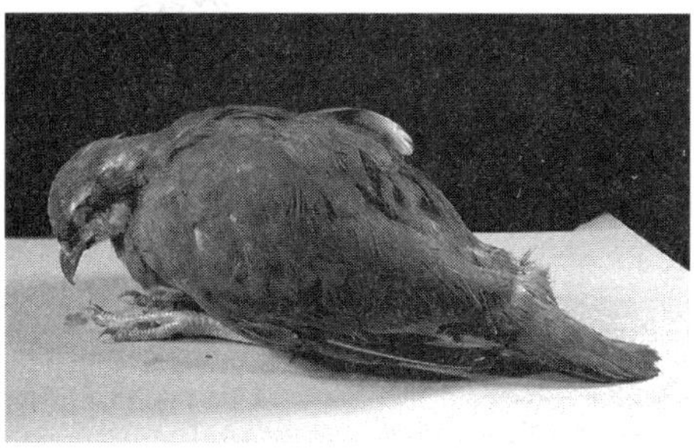

species), including Japanese Quail, "exotic" pheasants, "exotic" cranes, emus, Chukar Partridges, Pekin Ducklings, domestic pigeons and English Sparrows *Figure 15.2*.[22,46,55,60] Virus-induced disease also has been reported in some ducks, Whooping Cranes, finches, chickens and turkeys; affected birds, particularly neonates, have died acutely soon after exposure to the virus.[36] Many free-ranging Passeriformes, waterfowl, American Egrets, Snowy Egrets and White Ibis are susceptible to infection, but remain asymptomatic. They are thought to serve as reservoirs for the virus.[22,46,55,60] After being infected with EEE virus, New World finches (those that are indigenous to North America) develop a short

viremia and seroconvert while remaining asymptomatic. In comparison, English Sparrows, a European bird that has been introduced to North America, develop a prolonged, high-titered viremia.[36]

FIG 15.2
Clinical changes associated with EEE virus infections are most common in species of birds not native to North America like imported pheasants and cranes. Native species of birds that are infected with EEE virus usually remain clinically normal.

Response to experimental infection with EEE virus varies with the route of inoculation (intravenous, intramuscular or subcutaneous) and the species of infected bird. In some trials, even species of pheasants that are considered highly susceptible to infection and routinely develop clinical signs of disease may remain asymptomatic following experimental inoculation. Others may die within several days of being infected or develop mild incoordination five to twelve days after inoculation and then recover.[77] A summary of the outcome of experimental EEE virus infections in various species of birds is provided *Table 15.3*.

In one study, 50% of the English Sparrows infected with EEE virus died, while none of the birds infected with WEE virus demonstrated any clinical signs.[91] In another study, English Sparrows died from 17 to 48 hours after inoculation. The average period of viremia for EEE virus in experimentally infected English Sparrows was 2.2 days.[91] Cardinals experimentally in-

fected with EEE virus appeared in distress and vocalized incessantly prior to death 48 hours after inoculation.[55]

One crow inoculated intramuscularly with EEE virus remained asymptomatic, while another died acutely without showing clinical signs at five days after infection.[77] Two crows infected with EEE virus by oral inoculation remained asymptomatic but shed virus in their feces one to four days after infection.[54] An infected Red-shouldered Hawk remained asymptomatic and seroconverted, whereas a Great Horned Owl infected subcutaneously was depressed, reluctant to move and ataxic 48 hours after infection, had difficulty breathing 72 hours after infection and was clinically normal 6 days after inoculation.[77]

Some pigeons have died of natural and experimental EEE virus infections.[31] All 16 pigeons experimentally infected with EEE virus subcutaneously were viremic within 48 hours of inoculation. Four of the 16 birds died between 16 and 114 hours after infection. Virus-neutralizing antibodies were demonstrated in the pigeons that survived.[56] Doves infected subcutaneously remained asymptomatic but seroconverted.[57]

White Ibis, American Egrets and Snowy Egrets remained asymptomatic when inoculated subcutaneously with EEE virus or when bitten by infected mosquitoes. Even though these birds remained asymptomatic, they were susceptible to infection by small quantities of virus, suggesting a well-adapted host-parasite relationship.[55,57] In one study, all the wild-caught, Black-bellied

TABLE 15.3

Period of viremia and outcome of infection in some birds experimentally infected with EEE virus[2,55,67,77]

Species	Viremic Period	Clinical Outcome
Ibis	24 to 48 hours	asymptomatic
Egret	56 to 80 hours	asymptomatic
Snowy Egret	16 to 80 hours	asymptomatic
Purple Grackle	18 to 112 hours	asymptomatic
Cardinal	24 to 48 hours	death within 48 hours
English Sparrow	16 to 48 hours	died in 17 to 48 hours
Pheasant	by 48 hours	asymptomatic or death
Crow	24 to 72 hours	died 5 days after infection
Red-shouldered Hawk	by 24 hours	asymptomatic
Great Horned Owl	24 to 120 hours	ataxic and recovered
Pigeon	by 48 hours	asymptomatic or died 16 to 114 hours after infection
Dove	—	asymptomatic
Duck	24 to 48 hours	asymptomatic

Whistling Ducks inoculated with EEE virus developed viremia, remained clinically normal and seroconverted. Five of six ducks were viremic within 24 hours after inoculation, and the mean duration of viremia was two days. Antibody titers to EEE virus increased rapidly starting by day 6 after inoculation.[2]

EPIZOOTIOLOGY

It has been suggested that most birds are susceptible to alphaviruses, and that the prevalence of infection in a bird population is a function of behavioral and ecological factors and is not controlled by host susceptibility.[65,88] Generally, alphaviruses have geographically localized natural ranges. Cyclic outbreaks are dependent on specific regional insects that serve as biologic vectors and on vertebrate hosts that serve as reservoirs. Outbreaks are most common in the late summer when concentrations of biting insects (particularly mosquitoes and ticks) and young susceptible birds are the highest. Fall reduction in alphavirus activity coincides with decreased nesting of birds, reduced populations of young birds for feeding mosquitoes, the departure through migration of preferred birds for local mosquitoes and reduced mosquito activity with the onset of cooler weather. Additionally, the percentage of susceptible birds decreases in the late summer and early fall because more birds have been infected and developed immunity to a particular alphavirus increases.[45]

Eastern equine encephalitis virus infections were first confirmed in humans in the Massachusetts area in the 1930s, although epidemics were probably recognized in humans as early as 1831.[29,31,69] The virus was first reported in birds in 1938 when an outbreak occurred in pheasants in Connecticut.[98] Since the 1930s, periodic outbreaks of EEE have been reported in humans, horses and pheasants in the Caribbean, and in North, Central and South America.[13] Strains of EEE virus or antibodies to the virus have been recovered from the United States, Dominican Republic, Jamaica, Panama, Trinidad, British Guiana, Brazil, Argentina, Philippines, Czechoslovakia, Russia and Thailand.[3,12,63,88]

It is accepted that birds serve as the natural host for the EEE virus, playing various roles in its maintenance, transmission and dissemination.[46,56] In North America, virus activity is particularly prevalent on the east coast of the United States, and the virus is considered endemic in populations of birds that inhabit eastern swamps and coastal marshes *Figure 15.3*.[5,57] As is the case for other alphaviruses, the prevalence of EEE virus in certain areas of the United States is thought to be controlled by the regional distribution of mosquitoes that transmit this virus.[91] Eastern equine encephalitis virus has been implicated in disease from the Atlantic and Gulf coast states extending from New Hampshire and Quebec, Canada south to Texas and inland to Wisconsin and Michigan.[13,46,49,88]

FIG 15.3
Indigenous birds such as herons play various roles in the maintenance and dissemination of the EEE virus.

Over 90 species of birds have been found to be susceptible to EEE virus, including birds in the orders Ciconiiformes, Galliformes, Columbiformes, Gruiformes, Anseriformes, Strigiformes, Falconiformes, Psittaciformes and Passeriformes. Many free-ranging birds native to North America are highly susceptible to infection, resistant to disease, circulate virus in the blood in high concentrations for several days and are considered reservoirs of the virus.[55,60]

Three important factors govern the ability of free-ranging birds to serve as reservoirs for alphaviruses: the density and geographic distribution of the species, the species susceptibility to infection, and the likelihood that an infected bird will serve as a source of virus for an arthropod vector.[48,49] English Sparrows have been shown to develop long-term, high-level viremias and are thought to serve as reservoirs for EEE virus.[90] Reports on the duration of viremia following inoculation of waterfowl with alphaviruses vary. Some reports suggest a low-titered viremia of short duration; others document sufficient viremia for infected birds to serve as reservoirs.[2,67] Infected waterfowl that develop a low titered viremia of short duration would not be expected to serve a major role in dissemination of alphaviruses. The rapid death of infected emus may prevent them from serving as a reservoir for mosquitoes, thus making them of little importance in the transmission of EEE virus.[99]

The seasons of the year as they relate to the dynamics of alphavirus transmission include periods of overwintering of virus, spring viral amplification, summer viral activity and fall viral subsidence.[45] Seasonal epornitics of EEE virus have repeatedly been documented. These outbreaks typically occur in the late summer and early fall with subsidence of viral activity by October in the Northern Hemisphere.[43] In some areas, outbreaks may continue into November. In birds in a swamp in Maryland, WEE virus infections were most common in the mid-summer, and EEE virus infections did not peak until later in the season.[18] Of 3,020 free-ranging birds from Alabama that were sampled over a 12-month period, EEE virus was isolated from the blood of 42 birds and WEE virus from the blood of 23 birds. All the isolates were from birds trapped between September and December. Antibody surveys in these same birds indicated that there was minimal virus activity between January and July, and high virus activity between September and December.[89]

Several mechanisms for annual reintroduction of EEE virus into an area have been proposed, including return of the virus with spring migrants, maintenance of the virus in mosquitoes and latent viral infections in regional mammals, reptiles, amphibians or birds.[18] Substantial evidence suggests that migratory birds are probably not responsible for reintroducing the virus each spring. It has been well documented that southward-bound migratory birds may have either WEE or EEE viremias, whereas neither virus has been isolated from migratory birds returning from their wintering grounds.

In one study, the seroprevalence of antibodies to these viruses in summer residents leaving and returning to a study area remained essentially equal, suggesting the absence of virus activity in southern wintering areas.[92]

Neither EEE nor WEE virus could be isolated from any of 1,503 birds examined on return migration from Central and South America in the early spring. Serologically, the endemic bird population of Louisiana showed the highest antibody titers to EEE virus, with 22% of permanent residents and 15.5% of winter residents being seropositive. In Florida, 6.5% of permanent residents and 10% of winter residents tested had EEE virus antibodies. The presence of antibodies in winter residents of Florida would be expected because these birds migrate from Atlantic coastal areas where EEE virus activity is common. Likewise, coastal Louisiana is populated in the winter chiefly by migratory shore birds that originate from Arctic and subarctic regions. The low seroprevalence of EEE virus antibodies in these birds indicates that they experience minimal exposure to the virus during their summers in the Arctic, or during their wintering period in the southern United States.[57] It is most likely that EEE virus persists in freshwater swamps in the southeastern United States where endemic mosquitoes and mild winters allow for a low level of continued activity. Additionally, several species of birds have been shown to develop chronic viremias, and these birds could serve as a reservoir for the virus in the winter when mosquito activity subsides.[74,91]

While it is tempting to assume that an increase in the mosquito density in the spring would correlate with an increased incidence of viral transmission and disease, this assumption has been disproved. It has been suggested that an increase in the population of mosquitoes in combination with an increase in the population of infected and susceptible young birds that occurs in the summer and fall dictates the severity and longevity of an outbreak.[45,87] Eastern equine encephalitis virus is most frequently recovered from nestling birds in areas where the virus is endemic. The increased prevalence of infections in young birds is probably a result of their reduced immunity, coupled with the lack of feathers and inability to move, which make them easy targets for feeding mosquitoes.[45,46,49,90] Companion bird neonates raised in outdoor aviaries may provide a similarly attractive meal for infected mosquitoes.

<u>INCUBATION PERIOD</u> — Many birds infected with EEE virus develop a viremia from 24 to 72 hours after infection but remain asymptomatic. The incubation period of EEE virus in susceptible pheasants inoculated with virus by the intramuscular route averages 3.8 days. In pheasants, clinical signs may be noted 2 to 12 days following infection, depending on the route of inoculation. The incubation period in immunologically naive birds exposed to experimentally infected pheasants is 4.5 days.[51] Depending on the age of the birds, chickens experimentally infected with an EEE virus recovered from

Seroprevalence, or rate of alphavirus and flavivirus recovery, in various groups of birds

ALABAMA[89]

Free-ranging Piciformes, Passeriformes and Strigiformes
15% had EEE, and 13% had WEE virus antibodies in July
34% had EEE, and 34% had WEE virus antibodies in November and December

BRAZIL[82]

From 1.4% to 2.9% of free-ranging forest birds had EEE virus antibodies
During one year, EEE, WEE, SLE and Turlock viruses and antibodies to these viruses
were demonstrated in a group of free-ranging forest birds

FLORIDA[86]

360 free-ranging Ciconiiformes and Pelecaniformes
2% had EEE virus-neutralizing antibodies
7% had SLE virus-neutralizing antibodies
16% SLE virus seroprevelance in Pelecaniformes
5% SLE virus seroprevelance in Ciconiiformes
Seroprevalence higher in adults and sub-adults than in neonates

FLORIDA[28]

5 of 16 (31%) Ciconiiformes had EEE virus-neutralizing antibodies
1 of 5 (20%) Pelecaniformes had EEE virus-neutralizing antibodies
1 of 16 (6%) Ciconiiformes had antibodies to Highland J virus
None of 5 Pelecaniformes had antibodies to Highland J virus

FLORIDA[5,6]

4 of 85 (5%) Cattle Egrets had EEE virus antibodies
1 of 187 (0.5%) herons had SLE antibodies
2 of 187 (2%) herons had EEE antibodies
1 of 109 (1%) herons had WEE antibodies
none of 92 herons had EVE antibodies

KENTUCKY[64]

Birds tested during an SLE outbreak in people
35 of 241 (14%) free-ranging Passeriformes and Piciformes had SLE virus antibodies
37 of 122 (30%) chickens sampled had SLE virus antibodies
None of the free-ranging birds had HI antibodies to WEE or EEE viruses
6% of the chickens had EEE virus HI antibodies
2% of the chickens had WEE virus HI antibodies

LOUISIANA[55]

347 Ciconiiformes, Pelecaniformes, Gruiformes and Passeriformes tested
Overall, 12% of the birds had EEE virus-neutralizing antibodies
In some flocks, 75% of the ibis and night herons had EEE antibodies

MASSACHUSETTS - 2 year study [43]

748 free-ranging Passeriformes
Virus was not recovered from the blood of any of the birds
2% had EEE virus antibodies.
1,206 domestic turkeys, chickens and pheasants
3% had EEE virus antibodies

continued . . .

macaws died from 15 hours to 26 days after inoculation.[33]

AGE SUSCEPTIBILITY — The susceptibility of pheasants to EEE virus-induced disease varies with age and with the virulence of the infecting virus.[40] Pheasants develop resistance to disease from some strains of virus by four weeks of age, while other strains can cause high mortality in pheasants of any age. Pekin Ducks were found to develop natural resistance to EEE virus by 18 days of age.[22] Chickens appear to

MASSACHUSETTS[29]

Antibody survey during epidemic in humans
   47% of free-ranging birds had EEE virus antibodies.

MASSACHUSETTS[65]

8,417 free-ranging Piciformes and Passeriformes during a 12-year study
   EEE virus recovery from 8 birds of 7 species
   HJ virus recovery from 18 birds of 7 species
   VN antibodies to EEE virus in 2 of 3 species of Piciformes
   VN antibodies to EEE virus in 12 of 17 species of Passeriformes
   VN antibodies to HJ virus in 2 of 3 species of Piciformes
   VN antibodies to HJ virus in 14 of 17 species of Passeriformes

MEXICO[78]

Free-ranging Ciconiiformes, Columbiformes, Strigiformes and Passeriformes
   2 of 124 (1.6%) birds had EEE virus antibodies
   2 of 124 (1.6%) birds had WEE virus antibodies
   The birds that were seropositive were adults

MINNESOTA - 2 year study[71]

   68% of seronegative pigeons developed WEE virus antibodies during one year
   23% of the pigeons had antibodies to WEE virus the following year
   Seropositive pigeons were most commonly detected from August to October

NEW JERSEY[42]

Survey following EEE virus epornitic
   65% of chickens had EEE virus antibodies
   33% of free-ranging birds had EEE virus antibodies

NORTH DAKOTA[68]

342 immature and adult Piciformes, Passeriformes and Columbiformes tested
   Overall seroprevalence of SLE was 8%
   5% of nestlings had WEE virus antibodies
   17% of nestlings had SLE virus antibodies

SASKATCHEWAN[9]

471 free-ranging ducks
   11% seropositive for WEE virus
   16% had low levels of antibodies that were considered "suspicious"

TEXAS - 5 year study[48]

   WEE virus isolated from 234 of 3964 (6%) English Sparrows
   SLE virus was recovered from 15 of 3964 (0.4%) English Sparrows

WISCONSIN[20]

Epornitic in Whooping Cranes
   14 of 32 (44%) surviving cranes had EEE virus-neutralizing antibodies
   13 of 38 (34%) exposed Sandhill Cranes had EEE virus-neutralizing antibodies

develop age-related resistance to disease by three weeks of age.[50] Chukars were found to develop resistance to disease as they approached 12 weeks of age.[73] Whooping Cranes as old as 18 years died following a natural outbreak.[20]

SEROPREVALENCE — The seroprevalence of alphaviruses in birds varies from 0% to 100% depending on the species of birds being tested and their geographic location.[18,65] Additionally, from season to season and from year to year, fluctuations in the number of seropositive birds within a given population are common. The results of a group of alphavirus seroprevalence studies performed in the United States are provided *Table 15.4*.

The majority of seropositive birds are detected in the expected geographic range for each virus. For example, WEE virus activity is rare on the east coast of the United States, and consequentially, the number of birds on the east coast with antibodies to the virus is correspondingly low. By comparison, EEE virus activity is common on the east coast of the United States, and antibodies to the virus can be detected in over 75% of the birds in some flocks, indicating previous infections. In free-ranging birds and domestic poultry, the prevalence of EEE virus antibodies is generally higher in areas near fresh-water swamps.[43]

In a group of free-ranging birds in Alabama that included Piciformes, Passeriformes and Strigiformes, EEE and WEE antibodies were detected most commonly in permanent residents, followed in decreasing frequency by winter residents, summer residents and transient species. The prevalence of antibody titers was much higher in free-ranging adult birds than in neonates tested in the summer before the density of mosquitoes had increased. The seroprevalence rose sharply when immature birds were tested in the fall, a period when the number of mosquitoes and susceptible birds increased dramatically. These surveys indicated that there was minimal virus activity between January and July, and high virus activity between September and December.[89]

The number of birds with antibodies to EEE virus can vary dramatically in a region from year to year. In a multi-year study of ciconiiform and pelecaniform birds from salt-water and fresh-water environments in Florida, the seroprevalence of neutralizing antibodies to EEE virus was found to be cyclic; in some years, antibodies were not detected in any of the tested birds. A consistent finding was the more frequent detection of antibodies in adults and immature birds when compared to nestlings.[86]

Antibody surveys in Louisiana, New Jersey, Alabama and Massachusetts indicated that 13% to 54% of the passerines had neutralizing antibodies to EEE virus. The level of antibodies during a period when virus activity was considered minimal ranged from 13% to 22%. This level of seroprevalence was considered indicative of endemic infections allowing continued maintenance of the virus in a local population of susceptible birds. During an outbreak, the seroprevalence rate was as high as 54%, and virus was recovered from up to 11% of the free-ranging birds tested.[87,91] A similar finding has been reported in two other studies involving birds in Massachusetts. During a two-year study, EEE virus was not detected in the blood of any of 748 free-ranging Passeriformes, and only 17 (2%) of these birds and 35 of 1,206 (3%) domestic fowl had detectable levels of antibodies.[43] In another year when an epidemic of EEE occurred in humans, 47% of the free-ranging birds tested had EEE virus antibodies.[29]

The serologic response induced by EEE virus varies dramatically in degree and duration among naturally infected birds. In one study, Swamp Sparrows maintained antibody titers for up to three years, whereas antibody titers decreased rapidly in Chickadees. Previously infected birds in which antibody titers decrease may develop an increase in antibody titer when they are again exposed to the virus.[65]

If companion birds are as susceptible to alphavirus infections as free-ranging species are, psittacines in outdoor aviaries with natural exposure to mosquitoes should have a prevalence of antibodies that is at least as high as those demonstrated for birds that are permanent residents of the same geographic area. In a group of captive Psittaciformes from the eastern United States (Florida and Tennessee), up to 10% of the tested birds had EEE virus-neutralizing antibodies while none of the birds had WEE virus-neutralizing antibodies.[76a] This finding indicates that psittacine birds are susceptible to EEE virus infections *Figure 15.4*. The seroprevalence rate of companion birds in Chicago was 5% for EEE virus and 2% for WEE virus. Of particular interest was a subpopulation of Eclectus Parrots from Florida that had been experiencing high neonatal mortality. Sixty percent of the unaffected Eclectus Parrots from this flock were found to have EEE virus antibodies. This seroprevalence was higher than that described for other psittacine birds in the same collection and from other psittacine birds in Chicago and Tennessee.[76a]

TRANSMISSION — Alphaviruses are principally transmitted by mosquitoes, ticks and biting flies. An efficient insect-to-vertebrate-host transmission cycle requires that the susceptible host

FIG 15.4
Antibodies to EEE virus can be detected in asymptomatic psittacine birds from areas where virus activity is common.

produce a sufficient quantity of circulating virus to infect an insect that consumes its blood *Figure 15.5*.[45] Mosquitoes are the primary vector for EEE virus, and the natural cycle of infectivity involves mosquito transmission to birds, and bird transmission back to other mosquitoes. This virus is present in the blood stream of most birds within 24 hours after infection, and viremia can persist for two to five days or longer. The longevity of an infected mosquito is an important factor in alphavirus transmission. For a complete cycle of bird-to-mosquito-to-bird transmission to occur, a mosquito must be able to feed on an infected bird, survive for several days while the virus replicates, and then feed on a new susceptible host before dying.[46] Mosquitoes remain infected for their entire life, and as more mosquitoes are infected, the transmission rate to birds in an area increases proportionally.[45]

Because they share a habitat with large concentrations of mosquitoes, birds that inhabit aquatic environments are considered at greatest risk of becoming infected with alphaviruses and flaviviruses. These birds may then disseminate the virus southward as they migrate. Birds like non-native pheasants and emus that die soon after being infected are end hosts and are consid-

Alphaviruses are principally transmitted by mosquitoes, ticks and biting flies. Virus circulating in the blood of an infected bird is ingested when a mosquito feeds. The virus is then injected into new hosts with subsequent feedings.

ered of minimal importance in the transmission of alphaviruses.

These viruses also can be spread in other ways. In pheasants and ducks, EEE virus can be spread through direct contact without an arthropod vector.[22,47,60] In ducks, WEE virus infections can occur following ingestion of virus-contaminated water.[8] In experimentally infected pheasants, virus could be recovered from the feather quills for up to six days after infection and persisted in this site for 24 to 48 hours longer than in the blood *Figure 15.6*. Feather-picking

Because EEE virus is found in blood, a blood feather from an infected bird would be expected to contain virus; a susceptible bird could be infected by ingesting the virus when preening an infected mate.

and cannibalism were found to be responsible for the rapid dissemination of the virus among susceptible pheasants, but virus transmission did not occur in pheasants through contact with the aerosols of fecal material from infected birds.[77] Crows that ingested EEE virus-infected chicken embryos were

found to shed virus in their feces one to four days after infection.[54] Flocks of free-ranging starlings and eastern sparrows that consume EEE virus-infected carrion may serve as mechanical vectors for virus transmission.[77] Although virus is present in the feces and feathers of some infected birds, these sources of virus transmission are considered of little epizootiologic importance when compared to the likelihood of virus transmission by infected mosquitoes.[49]

PATHOLOGY

The necropsy findings associated with EEE virus infections vary widely among affected birds and in the same species of bird. The collective necropsy changes that have been associated with EEE virus infections in some bird species include a small pale spleen or splenomegaly, softening of the cerebral hemispheres, hepatomegaly, mottling of the spleen, distention of the kidneys, white streaking in the heart, ascites, mucoid enteritis and visceral gout *Figure 15.7*.

The microscopic lesions that have been reported in some birds affected by EEE virus include inflammation of the blood vessels in the brain, non-suppurative encephalitis, perivascular lymphocytic infiltrates in the brain, microgliosis, meningitis, neuronal degeneration, splenitis and myocardial necrosis.[51,73] Some affected birds develop lesions primarily in the brain; others primarily in the abdominal organs. In those birds with encephalitis, lesions tend to occur in the front portions of the brain area with a distribution characterized as "descending" encephalitis. This pattern differs from other causes of encephalitis in birds which are typically described as "ascending" encephalitis. Among 59 pheasants infected with EEE virus, 57 (97%) had encephalitis and 11 (19%) had hepatic necrosis, making these the most common lesions.[51] Crows inoculated with EEE virus developed lesions in the brain consisting of meningitis, glial cell proliferation, perivascular leukocyte infiltration and neuronal degeneration. Infected doves developed similar lesions in the brain. Severe liver necrosis in the absence of brain lesions was characteristic of EEE virus infections in cardinals, Red-winged Blackbirds, Whooping Cranes and sparrows.[20,55] Gouldian finches suspected of dying from an EEE virus infection had pale livers and kidneys with hemorrhage in the lungs. Severe liver necrosis was evident microscopically.[100]

PATHOGENESIS

In their natural host, many alphaviruses cause minimal disease because of evolutionary adaptations that have allowed for survival of the host and the virus. In unnatural hosts, many of these viruses can cause severe, life-threatening infections. Some of the most convincing support for the coevolution of EEE virus with species of birds native to the eastern United States where the virus is endemic, can be found in the resistance of native sparrows to disease, when contrasted with the fact that English Sparrows imported to the United States from Europe are highly susceptible to infection and may develop disease.[91]

In general, alphavirus replication occurs in the cells near the portal of entry or in regional lymphoid tissues. The virus is released from the site of initial replication, resulting in a primary viremia, followed by the dissemination

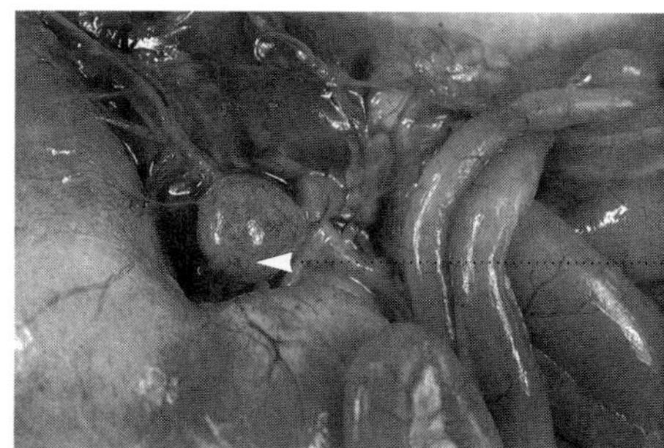

FIG 15.7
Necropsy findings suggestive of an EEE virus infection may include: a small pale spleen;

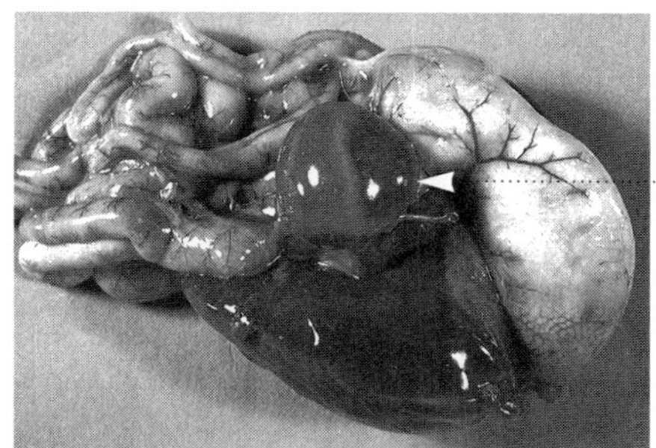

enlargement and mottling of the spleen;

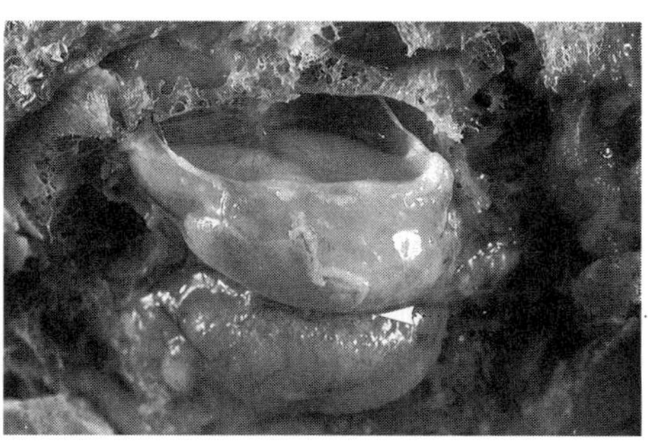

softening of the cerebral hemispheres (with or without an increased quantity of fluid);

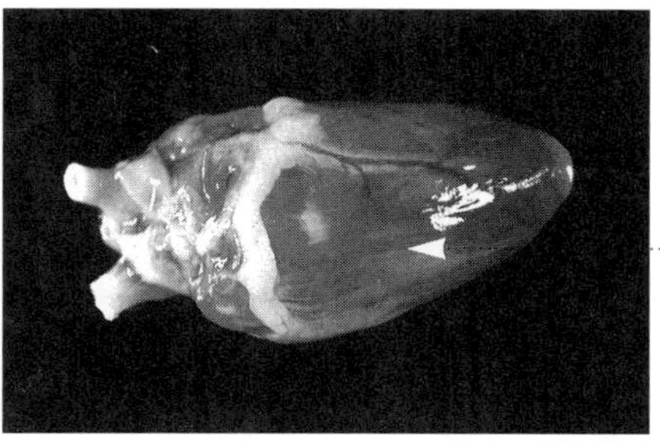

white streaking in the heart; or

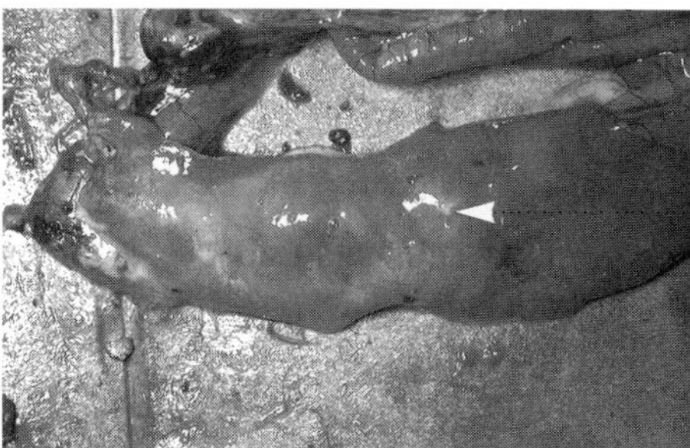

mucoid enteritis.

of the virus throughout the body. The virus then replicates in other tissues and the progeny virions that are released cause a secondary viremia. It is this secondary viremia that results in a sufficient quantity of circulating virus to infect feeding insects. Viremia is generally of short duration, lasting from several days up to a week *see Table 15.3.* Because alphaviruses have an affinity for red blood cells, they are transported throughout the body, including to the brain.[36]

In most avian species, the central nervous system is considered the primary target for EEE virus replication. However, not all species respond similarly to EEE virus infections. In cranes and some gallinaceous and passerine birds, the liver appears to be the primary organ affected by the EEE virus.[20,55] In emus, the virus principally affects the gastrointestinal tract, causing a severe enteritis *see Figure 15.11.*[97] No one yet knows why EEE virus has an affinity for the central nervous system in most species, but for the liver, spleen, heart or intestinal tract in others.

IMMUNITY

Birds that survive EEE virus infections become solidly immune and develop hemagglutination-inhibition (HI), complement-fixation and virus-neutralizing antibodies. The viremia that is associated with EEE virus infections ceases rapidly following the production of antibodies. Antibodies may be detected as soon as five days to two weeks after infection in gallinaceous and passerine birds.[58,91] Complement fixation and HI antibodies may persist for several weeks to months, while virus-neutralizing antibodies may persist for months to years following natural or experimental infections.[32,55,60,88]

In English Sparrows, the mean duration of viremia in immunologically naive birds was 2.4 days, but was only 0.4 days in birds that had pre-existing EEE virus antibodies.[91] Hemagglutination-inhibition antibody titers in chukars that survived natural infection ranged from 1:40 to 1:1280.[70] Virus-neutralizing antibody titers in pigeons that survived subcutaneous inoculation ranged from 1:40 to 1:1280.[56] Following a flock outbreak of EEE virus, 12 of 14 (86%) Whooping Cranes with neutralizing antibodies also had HI antibodies.[20] Following a natural EEE virus outbreak in pheasants, 20 survivors remained asymptomatic and developed neutralizing antibodies that persisted for over 16 months in some birds.[60] Some species of Passeriformes (Swamp Sparrows) were found to maintain virus-neutralizing antibody titers for over three years; others (Chickadees) had a rapid reversion to a seronegative status after infection. In a 12-year study, some birds that were seropositive when first tested were seronegative when evaluated years later. When tested subsequently, many of these seronegative birds had again developed antibodies to the virus, indicating re-exposure to an alphavirus.[65] A similar increase in antibodies with re-exposure to virus has been demonstrated in other studies.[88,92]

Birds that have antibodies to one alphavirus may have some increased resistance to disease induced by other alphaviruses.[44,50] In one study, 12 sparrows without and 15 sparrows with WEE virus (reported to be WEE but probably HJ virus) antibodies were experimentally infected with EEE virus. In the seronegative group, 50% died, but all of the birds with WEE virus antibodies survived. In a reciprocal experiment, nine sparrows without and eight sparrows with EEE virus antibodies were experimentally infected with WEE virus. In the sparrows with antibodies to EEE virus, the levels of viremia were lower and of shorter duration (0.8 days vs 2.4 days).[91]

Field and laboratory testing have indicated that EEE virus-neutralizing antibodies are transferred in the egg to nestling ibis, pigeons and chickens.[56,57,76] In one study, the level of HI antibodies transferred to a chick rarely exceeded 1:20 and did not correlate with the levels of similar antibodies in the hen. Maternally derived antibodies persisted in young pigeons for three to four weeks after hatching.[56,57]

Usually, viral-specific antibodies are responsible for preventing virus that is outside of a cell from binding to and entering a new cell. The removal of virus from within an infected cell is thought to be mediated by cytotoxic T-cells and macrophages. In contrast to this classic immunologic response, the removal of some alphaviruses from infected nerve cells in mice is mediated by antibodies, rather than by cytotoxic T-cells.[59]

DIAGNOSIS

Active or recent alphavirus infections can be diagnosed by demonstrating rising antibody levels in paired serum samples, using complement fixation, hemag-glutination-inhibition, agar-gel immuno-diffusion, enzyme-linked immuno-sorbent assay (ELISA) or virus-neutralizing assays. Screening chickens or quail for EEE virus antibodies is considered a more reliable indicator of virus activity in an area than is screening of free-ranging birds. In one study after an EEE virus outbreak in New Jersey, 65% of the chickens tested had EEE virus antibodies, while antibodies were detected in only 33% of free-ranging birds from the same area.[42] Alphaviruses are unstable outside of the host and are easiest to recover from fresh blood, brain, liver and spleen of recently infected birds.

CONTROL

Alphaviruses are enveloped and are not considered to be stable outside of the infected host. Most disinfectants would be expected to inactivate these viruses rapidly. Some inactivated EEE virus vaccines licensed for use in horses have been reported to be effective in preventing disease in pheasants, whooping cranes and emus.[16,93,96] Treatment of birds with hyperimmune serum was found to have little effect on the outcome of experimental or natural infections.[22]

WESTERN EQUINE ENCEPHALITIS VIRUS

Most WEE virus infections are asymptomatic. However, this virus has been associated with clinically recognizable disease in some infected pheasants, emus, chukars, English Sparrows, chickens and turkeys.[17] If present, clinical signs include ruffled feathers, somnolence, depression, weakness, incoordination, torticollis and paresis and paralysis.[73] This virus has been discussed as a cause of death in nestling English Sparrows,[17] but in one study, none of the sparrows infected with WEE virus developed clinical signs of disease.[91] Six of seven ducks inoculated with WEE virus remained clinically normal, and antibodies to the virus were detected starting 14 days after inoculation. Viremia was detected in only two of the ducks at three to four days after infection.[2] Ducks infected with WEE virus through consumption of contaminated drinking water sero-converted and developed neutralizing antibodies that persisted over a year.[8] When virus activity reaches an extremely high level in free-ranging birds, infections can occur in humans, horses and small mammals, all considered to be secondary hosts.

Epizootiology

Western equine encephalitis virus was first demonstrated in the brains of English Sparrows in California in 1931. This virus is endemic primarily west of the Mississippi River, with activity extending into Canada. The geographic prevalence of WEE in regional areas of the United States is thought to be controlled by regional distribution of the mosquito, *Culex tarsalis*, which is the primary vector for the virus.[87,91] Western equine encephalitis virus also has been documented in South America (Brazil, Uruguay, Argentina, Panama, Guyana), Mexico and on the island of Trinidad.[53,88] Previous reports of WEE virus along the east coast of the United States probably represent recoveries of the antigenically related, but distinct, Highlands J virus.[10,12,94]

The general epizootiologic principles discussed with EEE virus also apply to WEE virus. Western equine encephalitis virus is frequently recovered from asymptomatic Passeriformes in the Western United States, which are considered the primary reservoir for this virus. In one study, 17% of birds from 15 different species of free-ranging Passeriformes had antibodies to WEE virus. Virus recovery from English Sparrows between 55 and 330 days after infection suggests that these birds can be persistently infected.[74,75]

Nestling English Sparrows are considered to be the principal amplifying host for WEE virus in many areas. In one study, only 4 of 225 (2%) adult English Sparrows were viremic, whereas the virus was recovered from the blood of 53 of 205 (25.9%) nestlings tested during the same time period. Nine- to twelve-day-old nestlings were most likely to be infected, and virus was not detected in the feces of any of the nestlings.[49] A direct correlation between the amplification of WEE virus activity and the increase in numbers of nestling English Sparrows and mosquitoes has been reported by several investigators. It appears that *C. tarsalis*, the primary vector for WEE virus, preferentially feeds on nestling birds, accounting for the simultaneous increase and decrease of virus activity in both nestling birds and mosquitoes.[48,49] Cowbirds, Mourning Doves and pigeons are also commonly infected, but are considered to be of minimal importance in virus transmission in most areas.[18]

Epornitics of WEE typically occur in the late summer and early fall with subsidence of viral activity by October or November in the Northern Hemisphere.[41,45] In a study of birds in a swamp in Maryland, WEE (probably HJ) virus activity started in June and persisted to November. The peak of activity was during July and August. Eastern equine encephalitis virus activity occurred eight weeks later and was less intense. The previous year, EEE virus activity predominated in the swamp, suggesting alternating cycles of increased activity between these two viruses in areas where they coexist. Isolation of EEE virus from WEE virus immune birds (reported as WEE but probably HJ virus) and the detection of antibodies to both viruses in the same birds suggest that antibodies to one virus do not prevent an infection by the other.[18]

It is most likely that WEE virus persists through the winter in chronically infected birds. Isolation of WEE virus from a dove prior to the onset of mosquito activity was considered supportive evidence for chronic infections in this species. Also, the virus has been recovered from the blood of a few free-ranging sparrows during the winter.[9,46,74] Several studies have shown that infected birds can transport WEE virus southward during their fall migration, but no studies have documented that these birds transport the virus back into northern breeding grounds during the return spring migration.[18,45,92] Various small mammals (hares and squirrels) and reptiles that are susceptible to infection may also be involved in the overwintering of this virus.[45]

Western equine encephalitis virus has been isolated from many species of mosquitoes as well as from several bird mites and an assassin bug. In addition to arthropod transmission of WEE virus, experimental studies have shown that free-ranging ducks can be infected following the consumption of contaminated drinking water. Virus was recovered from the blood of infected ducks within 24 hours of ingesting contaminated water.[8] The gross and microscopic changes associated with WEE virus infections are similar to those described with EEE virus.

IMMUNITY

Antibodies to WEE virus have been demonstrated in more than 75 species of free-ranging birds, as well as from many domestic gallinaceous birds.[46] Antibodies are most commonly detected within two weeks after infection, and within several days after the cessation of viremia. Hemagglutination-inhibition antibodies are the first type of antibody produced in detectable concentrations; virus-neutralizing antibodies persist the longest.[58,91] Neutralizing antibodies produced following oral infections in ducks were found to persist in several of the birds for over a year after initial infection.[8] Re-exposure to the same virus usually elicits a rise in circulating antibodies; however, recovered birds probably develop lifelong immunity to disease.

The seroprevalence of WEE virus found in various studies in the United States is listed *Table 15.4*. Following a WEE outbreak, 50% of 23 adult sparrows tested in the fall and 75% of 116 sparrows tested in the spring were found to have protective antibodies to WEE.[41] During a two-year study in Minnesota, 68% of the pigeons developed WEE virus antibodies during the initial year, and 23% of the pigeons had antibodies the following year. Of 105 WEE virus-positive pigeons, 19 were positive in July, 41 were positive in August and 45 more were positive between September and October. The rise in seroconversion matched the increase in numbers of mosquitoes as the season progressed.[71] Antibodies to WEE virus were demonstrated in 11% of 471 blood samples from free-ranging ducks in Saskatchewan.[8]

Maternal transmission of virus-neutralizing antibody has been demonstrated by finding it in chicks less than two weeks old, in yolks and in embryos.[84] Maternally derived antibodies were found to persist for twelve weeks in doves and for four weeks in young chickens.[76] Maternally derived antibodies have been shown to be protective in doves and chickens, but not in English Sparrows.[56,76,84] Nestling English Sparrows derived from flocks with antibodies to WEE virus were shown to be as susceptible to infection as nestlings from flocks with no antibodies to the virus. However, young chickens from hens with WEE virus antibodies were resistant to infection. This virus was recovered from 59% of the nestling English Sparrows in flocks in which 56% of the adults had antibodies to the virus. The lack of protective immunity in nestling English Sparrows probably explains why infections are common in these birds, even though many of the adult birds have antibodies.[48] Antibodies to WEE virus also have been demonstrated in nestling pigeons and a heron chick as well as in adult pigeons, a Mourning Dove and a Northern Harrier.[84]

SPECIES-SPECIFIC ENCEPHALITIC DISEASES

AVIAN VIRAL SEROSITIS

An enveloped 55 to 62 nm viral particle with ultrastructural characteristics similar to those described for togaviruses was recovered from a group of young macaws and a Ring-necked Parakeet (ages ranged from 63 to 107 days) that died acutely over a two-week period. The disease was characterized by ascites. Based on the gross and microscopic changes, it was termed avian viral serositis. Gross changes varied with each bird, but collectively included distention of the abdomen with fluid, an enlarged, yellowish liver and congested edematous lungs *Figure 15.8*. The most consistent microscopic changes were lymphocytic proventriculitis, hepatic necrosis and serositis. The lymphoid organs, cardiac muscle and nervous tissue appear to be the primary target organs for this virus.[34]

A virus recovered from the pooled tissues of three of the affected young macaws was shown to be a strain of EEE virus. This virus induced characteristic clinical and gross changes in experimentally infected young chickens.[33] A group of five-day-old chickens inoculated intramuscularly with the virus became depressed and died within 15 hours after inoculation. In another group of 15-day-old chickens inoculated intramuscularly, clinical changes were generally less severe than those that occurred in younger chickens. Some of the older chicks developed depression and difficulty with breathing, then died within 48 hours. These birds had pale flaccid hearts, enlarged livers and pale kidneys. Other infected chickens developed ascites and died 5 to 26 days after exposure. Still others absorbed the ascitic fluid within two weeks of infection and recovered. When these last chickens were euthanized 40 days after infection, they had small hearts and chronic fibrotic myocarditis and epicarditis.[35]

Six of fifteen experimentally infected chickens died at some time following intramuscular inoculation. The nine chickens that survived developed antibodies to the virus. Some of the surviving chickens developed a chronic wasting type of disease and neurologic signs. By comparison, one of fifteen chicks inoculated orally died six days after infection, and only six of the remaining fourteen birds seroconverted. The fact that one of five chickens in contact with these birds also seroconverted indicates that natural transmission can occur. As would be expected, the EEE virus isolated from macaws also caused neurologic disease and death in experimentally infected mice and rats.[35]

In experimentally infected chicks, virus was recovered from most tissues and from cloacal and pharyngeal swabs early in the disease process, but not during the later phases of the disease. The recovery of the virus from the cloaca and pharyngeal regions implicates feces and respiratory secretions in the natural

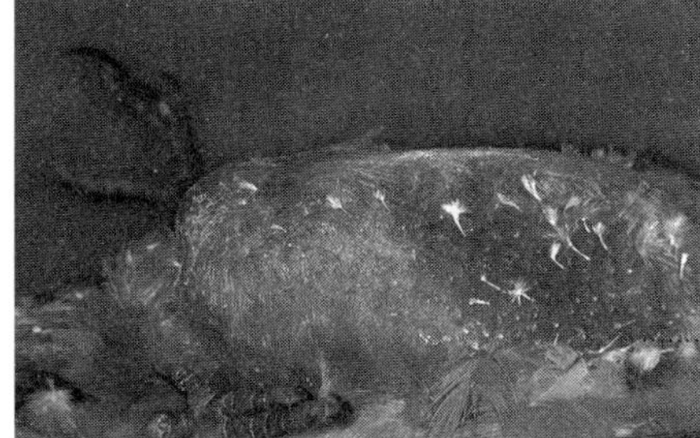

Avian viral serositis in psittacine birds has been shown to be caused by an EEE virus. Affected birds exhibit ascites;

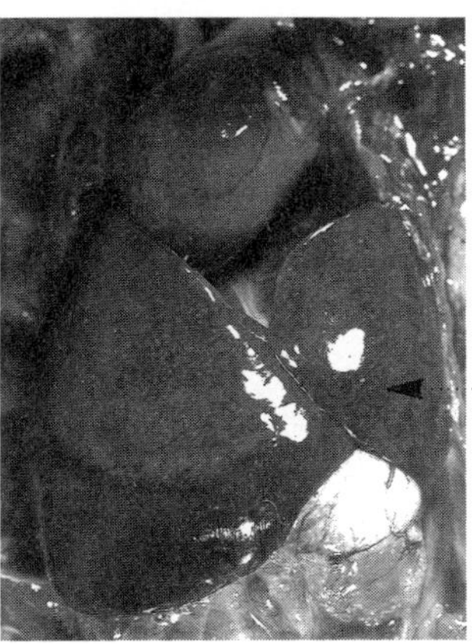

the liver is typically enlarged and yellowish.

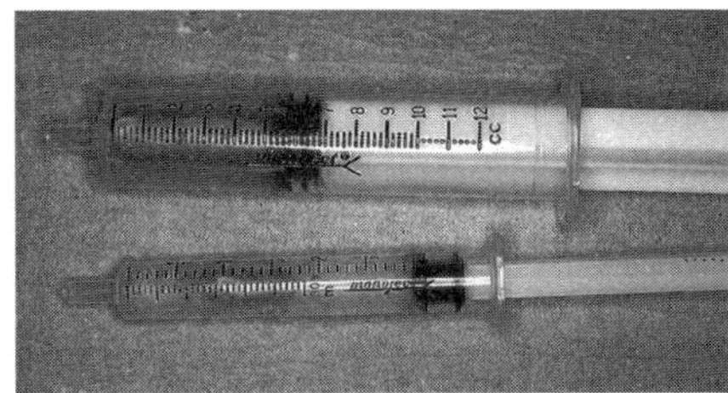

Fluid removed from the abdomen of a bird that died from avian viral serositis.

transmission of the virus. Antibodies to the virus were detected in the chickens that recovered from experimental infections, but not in the originally affected macaws, Ring-necked Parakeet or other psittacine birds that died from proventricular dilatation disease.[35]

In one study, two nestling Budgerigars died following oral inoculation with a psittacine isolate of EEE virus, while three Budgerigars that were in contact with the affected birds and fed with the same feeding utensils remained clinically normal.[33] In another experimental trial, EEE virus caused a transient diarrhea, polyuria and passage of undigested food in a Moluccan Cockatoo, an Eclectus Parrot and a Scarlet Macaw infected by intramuscular inoculation *Figure 15.9*. These infected birds seroconverted, recovered completely and remained clinically normal for more than 12 months after infection. An Eclectus Parrot and a Scarlet Macaw inoculated oral-

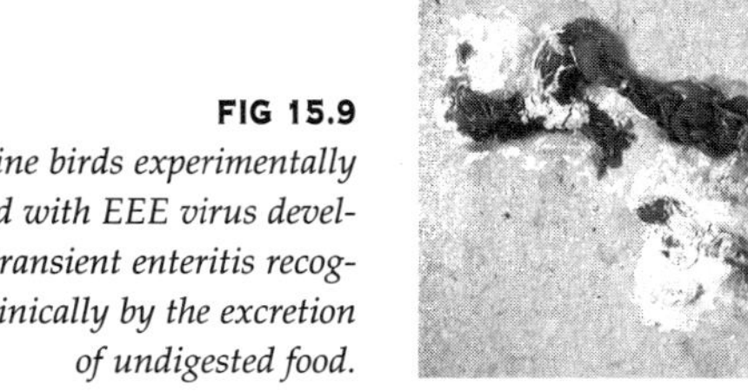

Psittacine birds experimentally infected with EEE virus developed a transient enteritis recognized clinically by the excretion of undigested food.

ly with the same preparation of EEE virus remained clinically normal and did not seroconvert, suggesting that they had not been infected when exposed to EEE virus by this route.[76a]

The outbreak of avian viral serositis described initially occurred in a group of birds that had been experiencing intermittent problems with proventricular dilatation disease (PDD). Many of the young psittacine birds that were exposed to the affected macaws developed PDD and died 4 to 18 months after the avian viral serositis outbreak. Additionally, many of the microscopic lesions in birds with naturally or experimentally induced avian viral serositis were similar to those described with PDD.

These findings have led to speculation that the EEE virus that causes avian viral serositis may also cause PDD.[34] However, antibodies to the EEE virus that caused avian viral serositis were not detected in a group of psittacine birds that died from PDD.[35] Moreover, seroprevalence studies have indicated that there is no significant difference in EEE virus-neutralizing antibodies in birds that died from PDD when compared to the population as a whole.[76a] Psittacine birds experimentally infected with EEE virus developed transient diarrhea, seroconverted and remained clinically normal; they did not develop clinical changes suggestive of PDD.[76a] Additionally, in a group of psittacine birds diagnosed with PDD, only 1 of 17 of the birds had antibodies to EEE, WEE and VEE viruses; another had antibodies to EEE and WEE viruses.[81] Collectively, these findings suggest that the EEE virus that causes avian viral serositis does not cause PDD.

ALPHAVIRUSES IN EMUS

Both EEE and WEE viruses have been documented in immunologically naive emus in the United States. Either of these viruses can cause high levels of morbidity, but mortality is most common with EEE virus. Emus infected with EEE virus can die suddenly without premonitory signs, or may die following a brief period of depression, anorexia, ataxia and hemorrhagic diarrhea *Figure 15.10*. Most emus that develop clinical signs of disease can be expected to die if not provided with aggressive supportive care. In an outbreak in Louisiana that started in late August, 14 of 24 (58%) emus in a flock died. In another outbreak involving a group of breeding emus that ranged in age from 20 to 36 months, 76% of the at-risk birds developed clinical signs, and 87% of affected birds died. Antibodies to the EEE virus were detected in two emus that recovered, but not in an unaffected bird.[95,97]

Emus of all ages have been shown to be susceptible to infection and disease; however, the virus is most commonly isolated from birds that are two to four years old. It is likely that this finding indicates the most common age of birds being maintained in captivity, rather than an age-related susceptibility to disease.[96] The rapid death of affected emus may prevent them from serving as an effective reservoir for mosquitoes, thus making them of little importance in the transmission of EEE virus.[99]

In emus, EEE virus causes a hemorrhagic enteritis, with hemorrhage also occurring in other abdominal organs *Figure 15.11*. Necrosis of the liver, spleen and intestinal mucosa are seen microscopically. Gross or microscopic changes have not been described in the central nervous system of affected emus. For diagnostic purposes, the virus can be isolated from the liver, spleen and intestines. Virus-neutralizing antibodies can be demonstrated in survivors.[97,99]

Some affected emus will favorably respond to supportive care that includes fluids, vitamins (particularly vitamin K) and assisted feeding; however, infections are best prevented in virus endemic areas through vaccination. Emus vaccinated with 1 ml of an inactivated vaccine licensed for use in horses are considered resistant to disease. In one study, 30% of emus seroconverted following the first vaccination, and 70% were seropositive when tested after receiving a booster. A virus neutralization titer of 1:20 is considered protective in emus, and many vaccinat-

FIG 15.10
Ratites infected with EEE virus can die suddenly or may die following a brief period of depression, anorexia, ataxia and hemorrhagic diarrhea.
photograph courtesy of Amy Raines

ed birds maintained a protective titer one year after vaccination.[96]

Currently, it is recommended that emus be vaccinated at three months of age, followed by a booster one month later. An additional booster should be given to adult birds after the breeding season in April and prior to the breeding season in October. These recommendations are considered preliminary, and

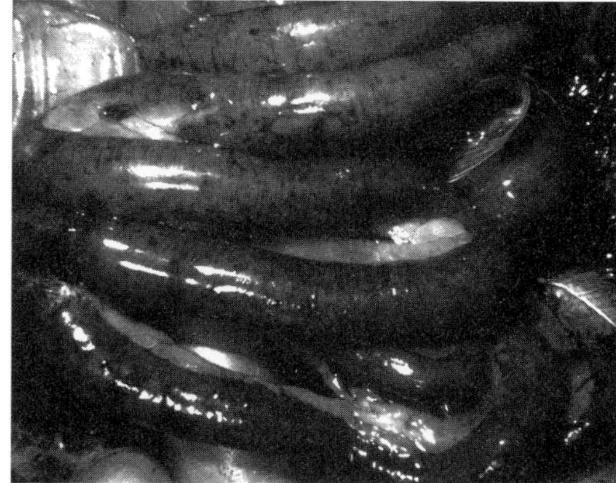

FIG 15.11
Hemorrhage of the intestinal tract is the most common gross change in emus infected with EEE virus.
photograph courtesy of Tom Tully

neither the effect of repeated vaccination nor the damage that vaccination may cause to muscle tissue intended for meat has been reported. Prior to vaccination, the insurer of an emu should be contacted to receive permission for the extra-label use of a vaccine licensed for use in horses.[96,97]

Epornitics of WEE virus occurred in emus in the southwestern United States in the summer of 1992, following unusually high levels of rain and mosquito activity during the spring. Interestingly, the outbreaks in emus occurred independently of an outbreak in horses or humans.[4,80]

Affected emus in various flocks ranged in age from eight weeks to three years;[80] in other studies birds as young as six weeks of age have been diagnosed with the disease.[96] The first signs of disease included lethargy and anorexia, followed by abnormal positioning of the head to one side or over the back. Some emus became acutely paralyzed but remained alert. Others died following a brief period of ataxia, recumbency and paddling. Some affected emus appeared totally normal 10 to 12 hours before dying; others developed clinical signs and died from 1 to 14 days later. In one study, many of the affected emus had elevated levels of aspartate aminotransferase (greater than 500) and elevated white blood cell counts (17,000 to 28,000).[4,80]

The morbidity rate during one outbreak was less than 20% of the at-risk birds, and many less severely affected emus responded to supportive care.[4,80] In another outbreak in several flocks in Texas, morbidity ranged from 15% to 50%, and 17 of 193 (9%) of the at-risk emus died. Some of the emus developed anorexia and depression of two to seven days' duration and recovered without therapy, whereas others with more severe clinical changes responded to administration of fluids and antibiotics.[4]

Reported necropsy findings in affected emus include accumulation of fluid around the heart, inflammation of the pericardium, and swelling and inflammation of the liver, kidneys and membranes covering the brain and spinal cord. The most common microscopic changes have been inflammation and necrosis in most tissues, including the heart, lungs, liver, spleen, kidneys, digestive tract, brain and meninges. Vasculitis with perivascular accumulation of lymphocytes and plasma cells can occur in most tissues, but has been noted less commonly in the brain of affected emus than in other avian species.[4,80]

Recovery from infection or virus activity within a flock can be demonstrated by detecting HI or virus-neutralizing antibodies.[4,80] In Texas, 10 of 13 emus from affected flocks had HI titers ranging from 1:20 to 1:1280. During an eight-month period, 35% of the serum samples from emus in Texas and adjacent states were found to have HI antibody titers that ranged from over 1:40 to 1:1280.[4]

Emus from affected flocks were vaccinated with 0.5 cc of an inactivated vaccine intramuscularly, and the outbreak stopped. It is currently recommended that emus in areas where WEE virus is endemic should be vaccinated at eight weeks of age, followed by a booster two to four weeks later. The birds should receive a booster at the end of the laying season in March and again in August prior to laying season.[80] The same cautions concerning vaccination for EEE virus also apply to WEE virus.

EEE Virus in Raptors

A Red-shouldered Hawk infected with EEE virus developed a low-titer viremia 24 hours after infection, then seroconverted. A Great Horned Owl infected subcutaneously was viremic from 24 to 120 hours afterwards, became depressed, reluctant to move and ataxic after 48 hours and had difficulty with breathing 72 hours after infection. The owl then recovered and was clinically normal within six days after infection. In contrast, another owl inoculated intracerebrally died within 34 hours. A species of hawk (*Circus* sp.) not native to North America developed clinical signs of disease and died eight days after inoculation.[77] Antibodies to EEE virus were detected in 1 of 16 Bald Eagles housed in an area during an epornitic in cranes, suggesting that the former are susceptible to natural infections. The seropositive eagle was considered clinically normal.[20]

ALPHAVIRUSES IN DUCKS

Ducks are susceptible to both natural and experimental EEE and WEE virus infections. Infections in most species are considered subclinical, although EEE virus was associated with death in domestic ducks on Long Island.[22] Antibodies to WEE virus were detected in 11% of 471 blood samples from free-ranging ducks in Saskatchewan.[8]

In an experimental infectivity study, all the wild-caught Black-bellied Whistling Ducks became viremic following inoculation with EEE or WEE viruses. Five of the six ducks had EEE viremia at one day after inoculation; the mean duration of viremia was two days. Two of seven WEE-infected ducks were viremic three to four days after infection and the mean duration of viremia was four days. Starting at day six after infection, antibody titers to EEE increased rapidly. All the EEE virus-inoculated ducks seroconverted. Antibodies were not detected to WEE until day 14 after infection, but six of seven inoculated ducks seroconverted. Infected ducks remained clinically normal, developed a viremia of short duration and had no gross or microscopic lesions suggestive of disease.[2]

Ducks also have been shown to be susceptible to WEE virus following oral inoculation, and can be infected through the ingestion of contaminated drinking water. Ducks inoculated orally developed detectable levels of virus-neutralizing antibodies by seven days post-ingestion, and some remained seropositive for over one year.[8] In an EEE virus outbreak, the virus was spread among susceptible ducklings by direct contact without an arthropod vector.[22]

An outbreak of encephalitis of undetermined cause killed 13 of 35 (37%) Muscovy Ducks in Canada. Clinical changes included recumbency, torticollis and opisthotonus. Pekin Ducks, geese and poultry maintained on the same farm remained clinically normal. No gross lesions were noted in the affected ducks; however, the accumulation of lymphocytes around the blood vessels in the brain, degeneration of the nerves and gliosis were considered indicative of a virus-induced encephalitis. Six Muscovy Ducks from an adjacent farm died with similar clinical and microscopic changes.[39]

ALPHAVIRUSES IN PHEASANTS

Eastern equine encephalitis virus was first recovered from pheasants during a

1938 epidemic in humans in Connecticut.[98] The clinical disease associated with EEE virus has been well characterized in pheasants, and is most common in pheasant species that are not native to North America. Highlands J virus, and possibly WEE virus, have been recovered from pheasants, but these viruses generally cause a less severe disease or lower levels of mortality than EEE virus.[27,41]

In 48 naturally and 113 experimentally infected pheasants, the clinical signs that occurred could be divided into two groups: systemic and neurologic. Systemic signs included droopiness, ruffled feathers, soiled vents and occasional respiratory distress. Neurologic signs included incoordination, unilateral or bilateral paralysis of the legs, droopy wings, tremors and torticollis.[51] During one EEE virus epornitic, disease was most common in Chinese Ring-necked, Impeyan, Mongolian and Tragapan Pheasants that ranged from two months to two years of age. Most of the affected pheasants were four to five months old; they were depressed, anorexic and frequently had profuse diarrhea. These clinical changes were followed by ataxia, trembling, paralysis and death.[27] A group of affected six-week-old pheasants was lethargic and had pasty-white diarrhea and paralysis of the neck. Affected pheasants generally died over a three-day period that started with depression and ended with paralysis prior to death.[60] Other pheasants may exhibit varying degrees of paresis and paralysis that can persist from weeks to months.[27,77] Some have been shown to recover completely following natural and experimental infections.

In the Northeastern United States, EEE virus outbreaks are most common in pheasants between August and October. Virus activity may persist into November in the Southeast.[27,60] The mortality rates associated with EEE virus infections in pheasants vary from 5% to 80% of the at-risk population, depending on the age of the birds when virus exposure occurs. A summary of the mortality associated with numerous outbreaks of EEE virus in pheasants is provided in *Table 15.5*. With most outbreaks, the mortality is highest in younger birds, and those less than 4 weeks of age suffer the highest mortality. However, in some reports, four- to five-month-old pheasants have been most commonly affected, and birds up to two years of age have died.[27,60] In one outbreak on a pheasant farm with birds that ranged from 10 to 17 weeks of age, only the 17-week-old birds developed clinical changes. In another outbreak, mortality reached 60% in one pen, with no mortality in an adjacent pen. A dead rat located outside the affected pen was found to have EEE virus. In still another study, 20 birds that remained asymptomatic during an outbreak were found to have detectable levels of neutralizing antibodies that persisted for 16 months.[60]

Response to experimental infection with EEE virus varies with the route of inoculation (intravenous, intramuscular or subcutaneous) and the species of infected pheasant. In some experimental trials, even species of pheasants that are considered highly susceptible to infection and routinely develop clinical signs of disease, have remained asymptomatic. In one trial, pheasants developed mild ataxia 2 to 12 days following injection with EEE virus. Some birds died within 90 hours of inoculation; others developed mild ataxia six to nine days after infection but recovered. Some pheasants infected orally also developed fatal infections. Pheasants inoculated subcutaneously developed 48 hours of viremia, while those inoculated intramuscularly were viremic for up to 72 hours.[77]

In another study, experimentally infected pheasants developed clinical signs from 1 to 7 days after infection (mean, 4.8 days), depending on the route of inoculation. The incubation period of EEE virus in susceptible pheasants inoculated with virus by the intramuscular route averages 3.8 days. The incubation period in pheasants in direct contact with experimentally infected birds is 4.5 days.[51]

Some experimental studies suggest that pheasants develop an age-related resistance to EEE virus-induced disease, but mature birds have been shown to die following naturally acquired infections. The levels of mortality associated with experimental infections in various ages of pheasants is provided *Table 15.6*. In one study, three-day-old pheasants were highly susceptible to disease following subcutaneous injection, while four-week-old pheasants were resistant to disease when inoculated by the same route. Pheasants of all ages were susceptible when inoculated intracerebral-

TABLE 15.5

Summary of EEE virus-induced mortality in various groups of pheasants[60,72]

Age (weeks)	# of pheasants	Mortality (%)	Duration of Outbreak
4	250	80	not reported
6	495	25	not reported
10	36	44	not reported
14	200	32	19 days
15	1300	31	57 days
16 - 18	10,682	90	not reported
26	350	43	17 days

ly. When a different strain of virus was used, even adult pheasants would die 120 to 132 hours after subcutaneous inoculation.[40] It appears that these supposed differences in age-related susceptibility to disease are controlled by the virulence of the virus and route of inoculation, and not by the age of the pheasant.

Eastern equine encephalitis virus has been shown to be transmitted in pheasants by mosquitoes as well as through direct contact where feather picking or cannibalism occurs. The virus has been shown to persist in the developing feathers of pheasants for up to six days, which is two days longer than the average period of viremia. In extensive experimental trails, it was shown that the virus was transmitted freely among pheasants in pens where cannibalism occurred, but transmission did not occur when feather picking and cannibalism were prevented by debeaking.[60,77] Virus was detected in the feces of pheasants infected intraperitoneally but not orally; however, virus transmission was not shown to occur through indirect contact with contaminated respiratory secretions or feces. It has been suggested that flocks of starlings, English Sparrows and carrion-eating birds could transmit the virus

TABLE 15.6

Mortality rates in pheasants by route of inoculation and by age[40]

Age (days)	# Exposed	% Mortality
Subcutaneous inoculation		
1	48	100
3	48	100
7	48	79
14	48	69
21	48	56
28	36	8
over 28	72	0
Intracerebrally		
84	36	100

among susceptible pheasants during epornitics.[77]

While the mortality rates associated with EEE virus in pheasants can be high, many affected birds die without developing gross lesions suggestive of disease. In 48 naturally and 113 experimentally infected pheasants, the only gross pathologic indication of disease was an occasional softening of the cerebral hemispheres.[51] In another outbreak, affected pheasants were reported to have small spleens.[27] Elongated white streaks have been reported in the heart of some infected pheasants *see Figure 15.7*.[73] The most common microscopic lesions associated with infection are vasculitis and patchy necrosis of affected tissues, particularly perivascular cuffing in the brain. Other changes include non-suppurative encephalitis, microgliosis, meningitis, neuronal degeneration and myocardial necrosis.[51,73,98] Microscopic changes tend to occur primarily in the front portion of the brain with a distribution characterized as "descending" encephalitis. This pattern differs from other avian encephalitides which typically cause "ascending" encephalitis.[51]

In pheasants, which are considered naturally susceptible to EEE, pathologic changes differed from the lesions found in chickens, which are considered relatively resistant to disease. Pheasants were found primarily to develop neurotropic disease, while chickens primarily developed myocardiotropic disease.[51] By comparison, Chukars were found to exhibit both neurologic and myocardial lesions following infection.[73] Some of the clinical and gross changes associated with EEE virus infections in pheasants also might be caused by botulism, Newcastle disease virus, pasteurellosis and salmonellosis.[51]

In one study, 69% of the vaccinated pheasants were resistant to subsequent challenge with EEE virus, while 70% of the unvaccinated birds died. This protection occurred even though the vaccinates did not develop detectable levels of neutralizing antibodies.[93] However, other studies have indicated that some inactivated vaccines are not effective in pheasants.[23]

ALPHAVIRUSES IN CHUKARS

Both EEE and WEE viruses have been reported in various flocks of pen-raised Chukars in Florida. It is likely that the WEE virus recovered from these birds was actually an isolate of Highlands J virus. These viruses caused clinical problems in five- to ten-week-old chukars that developed lethargy, ruffled feathers and an inability to stand one week after being moved from inside brooders to outside pens. Affected birds died 12 to 24 hours after developing clinical signs. Mortality rates ranged from 30% to 85%, depending on the age of the exposed Chukars. In one pen, many of the eight-week-old Chukars died, while all of the exposed 13-week-old birds remained clinically normal. This finding suggests that Chukars develop a natural resistance to the disease as they approach 12 weeks of age. Chukars that were inoculated subcutaneously with WEE virus at six weeks old developed lethargy in 24 hours, followed by prostration and death within 48 hours.[73]

An outbreak of EEE in Chukars from Maryland caused 40% mortality in eight-week-old birds, with death occurring 48 hours after clinical signs were first noted. The entire outbreak ran its course and stopped within seven days of the initial clinical signs. Hemagglutination-inhibition antibody titers that ranged from 1:40 to 1:1280 were detected in the surviving Chukars.[70]

Gross lesions that have been described in affected Chukars include white streaking in the heart, enlarged mottled spleen and distention of the kidneys with urates *see Figure 15.7*. Microscopic changes that have been described in the brain are similar to those described in pheasants and include perivascular accumulation of lymphocytes, gliosis and satellitosis. Necrosis of the heart with accumulation of lymphocytes also has been reported.[73]

ALPHAVIRUSES IN QUAIL

Eastern equine encephalitis virus caused high mortality in a flock of commercial quail that ranged in age from two to eight weeks. Affected birds developed a rapidly progressive disease characterized by depression, tremors and paralysis, and followed within hours by death. Egg production in the flock virtually stopped, and the few eggs that were produced had soft shells. The virus spread through the farm within five days of the initial case, and the outbreak subsided six days later. Substantial cannibalism of dead birds was considered important in the rapid transmission of the virus through the flock. Mortality rates varied from 40% to 90%, with the highest mortality occurring in the younger birds. Over 90,000 birds died during this outbreak. The most common gross finding was enteritis, and characteristic lesions were noted in the brain. Many affected quail that died had no gross lesions.[24]

ALPHAVIRUSES IN TURKEYS

Natural and experimental EEE and HJ virus infections have been demonstrated to cause increased mortality in young turkeys and an acute decrease in egg production in breeding birds; WEE virus was reported as a cause of depression, tremors and paralysis.[102] Naturally acquired EEE and HJ virus infections were associated with increased mortality in 10- to 28-day-old turkeys.[30] When two-week-old turkey poults were inoculated intramuscularly, 23% of the birds infected with EEE virus and 27% of birds infected with the HJ virus died. Clinical changes occurred from one to five days after infection and consisted of depression, anorexia and neurologic changes in some poults. The EEE virus-infected birds died from two to five days after infection, and the HJ virus-infected birds died from two to three days after infection. Survivors were clinically normal by day six.[38]

Turkeys in direct contact with the experimentally infected birds remained clinically normal and did not develop antibodies to the virus. The only gross lesions noted in the poults that died were atrophy of the thymus and bursa. Necrosis of the heart, kidney, pancreas, bursa, thymus and spleen was noted microscopically. No gross or microscopic lesions were detected in the brain of infected turkeys.[38] One group of naturally infected turkeys was found to have microscopic lesions in the brain. Their mortality rate was less than 5%.[85]

Both EEE and HJ viruses were recovered from the trachea of turkeys from flocks in which egg production had decreased up to 73%. Affected hens exhibited mild, transient lethargy. Eggs they produced were small and some had soft shells. The detection of antibodies to these viruses confirmed that previous infections had occurred in members of the affected flocks.[101] Within three days after breeding turkeys were experimentally infected with EEE or HJ viruses, egg production decreased to 20%. Egg production remained depressed for 15 days in birds infected with EEE virus and for 7 days in birds infected with HJ virus. These viruses could only be detected in the tissues of infected hens for one to two days, but were recovered from the eggs of these hens from two to five days after infection. Other viruses associated with decreased egg production in turkeys include paramyxovirus 3, avian influenza virus, avian encephalitis virus, Newcastle disease virus and rhinotracheitis virus.[37]

ALPHAVIRUSES IN CHICKENS

Two-week-old chickens were shown to be susceptible to experimental infection by either EEE or HJ viruses. The mortality rate in chicks inoculated with EEE virus was 80%, whereas only 7% of the HJ virus-infected birds died. Chickens experimentally infected with EEE virus recovered from turkeys may die within two to four days without developing clinical changes, or in three to seven days with lethargy, depression and paralysis. Birds infected with EEE virus were depressed and lethargic from one to six days after infection. After day six, survivors developed abdominal distention and decreased growth before they died. Two of five contact chickens developed clinical changes similar to those seen in older birds, and died 12 to 17 days post-exposure. Natural transmission of HJ virus was not noted. Surviving chickens that were necropsied 30 days after infection had necrotic lesions in the heart. The susceptibility of chickens to disease has been shown to decrease as they reach three weeks of age.[37] Necropsy changes include necrosis in heart, liver, thymus and bursa. Older chicks had evidence of ascites, pericardial effusion, and dilation of the heart.

ALPHAVIRUSES IN GUINEA FOWL

In southwestern France, a virus with morphological characteristics similar to a togavirus was recovered from the intestinal contents of guinea poults that died with clinical changes of anorexia, weight loss, recumbency and acute, severe diarrhea. The recovered virus caused enteritis in poults experimentally infected by the oral route and spread easily to uninoculated birds housed in the same enclosure. Gross lesions included loss of intestinal tone, distention of the ceca with yellowish foamy material, pancreatic necrosis, swollen kidneys and a distended gall bladder.[7]

EEE VIRUS IN CRANES

Over a seven-day period, 7 of 39 endangered, captive Whooping Cranes died from naturally acquired EEE virus infections. A group of 240 Sandhill Cranes that were adjacent to the Whooping Cranes remained clinically normal. Four of the seven Whooping Cranes died without clinical signs, and three others died following several days of lethargy, ataxia and leg and neck paresis. Affected birds ranged in age from over 1 to 18 years of age. Anemia, leukopenia, elevated levels of aspartate aminotransferase (2,763 to 6,660 IU/L), lactate dehydrogenase (7,767 to 11,998 IU/L) and uric acid (122 to 184 ml/dl) were common in the affected cranes. The apparent susceptibility of Whooping Cranes to EEE virus is particularly troublesome given that this virus is common in Florida, a targeted release site for reintroduction of the endangered crane.

Accumulation of clear yellow fluid in the air sacs, enlargement and reddishyellow mottling of the liver, splenomegaly, visceral gout and hyperemia of the small intestines were the characteristic findings at necropsy. Microscopically, inflammation and necrosis were evident in the liver, spleen, intestines, kidneys, adrenal glands, gonads and lungs, but not in the brain.[19,20]

Of the 32 Whooping Cranes that survived, 14 (44%) had EEE virus-neutralizing antibodies, as did 13 of 38 (34%) potentially exposed Sandhill Cranes. Finding antibody in these cranes indicated that they were infected; however, many of the exposed Whooping Cranes and all of the infected Sandhill Cranes remained clinically normal. Detectable levels of virus-neutralizing antibodies persisted in naturally infected cranes for over six months. Antibody studies using archival serum indicated that infections had occurred in cranes for years with no apparent mortality. The predisposing factors that induced the high mortality in this outbreak were not determined, nor was it determined why only the Whooping Cranes and not the related Sandhill Cranes were affected. Several of the cranes also had virus-neutralizing antibodies to Highlands J virus.[19,20]

Inactivated EEE vaccine given intramuscularly to Sandhill and Whooping Cranes induced virus-specific neutralizing antibodies. Birds that were seropositive because of a previous natural infection developed high titers of sustained duration following the initial vaccination. Immunologically naive vaccinates developed low virus-neutralizing antibody titers that decreased rapidly, but these titers increased in level and duration following administration of a booster. Cranes inoculated intramuscularly developed peak titers of 1:80 by 60 days following vaccination. Antibody was not detectable in most birds by 90 to 150 days post-vaccination, but titers increased rapidly when the cranes were revaccinated at 180 days. Intramuscular inoculations did not cause adverse reactions in cranes, even when birds were vaccinated weekly for five weeks. Vaccine administered intravenously did cause temporary incoordination 15 minutes after injection. Birds vaccinated intramuscularly developed higher and more persistent titers than birds vaccinated subcutaneously.[16]

ALPHAVIRUSES INFECTIONS IN HUMANS AND HORSES

Humans and horses are considered "accidental" hosts for EEE, WEE and HJ viruses. Because these viruses circulate in free-ranging birds, infections in humans and horses can occur when there is high virus activity in birds in combination with a large concentration of mosquitoes. Biting insects, not birds, are responsible for transmitting the virus to humans and horses. Many infected humans and horses remain asymptomatic, but a few may develop mild-to-severe encephalomyelitis. Eastern equine encephalitis virus is more likely to cause a severe encephalitis in humans and horses than is WEE virus.[9] Highland J virus has been associated with disease in pheasants, Chukars and turkeys but not in people or horses.[41]

There are frequent discussions as to whether individuals who have companion birds, aviary birds, ratites and horses may be at greater risk of being infected by EEE or WEE viruses. However, because these viruses are ubiquitous in free-ranging birds (particularly in swampy areas), people with companion or aviary birds are at no greater risk of infection than any other individual living in the same geographic area. Additionally, horses that are in contact with companion or aviary birds should be at no greater risk than horses exposed to free-ranging birds. Nor are birds at risk themselves - humans and horses play no role in the maintenance or transmission of these viruses.

VENEZUELAN EQUINE ENCEPHALITIS VIRUS

Venezuelan equine encephalitis (VEE) virus is also referred to as Everglades encephalitis virus (EVE). It has been documented in Venezuela, Trinidad, Colombia, Ecuador, Guatemala, El Salvador, Honduras, Nicaragua, Costa Rica, Panama, Peru, Brazil, Argentina, French Guiana, Mexico, southern Texas and southern Florida (the Everglades).[15,79] Birds experimentally infected with VEE develop a low grade viremia, remain asymptomatic and seroconvert.[14] Antibodies to the virus have been detected in free-ranging Whistling Ducks, Cattle Egrets, coots, turkeys, quail, doves, Killdeer, Cowbirds, and Barn Owls in Mexico.[1] No antibodies to this virus were detected in Ciconiiformes or Pelecaniformes during a study in Florida.[86] Antibodies to VEE occasionally have been detected in other birds (1 of 93 ibis, 2 of 8 Yellow-crowned Night Herons).[14] Humans develop a flu-like disease and rarely have neurologic signs.

In contrast to EEE and WEE viruses, birds are considered of little importance as a reservoir for VEE virus. Rodents and other small mammals are considered to be the principal reservoir for VEE virus, and mosquitoes that preferentially feed on rodents are the principal vector. In swampy areas, egrets and herons may function as an alternative reservoir.[36]

OTHER TOGAVIRUSES IN BIRDS

RUBELLA VIRUS

Antibodies to rubella virus, the virus that causes German measles in humans, were detected in 1% to 7% of the free-ranging pigeons in Europe.[21,36] It is equally possible that humans could serve as a source of virus for pigeons or that pigeons could serve as a source of virus for humans.[36]

OCKELBO VIRUS

Ockelbo virus-neutralizing antibodies were detected in free-ranging and captive Anseriformes, Galliformes and Passeriformes in Sweden. Of the tested birds, 26 of 324 (8%) were positive. The seroprevalence rate was highest in Passeriformes (27%), followed by Galliformes (6%) and Anseriformes (4%). In experimentally inoculated birds, antibody titers were higher and of longer duration in adult Passeriformes than in adult Anseriformes.[61,62]

SINDBIS VIRUS

Sindbis virus has been associated with death in some experimentally infected pigeons and chickens, but naturally infected birds remain asymptomatic.[52] In Europe, antibodies to Sindbis virus were detected in 9% of free-ranging Anseriformes and 0.7% of free-ranging Passeriformes.[66] In Czechoslovakia, 15% of free-ranging Anseriformes and 4% of free-ranging Passeriformes were shown to be seropositive.[25,26]

REFERENCES

1. Aguirre AA: Serologic survey for arboviruses and selected disease agents in wild mammals and birds from Mexico. Proc Assoc Zoo Vet, 1992, pp 12-15.

2. Aguirre AA, McLean RG, Cook RS: Experimental inoculation of three arboviruses in black-bellied whistling ducks (*Dendrocygna autumnalis*). J Wildl Dis 28:521-525, 1992.

3. Ananyan SA: Antibody against group A viruses in residents of certain areas of the USSR and preliminary results of investigations of new viruses. Acad Med Science USSR, 1963.

4. Ayers JR, Lester TL, Angulo AB: An epizootic attributable to western equine encephalitis virus infection in emus in Texas. J Am Vet Med Assoc 205:600-601, 1994.

5. Bigler WJ, Lassing E, Lewis AL, et al: Arbovirus surveillance in Florida: Wild vertebrate studies 1965-1974. J Wildl Dis 11:348-356, 1975.

6. Bigler WJ, Lassing E, Prather E, et al: Endemic eastern equine encephalomyelitis in Florida: A twenty-year analysis, 1955-1974. Am J Trop Med Hyg 25:884-890, 1976.

7. Brahem A, Demarquez N, Beyrie M, et al: A highly virulent togavirus-like agent associated with the fulminating disease of guinea fowl. Avian Dis 36:143-148, 1992.

8. Burton AN, Connell R, Rempel JG, et al: Studies on western equine encephalitis associated with wild ducks in Saskatchewan. Can J Microbiol 7:295-302, 1961.

9. Burton AN, McLintock JR, Spalatin J, et al: Western equine encephalitis in Saskatchewan birds and mammals, 1962-1963. Can J Microbiol 126:133-141, 1966.

10. Calisher CH, Karabatsos N, Lazuick JS, et al: Reevaluation of the western equine encephalitis antigenic complex of alphaviruses (family Togaviridae) as determined by neutralization tests. Am J Trop Med Hyg 38:447-452, 1988.

11. Casals J: Procedures for identification of arthropod-borne viruses. Bull World Hlth Org 24:723-734, 1961.

12. Casals J: Antigenic variants of eastern equine encephalitis virus. J Exper Med 119:547-565, 1964.

13. Chamberlain RW: Eastern equine encephalomyelitis. *In* Karabatsos N (ed): International Catalogue of Arboviruses 3rd ed. Fort Collins, American Society Tropical Medicine and Hygiene, 1985, pp 377-378.

14. Chamberlain RW, Kissling RE, Stamm DD, et al: Venezuelan equine encephalomyelitis in wild birds. Am J Hyg 63:261-273, 1956.

15. Chamberlain RW, Sudia WD, Coleman PH, et al: Venezuelan equine encephalitis virus from South Florida. Science 145:272-274, 1964.

16. Clark GG, Dein FJ, Crabbs CL, et al: Antibody response of sandhill and whooping cranes to an eastern equine encephalitis virus vaccine. J Wildl Dis 23:534-544, 1987.

17. Clark GG, Pretula HL, Jakubowsky T, et al: Arbovirus surveillance in Illinois in 1976. Mosquito News 37:389-395, 1977.

18. Dalrymple JM, Young OP, Eldridge BF, et al: Ecology of arboviruses in a Maryland freshwater swamp. III. Vertebrate hosts. Am J Epidemiol 96:129-140, 1972.

19. Dein FJ, Carpenter JW: Outbreak and control of eastern equine encephalitis virus in whooping cranes. Proc Assoc Avian Vet, 1985, pp 175-176.

20. Dein JF, Carpenter JW, Clark GG, et al: Mortality of captive whooping cranes caused by eastern equine encephalitis virus. J Am Vet Med Assoc 189:1006-1010, 1986.

21. Dorrestein GM: Viral infections in racing pigeons. Proc Assoc Avian Vet, 1992, pp 244-257.

22. Dougherty E, Price JI: Eastern encephalitis in white Pekin ducklings on Long Island. Avian Dis 4:247, 1960.

23. Eisner RJ, Nusbaum SR: Encephalitis vaccination of pheasants: A question of efficacy. J Am Vet Med Assoc 183:280-281, 1983.

24. Eleazer TH, Blalock HG, Warner JH, et al: Eastern equine encephalomyelitis outbreak in coturnix quail. Avian Dis 22:522-525, 1978.

25. Ernek E, Kozuch O, Nosek J, et al: Virus-neutralizing antibodies to

arboviruses in birds of the order Anseriformes in Czechoslovakia. Acta Virologica 19:349-353, 1975.

26. Ernek E, Kozuch O, Nosek J, et al: Arboviruses in birds captured in Slovakia. J Hyg Epid Micro Immuno 21:353-359, 1977.

27. Faddoul GP, Fellows GW: Clinical manifestations of eastern equine encephalomyelitis in pheasants. Avian Dis 9:530-535, 1965.

28. Favorite FG: Some evidence of local origin of EEE virus in Florida. Mosquito News 20:87-92, 1980.

29. Feemster RF, Wheeler RE, Daniels JB, et al: Field and laboratory studies on equine encephalitis. New Engl J Med 259:107-113, 1958.

30. Ficken MD, Wagers DP, Guy JS, et al: High mortality of domestic turkeys associated with Highlands J virus and eastern equine encephalitis virus infections. Avian Dis 37:585-590, 1993.

31. Fothergill LD, Dingle JH, Farber S, et al: Human encephalitis caused by the virus of the eastern variety of equine encephalomyelitis. N Engl J Med 219:411, 1938.

32. Froeschle JE, Reeves WL: Serologic epidemiology of western equine and St. Louis encephalomyelitis virus infection in California. II. Analysis of inapparent infections in residents of an endemic area. Am J Epid 81:44-51, 1965.

33. Gaskin JM: Questions and answers about psittacine proventricular dilatation disease and avian viral serositis. Proc Midwest Avian Research Exposition, 1992, pp 69-71.

34. Gaskin JM, Homer B, Eskelund K: Preliminary findings in avian viral serositis: A newly recognized syndrome of psittacine birds. J Assoc Avian Vet 5:27-34, 1991.

35. Gaskin JM, Homer BL, Eskelund KH: Some unofficial thoughts on avian viral serositis. Proc Assoc Avian Vet, 1991, pp 38-42.

36. Gerlach H: Viruses. In Ritchie BW, Harrison GJ, Harrison LR (eds): Avian Medicine: Principles and Application. Lake Worth, Wingers Publishing, 1994, pp 862-948.

37. Guy JS, Barnes HJ, Smith LG: Experimental infection of young broiler chickens with eastern equine encephalitis virus and Highlands J virus. Avian Dis 38:572-582, 1994.

38. Guy JS, Ficken MD, Barnes HJ, et al: Experimental infection of young turkeys with eastern equine encephalitis virus and Highlands J virus. Avian Dis 37:389-395, 1993.

39. Hanson J: Encephalitis in Muscovy ducks. Can Vet J 33:345, 1992.

40. Hanson RP, Vadlamudi PS, Trainer DO, et al: Comparison of the resistance of different-aged pheasants to eastern

encephalitis virus from different sources. Am J Vet Res 29:723-727, 1968.

41. Hayes CG, Wallis RC: Ecology of western equine encephalomyelitis in the eastern United States. Adv Virus Res 21:37-83, 1977.

42. Hayes RO, Beadle LD, Hess AD, et al: Entomological aspects of the 1959 outbreak of eastern encephalitis in New Jersey. Am J Trop Med Hyg 1961.

43. Hayes RO, Daniels JB, Anderson KS, et al: Detection of eastern encephalitis virus and antibody in wild and domestic birds in Massachusetts. Am J Hyg 75:183-189, 1962.

44. Hearn HJ: Cross-protection between Venezuelan equine encephalomyelitis virus. Proc Soc Exp Bio Med 107:607-610, 1981.

45. Hess AD, Hayes RO: Seasonal dynamics of western encephalitis virus. Am J Med Sci 253:333-348, 1967.

46. Hess AD, Holden P: The natural history of the arthropod-borne encephalitides in the United States. Ann N York Acad Sci 70:294-311, 1958.

47. Holden P: Transmission of eastern equine encephalomyelitis in ring-necked pheasants. Proc Soc Exp Biol Med 88:607-610, 1955.

48. Holden P, Francy DB, Mitchell CJ, et al: House sparrows, *Passer domesticus* (L), as hosts of arboviruses in Hale County, Texas. II. Laboratory studies with western equine encephalitis virus. Am J Trop Med Hyg 22:254-262, 1973.

49. Holden P, Hayes RO, Mitchell CJ, et al: House sparrows, *Passer domesticus* (L), as hosts of arboviruses in Hale County, Texas. I. Field studies 1965-1969. Am J Trop Med Hyg 22:244-253, 1973.

50. Ianconescu M: Arbovirus infections. In Calnek BW, et al (eds): Diseases of Poultry. Ames, Iowa State University Press, 1991, pp 674-679.

51. Jungherr EL, Helmboldt CF, Satriano SF, et al: Investigation of eastern equine encephalomyelitis III. Pathology in pheasants and incidental observations in feral birds. Am J Hyg 67:10-20, 1958.

52. Karabatsos N: International Catalogue of Arboviruses 1985 Including Certain Other Viruses of Vertebrates. Am Soc Trop Med Hyg, 1985.

53. Karstad L: Arboviruses. In Davis JW, et al (eds): Infectious and Parasitic Diseases of Wild Birds. Ames, Iowa State University Press, 1971, pp 17-21.

54. Karstad L, Spalatin J, Hannon RP: Experimental infections of wild birds with the viruses of equine encephalitis, Newcastle disease and vesicular stomatitis. J Infect Dis 105:188, 1959.

55. Kissling RE, Chamberlain RW, Sikes RK, et al: Studies on the North American arthropod-borne encephalitides. III. Eastern equine encephalitis in wild birds. Am J Hyg 60:251-265, 1954.

56. Kissling RE, Eidson ME, Stamm D: Transfer of maternal neutralizing antibodies against eastern equine encephalomyelitis virus in birds. J Infect Dis 95:179-181, 1954.

57. Kissling RE, Stamm DD, Chamberlain RW, et al: Birds as winter hosts for eastern and western equine encephalomyelitis viruses. Am J Hyg 66:42-47, 1957.

58. LaMotte LC, Crane GT, Shriner RB, et al: Use of adult chickens as arbovirus sentinels. I. Viremia and persistence of antibody in experimentally inoculated adult chickens. Am J Trop Med Hyg 16:348-356, 1967.

59. Levine B, Griffin DE: Persistence of viral RNA in mouse brains after recovery from acute alphavirus encephalitis. J Virol 66:6429-6435, 1992.

60. Luginbuhl RE, Satriano SF, Helmboldt CF, et al: Investigations of eastern equine encephalomyelitis. II. Outbreaks in Connecticut pheasants. Am J Hyg 67:4-9, 1958.

61. Lundstrom JO, Turell MJ, Niklasson B: Antibodies to ockelbo virus in three orders of birds (Anseriformes, Galliformes, and Passeriformes) in Sweden. J Wildl Dis 28:144-147, 1992.

62. Lundstrom JO, Turell MJ, Niklasson B: Viremia in three orders of birds (Anseriformes, Galliformes, and Passeriformes) inoculated with ockelbo virus. J Wildl Dis 29:189-195, 1993.

63. Mace DL, Ott RL, Cortez FS: Evidence of the presence of equine encephalomyelitis virus in Philippine animals. Bull US Army Med Dept 91:504, 1949.

64. Mack TM, Brown BF, Sudia WD, et al: Investigation of an epidemic of St. Louis encephalitis in Danville, Kentucky, 1964. J Med Ent 4:70-76, 1967.

65. Main AJ, Anderson KS, Maxfield HK, et al: Duration of alphavirus neutralizing antibody in naturally infected birds. Am J Trop Med Hyg 38:208-217, 1988.

66. McIntosh BM, McGillivray GM, Dickinson DB, et al: Ecological studies on Sindbis and west nile viruses in South Africa IV. Infection in a wild avian population. S African J Med Sci 33:105-112, 1968.

67. McLean RG, Bowen GS: Vetebrate hosts. In Monath TP (ed): St. Louis Encephalitis, Washington DC, Am Public Health Assoc, 1980, pp 381-450.

68. McLean RG, Shriner RB, Kirk LJ, et al: Western equine encephalitis in avian populations in North Dakota, 1975. J Wildl Dis 25:481-489, 1989.

69. Merrill M, TenBroeck C: The transmission of equine encephalomyelitis virus by *Aedes aegypti*. J Exp Med 62:687-695, 1935.

70. Moulthrop IM, Gordy BA: Eastern viral encephalomyelitis in chukar

(*Alectoris graeca*). Avian Dis 4:380-381, 1960.

71. Olson TA, Kennedy RC, Rueger ME, et al: Evaluation of activity of viral encephalitides in Minnesota through measurement of pigeon antibody response. Am J Trop Med Hyg 10:266-270, 1961.

72. Parikh GC, Colburn ZD, Larson DR: Eastern equine encephalitis outbreak on a South Dakota pheasant farm. Bacteriol Proceed 69:159, 1969.

73. Ranck FM, Gainer JH, Hanley JE, et al: Natural outbreak of eastern and western encephalitis in pen-raised chukars in Florida. Avian Dis 9:8-20, 1965.

74. Reeves WC, Hammon WM: Epidemiology of the arthropod-borne viral encephalitides in Kern County, California 1943-1952. Univ Calif Public Hlth 4:1-257, 1962.

75. Reeves WC, Hutson GA, Bellamy RE, et al: Chronic latent infections of birds with western equine encephalomyelitis virus. Proc Soc Exptl Biol Med 97:733, 1958.

76. Reeves WE, Sturgeon JM, French EM, et al: Transovarian transmission of neutralizing substances to western equine and St. Louis encephalitis viruses by avian hosts. J Infect Dis 95:168-178, 1954.

76a. Ritchie BW, et al: Unpublished data.

77. Satriano SF, Luginbuhl RE, Wallis RC, et al: Investigation of eastern equine encephalomyelitis. IV. Susceptibility and transmission studies with virus of pheasant origin. Am J Hyg 67:21-33, 1957.

78. Scherer WF, Campillo SJ, Macias DM, et al: Serologic survey for neutralizing antibodies to eastern equine and western equine encephalitis viruses in man, wild birds and swine in Southern Mexico during 1961. Am J Trop Med Hyg 15:211-218, 1966.

79. Scherer WF, Dickerman RW, Chia CW, et al: Venezuelan equine encephalitis virus in Veracruz, Mexico, and the use of hamsters as sentinels. Science 145:274-275, 1964.

80. Schwede GJ: Western encephalitis in emus: Symptoms and vaccine protocol. Proc Assoc Avian Vet, 1993, pp 299-300.

81. Shivaprasad HL: Diseases of the nervous system in pet birds. A review and report of diseases rarely documented. Proc Assoc Avian Vet, 1993, pp 213-222.

82. Shope RE, Andrade AH, Bensabath G, et al: The epidemiology of EEE, WEE, SLE and Turlock viruses, with special reference to birds in a tropical rain forest near Belem, Brazil. Am J Epidemiol 84:467-477, 1966.

83. Simpson DIH: Toga- and Flaviviruses. *In* McFerran JB, McNulty MS (eds): Virus Infections of Birds. New York, Elsevier Science Publishers, 1993, pp 231-241.

84. Sooter CA, Schaeffer M, Gorrie R, et al: Transovarian passage of antibodies following naturally acquired encephalitis infection in birds. J Infect Dis 95:165-167, 1954.

85. Spalatin JL, Karstad JR, Anderson JR, et al: Natural and experimental infections in Wisconsin turkeys with the virus of eastern encephalitis. Zoonoses Res 1:29-48, 1961.

86. Spalding MG, McLean RG, Burgess JH, et al: Arboviruses in water birds (Ciconiiformes, Pelecaniformes) from Florida. J Wildl Dis 30:216-221, 1994.

87. Stamm DD: Studies on the ecology of equine encephalomyelitis. Am J Publ Health 48:328-335, 1958.

88. Stamm DD: Relationship of birds and arboviruses. Auk 83:84-97, 1966.

89. Stamm DD: Arbovirus studies in birds in South Alabama, 1959-1960. Am J Epidemiol 87:127-137, 1968.

90. Stamm DD, Kissling RE: Influence of season on EEE infection of English sparrows. Proc Soc Exper Biol Med 92:73-78, 1956.

91. Stamm DD, Kissling RE: The influence of reciprocal immunity on eastern and western equine encephalomyelitis infection in horses and English sparrows. J Immunol 79:342-347, 1957.

92. Stamm DD, Newan RJ: Evidence of southward transport of arbovirus from the United States by migratory birds. Ann Microbiol 11:123-133, 1963.

93. Sussman G, Cohen D, Gerende JH: Preliminary report of equine encephalomyelitis vaccination studies in New Jersey pheasants during the 1956 epizootic. Proc 60th US Livestock Sanitary Assoc, 1956, pp 19-23.

94. Trent DW, Grant JA: A comparison of New World alphaviruses in the western equine encephalomyelitis complex by immunochemical and oligonucleotide fingerprint techniques. J Gen Virol 47:261-282, 1980.

95. Tully TN, Shane SM: Eastern equine encephalitis in emus. Proc Assoc Avian Vet, 1992, pp 316-317.

96. Tully TN, Shane SM: Multivalent equine encephalomyelitis vaccine to protect emus (*Dromaius novaehollandiae*). Proc Assoc Avian Vet, 1993, pp 297-298.

97. Tully TN, Shane SM, Poston RP, et al: Eastern equine encephalitis in a flock of emus (*Dromaius novaehollandiae*). Avian Dis 36:808-812, 1992.

98. Tyzzer EE, Sellards AW, Bennett BL: The occurrence in nature of equine encephalomyelitis in the ring-necked pheasant. Science 86:505-506, 1938.

99. Veazy RS, Vice CC, Cho D-Y, et al: Pathology of eastern equine encephalitis in emus (*Dromiceius novaehollandiae*). Vet Pathol 31:109-111, 1994.

100. Velasco MC: Eastern equine encephalomyelitis virus in a lady gouldian finch. J Assoc Avian Vet 6:227-228, 1992.

101. Wages DP, Ficken MD, Guy JS, et al: Egg-production drop in turkeys associated with alphaviruses: Eastern equine encephalitis virus and Highlands J virus. Avian Dis 37:1163-1166, 1993.

102. Woodring FR: Naturally occurring infection with equine encephalomyelitis virus in turkeys. J Am Vet Med Assoc 130:511-512, 1957.

CHAPTER 16

Other Avian Viruses

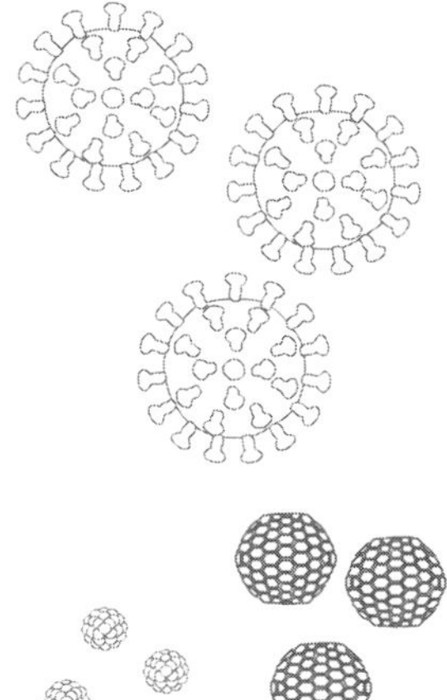

CORONAVIRIDAE

Coronaviridae currently contains only one genus, *Coronavirus*. These enveloped single-stranded RNA viruses can vary in size from 60 to 220 nm. Like other RNA viruses, replication occurs in the cytoplasm of infected cells. In general, these viruses are highly species-specific, which means that the coronaviruses that infect humans would not be expected to infect birds, and the coronaviruses that infect birds would not be expected to infect humans. These viruses replicate mostly in the cells that line the respiratory and gastrointestinal tracts, and are relatively easy to detect by electron microscopy in the feces of infected animals, including asymptomatic mammals and birds. Coronaviruses are generally associated with mild or asymptomatic infections in adults and severe diarrhea in young animals. Coronaviruses have been associated with disease in pigeons, pheasants and chickens (infectious bronchitis virus), and turkeys (bluecomb disease), and are suspected to be a cause of enteritis in ostriches, Japanese Quail, guinea fowl, pheasants and psittacines.[41,53,98,101] Antibodies to infectious bronchitis virus have been demonstrated in free-ranging Strigiformes and Passeriformes in Europe.[44]

CORONAVIRUS IN PSITTACINES

A virus morphologically similar to a coronavirus was recovered from three parrots. Adult Budgerigars and one-day-old chickens experimentally infected with the virus by the intramuscular, intraperitoneal or intraocular route developed anorexia, depression and greenish diarrhea within seven days after inoculation. Virus was recovered from the feces during episodes of diarrhea. One-day-old chickens inoculated intramuscularly died 10 to 13 days after inoculation. All of the Budgerigars survived, but had necrosis and hemorrhage of the liver and spleen when necropsied four weeks after inoculation. Experimentally infected quail remained clinically normal and did not seroconvert. The virus spread from the experimentally infected birds to chickens and Budgerigars in the same room. The virus was serologically distinct from infectious bronchitis virus, transmissible gastroenteritis virus (a coronavirus of pigs) and canine coronavirus.[53]

CORONAVIRUS IN OSTRICHES

A coronavirus was demonstrated in the intestinal contents of an 18-day-old ostrich chick that died following a one-week history of progressive weakness, anorexia and diarrhea *Figure 16.1*. Necropsy findings included a dilated, thin-walled proventriculus, "sticky, brown digesta" in the jejunum and necrotic-appearing kidneys. Microscopic changes included atrophy of the cells lining the intestinal tract and accumulation of necrotic debris. Eosinophilic inclusion bodies, which are not commonly seen with coronavirus, were

identified in the cytoplasm of enterocytes. Tibial chondrodysplasia was also noted. The bursa was depleted of lymphocytes. Other ostrich chicks from the farm had died from 3 to 12 days after hatching with similar clinical signs.[29] Antibodies to infectious bronchitis virus were detected by ELISA in 2% of 149 ostriches in Zimbabwe.[11]

CORONAVIRUS IN PIGEONS

A strain of infectious bronchitis virus (IBV) serologically related to a strain found in chickens was isolated from racing pigeons in Australia. Members of a flock of 150 racing pigeons developed an acute onset of disease characterized by ruffled feathers, difficulty with breathing and mucus accumulation at the commissure of the beak. Of the affected birds, 22 died from one day to two weeks after clinical signs were first noted in the flock. Other affected birds recovered over a two- to three-week period. Necropsy findings included ulcers in the esophagus and crop, fluid accumulation in the lower intestinal tract and mucus in the pharynx and trachea. A coronavirus was recovered from the tracheal mucosa and cloacal swabs of some of the affected birds; however,

many of the gross lesions were caused by concomitant trichomoniasis.

Antibodies to IBV were detected in the affected flock and in a nearby flock that had not experienced any clinical disease. It was theorized but not confirmed that the pigeons were exposed to the virus through contact with infected chickens. Experimentally infected four-week-old specific-pathogen-free chickens developed severe respiratory disease within four days following inoculation by intranasal, intraocular and oral routes, while eight-week-old pigeons remained clinically normal. Antibodies were detected in the experimentally infected chicks but not in the pigeons.[4]

CORONAVIRUS IN GALLINACEOUS BIRDS

Isolates of coronavirus from guinea fowl and pheasants have been shown to be serologically unrelated to chicken isolates.[41,98] Seven serologically distinct coronaviruses were isolated from pheasants in Great Britain.[41] Reduced egg production, poor egg shell quality and mild respiratory signs have been reported in pheasants infected with a coronavirus. Birds that die frequently have egg-related peritonitis, urolithiasis, visceral gout and enlarged kidneys *Figure 16.2*. Mortality can reach 40% in eight- to ten-week-old pheasant chicks.[41,77] Two-week-old pheasant chicks experimentally infected with coronavirus developed mild respiratory disease and seroconverted. Virus could not be recovered from the infected birds.[41]

An acute onset of decreased egg production, loss of shell pigmentation and abnormal shell quality was reported in a group of pheasants in England. Necropsy findings in affected hens were unremarkable. Two hens in the group of pheasants had hemagglutination-inhibition (HI) antibodies to IBV. The following year, hens developed egg-related problems accompanied by respiratory signs, which included head shaking and audible respiratory sounds. At this time, 80% of the pheasants tested had rising HI antibody titers to IBV. A coronavirus that was serologically related to IBV found in chickens was recovered from the affected pheasants. It was suspected that the virus entered the pheasant area through contact with an adjacent field that had been recently fertilized with chicken manure. An inactivated IBV vaccine administered intramuscularly induced HI antibody titers that ranged from 1:32 to 1:256, and the egg-related problems and respiratory disease decreased.[135]

In guinea fowl, IBV has been associated with enteritis, hepatitis and pancreatitis. Birds as young as three days of age may develop clinical changes that include anorexia, mild respiratory disease and polyuria, followed by death. Birds experimentally infected by the intranasal route develop similar clinical signs.[32] A serologically distinct coronavirus was recovered from Japanese Quail with respiratory signs.[101]

A coronavirus has been associated with acute enteritis in turkeys. Clinical changes include depression, loss of appetite and diarrhea. Turkey coronavirus can remain infectious in feces for months in areas where freezing temper-

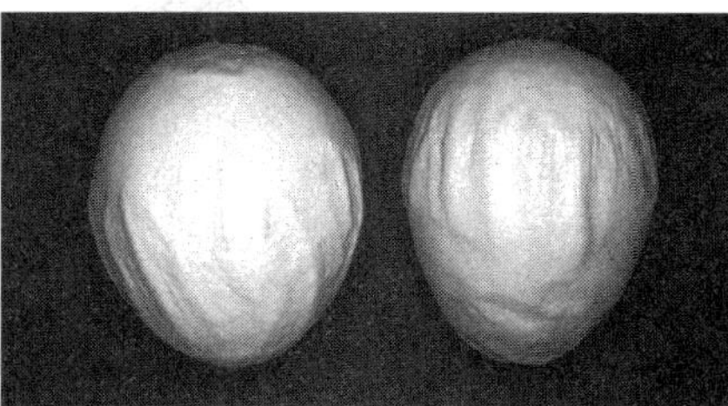

FIG 16.2

Abnormalities that have been associated with coronavirus infections in pheasants, chickens and turkeys include: decreased egg production or production of soft, abnormally formed eggs;

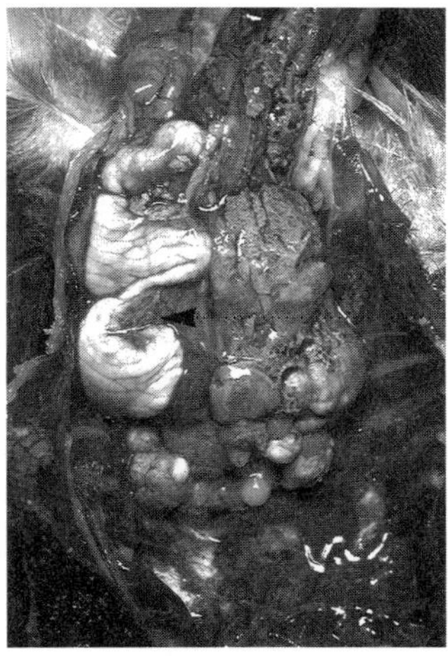

egg-related peritonitis; and

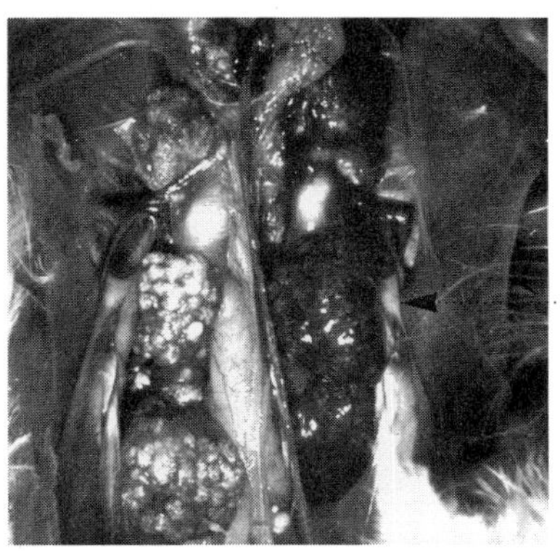

urolithiasis with enlarged kidneys.

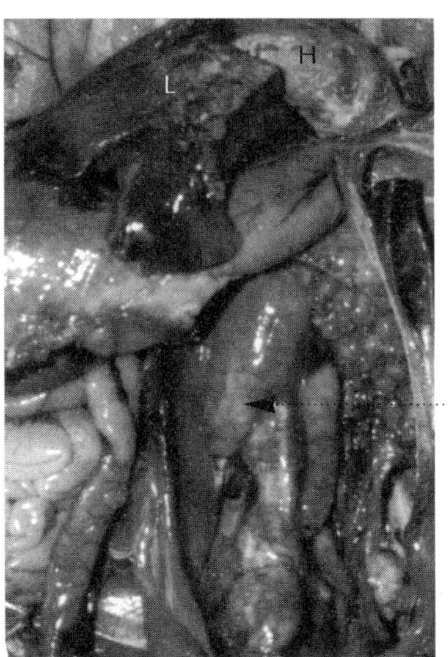

An adult pheasant hen infected with coronavirus had enlargement and inflammation of the kidneys (arrow) along with gout on the surface of the liver (L) and heart (H).

photographs courtesy of R.E. Gough and Jean Sander

atures are common. Pheasants, quail, chickens and seagulls have been shown to be resistant to turkey coronavirus infection.[102,103]

Infectious bronchitis virus commonly infects chickens maintained in high intensity commercial operations. Antigenically varied strains of the virus can cause disease of the respiratory tract, female reproductive tract (poor egg shell quality) or kidneys (gout, polyuria) *Figure 16.2*. The virus replicates in the cytoplasm of the cells lining the mucus membranes and does not induce the formation of inclusion bodies. Most affected chickens develop respiratory disease with lesions in the trachea. Clinical changes are characterized by sneezing, coughing and dyspnea. These changes may be noted within an entire flock within 24 to 72 hours. A decline in egg production is characteristic. If the lesions in the respiratory tract are not complicated by secondary invaders, most infected birds recover. However, high morbidity and mortality can occur in

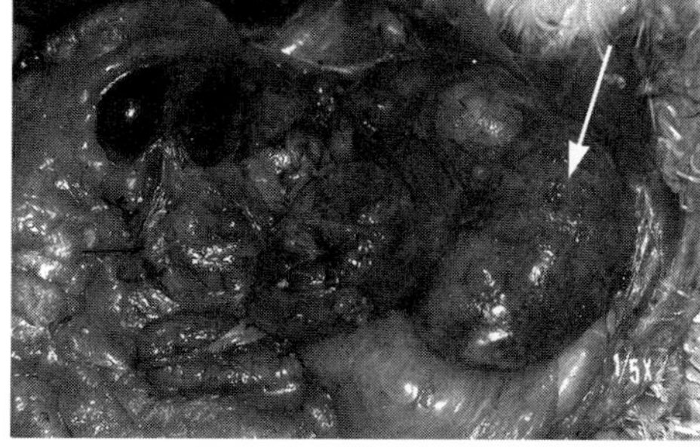

FIG 16.3

Infectious bronchitis virus has been shown to cause permanent proliferation of the wall of the oviduct (arrow), resulting in blockage and cystic changes in this organ. photograph courtesy of Jean Sander

flocks that are highly susceptible and maintained in non-hygienic, stressful, crowded conditions.

Infected birds shed virus in their feces and respiratory secretions. Virus is usually shed in highest concentration from the respiratory tract for five to seven days, and most birds stop shedding within several weeks after infection. Transmission occurs through ingestion of contaminated food or water or inhalation of contaminated aerosols. The virus has been shown to be present in the respiratory tract for 49 days, in the female reproductive tract (and eggs) for 50 days, in semen for two weeks and in the gastrointestinal tract for 20 weeks after infection.[2,17]

Gross changes in affected birds include thickening of the mucous membranes in the upper respiratory tract with accumulation of a catarrhal exudate. Dilatation of the intestinal tract with fluid and gas is a characteristic clinical sign. In some cases, IBV has been associated with rupture of a developing ovum and subsequent deposition of yolk in the abdomen.[128] Rarely, the kidneys may be congested. Some IBV strains can cause enlarged kidneys, with accumulation of uric acid in the kidney tubules and ureters. Infectious bronchitis virus has been shown to cause permanent proliferation of the wall of the oviduct, resulting in blockage and cystic changes in this organ *Figure 16.3*.[63]

Antibodies to the virus can be detected from 7 to 14 days after infection, and may persist for months. Recovered birds are considered immune to future infections. Until they wane three to four weeks posthatching, maternally derived antibodies protect a chick from disease. Virus can be recovered from respiratory excretions (tracheal swabs or transtracheal washes) or from the spleen, oviduct, kidney or feces of an infected bird. The virus can be recovered from the cecal tonsils or cloaca for several weeks after respiratory signs have resolved.[17,63,128]

Infectious bronchitis virus can be inactivated by heating to 50°C for 15 minutes, or with lipid solvents (chloroform and ether), detergents and most disinfectants.[31] The virus is stable at pH 3, which would be expected for a virus that must be ingested and pass through the acidic environment of the stomach to reach the intestinal tract before causing disease. Modified-live and inactivated vaccines are available for IBV.[31]

PICORNAVIRIDAE

The family Picornaviridae consists of five genera: *Enterovirus*, *Cardiovirus*, *Hepatovirus*, *Rhinovirus* and *Apthovirus*. Members of this family of viruses cause some of the most important diseases of humans and domestic mammals. In humans, a type of enterovirus is responsible for poliomyelitis. Hepatitis A is caused by a hepatovirus, and rhinoviruses are frequent causes of the "common cold." An apthovirus, the foot-and-mouth disease virus, can cause high mortality in cattle and other cloven-hoofed animals.

Several picornaviruses have been associated with disease in birds, including cockatoo enteritis virus, avian encephalomyelitis virus, duck hepatitis virus (types I and III) and turkey hepatitis virus. In one study, the enteroviruses from chickens and ducks were grouped into distinct serotypes that included duck hepatitis virus 1, duck hepatitis virus III, avian encephalomyelitis virus, avian nephritis virus and two other entero-like viruses isolated from chickens.[92]

It has been theorized that free-ranging birds living in and around cattle could serve as mechanical vectors in transmitting the apthovirus that causes foot-and-mouth disease. This virus has been recovered from birds in the Ukraine, and experimentally infected starlings remained asymptomatic but shed virus in their feces for 26 hours after infection.[25,83] Rhinoviruses and cardioviruses have not been reported in birds.

PICORNAVIRUSES IN COCKATOOS

A non-enveloped 35 nm virus with morphologic characteristics consistent with an enterovirus was recovered from the feces of free-ranging Galahs with intractable profuse mucoid diarrhea *Figure 16.4*. The birds rapidly became dehydrated and died from several days to three weeks after developing clinical signs. Sulphur-crested Cockatoos that were placed in direct contact with affected Galahs developed clinical signs of disease five days after exposure, and most of the Sulphur-crested Cockatoos died. All of the affected cockatoos had concurrent psittacine beak and feather disease (PBFD) virus infections. Normal Galahs, Galahs with PBFD, lovebirds with PBFD and specific-pathogen-free chickens inoculated with virus-laden feces that had been frozen remained normal. Because cockatoos exposed to birds with the disease developed characteristic clinical changes, it was suggested that the failure of virus transmission in the experimentally inoculated birds might have been caused by damage to the virus induced

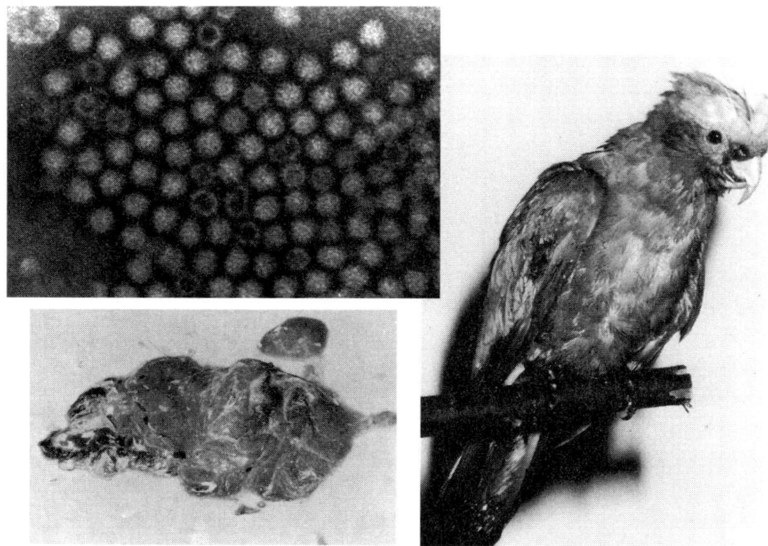

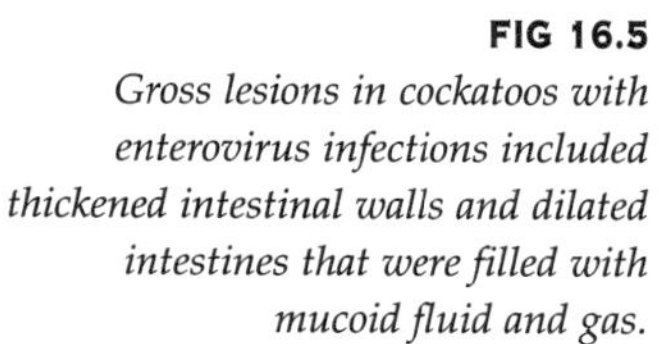

by freezing.[157] However, other studies have indicated that freezing increases the infectivity of enteroviruses recovered from chickens.[20]

Gross lesions in the affected cockatoos included thickened intestinal walls and dilated intestines that were filled with mucoid fluid and gas *Figure 16.5*. Microscopic changes were limited to the intestines and included elongated crypts, villous atrophy, chronic inflammation, epithelial cell hyperplasia and vacuolation of the enterocytes at the villous tips. Lesions could be detected throughout the intestinal tract but were most common in the duodenum and

proximal jejunum. Viral particles arranged in paracrystalline arrays were found within intracytoplasmic inclusion bodies in the enterocytes.

In another enterovirus-like outbreak in Australia, seven- to nine-week-old Sulphur-crested Cockatoos and Galah chicks developed profuse, yellow-green, mucoid diarrhea followed by anorexia, depression, weight loss and death. The clinical changes were noted two to seven days after the birds were removed from the wild. All the affected birds died over a one- to four-week period. Necropsy findings included a dilated duodenum and a jejunum filled with yellow-green mucoid fluid and gas. The walls of the intestinal tract were thickened. Of the affected cockatoos, 18 of 31 (58%) had particles in the feces that were morphologically suggestive of an enterovirus. In previous years, a disease with a similar clinical progression had been noted in 2% to 3% of the recently captured cockatoos.[94]

DUCK VIRUS HEPATITIS

CLINICAL FEATURES AND EPIZOOTIOLOGY

Hepatitis in ducks can be caused by at least two serologically distinct picornaviruses, as well as by a hepadnavirus and an astrovirus *Table 16.1*. The clinical progression varies with the type of virus. Currently, domestic ducks are considered the only natural host for duck hepatitis virus I (DHV I) and duck hepatitis virus III (DHV III) infections. However, turkeys, chickens, quail, Mallard Ducks, pheasants, guinea fowl and goslings are susceptible to experimental infections. High mortality occurs in experimentally infected pheasants, geese and guinea fowl.[57] Duck hepatitis virus infections have not been documented in free-ranging ducks as of 1989.[155]

TABLE 16.1

Viruses associated with hepatitis in ducks

Duck hepatitis I — Picornavirus
Duck Astrovirus — Astrovirus
    Formerly duck hepatitis II
Duck hepatitis III — Picornavirus
Duck hepatitis B — Hepadnavirus

Duck hepatitis virus I primarily affects ducklings less than six weeks of age, is distributed worldwide and can cause high levels of mortality. The incubation period is less than 24 hours following natural or experimental infection, and affected ducklings usually die within several hours of developing clinical signs that include progressive lethargy over a several-hour period, unwillingness to move, ataxia, loss of balance and recumbency prior to death. The virus can move rapidly through a flock with the majority of exposed ducklings dying within two to four days after the first clinical signs are noted.[155] Mortality is highest (may reach 100%) in birds less than one week of age, decreases to 50% in three- to four-week-old ducklings, and reduces to a point where infections in most birds greater than five to six weeks of age are subclinical.[142]

Duck hepatitis virus III causes clinical changes similar to but less severe than DHV I and with lower levels of morbidity and mortality. This virus has been documented only in domestic ducks in the United States. The mortality associated with DHV III varies from 10% to 50%, depending on the age of the infected duck.[45] The highest mortality occurs in ducklings less than one week of age and resistance to the disease begins by two weeks of age.[142]

Some researchers have induced infections by inoculating ducks orally, while others have been unable to infect birds by this route.[48,104,144] The capacity to infect birds by intratracheal inoculation has led some to conclude that this is the natural route of virus entry to ducklings.[104,144] Viremia occurs following primary infection, and virus can be detected in the blood, visceral organs (particularly liver) and feces.[155] Some ducks that recover have been found to shed virus in their feces for up to eight weeks.[109] Contaminated equipment, supplies and clothing may be involved in the indirect transmission of these viruses. In one study, the virus was shown to persist in naturally contaminated brooders for ten weeks.[156] Vertical transmission has not been shown to occur; however insufficient studies have been performed to eliminate the possibility of this method of transmission.

TABLE 16.2

Comparison of duck hepatitis viruses

DHV type I
    Ducklings less than 6 weeks of age
    Clinical signs include weakness, ataxia and
        loss of balance
    Worldwide distribution
    Rapid infection, up to 100% mortality in birds
        less than one week old
    Hemorrhagic lesions in the liver of young
        ducklings are highly suggestive

DHV type II
    Loose droppings, polyuria, ataxia and
        convulsions prior to death
    Only documented in the United Kingdom
    Mortality generally less than 20%

DHV type III
    Ducklings less than 2 weeks of age
    Clinical changes similar to DHV I but
        less severe
    Only documented in the United States
    Mortality varies from 10% to 50%

Duck hepatitis B Virus
    Subclinical infections

Because of the durability of this virus when outside of the host, these viruses become enzootic and cycle through successive generations of young ducks housed in contaminated areas. In one study, DHV I was recovered from 4 of 45 (9%) ducks presented for necropsy over a one-year study period in Great Britain. Affected ducks were from flocks experiencing high mortality and severe liver disease.[38]

PATHOLOGY, IMMUNITY AND DIAGNOSIS

The primary gross lesion associated with DHV is an enlarged, congested liver with reddish-brown mottling and hemorrhage *Figure 16.6*. Severe necrosis that may involve most of the liver (DHV I) is noted microscopically. The gross and microscopic lesions noted with DHV III are similar but less severe.[26,142] Hemorrhagic lesions in the liver of young ducklings are highly sug-

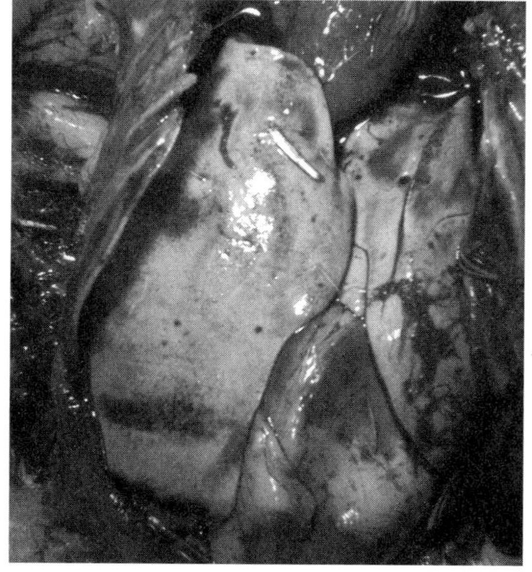

FIG 16.6

The primary gross lesion associated with duck hepatitis virus is an enlarged, congested liver with reddish-brown mottling and hemorrhage.
photograph courtesy of Jean Sander

gestive. Virus isolation from the liver is the best way to confirm an infection.

A fatty kidney syndrome and pancreatic necrosis has also been associated with

DHV I in Pekin Ducklings up to six weeks of age in Britain. All the affected ducklings had been vaccinated with an attenuated-live DHV I vaccine at one day of age. Gross lesions included pale, swollen kidneys and swollen mottled livers and spleens. Microscopically, lesions included lipid accumulation in the renal tubules, bile duct hyperplasia, necrotizing hepatitis and liver hemorrhage.[28]

Virus-neutralizing antibodies can be detected as soon as one week after ducklings are infected by the intratracheal, subcutaneous or intramuscular route.[144] Duck hepatitis viruses cause the most severe problems in ducklings that do not have high levels of maternally derived antibodies. High levels of maternal antibodies provide protection from disease until they begin to wane at three to four weeks post-hatching. By this time the ducklings will be resistant to DHV III and partially resistant to DHV I. Concomitant infections with *Chlamydia* sp. or influenza A virus have been shown to decrease the age-related resistance to DHV.[14,39] Salmonellosis and aflatoxocosis can cause clinical changes in ducklings similar to those caused by DHV.[155]

CONTROL

Duck hepatitis virus is nonenveloped, is relatively stable in the environment and has been shown to remain infectious for two years when maintained at 4°C. The virus is resistant to lipid solvents, acid (pH 3), 2% Lysol, 0.1% formalin, 15% creolin and 20% anhydrous sodium carbonate, and is heat stable at 50°C (one hour), 56°C (30 minutes to 23 hours) and 62°C (30 minutes).[20,156] These viruses can be inactivated by many standard disinfectants, including 2% sodium hypochlorite (three hours), 5% phenol, 3% chloramine, 0.2 to 1% formalin (two hours) and 2% caustic soda (two hours).[156]

Hatchlings that do not have maternally derived antibodies can be provided protection from the disease by giving them hyperimmune serum (IM) or better yet, β-propiolactone-treated yolk (PO) from the eggs of vaccinated hens.[110] Both attenuated-live and inactivated vaccines have been developed for DHV I. The best levels of antibodies are achieved in birds vaccinated with modified-live virus at twelve weeks of age, followed by inactivated virus six weeks later.[156]

Vaccinated hens pass antibodies to their embryos that provide protection from the disease for two to three weeks post-hatching. An attenuated-live virus vaccine has been shown to protect seronegative ducklings as soon as three days following intramuscular injection.[55,56] Virus-neutralizing antibodies were detected four days post-vaccination in two-day-old seronegative ducklings.[21] Because of the rapid immune response that occurs with this vaccine, it may by advantageous to vaccinate early in an outbreak;[156] however, care should be exercised to prevent the vaccination procedure from hastening the transmission of the virus through the flock. Maintaining a vaccinated flock, eliminating overcrowded conditions, and practicing sound hygiene to reduce the accumulation of the virus in the environment (particularly with ducklings less than six

weeks of age) are the best methods to prevent DHV-induced disease.[143]

AVIAN ENCEPHALOMYELITIS VIRUS

Avian encephalomyelitis virus causes a neurologic disease in pheasants, quail, chickens and turkeys. Experimental infections have been documented in ducks, pigeons and guinea fowl.[12] This disease, which is caused by an enterovirus, should not be confused with eastern or western equine encephalitis, which are caused by togaviruses.

Avian encephalomyelitis virus infections are most severe in young (two- to four-week-old) birds that have incompletely developed immune systems. Young chicks that are infected in the oviduct or soon after hatching develop neurologic signs that include progressive ataxia, reluctance to move and tremors of the head and neck. Birds may die acutely or within several hours of developing clinical signs. Survivors may have permanent neurologic and ocular changes that include enlarged eyes, lens opacities and fixed-appearing pupils.[12] Chicks greater than four weeks old generally develop a mild disease or remain asymptomatic. A decrease in egg production may occur in older birds that are infected.[13]

Avian encephalomyelitis virus has a worldwide distribution but has not been documented in free-ranging birds. In one study, antibodies to the virus were detected in partridges and pheasants, but not in sparrows, starlings, pigeons, finches, ducks or doves.[146] Antibodies to avian encephalomyelitis virus were detected in 22 (15%) of 149 ostriches in Zimbabwe using an enzyme-linked immunosorbent assay (ELISA). Of nine farms evaluated, four had no seropositive birds; the other five had seropositive rates ranging from 5% to 50%.[11]

The virus is shed in the feces from 5 to 21 days after infection and can remain infectious in litter for months to years. Younger birds shed the virus for the longest periods.[152] Because of the stability of the virus, contaminated fomites should be considered in the transmission of the virus among flocks. This virus is transmitted primarily from viremic hens to their embryos.[12] The incubation period for transovarial infections is one to seven days. Horizontal transmission results in an eleven-day incubation period.[13]

PATHOLOGY, PATHOGENESIS, IMMUNITY AND DIAGNOSIS

Affected chicks generally have no gross lesions.[12] A perivascular accumulation of lymphocytes in the brain and spinal cord is common microscopically. Virus is ingested, replicates in the intestinal tract and is shed in the feces. Because of its stability to acids, the virus is able to pass through the proventriculus (pH of 2). The central nervous system signs occur secondary to replication of the virus in the intestinal tract and the resulting viremia.[96]

Antibodies to the virus can be detected using ELISA, virus neutralization or agar-gel immunodiffusion assays. Maternally derived antibodies reduce the severity of disease in a flock, but do not prevent enteric infections.[13,92,151,152] Young birds without maternally derived antibodies develop clinical signs, whereas those with maternally derived antibodies are temporarily resistant to disease. An age-related resistance occurs as chicks approach and pass six weeks of age. The production of virus-neutralizing antibodies correlates with a cessation of viremia. These antibodies can be detected as early as 11 days after infection.[152] Antibodies to the virus can be detected throughout a bird's life, possibly as a result of frequent exposure to the virus.116 Recovered birds are considered immune to the disease. Infections can be diagnosed by isolating the virus from the brain, liver or pancreas, or by demonstrating a rising antibody titer.

CONTROL

Avian encephalomyelitis virus is resistant to extreme environmental conditions, and has been shown to be stable to lipid solvents, acid pH and 62°C (several hours). The virus was found to be inactivated within 18 hours when exposed to formaldehyde gas.[60] Vaccination of breeding flocks to induce maternal antibodies is used to control the disease in chickens and turkeys.[23]

OTHER AVIAN ENTEROVIRUSES

In two studies, enterovirus-like particles were detected by electron microscopy in the intestines of 2 of 160 and 1 of 192 ostriches.[130,131] Other enteroviruses have been associated with diarrhea and growth retardation in domestic poultry. In companion and aviary birds, it is likely that these viruses are causing similar problems that have yet to be documented.[92]

An enterovirus that is serologically distinct from avian encephalomyelitis virus, duck hepatitis viruses and other enteroviruses has been associated with renal disease in young chickens and turkeys. This virus is referred to as avian nephritis virus. Affected chicks may remain asymptomatic or exhibit reduced weight gains.

BIRNAVIRIDAE

The Birnaviridae family contains a single genus, *Birnavirus*. These 60 nm virus particles contain a double stranded RNA that is bisegmented. As is typical for RNA viruses, replication occurs in the cytoplasm of an infected cell. Strains of *Birnavirus* have been recovered from fish, birds and mollusk.

INFECTIOUS BURSAL DISEASE VIRUS

Infectious bursal disease (IBD) virus has been shown to infect ostriches, ducks, pheasants, chickens and turkeys, and is of major importance in all poultry-producing regions of the world. Because IBD virus was first recovered from birds in Gumboro, Delaware, USA, it is frequently called Gumboro disease. The birnavirus that causes IBD is highly infectious in young chickens and causes severe damage to the bursa, resulting in suppression of the immune system. Once immunosuppressed, the young birds are increasingly susceptible to both bacterial and other viral-induced diseases. Clinical changes, characterized by anorexia and watery, bile-tinged diarrhea, are most common in chicks three to six weeks of age as the maternally derived antibodies decrease. Subclinical infections are common in birds less than two weeks

old. Many infections are subclinical but still may cause immunosuppression that allows other pathogens to cause disease. Mortality rates of 2% to 80% have been reported in infected pheasants.[79] The bursa is critical for virus replication, and infections do not occur in chickens in which the bursa is removed. The incubation period in chickens is three to seven days.[51]

Both virus and antibodies to the virus have been demonstrated in ducks.[88] Antibodies to the virus have been detected in Passeriformes in Africa[99] and in Anseriformes in Australia.[154] Rising IBD virus antibodies were demonstrated in a group of three- to five-week-old ducklings with respiratory disease.[87] Antibodies to infectious bursal disease were detected in 22 (15%) of 149 ostriches in Zimbabwe. On nine farms, the seroprevalence rate varied from 0 to 60%.[11] The virus has been recovered from young ostriches in the United States, but its clinical effects have not been confirmed. Pigeons have not been shown to be susceptible experimentally.[149]

Transmission occurs primarily through direct contact with contaminated feces, although the virus also can be found in ocular and respiratory secretions. Food, litter or water contaminated with virus-laden feces can also serve as a source of virus exposure, because the virus is stable outside of the host. Infected chickens shed the virus in their feces for up to two weeks. It is theorized that infected free-ranging birds (particularly pheasants) could serve as a transport host for captive birds susceptible to infection. Rodents and insects should also be considered as mechanical vectors.[88]

Swelling of the bursa occurs early during the infectious process, followed by atrophy as the disease progresses. Virus is present in the bursa, thymus, spleen, kidney and cecal tonsils two to eight days after infection, and can be recovered from these organs as well as in the feces.[27] Antibodies can be demonstrated using various assays including agar-gel immunodiffusion, virus-neutralizing or ELISA. Virus-neutralizing antibodies persist for a longer period than precipitating antibodies (those detected by agar-gel immunodiffusion) do.

Infectious bursal disease virus is stable when heated to 56°C (five hours) or 60°C (30 minutes), and is resistant to lipid solvents and pH 2. This virus is stable in 0.5% formalin for six hours. The virus can be inactivated by pH 12, 70°C (30 minutes), 2% chloramine and glutaraldehyde.[95,112] The stability of IBD virus creates a problem in infected flocks where facilities can become contaminated and remain so for prolonged periods. Virus has been shown to remain infectious for 122 days in a chicken house and for 52 days in food and water.[87]

Maternally derived immunity is critical in protecting chicks from infection during their first several weeks after hatching when they are most susceptible.[81] Attenuated-live and inactivated vaccines are available for use in chickens. Breeder chickens are vaccinated prior to laying and a booster is given several months later. An effective recombinant vaccine has been developed for IBD virus but it is not currently available.[5] Many breeders are vaccinating ostriches with products designed and tested for use in chickens. If it is determined that a flock of ostriches should be vaccinated, it is important that an inactivated vaccine, and not an attenuated-live virus vaccine, be used.

FLAVIVIRIDAE

Historically, the term arbovirus has been used — without regard for taxonomic classification — to describe any virus that is transmitted in nature by an arthropod vector. Hundreds of viruses, principally within the families Togaviridae, Flaviviridae, Reoviridae, Rhabodoviridae and Bunyaviridae, have been shown to be transmitted from host to host by arthropods.[65] Many of the viruses within the Flaviviridae family can cause encephalitis in humans and mammals, and birds can play various roles in the maintenance, transmission and dissemination of this arthropod-vectored virus *Table 16.3.*

Flaviviridae contains two groups of viruses: the flaviviruses and hepatitis C virus. These 40 to 60 nm diameter viruses replicate in the cytoplasm of infected cells. Flaviviruses have been shown to infect many species of mammals and a few of these viruses have been documented in birds. The most common flaviviruses in birds are St. Louis encephalitis virus, Japanese encephalitis virus, Murray Valley encephalitis virus and turkey meningoencephalitis virus. Each of these viruses and others that occasionally have been described are serologically distinct and have geographically restricted ranges. The only one considered of importance in birds is turkey meningoencephalitis virus. Hepatitis C virus can infect humans and other primates.

St. Louis Encephalitis Virus

St. Louis encephalitis (SLE) virus has been recovered from many avian species, and virus-neutralizing antibodies have been reported in most common domestic birds and in some 60 species of free-ranging birds.[105,140] Birds are considered to be the natural reservoir for SLE virus, but after experimental infections, the levels of viremia that developed were low and somewhat erratic in comparison to other arthropod-vectored viruses.[15] Most birds infected with SLE virus remain asymptomatic and seroconvert, but a few deaths have been associated with infections in doves, pigeons, blackbirds, ducks, sparrows and chickens.[16,78,132]

All wild-caught Black-bellied Whistling Ducks inoculated with SLE virus developed viremia, remained asymptomatic and seroconverted. The mean duration of viremia was six days, and antibodies were detected starting on day six after infection. None of the birds had gross or histologic lesions suggestive of disease.[1] During an outbreak in Texas, SLE virus was recovered from a free-ranging Blue Jay, a Mockingbird, a domestic goose and a pigeon.[80] As with alphaviruses, nestling birds are considered the most likely host for amplifying SLE virus, because nestlings are easy targets for mosquitoes.[82]

Antibodies to SLE virus are common in free-ranging birds in the United States,

TABLE 16.3

Some characteristics of flaviviruses recovered from free-ranging birds[132]

St Louis encephalitis (SLE) virus (USA, Trinidad)
  Isolates and antibodies in birds
  Asymptomatic to death

Japanese encephalitis (JEE) virus (Asia and portions of Europe)
  Isolates from birds and mosquitoes

West Nile virus (Africa)
  Isolates from migrant birds, pigeons and mosquitoes
  Asymptomatic to death

Ilheus viruses (Central and South America)
  Isolates from birds and mosquitoes

Alfuy (Queensland)
  Isolates from birds and mosquitoes, and antibodies in birds

Kunjin (Queensland)
  Isolates from birds and mosquitoes

Murray Valley encephalitis (Australia and New Guinea)
  Isolates from birds and mosquitoes, antibodies common in birds

Rocio (Brazil)
  Isolates and antibodies in birds

Uganda S (Uganda)
  One isolate from a bird

Usutu (South Africa)
  Isolates from birds

Hypr (Czechoslovakia)
  Isolates from ticks,
  antibodies in passerine, gallinaceous and predatory birds

Kumlinge (Finland)
  Isolates in birds and ticks

Kyasanur Forest disease (India)
  Occasional isolates in birds and ticks

Louping Ill (Scotland, Great Britain)
  Isolates from birds and ticks
  Asymptomatic to death (grouse)

Russian Spring Summer encephalitis
  Isolates from ticks and birds

Cacipacore (Brazil)
  Isolates and antibodies from birds

Turkey meningoencephalitis (Israel, South Africa)
  No arthropod vector defined
  Paralysis and death

particularly west of the Mississippi. Virus activity also has been documented in the eastern United States (New Jersey, Kentucky, Tennessee, Florida), and in Trinidad, Panama, Haiti, Ecuador and the Caribbean region.[140] Antibodies to SLE virus were detected in free-ranging turkeys, herons and doves in Mexico.[1] Outbreaks of SLE in humans and birds have been reported in Missouri, Ohio, Indiana, Kentucky Louisiana and Florida.[52,140] Both western equine encephalitis (WEE) virus and SLE virus have similar geographic distributions, but WEE virus infections tend to occur in the midsummer, whereas SLE virus outbreaks are more common during the late summer and early fall.[52]

Several studies have indicated that from 3% to 56% of the free-ranging birds in endemic areas can have antibodies to the virus.[89] Neutralizing antibodies to SLE virus have been found to persist longer than hemagluttination-inhibition (HI) antibodies in the avian species tested.[106] During an outbreak in Kentucky, 35 of 241 (14%) free-ranging birds and 37 of 122 (30%) chickens were seropositive.[84] In another study, 17% of nestling and 5% of adult free-ranging Piciformes, Passeriformes and doves tested had SLE antibodies.[89-91] Virus-neutralizing antibodies were detected in nestling Ciconiiformes and Pelecaniformes from salt water and fresh water environments in Florida. The sero-prevalence was higher in adults and juveniles than in nestlings, with 16% of the Pelecaniformes and 5% of Ciconiiformes being seropositive. Antibodies were detected in nestling egrets at 8 to 16 days of age, suggesting either the maternal transmission of antibodies or induction of an active infection soon after hatching. During some years, antibody titers to SLE virus were more common than in other years when antibodies to this virus were not detected in any of the birds sampled.[136]

Virus-specific HI antibodies have been shown to be maternally transmitted in chickens, pigeons and doves. Most young chickens with maternally derived antibodies are considered seronegative by four weeks post-hatching, whereas antibodies may persist for up to 12 weeks in doves.[9,134] Antibodies to the virus were not detected in House Sparrows 14 to 18 days after hatching.[82] With few exceptions, the SLE hemagglutination-inhibition antibody titers in chicks 24 hours after hatching correlated well with maternal antibody levels.[9] Transovarian transfer of SLE HI antibody has also been shown to occur in pigeons.[134]

JAPANESE ENCEPHALITIS VIRUS

The Japanese encephalitis (JE) virus has a widespread distribution in Asia and has been documented in portions of Europe.[140] Virus isolations were more common from Black-crowned Night Herons in Asia than from two other birds in the same area.[119] Following JE virus inoculation, young chicks were found to be viremic more often than older birds, and young birds also were considered to be more susceptible to infection as measured by the production of virus-neutralizing antibodies. The duration of viremia in chicks two weeks of age was longer (four days) than in 1.5- to 7-month-old birds (one to two days).[10] Experimentally infected birds remained asymptomatic. Hemagglutination-inhibition (HI) antibodies were detected within 7 to 14 days, with a maximum titer occurring 14 to 21 days after infection. Neutralizing antibodies did not reach significant levels until 4 to 8 weeks after inoculation. In one study, HI antibodies decreased within two to three months, and virus-neutralizing antibodies persisted up to nine months. Neutralizing antibodies were detected in some birds that were not viremic.[10] In one group of birds, 48 of 82 birds were found to develop HI antibodies following infection, although only 11 of the 82 birds were found to be viremic. Because HI antibodies appeared uniformly after infection and persisted longer than viremia, detecting these antibodies was considered to reflect the time and incidence of natural infections more accurately than virus isolation studies.[10]

Several studies have shown that non-specific substances that will neutralize Japanese encephalitis, St. Louis encephalitis and western equine encephalitis viruses occur in blood collected from birds that have died from lethal injuries. For this reason, it is important that blood for virus-neutralizing assays be collected from uninjured birds by jugular venipuncture.[47,49,120]

Louping Ill Virus

Louping ill virus, which is transmitted by ticks, has been demonstrated in cattle, horses, pigs, grouse and humans in Great Britain and Ireland. Some species of grouse like Willow Grouse, Red Grouse and Rock Ptarmigan are particularly susceptible and frequently develop central nervous system signs following infection. The birds that naturally inhabit mainland and tundra appear to be more susceptible than upland species. Various species of pheasants also have been shown to be susceptible to infection.[107]

Turkey Meningoencephalitis Virus

Turkey meningoencephalitis virus has been shown to cause paralysis and decreased egg production in naturally and experimentally infected turkeys. Currently, the virus has been documented only in Israel and South Africa.[3,58] In natural infections, turkeys older than ten weeks are affected and develop clinical signs starting five to eight days post-exposure.[115] Recently hatched turkeys are highly susceptible to experimental infection. Mortality rates are generally 10% to 30% but can reach 80% in some flocks.[58] The virus also has been shown to infect turkeys and Japanese Quail (experimental). Ducks, chickens, geese and pigeons have been shown to be refractory to infection.[59,73] Virus can be recovered from blood from one to eight days after infection, and antibodies to the virus can be demonstrated using hemagglutination-inhibition or virus-neutralization assays. An attenuated-live virus vaccine has been reported to be effective in preventing the disease.[58]

Astrovirus

A duck astrovirus formerly called duck hepatitis virus II has been shown to be the cause of acute, fatal hepatitis in domestic ducklings in England. Affected ducklings developed polydipsia, polyuria and diarrhea for one to four days. Ducklings died within several hours of becoming ataxic. Mortality varied from 10% to 50% and was highest in ducklings less than two weeks old. Older ducks were refractory to disease. Natural outbreaks have been limited to commercial meat ducks raised in open settings. The fact that the virus occurs in ducks raised in fields suggests that free-ranging birds may be the source of this virus;[39,40] however, it has been documented only in domestic ducks in England.

The duck astrovirus appears to be antigenically distinct from those that infect chickens and turkeys.[43] The virus is shed in the feces for up to one week after infection, and ingestion is considered the major route of exposure. An enlarged, pale liver and spleen with hemorrhages and swollen congested kidneys are characteristic. Hepatic necrosis and bile duct hyperplasia are characteristic microscopic changes. Survivors are immune to subsequent infection, but only low levels of antibodies can be detected following natural or experimental infection. Infections are most easily diagnosed by electron microscopic examination of the liver for characteristic virus particles. Hyperimmune serum and an attenuated-live virus vaccine have been shown to be protective.[43]

An astrovirus has been observed in conjunction with diarrhea in young turkeys in the United States and Europe.[93,114] In two studies, astroviruses were detected by electron microscopy in the intestines of 2 of 160 (1.3%) and 2 of 192 (1%) ostriches.[130,131]

HEPADNAVIRIDAE

In mammals, highly host-specific hepadnaviruses have been shown to be associated with acute hepatitis, chronic hepatitis, cirrhosis, liver cancer and aplastic anemia. In birds, two antigenically distinct hepadnaviruses have been recovered from free-ranging Grey Herons in Germany and from domestic and free-ranging ducks throughout the world.[137] A virus antigenically related to duck hepatitis B virus (DHBV) was recovered from domestic geese, and it has been theorized that these birds were infected through the use of a contaminated vaccine of duck origin.[138] The hepadnaviruses that infect mammals are substantially different from the hepadnaviruses that infect birds; thus, there is no potential for cross-species transmission.[24,86]

Hepadnavirus has been recovered from the blood of free-ranging Mallard Ducks and a free-ranging duck species in Australia.[24,75,86] The virus is considered endemic in most flocks of domestic ducks. All species of ducks and geese that have been tested, except for Muscovy Ducks, have been shown to be susceptible to experimental infection.[85] The hepadnavirus that infects herons is considered endemic in Grey Herons in Germany. Ducks and geese were not susceptible to experimental infection with the virus recovered from herons. This strictly defined host-specificity is characteristic for the hepadnaviruses.[139]

Although hepadnaviruses have been associated with hepatitis and liver cancer in mammals, ducks naturally infected with hepadnavirus remain asymptomatic, and studies in ducks have failed to demonstrate a correlation between infection and liver cancer.[61] Mild hepatitis has been noted in some experimentally infected ducklings;[62] however, in another study, there was no difference in liver enzyme activity between infected and non-infected birds.[127]

Experimentally infected ducklings become viremic within several days to several weeks after inoculation and usually develop chronic infections. Young ducklings and goslings less than six weeks old are most susceptible to experimental infection. Older birds may develop a transient viremia and eliminate the virus.[46,62,127]

The primary route of natural transmission is from the chronically infected hen to embryos. Eggs from chronically infected hens contain high concentrations of the virus. In one study, virtually all of the ducklings from infected hens had persistent viremia.[127] Experimental infections can be induced through various parenteral routes. But horizontal transmission is probably of little importance in the natural transmission of this virus. Biting insects have not yet been demonstrated to transmit the virus.

Ducklings with congenitally derived infections are persistently infected and immunotolerant. Older ducklings infected with the virus do seroconvert.[46,148] Free-ranging ducks have been shown to have higher levels of virus in the blood stream and to develop higher levels of antibodies than do domestic ducks.[75] Antibodies were detected in 20% to 50% of free-ranging Grey Herons in Germany.[137,139] This finding suggests that the birds were infected after hatching or that congenitally infected heron chicks are not immunotolerant.

Chronically infected ducks circulate high levels of the virus in the blood, and envelope proteins of the virus can be detected by using an ELISA.[76] Hepadnavirus nucleic acid can be demonstrated in the blood or liver using viral-specific DNA probes.[19] In one study, percutaneous biopsy of the liver was used to obtain samples for diagnosing duck hepatitis B virus.[147]

Duck hepatitis B virus in captive ducks can be prevented by detecting and removing birds that are chronically infected. It is likely that a vaccine would be effective in preventing infections; however, because the virus has not been associated with a specific disease, a vaccine has not been developed.[148]

PARVOVIRIDAE

Parvoviruses are 18 to 26 nm nonenveloped virions that replicate in the nucleus of rapidly dividing cells and produce large intranuclear inclusion bodies. Depending on the genus of virus, replication may occur autonomously or a helper virus (usually an adenovirus or herpesvirus) may be required. The progression of disease is linked with its requirement for rapidly growing cells in which to replicate. The rapidly dividing cells that line the intestinal tract are the principal sites of parvovirus replication and disease.

DERZSY'S DISEASE
A parvovirus with a relatively restricted host range has been shown to cause hepatitis and high mortality in goslings

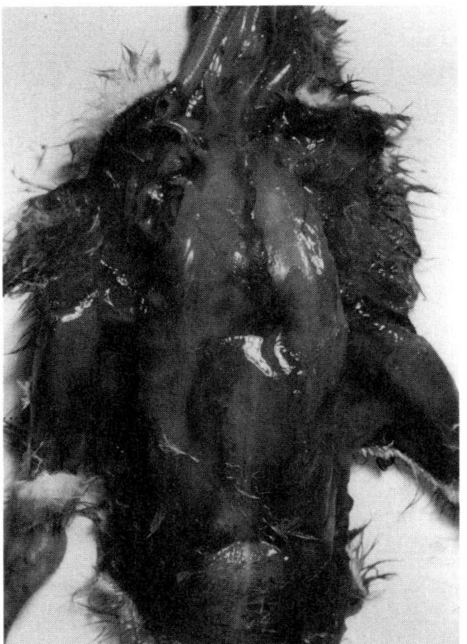

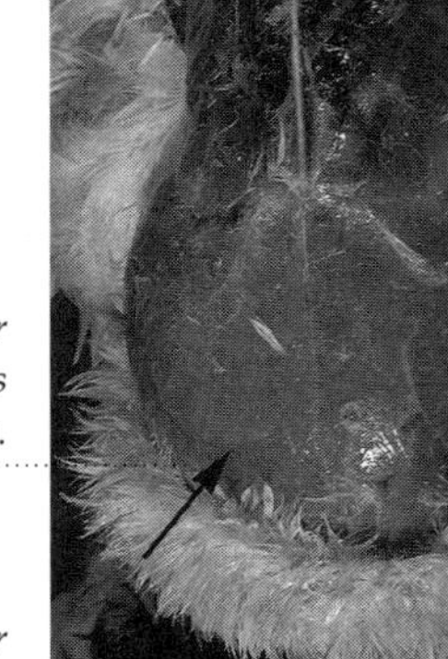

less than four weeks old. The disease has been called goose hepatitis, goose plague, infectious myocarditis, hepatonephritis-ascites, viral enteritis and goose influenza. The commonly used name is Derzsy's disease. This disease was first confirmed in geese in Hungary in 1964, and has subsequently been demonstrated in most European and Asian countries where goose farming is common.[22] The virus has been shown to naturally infect Muscovy Ducks and domestic breeds of geese. A group of Canada and Snow goslings from a zoological park that were hatched in a research laboratory working with this parvovirus developed

characteristic disease post-hatching.[54,121,122] This finding suggests that free-ranging geese could be susceptible to infection and be responsible for widespread dissemination of the virus; however, there have been no reports of this virus, or of antibodies to the virus, in free-ranging ducks or geese. Mute Swans can be infected experimentally.[32]

The clinical changes associated with the disease vary with the age of the infected bird. Recently hatched goslings that do not have maternally derived antibodies are the most susceptible. Experimentally, mortality has been shown to decrease from 100% to 0% as goslings exposed to the virus approach and reach 34 days of age.[122,123] Embryos infected in the oviduct may die before or soon after hatching.[22,68] Acute infections, characterized by several days of depression, anorexia, mild respiratory disease, hemorrhage under the skin, diarrhea, abdominal distention and ataxia followed by death are common in birds less than two weeks of age *Figure 16.7*. Initially, it was believed that this disease was caused by a reovirus.[22,42,122] Infections can appear explosive, with the majority of susceptible, naturally exposed goslings developing clinical signs and dying three to ten days after an initial case is noted. Epornitics generally run a three-week course.[22]

Goslings infected after they reach several weeks of age develop clinical changes similar to those that occur in young birds, but of longer duration and sometimes accompanied by ascites. Survivors exhibit retarded growth and may have poor feather development on the neck and thorax. Some surviving goslings have been described as being "naked" after infection. Older goslings develop subclinical infections. Secondary aspergillosis or salmonellosis may alter the clinical changes and make diagnosis more difficult.[42,66,124] A parvovirus serologically related to the one that affects goslings was recovered from Muscovy Ducks in Japan. Affected ducks experienced weakness, poor feathering and high mortality.[141]

EXPERIMENTAL INFECTIONS

One- to four-day-old goslings inoculated intramuscularly developed an acute illness characterized by anorexia and polydipsia three to six days, and death four to seven days, after infection. Some goslings infected at two weeks of age developed anorexia, mild conjunctivitis, a nasal discharge and necrotic lesions in the mouth; death occurred ten days after inoculation. Others developed a more protracted disease associated with conjunctivitis, lesions in the oral cavity, nasal discharge, feather abnormalities, poor weight gains, hyperemia of the skin and swollen uropygial glands; some died two to ten weeks after infection. Older goslings that recovered grew new feathers but remained stunted. Disease developed in contact controls several days later than in the experimentally infected goslings, suggesting a rapid excretion of the virus after infection. By four weeks of age, goslings infected by the intramuscular route remained asymptomatic, while those inoculated intravenously developed clinical changes similar to those described in younger birds.[122,123] When a group of goslings, ducklings and chickens were infected intramuscularly, the goslings died five to ten days after inoculation, while the ducks and chickens remained clinically normal and did not seroconvert.[42] In another study, 72% of three-day-old Muscovy Ducklings infected intramuscularly died by 12 days after infection.[141]

EPIZOOTIOLOGY AND PATHOLOGY

Derzsy's disease virus is transmitted principally through ingestion of contaminated feces, food or drinking water. Infected hens can transmit the virus to their embryos.[22,122,123] Virus is shed in the feces soon after infection and can continue for up to six months.[32] Recovered birds are considered to be latently infected and can disseminate the virus through a flock.

The most common gross lesions noted in young birds that die acutely are an enlarged heart and liver with grayish-white areas of necrosis. Hemorrhage may also be noted in the intestinal tract, pancreas and bursa. Chronically infected birds may develop hyperemia of the intestinal mucosa, ascites, and accumulation of a fibrinous material on the heart, liver and intestines *Figure 16.8*.[37] Hepatic necrosis, pancreatic necrosis, serofibrinous enteritis, pericarditis, perihepatitis and pulmonary edema may be seen microscopically. Intranuclear inclusion bodies have been reported most commonly in the liver, but may also be noted in the spleen, heart, intestines, thymus and thyroid gland. In one report, degeneration of the cardiac, skeletal and smooth muscles was common.[97] Another investigator described vacuolation of hepatocytes and hepatic necrosis with no change in the muscles.[121,122]

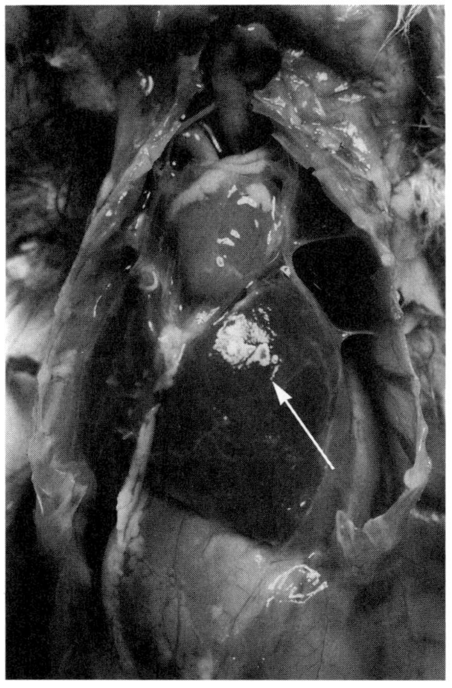

Late in the breeding season, the hen transfers decreasing levels of antibodies to her embryos, and young birds are increasingly susceptible to infection.[70] Older birds (greater than four weeks) develop a resistance to natural disease that is independent of antibodies *Figure 16.9*.[124] Recovered birds are considered to be persistently infected.[68]

Parvovirus infections can be diagnosed by demonstrating an increase in virus-neutralizing antibodies or by recovering the virus from the feces, liver, pancreas or heart. The demonstration of antibodies in birds greater than four to six weeks of age indicates that an active infection has occurred, because maternally derived antibodies are not detectable after three weeks of age. Other diseases that can cause similar clinical or gross changes include reovirus infection, nephroenteritis, chlamydiosis, aspergillosis and salmonellosis.[32,68]

IMMUNITY AND DIAGNOSIS

Agar-gel immunodiffusion, or the more sensitive virus-neutralization, assay can be used to detect antibodies. Naturally and experimentally infected goslings and ducklings develop antibodies to the virus five to ten days after infection; neutralizing antibodies may persist for over 43 months.[42,66] High levels of maternally transmitted neutralizing antibodies protect chicks from

CONTROL

Parvoviruses are resistant to many common disinfectants and to a pH range of 3 to 9. They are heat stable at 37°C for one hour, 56°C for up to three hours, and 60°C for up to 30 minutes.[42,69] Formalin and oxidizing agents (sodium hypochlorite and stabilized chlorine dioxide) will inactivate the parvoviruses that have been studied. A solution of 0.5% formaldehyde destroys the virus in 15 minutes.

Parvoviral disease is best prevented in geese and ducks through vaccination.[54] Geese injected with an attenuated-live virus vaccine will seroconvert within three weeks post-vaccination.[36] Intra-

disease during the first several weeks of life. Antibody levels begin to wane in chicks by two weeks, and are undetectable by three weeks post-hatching.[66]

muscular vaccination four to six weeks prior to egg-laying will provide protective levels of antibodies to the chicks hatched early in the breeding season. A booster vaccination may be needed late in the breeding season as the antibody levels in the hens decrease.[68] The flock should be revaccinated annually. Inactivated vaccines have not been shown to produce sufficient or sustained antibody levels.[125]

Goslings that do not have maternally derived antibodies can be provided with temporary immunity by injecting them intramuscularly or subcutaneously with hyperimmune serum (1 ml) at one day of age.[54,122] Injecting the yolk of a developing egg with hyperimmune serum has been shown to provide chicks with protection from even intravenous challenge.[124,125]

PARVOVIRUS IN OTHER BIRDS

A virus that resembled parvovirus was described in adult canaries with myocarditis and encephalitis. Clinical changes ranged from acute death with no premonitory signs to death following a brief period of depression and weight loss. Some affected canaries had no gross lesions; others had an enlarged spleen. Intranuclear inclusion bodies were detected in the brain and in the heart, where they were associated with myocarditis.[50]

A virus with morphologic characteristics similar to parvovirus was recovered from the feces of cockatoos experiencing anorexia and intermittent diarrhea. The affected birds had concomitant psittacine beak and feather disease virus infections.[111]

Parvovirus-like particles were detected in the liver of a pigeon with a concomitant adenovirus infection.[30] Similar particles have been seen in pigeons with enteritis.[7]

A parvovirus serologically unrelated to the virus found in geese has been associated with a stunting syndrome in young broiler chickens. Experimentally infected young chickens develop a mucoid, gaseous, yellow diarrhea three to five days after inoculation, exhibit poor weight gains and fail to thrive. Parvovirus infections have also been reported in domestic turkeys.[67,71]

RHABDOVIRIDAE

The family Rhabdoviridae consists of two genera: *Lyssavirus*, which contains the virus that causes rabies, and *Vesiculovirus*, which causes blistering diseases in mammals and experimentally in birds.

Birds, like all warm-blooded animals, are susceptible to the rabies virus. However, infected birds generally remain asymptomatic and are not considered a natural reservoir for this virus. Naturally acquired rabies infections (as documented by virus recovery or detection of antibodies to the virus) have been documented in Anseriformes, Falconiformes, Galliformes, Strigiformes and Passeriformes. Experimental infections have been documented in these birds as well as in Ciconiiformes, Galliformes, Columbiformes and canaries.[35,74,100,150] Infections can occur when birds are bitten by infected prey or from ingestion of infected prey or carrion. Species susceptibility appears to vary, with predatory and scavenging birds being more susceptible to natural infection than are pigeons and chickens.[150]

No clinical changes associated with rabies have been seen in naturally infected birds, and most experimentally infected birds remain asymptomatic or recover. Some experimentally infected chickens became excited (aggressive, increased vocalization, wing fluttering), followed by anorexia, depression, paralysis and death one to three days later. Many of the affected chickens seroconverted and recovered. The incubation period in experimentally infected chickens that did develop clinical signs ranged from 12 days to 12 months.[126] Clinical infections may last from 2 to 42 days, with birds developing behavioral changes and progressive neurologic signs.[32]

EXPERIMENTAL INFECTIONS

Experimental infections can be induced by injecting the virus or by feeding birds rabid prey.[6,8,18] A Great Horned Owl was experimentally infected with rabies by feeding the animal an infected skunk. The owl remained clinically normal and seroconverted 27 days after ingesting the prey. The virus was isolated from oral swabs (saliva), corneal smears, the brain and eyes. Viral antigen was demonstrated by fluorescent antibody staining for 52 days after infection in pharyngeal mucosa and for 86 days after infection in corneal smears of an owl that ingested infected prey.[64] In one study, chickens bitten by rabid dogs were shown to be infected and developed aggressive behavior and neurologic signs.[108] Experimentally infected owls and crows remained asymptomatic but

seroconverted following oral inoculation.[35,118] Virus was recovered from six of nine crows fed rabid mice.[33]

Chickens inoculated intracranially with rabies virus rapidly produced central nervous system-bound antibodies. The self-limiting nature of the virus in avian species is thought to be due to a rapid production of antibodies.[126] This antibody response appears to have little effect on the replication of the virus during the primary infection, but interferes with subsequent spread of the virus within the host.[64]

EPIZOOTIOLOGY

In one study, rabies virus was recovered from 25 of 44 free-ranging raptors and scavengers from an area in Germany experiencing a rabies epizootic.[100] Virus isolation has also been reported from Common Buzzards, Goshawks, ducks, Red Kites and a Barn Owl in Europe.[32] Some reports indicate that free-ranging populations of raptors and other carrion-feeding birds in United States frequently have antibodies to rabies virus.[34] Other studies indicate that antibodies to rabies virus are rare in these species of birds.[129] In one serologic survey of 343 free-ranging birds representing 22 avian species in the United States, antibodies to the virus were detected in 23% of 65 predatory and 2.9% of 278 nonpredatory birds.[34] In another study, 23% of raptors and 8% of the non-predatory scavengers, including starlings, crows and ravens, had rabies virus antibodies.[35] Sixty of 332 crows were found to have rabies virus antibodies.[117] These findings suggest that viral exposure occurs through contact with infected prey species.[34] However, in another study in the United States, rabies virus antibodies were not detected in any of 53 free-ranging raptors. It was theorized that the difference in reported seroprevalence of infections in raptors was due to the use of a 1:2, instead of a 1:16, dilution of serum as a significant titer.[129]

Rabies virus can be detected in the oral secretions of experimentally infected birds. However, birds have never been shown to transmit this virus to humans or other mammals and are considered of minimal importance in the epizootiology of rabies. Because raptors can be infected by this virus, those that have been recently removed from the wild should be handled with care. Any person intending to work with free-ranging wildlife, including raptors, should be vaccinated for rabies.

OTHER RHABDOVIRUSES

Vesicular stomatitis virus (VSV) can occur in free-ranging turkeys during epizootics in mammals. In a year when VSV was epidemic in cattle and horses, 35% of free-ranging turkeys were found to have antibodies to the virus.[145] Neutralizing antibodies to the virus have also been documented in free-ranging vultures in South America.[145] Clinical changes have not been described in naturally infected birds; however, experimentally infected geese developed vesicular lesions on the tongue and feet, and experimentally infected chickens and ducks developed vesicular lesions on the tongue.[133]

Flanders virus has been recovered from the blood of asymptomatic free-ranging Passeriformes in the United States.[72,153] In one study, the virus was recovered from 213 of 1,333 (16%) free-ranging birds, and antibodies to the virus were detected in adult and juvenile birds.[113]

REFERENCES

1. Aguirre AA, McLean RG, Cook RS: Experimental inoculation of three arboviruses in black-bellied whistling ducks (*Dendrocygna autumnalis*). J Wildl Dis 28:521-525, 1992.

2. Alexander DJ, Gough RE: Isolation of avian infectious bronchitis virus from experimentally infected chickens. Res Vet Sci 23:344-347, 1977.

3. Barnard BJH, Buys SB, du Prees JH, et al: Turkey meningo-encephalitis in South Africa. Onderstepoort J Vet Res 47:89-94, 1980.

4. Barr DA, Reece RL, O'Rourke D, et al: Isolation of infectious bronchitis virus from a flock of racing pigeons. Aust Vet J 65:228, 1988.

5. Bayliss CD, Peters RW, Cook JKA: A recombinant fowlpox virus that expresses the VP2 antigen of infectious bursal disease virus induces protection against mortality caused by the virus. Arch Virol 120:193-205, 1991.

6. Bell JF, Moore GJ: Susceptibility of carnivores to rabies virus administered orally. Am J Epid 93:176-182, 1971.

7. Bergmann V, Kiupel H: Einschlußkorperchen enteritis bei Tauben, hervorgerufen durch Adeno- und Parvovirus. Archiv Exper Vet Med 36:445-453, 1982.

8. Blancou J, Samudo A: Echec de l'infection rabique experimentale de la buse variable. Revue de Med Vet 130:427, 1979.

9. Bond JO, Lewis FY, Jennings WL, et al: Transovarian transmission of hemagglutination-inhibition antibody to St. Louis encephalitis virus in chickens. Am J Trop Med Hyg 14:1085-1089, 1965.

10. Buescher EL, Scherer WF, Rosenberg MZ, et al: Immunologic studies of Japanese encephalitis virus in Japan. J Immunol 83:605-613, 1959.

11. Cadman HF, Kelly PJ, Zhou R, et al: A serosurvey using enzyme-linked immunosorbent assay for antibodies against poultry pathogens in ostriches (*Struthio camelus*) from Zimbabwe. Avian Dis 38:621-625, 1994.

12. Calnek BW, Luginbuhl RE, Helmboldt CF: Avian encephalomyelitis. *In* Calnek BW, et al (eds): Diseases of Poultry. Ames, Iowa State University Press, 1991, pp 520-531.

13. Calnek BW, Taylor PJ, Sevoiam M: Studies on avian encephalomyelitis. IV. Epizootiology. Avian Dis 4:325-347, 1960.

14. Chalmers WSK, Woolcock PR: The effect of animal sera on duck hepatitis virus. Avian Pathol 13:727-732, 1984.

15. Chamberlain RW, Kissling RE, Stamm DD, et al: Virus of St. Louis encephalitis in three species of wild birds. Am J Hyg 65:110-118, 1957.

16. Clark GG, Pretula HL, Jakubowsky T, et al: Arbovirus surveillence in Illinois in 1976. Mosquito News 37:389-395, 1977.

17. Cook JKA: Recovery of infectious bronchitis virus from eggs and chicks produced by experimentally infected hens. J Comp Path 81:203-211, 1971.

18. Correa-Giron RA, Sulkin ES: The infectivity and pathogenesis of rabies virus administered orally. Am J Epid 91:203-215, 1970.

19. Cova L, Hantz O, Arliaud-Gassin M, et al: Comparative study of DHV DNA levels and endogenous DNA polymerase activity in naturally infected ducklings in France. J Virol Meth 10:251-260, 1985.

20. Davis D: Temperature and pH stability of duck hepatitis virus. Avian Pathol 16:21-30, 1987.

21. Davis D, Hannant D: Fractionation of neutralizing antibodies in serum of ducklings vaccinated with live duck hepatitis virus vaccine. Res Vet Sci 43:276-277, 1987.

22. Derzsy D: A viral disease of goslings. I. Epidemiological, clinical, pathological and aetiological studies. Acta Vet Acad Sci Hung 17:443-448, 1967.

23. Deshmukh DR, Larsen CT, Dutta SK, et al: Characterization of pathogenic filtrate and viruses isolated from turkeys with bluecomb. Am J Vet Res 30:1019-1025, 1969.

24. Dixon RJ, Jones NF, Campbell M, et al: The Australian maned duck hepatitis B virus. *In* Marion PL, Schaller H (eds): Hepatitis B Viruses. Cold Spring Harbor, New York, Cold Spring Harbor Laboratory, 1989.

25. Eccles MA: The role of birds in the spread of foot-and-mouth disease. Bull Off Int Epiz 18:118-148, 1939.

26. Fabricant J, Rickard CG, Levine PP: The pathology of duck virus hepatitis. Avian Dis 1:256-275, 1957.

27. Fadly AM, Nazerian K: Pathogenesis of infectious bursal disease in chickens infected with virus at various ages. Avian Dis 27:714-723, 1983.

28. Farmer H, Chalmers WSK, Woolcock PR: The duck fatty kidney syndrome: An aspect of duck viral hepatitis. Avian Pathol 16:277-236, 1987.

29. Frank RK, Carpenter JW: Coronaviral enteritis in ostrich (*Struthio camelus*) chicks. J Zoo Wildlife Med 23:103-107, 1992.

30. Fuchs A, Weissenbock H: Inclusion body hepatitis in psittacine birds and pigeons: Comparative histological and ultrastructural findings. Proc Europ Conf Avian Med Surg, 1993, pp 552-557.

31. Gelb J: Infectious bronchitis. *In* Purchase HG, et al (eds): A Laboratory Manual for the Isolation and Identification of Avian Pathogens 3rd ed. Dubuque, Kendall/Hunt Publishing Co, 1989, pp 124-127.

32. Gerlach H: Viruses. *In* Ritchie BW, Harrison GJ, Harrison LR (eds): Avian Medicine: Principles and Application. Lake Worth, Wingers Publishing, 1994, pp 862-948.

33. Gough PM: Oral transmission of rabies virus in wildlife. Federal Research in Progress Database November: 0039219, 1983.

34. Gough PM, Dierks RE: Passive haemagglutination test for antibodies against rabies virus. Bull World Hlth Org 45:471-475, 1971.

35. Gough PM, Jorgenson RD: Rabies antibodies in sera of wild birds. J Wildl Dis 12:392-395, 1976.

36. Gough RE: Application of the agar gel precipitin and virus neutralization tests to the serological study of goose parvovirus. Avian Pathol 13:501-509, 1984.

37. Gough RE: Goose parvovirus infection. *In* Calnek BW, et al (eds): Diseases of Poultry. Ames, Iowa State University Press, 1991, pp 684-690.

38. Gough RE, Alexander DJ, Collins MS, et al: Routine virus isolation or detection in the diagnosis of diseases in birds. Avian Pathol 17:893-907, 1988.

39. Gough RE, Borland ED, Keymer IF, et al: An outbreak of duck hepatitis type II in commercial ducks. Avian Pathol 14:227-236, 1985.

40. Gough RE, Collins MS, Borland DE, et al: Astrovirus-like particles associated with hepatitis in ducklings. Vet Rec 114:279, 1984.

41. Gough RE, et al: Pheasant coronavirus in Great Britain. Proc World Vet Poult Assoc, 1991, pp 308-314.

42. Gough RE, Spackman D, Collins MS: Isolation and characterization of a parvovirus from goslings. Vet Rec 108:399-400, 1981.

43. Gough RE, Stuart JC: Astroviruses in ducks (duck virus hepatitis virus type II). *In* McFerran JB, McNulty MS (eds): Virus Infection of Birds. New York, Elsevier Science Publishers, 1993, pp 505-508.

44. Gylstorff I: Coronavirus. *In* Gylstorff I, Grimm F (eds): Vogelkrankheiten. Stuggart, Verlan Eugen Ulmer, 1987, p 263.

45. Haider SA, Calnek BW: In vitro isolation, propagation and characterization of duck hepatitis virus type III. Avian Dis 23:715-729, 1979.

46. Halpern MS, Mason WS, Coates L, et al: Humoral immune responsiveness in duck hepatitis B virus-infected ducks. J Virol 61:916-920, 1987.

47. Hammon WM, Sather GE, McClure HE: Serologic survey of Japanese encephalitis virus infection in birds in Japan. Am J Hyg 67:118-133, 1958.

48. Hanson LE, Tripathy DN: Oral immunization of ducklings with attenuated duck hepatitis virus. Dev Biol Stand 33:357-363, 1976.

49. Hardy JL, Scherer WF, Warner DW: Arbovirus neutralizing substances in avian plasmas. II. Studies on their mechanism of release and specificity after collection of plasmas by shooting and cardiac puncture in birds. Am J Trop Med Hyg 12:867-875, 1964.

50. Helfer DH, et al: Myocarditis-encephalopathy in canaries. Proc 30th West Poultry Dis Conf, 1981, p 92.

51. Helmboldt CF, Garner E: Experimentally induced Gumboro disease (IBD). Avian Dis 8:561-575, 1964.

52. Hess AD, Holden P: The natural history of the arthropod-borne encephalitides in the United States. Ann N York Acad Sci 70:294-311, 1958.

53. Hirai K, Hitchner SB, Calnek BW: Characterization of a new coronavirus-like agent isolated from parrots. Avian Dis 23:515-525, 1978.

54. Hoekstra J, Smit T, Van Brakel C: Observations on host range and control of goose virus hepatitis. Avian Pathol 2:169-178, 1973.

55. Hwang J: Early induction of resistance in ducklings against virus hepatitis. Am J Vet Res 32:2095-2097, 1971.

56. Hwang J: Active immunization against duck hepatitis virus. Am J Vet Res 31:805-807, 1972.

57. Hwang J: Susceptibility of poultry to duck hepatitis viral infection. Am J Vet Res 35:477-479, 1974.

58. Ianconescu M: Arbovirus infections. *In* Calnek BW, et al (eds): Diseases of Poultry. Ames, Iowa State University Press, 1991, pp 674-679.

59. Ianconescu M, Aharonovici A, Samberg Y: The Japanese quail as an experimental host for turkey meningoencephalitis virus. Refu Vet 31:100-108, 1974.

60. Ide PR: The sensitivity of some avian viruses to formaldehyde fumigation. Can J Comp Med 43:211-216, 1979.

61. Imazeki F, Yaginuma K, Omata M, et al: Integrated structures of duck hepatitis B virus DNA in hepatocellular carcinoma. J Virol 62:861-865, 1988.

62. Jilbert AR, Wu TT, England JM, et al: Rapid resolution of duck hepatitis B virus infections occur after massive hepatocellular involvement. J Virol 66:1377-1388, 1992.

63. Jones RC, Jordon FTW: Persistence of virus in the tissues and development of the oviduct in the fowl following infection at day old with infectious bronchitis virus. Res Vet Sci 13:52-60, 1972.

64. Jorgenson RD, Gough PM: Experimental rabies in a great horned owl. J Wildl Dis 12:444-447, 1976.

65. Karabatsos N: International Catalogue of Arboviruses 1985 Including Certain Other Viruses of Vertebrates. Am Soc Trop Med Hyg, 1985.

66. Kisary J: Immunological aspects of Derzsy's disease in goslings. Avian Pathol 6:327-334, 1977.

67. Kisary J: Indirect immunofluorescence as a diagnostic tool for parvovirus infection of chickens. Avian Pathol 14:269-273, 1985.

68. Kisary J: Derzsy's disease of geese. *In* McFerran JB, McNulty MS (eds): Virus Infections of Birds. New York, Elsevier Science Publishers, 1993, pp 157-162.

69. Kisary J, Derzsy D: Viral disease of goslings. IV. Characterization of the causal agent in tissue culture system. Acta Vet Acad Sci Hung 24:287-292, 1974.

70. Kisary J, Derzsy D, Horvath E: Jaz unllibianfluenza elleni szikimmunitas

jarvanytani jelentosege. Magyar Alla-torrosök Lapsa 30:721-723, 1975.

71. Kisary J, Nagy B, Bitay Z: Presence of parvoviruses in the intestines of chickens showing stunting syndrome. Avian Pathol 13:339-343, 1984.

72. Kokernot RH, Hayes J, Will RL, et al: Arbovirus studies in the Ohio-Mississippi Basin 1964-1967. III Flanders Virus. Am J Trop Med Hyg 18:762-767, 1969.

73. Komarov A, Kalmar E: A hitherto undescribed disease - turkey meningo-encephalitis. Vet Rec 72:257-261, 1960.

74. Krans R, Clairmont P: Über experimentelle lyssa bei vogeln. Zeitschr Hyg Infektionskraukh 34:1-30, 1900.

75. Lambert V, Cora L, Chevallier P, et al: Natural and experimental infection of wild mallard ducks with duck hepatitis B virus. J Med Virol 72:417-420, 1991.

76. Lambert V, Fernholz D, Sprengel R, et al: Virus-neutralizing monoclonal antibody to a conserved epitope on the duck hepatitis B virus pre-s protein. J Virol 64:1290-1297, 1990.

77. Lister SA, et al: Outbreaks of nephritis in pheasants (*Phasianus colchinus*) with a possible coronavirus etiology. Vet Rec 117:612-613, 1985.

78. Lord RD, Works TH, Coleman PH, et al: Virological studies of avian hosts in the Houston epidemic of St. Louis encephalitis in 1964. Am J Trop Med Hgy 22:662-671, 1973.

79. Louzis C, Gillet JP, Irgens K, et al: La maladie de Gumboro: apparition chez le fiasan d'elerage. Bull Mens Soc Prat Fr 63:785-789, 1979.

80. Luby JP, Miller G, Gardner P, et al: The epidemiology of St. Louis encephalitis in Houston, Texas, 1964. Am J. Epidemiol 86:584-597, 1967.

81. Lucio B, Hitchner SB: Infectious bursal disease emulsified vaccine: Effect upon neutralizing antibody levels in the dam and subsequent protection of the progeny. Avian Dis 23:466-478, 1979.

82. Ludwig GV, Cook RS, McLean RG, et al: Viremic enhancement due to transovarially acquired antibodies to St. Louis encephalitis virus in birds. J Wildl Dis 22:326-334, 1986.

83. Lvov DK, Iljicev VD: Avian Migrations and Transport of Infectious Agents. Moskva, Nauka, 1979.

84. Mack TM, Brown BF, Sudia WD, et al: Investigation of an epidemic of St. Louis encephalitis in Danville, Kentucky, 1964. J Med Ent 4:70-76, 1967.

85. Marion PL, Cullen JM, Azcarraga RR, et al: Experimental transmission of duck hepatitis B virus to Pekin ducks and to domestic geese. Hepatology 7:724-731, 1987.

86. Mason WS, Taylor JM, Seal G, et al: An HBV-like virus of domestic ducks. *In* Szmuness W, et al (eds): Viral Hepati-tis. New York, Franklin Institute Press, 1981, pp 107-118.

87. McFerran JB, McNulty MS (eds): Virus Infections of Birds. London, Elsevier Science Publishers, 1993.

88. McFerran JB, McNulty MS, McKillup ER, et al: Isolation and serological studies with infectious bursal disease viruses from fowl, turkeys, and ducks: Demonstration of a second serotype. Avian Pathol 9:395-404, 1980.

89. McLean RG, Bowen GS: Vertebrate hosts. *In* Monath TP (ed): St. Louis Encephalitis. Washington DC, Am Public Health Assoc, 1980, pp 381-450.

90. McLean RG, Francy DB, Campos EG: Experimental studies of St. Louis encephalitis virus in vertebrates. J Wildl Dis 21:85-93, 1985.

91. McLean RG, Scott TW: Avian hosts of St. Louis encephalitis virus. Proc 8th Bird Control Seminar, 1979, pp 143-155.

92. McNulty MS, Conner TJ, McNeilly F, et al: Biological characterisation of avian enteroviruses and enterovirus-like viruses. Avian Pathol 19:75-87, 1990.

93. McNulty MS, Curran WL, McFerran JB: Detection of astroviruses in turkey faeces by direct electron microscopy. Vet Rec 106:561, 1980.

94. McOrist S, Madill D, Adamson M, et al: Viral enteritis in cockatoos (*Cacatua* sp). Avian Pathol 20:531-539, 1991.

95. Meulemans G, Halen P: Efficacy of some disinfectants against infectious bursal disease virus and avian reovirus. Vet Rec 111:412-413, 1982.

96. Miyamae T: Distribution and persistence of avian encephalomyelitis viral antigens in internal organs of naturally infected chickens. Am J Vet Res 35:1229-1234, 1974.

97. Nagy Z, Derzsy D: Viral disease of goslings. II. Microscopic lesions. Acta Vet Acad Sci Hung 18:3-18, 1968.

98. Nair M, et al: Studies on broiler's IBV and IB-like virus from guinea fowl. Proc World Vet Poult Assoc, 1991, pp 302-307.

99. Nawathe DR, Onunkwo O, Smith IN: Serological evidence of infection with the virus of infectious bursal disease in wild and domestic birds in Nigeria. Vet Rec 102:144, 1978.

100. Paarmann E: Ein bietrag zur lyssa der vogel. Z Hyg Infektionskrankh 141:103-109, 1955.

101. Pascucci S, et al: Characterization of a coronavirus-like agent isolated from coturnix quail. Proc VIII Intern Cong WVPA, 1985, p 52.

102. Pomeroy KA, Nagaraja KV: Viral enteric infections. Coronaviral enteritis of turkeys (bluecomb disease). *In* Calnek BW, et al (eds): Diseases of Poultry 9th ed. Ames, Iowa State University Press, 1991, pp 619-627.

103. Pomeroy KA, Patel BL, Larsen CT, et al: Combined immunofluorescence and transmission electron microscopic studies of sequential intestinal samples from turkey embryos and poult infected with turkey enteritis coronavirus. Am J Vet Res 39:1348-1354, 1978.

104. Priz NN: Comparative study of virus hepatitis in animals (dogs and ducks) using different routes of inoculation. Vopr Virusol 6:696-700, 1973.

105. Razenhoter ER, Alexander ER, Beadle LD, et al: St. Louis encephalitis in Calvert City, Kentucky, 1955: An epidemiologic study. Am J Hyg 65:174-181, 1957.

106. Reeves WE, Sturgeon JM, French EM, et al: Transovarian transmission of neutralizing substances to western equine and St. Louis encephalitis viruses by avian hosts. J Infect Dis 95:168-178, 1954.

107. Reid HW, et al: Experimental louping-ill virus infection of black grouse (*Tetrao tetrix*). Arch Virology 78, 1983.

108. Remlinger P, Bailly J: La rage du coq. Ann Inst Pasteur 49:153-167, 1929.

109. Reuss U: Virusbiologische untersuchungen bei der enternhepatitis. Zentralbl Vet 6:808-815, 1959.

110. Rispens BH: Some aspects of control of infectious hepatitis in ducklings. Avian Dis 13:417-426, 1969.

111. Ritchie BW, Niagro FD, Latimer KS, et al: Hemagglutination by psittacine beak and feather disease virus and use of hemagglutination-inhibition for detection of antibodies against the virus. Am J Vet Res 52:1810-1815, 1991.

112. Rosenberger JK: Infectious bursal disease. *In* Purchase HG, et al (eds): A Laboratory Manual for the Isolation and Identification of Avian Pathogens 3rd ed. Dubuque, Kendall/Hunt Publishing Co, 1989, pp 165-166.

113. Rowley WA, Hunt GJ, Dorsey DC: Flanders virus activity in Iowa, USA. J Med Entomol 20:409-413, 1983.

114. Saif LJ, Saif YM, Theil KW: Enteric viruses in diarrheic turkey poults. Avian Dis 29:798-811, 1985.

115. Samberg Y, Ianconescu M, Hornstein K: Epizootiological aspects of turkey meningo-encephalitis. Refu Vet 29:103-110, 1972.

116. Sato G, Choi WP: A note on duration of neutralizing antibody against avian encephalomyelitis virus in chickens. Jap J Vet Res 20:45-49, 1972.

117. Schaefer JM: The common crow as a sentinel species of rabies in wildlife populations. Iowa State University, 1983.

118. Schafer W: Vergleichende sero-immunologische Untersuchungen über die viren der influenxa und klassichen Geflugelpest. Z Naturforsch 10:81-91, 1955.

119. Scherer WF, Bueshcer EL, Flemings MB, et al: Ecologic studies of Japanese

encephalitis virus in Japan. Am J Trop Med Hyg 8:665-677, 1959.

120. Scherer WF, Hardy JL, Gresser I, et al: Arbovirus-neutralizing substances in avian plasmas. Their higher prevalence after collection of birds by shooting and cardiac puncture than after netting and jugular venipuncture. Am J Trop Med Hyg 13:859-866, 1964.

121. Schettler CH: Goose virus hepatitis in the Canada goose and snow goose. J Wildl Dis 7:147-148, 1971.

122. Schettler CH: Isolation of a highly pathogenic virus from geese with hepatitis. Avian Dis 15:323-325, 1971.

123. Schettler CH: Virus hepatitis of geese. II. Host range of goose hepatitis virus. Avian Dis 15:809-823, 1971.

124. Schettler CH: Virus hepatitis of geese 3: Properties of the causal agent. Avian Pathol 2:179-193, 1973.

125. Schettler CH: Large scale investigation on the serological response of breeder geese and Muscovy ducks to different types of live attenuated and inactivated goose hepatitis virus strains GHV-SHM 319. Proc Intern Conf Breeding Geese Produc, 1979, pp 197-202.

126. Schneider LG, Burtscher H: Untersuchungen über die Pathogenese der Tollwut bei Huhnern nach intracerebraler Infektion. Infektion Zentbl Vet Med B14:598-624, 1967.

127. Schodel F, Weimer T, Fernholz D, et al: The biology of avian hepatitis B viruses. *In* MacLachlan A (ed): Molecular Biology of the Hepatitis B Virus. Cleveland, CRC Press, 1991, pp 53-80.

128. Sevoian M, Levine PP: Effects of infectious bronchitis on the reproductive tracts, egg production, and egg quality of laying chickens. Avian Dis 1:136-164, 1957.

129. Shannon LM, Poulton JL, Emmons RW, et al: Serological survey for rabies antibodies in raptors from California. J Wildl Dis 24:264-267, 1988.

130. Shivaprasad HL: Neonatal mortality in ostriches: An overview of possible causes. Proc Assoc Avian Vet, 1993, pp 282-285.

131. Shivaprasad HL, Woolcock PR, Chin RP, et al: Identification of viruses from the intestines of ostriches. Proc Assoc Avian Vet, 1994, p 442.

132. Simpson DIH: Toga- and Flaviviruses. *In* McFerran JB, McNulty MS (eds): Virus Infections of Birds. New York,

Elsevier Science Publishers, 1993, pp 231-241.

133. Skinner HH: Infection of chickens and chick embryos with the viruses of foot-and-mouth disease and of vesicular stomatitis. Nature 174:1052-1053, 1954.

134. Sooter CA, Schaeffer M, Gorrie R, et al: Transovarian passage of antibodies following naturally acquired encephalitis infection in birds. J Infect Dis 95:165-167, 1954.

135. Spackman D, Cameron IRD: Isolation of infectious bronchitis virus from pheasants. Vet Rec 113:354-355, 1983.

136. Spalding MG, McLean RG, Burgess JH, et al: Arboviruses in water birds (Ciconiiformes, Pelecaniformes) from Florida. J Wildl Dis 30:216-221, 1994.

137. Sprengel R, Kaleta EF, Will H: Isolation and characterization of a hepatitis B virus endemic in herons. J Virol 62:3832-3839, 1988.

138. Sprengel R, Schneider R, Marion R, et al: Comparative sequence analysis of defective and infectious avian hepadnaviruses. Nucl Ac Res 19:4289, 1991.

139. Sprengel R, Will H: Duck hepatitis B virus. *In* Darai G (ed): Virus Disease in Laboratory and Captive Animals. Boston, Martinus Nijhoff Publishing, 1988, pp 363-386.

140. Stamm DD: Relationship of birds and arboviruses. Auk 83:84-97, 1966.

141. Takehara K, Hyakutake K, Imamura T, et al: Isolation, identification, and plaque titration of parvovirus from muscovy ducks in Japan. Avian Dis 38:810-815, 1994.

142. Toth TE: Studies of an agent causing mortality among ducklings immune to duck virus hepatitis. Avian Dis 13:834-846, 1969.

143. Toth TE: Alum-precipitated and sodium hydroxide conjugated vaccines for duck virus hepatitis: Immunologic and serologic response of susceptible and low-level parentally immune white Pekin ducklings. Avian Dis 13:249-259, 1972.

144. Toth TE, Norcross NL: Humoral immune response of the duck to duck hepatitis virus: Virus-neutralizing versus virus-precipitating antibodies. Avian Dis 25:17-28, 1981.

145. Trainer DO, Glazener WC, Hanson RP, et al: Infectious disease exposure in a wild turkey population. Avian Dis 12:208-214, 1968.

146. Van Steenis G: Survey of various avian species for neutralizing antibody and susceptibility to avian encephalomyelitis virus. Res Vet Sci 12:308-311, 1971.

147. Varagona G, Ellis LA, Moore D, et al: A percutaneous liver biopsy technique in ducks (*Anas platyrhynchos*) experimentally infected with duck hepatitis B virus. Lab Animal 24:254-257, 1991.

148. Vickery K, Freiman JS, Dixon RJ, et al: Immunity in Pekin ducks experimentally and naturally infected with duck hepatitis B virus. J Med Virol 28:231-236, 1989.

149. Vindevogel H: Resistance du pigeon aux virus de la maladie de Gumboro. Ann Med Vet 123:285-286, 1979.

150. Von Lote J: Beitrange zue kenntris der experimentellen lyssa der vogel. Zentr Bakterviol Parasitenk 35:741-744, 1904.

151. Westbury HA, Sinkovic B: The pathogenesis of infectious avian encephalomyelitis. II. The effect of immunosuppression on the disease. Aust Vet J 54:72-75, 1978.

152. Westbury HA, Sinkovic B: The pathogenesis of infectious avian encephalomyelitis. IV. The effect of maternal antibody on the development of disease. Aust Vet J 54:81-85, 1978.

153. Whitney E: Flanders strain, an arbovirus newly isolated from mosquitoes and birds of New York State. Am J Trop Med Hyg 13:123-131, 1964.

154. Wilcox GE, Flower RLP, Baxendale W, et al: Serological survey of wild birds in Australia for the prevalence of antibodies to egg drop syndrome 1976 and infectious bursal disease viruses. Avian Pathol 12:135-139, 1983.

155. Woolcock PR: Duck virus hepatitis. *In* Purchase HG, et al (eds): A Laboratory Manual for the Isolation and Identification of Avian Pathogens 3rd ed. Dubuque, Kendall/Hunt Publishing Co, 1989, pp 152-155.

156. Woolcock PR: Duck hepatitis virus type I: Studies with inactivated vaccines in breeder ducks. Avian Pathol 20:509-522, 1991.

157. Wylie SL, Pass DA: Investigations of an enteric infection of cockatoos caused by an enterovirus-like agent. Aust Vet J 66:321-324, 1989.

Proventricular Dilatation Disease

Christopher R. Gregory, DVM

A fatal disease associated with progressive weight loss despite a ravenous appetite was first described in macaws in the late 1970s in the United States. Because of the clinical changes and the affected species of bird, the syndrome was initially called macaw wasting syndrome. The disease is associated with damage to the nerves that supply the gastrointestinal tract, causing a partial or generalized ileus. Names used in the literature to describe this syndrome include neuropathic gastric dilatation (NGD), proventricular dilatation syndrome (PDS), proventricular dilatation disease (PDD), myenteric ganglioneuritis, psittacine wasting syndrome, proventricular hypertrophy and infiltrative splanchnic neuropathy. The most recent name suggested for the disease is lymphoplasmacytic ganglioneuritis and encephalomyelitis, a comprehensive term which most accurately describes the microscopic changes in affected birds.[7,8,10,13] In this chapter proventricular dilatation disease (PDD) is being used to describe a group of changes recognized by dysfunction and distention of the ventriculus and proventriculus, and the accumulation of lymphocytes and plasma cells adjacent to the nerves that infiltrate these organs.

Histologic changes in affected birds and epizootiologic evidence suggests that PDD has a viral etiology; however,

a direct association between a virus and the disease has not been confirmed. Some evidence indicates that the disease is caused by a highly infectious organism; other data suggest that the disease occurs randomly and is not caused by a readily transmissible agent. Multiple cases of PDD can occur in the same flock, and mates or siblings can develop the disease following the death of an affected bird — facts that support the theory that this disease is caused by an infectious agent. However, many birds exposed to affected individuals remain asymptomatic, suggesting that the etiologic agent may be of low transmissibility or that some infected birds survive. It has been suggested that the clinical changes associated with the disease occur after a virus that causes the disease has been eliminated, making the detection of the etiologic agent difficult.[8] However, the microscopic changes that occur with the disease suggest that an active inflammatory response is occurring.

Adenovirus-like particles have been demonstrated within intranuclear inclusion bodies in the cells lining the kidneys of one affected bird.[15] Paramyxovirus-like viral particles have been demonstrated within inclusion bodies located in the neural cells of the spinal cord and in visceral nerve ganglia.[2,11,22] Similar inclusion bodies have been described in the nerves of pigeons with paramyxovirus infections.[10] Birds with PDD have been shown to lack detectable levels of antibodies to paramyxovirus (serotypes 1, 2, 3, 4, 6 and 7), Pacheco's disease virus, avian polyomavirus and avian encephalitis virus.[16,28]

An eastern equine encephalitis (EEE) virus was recovered from neonates with abdominal distention from an aviary with a history of PDD.[7,8,11] However, EEE virus occurs primarily in the eastern portion of the United States, and PDD has been shown to occur throughout North America and Europe. Additionally, only 2 of 17 birds from a California aviary with a history of PDD had antibodies to EEE virus,[25] and serologic surveys indicate EEE antibodies are detected with a similar prevalence in birds with and without PDD.[24a] These findings suggest that PDD is not caused by currently known strains of EEE virus.

Because an infectious agent has not been recovered from affected birds, it has been suggested that the observed changes in the nerves may be a result of an autoimmune process.[11] However, most autoimmune diseases that affect nerves cause demyelination, which has not been reported in birds with PDD.

CLINICAL FEATURES

Clinical changes suggestive of PDD in mature birds include depression, sustained regurgitation, passage of undigested food, impaction of the crop, abdominal distention, progressive weight loss (weeks to months), and central nervous system signs *Figure 17.1*. Diarrhea or scant feces has been reported in rare cases. Some birds develop only gastrointestinal or central nervous system signs; others develop a combination of both.[5,13,20,23] Of 89 birds described in the literature with confirmed PDD, 86.5% had one or more of the most common clinical signs including depression, weight loss, regurgitation or passage of undigested food.[13] In one affected Goffin's Cockatoo, the only clinical sign was progressive seizures.[1] A mature, wild-caught Nanday Conure with PDD developed difficulty in walking, and progressively lost control of both legs over a three-month period.[23] Clinical changes suggestive of the syndrome in young birds include frequent regurgitation, failure to properly gain weight and a reversion of weaned birds to begging behavior.[7]

Proventricular dilatation disease is invariably fatal; some reports describe the disease as having a protracted course that can persist for months, while others document death within five days of developing clinical signs.[8,23,28] In one study, affected birds died over a period that ranged from six weeks to four months after diagnosis.[26] Birds with the more chronic form of the syndrome will gradually lose weight and pass increasing quantities of poorly digested food. The excrement from these birds is frequently odoriferous and of increased volume because of poor absorption of nutrients. Intact seeds may be noted in the excrement of birds on a seed-based diet. Affected birds that are fed formulated diets may have only voluminous stools, and it is more difficult to discern that the food is being poorly digested. It should be noted that weight loss and regurgitation

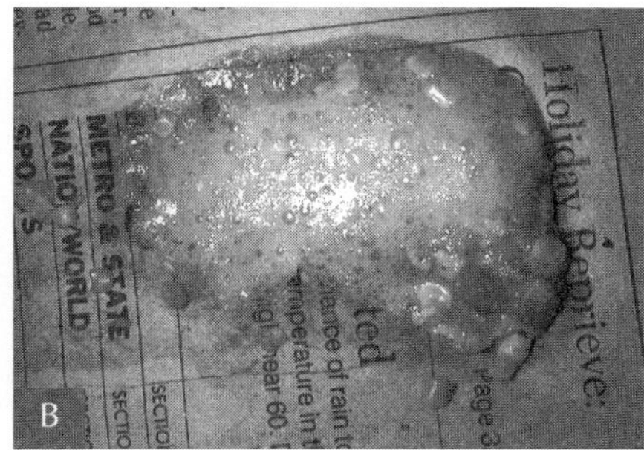

FIG 17.1
Clinical changes suggestive of PDD in mature birds include:
A depression (also note the promient keel indicating severe weight loss); and B sustained regurgitation.

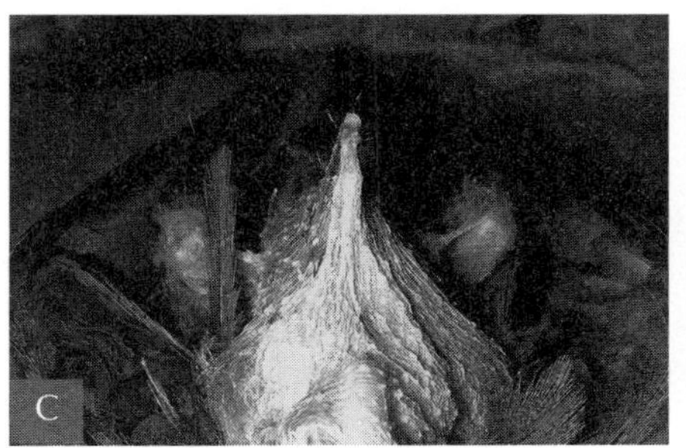

C Progressive weight loss may occur over several weeks to months. This can be detected by a loss of pectoral muscle mass making the keel more prominent. Here, the bird is lying on its back and the view is from the neck toward the tail. D Central nervous system signs that include ataxia, tremors and seizures may occur in some birds, particularly during the end stages of the disease.

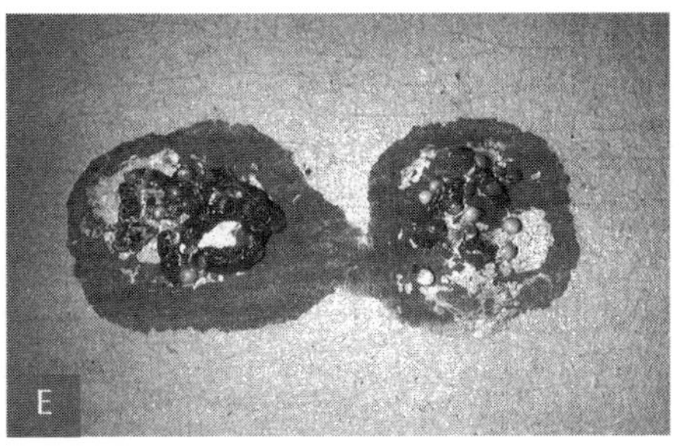

E Intact seeds may be noted in the excrement of birds on a seed-based diet. F Affected birds fed formulated diets may have only voluminous stool, which is difficult to discern.

can be induced by any extra-luminal or intraluminal mass that restricts passage of ingesta through the upper alimentary tract. Weight loss, regurgitation and passage of undigested food may occur in birds with enteritis or pancreatitis.

Changes in the complete blood count or serum chemistries of affected birds are inconsistent, possibly due to the effects of concomitant bacterial or fungal infections in the damaged gastrointestinal tract.[13] Hypoproteinemia was the only consistent change in three birds with PDD.[26] In other reports, affected birds had heterophilia, monocytosis, hypoglycemia and anemia.[23]

It has been suggested that increased creatinine kinase (CPK) activity may be an indication of PDD.[17] However, not all birds with microscopic changes consistent with PDD have elevated CPK activity.[26] In mammals, CPK activity transiently increases when active damage is occurring to skeletal muscle or to the central nervous system. Because the skeletal muscle change associated with PDD is atrophy and not necrosis, it is likely that any increased CPK activity associated with this disease is caused by continuing damage to the central nervous system.

Birds in which microscopic changes suggestive of PDD have been described

African Grey Parrots	Amazon parrots
Brotegeris spp.	Budgerigars
Buffalo Weaver	Canada Geese
Cockatiels	Cockatoos
Conures	Eclectus Parrots
Hawk-headed Parrots	King Parakeet
Lovebirds	Macaws
Meyer's Parrots	*Poicephalus* spp.
Pionus Parrots	*Psittacula* spp.
Quaker Parakeets	Red-bellied Parrots
Rock Pebbler	Senegal Parrots
Spoonbills	Thick-billed Parrot
Toucans	Vasa Parrot

EPIZOOTIOLOGY

Proventricular dilatation disease was first discussed in the late 1970s in imported psittacine birds in the United States and Germany.[10,11] Subsequently, an epornitic of PDD has been occurring in imported and captive bred psittacine birds in North America and Europe, probably as a result of the widespread shipment of companion birds.[10,11,16,23,24,27,28] Clinical or microscopic changes characteristic of the disease have been described in over 50 species of psittacine birds, as well as toucans, spoonbills, a Buffalo Weaver and free-ranging Canada Geese *Table 17.1*.[4,13,24c] The description of PDD in multiple families of birds would suggest that its cause is not restricted to distinct hosts *Figure 17.2*. There is no reference to spontaneous disease in free-ranging psittacine birds; however, there is every reason to assume that these birds would be susceptible. Given the severe nature of PDD and its apparent ability to affect a wide range of bird species, the importation of psittacine birds or their eggs into any region with indigenous Psittaciformes must be considered extremely dangerous.

It is likely that this disease can occur in any psittacine bird of any age; however, adult birds appear to be more commonly affected *Figure 17.3*. In a group of 127 birds that had histologic lesions suggestive of PDD, 43.3% were macaws, 21.3% were cockatoos, 12.6% were conures, 18.1% were African Grey Parrots, and 4.7% were other psittacines (Eclectus Parrots, Thick-billed Parrot, Blue-fronted Amazon Parrot). In a large survey of over 10,000 companion and free-ranging birds evaluated histologically over a ten-year period, 1.2% of the birds had lesions suggestive of PDD.[12] In one study, PDD was diagnosed in birds ranging from 10 weeks to 17 years of age, with a mean age of 3.8 years.[12] The average age of a group of 44 affected psittacine birds was 4.6 years.[6]

Proventricular dilatation disease can occur in any aviary despite excellent hygiene, valid quarantine procedures and the absence of new additions to the flock. In some aviaries, numerous cases of PDD will occur simultaneously. In others, several affected birds may die and the problem seemingly resolves, only to reappear one to two years later.[8] In a group of mixed species, wild-caught psittacine birds exposed to a bird confirmed by histopathology to have PDD, signs of disease in the other

FIG 17.2

Throughout the late 1970s and mid 1980s, PDD was considered a disease of psittacine birds. However, microscopic lesions suggestive of the disease have been reported in geese, toucans and spoonbills, indicating that many avian species may be susceptible to the disease.

birds were not noted for up to a year following exposure.[1] In other cases, a single bird in a breeding pair may die, with no subsequent losses in the aviary even four to five years later. It is common for many birds exposed directly or indirectly to an affected bird to remain asymptomatic. These variances in the occurrence of clinical changes in exposed birds would suggest that some are resistant to the disease or are able to mount an effective defense.

INCUBATION AND TRANSMISSION —

Because the etiology of PDD has not been confirmed, the incubation period or potential routes by which a bird may be exposed to the reputed infectious agent have not been confirmed. It has been proposed that the incubation period for PDD may be as long as four years; however, the occurrence of acute outbreaks in multiple birds at the same time suggests a much shorter incubation period.[23,26] Microscopic lesions suggestive of the disease have been detected in birds as young as ten weeks of age.[12] Young birds may die within a week of developing suggestive clinical changes, while older birds can slowly deteriorate over a period of months to years.

During one outbreak, there was a two-week interval between the time that the first bird developed clinical signs and the occurrence of additional cases. Five of ten exposed psittacine birds (macaws and cockatoos) developed clinical signs that included acute depression, weight loss, regurgitation or abnormal digestion over a three-week period. An exposed Green-winged Macaw remained clinically normal, as did two Red-lored Amazon Parrots and two Senegal Parrots.[23]

Proventricular dilatation disease was diagnosed in a mature, wild-caught, Nanday Conure that developed neurologic signs after being isolated from direct contact with other birds for seven months.[23] An affected hand-raised African Grey Parrot had no direct contact with other birds for four years.[26] A captive-raised male Moluccan Cockatoo with a one-year history of passing undigested food was found by biopsy of the ventriculus to have histologic lesions suggestive of PDD. Radiographically, this bird's parents and a clutchmate had an enlarged proventriculus and ventriculus; however, proventricular biopsies of these birds were normal.[1] The random occurrence of this disease in established aviaries or single bird households could indicate that PDD is not caused by an infectious agent, or that it is a slow progressive disease, has a long incubation period, can be transmitted by a mechanical or biologic vector or can occur following activation of a latent infection.[26] The repeated occurrence of PDD in neonates from the same pair of adults suggests that a carrier state may exist.[8]

PATHOLOGY

Gross changes associated with PDD include emaciation, ventricular ulcers, proventricular dilatation, ventricular dilatation, flaccid crop and undigested food in the dilated lower gastrointestinal tract *Figure 17.4*.[16] The proventriculi in emaciated Canada Geese with lesions suggestive of PDD were described as being five times the normal

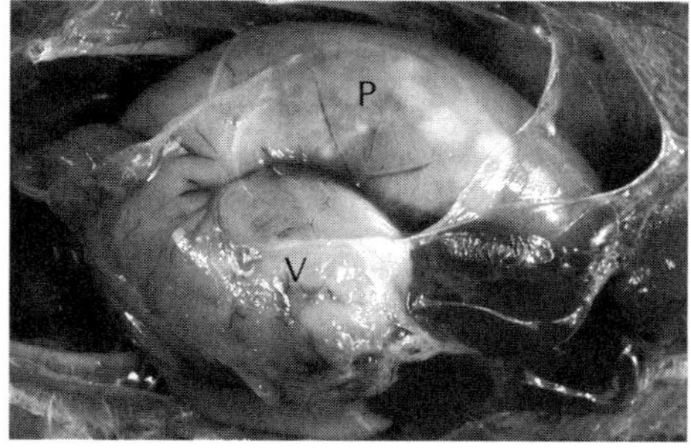

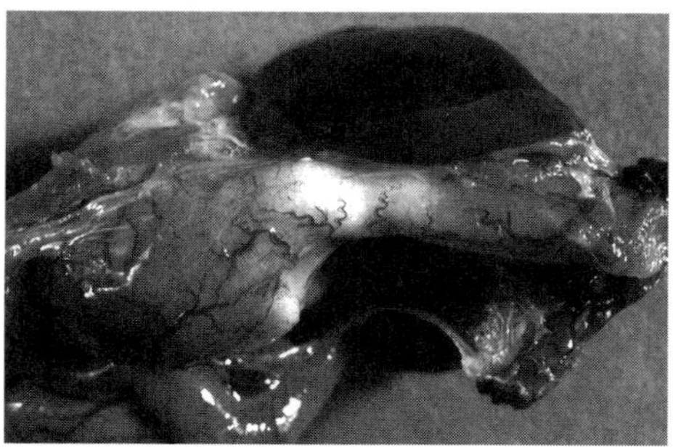

FIG 17.4

The most common gross changes associated with PDD include emaciation and accumulation of food in a dilated proventriculus (P) and ventriculus (V). The ventriculus is typically flaccid.

At necropsy, congestion of the blood vessels to the proventriculus may be the only abnormality noted in some birds with microscopic lesions consistent with PDD.

size.[4] These lesions are suggestive of PDD, but confirming the diagnosis requires the histologic detection of an accumulation of lymphocytes and plasma cells in the nerves of the gastrointestinal tract, brain or spinal cord.[9,13,23]

Some birds with histologic lesions consistent with PDD die acutely with or without neurologic signs, and the only gross lesions in these birds may be congestion of the blood vessels to the proventriculus. Gross lesions in one Umbrella Cockatoo were limited to dilatation of the duodenum and proximal jejunum.[18]

As an example of the importance of microscopic examination of the proventriculus, ventriculus and brain to confirm PDD, only 34 of 55 (62%) of the birds suspected to have PDD based on gross findings were confirmed to have the disease microscopically.[10] In another study, 16 of 421 (3.8%) psittacine birds examined had dilatation of the proventriculus, and only 4 of these 16 suspect birds had microscopic changes consistent with PDD.[3] Of 89 birds in the literature with lymphoplasmacytic infiltrates in the proventriculus, 80 (90%) had dilatation of the proventriculus.[13] In a group of 15 birds with lymphoplasmacytic infiltrates in the proventriculus, only 10 (67%) had proventricular dilatation. The distribution and prevalence of lymphoplasmacytic infiltrates in birds diagnosed with PDD are listed *Table 17.2*.

Accumulation of lymphoplasmacytic infiltrates in nerves is characteristic of the disease. Other histologic changes that have been reported in some birds include myocarditis, serositis, ventriculitis, degeneration of the lining of the ventriculus, smooth muscle degeneration and accumulation of lymphocytes in affected tissues.[9,16,23] Intranuclear and intracytoplasmic eosinophilic inclusion bodies have been identified in nerve cells of intestinal ganglia; however, their importance in the disease remains unconfirmed.[9,16,19,23]

PATHOGENESIS

Whatever its etiology, PDD clearly is associated with damage to the nerves supplying the proventriculus, ventriculus and, in some cases, the duodenum. This impedes an affected bird's ability to properly digest food. The tremors and incoordination associated with PDD are caused by damage that occurs in the brain and spinal cord. The progression of PDD varies widely among affected individuals, probably as a result of differences in the speed with which damage to the nerves occurs.

As food is swallowed by a psittacine bird, it goes down the upper portion of the esophagus and into the crop. Food then exits the crop, enters the lower portion of the esophagus and travels into the proventriculus, where the food is mixed with stomach acids and digestive enzymes. The food then passes from the proventriculus to the thick-walled, muscular ventriculus where the food is crushed, allowing further mixing with digestive secretions. The food passes from the ventriculus into the intestinal tract where further digestion takes place and nutrients are absorbed into the body. If food is not passing correctly from the ventriculus into the intestines, it accumulates in the proventriculus, esophagus and crop, causing them to dilate. Accumulated food may be expelled by vomiting. If the ventriculus does not properly crush ingested food, whole portions of the diet may be passed in the feces. Thus, the primary

TABLE 17.2

Prevalence of lymphoplasmacytic infiltrates in birds with PDD[12,13,16]

	7 birds	15 birds	Mixed Number
Proventriculus	100%	100%	—
Ventriculus	66%	100%	—
Proventriculus/Ventriculus	—	—	100%
Small intestines	—	—	78%
Crop	—	67%	22%
Heart	28%	73%	—
Brain or spinal cord	—	67%	70%
Pons, Medulla, Midbrain	100%	—	
Cerebrum	47%	—	—
Cerebellum	16%	—	—

clinical changes noted with PDD (regurgitation, weight loss, passing undigested food) can be attributed to failure of the proventriculus and ventriculus to properly contract. Affected birds continue to have ravenous appetites but the food that passes through the gastrointestinal tract is improperly digested, nutrients are poorly absorbed and the bird slowly starves due to insufficient consumption of energy.

The passage of food and water through the gastrointestinal tract ensures that bacteria, fungi and other microorganisms found in the lumen are constantly being removed. This ongoing "flushing" of the gastrointestinal tract helps prevent infectious agents from accumulating in the lumen and colonizing the intestines in large numbers *see Figure 3.16*. When the gastrointestinal tract is not emptying properly, bacteria and other organisms accumulate and are afforded a greater opportunity to colonize the gastrointestinal tract and cause disease. In some birds with PDD, the severely dilated, thin-walled proventriculus may rupture, resulting in the movement of impacted food into the abdominal cavity causing peritonitis.

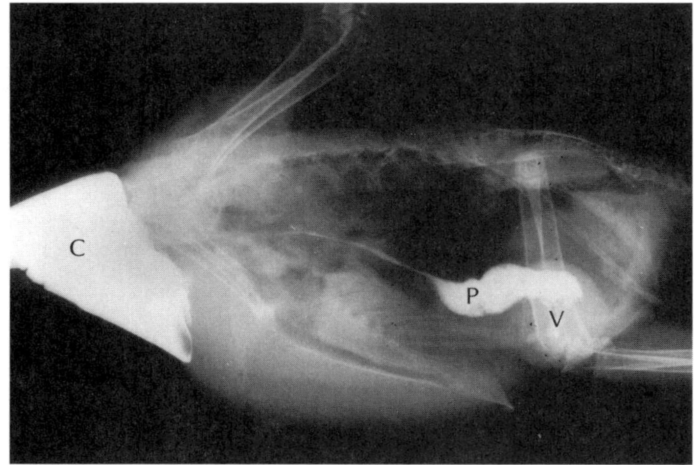

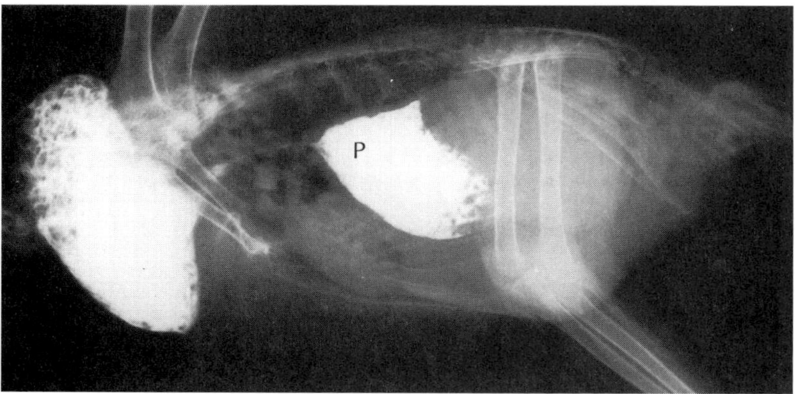

FIG 17.5

top Contrast radiograph of a normal bird 20 minutes after receiving barium. Note that the barium has already passed through the proventriculus (P) and into the ventriculus (V). The crop (C) is filled with barium.

bottom Contrast radiograph of a bird with PDD three hours after receiving barium. The proventriculus is severely dilated and the gastric emptying time is extremely slow.

DIAGNOSIS

Progressive weight loss, regurgitation and passing of undigested food, with or without neurologic signs, are suggestive of PDD. This disease should also be suspected in any bird with neurologic signs that include ataxia, head tremors or seizures. Contrast radiographs of affected adult birds may indicate a rapid gastrointestinal transit time early in the disease process, followed by a slowed transit time (chronic disease) and gaseous distention of the intestines as the ventriculus becomes increasingly dysfunctional and the proventriculus and intestines dilate *Figure 17.5*.[15] However, any process that causes partial blockage of the intestinal tract or maldigestion — that prevents the passage of food — including fungal proventriculitis, megabacteriosis, parasitic enteritis, foreign bodies, neoplasia, heavy metal toxicosis, bacterial enteritis, papillomatosis of the stomach or any intraluminal or extraluminal mass can cause similar clinical, radiographic or gross changes. The proventriculus of neonates is normally dilated, a condition which should not be misinterpreted as PDD *Figure 17.6*. The proventriculus of a neonate attains its adult tone and size as the bird enters and completes the weaning period.

Currently, diagnosis of PDD requires the microscopic demonstration of accumulated lymphocytes and plasma cells in association with the nerves of the gastrointestinal tract or central nervous system. In most cases, a post mortem diagnosis is rendered when a complete set of tissues (including proventriculus, ventriculus, brain and spinal cord) are examined microscopically. It may be possible to obtain a diagnosis before death by submitting a biopsy of the crop or ventriculus for microscopic evaluation *Figure 17.7*. Biopsy of the crop is simple, relatively noninvasive and may have some diagnostic merit. In one study, PDD was reliably diagnosed in 66% of crop biopsies collected from positive birds.[6] Other studies have indicated that a sample of the crop collected at necropsy was 76% accurate in detecting birds that died from PDD.[14] Thus, a positive crop biopsy in a bird with suggestive clinical changes is of diagnostic value, but a negative crop biopsy in a bird with suggestive clinical changes does not rule-out PDD. When the etiology of PDD is confirmed, a screening test to determine whether exposed birds are

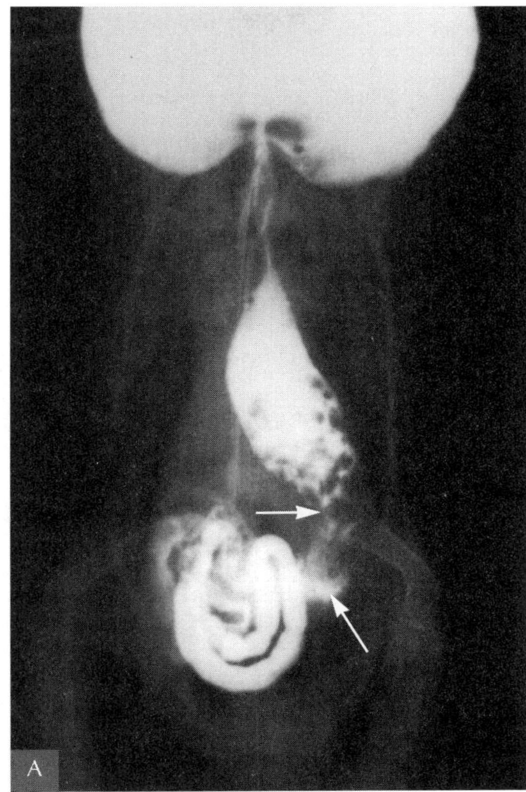

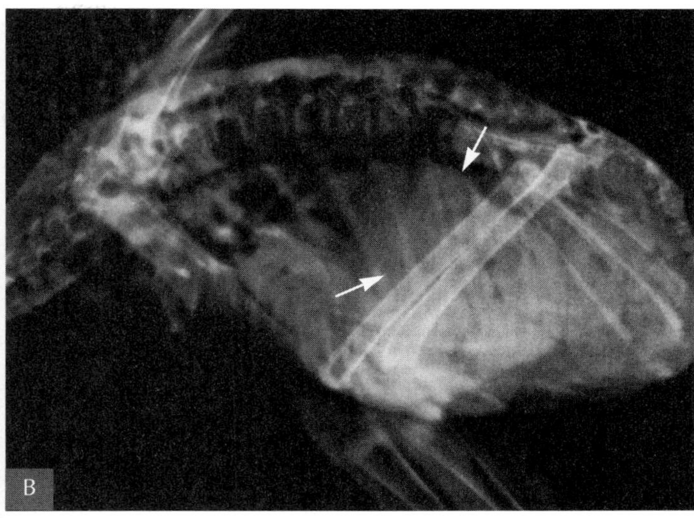

FIG 17.6

A It should be noted that any process that causes partial blockage of the intestinal tract or maldigestion can cause clinical or radiographic changes suggestive of PDD. In this case, a tumor in the ventriculus (arrows) was blocking the passage of food, causing regurgitation and dilatation of the proventriculus.

B The proventriculus (arrows) of neonates is normally dilated, a condition which should not be misinterpreted as PDD.

infected may be of value in controlling the disease.

It should be noted that microscopic lesions suggestive of PDD have been described in birds that die from other causes and have absolutely no gross or clinical signs of the disease. This finding may indicate more than one cause for the accumulation of lymphoplasmacytic infiltrates in the nerves of the gastrointestinal tract. All clinical, necropsy and microscopic findings should be considered in diagnosing PDD.

CONTROL

Definitive control or prevention of PDD will require confirmation of the etiology. Mates, offspring or clutchmates of birds that are diagnosed histologically with PDD should be considered at extra risk of developing the disease; however, they should not be euthanized. Many of the birds that are directly exposed to those with PDD never develop the disease. Until the cause of PDD is confirmed, and an appropriate screening test is developed, it would be prudent to place exposed birds in single-bird households where they have no direct or indirect exposure to other birds.

Provided with an easily digested, high-energy diet, a stress-free environment and treatment for secondary bacterial or fungal infections, affected companion birds can survive for months or years. Any bird with the disease that is being treated should be placed in strict isolation with no direct or indirect contact with other birds. Some birds with clinical changes suggestive of PDD have been reported to recover when provided supportive care. However, a

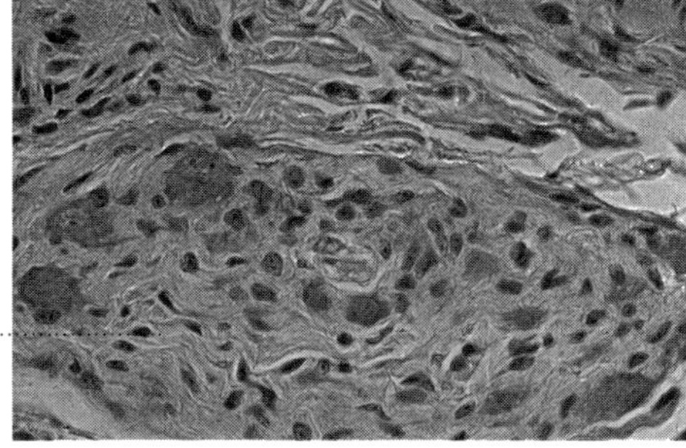

Biopsy of the crop may be a simple, relatively noninvasive method to confirm a diagnosis of PDD. Microscopic appearance of a nerve from a normal crop.

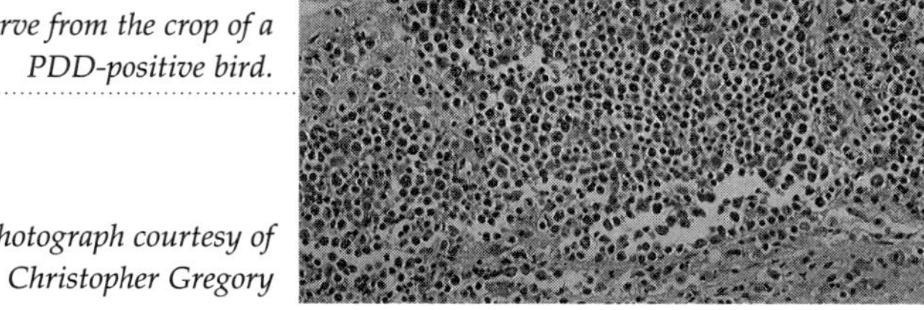

Lymphocytes have accumulated in the nerve from the crop of a PDD-positive bird.

photograph courtesy of Christopher Gregory

positive diagnosis of this disease requires the demonstration of microscopic lesions in the nerves, and none of the reported recoveries have been in birds confirmed to have PDD.[21]

Some clinicians believe that birds with clinical and radiographic signs consistent with the disease respond favorably to treatment with interferon. While there are no scientific data to support these clinical observations, the treatment regime that has been loosely discussed is 30 units of interferon daily for five days, followed by 30 units twice a week for two weeks, then 30 units once a week for an additional two weeks.[24b]

REFERENCES

1. Bond MW, Downs D, Wolf S: Screening for psittacine proventricular dilatation syndrome. Proc Assoc Avian Vet, 1993, pp 92-97.

2. Busche R, Frese K, Weingarten M: Zur pathologie des macaw wasting-syndromes. Int Symp Zoo Animal, 1985, pp 325-329.

3. Clark FD: Proventricular dilatation syndrome in large psittacine birds. Avian Dis 28:813-815, 1984.

4. Daoust PY, Julian RJ, Yason CV, et al: Proventricular impaction associated with nonsuppurative encephalitis and ganglioneuritis in two Canada Geese. J Wildl Dis 27:513-517, 1991.

5. Degernes LA, Flammer K: Proventricular dilation syndrome in a green-wing macaw. Proc Assoc Avian Vet, 1991, pp 45-49.

6. Doolen M: A low risk diagnosis for neuropathic gastric dilatation. Proc Assoc Avian Vet, 1994, pp 193-196.

7. Gaskin JM: Questions and answers about psittacine proventricular dilatation disease and avian viral serositis. Proc Midwest Avian Research Exposition, 1992, pp 69-71.

8. Gaskin JM, Homer B, Eskelund K: Preliminary findings in avian viral serositis: A newly recognized syndrome of psittacine birds. J Assoc Avian Vet 5:27-34, 1991.

9. Gerlach H: Viruses. *In* Ritchie BW, Harrison GJ, Harrison LR (eds): Avian Medicine: Principles and Application. Lake Worth, Wingers Publishing, 1994, pp 862-948.

10. Gerlach S: Macaw wasting disease: A 4 year study on clinical case history, epizootiology, analysis of species, diagnosis, microbiological and virological results. Proc Europ Assoc Avian Vet, 1991, pp 273-281.

11. Graham DL: Infiltrative splanchnic neuropathy, a component of the "wasting macaw" complex. Proc Assoc Avian Vet, 1984, p 275.

12. Graham DL: Wasting/proventricular dilation disease: A pathologist's view. Proc Assoc Avian Vet, 1991, pp 43-44.

13. Gregory CR, Latimer KS, Niagro FD, et al: A review of proventricular dilation syndrome. J Assoc Avian Vet 8:69-75, 1994.

14. Gregory CR, Latimer KS, Ritchie BW, et al: Evaluation of crop biopsy for the diagnosis of proventricular dilatation syndrome in psittacine birds. J Vet Diag Invest Submitted for publication 1995.

15. Heldstab A, Morgenstern D, Ruedi A, et al: Pathologie einer endemieartig verlaufenden neuritis im magen/darmberich bei großpapageien. Int Symp Zoo Animal, 1985, pp 317-324.

16. Hughes E: The pathology of myenteric ganglioneuritis, psittacine encephlomyelitis, proventricular dilatation of psittacines, and macaw wasting syndrome. Proc 33rd West Poult Dis Conf, 1984, pp 85-87.

17. Jenkins T: Creatinine kinase as a diagnostic indicator of splanchnic neuropathy. J Assoc Avian Vet 5:49, 1991.

18. Joyner KL, Kock N, Styles D: Encephalitis, proventricular and ventricular myositis and myenteric ganglioneuritis in an umbrella cockatoo. Avian Dis 33:379-381, 1989.

19. Lowenstine LJ: Emerging viral diseases of psittacine birds. *In* Kirk RW (ed): Current Veterinary Therapy IX. Philadelphia, WB Saunders Co, 1986, pp 705-710.

20. Lutz ME, Wilson RB: Psittacine proventricular dilatation syndrome in an umbrella cockatoo. J Am Vet Med Assoc 198:1962-1963, 1991.

21. Malley D: Case report: A study of a Moluccan cockatoo with proventricular dilatation. Proc Europ Assoc Avian Vet, 1991, pp 271-272.

22. Mannl A, Gerlach H, Leipold R: Neuropathic gastric dilatation in Psittaciformes. Avian Dis 31:214-221, 1987.

23. Phalen DN: An outbreak of psittacine proventricular dilatation syndrome (PPDS) in a private collection of birds and an atypical form of PDS in a nanday conure. Proc Assoc Avian Vet, 1986, pp 27-34.

24. Ridgway RA, Gallerstein GA: Proventricular dilatation in psittacines. Proc Assoc Avian Vet, 1983, pp 228-230.

24a. Ritchie BW, et al: Unpublished data.

24b. Ritchie BW: Personal communication.

24c. Schmidt R: Personal communication.

25. Shivaprasad HL: Diseases of the nervous system in pet birds: A review and report of diseases rarely documented. Proc Assoc Avian Vet, 1993, pp 213-222.

26. Suedmeyer WK: Diagnosis and clinical progression of three cases of proventricular dilatation syndrome. J Assoc Avian Vet 6:160-163, 1992.

27. Turner R: Macaw fading or wasting syndrome. Proc 33rd West Poult Dis Conf, 1984, pp 87-88.

28. Woerpel RW, Rosskopf WJ: Proventricular dilatation and wasting syndrome: Myenteric ganglioneuritis and encephalomyelitis of psittacines: An update. Proc Assoc Avian Vet, 1984, pp 25-28.

ADENOVIRUS IN PIGEONS

FAMILY - ADENOVIRIDAE

PRIMARY CLINICAL FEATURES
- Widely varied species susceptibility to disease
- Acute death with no premonitory signs
- Depression, anorexia, dyspnea, a crouched stance, polydipsia, polyuria, and slimy green diarrhea
- Infections are most common in 2- to 4-month-old pigeons
- Infected young may die within 48 hours of clinical signs

INCUBATION PERIOD
- Mortality is highest 3-4 days after infection

INFECTED SURVIVORS
- Chickens are considered to be latently infected
- Undetermined in pigeons, but survivors should be considered latently infected

COMMON PATHOLOGY
- Enlarged liver and spleen
- Yellowish to copper-gold, patchy discoloration of the liver
- Hyperemia of the lining of the intestinal tract
- Greenish, mucoid fluid in the crop and intestinal tract

HISTOPATHOLOGY
- Intranuclear inclusions in the liver and intestines
- Similar to those caused by pigeon herpesvirus

DIFFERENTIAL DIAGNOSIS
- Bacterial hepatitis or enteritis, chlamydiosis, salmonellosis, reovirus, pigeon herpesvirus

DIAGNOSIS - IN THE LIVING BIRD
- Demonstration of virus in feces by electron microscopy
- Isolation of virus from feces
- Birds with antibodies considered latently infected
- Viral-specific DNA probes (theoretical, unproven)

DIAGNOSIS - POST MORTEM
- Detection of suggestive microscopic changes
- Virus isolation from liver or intestines
- Viral-specific DNA probes

TRANSMISSION
- Shed in high concentration in the feces
- Primarily fecal/oral
- Free-ranging pigeons can be a source of virus for captive birds

PREVENTION
- In chickens, maternal antibodies provide temporary (up to 3 weeks) protection
- Vaccines available for turkeys, pheasants and ducks *Appendix V*
- No vaccines available for pigeons

CONTROL
- Adenoviruses are stable outside the host and resistant to many disinfectants
- Inactivated by exposure for 1 hour to formalin, aldehydes and iodophors

TREATMENT
- No specific therapy
- Supportive care, fluids, assisted feeding and broad-spectrum antibiotics to prevent secondary infections

ZOONOTIC POTENTIAL
- None noted nor expected

ADENOVIRUS IN PSITTACINES

FAMILY - ADENOVIRIDAE

PRIMARY CLINICAL FEATURES
- Widely varied species susceptibility to disease
- Acute death with no premonitory signs
- Depression, anorexia, yellowish-green diarrhea and death

INCUBATION PERIOD
- Undetermined in psittacines
- In chickens, 24-48 hours; in quail, 2-4 days

INFECTED SURVIVORS
- Undetermined in psittacine birds
- Chickens considered to be latently infected

COMMON PATHOLOGY
- Enlarged, friable, discolored liver with mottling and sub-capsular hemorrhages
- Congestion of the intestines

HISTOPATHOLOGY
- Necrosis in the liver, spleen, pancreas and/or intestines
- Intranuclear inclusions in liver, spleen, pancreas and/or intestines
- Inclusion bodies similar to those caused by polyomavirus, herpesvirus or PBFD virus
- Inclusion bodies may be present in asymptomatic birds

DIFFERENTIAL DIAGNOSIS
- Bacterial hepatitis or enteritis, chlamydiosis, salmonellosis, polyomavirus, reovirus, Pacheco's disease virus, paramyxovirus, influenza A virus

DIAGNOSIS - IN THE LIVING BIRD
- Demonstration of virus in feces or pharyngeal secretions by electron microscopy
- Isolation of virus from feces or pharyngeal secretions
- In chickens, neutralizing antibodies present 1-2 weeks after infection (birds with antibodies are considered latently infected)

DIAGNOSIS - POST MORTEM
- Detection of suggestive microscopic changes
- Virus isolation from liver, intestines, kidney or pancreas
- Viral-specific DNA probes

TRANSMISSION
- Some strains transmitted vertically and horizontally; only horizontal routes confirmed in psittacines
- In chickens, virus shed in the feces (up to 20 days) or respiratory secretions (up to 4 weeks)
- Latently infected chickens can intermittently shed virus
- High levels of antibodies associated with low levels of viral shedding; low levels of antibodies — high levels of viral shedding
- Free-ranging pigeons and waterfowl — potential reservoirs

PREVENTION
- Vaccines available for turkeys, pheasants and ducks; none for psittacines *Appendix V*

CONTROL
- Stable outside the host and resistant to many disinfectants
- Inactivated 1 hour of contact with formalin, aldehydes and iodophors

TREATMENT
- Supportive care, fluids, assisted feeding and broad-spectrum antibiotics to prevent secondary infections

ZOONOTIC POTENTIAL
- None noted nor expected

AVIAN LEUKOSIS/SARCOMA VIRUS

FAMILY - RETROVIRIDAE

PRIMARY CLINICAL FEATURES
- Widely varied species susceptibility to disease
- Many infected birds are asymptomatic
- May cause cancers, chronic weight loss, neurologic disease, damage to the immune system and persistent infections

CHICKENS
- Most common neoplasia is lymphoid leukosis
- Most asymptomatic, some develop tumors over a 16-week period
- Most common in chickens 4-9 months of age
- Some chickens have natural resistance
- Neoplasia in circulating blood cells or solid tumors in organs
- Decreased egg production, reduced hatchability and stunting
- Abdominal masses or enlargement of the liver may cause dyspnea
- An increased number of heterophils, lymphocytes and monocytes may be noted
- Females more susceptible to disease than males
- Young birds develop resistance to the disease as their cloacal bursa matures

OTHER BIRDS
- Clinical signs depend on the location of tumors and presence of secondary infections
- Affected birds may die acutely with no premonitory signs, die following a brief period of depression and diarrhea, or develop chronic problems associated with the formation of proliferative masses in numerous tissues
- Leukocytosis, heterophilia, lymphocytosis, monocytosis and a partially regenerative anemia
- Predominance of mature lymphocytes that may have scalloped borders
- Activities of some serum enzymes may be increased if the liver is affected

BUDGERIGARS
- Renal tumors with lameness and abdominal enlargement
- Tumors were most common in birds 3-6 years of age

CANARIES
- Die suddenly, die following a brief period of dyspnea, or develop chronic weight loss and respiratory problems
- Subcutaneous masses on the head and neck
- Abdominal swelling

INCUBATION PERIOD
- In chickens - varies with the strain of virus and the type of neoplasia induced

INFECTED SURVIVORS
- In chickens, persistent infections are common

COMMON PATHOLOGY
- Enlarged friable livers and spleens
- Renal enlargement
- Cream- to grayish-colored masses in abdominal organs and lungs, thymus and bone marrow

HISTOPATHOLOGY
- Accumulation of immature lymphocytes in the liver and spleen
- Sheets of lymphocytes with a high mitotic index

DIFFERENTIAL DIAGNOSIS
- Marek's disease virus

DIAGNOSIS - IN THE LIVING BIRD
- Virus recovery from plasma, serum, feces, oviduct swabs, pharyngeal secretions, feather pulp or albumen of freshly laid eggs
- Detection of antibodies using ELISA or VN assays
- Chicks infected in the oviduct or immediately post-hatching are immunotolerant

DIAGNOSIS - POST MORTEM
- Microscopic examination of affected tissues for accumulation of lymphocytes
- Virus isolation from liver, spleen, kidney or distinct tumors

TRANSMISSION
- Endogenously transmitted strains are spread from infected parent to offspring directly in the transferred genetic material
- Exogenously transmitted strains are spread through direct or indirect contact with contaminated blood, saliva, respiratory secretions, semen or feces
- Shed from the oviduct of an infected hen and can enter the albumen of the developing egg
- Infected males can infect hens during copulation

PREVENTION AND CONTROL
- Virions are unstable when outside the bird
- Most disinfectants would inactivate the virus
- No vaccines for ALSV
- Prevention through excellent hygiene and reducing exposure to the virus
- Identify and remove infected hens from the breeding flock
- Selective breeding to produce strains of chickens that are resistant
- Mibolerone induces resistance in chicks

TREATMENT
- Steroids - prednisolone
- Expected survival rate less than 1 year

ZOONOTIC POTENTIAL
- None noted nor expected

AMAZON TRACHEITIS VIRUS

FAMILY - HERPESVIRIDAE

PRIMARY CLINICAL FEATURES
- Rapid death following a brief period of severe dyspnea
- Chronic upper respiratory disease

INCUBATION
- Experimental - 3-4 days

INFECTED SURVIVORS
- Considered to be latently infected (not confirmed)
- Considered resistant to disease

COMMON PATHOLOGY
- Diphtheritic membranes in the trachea

HISTOPATHOLOGY
- Necrosis of cells lining the respiratory tract, pharynx, larynx and esophagus or crop
- Intranuclear inclusion bodies may be difficult to detect

DIFFERENTIAL DIAGNOSIS
- Bacterial sinusitis, rhinitis or tracheitis; chlamydiosis, paramyxoviruses (Newcastle disease virus), influenza A virus, aspergillosis, trichomoniasis, tracheal mites, vitamin A deficiency, diphtheritic form of poxvirus

DIAGNOSIS - IN THE LIVING BIRD
- Virus isolation from tracheal swabs

DIAGNOSIS - POST MORTEM
- Microscopic examination of tissues for intranuclear inclusions
- Virus isolation from trachea or lungs

TRANSMISSION
- Not confirmed; probably through respiratory secretions

PREVENTION
- Efficacy of inactivated ILT virus vaccines in psittacine birds is undetermined
- Attenuated-live ILT virus vaccines should not be used in psittacine birds

TREATMENT
- The effectiveness of acyclovir in preventing disease is undetermined

ZOONOTIC POTENTIAL
- None noted nor expected

CANARY POXVIRUS

FAMILY - POXVIRIDAE

PRIMARY CLINICAL FEATURES
- Considered enzootic in canaries
- Cutaneous, diphtheritic and septicemic forms of the disease
- Septicemic form is most common
- Acute onset of dyspnea, followed by death
- Most canaries (>70%) die shortly after developing clinical signs
- Recovered canaries may develop tumors in the lungs
- Wart-like lesions on the eyelids and around the beak
- Diphtheritic form - necrotic lesions in the trachea, larynx and oral cavity

INCUBATION PERIOD
- 4 days to 3 weeks; infections rapidly spread through a susceptible population

COMMON PATHOLOGY
- Congestion in the lungs
- Hyperemia in the trachea

HISTOPATHOLOGY
- Inflammation of the bronchi
- Bollinger bodies in cells of the skin, bronchi and lungs

DIAGNOSIS - IN THE LIVING BIRD
- Virus isolation from vesicles, pharyngeal swabs or feces
- Demonstration in lesions by cytology or electron microscopy
- AGID, VN, ELISA or HI assays to detect antibodies

DIAGNOSIS - POST MORTEM
- Microscopic examination of tissues for intracytoplasmic inclusions
- Virus isolation from skin lesions, mucosal lesions or lung

TRANSMISSION
- A break in the bird's epithelium is necessary
- Direct contact of virus from an infected bird or indirect contact with a contaminated object or insect
- Mosquitoes and mites serve as the primary mechanical vectors; epornitics are common in the spring and fall when mosquitoes are most prevalent
- Affected birds with lesions or scabs are most infectious
- Infected free-ranging Passeriformes may serve as a source of virus for captive canaries

PREVENTION
- Attenuated-live vaccine provides temporary immunity *Appendix V*
- Vaccinate all canaries and finches
- Maximum immunity develops 3-4 weeks after vaccination

CONTROL
- Virions stable when outside of bird
- Inactivated with 1% potassium hydroxide, heating to 50°C for 30 min or to 60°C for 8 min, steam, 2% NaOH and 5% phenol
- Isolate infected birds and directly exposed birds
- Destroy contaminated wooden perches or nest boxes
- Sterilize contaminated nets, towels, clothing, holding containers, equipment, supplies
- Use biologically secure shipping containers

TREATMENT
- Rarely effective

ZOONOTIC POTENTIAL
- None noted nor expected

DERZSY'S DISEASE

FAMILY - PARVOVIRIDAE

PRIMARY CLINICAL FEATURES
- Vary with the age and immune status of the goose
- Embryos infected in utero may die before or soon after hatching
- Hatchlings - depression, anorexia, respiratory disease, diarrhea, ataxia, death
- Survivors exhibit retarded growth and poor feather development on the neck and thorax
- Goslings resistant by 4-6 weeks of age

INCUBATION PERIOD
- Experimental in goslings - 3-6 days

INFECTED SURVIVORS
- Considered resistant to disease
- Considered to be persistently infected

COMMON PATHOLOGY
- Acute infections - enlarged heart and liver with grayish-white areas of necrosis, hemorrhage in the intestinal tract, pancreas and bursa
- Chronic infections - hyperemia of the intestinal mucosa, ascites and accumulation of a fibrinous material on the heart, liver and in the intestines

HISTOPATHOLOGY
- Hepatic necrosis, pancreatic necrosis, serofibrinous enteritis, pericarditis, perihepatitis and pulmonary edema

DIFFERENTIAL DIAGNOSIS
- Reovirus, chlamydiosis, influenza A virus, salmonellosis, aflatoxicosis, aspergillosis, nephroenteritis syndrome

DIAGNOSIS - IN THE LIVING BIRD
- Previous infection - demonstration of antibodies in birds over 4 weeks old
- Active infection - demonstration of rising antibody titers
- Isolation of virus from feces
- Demonstration of virus in feces by electron microscopy

DIAGNOSIS - POST MORTEM
- Isolation of virus from intestines, liver, pancreas or heart
- Intranuclear inclusions in liver, spleen or intestines

TRANSMISSION
- Excreted in feces
- Ingestion of contaminated litter, food and water
- Vertical transmission

PREVENTION
- Vaccinate flock with attenuated-live virus vaccine *Appendix V*
- Administration (IM or orally) of hyperimmune serum to antibody negative 1-day-old goslings

CONTROL
- Virions stable when outside the bird
- Inactivated by sodium hypochlorite, glutaraldehyde and stabilized chlorine dioxide

TREATMENT
- Supportive care
- Recovered birds considered to be persistently infected

ZOONOTIC POTENTIAL
- None noted nor expected

DUCK PLAGUE VIRUS

FAMILY - HERPESVIRIDAE

PRIMARY CLINICAL FEATURES
- Wide species variability from asymptomatic to high mortality
- Acute death with no premonitory signs
- Death following depression, anorexia, ocular discharge, nasal discharge, diarrhea or central nervous system signs
- Death 1-10 days after clinical signs (1-4 days most common)

INCUBATION
- 3-12 days (3-7 days most common)

INFECTED SURVIVORS
- Latently infected
- In most cases considered resistant to disease
- May intermittently develop erosions on the mucosa of the oral cavity, particularly under the tongue

COMMON PATHOLOGY
- Hemorrhage throughout the body, particularly in the alimentary tract and liver
- Diphtheritic membranes on the mucosa of the pharynx, esophagus, intestinal tract and cloaca

HISTOPATHOLOGY
- Necrosis of the liver, spleen, esophagus and intestines
- Intranuclear inclusions in the liver, esophagus and intestines

DIFFERENTIAL DIAGNOSIS
- Toxins, duck hepatitis virus, pasteurellosis, poxvirus (diphtheritic form), paramyxovirus

DIAGNOSIS - IN THE LIVING BIRD
- Virus isolation from feces or pharyngeal swabs
- Increase in virus-neutralizing antibody levels
- Electron microscopic demonstration of virus in feces or oral secretions

DIAGNOSIS - POST MORTEM
- Microscopic examination of liver, esophagus or intestines for intranuclear inclusions
- Virus isolation from the liver, spleen or intestines

TRANSMISSION
- Shed in feces and pharyngeal secretions
- Shedding may start within 2-3 days after infection
- Ingestion of contaminated water or food
- Latently infected birds shed virus intermittently

CONTROL
- Virions relatively stable in pond water
- Most disinfectants will inactivate the virus
- Self-limiting in closed, captive populations of waterfowl
- Prevent exposure of captive birds to free-ranging waterfowl
- Attenuated-live virus vaccine

ZOONOTIC POTENTIAL
- None noted nor expected

EASTERN EQUINE ENCEPHALITIS VIRUS

FAMILY - TOGAVIRIDAE

PRIMARY CLINICAL FEATURES

GENERAL CONSIDERATIONS
- Natural susceptibility varies among birds
- All species are considered susceptible to experimental infections
- Birds native to North America are generally susceptible to infection but resistant to disease
- Birds not native to North America may develop encephalitis or enteritis
- Epornitics are most common in the summer and fall
- Depending on the species, birds may remain asymptomatic, die suddenly with no signs of illness or die following a brief period of severe enteritis or neurologic disease
- Depression, anorexia, weakness, profuse diarrhea
- Central nervous system abnormalities
- Some affected birds recover completely, while others die within 48 hours
- Age-related resistance in some species
- Psittacine neonates - serositis, ascites and death
- Experimental infections in Budgerigars - fatal
- Experimental infections in other psittacines - transient diarrhea and recovery

PHEASANTS
- Systemic or neurologic changes
- Systemic signs include droopiness, ruffled feathers, soiled vents and occasional respiratory distress
- Neurologic signs include incoordination, paralysis of the legs, droopy wings, tremors and head tilt
- Mortality rate varies from 5%-80%
- Some with severe signs may recover

EMUS
- High mortality
- Die suddenly without premonitory signs
- Depression, anorexia, ataxia and hemorrhagic diarrhea followed by death

WHOOPING CRANES
- Die suddenly without premonitory signs
- Lethargy, ataxia and leg and neck paresis followed by death in several days
- Anemia, leukopenia, elevated AST, LDH, GGT and uric acid

INCUBATION PERIOD
- Pheasants - experimental, 1-12 days
- Chickens - experimental, 15 hours to 26 days

INFECTED SURVIVORS
- Considered resistant to disease

COMMON PATHOLOGY
- Varies among affected birds
- Small pale spleen or splenomegaly with mottling
- Softening of the cerebral hemispheres
- Enlarged, mottled liver, ascites
- Distention of the kidneys
- Emus - hemorrhagic enteritis with hemorrhage in other abdominal organs

HISTOPATHOLOGY
- Some species develop lesions primarily in the brain
- Inflammation of the blood vessels in the brain
- Non-suppurative encephalitis
- Perivascular lymphocytic infiltrates in the brain
- Some species develop lesions primarily in the abdominal organs
- Hepatic necrosis
- Myocardial necrosis
- Emus - hemorrhage in the intestines and abdominal organs
- Necrosis of the liver, spleen and intestinal mucosa

DIFFERENTIAL DIAGNOSIS
- Paramyxovirus, bacterial encephalitis, other alphaviruses, botulism, pasteurellosis and salmonellosis

DIAGNOSIS - IN THE LIVING BIRD
- Detection of antibodies using HI, CF, ELISA or VN assays
- Antibodies may be detected as soon as 5 days to 2 weeks after infection
- Complement-fixation and HI antibodies may persist for several weeks to months
- Virus-neutralizing antibodies may persist for months to years
- Virus isolation from fresh blood

DIAGNOSIS - POST MORTEM
- Microscopic examination of affected tissues for suggestive changes
- Virus isolation from fresh, liver, spleen or intestines

TRANSMISSION
- Primarily transmitted by mosquitoes
- Free-ranging birds considered to be a reservoir
- Virus is present in the blood of most birds from 1-5 days after infection
- Mosquitoes remain infected for their entire life
- Pheasants, ducks and crows - direct transmission without arthropod vector

PREVENTION
- Some inactivated EEE virus vaccines are effective in birds
- Emus - Solvay-Triple E has been shown to be effective
- Vaccinate at 3 months of age and boost 2-4 weeks later
- Vaccinate adults after the breeding season in April and prior to the breeding season in October
- Obtain approval from insurer prior to vaccination

CONTROL
- Virions unstable when outside the host
- Most disinfectants rapidly inactivate the virus

TREATMENT
- Supportive care - fluids, assisted feeding, antibiotics

ZOONOTIC POTENTIAL
- Infected mosquitoes can transmit virus to humans
- Many infected humans remain asymptomatic, but a few develop mild-to-severe encephalomyelitis
- Aviculturists are at no greater risk than anyone else exposed to mosquitoes

EGG DROP SYNDROME

FAMILY - ADENOVIRIDAE

PRIMARY CLINICAL FEATURES
- Decrease in egg quality and reduced hatchability
- Thin-shelled, soft-shelled, shell-less or poorly pigmented eggs

INCUBATION PERIOD
- Experimental incubation period in chickens is 7-17 days

INFECTED SURVIVORS
- In chickens, recovered hens are latently infected
- Virus persists in the infundibulum
- Virus dormant except during the laying season

COMMON PATHOLOGY
- Edema and accumulation of mucoid exudates in oviduct

HISTOPATHOLOGY
- Intranuclear inclusions are transiently present in the oviduct

DIFFERENTIAL DIAGNOSIS
- Infectious bronchitis virus, Newcastle disease virus and influenza A

DIAGNOSIS - IN THE LIVING BIRD
- Highest concentration of virus in the uterus when abnormal eggs are formed
- Antibody titers increase during periods when the virus is active
- Antibodies present 5 days after infection and can persist for up to 6 months
- Virus isolation from cloaca or shell gland when clinical changes are first noted

DIAGNOSIS - POST MORTEM
- Microscopic examination of oviduct for intranuclear inclusions

TRANSMISSION
- Vertical, or direct or indirect contact with contaminated droppings or eggs
- Infected birds shed virus in excrement for up to 2 weeks after infection
- Vertically infected chicks can shed virus immediately after hatching
- Infected free-ranging waterfowl can contaminate ponds

PREVENTION
- Prevent infected free-ranging birds from contaminating ponds
- Inactivated vaccines are effective in chickens, probably effective in ducks *Appendix V*

CONTROL
- Stable outside the host and resistant to many disinfectants
- Inactivated by exposure for 1 hour to formalin, aldehydes and iodophors

TREATMENT
- Supportive care, fluids, assisted feeding and broad-spectrum antibiotics to prevent secondary infections

ZOONOTIC POTENTIAL
- None noted nor expected

FALCON HERPESVIRUS

FAMILY - HERPESVIRIDAE

PRIMARY CLINICAL FEATURES
- Death with no premonitory signs
- Death 24-48 hours after onset of depression and anorexia
- Considered 100% fatal

INCUBATION
- Death 4-6 days after infection (experimental)

COMMON PATHOLOGY
- Hepatomegaly and splenomegaly
- Hyperemia of the small intestines

HISTOPATHOLOGY
- Necrosis in the liver, spleen, intestines and bone marrow
- Intranuclear inclusions in the liver, spleen, intestines and bone marrow

DIFFERENTIAL DIAGNOSIS
- Toxicosis

DIAGNOSIS - POST MORTEM
- Virus isolation from liver, spleen or bone marrow

TRANSMISSION
- Direct contact suspected
- Ingestion of contaminated prey

CONTROL
- Virus would be expected to be susceptible to most virucidal disinfectants

TREATMENT
- No reports on the ability of acyclovir to protect falcons from disease

ZOONOTIC POTENTIAL
- None noted nor expected

INFECTIOUS LARYNGOTRACHEITIS VIRUS

FAMILY - HERPESVIRIDAE

PRIMARY CLINICAL FEATURES
- Coughing, sneezing, dyspnea, bloody discharge from trachea
- Acute death, high mortality

INCUBATION
- Natural - 3-12 days (3-6 days most common)

INFECTED SURVIVORS
- Latently infected
- Resistant to disease

COMMON PATHOLOGY
- Thickening and hemorrhage of the tracheal mucosa
- Accumulation of necrotic debris in the lumen of the trachea
- Blood-tinged mucus in the larynx, trachea or bronchi
- Yellow necrotic plaques on the mucosa of the oropharynx

HISTOPATHOLOGY
- Necrosis of tracheal epithelial cells
- Intranuclear inclusion bodies in tracheal epithelial cells (early in the disease process)

DIFFERENTIAL DIAGNOSIS
- *Hemophilus* sp. (infectious coryza), mycoplasmosis, poxvirus (diphtheritic form), tracheal mites, aspergillosis, bacterial tracheitis, infectious bronchitis virus, vitamin A deficiency, trichomoniasis, candidiasis

DIAGNOSIS - IN THE LIVING BIRD
- Demonstration of intranuclear inclusions in cells collected from the pharynx or trachea
- Virus isolation from respiratory excretions (pharyngeal or tracheal swabs)
- Virus-neutralizing antibodies develop within 4 days of infection

DIAGNOSIS - POST MORTEM
- Microscopic examination of trachea for intranuclear inclusions
- Virus isolation from trachea

TRANSMISSION
- Primarily by inhalation of contaminated respiratory secretions
- Contaminated fomites (virus is relatively stable outside of host)

CONTROL
- Virions unusually stable for a herpesvirus
- Susceptible to sunlight, 5% phenol, formalin, sodium hypochlorite and iodophors
- Vaccination of chickens with subunit, inactivated or attenuated-live vaccine *Appendix V*
- Attenuated-live vaccine causes disease in pheasants

TREATMENT
- Supportive care
- Acyclovir is ineffective

ZOONOTIC POTENTIAL
- None noted nor expected

INFECTIOUS BURSAL DISEASE VIRUS

FAMILY - BIRNAVIRIDAE

PRIMARY CLINICAL FEATURES
- Highly infectious in young chickens
- Severe necrosis of the bursa causes suppression of the immune system
- Anorexia and watery, bile-tinged diarrhea
- Most common in chicks 3-6 weeks of age as maternally derived antibodies decrease
- Panleukopenia, increased activity of LDH and hypergamma-globulinemia
- Mortality rates of 2%-80% have been reported in pheasants

INCUBATION PERIOD
- Chickens - experimental, 3-7 days

INFECTED SURVIVORS
- May have permanently damaged immune system

COMMON PATHOLOGY
- Swelling of the bursa, followed by atrophy as the disease progresses

HISTOPATHOLOGY
- Necrosis of B-lymphocytes in the bursa and spleen
- Infiltration of heterophils in the bursa

DIAGNOSIS - IN THE LIVING BIRD
- Virus recovery from the feces for 2 weeks after infection
- Detection of VN antibodies

DIAGNOSIS - POST MORTEM
- Microscopic examination of bursa
- Recovery of virus from the bursa, thymus, spleen, kidney or cecal tonsils

TRANSMISSION
- Chickens shed virus in feces for up to 2 weeks
- Direct contact with contaminated feces, ocular and respiratory secretions
- Indirect contact with contaminated food, litter or water
- Theoretically, free-ranging birds, rodents and arthropods could be mechanical vectors

PREVENTION
- Maternally derived antibodies protect young chicks from disease
- Attenuated-live and inactivated vaccines *Appendix V*
- Breeder birds are vaccinated prior to laying and a booster is given several months later

CONTROL
- Virus is stable outside the bird
- Virus remains infectious for 122 days in litter and 52 days in food and water
- Virus is inactivated by pH 12, 70°C (30 min), 2% chloramine and glutaraldehyde

ZOONOTIC POTENTIAL
- None noted nor expected

INFLUENZA A VIRUS

FAMILY - ORTHOMYXOVIRIDAE

PRIMARY CLINICAL FEATURES
- Most species of birds of all ages are considered susceptible
- Morbidity and mortality (if any) vary widely among species
- Infections range from asymptomatic to severe respiratory disease and death
- Decreased egg production, diarrhea or edema of the head and neck
- Virulent strains can cause lymphopenia and bleeding disorders
- Clinical changes may be complicated by concomitant disease processes

INCUBATION
- Period varies from several hours to several days depending on the virulence of the subtype and the susceptibility of the host

INFECTED SURVIVORS
- Temporarily resistant to the same subtype of virus, but susceptible to differing subtypes of the virus

COMMON PATHOLOGY
- May die soon after infection with no gross lesions
- Cloudy air sacs and accumulation of necrotic debris in the sinuses, air sacs and trachea
- Hemorrhage and congestion in the brain, lungs and intestinal tract

HISTOPATHOLOGY
- Catarrhal or hemorrhagic tracheitis, hemorrhagic enteritis, air sacculitis, sinusitis and pneumonia

DIFFERENTIAL DIAGNOSIS
- Bacterial or fungal enteritis, tracheitis or pneumonia, chlamydiosis, mycoplasmosis, paramyxovirus, reovirus and some herpesviruses

DIAGNOSIS - IN THE LIVING BIRD
- Virus isolation from cloacal or tracheal swabs, contaminated pond or lake water
- Detection of antibodies by agar-gel immunodiffusion assay
- Detection of subtype specific antibodies by ELISA or VN assays

DIAGNOSIS - POST MORTEM
- Virus isolation from trachea, lung, air sacs, sinuses, spleen, liver, intestines and cloaca

TRANSMISSION
- Feces, respiratory secretions, ocular secretions
- Ducks shed virus from trachea (1-3 days) and cloaca (3-10 days) after infection; other birds may shed from 5 days to 4 weeks
- Free-ranging birds (waterfowl, shore birds) may serve as reservoir
- Contaminated lake or pond water considered important
- Humans could serve as mechanical vectors
- Vertical transmission may occur in some species

PREVENTION/CONTROL
- Stable in lake water from several to over 200 days depending on temperature
- Inactivated by heat, sunlight and most detergents or disinfectants
- Prevent free-ranging birds from entering aviaries or ponds
- Drain, clean and disinfect contaminated ponds or lakes
- Inactivated vaccine available *Appendix V*

TREATMENT
- Supportive care
- Maintain treated birds in isolation for 3-4 weeks after clinical signs resolve

ZOONOTIC POTENTIAL
- Humans infected through contact with birds (mainly poultry, turkey and waterfowl)
- Theoretically possible for humans to infect companion birds

MARBLE SPLEEN DISEASE

FAMILY - ADENOVIRIDAE

PRIMARY CLINICAL FEATURES
- Serologic surveys indicate many infections are subclinical
- Acute death with no clinical signs
- Anorexia, depression, dyspnea, diarrhea, vomiting and death
- Disease most common in 3- to 8-month-old pheasants

INCUBATION PERIOD
- In chickens, 24-48 hours
- In quail, 2-4 days

INFECTED SURVIVORS
- Considered to be latently infected
- Considered resistant to disease

COMMON PATHOLOGY
- Enlarged, mottled spleen (3 times normal size) with gray necrotic areas highlighted by red congested areas
- Edematous lungs

HISTOPATHOLOGY
- Necrosis of the spleen and liver
- Intranuclear inclusions in the spleen, liver, lungs, bursa, bone marrow or proventriculus

DIAGNOSIS - IN THE LIVING BIRD
- Theoretically, DNA probes should detect shedders
- Virus isolation from feces
- Antibodies detected by AGID assay by 2 weeks after infection

DIAGNOSIS - POST MORTEM
- Microscopic examination of tissues for intranuclear inclusions
- Viral-specific DNA probes
- Viral detection using AGID assay and viral-specific antibodies
- Virus isolation from spleen or liver

TRANSMISSION
- Contaminated feces involved in direct and indirect transmission
- Virus can remain infectious in contaminated litter for months
- Not egg transmitted

PREVENTION
- Adequate sanitation helps prevent transmission through a flock
- An attenuated-live virus vaccine is available *Appendix V*

CONTROL
- Adenoviruses are stable outside the host and resistant to many disinfectants
- Inactivated by exposure for 1 hour to formalin, aldehydes and iodophors

TREATMENT
- Supportive care, fluids, assisted feeding and broad-spectrum antibiotics to prevent secondary infections

ZOONOTIC POTENTIAL
- None noted nor expected

MAREK'S DISEASE VIRUS

FAMILY - HERPESVIRIDAE

PRIMARY CLINICAL FEATURES
- Partial or complete paralysis

INCUBATION
- 2 weeks to several months (3-4 weeks most common)

INFECTED SURVIVORS
- Latently infected

COMMON PATHOLOGY
- Enlarged, edematous and grayish peripheral nerves
- Grayish nodules in the ovary, kidney, spleen, skin, skeletal muscle or liver
- Atrophy of the bursa and thymus

HISTOPATHOLOGY
- Accumulation of neoplastic lymphocytes in the peripheral nerves or various organs

DIFFERENTIAL DIAGNOSIS
- Reticuloendotheliosis, lymphoid neoplasias

DIAGNOSIS - IN THE LIVING BIRD
- Virus isolation from white blood cells
- Virus detection in feather pulp using an AGID assay
- Detection of antibodies using AGID, IFA, VN or ELISA assays

DIAGNOSIS - POST MORTEM
- Virus isolation from liver, spleen, kidneys or viral-induced tumors

TRANSMISSION
- Primarily through direct contact with contaminated feather dust
- Contaminated fomites (virus is relatively stable outside of host)

CONTROL AND PREVENTION
- Virions unusually stable for a herpesvirus
- Susceptible to most disinfectants
- Vaccination of chickens with inactivated or attenuated-live vaccine *Appendix V*
- Safety and immunogenicity of vaccines has been defined only in gallinaceous birds

TREATMENT
- Acyclovir reduces severity of disease

ZOONOTIC POTENTIAL
- None noted nor expected

NEWCASTLE DISEASE VIRUS

FAMILY - PARAMYXOVIRIDAE

PRIMARY CLINICAL FEATURES
- Widely varied species susceptibility to disease
- Birds may remain asymptomatic, develop disease and recover, die acutely or die following a protracted illness
- Respiratory signs - ocular and nasal discharge, conjunctivitis, rhinitis, sneezing, coughing, dyspnea, bluish discoloration of facial appendages
- Gastrointestinal signs - anorexia, diarrhea
- Neurologic signs - ataxia, torticollis, opisthotonos, convulsions, circling, tremors and paralysis of the wings and legs

INCUBATION PERIOD
- 3-28 days depending on susceptibility of host and virulence of the virus strain
- 3 days to 2 weeks in experimentally infected psittacines; 5-16 days in naturally infected psittacines
- 28 days in experimentally infected mynahs

NEWCASTLE DISEASE VIRUS *CONT.*

- 2-15 days in experimentally infected chickens

INFECTED SURVIVORS
- May develop intermittent neurologic signs
- Chronic infections in free-ranging Anseriformes, Psittaciformes, Strigiformes and Passeriformes

COMMON PATHOLOGY
- Varies with virus strain and species of bird
- Either no gross lesions or varying combinations of cardiomegaly, hepatomegaly, splenomegaly, mucus in the upper respiratory tract, hemorrhage in the trachea, hemorrhage of the ovaries, hyperemia of the brain and hemorrhage and edema in the respiratory and gastrointestinal tracts
- Gallinaceous species - inflammation and hemorrhage of lymphoid tissue in the jejunum are characteristic of VVND

HISTOPATHOLOGY
- Hemorrhage, necrosis and perivascular cuffing in the brain
- Rarely, intranuclear or intracytoplasmic inclusions in the brain
- Microscopic lesions are rare in birds with chronic CNS signs

DIFFERENTIAL DIAGNOSIS
- Chlamydiosis, salmonellosis, encephalomalacia, heavy metal toxicity, calcium deficiencies, bacterial hepatitis and enteritis

DIAGNOSIS - IN THE LIVING BIRD
- Electron microscopic detection of virus in feces
- Virus isolation from feces or pharyngeal swabs
- Increase in antibody titer demonstrated in paired serum samples
- Antibodies detected from 4-14 days after infection using HI, AGID or ELISA assays
- Antibodies to PMV-3 may cross react with PMV-1

DIAGNOSIS - POST MORTEM
- Virus isolation: intestines, trachea, lung, spleen, liver, brain
- Microscopic examination of brain for inclusions

TRANSMISSION
- Shed primarily in respiratory secretions and feces
- Some chronically infected birds can shed virus for over a year
- Insects, rodents and humans may serve as mechanical vectors
- Rarely, persistently infected migratory birds may transmit the virus
- Vertical transmission is possible, but is unlikely
- Incubator contamination from eggs covered with virus-laden feces

PREVENTION
- Vaccines available for gallinaceous species
- NDV is a reportable disease in many countries
- Avoid birds that may have illegally entered a country
- Combination of inactivated and attenuated live-virus vaccines successful in immunizing aviary birds (Germany)
- Inactivated vaccines available *Appendix V*

CONTROL
- Virions relatively stable (at 50°C for 19 weeks or 28°C for 4 weeks)
- Inactivated by high temperatures (56°C), sunlight, detergents, chloramine (1%), sodium hypochlorite, lysol, phenol and 2% formalin

TREATMENT
- Supportive care
- Hyperimmune serum (2 ml/kg body weight IM)

ZOONOTIC POTENTIAL
- Can cause conjunctivitis in humans
- Antibodies to mumps virus can cross-react with NDV, causing false-positive results

OWL HERPESVIRUS

FAMILY - HERPESVIRIDAE

PRIMARY CLINICAL FEATURES
- Death with no premonitory signs
- Death 2-5 days after onset of depression and anorexia
- Occasionally, small yellow nodules on the pharyngeal mucosa

INCUBATION
- Death 3-9 days after infection (experimental)

INFECTED SURVIVORS
- Considered to be latently infected

COMMON PATHOLOGY
- Hepatomegaly and splenomegaly

HISTOPATHOLOGY
- Necrosis in the liver, spleen and bone marrow
- Intranuclear inclusions in the liver, spleen and bone marrow

DIFFERENTIAL DIAGNOSIS
- Toxicosis

DIAGNOSIS - IN LIVING BIRD
- Survivors develop virus-neutralizing antibodies

DIAGNOSIS - POST MORTEM
- Virus isolation from bone marrow

TRANSMISSION
- Virus excreted in pharyngeal secretions and urine
- Ingestion of virus is considered the principal route of exposure
- Experimental infections following inhalation of contaminated aerosols

CONTROL
- Virus is expected to be susceptible to most virucidal disinfectants

TREATMENT
- No reports on the ability of acyclovir to protect owls from disease

ZOONOTIC POTENTIAL
- None noted nor expected

PACHECO'S DISEASE VIRUS

FAMILY - HERPESVIRIDAE

PRIMARY CLINICAL FEATURES
- Widely varied species susceptibility to disease
- Acute death with no premonitory signs
- Asymptomatic or greenish-yellow diarrhea, regurgitation, central nervous system signs
- Elevated serum enzymes, lipemia, leukopenia, anemia may occur

INCUBATION PERIOD
- Reported range of 3-14 days; 5-7 days most common
- 6-14 days in Quakers inoculated orally

INFECTED SURVIVORS
- Considered to be latently infected
- Considered resistant to disease

COMMON PATHOLOGY
- Enlarged, hemorrhagic, reddish to yellow-brown liver
- Enlarged spleen and congestion of blood in abdominal organs
- Hemorrhage on the surface of various organs

HISTOPATHOLOGY
- Necrosis of the liver, spleen, kidneys, other tissues
- Intranuclear inclusion bodies (liver, spleen, kidney, other tissues)

DIFFERENTIAL DIAGNOSIS
- Bacterial hepatitis, chlamydiosis, salmonellosis, heavy metal toxicity, avian polyomavirus, reovirus, adenovirus

DIAGNOSIS - IN THE LIVING BIRD
- Birds that are shedding virus should be detectable using DNA probes
- Virus isolation from feces or pharyngeal swabs
- Previously infected birds (carriers) - Detection of antibodies

DIAGNOSIS - POST MORTEM
- Microscopic examination of tissues for intranuclear inclusions
- Viral-specific DNA probes
- Virus isolation from liver, spleen, brain or kidney

TRANSMISSION
- Shed in feces and pharyngeal secretions several days before death
- Enters new bird when ingested
- Latently infected birds shed virus intermittently

PREVENTION
- Inactivated vaccine available *Appendix V*

CONTROL
- Virions unstable when outside of bird
- Most disinfectants inactivate the virus

TREATMENT
- Acyclovir - 80 mg/kg orally, every 8 hours

ZOONOTIC POTENTIAL
- None noted nor expected

PAPILLOMATOSIS

FAMILY - UNDETERMINED

PRIMARY CLINICAL FEATURES
- Suspected, but unproven, to be caused by a virus
- Most birds are asymptomatic
- Cloacal masses - tenesmus, putrid-smelling feces, infertility, recurrent enteritis, hematochezia, recurrent prolapses, accumulation of droppings around the vent
- Oral or esophageal masses - halitosis, dysphagia, dyspnea or wheezing
- Intestinal tract masses - anorexia, chronic weight loss or vomiting; may mimic proventricular dilatation disease

INCUBATION
- Cause of papillomatosis and incubation period are unknown

COMMON PATHOLOGY
- Proliferative masses on mucosa of the oral cavity, gastrointestinal tract or cloaca
- Cancer of the liver and pancreas are common in affected birds

HISTOPATHOLOGY
- Proliferations of epithelial cells on a well vascularized, fibrous stalk
- Classic histopathologic description involves acanthosis and hyperkeratosis

DIFFERENTIAL DIAGNOSIS
- Neoplasms, oral masses, vitamin A deficiency, poxvirus

DIAGNOSIS - IN THE LIVING BIRD
- 5% acetic acid will cause suspect lesions to appear white
- Microscopic examination of suspect lesions for characteristic changes
- Liver biopsy if cancer is suspected

DIAGNOSIS - POST MORTEM
- Microscopic examination of suspect lesions for characteristic changes
- Microscopic examination of the liver and pancreas for neoplastic changes

TRANSMISSION
- Unknown; if infectious agent is involved, it appears to be of low transmissibility
- Chicks from parents with cloacal papillomatosis appear to remain clinically unaffected if the eggs are artificially incubated

PREVENTION
- Restrict contact between affected and unaffected birds
- Examine all birds during the quarantine period to prevent introduction of affected birds

TREATMENT
- Oral or esophageal lesions can be surgically or radiosurgically removed
- Cloacal lesions are best removed with silver nitrate cauterization

ZOONOTIC POTENTIAL
- None noted nor expected

PAPILLOMAVIRUS AND PAPILLOMAS

FAMILY - PAPOVAVIRIDAE

PRIMARY CLINICAL FEATURES
- Formation of benign skin tumors (warts) in some birds

INCUBATION
- In mammals, the incubation period varies from weeks to months
- An incubation period for birds has not been confirmed

COMMON PATHOLOGY
- Proliferative skin masses particularly on the face, legs and feet

HISTOPATHOLOGY
- Long thin folds of hyperkeratotic epidermis accompanied by acanthotic parakeratosis

DIFFERENTIAL DIAGNOSIS
- Proliferative skin lesions induced by herpesvirus or poxvirus

DIAGNOSIS
- Electron microscopic examination of affected tissue to demonstrate virus particles

TRANSMISSION
- In mammals, direct skin-to-skin contact
- Unproven in birds, but direct skin-to-skin contact most likely

PREVENTION
- Avoid contact between infected and susceptible birds

TREATMENT
- Removal of affected tissue, if necessary

CONTROL
- Mammalian papillomaviruses are stable outside of the host
- A similar stability would be expected with avian papillomavirus

ZOONOTIC POTENTIAL
- None described nor expected
- Papillomaviruses are highly host-specific

PICORNAVIRUS

FAMILY - PICORNAVIRIDAE

PRIMARY CLINICAL FEATURES
- Cockatoos - yellowish-green, mucoid diarrhea, death in 3 days to 4 weeks

DUCKS - DHV I
- Mortality of 100% in ducklings less than 1 week old
- Asymptomatic in ducklings greater than 6 weeks old
- Progressive lethargy, unwillingness to move, ataxia, recumbency and death
- May be complicated by chlamydia or influenza A virus infections

DUCKS - DHV III
- Mortality from 10% to 50% in birds less than 2 weeks of age
- Asymptomatic in ducklings greater than 2 weeks of age
- Progressive lethargy, unwillingness to move, ataxia, recumbency and death

INCUBATION PERIOD
- Cockatoos - experimental, 5 days
- Ducks - less than 24 hours

INFECTED SURVIVORS
- Ducks - resistant to disease

COMMON PATHOLOGY
- Cockatoos - thickened intestinal walls, dilated intestines filled with mucoid fluid and gas
- Ducks - enlarged, congested liver with reddish-brown mottling and hemorrhage

HISTOPATHOLOGY
- Cockatoos - elongated crypts, villous atrophy, vacuolation of the enterocytes; intracytoplasmic inclusions in enterocytes
- Ducks - severe hepatic necrosis

DIFFERENTIAL DIAGNOSIS
- Ducks - chlamydiosis, influenza A virus, salmonellosis and aflatoxicosis

DIAGNOSIS - IN THE LIVING BIRD
- Cockatoos - virus particles in feces by electron microscopy
- Ducks - VN antibodies detectable 1 week after infection

DIAGNOSIS - POST MORTEM
- Cockatoos - microscopic examination of intestines for intracytoplasmic inclusions
- Ducks - virus isolation from the liver

TRANSMISSION
- Cockatoos - shed in feces; birds in direct contact develop disease
- Ducks - virus shed in feces from 1 day to 8 weeks after infection
- Contaminated litter, equipment, clothing, brooders important
- Vertical transmission has not been confirmed

PREVENTION
- Ducks - hyperimmune serum, inactivated and attenuated-live virus vaccines *Appendix V*
- Careful vaccination in face of outbreak will decrease mortality

CONTROL
- Virions stable when outside the bird
- Inactivated by 2% sodium hypochlorite (3 hours), 5% phenol, 3% chloramine

TREATMENT
- Supportive care
- Birds with clinical signs should be isolated for at least 8 weeks

ZOONOTIC POTENTIAL
- None noted nor expected

PIGEON HERPESVIRUS

FAMILY - HERPESVIRIDAE

PRIMARY CLINICAL FEATURES
- Conjunctivitis, rhinitis, dyspnea
- Ulcerative lesions on the mucosa of the larynx, pharynx or oral cavity
- Diarrhea, vomiting
- Neurologic signs

INCUBATION PERIOD
- Experimental - 1 to 3 days

INFECTED SURVIVORS
- Latently infected
- Considered resistant to disease unless immunosuppressed

COMMON PATHOLOGY
- Ulcers on the mucosa of the alimentary tract, larynx and pharynx
- Hepatomegaly, splenomegaly
- Congestion of the intestines
- Formation of sialoliths (unproven to be caused by PHV-1)

HISTOPATHOLOGY
- Necrosis of the liver, spleen, pancreas, kidneys, lungs
- Intranuclear inclusion bodies in pharyngeal mucosa, liver, spleen, pancreas and tracheal mucosa

DIFFERENTIAL DIAGNOSIS
- Chlamydiosis, salmonellosis, trichomoniasis, poxvirus (diphtheritic form), paramyxovirus (PMV-1), adenovirus

DIAGNOSIS - IN THE LIVING BIRD
- Virus-neutralizing antibodies detectable within a week of infection (inconsistent in carriers)
- Virus isolation from feces or respiratory secretions (pharyngeal swabs)
- Inclusion bodies in impression smears from ulcerative lesions

DIAGNOSIS - POST MORTEM
- Microscopic examination of tissues for intranuclear inclusions
- Virus isolation from pharynx, liver, spleen or pancreas

TRANSMISSION
- Shed in feces and respiratory secretions
- Contaminated secretions fed to squabs
- Egg transmission theoretically possible but unproven
- Experimentally infected birds shed virus in 1-3 days

PREVENTION
- Experimental vaccine prevents disease but not latent infections

CONTROL
- Virions unstable outside of an infected pigeon
- Virus considered susceptible to most disinfectants

TREATMENT
- Acyclovir is ineffective
- Trisodium phosphonoformate does not prevent latent infections

ZOONOTIC POTENTIAL
- None noted nor expected

PIGEON POXVIRUS

FAMILY - POXVIRIDAE

PRIMARY CLINICAL FEATURES
- Cutaneous form - small blister-like areas that progressively enlarge, ulcerate and scab
- Large wart-like proliferative growths on the skin, eyelids, commisure of the beak, feet and legs
- Lesions are common around the cloaca and umbilicus of unfeathered young
- Uncomplicated lesions resolve in 3-4 weeks
- Diphtheritic form - thickened, yellowish, necrotic plaques in mouth, throat, crop or trachea
- Accumulation of necrotic debris in trachea may cause asphyxiation
- Skin tumors
- Mortality highest in young pigeons, and may approach 50%

INCUBATION PERIOD
- 7-9 days

INFECTED SURVIVORS
- Some birds are considered to be persistently infected
- Previously infected pigeons develop lesions when given immunosuppressive drugs

COMMON PATHOLOGY
- Many gross lesions are visible on the skin or oral mucosa
- Fibrinonecrotic lesions in the gastrointestinal tract
- Ulceration of the nasal, tracheal and bronchial mucosa

HISTOPATHOLOGY
- Necrosis of cells lining the mouth, esophagus, crop or trachea
- Bollinger bodies seen in skin or mucosa of the sinuses, trachea, crop, esophagus or throat

DIFFERENTIAL DIAGNOSIS
- Cutaneous form - trauma, *Trichophyton* sp. and bacterial dermatitis
- Diphtheritic form - candidiasis, hypovitaminosis A, aspergillosis, trichomoniasis or herpesvirus

DIAGNOSIS - IN THE LIVING BIRD
- Virus isolation from vesicles, necrotic mucosa, pharyngeal swabs or feces
- Demonstration of poxvirus in lesions by cytology or electron microscopy
- AGID, VN, ELISA or HI assays to detect antibodies

DIAGNOSIS - POST MORTEM
- Microscopic examination of tissues for intracytoplasmic inclusions
- Virus isolation from skin lesions, mucosal lesions or liver

TRANSMISSION
- A break in the epithelium is necessary for infection
- Direct contact with an infected bird or indirect contact with a contaminated object or insect
- Mosquitoes and mites serve as the primary mechanical vectors
- Epornitics common in the spring and fall when mosquitoes are most prevalent
- Affected birds most infectious when lesions or scabs are present
- Infected free-ranging pigeons and doves may serve as a source of virus for captive birds

PREVENTION
- Attenuated-live virus vaccine available *Appendix V*
- Vaccinate pigeons in high risk areas at 4-6 weeks of age
- Vaccinate pigeons 1 month before the race or show season

CONTROL
- Virions stable when outside of bird
- Inactivated with 1% potassium hydroxide, heating to 50°C for 30 minutes, heating to 60°C for 8 minutes, steam, 2% NaOH and 5% phenol
- Isolate infected birds and those birds directly exposed
- Reduce overcrowding
- Thoroughly clean and disinfect the loft frequently
- Destroy contaminated wooden perches or nest boxes
- Sterilize contaminated nets, towels, clothing, holding containers, equipment, supplies
- Use biologically secure shipping containers

TREATMENT
- Keep uncomplicated lesions clean and do not disturb
- Gently remove infected tissue
- Thoroughly cleanse open wounds and apply topical antimicrobial agents
- Use systemic antibiotics in birds with lesions in the respiratory or gastrointestinal tracts
- Treat secondary candidiasis or trichomoniasis
- Supplement vitamin A (10,000-25,000 IU/300g body weight IM) and vitamin C

ZOONOTIC POTENTIAL
- None noted nor expected

POLYOMAVIRUS

FAMILY - PAPOVAVIRIDAE

PRIMARY CLINICAL FEATURES
BUDGERIGARS

- Sudden death in birds less than 15 days of age
- Older Budgerigars - Abdominal distention, hemorrhage under the skin and reduced formation of down and contour feathers
- 10% of affected neonates - ataxia and tremors of the head and neck
- Decreased egg hatchability, early embryonic death and high hatchling mortality

PRIMARY CLINICAL FEATURES
NONBUDGERIGAR PSITTACINE BIRDS

- Subclinical infections are most common
- Peracute death with no clinical signs in young birds
- Acute death 12-48 hours after depression, delayed crop emptying, regurgitation, diarrhea, subcutaneous hemorrhage
- Bleeding from injection sites or feather follicles
- Hematuria in Eclectus Parrots
- Feather abnormalities are rare

INCUBATION

- Experimentally infected Budgerigar neonates died 11 days after inoculation
- Experimentally infected nonbudgerigar psittacines shed virus in feces starting from 2-7 days after inoculation
- Experimentally infected chickens can seroconvert within three days of exposure

INFECTED SURVIVORS

- Budgerigars may develop feather abnormalities or neurologic signs; considered to be latently infected
- Most infections in nonbudgerigar psittacines are subclinical, and infected birds develop an antibody response; latent infections are suspected but not confirmed

COMMON PATHOLOGY

- In Budgerigars - hydropericardium, cardiomegaly, hepatomegaly, ascites, splenomegaly, hemorrhage of the intestines and heart, subcutaneous hemorrhage
- In nonbudgerigar psittacines - hepatomegaly, splenomegaly, pale swollen kidneys, pale cardiac and skeletal muscles, subcutaneous hemorrhage of the intestines, liver and heart

HISTOPATHOLOGY

- In Budgerigars - necrosis in the heart, liver and kidneys; intranuclear inclusion bodies in most tissues, particularly the feather follicles, kidney, liver, heart and spleen
- In nonbudgerigar psittacines - Massive liver necrosis and hemorrhage throughout the body are characteristic Intranuclear inclusions bodies may occur in the liver, spleen and kidneys

DIFFERENTIAL DIAGNOSIS

- Feather lesions - PBFD virus, adenovirus, endocrine abnormalities, bacterial infections, fungal infections, traumatic injuries and drug reactions (cephalosporins and penicillins)
- Systemic lesions - Chlamydiosis, liver disease, clotting disorders, bacterial septicemia, Pacheco's disease virus, adenovirus and reovirus

DIAGNOSIS - IN THE LIVING BIRD

- Shedders - viral-specific DNA probes
- DNA probe detection of virus in excrement, microscopic evaluation of liver biopsies, demonstration of a four-fold increase in antibody titer in paired serum samples; infected birds may have increased activities of LDH, AST and AP; hematuria may occur in Eclectus Parrots
- Exposed birds - DNA probe detection of virus in excrement or demonstration of a four-fold increase in antibody titer in paired serum samples

DIAGNOSIS - POST MORTEM

- Demonstration of virus particles by electron microscopy, isolation of the virus in cell culture, specialized staining of suspect lesions using viral-specific antibodies or the detection of viral nucleic acid using polyomavirus-specific DNA probes

TRANSMISSION

- Budgerigars - intranasal (experimental); exposure to contaminated aerosols and direct exposure to contaminated feces, crop excretions, feather dust, urates and respiratory secretions suspected in natural cases; vertical transmission
- Nonbudgerigar psittacines - orally, intramuscularly and intravenously (experimental); virus present in excrement; vertical transmission not documented
- Chickens - intramuscularly and intravenously (experimental)

PREVENTION

- Vaccination - inactivated vaccine available *Appendix V*
- Sound hygienic practices, maintaining closed aviaries, preventing visitors from entering avian nurseries and attempting to identify and isolate subclinical shedders using viral-specific DNA probes *Appendix V*

TREATMENT

- Supportive care - unsubstantiated reports that interferon and other immunostimulants may be helpful in some early cases

CONTROL

- Polyomavirus is environmentally stable and resistant to inactivation by chlorhexidine, and partially resistant to some iodine containing disinfectants and quaternary ammonium; inactivated by Clorox, stabilized chlorine dioxide, phenol and ethanol

ZOONOTIC POTENTIAL

- None demonstrated nor expected

POXVIRUS IN PSITTACINES

FAMILY - POXVIRIDAE

PRIMARY CLINICAL FEATURES

GENERAL CONSIDERATIONS
- Widely varied species susceptibility to disease
- Serous ocular discharge, rhinitis and conjunctivitis followed by ulcerations on the eyelids
- Dry crusty lesions on the lid margins and corners of the eye
- Lesions most severe when secondary infections are present
- Distortions in the margins of the eyelids and loss of filoplumes around the eyes

AMAZON PARROTS
- Severe upper respiratory tract disease
- Diphtheritic lesions on the oral, pharyngeal, esophageal or crop mucosa
- Respiratory distress
- Depression, anorexia, diarrhea and bloody stools
- Mortality rates are highest when diphtheritic lesions are secondarily infected

LOVEBIRDS
- Cutaneous or diphtheritic forms of poxvirus
- 75% morbidity and mortality in highly susceptible populations

INCUBATION PERIOD
- 7 days - experimental in Amazon parrots and conures
- 10-14 days - natural infections in Amazon parrots

INFECTED SURVIVORS
- Chickens and lovebirds considered to be persistently infected
- Considered resistant to disease by same strain of virus for 6 months to 1 year

COMMON PATHOLOGY
- Many gross lesions are visible on the skin or oral mucosa
- Fibrinonecrotic lesions in the gastrointestinal tract
- Ulceration of the nasal, tracheal and bronchial mucosa

HISTOPATHOLOGY
- Necrosis of cells lining the mouth, esophagus, crop or trachea
- Bollinger bodies in skin or mucosa of the sinuses, trachea, crop, esophagus or throat

DIFFERENTIAL DIAGNOSIS
- Cutaneous form - trauma, *Trichophyton* sp., knemidokoptes mites, papillomavirus and bacterial infection
- Diphtheritic form - candidiasis, hypovitaminosis A, aspergillosis, trichomoniasis or herpesvirus

DIAGNOSIS - IN THE LIVING BIRD
- Virus isolation from vesicles, necrotic mucosa, pharyngeal swabs or feces
- Demonstration of poxvirus in lesions by cytology or electron microscopy
- AGID, VN, ELISA or HI assays to detect antibodies

DIAGNOSIS - POST MORTEM
- Microscopic examination of tissues for intracytoplasmic inclusions
- Virus isolation from skin lesions, mucosal lesions or liver

TRANSMISSION
- A break in the epithelium is necessary for infection to occur
- Direct contact with an infected bird or indirect contact with a contaminated object or insect
- Mosquitoes and mites serve as the primary mechanical vectors
- Epornitics common in the spring and fall when mosquitoes are most prevalent
- Affected birds most infectious when lesions or scabs are present
- Infected free-ranging birds may serve as source of virus for some species

PREVENTION
- Poxvirus infections in psittacines can be prevented through vaccination *Appendix V*

CONTROL
- Virions stable when outside of bird
- Inactivated with 1% potassium hydroxide, temperatures of 50°C for 30 minutes or 60°C for 8 minutes, steam, 2% NaOH and 5% phenol
- Isolate infected birds and those birds to which they are directly exposed
- Destroy contaminated wooden perches or nest boxes
- Sterilize contaminated nets, towels, clothing, holding containers, equipment, supplies
- Use biologically secure shipping containers

TREATMENT
- Keep uncomplicated lesions clean and do not disturb
- Gently remove infected tissue
- Thoroughly cleanse open wounds and apply topical antimicrobial agents
- Use systemic antibiotics for birds with lesions in the respiratory or gastrointestinal tract
- Treat secondary candidiasis or trichomoniasis
- Supplement vitamin A (10,000-25,000 IU/300g body weight IM) and vitamin C

ZOONOTIC POTENTIAL
- None noted nor expected

PIGEON PARAMYXOVIRUS-1

FAMILY - PARAMYXOVIRIDAE

PRIMARY CLINICAL FEATURES
- Incoordination, tremors of the head and wings, circling, paralysis, head shaking and torticollis
- Polydipsia, polyuria and diarrhea (may be watery or hemorrhagic)
- Clinical signs in 20% to 80% of exposed pigeons
- Mortality in up to 90% of young pigeons; older birds typically recover in 3-4 weeks

INCUBATION PERIOD
- 5 days to over a month (natural); 5-18 days (experimental)

INFECTED SURVIVORS
- Undetermined status

COMMON PATHOLOGY
- Gross lesions may or may not occur
- Hyperemia of the brain and abdominal organs, catarrhal enteritis, enlarged kidneys and hemorrhage of the pancreas

HISTOPATHOLOGY
- Inflammation of the intestines is common during early infection
- Hepatitis and accumulation of lymphocytes and plasma cells in the kidneys
- Non-suppurative encephalitis, hepatic necrosis and pancreatitis

DIFFERENTIAL DIAGNOSIS
- Chlamydiosis, salmonellosis, Newcastle disease virus, pigeon herpesvirus, heavy metal toxicity, organophosphate poisoning, bacterial hepatitis, bacterial enteritis

DIAGNOSIS - IN THE LIVING BIRD
- Electron microscopic detection of virus in feces
- Virus isolation from feces or pharyngeal swabs
- Increase in antibody titer using paired serum samples

DIAGNOSIS - POST MORTEM
- Virus isolation from intestines, trachea, lung, spleen, liver and brain

TRANSMISSION
- Ingestion/inhalation of contaminated respiratory secretions or feces
- Virus excreted in laryngeal secretions from 2-9 days after infection and in feces 2-14 days after infection
- Shedding may continue for up to a month
- Free-ranging pigeons can introduce virus to a loft

PREVENTION
- Vaccination prevents disease but not infection or shedding
- Both attenuated live-virus and inactivated vaccines have been tested *Appendix V*
- Vaccinate adults 4 months prior to the breeding, racing or show season; squabs at 3-4 weeks of age; new pigeons during the quarantine period

CONTROL
- Virions are relatively stable outside of a bird
- Stable at 50°C for 19 weeks and 27°C or 40°C for 4 weeks
- Inactivated by high temperatures (56°C), sunlight, detergents, chloramine (1%), sodium hypochlorite, lysol, phenol and 2% formalin
- Isolate pigeons with clinical signs
- Quarantine new birds or those that leave the aviary for 6-8 weeks
- Prevent free-ranging pigeons from entering the loft
- Use biologically secure shipping containers

TREATMENT
- Supportive care, fluids and antibiotics to prevent secondary infections

ZOONOTIC POTENTIAL
- None reported nor expected

PROVENTRICULAR DILATATION DISEASE

FAMILY - UNDETERMINED

PRIMARY CLINICAL FEATURES
- Etiology not confirmed
- Gastrointestinal signs, neurologic signs or a combination
- Progressive weight loss despite a good appetite
- Passage of poorly digested food, regurgitation
- Central nervous system signs including ataxia and seizures
- Increased CPK activity may be an indication of PDD
- Hypoproteinemia, hypoglycemia, heterophilia and anemia
- Any age bird susceptible; however, adult birds appear to be more commonly affected
- Diagnosed in hand-raised and wild-caught psittacines

INCUBATION PERIOD
- Not confirmed
- Clinical observations suggest several weeks to 4 years

INFECTED SURVIVORS
- Unknown status
- Many exposed birds remain asymptomatic

COMMON PATHOLOGY
- Emaciation
- Dilation of the proventriculus and/or ventriculus
- Flaccid, ingesta-filled crop
- Dilatation of the intestines with undigested food and gas

HISTOPATHOLOGY
- Accumulation of lymphocytes and plasma cells in the nerves of the gastrointestinal tract, brain or spinal cord

DIFFERENTIAL DIAGNOSIS
- Fungal proventriculitis, megabacteriosis, parasitic enteritis, foreign bodies, neoplasia, heavy metal toxicosis, bacterial enteritis, papillomatosis of the proventriculus/ventriculus or any intraluminal or extra-luminal mass that prevents the passage of food
- Proventriculus of neonates is normally dilated

DIAGNOSIS - IN THE LIVING BIRD
- Acute phase - suspected in birds with rapid gastrointestinal transit time
- Chronic phase - suspected in birds with slowed gastrointestinal transit time
- Accumulation of lymphocytes and plasma cells in association with the nerves of the gastrointestinal tract or central nervous system
- Crop biopsy diagnostic in some cases

DIAGNOSIS - POST MORTEM
- Accumulation of lymphocytes and plasma cells in association with the nerves of the gastrointestinal tract or central nervous system

TRANSMISSION
- Not confirmed - suspect direct and indirect transmission

CONTROL AND PREVENTION
- Avoid direct or indirect contact with affected or exposed birds

TREATMENT
- Supportive care, assisted alimentation, broad-spectrum antibiotics as needed
- Clinical observations suggest that interferon may be helpful in some cases

ZOONOTIC POTENTIAL
- Undetermined

PSITTACINE BEAK AND FEATHER DISEASE

FAMILY - CIRCOVIRIDAE

PRIMARY CLINICAL FEATURES
- Peracute infection - pneumonia, enteritis, rapid weight loss and death
- Acute infection - depression followed by necrosis, fractures, bending, hemorrhage or premature shedding of developing feathers
- Chronic infection - progressive appearance of abnormally developed feathers during each successive molt
- Feather changes include retention of feather sheaths, hemorrhage within the pulp cavity, fractures of the feather shaft, short clubbed feathers, deformed curled feathers, stress lines within vanes and circumferential constrictions
- Beak pathology - if present, includes progressive beak elongation with transverse or longitudinal fractures, palatine necrosis and ulcerations in the mouth

INCUBATION PERIOD
- Minimum experimental incubation period is 21-25 days, depending on species
- Maximum documented incubation period is 18 months, but suspected to be years

INFECTED SURVIVORS
- Considered resistant to disease

COMMON PATHOLOGY
- Gross feather and beak changes listed under clinical features
- In young birds, the cloacal bursa and thymus may be small

HISTOPATHOLOGY
- Necrosis and swelling of cells lining the developing feather
- Accumulation of plasma cells, lymphocytes, macrophages and heterophils in the feather pulp
- Necrosis in the bursa and thymus
- Intranuclear inclusions in epithelial cells
- Intracytoplasmic inclusions within macrophages

DIFFERENTIAL DIAGNOSIS
- Avian polyomavirus, adenovirus, trauma, bacterial folliculitis, fungal folliculitis, septicemias, malnutrition, endocrine abnormalities, reaction to antibiotics

DIAGNOSIS - IN THE LIVING BIRD
- Active or subclinical infections detected using DNA probes on blood
- Active infections - demonstration of intracytoplasmic inclusions in diseased feathers
- DNA probe detection of virus in feces or crop secretions
- Detection of HI antibodies indicate previous infection

DIAGNOSIS - POST MORTEM
- Microscopic examination of tissues for characteristic intracytoplasmic inclusions
- Viral-specific DNA probes or viral-specific antibodies to confirm suspect lesions

TRANSMISSION
- Shed in feces, crop secretions and feather dust
- Contact with virus-contaminated materials or surfaces
- Affected hens may pass the virus to the developing egg

PREVENTION
- Establishing and maintaining negative flocks using DNA probe testing
- Avoid direct or indirect contact with infected birds
- Use shipping containers that prevent contact with the PBFD virus during transport *Appendix V*

CONTROL
- Virions in feces or feather dust are suspected to be extremely durable
- Sodium hypochlorite, stabilized chlorine dioxide or glutaraldehyde are probably the best disinfectants
- Virus contamination of an environment can be determined using DNA probe testing

TREATMENT
- A treatment that eliminates the virus nucleic acid from the blood has not been reported
- Supportive care and broad-spectrum antibiotics may improve overall condition
- Affected birds must not be in direct or indirect contact with susceptible birds

ZOONOTIC POTENTIAL
- None noted nor expected

REOVIRUS

FAMILY - REOVIRIDAE

PRIMARY CLINICAL FEATURES
- Morbidity and mortality highest with concomitant infections
- Galliformes - most asymptomatic or arthritis/tenosynovitis, diarrhea, stunting, osteoporosis, feather abnormalities
- Chickens - experimental rarely as severe as natural infections
- Psittacines - asymptomatic or weight loss, diarrhea, dyspnea, ascites, subcutaneous hemorrhage, edema of the head and legs
- Lories and lorikeets - sudden death in fledglings
- Anemia, hypoproteinemia, hypoalbuminemia and hyperglobulinemia
- Leukocytosis early, followed by leukopenia
- Increases in AST and LDH activities terminally
- Pigeons - diarrhea alone or in conjunction with dyspnea

INCUBATION PERIOD
- Reported range of 2-9 days in most species

INFECTED SURVIVORS
- In chickens, some survivors considered to be latently infected
- Latently infected hens can transmit virus to their developing eggs

COMMON PATHOLOGY
- Pale yellow mottling of an enlarged liver and spleen, ascites, subcutaneous hemorrhage; hyperemia of the intestines
- Chickens - swollen hocks

HISTOPATHOLOGY
- Disseminated or focal necrosis of the liver
- Tenosynovitis common in chickens, turkeys
- Intracytoplasmic inclusions occasionally seen in liver

DIFFERENTIAL DIAGNOSIS
- Avian polyomavirus, adenovirus, Pacheco's disease virus, bacterial hepatitis, chlamydiosis, salmonellosis

DIAGNOSIS - IN THE LIVING BIRD
- Virus in feces or joint fluid by electron microscopy
- Virus isolation from feces, pharyngeal swabs, or swollen joints
- Chickens - antibodies within 17 days; persist over 6 months
- Psittacines - precipitating antibodies within 7 to 21 days, persist 2 to 6 weeks

DIAGNOSIS - POST MORTEM
- Examination of liver for necrosis and intracytoplasmic inclusions
- Virus isolation from liver, intestinal tract, joint fluid or lung

TRANSMISSION
- Shed principally in feces
- Chickens - experimentally by oral, intratracheal or IM route
- Egg transmission in ducks, geese, turkeys and chickens
- Maternal antibodies protect young chicks from infection

PREVENTION
- Galliformes - inactivated and attenuated-live virus vaccines
- Psittacines - experimental vaccine, chicken vaccine not effective
- Ducks - attenuated-live chicken vaccine protective in some ducks
- Goslings - hyperimmune serum provides temporary protection

CONTROL
- Stable in organic materials and respiratory secretions
- Inactivated by prolonged contact (hours) with phenols, aldehydes, heating to 70°C, 70% ethanol, halides, formalin and iodine

TREATMENT
- Supportive care, fluids and broad-spectrum antibiotics

ZOONOTIC POTENTIAL
- None noted nor expected

RETICULOENDOTHELIOSIS VIRUS

FAMILY - RETROVIRIDAE

PRIMARY CLINICAL FEATURES
- Widely varied species susceptibility to disease
- Mortality varies with the species from 2% to 100%
- Immunosuppressive, predisposing infected birds to secondary infections
- Chickens - many are asymptomatic; rapid formation of neoplasias; immunodepression; failure to thrive in young birds; formation of lymphoid tumors; abnormal feather growth
- Pheasants - tumors of the head and in the mouth; distention of the infraorbital sinuses and ulcerative lesions in the mouth; many infected pheasants die by 6-12 months of age
- Turkeys - may die without clinical signs or may develop lameness and diarrhea

INCUBATION PERIOD
- Chickens - experimental, 2-3 weeks
- Turkeys - experimental, 8-11 weeks

INFECTED SURVIVORS
- Some recovered birds are considered persistently infected

COMMON PATHOLOGY
- Decrease in size of the bursa, spleen and thymus
- Reduced weight gain in young birds
- Liver and spleen are enlarged and contain multiple yellowish-to grayish-white nodules
- Tumors in the kidney, bursa, gonads and the digestive tract

HISTOPATHOLOGY
- Proliferation of reticular and endothelial cells that line capillaries of affected tissues

DIFFERENTIAL DIAGNOSIS
- Pheasants - masses on head and in mouth are similar to poxvirus
- Enlargement of the peripheral nerves similar to Marek's disease virus
- Lymphoid leukosis and other neoplasias

DIAGNOSIS - IN THE LIVING BIRD
- Virus isolation from tumors or blood (if bird is viremic)
- Detection of antibodies in serum or yolks using IFA, AGID, ELISA or VN assays
- Antibodies detectable 2 weeks after infection unless bird is immunotolerant; turkeys infected in the oviduct or soon after hatch are immunotolerant
- Antibodies may persist for 35 weeks in some birds

DIAGNOSIS - POST MORTEM
- Microscopic examination of affected tissues
- Virus isolation from liver, spleen, kidney, ovary or oviduct

TRANSMISSION
- Horizontal - direct contact with pharyngeal secretions, feces, litter or during copulation
- Vertical - common in ducks, but considered rare in other species
- Domestic fowl - use of virus-contaminated vaccines
- Mosquitoes and other blood-feeding arthropods can serve as mechanical vectors

PREVENTION AND CONTROL
- Virions unstable when outside the bird
- Most disinfectants inactivate the virus

ZOONOTIC POTENTIAL
- None noted nor expected

ROTAVIRUS

FAMILY - REOVIRIDAE

PRIMARY CLINICAL FEATURES
- Virus common in feces of asymptomatic gallinaceous birds
- When they occur, diarrhea and weight loss are the most common clinical changes
- Young birds are most severely affected

INCUBATION PERIOD
- Reported range of 1 to 5 days in chickens and turkeys

INFECTED SURVIVORS
- Latent infections not demonstrated

COMMON PATHOLOGY
- Dilated intestinal loops filled with watery material and gas

HISTOPATHOLOGY
- Necrosis of the enterocytes

DIFFERENTIAL DIAGNOSIS
- Any bacterial, viral, parasitic or environmental cause of diarrhea

DIAGNOSIS
- Virus in feces by electron microscopy
- Virus isolation from feces
- Antibodies detected within 4 to 6 days in chickens and turkeys

TRANSMISSION
- Virus excreted in feces from 1 to 5 days after infection
- Virus accumulation in enclosures or ponds
- Cross-facility transmission likely in contaminated aviaries with poor sanitation
- Maternal antibodies do not protect young chicks from infection

PREVENTION
- Vaccines not available
- Hygiene to reduce transmission

CONTROL
- Stable in feces
- Iodophors most effective disinfectant

TREATMENT
- Supportive care, fluids and broad-spectrum antibiotics to prevent secondary infections

ZOONOTIC POTENTIAL
- None noted nor expected

WESTERN EQUINE ENCEPHALITIS VIRUS

FAMILY - TOGAVIRIDAE

PRIMARY CLINICAL FEATURES
- Most infections are asymptomatic
- WEE virus infections generally less severe and result in lower mortality than EEE virus infections
- If present, clinical signs include ruffled feathers, depression, weakness, incoordination, head tilt, paresis and paralysis
- Epornitics are most common in the summer and fall

EMUS
- Lethargy and anorexia, followed by torticollis, ataxia, paralysis and death
- Elevated activities of AST and high white blood cell count
- Morbidity from 15% to 50%, mortality 10% to 15%

INCUBATION PERIOD
- Varies with the species
- Viremia 1 to 4 days after infection

INFECTED SURVIVORS
- Considered resistant to disease

WESTERN EQUINE ENCEPHALITIS VIRUS

COMMON PATHOLOGY
- Varies among affected birds
- Small pale spleen or splenomegaly with mottling
- Softening of the cerebral hemispheres
- Enlarged, mottled liver, ascites
- Distention of the kidneys

EMUS
- Accumulation of fluid around the heart
- Swelling of the liver, kidneys and membranes covering the brain and spinal cord

HISTOPATHOLOGY
- Some species develop lesions primarily in the brain
- Inflammation of the blood vessels in the brain
- Non-suppurative encephalitis
- Perivascular lymphocytic infiltrates in the brain
- Some species develop lesions primarily in the abdominal organs
- Hepatic and myocardial necrosis

EMUS
- Inflammation and necrosis of most tissues, including the heart, lungs, liver, spleen, kidneys, digestive tract and brain
- Vasculitis with perivascular accumulation of lymphocytes

DIFFERENTIAL DIAGNOSIS
- Paramyxovirus, bacterial encephalitis, other alphaviruses, botulism, pasteurellosis and salmonellosis

DIAGNOSIS - IN THE LIVING BIRD
- Antibodies detected within 2 weeks after infection
- CF and HI antibodies may persist for several weeks to months
- Virus-neutralizing antibodies may persist for months to years
- Virus isolation from fresh blood

DIAGNOSIS - POST MORTEM
- Microscopic examination of affected tissues
- Virus isolation from fresh, liver, spleen or intestines

TRANSMISSION
- Primarily transmitted by mosquitoes
- Free-ranging birds, particularly nestling English Sparrows are considered reservoirs
- Viremia in most birds from 1 to 5 days after infection
- Ducks can be infected by ingestion of virus-contaminated water

PREVENTION
- Some inactivated WEE virus vaccines are effective in birds

EMUS
- Solvay-Triple E has been shown to be effective
- Vaccinate at 8 weeks of age and boost 2 to 4 weeks later
- Vaccinate adults after the breeding season in April and prior to the breeding season in October
- Obtain approval from insurer prior to vaccination

CONTROL
- Virions unstable when outside the host
- Most disinfectants rapidly inactivate the virus

TREATMENT
- Supportive care - fluids, assisted feeding, antibiotics

ZOONOTIC POTENTIAL
- Infected mosquitoes can transmit virus to humans (aviculturists are at no greater risk than anyone exposed to mosquitoes)
- Many infected humans remain asymptomatic, but a few develop mild to severe encephalomyelitis

CHARACTERISTICS OF FAMILIES OF VIRUSES THAT INFECT ANIMALS

	Size	Envelope	Nucleic acid	Inclusions
Circoviridae				
PBFDV	14-17 nm	nonenveloped	ssDNA	IN and IC
CAV	18-26 nm	nonenveloped	ssDNA	IN, inconsistent
Parvoviridae	18-26 nm	nonenveloped	ssDNA	IN
Hepadnaviridae	46-48 nm	enveloped	dsDNA	—
Papovaviridae				
Polyomavirus	45 nm	nonenveloped	dsDNA	IN
Papillomavirus	50-55 nm	nonenveloped	dsDNA	IN
Adenoviridae	70-90 nm	nonenveloped	dsDNA	IN contain virus IC do not contain virus
Herpesviridae	120-200 nm	enveloped	dsDNA	IN
Iridoviridae	125-300 nm	enveloped	dsDNA	—
Poxviridae	300-450 X 170-260 nm	enveloped	dsDNA	IC, IN uncommon
Astroviridae	27-30 nm	nonenveloped	ssRNA	—
Picornaviridae	28-30 nm	nonenveloped	ssRNA	IC in cockatoos
Caliciviridae	35-40 nm	nonenveloped	ssRNA	—
Torovirus	35 X 170 nm	enveloped	ssRNA	—
Flaviviridae	40-60 nm	enveloped	ssRNA	—
Birnaviridae	60 nm	nonenveloped	dsRNA	—
Togaviridae	60-70 nm	enveloped	ssRNA	—
Reoviridae	60-80 nm	nonenveloped	dsRNA	IC
Retroviridae	80-130 nm	enveloped	ssRNA	—
Rhabdoviridae	70-85 X 130-380	enveloped	ssRNA	IC
Coronaviridae	60-220 nm	enveloped	ssRNA	—
Bunyaviridae	90-120 nm	enveloped	ssRNA	—
Orthomyxoviridae	90-120 nm	enveloped	ssRNA	—
Arenaviridae	50-300 nm (110-130)	enveloped	ssRNA	—
Paramyxoviridae	150-300 nm	enveloped	ssRNA	IN and IC
Filoviridae	80 x 790 to 14,000 nm	enveloped	ssRNA	—

listed in order of increasing size according to nucleic acid
ds = double stranded, ss = single stranded, IN = intranuclear, IC = intracytoplasmic

COMMON NAMES OF DISEASES CAUSED BY FAMILIES OF VIRUSES THAT INFECT BIRDS

ADENOVIRIDAE

CELO
Chicken splenomegaly
Duck hepatitis
Egg drop syndrome in chickens
Egg drop syndrome in ducks
Hemorrhagic enteritis in turkeys
Inclusion body disease
Inclusion body hepatitis in chickens
Marble spleen disease in pheasants
Quail bronchitis
Turkey viral hepatitis

ARENAVIRIDAE

None confirmed in birds

ASTROVIRIDAE

Duck hepatitis type II

BIRNAVIRIDAE

Infectious bursal disease
(Gumboro disease)

BUNYAVIRIDAE

None confirmed in birds

CALICIVIRIDAE

None confirmed in birds

CIRCOVIRIDAE

Chicken anemia virus
Pigeon circovirus
Psittacine beak and feather disease

CORONAVIRIDAE

Bluecomb in turkeys
Infectious bronchitis virus

FILOVIRIDAE

None confirmed in birds

FLAVIVIRIDAE

Alfuy
Cacipacore
Hypr
Ilheus virus
Japanese encephalitis virus
Kumlinge
Kunjin
Kyasanur Forest disease
Louping Ill
Murray Valley encephalitis
Rocio
Russian Spring Summer encephalitis
St. Louis encephalitis virus
Turkey meningoencephalitis virus
Uganda S
Usutu
West Nile virus

HERPESVIRIDAE

Amazon tracheitis
Crane hepatitis
Duck plague or duck enteritis
Hepatosplenitis (owls, falcons)
Inclusion body disease (falcons)
Inclusion body hepatitis of pigeons
(pigeon herpesvirus 1)
Infectious esophagitis
(pigeon herpesvirus 1)
Infectious laryngotracheitis
Marek's disease
Pacheco's disease
Pigeon herpes encephalitis virus

HEPADNAVIRIDAE

Duck hepatitis B virus
Heron hepatitis B virus

IRIDOVIRIDAE

None confirmed in birds

ORTHOMYXOVIRIDAE

Fowl pest (influenza A virus)
Fowl plague (influenza A virus)

PAPOVAVIRIDAE

Polyomavirus
Avian polyomavirus
Budgerigar fledgling disease
Papillomavirus

PARAMYXOVIRIDAE

Paramyxovirus-1
Newcastle disease
Avian distemper
Avian pest
Fowl pest
Lentogenic Newcastle disease
Mesogenic Newcastle disease
Ranikhet disease
Velogenic viscerotropic Newcastle
disease (Asiatic Newcastle disease)
Velogenic, neurotropic Newcastle
disease (American Newcastle disease
or pneumoencephalitis)
Kunitachi virus (PMV-5)
Pneumovirus
Swollen head syndrome in chickens
Turkey rhinotracheitis

PARVOVIRIDAE

Parvoviruses
In geese called:
Derzsy's disease
Goose hepatitis
Goose influenza
Goose plague
Hepatonephritis-ascites
Infectious myocarditis
Viral enteritis

PICORNAVIRIDAE

Avian encephalomyelitis
Avian nephritis virus
Cockatoo enteritis virus
Duck fatty kidney syndrome (sequelae to
duck hepatitis virus infection)
Duck hepatitis (types I and III, frequently
referred to as duck hepatitis-A virus)
Turkey hepatitis virus

POXVIRIDAE

Kikuth's virus (canary poxvirus)

REOVIRIDAE

Rotaviral enteritis
Viral arthritis

RETROVIRIDAE

Duck infectious anemia
Leukosis/Sarcoma viruses
Reticuloendotheliosis
Spleen necrosis virus

RHABDOVIRIDAE

Rabies

TOGAVIRIDAE

Avian viral serositis
Eastern equine encephalitis
Highlands J virus
Venezuelan equine encephalitis
(Everglades viral encephalitis)
Western equine encephalitis

TOROVIRUS

None confirmed in birds

ABDOMINAL CAVITY
The area within the body that contains the proventriculus, ventriculus, liver, kidneys, pancreas, spleen, testes, ovaries and intestines.

ACCIPITRIDAE
Family of birds within the Falconiformes order containing kites, hawks and eagles.

ACTIVE IMMUNIZATION
The immune response that should occur following vaccination. This is differentiated from passive immunization, which is the immunity that a hen transfers to her developing chicks through the yolk.

ACUTE
A disease characterized by a rapid onset and generally a short duration of clinical signs.

ADJUVANT
A substance that is added to an inactivated vaccine to increase the degree and duration of the immunologic response. Common adjuvants include mineral oils and aluminum hydroxide gel.

AEROSOL
Floating in air. With respect to viruses, a mist that carries fluids contaminated with virus particles. The discharge produced when a bird coughs or sneezes is considered an aerosol as is the mist that is produced when washing the floor of an aviary.

AGAR-GEL IMMUNODIFFUSION ASSAY
An assay system that can be used to detect a virus or antibodies to a virus, depending on how it is constructed.

AGID
See agar-gel immunodiffusion assay.

ALAUDIDAE
The family of birds within the Passeriformes order containing larks.

ALCIDAE
The family of pelagic birds within the Charadriformes order containing murres, guillemots, puffins and auks.

ALIMENTARY TRACT
The tube through which the intake, processing and absorption of food takes place. The digestive tract from the mouth to the cloaca.

ALLOPREENING
The preening of one bird by another.

ALSV
Avian leukosis/sarcoma virus.

ALTRICIAL
Young birds that are hatched with their eyes closed, virtually naked and require maternal care for feeding during the early stages of development. Psittacine birds produce altricial young. By comparison, precocial young are hatched with their eyes open, covered with down and are capable of eating without assistance from the hen. Ducks produce precocial young.

ALULA
The bird equivalent of the thumb.

ANAMNESTIC
The immune response that occurs the second time a bird is exposed to an antigen.

ANATIDAE
The family of birds in the Anseriformes order containing swans, geese and ducks. Frequently referred to as waterfowl.

ANEMIA
A reduced number of circulating red blood cells.

ANSERIFORMES
The order of birds containing ducks, swans and geese. Frequently referred to as waterfowl.

ANTEMORTEM
Prior to death. With respect to diagnostic tests it applies to those that are used in living birds (eg, evaluation of blood components). By comparison, postmortem diagnostic tests are those that are performed after a bird has died (eg, necropsy).

ANTERIOR UVEITIS
Inflammation of the structures in the front portion of the eye.

ANTIBODY
The specific protein that the immune system produces when stimulated by an antigen (foreign substance or molecule that enters the body). Generally, the antibody is specific to the antigen.

ANTIGEN
A substance or molecule that stimulates the immune system, and reacts specifically with its products (and cells).

ANTIGENIC
Substances or molecules that are able to stimulate an immune response.

ANTIGENIC DIFFERENCES
Substances or molecules that are varied sufficiently to induce unique antibodies. For example, many influenza viruses are sufficiently different that they stimulate the production of specific antibodies that do not react with other related but different strains of influenza virus.

ANTIGENIC VARIANCE
An alteration in the proteins on the surface of a virus that cause it to appear different to the immune system.

ANTIGENICALLY RELATED
Organisms that induce antibodies that have similar activity. For example, the antibodies produced in response to some strains of paramyxovirus are sufficiently similar that they will cross react with other different but related strains of paramyxovirus.

ANTIGENICITY
The degree to which a substance or molecule is able to stimulate an immune response. Highly antigenic compounds should stimulate the most pronounced immune response.

ANTIMICROBIALS
Agents that destroy or suppress the development or growth of microorganisms (eg, viruses, bacteria, fungi, protozoa).

ANTISEPTIC
An agent that can be used as directed to reduce the microbial population found on skin. Agents classified as antiseptics may or may not destroy microorganisms.

ANTISERUM
Serum that contains antibodies to a specific antigen (protein). This serum can be produced when a bird naturally survives an infection or through vaccination.

AP
Alkaline phosphatase.

APATHOGENIC
An organism that is able to infect a bird but does not cause disease.

APODIDAE
The family of birds within the Passeriformes order containing swifts.

APV
Avian polyomavirus.

ARBOVIRUS
Historically, the term arbovirus has been used without regard for taxonomic classification to collectively describe all of the viruses transmitted in nature by an arthropod (eg, mosquitoes and ticks).

ARTHRITIS
Inflammation of the joint.

ARTHROPODS
Members of the phylum Arthropoda that includes insects, crustaceans and arachnids (spiders).

ARTIFICIAL INCUBATION
The use of an incubator in place of a hen to provide the necessary condi-

tions for the development of a fertile egg.

ASPHYXIATION
Insufficient intake of oxygen that results in death.

AST
Aspartate aminotransferase.

ASYMPTOMATIC
A lack of clinical signs. A bird that is infected by an organism but remains clinically normal is considered to be asymptomatic.

ATAXIA
A lack of coordination resulting in irregular body movements.

ATHEROSCLEROSIS
Thickening and loss of elasticity of the arterial walls caused by an accumulation of plaques containing cholesterol and fat.

ATROPHY
A wasting away or decrease in size of an organ or tissue.

ATTENUATION
The process of reducing the ability of an organism to cause disease. With respect to viruses, attenuation is typically accomplished by repeated growth in cell culture. Vaccines containing attenuated strains or virus are frequently referred to as modified live.

ATV
Amazon tracheitis virus.

AUTOGENOUS
Derived from self. For example, autogenous vaccines are produced by processing infected tissues from a bird and using this material to inoculate said bird.

AUTOIMMUNE
Disease process in which the body produces an immune response to its own tissues.

AVASCULAR NECROSIS
Avascular refers to a lack of blood flow to an area. Avascular necrosis is the death of tissues caused by a lack of blood supply to the area.

AVES
The taxon (Class) that contains all birds.

AVIRULENT
A term used to describe an organism that causes minimal or no disease in an infected bird.

AVS
Avian viral serositis.

B-LYMPHOCYTES
A subgroup of lymphocytes (a type of white blood cells) that responds to infectious organisms by producing antibodies.

BENIGN
With respect to tumors, those that are not likely to be invasive or recur following therapy.

BFD
Budgerigar fledgling disease.

BFDV
Budgerigar fledgling disease virus or avian polyomavirus.

BILATERAL
Occurring roughly equally on both sides of the body.

BINARY FISSION
The process of replication that results in the splitting of a microorganism into two new organisms.

BIOLOGIC VECTORS
A living being that is infected with a virus that subsequently transmits that virus to a noninfected, susceptible individual. A virus must be capable of replicating in the infected host for the host to be considered a biologic vector. Mosquitoes and ticks are two of the most common biologic vectors of viruses.

BIOPSY
Collection of a small piece of tissue from a living bird that is used for microscopic examination.

BLAST CELLS
An immature stage of a developing cell.

BLEPHARITIS
Inflammation of the eyelids.

BONE MARROW
The soft material that fills the interior of a bone where blood cells are manufactured.

BOOSTER
The additional dose of a vaccine that is used to enhance the immune response induced by a previous series of injections. Booster vaccines are usually given on an annual basis.

BRONCHI
The tubes that carry air from the trachea to the lungs.

BURSA OF FABRICIUS
A sac-like organ found in the cloaca of young birds. Referred to as the cloacal bursa or simply bursa, it is the portion of the immune system where young lymphocytes released from the bone marrow mature into B-type lymphocytes.

BURSAL ATROPHY
A decrease in the size or number of cells present in the bursa of Fabricius. The bursa undergoes a natural process of atrophy as a bird matures.

CAA
Chicken anemia agent.

CANDIDIASIS
An infection caused by *Candida* sp.

CANTHI
The angle at either corner of the eye formed by the eyelids.

CAPILLARIES
The smallest of the blood vessels that course through tissues and connect the smallest of the arteries to the smallest of the veins.

CAPSID
The protein covering around the nucleic acid of a virus particle.

CARCINOGEN
Any substance or agent that promotes the formation of cancer.

CARCINOMAS
A malignant tumor that originates from epithelial tissue.

CARDIAC
Pertaining to the heart.

CARRIER
A bird that is infected with an organism yet manifest no discernible clinical signs of disease. These birds are potentially capable of spreading the organism to other susceptible birds.

CATARRHAL
Inflammation of the membranes lining the intestinal or respiratory tract, typically with the production of excess mucus.

CAV
Chicken anemia virus.

CELL
The simplest, fundamental unit of structure for all living organisms. Collections of cells form tissues, and collections of tissues form organs.

CELL CULTURE
A plate of cells grown in the laboratory.

CERTHIIDAE
The family of birds within the Passeriformes order containing creepers.

CF
See complement fixation assay.

CHALLENGE
A term used in virology to indicate experimental exposure of a bird to a virus.

CHARADRIIFORMES
The order of birds containing shore-birds, plovers, sandpipers, gulls, terns and puffins.

CHEMOSIS
Edema of the conjunctiva.

CHLAMYDIOSIS
The disease caused by *Chlamydia* sp.

CHRONIC
With respect to infections, those that are characterized by a long, frequently progressive duration of disease.

CICONIIFORMES
The order of birds containing herons and egrets.

CILIA
Tiny hairlike processes that project from the surface of some mucus membranes, particularly those that line the nasal passages and trachea.

CLINICAL SIGNS
Detectable abnormalities indicating that a disease process is occurring. Clinical signs may include depression, diarrhea, sneezing, coughing and weight loss.

CLOACA
The cavity at the bottom end of a bird where the digestive, urinary and reproductive tracts exit the body.

CLOACAL BURSA
See bursa of Fabricius.

CLOSED AVIARY
An aviary that does not add new birds or allow visitors.

CM
Centimeter. A centimeter is 0.01 of a meter.

CNS
Central nervous system. Clinical changes associated with diseases of the central nervous system may include ataxia, convulsions, opisthotonos, tremors, paresis and paralysis.

COLUMBIFORMES
The order of birds containing doves and pigeons.

COMMISSURE
The coming together of two structures like the upper and lower lips or beaks.

COMPLEMENT FIXATION ASSAY
An assay used to determine the concentration of antibodies in the blood.

CONCOMITANT
Disease processes or infections by several organisms at the same time.

CONGENITAL
Passed genetically from parent to offspring.

CONGESTION
The accumulation of an excess quantity of blood or fluids in a tissue or organ. Congested organs often appear hyperemic and enlarged.

CONJUNCTIVA
The mucous membranes that line the eyelids.

CONJUNCTIVITIS
Inflammation of the mucous membranes that line the eyelids.

CONSPECIFIC
A member of the same species.

CONTACT CONTROLS
A term used to denote birds that are placed in direct contact with those that have been infected by a specific organism. Contact controls are used to determine if an infectious agent passes directly from an infected to a susceptible bird.

CONVALESCENCE
The period of recovery from a disease.

CONVULSION
Usually violent, involuntary contraction with subsequent relaxation of the muscles.

COPROPHAGY
The ingestion of excrement.

COPULATION
Sexual intercourse between a male and female.

CORNEAL CRYSTALLIZATION
Formation of crystals on the cornea (outer surface of the eye).

CORVIDAE
The family of birds within the Passeriformes order containing jays, magpies and crows.

CORYZA
An acute inflammation of the membranes that line the upper respiratory tract, resulting in a profuse discharge of mucus.

CPK
Creatine phosphokinase.

CROP
A sac-like structure in the esophagus of some birds that is used for the temporary storage of ingested food before it passes into the proventriculus.

CROSS REACTION
The binding of antibodies produced against one virus to another virus or organism. Cross reaction of antibodies can cause false-positive test results.

CRYOTHERAPY OR CRYOSURGERY
Exposure of tissues to extremely cold temperatures resulting in a localized destruction or the treated area. Used to remove some types of tumors (warts) from the skin.

***CRYPTOSPORIDIUM* SP.**
A genus of protozoan parasites that can infect the intestinal tract of birds. Infection in birds with a functional immune system is usually mild and self-limiting, while infection in birds with a damaged immune system can be severe and life-threatening.

CULTURE AND SENSITIVITY
Plates of specialized prepared growth medium are used for isolating bacteria from tissue samples (eg, blood, feces). This process is called culturing. Once a bacteria has been isolated its sensitivity to destruction by various antibiotics can be determined.

CUTANEOUS
Pertaining to the skin.

CYTOMEGALY, CYTOMEGALIC
Enlargement of a cell.

CYTOPLASM
The area within a cell that surrounds the nucleus.

DEV
Duck enteritis virus.

DHBV
Duck hepatitis B virus.

DHV
Duck hepatitis virus.

DIFFERENTIAL DIAGNOSIS
A list of disease processes that can cause similar changes.

DIGESTIVE TRACT
See alimentary tract.

DIPHTHERITIC
Descriptive of an inflammatory disease characterized by the formation of sheets of necrotic debris (pseudomembrane) on a mucous membrane.

DISEASE
Abnormal changes to a bird's tissues (tissues are a collection of cells).

DISINFECTANT
An agent that will destroy many of the disease-causing microorganisms present on the surface of an inanimate object. These agents may or may not be effective against all viruses, *Mycobacterium* sp, protozoa, bacteria or heat-resistant bacterial spores.

DNA
Deoxyribonucleic acid. The genetic material contained in most microorganisms and all higher forms of life. Some viruses contain DNA as their genetic material whereas others contain RNA.

DNA PROBES
Specially designed segments of DNA that are designed to detect specific sequences of nucleic acid in tissues, body fluids, secretions or excretions.

DORMANCY
The inactive period when a virus or other infectious agent is outside of a bird's body.

DPV
Duck plague virus.

DYSPHAGIA
Difficulty or inability to swallow.

DYSPNEA
Difficult breathing.

DYSTOCIA
Difficult labor. Sometimes used with birds to describe difficulty in laying an egg.

DYSTROPHIC
An abnormality that is caused by an alteration in normal metabolism.

EDEMA, EDEMATOUS
Accumulation of an excess amount of fluids in the body tissues.

EDS
Egg drop syndrome.

EEE
Eastern equine encephalitis.

ELISA
See enzyme-linked immunosorbent assay.

ENCEPHALITIS
Inflammation of the brain.

ENCEPHALOMYELITIS
Inflammation of the brain.

ENDEMIC
The continuous presence of an infectious disease-causing organism in a specific population of animals or in a particular geographical area.

ENDOCYTOSIS
The process of a cell taking in material that is present in its surrounding environment. Endocytosis can be viewed as a cell "swallowing" the material that is attached to its surface.

ENDOGENOUS
Produced or arising from within an organism.

ENDOTHELIUM, ENDOTHELIAL
Cells that line the inside of blood and lymphatic vessels.

ENTERIC VIRUSES
Viruses that infect the intestinal tract.

ENTEROCYTES
The cells than line the inside of the intestines.

ENTERITIS
Inflammation of the intestinal tract.

ENVELOPE
The fatty-protein coat that surrounds the capsid of some types of viruses.

ENZOOTIC
The occurrence of a microorganism in a population of animals at all times, usually at a relatively low level of activity.

ENZYME-LINKED IMMUNOSORBENT ASSAY
A diagnostic test that, depending on its configuration, can be used to detect the presence of a virus or antibodies to a virus.

ENZYMES
Complex proteins that are capable of inducing chemical changes in other substances without being changed themselves.

EPICARDITIS
Inflammation of the outermost layer of the heart.

EPIDEMIC
The occurrence of a disease in numerous people at the same time in the same geographical area.

EPITHELIUM, EPITHELIAL
The layer of cells forming the skin and the lining of internal organs including the alimentary tract, respiratory tract and urinary tract.

EPITOPE
The specific site on an antigen to which antibodies are produced.

EPIZOOTIC
The sudden occurrence and rapid spread of a disease in multiple types of animals in the same geographical area at the same time.

EPIZOOTIOLOGY
The study of the frequency and distribution of infectious diseases among animals. This science attempts to define how an infectious agent interacts with host and external environmental factors to cause a disease process.

EPORNITIC
The occurrence of a disease in numerous birds at the same.

EROSION
An area of cells on the surface of the skin or mucus membrane that are destroyed by a physical or inflammatory process.

ERYTHROID LEUKOSIS (ERYTHROBLASTOSIS)
A type of cancer involving the proliferation of immature red blood cells.

ESOPHAGUS
The tube that connects the oral cavity to the proventriculus (stomach).

ESTRILDIDAE
The family of birds within the Passeriformes order containing some finches, waxbills, nuns, manikins and Java rice sparrows.

ETIOLOGIC AGENT
The cause of a disease process.

ETIOLOGY
Cause of, or study of the cause of disease.

EVE
Everglades viral encephalitis.

EXCRETIONS
The waste products of digestion and metabolism (eg, feces, urine) that are released from the body.

EXHALATION, EXHALE
The process of breathing outward.

EXOGENOUS
Originating outside of an animal.

EXTRA-LUMINAL MASSES
Masses that are located outside of the lumen of a hollow organ. These masses may press on an organ causing partial or complete obstruction.

EXTRACELLULAR
The space outside of a cell.

FALCONIDAE
The family of birds within the Falconiformes order containing caracaras and falcons.

FALCONIFORMES
The order of birds containing hawks, falcons and eagles.

FALSE-POSITIVE
A test result that indicates that a bird is infected with a particular organism when it is not really infected.

FAMILY
The subgrouping of birds within an Order; all Families end with the suffix: idae.

FEATHER DUSTER
A term used to describe budgerigars in which the feathers have grown beyond their normal length. The cause of the syndrome has not been confirmed.

FEATHER EPITHELIUM
The epithelial cells that line the inside of a developing feather.

FECAL-ORAL TRANSMISSION
The passage of a virus from the contaminated feces of one bird to the oral cavity of another bird.

FECOBEZOAR
An accumulation of feces around the vent; commonly referred to as "pasted" vent.

FHV
Falcon herpesvirus.

FIBRIN
A whitish, filamentous protein that forms into a gel when blood clots.

FLORA
The microorganisms that are normally found in the body (eg, alimentary and respiratory tracts) or on the skin of a bird.

FOCI, FOCUS
The site where an infection initiates.

FOLLICULAR EPITHELIUM
The epithelial cells that line the feather follicle.

FOMITE
An inanimate object on which a virus can be transported from one bird to another. Fomites can be contaminated with virus that is present in aerosolized feather dust, feces, urine, saliva, or aerosolized respiratory secretions (generated when an infected animal coughs or sneezes).

FOREIGN
Infectious organisms that are not normally found in a particular geographical area. With respect to the immune system, any substance that comes in contact with or enters the body and induces an immune response.

FRIABLE
Easily torn or pulverized.

FRINGILLIDAE
The family of birds within the Passeriformes order containing the green singing finch, chaffinches, bramblings and the European gold finch.

FULMINATING
A sudden and severe onset of disease.

GALLIFORMES
The order of birds containing chickens, quail, pheasants, guinea fowl and Chukars.

GALLINACEOUS
Pertaining to birds within the order Galliformes.

GASTROINTESTINAL TRACT
The stomach and intestines.

GAVIIFORMES
The order of birds that contains the loons.

GENE
A segment of DNA that represents the basic unit of heredity that is transferred from a parent to its offspring.

GENETIC
Pertaining to reproduction and the transfer of hereditary information from parent to offspring.

GENITAL TRACT
The reproductive system including the copulatory organs, ovary and testis.

GENITOURINARY TRACT
The combined reproductive (copulatory organs, ovary, testis) and urinary (kidneys, ureters) systems.

GENOME
With respect to viruses, the complete DNA or RNA sequence necessary for a virus to replicate.

GENUS
The subgrouping of closely related birds within a Family.

GERM-FREE
The absence of microorganisms.

GERMICIDE
An agent that when used as directed will kill a specific group of organisms listed on the label.

GLOBULINS
The type of protein that composes antibodies. These proteins are present in the serum or plasma component of the blood.

GLOMERULOPATHY
Pathologic changes in the glomeruli (vascular plexus) of the kidneys.

GRANULOMA, GRANULOMATOUS
A nodule composed of white blood cells and the dead and dying cells they are attempting to "wall off."

GROSS
Visible to the naked eye. Gross lesions are those that are visible to the eye without the aid of magnification.

GRUIFORMES
The order of birds containing the cranes and rails.

GRURIDAE
The family of birds within the Gruiformes order containing cranes.

HALF-LIFE
The amount of time required for the activity of a substance or compound to be reduced by 50%.

HALITOSIS
Malodorous breath.

HARD PALATE
The plate of bone that separates the roof of the mouth from the sinuses.

HEMAGGLUTINATION
The clumping of red blood cells.

HEMAGGLUTINATION-INHIBITION ASSAY
A type of test used to detect antibodies to some viruses and other microorganisms. Some viruses will cause red blood cells to clump together (hemagglutination). When antibodies to these viruses are incubated with the virus, the antibodies bind to the virus particles causing them to lose their ability to agglutinate red blood cells.

HEMANGIOMA
A benign tumor of blood vessels.

HEMATOCHEZIA
Passage of excrement containing blood.

HEMOLYSIS
The destruction of red blood cells.

HEMORRHAGE
The abnormal leakage of blood from the vessels.

HEMORRHAGIC
Used to define tissues in which bleeding has occurred.

HEPATIC
Pertaining to the liver.

HEPATIC NECROSIS
Death of the cells within the liver.

HEPATITIS
Inflammation of the liver.

HEPATOCYTES
The specific cells that form the liver.

HETEROLOGOUS
Tissues, cells or blood obtained from a different bird. Homologous is tissue, cells or blood obtained from the same bird. A person that receives a homologous blood transfusion would be receiving their own blood. A person that receives a heterologous blood transfusion is receiving blood from any other individual. Homologous antibodies are those that are produced against a specific antigen. Heterologous antibodies are those that are produced against a different antigen to the one being evaluated.

HETEROPHIL
A type of white blood cell found in birds.

HEV
Hemorrhagic enteritis virus.

HI
See hemagglutination-inhibition assay.

HISTOLOGY, HISTOLOGIC
The study of the microscopic anatomy of tissues and organs.

HISTOPATHOLOGY
The microscopic changes that occur in diseased tissues.

HJ
Highlands J virus.

HOMOGENIZE
To render a uniform quality and consistency throughout. Tissues are homogenized by placing them in a high speed blender.

HOMOLOGOUS
See heterologous.

HORIZONTAL TRANSMISSION
Transfer of an infectious agent from one bird to another through inhalation, ingestion, insect or animal bites (injection); or skin-to-skin contact (eg, breeding, preening).

HOST
The animal that provides nourishment to an infectious agent (eg, virus, bacteria, parasite).

HOST-SPECIFIC
An infectious agent that only attacks a specific type of animals. For example, Pacheco's disease virus has been shown to infect psittacine birds and not chickens, and is host specific for psittacines.

HOST-ADAPTED
An infectious agent that has adapted to living in a particular host. Generally, host adapted viruses cause minimal disease in the species of bird to which they have adapted, but may cause severe disease in other species. (see also non-host adapted).

HYBRIDIZATION
The process of binding together two complimentary strands of nucleic acid (DNA or RNA).

HYPEREMIA
An increased quantity of blood in a particular part of the body that causes the engorged tissue to appear reddened.

HYPERGLOBULINEMIA
An increase in the concentration of globulins (the protein antibodies are made from) in the blood. Can indicate that the bird's immune system has been responding to some type of infectious agent.

HYPERKERATOSIS
An increase in growth in the outer layer of the skin.

HYPERPLASIA
An increase in the number of cells that are present in a tissue.

HYPOALBUMINEMIA
A decrease in the concentration of albumin in the blood.

HYPOPROTEINEMIA
A decrease in the concentration of protein in the blood.

IATROGENIC
Problems that are induced by the action or through contact with a doctor or hospital staff.

IBD
Infectious bursal disease.

IBH
Inclusion body hepatitis.

IBV
Infectious bronchitis virus.

ICTERIDAE
The family of birds within the Passeriformes order containing blackbirds and orioles.

IDIOPATHIC
A term used to describe problems of unknown cause.

IF
See indirect fluorescent assay.

ILEUS
Obstruction of the intestines that prevents the passage of ingesta resulting from inhibition of bowel motility.

ILT
Infectious laryngotracheitis virus

IMMUNE SYSTEM
The systems and cells that are responsible for protecting a bird from infectious agents.

IMMUNIZATION
Administration of an antigen intended to induce an immune response (see vaccination).

IMMUNOCOMPETENCE
The ability to develop an immune response following stimulation by an antigen (foreign protein).

IMMUNOCOMPROMISED
Having an immune system with a reduced ability to defend the body against infectious agents. The immune system can be weakened by many types of medications, malnutrition or concomitant infections.

IMMUNODEFICIENT
See immunocompromised.

IMMUNOGENICITY
The capacity of an organism or substance to stimulate the immune system, particularly an antibody response.

IMMUNOLOGICALLY NAIVE
A bird that has not been previously exposed to a particular infectious organism, and thus has no prior immunity to that organism.

IMMUNOLOGICALLY PRIVILEGED SITE
A part of the body where antibodies do not reach.

IMMUNOSTIMULANT
A substance or compound that stimulates the immune system.

IMMUNOSUPPRESSED
See immunocompromised.

IMMUNOTOLERANT
A bird that has been exposed to an antigen (ie, an infectious agent) in such a way that the immune system does not recognize the organism as an invader. Immunotolerant birds do not develop an antibody response to the organisms for which they are tolerant.

IN SITU HYBRIDIZATION
The process of using nucleic acid probes to detect viral-specific nucleic acid sequences in tissues.

IN UTERO
Occurring during the development of the egg in the oviduct.

INACTIVATE, INACTIVATED
Render inactive or unable to replicate. A virus that has been damaged sufficiently to prevent its replication. Vaccines containing inactivated virus are commonly referred to as "killed" vaccines.

INBREEDING
The mating of closely related birds such as a brother to a sister or a cousin to an aunt.

INCIDENCE
The frequency with which a disease occurs within a specific population of birds over a given period of time. For example, the number of cases of avian polyomavirus infection that are detected in a nursery during a six-month period.

INCLUSION BODIES
Visible accumulations of virus particles, proteins, viral components or damaged cellular debris in the nucleus or cytoplasm of some cells infected with certain viruses.

INCOORDINATION
Inability of the muscular system to function in a normal rhythmic pattern resulting in stumbling, shaking or tremors.

INCUBATION PERIOD
The interval between exposure to an infectious agent and the first appearance of clinical signs.

INDEX CASE
The first case. The first bird in a nursery to die from a psittacine beak and feather disease virus infection would be considered the index case.

INDIGENOUS
Native to a particular geographical area.

INDIRECT FLUORESCENT ASSAY
A diagnostic test that depending on its configuration can be used to detect the presence of a virus in tissues or antibodies to a virus in serum.

INFECTION
A condition that occurs when a virus or other microorganism invades and replicates in the body.

INFECTIOUS AGENT
An organism that is capable of being transmitted from one bird to another.

INFLAMMATION, INFLAMMATORY RESPONSE
The reaction of tissues to injury. The response is designed to destroy and contain the injurious agent as well as to facilitate repair of the damaged tissue.

INFUNDIBULUM
The funnel-shaped end of the oviduct that surrounds and catches an ovum as it is released from the ovary.

INGESTA
Food and drink that have been taken into the body through the mouth.

INGESTED
To enter the body through swallowing.

INGLUVEITIS (INGLUVITIS)
Inflammation of the crop.

INHALATION
The act of taking air, and the substances and organisms it contains, into the lungs.

INSPIRATION
See inhalation.

INTESTINAL EPITHELIAL CELLS
The cells that line the lumen (interior surface) of the intestines.

INTESTINAL TRACT
The portion of the alimentary tract that extends from the end of the ventriculus to the cloaca.

INTRALUMINAL MASSES
Masses contained within the lumen of the intestinal tract.

INTRACELLULAR
Contained within a cell. Viruses are considered intracellular parasites because they " live" inside of an infected cell.

INTRACEREBRAL
Within the brain. Intracerebral injections are those that are made into the brain.

INTRACYTOPLASMIC INCLUSION BODIES
Inclusion bodies that are present within the cytoplasm, as apposed to the nucleus, of a cell.

INTRAMUSCULAR
Within the muscle. Intramuscular injections are those that are made into a muscle.

INTRANASAL
Within the nostrils. Medications that are administered intranasally are those that are placed in the nostrils.

INTRANUCLEAR INCLUSION BODIES
Inclusion bodies that are present within the nucleus, as apposed to the cytoplasm, of a cell.

INTRAOCULAR
Within the eye. Intraocular infections are those that occur within the eye.

INTRAPERITONEAL
Within the peritoneal cavity. Intraperitoneal injections are those that are made into the peritoneal cavity (abdomen).

INTRAVENOUS
Within a vein. Medications that are given intravenously are injected directly into a vein.

ISOLATED
See quarantine.

ISOLATION
The recovery of a virus from infected tissues. Or, the physical separation of an infected bird from other members of the flock.

JEV
Japanese encephalitis virus.

KARYOMEGALY
Abnormal enlargement of the nucleus of a cell.

KERATITIS
Inflammation of the cornea (the transparent structure that covers the front of the eye).

LABILE
Unstable. Easily altered or destroyed.

LARIDAE
The family of birds within the Charadriiformes order containing gulls.

LARYNGOTRACHEITIS
Inflammation of the larynx and trachea.

LARYNX
The enlarged upper end of the trachea that impedes the entrance of food or water into the trachea.

LATENT
State of inactivity. A dormant infection that has the potential to be become active.

LDH
Lactate dehydrogenase.

LENTOGENIC
Strain of Newcastle disease virus that causes a mild or inapparent infection in chickens.

LESION
The location of injury or dysfunction in a tissue.

LETHARGIC
Drowsiness. A lack of energy or enthusiasm.

LINEBREEDING
The breeding of closely related family members to each other (see inbreeding)

LIPEMIA
An increased concentration of lipids (fat) in the blood.

LUMEN
The hollow space within a tube such as the interior of a blood vessel or the intestines.

LYMPHOBLAST, LYMPHOBLASTIC
An immature blood cell that matures into a lymphocyte.

LYMPHOCYTE
A type of white blood cell. Lymphocytes are divided into B-cells that produce antibodies and T-cells that are involved in destroying virus infected cells and regulating the immune response. T-cells are part of the cell-mediated immunity as opposed to the humoral or antibody-mediated immunity.

LYMPHOID LEUKOSIS
The abnormal accumulation of immature lymphocytes in tissue.

LYMPHOID TUMOR
A mass containing neoplastic (cancerous) lymphocytes.

LYSE, LYSIS
Rupture or disintegration of a cell.

MACAW WASTING DISEASE
See PDD.

MACROMOLECULES
Complex molecules that are made of multiple components.

MACROPHAGE
A type of white blood cell that is responsible for engulfing microorganisms that enter the body. Once activat-

ed, macrophages send out a chemical signal that stimulates other portions of the immune system.

MALABSORPTION
The improper absorption of nutrients from the intestinal tract.

MALAISE
Discomfort or uneasiness.

MASTICATION
Chewing of food.

MATERNALLY DERIVED ANTIBODIES
Antibodies that are passed from the hen to her developing eggs in the yolk.

MDV
Marek's disease virus.

MECHANICAL VECTOR
Any object that is contaminated with a virus that functions to carry that virus or other infectious agent from an infected to a noninfected, susceptible bird. Mechanical vectors may include contaminated gloves, nets, feeding utensils, nest boxes and surgical instruments.

MECONIUM
The initial feces passed by a newborn consisting of mucus and bile.

MELEAGRIDIDAE
The family of birds within the Galliformes order containing turkeys.

MESOGENIC
Strain of Newcastle disease virus that causes mild to severe disease in chickens. These strains of virus may cause disease in other birds.

METABOLIC PATHWAYS
The chemical processes in the body that are responsible for turning ingested protein, fats and carbohydrates into nutrients that can be used by the cell, and for processing and removing damaged cells and their by-products from the body.

METASTASIS, METASTATIC
Transfer of a disease process from one area of the body to another. Is most commonly used to refer to dangerous cancers that move from an initial site to other organs.

MICROBIAL FLORA
The total population of microorganisms found in a portion of the body. For example, the normal microbial flora of the intestinal tract may contain various bacteria, fungi and protozoa.

MICROBIAL PATHOGENS
Microorganisms that have the potential to cause disease.

MICROORGANISM
Tiny, simple forms of life that cannot be visualized without magnification. Included in this group are bacteria, viruses and fungi.

MIMIDAE
The family of birds within the Passeriformes order containing mockingbirds and thrashers.

MITOTIC FIGURES
The microscopic changes that are noted when a cell is dividing. Cancer-causing agents frequently increase the rate of cell division, resulting in an unusually high number of mitotic figures.

MITOTIC INDEX
A grading scale used to determine the rate of cell division. A high mitotic index means that a large number of cells are dividing.

MODIFIED LIVE
See attenuation.

MOLECULAR BIOLOGY
The study of the development and function of the physical and chemical characteristics of a cell. Used frequently to refer to the study of nucleic acid and its relationship to the cell.

MONOCLONAL ANTIBODY
A type of pure antibody that has been produced in a laboratory in such a way that it only reacts with one specific epitope. This is in contrast to most antibody preparations which are polyclonal; contain a mixture of different antibodies, each of which reacts with a specific antigen.

MONOCYTE
A type of white blood cell that is responsible for consuming foreign material and organisms that enter the body. Monocytes are found circulating in the blood while macrophages are most commonly found in tissues.

MORBIDITY
A state of disease. The number of individuals within a given population that are diseased.

MORPHOLOGY, MORPHOLOGICAL
The study of the form and structure of an organism. Two organisms that appear similar are said to be morphologically related.

MORTALITY
Pertaining to death. The number of individuals within a given population that die from a specific cause.

MSD
Marble spleen disease.

MUCOID EXUDATE
Discharge from the body containing mucus.

MUCOPURULENT
A discharge that contains mucus and pus (degenerate cells and debris).

MUCOSA
A mucous membrane that lines the interior of the intestinal tract or nasal passages.

MUCOSAL
Pertaining to a mucous membrane.

MUCOUS MEMBRANE
See mucosa.

MUCUS
A viscid, slimy secretion found on the surface of mucous membranes.

MUTANT, MUTATED
A variation in genetic structure from the normal. A virus that has mutated has a different nucleic acid sequence than its precursor.

MYELOID LEUKOSIS (MYELOBLASTOSIS)
A neoplastic condition characterized by the appearance of immature myeloid cells in the blood.

MYOCARDITIS
Inflammation of the muscle (myocardium) in the heart.

MYOCARDIUM
Heart muscle.

NAIVE
With respect to the immune system, an animal that has not previously been exposed to a particular microorganism.

NANOMETER (NM)
A nanometer is 0.000000001 (one-billionth) of a meter.

NDV
Newcastle disease virus.

NECROPSY
The examination and dissection of a body in an attempt to determine the cause of death. An animal equivalent of an autopsy.

NECROSIS, NECROTIZING, NECROTIC
The death and decay of cells in a particular area of tissue.

NGD
Neuropathic gastric dilatation. See PDD.

NEOPLASIA
The formation of neoplasms. Cancer. The abnormal progressive multiplication of cells.

NEOPLASM
The tumor (mass) formed by the abnormal progressive multiplication of cells.

NEOPLASTIC
Pertaining to neoplasia. Having the appearance of new, abnormal tissue.

NEUROLOGIC
Pertaining to the nervous system.

neutralizing, neutralization
With respect to specific antibodies, those that are able to bind to viruses and render them noninfectious.

NICTITATING MEMBRANE
The membrane under the eyelids that covers the eye during a blink response.

NM
Nanometer.

NON-HOST ADAPTED
An infectious agent that has not adapted to living in a particular host. Generally, non-host adapted viruses cause severe disease in an infected bird of a susceptible species. (see also host-adapted).

NONPATHOGENIC
An organism that infects a bird but does not cause disease.

NOSOCOMIAL INFECTIONS
Infections derived from an animal's visit to a hospital.

NUCLEAR MEMBRANE
The membrane that surrounds the nucleus and separates it from the cytoplasm.

NUCLEIC ACID
The DNA and RNA that are responsible for transfer of genetic material and control of a cell's functions.

NUCLEIC ACID PROBES
Synthetically produced sequences of DNA that are designed to bind to a complimentary strand of nucleic acid in a sample.

NUCLEUS
The portion of a eukaryotic cell that contains DNA and dictates the development, metabolism and function of a cell.

OHV
Owl herpesvirus.

ONCOGENIC
Having the capacity to cause the formation of tumors (cancer). A drug that causes cancer is described as oncogenic.

OPISTHOTONOS
A clinical sign indicating neurologic disease in which the top of the head is bent over and approaches the back.

OPPORTUNISTIC
An organism that does not normally cause disease, but can if the host is

weakened or has a damaged immune system

ORAL CAVITY
The mouth.

ORDER
The broadest subgrouping of birds. All orders end with the suffix: iformes.

ORGAN
A collection of tissues that serve a particular function in the body (eg, heart, liver, brain, kidney).

ORGANIC
Pertaining to substances derived from living tissue (eg, soil, food, feces, blood, bedding material, mucus).

ORONASAL
Pertaining to the mouth and nose.

OROPHARYNGITIS
Inflammation of the mouth and pharynx.

OROPHARYNX
The portion of the pharynx between the palate and the glottis.

OSTEOPETROSIS
Excessive calcification of the bones causing them to be brittle and easily fractured.

OVIDUCT
The tube through which an ovum passes from the ovary to the cloaca and develops into an egg.

PALATE, PALATINE
The structure separating the mouth from the nasal cavity.

PALATINE NECROSIS
Death and decay of portions of the tissue that form the palate.

PALLOR
Paleness. A lack of color.

PANDEMIC
The occurrence of disease in a large portion of humans in a geographical area, or an increase in the number or cases of a disease in multiple places within the world at the same time.

PANOPHTHALMIA
Inflammation of all the structures or tissues that form the eye.

PANSYSTEMIC
Involvement of most of the organ systems at the same time.

PANTROPIC
Having an affinity for multiple organ systems. A virus that has a tendency to infect several different organs is described as pantropic.

PANZOOTIC
The occurrence of disease in a large portion of many different species of

animals in a geographical area, or an increase in the number or cases of a disease in multiple places within the world at the same time.

PAPILLOMATOSIS
A term used to describe the papilloma-like changes that have been described along various portions of a bird's alimentary tract including the oral cavity, proventriculus and cloaca. The cause of papillomatosis is unknown.

PAPULE
A raised, circumscribed area of skin.

PARULIDAE
The family of birds within the Passeriformes order containing warblers.

PARALYSIS
Damage to the nervous system that results in loss of sensation or voluntary motion.

PARESIS
Partial or incomplete loss of sensation of voluntary motion. Weakness.

PASSERIFORMES
The order of birds containing finches, canaries, sparrows and mynahs and their relatives, collectively referred to as perching birds.

PASSIVE IMMUNIZATION
The passage of antibodies from the hen to her chicks in the yolk.

PATHOGEN, PATHOGENIC
An organism or substance that is capable of causing disease.

PATHOGENESIS
The origination and development of a disease process.

PATHOGENICITY
The capacity of an organism to cause disease in a host. A highly pathogenic organism is likely to cause a severe disease.

PATHOGNOMONIC
A change that is characteristic or peculiar to a specific disease process.

PATHOLOGY
Study of the changes in structure and function associated with disease.

PATHOTYPE
A classification system used for some microorganisms that divides them into groups based on the type of pathologic changes they cause.

PBFD
Psittacine beak and feather disease.

PCV
Porcine circovirus.

PDD
Proventricular dilatation disease.

PDS
Proventricular dilatation syndrome.

PDV
Pacheco's disease virus.

PELAGIC
Living in the open sea.

PELECANIDAE
The family of birds within the Pelecaniformes order containing pelicans.

PELECANIFORMES
The order of birds containing pelicans, tropicbirds, frigates, gannets, and cormorants.

PERACUTE
A disease characterized by an extremely rapid onset and usually severe clinical signs.

PERCUTANEOUS
Through the skin.

PERICARDITIS
Inflammation of the covering (pericardium) of the heart.

PERIOCULAR
Located around the eye.

PERMEABLE
Allowing the passage of fluids or air.

PETECHIAE, PETECHIAL
Small, purplish spots indicative of hemorrhage.

PH
A scale used in chemistry to define the degree of acidity or alkalinity of a substance.

PHAETHONTIDAE
The family of birds within the Pelecaniformes order containing tropicbirds.

PHAGOCYTES, PHAGOCYTIC
Cells with the ability to ingest and destroy cellular debris, bacteria, viruses, dust particles, etc.

PHARYNX, PHARYNGEAL
The area in the throat that controls the passage of air from the nasal passage to the trachea and the passage of food from the mouth to the esophagus.

PHASIANIDAE
The family of birds within the Galliformes order containing quail, partridges and pheasants.

PHV
Pigeon herpesvirus.

PHYSIOLOGY, PHYSIOLOGIC
The study of the chemical and physical processes that control the functions of a living organism.

PICIDAE
The family of birds within the Piciformes order containing woodpeckers.

PICIFORMES
The order of birds containing toucans and woodpeckers.

PLEOMORPHIC
Having many shapes.

PLOCEIDAE
The family of birds within the Passeriformes order containing the weavers.

PMV
Paramyxovirus.

PO
per os, orally.

PODICIPEDIFORMES
The order of birds containing the grebes.

POLYDIPSIA
An increase in the quantity of water consumed.

POLYURIA
An increase in the quantity of urine produced.

POSTMORTEM
After death. With respect to diagnostic tests it applies to those that are used in dead birds (eg, necropsy, histopathology). By comparison, antemortem diagnostic tests are those that are performed in a living bird (eg, evaluation of blood components).

POULT
Young fowl. Frequently used to describe young turkeys.

PRECIPITATING ANTIBODIES
A group of antibodies that bind with an antigen (protein) causing the bound product to settle out of solution.

PRECOCIAL
See altricial.

PREMONITORY
Providing a warning of a disease process through the induction of an early clinical sign.

PREVALENCE
The number of cases of disease present in a specific population of birds at a specific point in time.

PREVALENT
An infection or disease that occurs frequently.

PRIMARY FEATHERS
The large flight feathers that originate from the area of the carpus and metacarpus.

PRIMARY INFECTION
When more than one microorganism is involved in a disease, the primary infectious agent is the one that initiated the original abnormal changes.

PRIMARY VIRAL PATHOGEN
A virus that is capable of causing disease in a healthy bird without the assistance of other infectious agents.

PROBE
See nucleic acid probe.

PROCELLARIIDAE
The family of birds within the Procellariiformes order containing shearwaters and petrels.

PROCELLARIIFORMES
The order of birds containing the shearwaters, albatross, petrels and fulmars.

PROGENY
Offspring.

PROPHYLACTIC
An agent that aids in the prevention of an infection.

PROVENTRICULUS
The glandular stomach in a bird that produces acids and digestive enzymes (pepsin).

PSEUDOMEMBRANOUS
Descriptive of the accumulation of a sheet of necrotic cells and debris on a mucous membrane (as in diphtheritic).

PSITTACIDAE
The family of birds within the Psittaciformes order containing parrots.

PSITTACIFORMES
The order of birds that contains parrots, cockatoos, macaws and their relatives.

PTERYLAE
The tracts on a bird from which feathers originate.

PULMONARY
Pertaining to or involving the lungs.

PUS
An accumulation of dead cells, debris and inflammatory cells (mostly heterophils and macrophages).

QBV
Quail bronchitis virus.

QUARANTINE
The physical separation of individuals from each other.

RADIOIMMUNOASSAY
A sensitive method of determining the concentration of proteins (particularly antibodies and hormones) in the blood using radioactively-labeled reagents.

RADIOSURGERY
A surgical technique that uses electromagnetic waves to cut tissue. Commonly referred to as electrocautery.

RECEPTOR
The location on a cell to which a virus or other substance attaches.

RELAPSE
Recurrence of a disease after apparent recovery.

RENAL
Pertaining to the kidneys.

REPLICATE, REPLICATION
The process used by a virus to produce progeny.

RESERVOIR
Any source of an infectious agent. An animal that harbors an infectious agent and transmits it to other susceptible animals.

RESISTANT
With respect to viruses, an animal that is not susceptible to infection.

RESPIRATORY TRACT
The organs involved in the intake and processing of air including the nasal passages, trachea, bronchi, lungs and air sacs.

REV
Reticuloendothelial virus.

RHINITIS
Inflammation of the nasal mucosal.

RIA
See radioimmunoassay.

RNA
Ribonucleic acid.

SANITIZER
An agent that reduces microbial contamination on the surface of an object to an acceptable level. Sanitizers must not leave a harmful residue.

SECONDARY FEATHERS
The large flight feathers that originate from the area of the radius and ulna.

SECONDARY PATHOGENS
Infectious agents that do not routinely cause disease in a healthy bird, but are capable of causing disease in a bird that is weakened or has a damaged immune system.

SECRETION
Substances, like saliva, produced by glandular organs.

SENSITIVITY
With respect to a diagnostic test, the ability to correctly detect small quantities of a virus in a sample. The statistical probability that a diagnostic test will correctly identify a diseased individual. (see specificity)

SENTINEL
A susceptible bird that is placed with those in quarantine to determine if infectious agents are present.

SEQUENCING
The process of determining the bases (building blocks) that form a segment of nucleic acid.

SEROCONVERT
The development of antibodies in response to an infection or vaccination.

SEROLOGICALLY RELATED
A term used to indicate that a group of organism induce cross-reactive antibodies indicating that they have a similar antigenic structure.

SEROPREVALANCE
The number of birds in a study population that have antibodies to a particular organism.

SEROTYPES
A classification used to separate organisms based on the reactivity of the antibodies they illicit. Related strains of a virus that induce similar antibodies would be placed in the same serotype, while viruses that induce different antibodies would be placed in different serotypes.

SEROUS
Descriptive of a clear, fluid discharge.

SERUM
The watery portion of the blood. The serum is a collection of minerals, water, nutrients and antibodies.

SERUM ELECTROPHORESIS
A process used to separate the proteins found in serum by subjecting the liquid to an electric current.

SERUM ENZYMES
Enzymes produced by various organs that are found circulating in the serum. Many organs release increased quantities of certain enzymes when they are damaged and detection of these enzymes can be diagnostic.

SGOT
Serum glutamic-oxaloacetic transaminase. Now called AST or ASAT (aspartate aminotransferase).

SIALOLITHS
Masses of accumulated saliva described in the salivary glands of some pigeons and other birds.

SINUS
The passages in the head through which air is transferred when breathing.

SINUSITIS
Inflammation of the sinus passages.

SLE
St. Louis encephalitis.

SPECIES
Specific type of bird such as Red-tailed Hawk (*Buteo jamaicensis*) or St. Lucia Amazon Parrot (*Amazona versicolor*).

SPECIFIC-PATHOGEN-FREE
A bird that has been tested and shown to be free of a specific group of infectious agents.

SPECIFICITY
With respect to a diagnostic test, the ability to differentiate between closely related viruses. The statistical probability that a diagnostic test will properly indicate that a bird is not diseased.

SPF
See specific-pathogen-free.

SPHENISCIDAE
The family of birds within the Sphenisciformes order containing penguins.

SPHENISCIFORMES
The order of birds containing penguins.

STASIS
Stagnation of the normal flow of fluids.

STERILANT
An agent that destroys all microbial organisms including heat-resistant, bacterial spores. Sterilization is usually achieved by autoclaving. Solutions that contain chlorine or glutaraldehyde are frequently labeled as chemical sterilants but may not destroy all agents.

STRUTHIONIDAE
The family of birds within the Struthioniformes order containing ostriches.

STRUTHIONIFORMES
The order of birds containing ostriches.

SUBCLINICAL
An infection or abnormality that occurs without causing clinical signs. See asymptomatic.

SUBCUTANEOUS
Under the skin.

SUPPORTIVE ALIMENTATION
The process of assisted feeding.

STURNIDAE
The family of birds within the Passeriformes order containing starlings.

SUSCEPTIBLE
A bird that can be infected by a virus or other microorganism.

SYMPTOMATIC
The occurrence of clinical signs in conjunction with a disease.

SYNDROME
A group of clinical signs or changes that characterize a particular disease.

SYSTEMIC
Occurring throughout the body as opposed to any particular organ.

T-LYMPHOCYTES
A subgroup of lymphocytes (a type of white blood cells) that seeks out and destroys virus infected cells and regulates the response of other immune system cells.

TAXONOMY
The science of classifying the world's biota, whereby all life is grouped into a series of every more specific categories that uniquely identify each type by its Kingdom, Phylum, Class, Order, Family, Genus and species.

TENESMUS
Straining to defecate or urinate.

TENOSYNOVITIS
Inflammation of the tendon sheaths and synovial membranes (lining) of a joint.

TETRAONIDAE
The family of birds within the Galliformes order containing some grouse.

THERAPEUTIC AGENT
A medication designed to correct the cause or relieve the clinical signs of disease.

TID
Every eight hours.

TISSUES
A group of a similar type of cells which act together to perform a particular function. For example, muscle tissue is made of muscle cells that function together to cause the body to move.

TITER
A measure of the level of antibody found in a bird's serum.

TMEV
Turkey meningoencephalitis virus.

TORTICOLLIS
A clinical sign of disease in the nervous system in which the head is tilted to one side.

TRANSFORM, TRANSFORMATION
The change that a cell undergoes as it becomes cancerous. The changing of one type of tissue into another.

TRANSMISSIBILITY
The ease with which an infectious agent is transferred from one animal to another. A highly transmissible organism is easily transferred between susceptible animals.

TREMORS
Involuntary movement of parts of the body. Uncontrolled quivering.

TRICHOMONIASIS
An infection caused by *Trichomonas* sp.

TURDIDAE
The family of birds within the Passeriformes order containing thrushes, solitaires and bluebirds.

UNICELLULAR
Composed of only one cell.

UNILATERAL
Occurring on only one side.

URETERS
The tubes that carry urates and urine from the kidneys to the cloaca.

URINARY TRACT
The system that processes and removes urates and urine from the body including the kidneys and ureters.

USDA
United States Department of Agriculture.

VACCINATE
An animal that has been vaccinated.

VACCINATION
Administration of an antigen intended to induce an immune response (see immunization).

VARIANT
An organism that is different in appearance or structure from its predecessor.

VASCULITIS
Inflammation of the blood vessels.

VECTOR
An object or living being that serves to carry a virus from an infected to a noninfected, susceptible bird. Vectors may include contaminated gloves, nets, feeding utensils, surgical instruments or insects.

VEE
Venezuelan equine encephalitis.

VELOGENIC NEUROTROPIC (VNND)
Strain of Newcastle disease virus that causes severe disease with high mortality in chickens. These strains do not cause hemorrhagic lesions in the gastrointestinal tract. The clinical changes induced in other avian species vary.

VELOGENIC VISCEROTROPIC (VVND)
Strain of Newcastle disease virus that causes severe disease with high mortality in chickens. These strains cause hemorrhagic lesions in the gastrointestinal tract. The clinical changes induced in other avian species vary.

VENTRICULUS
The stomach in birds that is responsible for grinding food. The gizzard.

VESICLES, VESICULAR
Small blisters. Pertaining to the formation of small blisters.

VIREMIC
The presence of virus in the blood.

VIRION
A complete infectious virus particle.

VIROLOGY
The study of viruses.

VIRUCIDAL
Descriptive of a process or substance that inactivates (kills) a virus.

VIRULENCE
Relative ability of a virus to cause disease in an infected animal. A particular virus that is capable of producing disease is said to be virulent. A member of the same group of viruses that does not produce disease would be classified as avirulent.

VIRURIA
The presence of virus in the urine.

VIRUS NEUTRALIZING ASSAY
A diagnostic test used to detect the presence of neutralizing antibodies to a particular virus.

VIRUS PARTICLE
A complete virus that may or may not be infectious.

VISCERA
The internal organs, particularly those in the abdominal area (eg, proventriculus, liver, pancreas, spleen).

VN
Virus-neutralizing.

VNND
Velogenic neurotropic Newcastle disease virus.

VSV
Vesicular stomatitis virus.

VVND
Velogenic viscerotropic Newcastle disease virus.

WATERFOWL
A common term used to describe ducks, swans and geese.

WEE
Western equine encephalitis.

WHITE BLOOD CELLS
The cells in the blood (including heterophils, macrophages, monocytes, T-lymphocytes and B-lymphocytes) responsible for protecting the body against infectious organisms.

WILD-TYPE
As appears in nature.

ZOONOTIC
An infectious agent that can be transmitted from animals to man.

PRODUCTS AND SERVICES IN TEXT

Product	Supplier
DNA probe testing for PBFD virus DNA probe testing for polyomavirus	Avian Research Associates (ARA) Suite 101 TechneCenter Dr Milford, OH 45150 513-248-4700 VetGen Europe PO Box 60 Winchester, Hants UK SO23 9XN 011-44-962-880376
Acyclovir	Burroughs Welcome Co 3030 Cornwallis Rd Research Triangle Park, NC 27709 800-443-6763
Acemannan	Carrington Veterinary Medical Division PO Box 569003 Dallas, TX 75356-9003 214-717-5009
Biosecure shipping containers	Horizon Micro-Environments, Inc 1081 Industrial Dr Watkinsville, GA 30677 800-443-2498
Pacheco's disease virus antibody assay	Texas Veterinary Medical Diagnostic Lab PO Drawer 3040 College Station, TX 77841-3040 409-845-3414
Equimune	Vetrepharm, Inc 119 Rowe Rd Athens, GA 30601

SOURCES OF VACCINES FOR AVIAN VIRUSES

AVIAN ENCEPHALOMYELITIS VIRUS

Intervet	Maine Biological
Sanofi	Schering-Plough
Solvay	Sterwin
Tri Bio	Vineland

CANARY POXVIRUS

Biomune

INFLUENZA A VACCINE

Maine Biological

INFECTIOUS BRONCHITIS VIRUS

Biomune	Intervet
Maine Biological	Select
Solvay	Sterwin
Tri Bio	Vineland

INFECTIOUS BURSAL DISEASE

Biomune	Intervet
Maine Biological	Select
Solvay	Sterwin
Tri Bio	Vineland

INFECTIOUS LARYNGOTRACHEITIS VIRUS

Intervet	Sanofi
Schering-Plough	Solvay
Sterwin	Tri Bio
Vineland	

MARBLE SPLEEN DISEASE VIRUS

Tri Bio

MAREK'S DISEASE VIRUS

Intervet	Sanofi
Select	Solvay
Tri Bio	Vineland

NEWCASTLE DISEASE VIRUS

Biomune	Intervet
Maine Biological	Sanofi
Select	Solvay
Sterwin	Tri Bio
Vineland	

PACHECO'S DISEASE VIRUS

Biomune

PARAMYXOVIRUS (PMV-3)

Maine Biological

PIGEON PARAMYXOVIRUS (PMV-1)

Tradewinds

PIGEON POXVIRUS

Sterwin	Vineland

POLYOMAVIRUS

Biomune

POXVIRUS

Biomune	Intervet
KeeVet	Maine Biological
Sanofi	Schering-Plough
Select	Solvay
Sterwin	Tri Bio
Vineland	

REOVIRUS

Biomune	Intervet
KeeVet	Maine Biological
Sanofi	Schering-Plough
Select	Solvay
Sterwin	Vineland

AVIAN VACCINE PRODUCERS

Biomune Company
8906 Rosehill Rd
Lenexa, KS 66215
800-846-0230

Intervet, Inc
405 State St
PO Box 318
Millsboro, DE 19966
800-441-8272

KeeVet Laboratories
PO Box 1706
1900 Coleman Rd
Anniston, AL 36202
205-835-0425

Maine Biological Laboratories
PO Box 255
Waterville, ME 04903-0255
207-873-3989

Sanofi Animal Health, Inc
7101 College Blvd
Overland Park, KS 66210
800-538-2382

Schering-Plough Animal Health
PO Box 529
Galloping Hill Rd
Kenilworth, NJ 07033
800-648-2118

Select Laboratories, Inc
1168 Airport Parkway
PO Box 2497
Gainesville, GA 30501
404-536-8787

Solvay Animal Health
1201 Northland Dr
Mendota Heights, MN 55120-1139
612-681-9556

Sterwin Laboratories, Inc
Route 113
PO Box 537
Millsboro, DE 19966-0537
800-633-0462

Tradewinds
2339 Tacoma Avenue South
Tacoma, WA 98401
206-272-4887

Tri Bio Laboratories, Inc
1400 Fox Hill Rd
State College, PA 16803
800-677-6639

Vineland Laboratories
2285 East Landis Ave
Vineland, NJ 08360
800-225-0270

SELECTED DISTRIBUTORS OF DISINFECTANTS

Class of disinfectant	Distributor
Chlorhexidine gluconates	Bio-Ceutic Division 2621 North Belt Highway St Joseph, MO 64506 800-325-9167 Fort Dodge Laboratories 800 5th St NW PO Box 518 Fort Dodge, IA 50501 515-955-4600
Glutaraldehyde	Johnson & Johnson Medical Arlington, TX 76004-3130
Iodines	The Butler Company 5000 Bradenton Ave Dublin, OH 43017-0753 614-761-9095
Phenol	Veterinary Products Lab PO Box 34820 Phoenix, AZ 85067-4820
Quaternary Ammonium	Huntington Labs, Inc 970 East Tipton St Huntington, IN 46750 800-537-5724 KenVet Animal Care Prod. 7th and Orange St Ashland, OH 44805 800-338-7953 Stearns Packaging Co. 4200 Sycamore Ave PO Box 3216 Madison, WI 53704 608-246-5150 The UpJohn Company 7000 Portage Rd Kalamazoo, MI 49001 616-323-4000
Stabilized chlorine dioxide	Oxyfresh, USA, Inc PO Box 3723 Spokane, WA 99220 800-999-9551 ext 105270

parvovirus and 430
PBFD virus and 232
picornavirus and 417
pigeon PMV-1 and 274
PMV and 277, 278, 280
PMV-2 and 266
poxvirus and 292, 295, 301, 303, 304, 306, 308
rabies virus and 434
reovirus and 335, 339
reticuloendotheliosis viruses and 373
rotavirus and 347
rubella virus and 409
sindbis virus and 409
SLE and 425
VEE virus and 408
virus damage of 37, 54, 267
virus transmission and 49
VSV and 435
WEE virus and 394, 395
French moult 139
   PBFD virus and 223, 226
   polyomavirus and 138, 223
Fringillidae
   papillomas and 128
Fumes
   damage from (F) 62, 63
Fungi 1, 83
   adenovirus and 319
   Amazon tracheitis virus and 186
   as normal flora 53, 62, 64, 75
   cellular effects of 41
   damage from 44
   destruction of 71
   disease and 31
   disinfectants and 116
   ILT and 198
   immune system and 39
   immunodeficiency and 78
   immunosuppression and 36, 46
   neonates and 49
   normal flora and 64, 75
   owl herpesvirus and 206
   PBFD virus and 238
   PDD and 446
   pigeon herpesvirus and 190
   poxvirus and 296
   protection from 60
   structure of (F) 5
   susceptibility to 78

Galahs
   PBFD virus and 223, 229, 232, 236
Galliformes
   see also Chickens, Turkeys, Pheasants
   adenovirus and 320
   coronavirus and 414
   EEE virus and 384, 392
   herpesvirus and (T) 172
   ILT and (F) 198

MDV and (F) 203
NDV and 262
ockelbo virus and 409
PMV-1 and 261
PMV-3 and 268
polyomavirus and 142, 143
poxvirus and 292, 306
reovirus and (F) 335, 341, 344
vesiculovirus and 433
virus susceptibility of (T) 2
WEE virus and 395
Gammaherpesvirinae
   classification of (F) 15
Gastrointestinal tract (F) 65
   see also Alimentary tract
   adenovirus and 319
   antibodies and 72
   coronavirus and 413
   defense mechanisms of 60, 63
   EEE virus and 392
   falcon herpesvirus and 205
   IBV and 416
   immunodeficiency and 78
   influenza A virus and 353, 357, 362
   movement of (F) 66
   NDV and 259, 274, 279
   owl herpesvirus and 206
   papillomas and 131
   papillomatosis and (F) 134
   PDD and 439, 444
   persistent infections and 82
   PMV-1 and 256, (F) 259, 263
   poxvirus and 290, 294, 296, 298
   reovirus and 335, 339
   rotavirus and 347
   vaccination and 122
   virus transmission and 51, 52
Geese
   adenovirus and 313, 316-317, 329
   duck plague virus and 207, (T) 209, 210-211, 213
   EDS-76 virus and 330
   hepadnavirus and 428
   hepatitis and 430
   herpesvirus and (T) 172
   immune response and 73
   influenza A virus and 353, 354, 356, 360
   influenza and 430
   lymphoid leukosis and 368
   MDV and 201
   parvovirus and (F) 430, (F) 432, 433
   PDD AND (F) 442, 444
   PMV and 268, (T) 254, 278
   PMV-2 and 268
   PMV-4 and 271
   PMV-6 and 272
   PMV-8 and 272
   pneumovirus and 281
   poxvirus and (T) 286, (T) 292, 304-305

reovirus and 335, 338, 344, 346
reticuloendotheliosis viruses and 372, 374-375, (F) 376
SLE and 425
TMV and 427
VSV and 435
Genetic malformations 46
Genetics
   disease resistance and 59
   immune system and 74
   immunodeficiency and 78
   vaccination and 125
Genitourinary tract
   defense mechanisms of 57
   papillomas and 128
   persistent infections and 82
   virus entrance and 54, 57
   virus transmission and 47
Genome 22, 29
   falcon herpesvirus and 205
   PDV and 192, 205
   pigeon herpesvirus and 192, 205
   poxvirus and 293
   retroviruses and 365
Genus 13
Germany
   adenovirus and 316
   MSD virus and 322
   papillomas and 128
   PDV and 176
   PMV-3 and 270
   polyomavirus and 142
   stork herpesvirus and 217
Germicide (T) 110
Globulins 46
Gloves
   transmission and 49, 295
Glutaraldehyde 110, 115, 117
Gonads
   EEE virus and 407
   lymphoid leukosis and 370
   reticuloendotheliosis viruses and 373, 374
   tumors and 369
Goose plague 430
Goslings
   DHV and 418
   duck plague virus and 208
   picornavirus and 418
   pigeon herpesvirus and 191
Gout
   coronavirus and 414, (F) 415
   EEE virus and 390, 407
   IBV and 416

Great Britain
   DHV and 420
   duck plague virus and 209, 210
   PMV-7 and 272
Green finches

AVIAN VIRUSES: FUNCTION AND CONTROL

525